FRACTURE MECHANICS: SEVENTEENTH VOLUME

Seventeenth National Symposium
on Fracture Mechanics
sponsored by
ASTM Committee E-24
on Fracture Testing
Albany, New York, 7–9 August 1984

ASTM SPECIAL TECHNICAL PUBLICATION 905
J. H. Underwood, U.S. Army Armament
Research & Development Center, R. Chait, U.S.
Army Materials & Mechanics Research Center,
C. W. Smith, Virginia Polytechnic Institute &
State University, D. P. Wilhem, Northrop
Aircraft, W. A. Andrews, General Electric
Company, and J. C. Newman, NASA Langley
Research Center, editors

ASTM Publication Code Number (PCN)
04-905000-30

ASTM 1916 Race Street, Philadelphia, Pa. 19103

Library of Congress Cataloging-in-Publication Data

National Symposium on Fracture Mechanics (17th:
1984: Albany, N.Y.)
Fracture mechanics.

(ASTM special technical publication; 905)
"ASTM publication code number (PCN) 04-905000-30."
Includes bibliographies and index.
1. Fracture mechanics—Congresses. I. Underwood,
John H. II. ASTM Committee E-24 on Fracture Testing.
III. Title. IV. Series.
TA409.N38 1984 620.1′126 86-8000
ISBN 0-8031-0472-3

Library of Congress Catalog Card Number: 86-8000

NOTE

The Society is not responsible, as a body, for the statements and opinions advanced in this publication.

Printed in Baltimore, Md.
July 1986

Dedication

This publication is dedicated to the following group of individuals and their pioneering work in fracture testing:

William F. Brown, Jr.
James E. Campbell
Roy H. Chirstensen
John Hodge
George R. Irwin
Joseph M. Krafft
William T. Lankford
John R. Low, Jr.
Richard A. Rawe
John E. Srawley
Henry J. Stremba
Charles F. Tiffany

Their important contributions were central to the ASTM Special Committee on Fracture Testing of High Strength Sheet Materials, forerunner of Committee E-24 on Fracture Testing.

As a tribute to the founders of ASTM Committee E-24 and to the series of symposia which they helped to establish, the poem on the following page was offered as a special presentation at the Albany meeting.

The 17th Symposium on Fracture

At first a Committee, called E-24,
Studied aspects of fracture not known before;
And Irwin suggested the very best way
Was to write all the terms as functions of *K*.

This worked for bodies whilst still elastic,
But needed correction as the stresses turned plastic;
Till Rice and some others showed us the way
To express all the terms by the integral *J*.

And presently users were nothing loath
To use *dJ* for stable crack growth;
So fracture was thought to be well understood
At the Albany meeting of John Underwood.

But then the Symposium, in second day session,
Was taught a quite salutary lesson;
As the crucial question was faced by John Srawley
That sometimes *J* would serve us but poorly.

But if these complexities seem to confuse us,
Just follow the founders' advice on consensus
And study the problem until a year older,
Then tell us next time in the Conference at Boulder.

Dedicated to those founding members
of the original Committee, whom
it was my good fortune to know.

Cerdic Renrut
9 August 1984

Foreword

The Seventeenth National Symposium on Fracture Mechanics was held on 7–9 August 1984 in Albany, New York. ASTM Committee E-24 on Fracture Testing was the sponsor. J. H. Underwood, U.S. Army Armament Research & Development Center, served as symposium chairman and co-editor of this publication. R. Chait, U.S. Army Materials & Mechanics Research Center, C. W. Smith, Virginia Polytechnic Institute & State University, D. P. Wilhem, Northrop Aircraft, W. A. Andrews, General Electric Company, and J. C. Newman, NASA Langley Research Center, served as symposium co-chairmen and co-editors of this publication.

Related ASTM Publications

Fracture Mechanics: Sixteenth Symposium, STP 868 (1985), 04-868000-30

Fracture Mechanics: Fifteenth Symposium, STP 833 (1984), 04-833000-30

Fracture Mechanics: Fourteenth Symposium—Volume I: Theory and Analysis, STP 791 (1983), 04-791001-30

Fracture Mechanics: Fourteenth Symposium—Volume II: Testing and Applications, STP 791 (1983), 04-791002-30

Fracture Mechanics (Thirteenth Conference), STP 743 (1981), 04-743000-30

Fracture Mechanics (Twelfth Conference), STP 700 (1980), 04-700000-30

A Note of Appreciation to Reviewers

The quality of the papers that appear in this publication reflects not only the obvious efforts of the authors but also the unheralded, though essential, work of the reviewers. On behalf of ASTM we acknowledge with appreciation their dedication to high professional standards and their sacrifice of time and effort.

ASTM Committee on Publications

ASTM Editorial Staff

Contents

Subcritical Crack Growth

Fracture Testing

Introduction

This volume and the Seventeenth National ASTM Symposium on Fracture Mechanics on which it is based are part of a continuing series. These symposia have become clearly the most prestigious in the field of fracture. As such, they are the focus and forum for quality work in all areas of the field, and this is the important purpose of the symposium and volume.

If the field can be divided into testing and analysis, the former has been and continues to be the more emphasized in this symposium series. This is appropriate, considering the sponsor, ASTM Committee E-24 on Fracture Testing. Nevertheless, analysis is a required part of any test, and much of the work reported here is primarily analysis.

At least four general topics or categories of work frequently occur in the papers: ductile fracture, test method development, surface cracks and crack shape effects, and high temperature and loading rate effects. The prevalence of these four categories attests to the basic practical nature of the field of fracture and of those who work in it. Each of these categories defines an area of important current concern in the design and use of load-carrying components and structures. It is the hope and belief of all those involved that this symposium and volume have contributed to these and other important areas in the field of fracture.

The National Symposium on Fracture Mechanics is often the occasion at which ASTM awards are presented to recognize the achievements of current investigators. At the Seventeenth Symposium two awards were presented. The ASTM Committee E-24 Irwin Medal was presented by Dr. Irwin to Mr. John G. Merkle, Martin Marietta Energy Systems, for his outstanding work in the field of fracture mechanics. The ASTM Award of Merit and honorary title of Fellow were given to Mr. David P. Wilhem, Northrup Corporation, for his distinguished service and leadership in Committee E-24. Dr. J. Gilbert Kaufman, Arco Metals, past chairman of E-24, made the presentation to Mr. Wilhem.

Wc takc this opportunity to thank two groups who deserve a significant share of credit for this symposium. The first is the combined support staff of all of us listed below. The administrative and clerical work of this whole group was essential to the task and is greatly appreciated. The second group is made up of those behind-the-scenes people whose work is nonetheless critical.

In particular, we thank Professor Ray Eisenstadt of Union College for his help in administering the symposium, Mr. Jim Gallivan of the Army Materials and Mechanics Research Center for financial support, the late Dr. Fred

Schmeideshoff of the Army Research Office for his help in organizing the symposium, and Professor Jerry Swedlow for his continuing support and sound advice during the entire process.

John H. Underwood

Army Armament Research & Development Center, Watervliet, New York; chairman and co-editor.

Richard Chait

Army Materials and Mechanics Research Center, Watertown, Massachusetts; co-chairman and co-editor.

C. William Smith

Virginia Polytechnic Institute & State University, Blacksburg, Virginia; co-chairman and co-editor.

David P. Wilhem

Northrop Aircraft, Hawthorne, California; co-chairman and co-editor.

Wayne R. Andrews

General Electric Company, Schenectady, New York; co-chairman and co-editor.

James C. Newman

NASA Langley Research Center, Hampton, Virginia; co-chairman and co-editor.

Applications

P. D. Hilton,[1] *R. A. Mayville,*[1] *and D. C. Peirce*[1]

An Application of Fracture Mechanics to a Ship Controllable Pitch Propeller Crank Ring

REFERENCE: Hilton, P. D., Mayville, R. A., and Peirce, D. C., **"An Application of Fracture Mechanics to a Ship Controllable Pitch Propeller Crank Ring,"** *Fracture Mechanics: Seventeenth Volume, ASTM STP 905,* J. H. Underwood, R. Chait, C. W. Smith, D. P. Wilhem, W. A. Andrews, and J. C. Newman, Eds., American Society for Testing and Materials, Philadelphia, 1986, pp. 5–21.

ABSTRACT: A fracture mechanics analysis was conducted to establish the fracture toughness of a controllable pitch propeller crank ring material required to prevent a fracture mode in which loss of a propeller blade occurs. Loss of the propeller was assumed to be prevented if fracture instability could not occur before the fatigue crack grew to a size beyond which crack growth would proceed radially through the flange of the crank ring and not around the circumference. The fracture analysis was conducted by modeling the cracked crank ring as a plate with a part-through crack in bending. Numerical solutions for part-through cracks in bending were combined with results for large crack length-to-plate width geometries for through cracks in bending to determine K_I for the large crack size of interest. Values of K_I with plastically adjusted crack lengths were converted to values of J_I and crack driving force curves were generated. Estimates of the plastic collapse moment for the crank ring were made as an alternative method of determining fracture conditions. The results of the analysis are a minimum acceptable value of yield strength and curves of yield strength versus minimum acceptable values of J_{IR} and T_{mat} at a crack extension of 1.27 mm as determined by a *J-R* curve test.

KEY WORDS: fracture mechanics, application, bending, ship component

Controllable pitch propellers are commonly found in current ship propulsion systems. They are used for both small vessels and large ships with power as great as 40 000 hp. All controllable pitch propellers require some mechanism to rotate the propeller blades. In the study described in this paper, rotation is brought about by a crank ring to which the propeller blade is attached by several bolts. An illustration of such a crank ring is shown in Fig. 1. Not

[1]Arthur D. Little, Inc., Acorn Park, Cambridge, MA 02140.

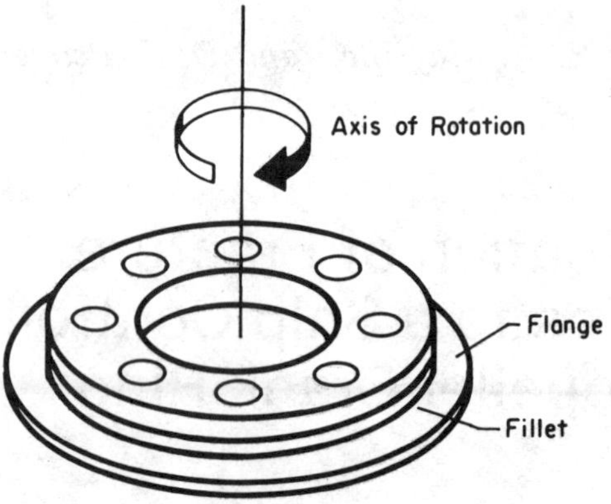

FIG. 1—*A controllable pitch propeller crank ring.*

shown in the figure is the protrusion from the underside of the ring to which a mechanical "crank" is attached for the purpose of rotating the crank ring.

On installation, the crank ring is set over a central post which is attached to the propeller hub. Next, a narrow bearing ring is threaded into the hub body over the crank ring so that if the crank ring were lifted, its flange would contact the underside of the bearing ring. Finally, the propeller blade is bolted to the crank ring. Thus, under the action of centrifugal and hydrodynamic loads during operation of the propeller, significant pressure loads are transferred between the underside of the bearing ring and the upper surface of the crank ring flange. This in turn causes cyclic stresses in the fillet at the point where the flange meets the main crank ring body (Fig. 1). These high stresses can lead to cracking at the fillet [*1*], and there is a possibility that the crank ring can fracture. In fact, one can imagine a scenario in which rapid fracture from a fillet crack could proceed around the circumference of the crank ring and lead to separation of the propeller blade from the hub body.

The objective of the investigation described in this paper was to establish through analysis the material fracture toughness for a particular crank ring such that, in the unlikely event that a fatigue crack does initiate, a fracture mode leading to loss of the propeller would be avoided. Periodic inspection of crank rings is generally not conducted, so that in this scenario some other incident, such as excessive deformation, must occur to make the failure detectable. There has been one reported failure incident in which a fillet fatigue crack initially propagated around the circumference of the crank ring but eventually propagated and broke through the flange. The severed piece of the crank ring then prevented rotation on the next attempt at pitch control and this led to the discovery of the fracture. It is not clear that fatigue cracks in all crank rings will proceed in this manner and, in fact, results of our analysis, presented below, show that there is a significant driving force for continued circumferential fatigue crack growth. Nevertheless, based on limited evidence, it has been assumed that the crack will propagate initially around the circumference and then through the flange, provided the mode of fracture is by fatigue and not by rapid brittle or ductile fracture. Furthermore, loss of the propeller is assumed to be prevented if fracture instability cannot occur before the fatigue crack grows to a size beyond which crack growth by any mode would proceed approximately radially through the flange and not around the circumference.

The first problem in establishing the required crank ring toughness is to choose a crack size and geometry from which fracture instability would proceed through the flange. Guaranteeing that fracture instability will not occur prior to attaining this crack size is then equivalent to finding the conditions—material toughness—required for instability to occur at this crack size; this assumes that smaller crack sizes are less severe.

Crank Ring and Crack Geometry

A cross section of the single crank ring geometry analyzed in this investigation is shown in Fig. 2. A full-scale laboratory test was performed for this crank ring resulting in a fatigue crack, the geometry of which was used in our analyses and is shown in Fig. 3. The crack had several initiation sites located in the fillet on the thrust side of the blade and at discovery extended about 85° around the circumference. At its midpoint the crack was inclined approximately 45° to the vertical. The crack front extended part way into the flange and to within about 8.9 mm of the crank ring bottom at the center of the

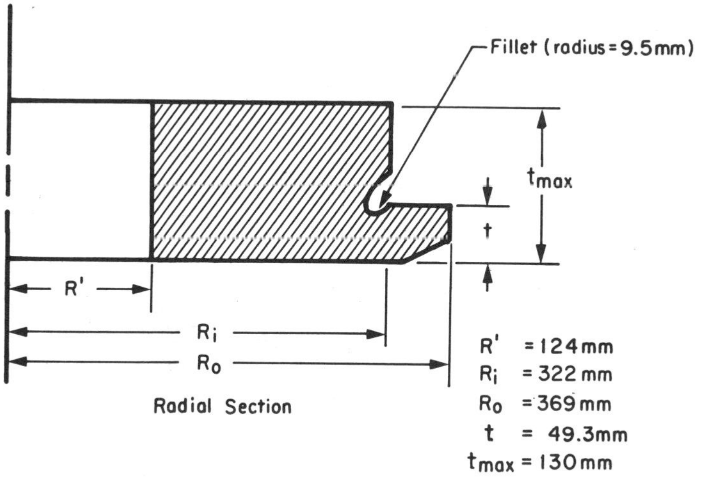

FIG. 2—*Geometry of the crank ring.*

crack front. The crack front in its plan view was not parallel to the bending axis but was instead curved somewhat (Fig. 3). The geometry of the crack in its plane is unknown. Indications are that it grew more to a trapezoidal shape than to an elliptical shape. The location, geometry, and orientation of this crack are consistent with the tensile stresses developed in the crank ring fillet by the hydrodynamically induced bending loads on the propeller.

No calculations were performed to establish the radial extent of the crack in the flange required to ensure that fracture would proceed through the flange and not turn in the circumferential direction. Instead, it was assumed that if the radial extent of the crack is half-way through the flange, then further crack growth will also be radial.

Calculation of Loads

Loads on the crank ring arise from centrifugal forces due to propeller hub rotation and from bending forces due to hydrodynamic pressure on the pro-

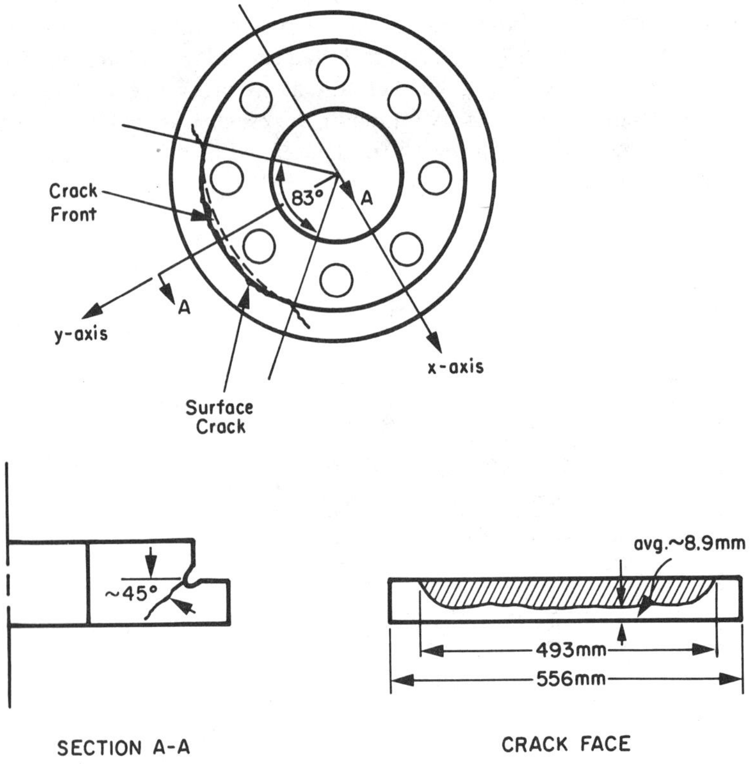

FIG. 3—*Crack geometry from laboratory-tested crank ring.*

peller blades. Measurements made with strain gages in the fillet of the crank ring during ship operation indicate that these forces result in loads on the flange which can be modeled as the sum of a uniformly distributed load and a load whose magnitude varies linearly with respect to the y-axis (Fig. 3).

Two methods were used to estimate the magnitude of the crank ring flange loads on the thrust side of the blade: strength of materials calculations based on strain gage measurements and finite element analysis. In both cases it was assumed that the flange is subjected to a normal line load along its periphery.

Strain gage measurements made on a crank ring with the geometry shown in Fig. 2 in a 35 000 hp ship indicate that the most severe radial tensile stress in the crank ring fillet is approximately 503 MPa. This stress, σ, is related to the nominal bending moment, m, per unit circumferential length at the fillet by

$$\sigma = K_t 6m/t^2$$

Where t is the flange thickness, 49.3 mm from Fig. 2, and K_t is a stress concentration factor, estimated from Peterson [2] to be 1.5. The load per unit length on the outer flange circumference causing the bending moment is $p = m\ (R_i/R_o)/(R_o - R_i)$, where R_i and R_o are the inner and outer radii of the crank ring flange; $R_o = 369$ mm and $R_i = 322$ mm. Therefore the maximum flange load estimated from the strain gage readings is

$$p = 2540 \text{ N/mm}$$

Estimates for load distribution along the flange based on strain gage readings are probably upper bound predictions, because the strain gage readings were made in the fillet directly adjacent to the bolts that connect the propeller blade to the crank ring. Stress in this region may be influenced by the local stress concentration effect of the nearest bolt. On the other hand, the load distribution along the flange will not be as significantly influenced by load concentrations associated with individual bolts, because the distance from the bolts to the load transfer region is larger than that from the bolts to the fillet where the strain gage was located.

A simple finite element model of the crank ring was used as an alternative method to obtain crank ring load distribution estimates. The crank ring, modeled by a ten-element, three-dimensional mesh (Fig. 4), was subjected to a combination of axial load and bending moment to simulate the centrifugal and hydrodynamically induced bending loads. This loading is obtained by the superposition of two solutions, symmetric and skew symmetric with respect to an axis, x, that passes through the center of the crank ring and is parallel to the neutral axis. The inner circumference of the flange is held fixed, modeling its interaction with the stiffer central portion of the crank ring. Displacements are prescribed at the upper edge of the outer circumference on the as-

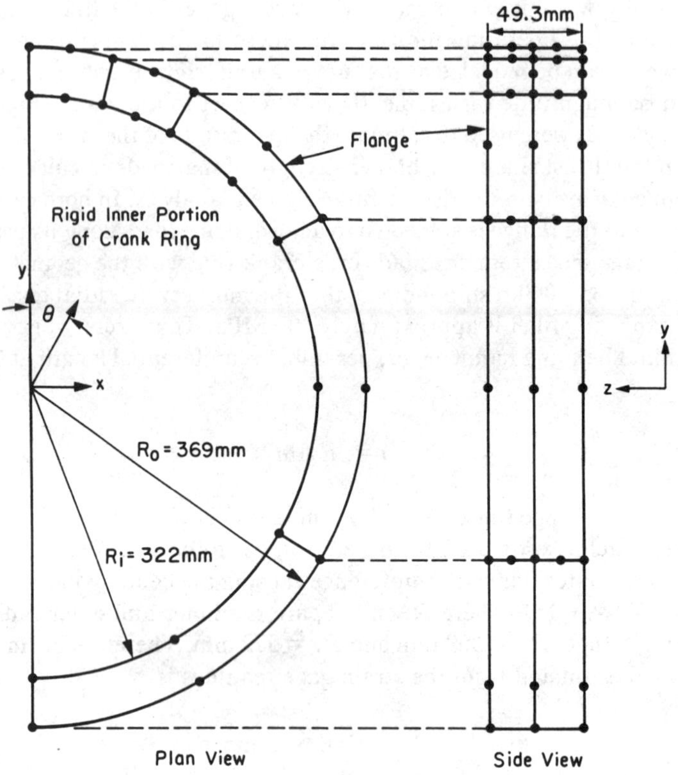

FIG. 4—*Finite element grid pattern.*

sumption of a rigid bearing ring. Two calculations are carried out: the first prescribes a constant downward displacement on the outer circumference (modeling the centrifugal load), while the second prescribes a set of displacements varying linearly in y (modeling the hydrodynamic load). These two solutions are superposed in such a way that the nodal reaction forces at the outer edge of the flange balance the centrifugal load and hydrodynamic bending moment. These loads were obtained from the same example used in the previous strain gage calculations.

The maximum load obtained by finite element analysis was approximately 2100 N/mm. This differs from the maximum load derived from the strain gage readings by 17%. Both of the methods used to estimate the load distribution along the circumference of the crank ring are approximate, and it is difficult to establish which of the estimates is more accurate. Since the finite element analysis ensures that force and moment equilibrium are satisfied and avoids the complications associated with load (or stress) concentrations, the finite element analysis results are used to perform the fracture mechanics analysis to be described later.

A finite element calculation was also performed to quantify the load redistribution which occurs in the presence of a crack. The mesh and loading were exactly the same as for the uncracked case, except nodes on the inner flange circumference were released to simulate a crack that extends 90° around the circumference and 75% through the flange thickness. The center of the crack was symmetric with respect to the y-axis.

The load distributions for the uncracked and cracked crank ring models, as calculated by the finite element analysis, are shown in Fig. 5. There is a substantial redistribution in load in the presence of a crack. In particular, the load at the intersection of the y-axis with the flange periphery, which is the location of maximum load in the uncracked crank ring, shows a reduction in load to about 700 N/mm from 2100 N/mm. Figure 5 also shows that the maximum load in the cracked crank ring occurs at about 50° from the y-axis. This is in part due to the coarseness of the mesh used and the presence of the crack tip at this point as well as the reduction in load with y which must occur because of the bending nature of the problem. This large load near the crack tip suggests that there may still be a significant driving force for circumferential crack growth.

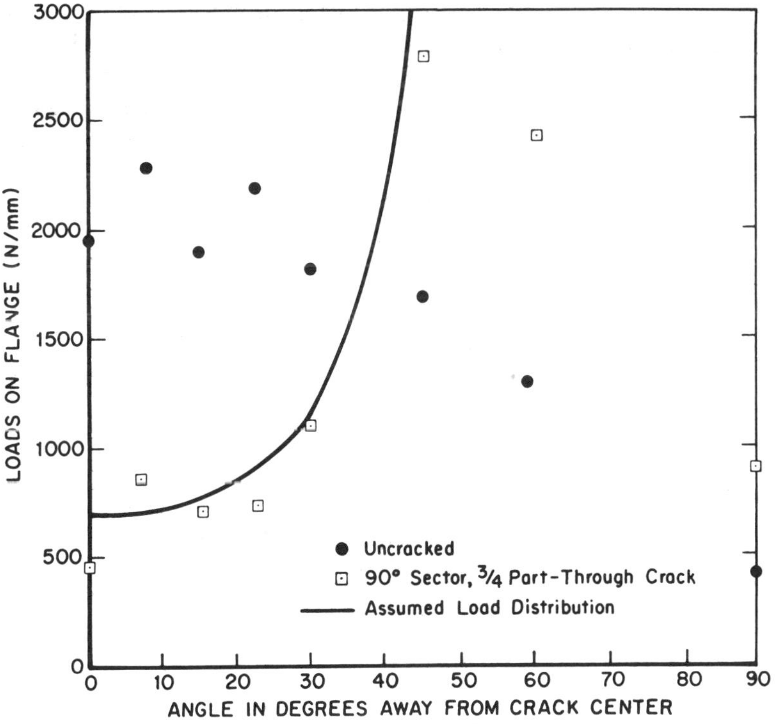

FIG. 5—*Crank ring flange load distributions.*

results for K_I at the surface in terms of the nondimensional parameter F defined by[2]

$$F = K_I/\sigma_b\sqrt{\pi c}$$

where σ_b is the nominal outer fiber bending stress equal to $6m/t^2$ and m is the applied bending moment per unit width. The assumed crank ring crack is characterized by the ratios $a/c = 0.16$, $a/t = 0.82$, and $c/W = 0.89$ (Fig. 3). Newman and Raju's results closest to this case are for $a/c = 0.2$, $a/t = 0.8$ and $c/W = 0.8$, which give a value of $F = 0.49$. This value is certainly low because F increases sharply as c/W approaches unity and there is a considerable difference between $c/W = 0.8$ and $c/W = 0.89$.

Results by Boduroglu and Erdogan [5], who have recently published K_I solutions for plates of finite width containing through cracks and loaded in remote bending, are used to quantify the effect of a greater crack length-to-width ratio for the plate containing a part-through crack. Results for the through crack case are given for values of c/W very close to unity. The approach in using these results is to assume that the effect of finite width for the through crack geometry is the same as the finite width effect for the part-through crack geometry. Table 1 lists the factors F for a number of c/W values for part-through cracks with $a/c = 0.2$ and $a/t = 0.8$ as obtained from Ref *4* and for through cracks with $c/t = 4$ [$=(a/t)/(a/c)$] as obtained from Ref *5*. The ratio of F-factors for the two cases is also listed.

The results in Table 1 indicate that for $a/c = 0.2$ and $a/t = 0.8$ the factor F for the part-through crack geometry is approximately one half of the factor for the through crack geometry. Therefore the approach taken to obtain a value of F for the cracked crank ring geometry is to obtain a value of F from the through crack analysis for a geometry close to the crank ring geometry and to multiply it by 0.5. The value of F for a through crack geometry with $c/W = 0.89$ and $c/t = 5$ is $F = 1.6$ [*6*] so that for the part-through crack $F = 0.5(1.6) = 0.8$. This is the value used in the fracture mechanics analysis.

TABLE 1—*Effect of finite width on* K_I *at the surface for plates loaded in remote bending containing part-through and through cracks.*

Where $F = K_I/\sigma_b\sqrt{\pi c}$, $a/c = 0.2$, $a/t = 0.8$, $c/t = 4$:

c/W	W/t	F_{pt} (Part-Through Crack [*4*])	F_t (Through Crack [*5*])	F_{pt}/F_t
0.2	20	0.33	0.70	0.47
0.4	10	0.35	0.73	0.48
0.6	6.7	0.40	0.83	0.48
0.8	5	0.49	1.13	0.43

[2]This nondimensional factor differs from that used by Newman and Raju [*4*].

Using the dimensions and loading for the cracked crank ring, $m = 88.9$ N-m/mm, $t = 49.3$ mm, and $c = 246$ mm, the value of K_I at the free surface, without correction for plasticity, is equal to 158 MPa $\sqrt{m}$. One notes immediately that, according to this analysis, a very tough material is needed to avoid loss of the propeller in the presence of the assumed fatigue crack.

The fracture mechanics analysis of the crank ring will account for a certain amount of stable crack growth, so it is necessary to quantify the dependence of F on crack length. This is done by employing the dependence of F on crack length for the through crack and multiplying by one half. The variation of F with c for the crank ring crack dimensions, $c/t = 5$, $W/t = [(c/t)/(c/W)] = 5.64$, is approximately [6]

$$F_t = 2.73c - 24.9$$

Multiplying this expression by one half provides the relation to be used in the fracture mechanics assessment of the cracked crank ring:

$$F_{pt} = 1.37c - 12.4 \tag{1}$$

Newman and Raju's results can also be used to estimate the stress intensity factor at the bottom of the crank ring crack. For $a/c = 0.2$ and $a/t = 0.8$ the stress intensity factor at the bottom of the crack is approximately one half of the value at the surface for $0.2 < c/W < 0.8$; data for $c/W > 0.8$ were not given. Finite element analyses of an elliptical part-through crack in bending, which model the close proximity of the vertical flange loads to the crack plane and the short moment arm in comparison to the plate width and surface crack length, also indicate that K_I at the deepest point of the crack is about 50% of K_I at the surface [3]. This would appear to contradict the possibility of a fatigue crack in the crank ring growing to the shape shown in Fig. 3. Without attempting to explain this apparent contradiction, it is noted that crack propagation from the top part of the crack is the fracture that would lead to loss of the propeller and is therefore of greatest interest in this analysis.

The large K_I value calculated earlier for the crank ring crack shows that tough materials must be used to avoid loss of the propeller according to this methodology. This implies that the material will be at or near its upper shelf behavior, that it can experience stable tearing, and that elastic-plastic fracture mechanics techniques are necessary to quantify its resistance to fracture (at least to characterize the material toughness with small specimens). Consequently, crack driving force curves are calculated in terms of J_I, since the material's fracture resistance is expected to be expressed in terms of J_I-R curves.

The crack driving force curve is the relation between J_I and crack length or crack extension and is estimated from values of K_I through the relation

$$J_I = K_I^2/E' \tag{2}$$

where E' is the effective elastic modulus equal to E for plane stress and $E/(1-\nu^2)$ for plane strain: E is Young's modulus and ν is Poisson's ratio. Plasticity is accounted for in the analysis by making a crack length plastic zone correction to K_I before converting to J_I:

$$c_{eff} = c + (1/6\pi)(K_I/\sigma_o)^2$$

where K_I on the right-hand side of the equation is calculated using the original crack length, c. Two-dimensional, plane strain conditions are assumed to prevail near the flange surface. The crack-driving force relation with the plastic zone correction is then given by

$$J_I = F^2\,(c_{eff})\,\sigma_b^2\pi c_{eff}/E' \qquad (3)$$

where Eq 1 is used to calculate F, and the applied load or stress σ_b, is assumed to be constant (load control).

Figure 8 shows a plot of J_I versus crack extension for the crack geometry of Fig. 3 and a material with yield strength equal to 690 MPa. The J_I-R curve for a Ni-Cr-Mo steel with σ_o = 690 MPa is shown for comparison.

An alternative driving force for unstable fracture is the attainment of the plastic collapse load. A lower bound to the limit moment for a plate with a part-through crack in bending is obtained by calculating the moment which arises when the axial stress over the entire net section is equal to the yield

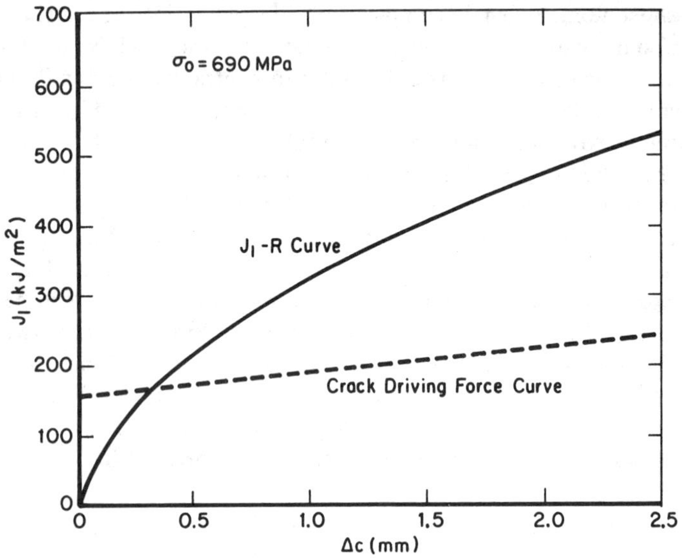

FIG. 8—*Crack-driving force curve for the cracked crank ring in comparison to the* J_I-R *curve for a 690 MPa yield strength Ni-Mo-Cr steel.*

strength [7]. The idealized cross-sectional geometry shown in Fig. 9 was used to perform this calculation. The neutral axis for this section is essentially at the lower crack front, and the limit moment per unit width ($2W = 556$ mm) is given by

$$m_{\rm lim} = 0.17\sigma_o \tag{4}$$

where the units for $m_{\rm lim}$ and σ_o are N-m/mm and MPa. Therefore, for $\sigma_o =$ 690 MPa, $m_{\rm lim} = 117$ N-m/mm, which is greater than the applied moment assumed for the crank ring in this analysis: 88.9 N-m/mm. The actual collapse moment would be larger because the crack is probably smaller than the idealization shown in Fig. 9, the material will harden, and the cracked geometry induces some constraint to plastic deformation.

Strength and Toughness Requirements for the Crank Ring

It is now possible to determine the strength and toughness of the crack ring material required to prevent failure from occurring under the assumptions of this investigation. In the section on loads, it was determined that the effective bending moment per unit width on the cracked section shown in Fig. 3 is approximately equal to 88.9 N-m/mm.

A lower bound estimate of yield strength necessary to prevent collapse from occurring can be calculated from Eq 4. In this case:

$$\sigma_o = m_{\rm lim}/0.17 = 523 \text{ MPa}$$

Therefore the minimum yield strength for the crank ring material should be greater than 523 MPa.

Two approaches are taken to specify the crank ring material toughness to avoid ductile tearing instability. Both are based on the assumption that the material does not fail by a cleavage mechanism of fracture for the tempera-

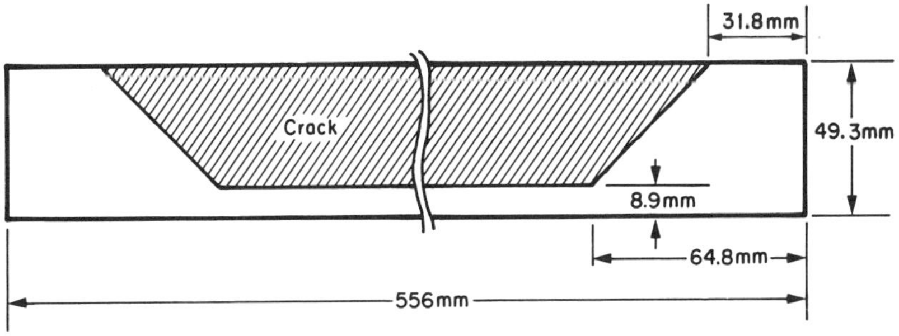

FIG. 9—*Idealized crank ring crack geometry used for collapse moment calculation.*

ture and loading rates characteristic of the crank ring. This can be accomplished by requiring the minimum upper shelf temperature, say, as determined by Charpy tests, to be below the operating temperature.

In the first approach to specifying required toughness, no crack extension by tearing is permitted in the engineering sense; that is,

$$J_I\ (\Delta c = 0) < J_{Ic}$$

where J_{Ic} is determined in accordance with a procedure such as ASTM Test for J_{Ic}, A Measure of Fracture Toughness (E 813).

Equation 3 is used to calculate $J_I\ (\Delta c = 0)$ for an arbitrary yield strength and this becomes the specified minimum value of J_{Ic} for the crank ring material. The required value of J_{Ic} for a yield strength of 690 MPa, according to this procedure, is 159 kJ/m^2. The value of J_{Ic} for the 690 MPa yield strength material whose J_I-R curve is shown in Fig. 8 is 131 kJ/m^2. Therefore a specification permitting no crack extension by tearing in the engineering sense would eliminate this steel as a candidate material for the crank ring. Again, this is based on the assumption that a large fatigue crack could develop in the crank ring.

A fracture criterion based on J_{Ic} alone for a ductile material is conservative, because it does not take advantage of the increase in resistance to ductile crack extension which generally accompanies small amounts of tearing. A more realistic approach is to specify toughness so that tearing will arrest and not become unstable. This is accomplished by requiring that the J_I-Δc and the J_I-R curves intersect and that at the point of intersection the slope of the J_I-Δc curve is less than the slope of the J_I-R curve; in other words, $T_{app\ell}$ is less than T_{mat} where $T = (E/\sigma_o^2)\ dJ/dc$. Such a procedure is illustrated in Fig. 8. Quantifying this criterion is difficult, because the J_I-R curve requires at least two parameters to be represented. This means technically that there are an infinite number of J_I-R curves which intersect the J_I-Δc curve.

A practical implementation of this approach is to specify a minimum value of J_I-R, for a certain amount of crack extension, which is greater than the value of J_I for the same amount of crack extension for the yield strength in question. A value of $\Delta c = 1.27$ mm (0.050 in.) has been chosen for this investigation. This amount of crack extension has a negligible effect on the collapse moment and is within the range of crack extension investigated in the determination of J_I-R values in accordance with ASTM E 813.

The required value of J_I-R for a yield strength of 690 MPa according to this criterion as obtained from Fig. 8 is 194 kJ/m^2. An additional requirement is that T_{mat} at $\Delta c = 1.27$ mm be greater than $T_{app\ell}$ at $\Delta c = 1.27$ mm for the yield strength in question. The minimum allowable value of T_{mat} for $\sigma_o = 690$ MPa is 13.4. As a comparison, the values of J_I-R ($\Delta c = 1.27$ mm) and T_{mat} ($\Delta c = 1.27$ mm) for the steel whose J_I-R curve is shown in Fig. 8 are, respectively, 368 kJ/m^2 and 68.2. Thus this steel would be considered suitable for the crank ring.

Application to a Full-Scale Laboratory Test

The basis for the fatigue crack geometry used in this analysis was the crack that occurred in a controllable pitch propeller crank ring assembly. The crank ring was made of 4150H steel which has a quoted yield strength of almost 759 MPa. This yield strength is greater than the 523 MPa value required to avoid plastic collapse. The corresponding required value of J_I-R (Δc = 1.27 mm) is calculated according to the method described in the previous section to be

$$J_I\text{-}R\ (\Delta c = 1.27\ \text{mm}) = 175\ \text{kJ/m}^2$$

The toughness data generated for the 4150H steel show that it is not in the upper shelf at room temperature, which was the temperature for the full-scale laboratory test. The Charpy energy at room temperature is quoted as ranging from 8 to 15 J. The value of K_I at fracture varied from 60 to 123 MPa$\sqrt{\text{m}}$. Two of the tests provided valid K_{Ic} values, 60 and 82 MPa$\sqrt{\text{m}}$, in accordance with ASTM Test for Plane-Strain Fracture Toughness of Metallic Materials (E 399); the other tests provided invalid values because either too much plasticity or crack growth occurred—it was not determined which—or the specimen dimensions did not satisfy the plane strain requirements. In any case, the criterion proposed in the analysis of this paper, that the fracture mode be ductile tearing, was violated.

The range of critical J_I values converted from the K_c values is 16 to 68 kJ/m^2, all of which are considerably lower than the required value of J_I-R = 175 kJ/m^2. Therefore the methodology developed in this investigation predicts that the 4150H is not a suitable crank ring material. The fact that the full-scale laboratory tested 4150H crack ring did not experience unstable fracture shows that the analysis is conservative. The degree of conservatism on the required J_I-R value in this case is greater than a factor of two.

This degree of conservatism arises because of the many assumptions made in the analysis. Loads calculated using finite element analysis for an idealized crack geometry are undoubtedly too high. Since J_I is proportional to the load squared, a decrease in load will cause a substantial decrease in required J_I.

Summary and Conclusions

The objective of the investigation reported in this paper was to set material toughness requirements to avoid loss of a ship propeller blade from a controllable pitch propeller crank ring that has a fatigue crack. Loss of the propeller was assumed to be prevented if fracture instability could not occur before the fatigue crack grew to a size beyond which crack growth would proceed radially through the crank ring flange and not around its circumference. Choice of this crack size and geometry was based on a full-scale crank ring laboratory test in which a large fatigue crack occurred.

Kiyoshi Tanaka,[1] *Mitsuo Sato,*[1] *Tadashi Ishikawa,*[1] *and Hironori Takashima*[1]

A New Wide Plate Arrest Test (SCA Test) on Weld Joints of Steels for Low Temperature Application

REFERENCE: Tanaka, K., Sato, M., Ishikawa, T., and Takashima, H., **"A New Wide Plate Arrest Test (SCA Test) on Weld Joints of Steels for Low Temperature Application,"** *Fracture Mechanics: Seventeenth Volume, ASTM STP 905,* J. H. Underwood, R. Chait, C. W. Smith, D. P. Wilhem, W. A. Andrews, and J. C. Newman, Eds., American Society for Testing and Materials, Philadelphia, 1986, pp. 22–40.

ABSTRACT: As a realistic and practical criterion for brittle fracture arrest in low temperature storage tanks, the present authors propose the short crack arrest (SCA) concept. This aims at arresting brittle fracture before propagation to a catastrophe. In order to investigate the capability of steel materials from the viewpoint of this concept, two types of wide plate tests, the SCA tests on weld heat-affected zone (HAZ) and base metal, were developed. These two test methods were designed to simulate the run-and-arrest phenomenon of brittle fracture in actual storage tanks.

The two types of tests, together with the compact crack arrest (CCA) test, were applied to base metals and welded joints of a wide range of steels for low temperature application. Efforts were made to clarify the effects of base metal chemical compositions and conditions for welding on the weld HAZ characteristics.

The results revealed that addition of about 3% nickel to the base metal significantly improved the weld HAZ arrest toughness. It was also shown that the SCA capability of steels could be predicted from the K_a values obtained by the CCA test.

KEY WORDS: low temperature, fracture toughness, crack arrest, brittle fracture propagation, welded joint, weld heat-affected zone, storage tank

In Japan, in 1968, there was a catastrophic fracture accident in a spherical tank with a diameter of 16.2 m during hydrotest [*1*]. The tank, made of an 800 MPa high tensile strength steel, collapsed after fast fracture in three fourths of the circumference. The fracture surface consisted of shear fracture in the base metal and brittle fracture along the heat-affected zone (HAZ) of a

[1]Senior Research Engineer, Research Engineer, Research Engineer, and Chief Research Engineer, respectively, R&D Laboratories II, Nippon Steel Corporation, Sagamihara, Japan.

vertical weld joint with a length of 6.2 m. It is believed that a weld hydrogen crack, with a length of 60 mm and a depth of 13 mm, at the toe of the weld was the cause of the brittle fracture initiation. The brittle fracture propagated in the HAZ along the weld joint and penetrated the entire weld length. Although the brittle fracture was arrested by the neighboring base plates, ductile fracture took place and continued to propagate in an unstable manner until the tank collapsed.

In 1977 a disaster occurred in the Middle East in which a gas liquefaction and storage plant completely collapsed [*2*]. It is believed that the cause of the disaster was brittle fracture in a liquefied propane gas (LPG) storage tank. In this case, the brittle fracture propagated in the base plates. After this accident, Cuperus [*3*] questioned the safety of storage tanks for liquefied light hydrocarbons and proposed the double integrity principle.

From these experiences and other examples of failure accidents, the present authors recognized the necessity of a comprehensive investigation of the performance of steel plates and their weld joints in terms of brittle fracture initiation and arrest:

1. Stringent evaluation of fracture initiation toughness of weld HAZs as well as base metals.
2. Realistic evaluation of fracture arrest toughness by means of suitable simulation methods of the brittle fracture run-arrest phenomenon and a convenient test method for arrest toughness.

This report describes the results of investigations made by the authors for establishing these evaluation methods and for collecting a body of data for material selection.

Fracture Initiation Characteristics of Welds

It is well known that welded joints are heterogeneous in terms of the microstructure and show a large scatter of brittle fracture toughness. Figure 1 shows examples of zones in a welded joint where initiation points of brittle fractures were found after microscopic investigation on specimens which presented low initiation toughness. The reasons for, and the amount of, the embrittlement for each of the zones are different depending on the chemical compositions of the base metals, the weld thermal cycles, and the hot strain during welding. It is therefore quite important to investigate the fracture initiation toughness of the welded joint carefully by using many specimens having their crack tips at various points in the welded joint. It is also important to apply two kinds of specimen geometries having different notching and cracking directions (i.e., side-cracked and face-cracked specimens). The notch tip in the former specimen can sample a wide range of microstructures but without a long contact with the same microstructure, whereas that in the latter can sample a specific microstructure with a long contact. This difference pro-

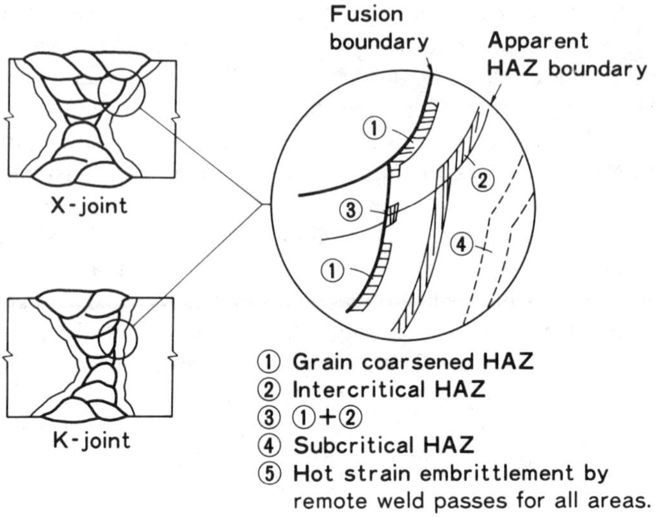

FIG. 1—*Embrittled regions in weld joints.*

duces a significant contrast in the results (i.e., a wider scatter and a lower minimum value of fracture toughness for the face-cracked specimens).

In order to simplify the investigation, the present authors proposed the application of the fatigue crack tip opening displacement (CTOD) test method [*4,5*], the basis of which comes from the fatigue wide plate test by Nibbering et al [*6*] and the fatigue fracture toughness test by Yokobori et al [*7*]. The characteristics of this method, when compared with the ordinary fracture toughness test method, are as follows:

1. Specimens are cyclically loaded at the test temperature, and the fatigue crack develops scanning the various microstructures in the crack line.
2. When the toughness, or the critical CTOD, δ_c, of a microstructure is lower than the working CTOD, or applied CTOD, brittle fracture takes place.
3. With several specimens, the minimum δ_c value of the welded joint tested can be determined. The reliability of this value is high because microstructures in the area where the fatigue crack passed are all tested by each of the cyclic loads.

When the crack line is misaligned a little from the weld line (Fig. 2) the reliability of the test results increases further. It was found, however, that a small cyclic range of the stress intensity factor, ΔK, preferably smaller than 30 MPa$\sqrt{m}$, has to be maintained during the whole period of the low temperature fatigue loading in order to achieve results consistent with the monotonic CTOD test [*5*].

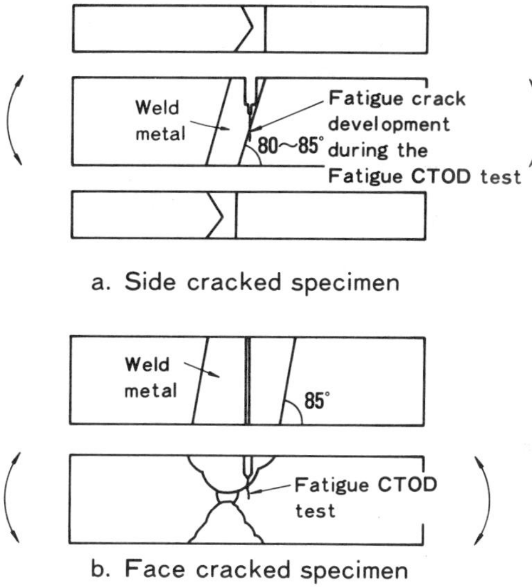

FIG. 2—*Fatigue CTOD test on weld joints.*

Table 1 lists examples of the monotonic and fatigue CTOD test results for weld HAZs for three kinds of low temperature steels. Numerous monotonic CTOD tests were conducted on each of the weld joints. The results are shown in the form of the coefficients in the Weibull distribution equations; these fit well the test results. For comparison, δ_c values for 2% cumulative probability were taken as the minimum δ_c values of the weld joints. The fatigue CTOD test showed results conforming to those for the monotonic CTOD test. This fact indicates the possibility of economization of fracture toughness tests by means of the new test method without losing, or instead with increasing, reliability. Fracture initiation points in the fatigue CTOD test were also the grain-coarsened HAZ close to the fusion line. Moreover, it is surprising to find that the low-carbon grain-refined steels showed δ_c values of 0.04 mm or lower when some unfavorable combinations of metallurgical and testing conditions were met, since these steels have showed satisfactory performance in all conventional tests and notched wide plate tests as well as in actual application to LPG storage tanks for more than 20 years.

Significance of Local Brittle Zones

In small-scale tests for fracture initiation toughness, such as the CTOD and K_{Ic} tests, the first incident which leaves a discontinuity in the load versus displacement record is taken as the critical point for investigation. With these tests, however, it is difficult to clarify the significance of the incident with regard to the total safety of the structures. In order to assess the low δ_c values

TABLE 1—*Monotonic and fatigue CTOD tests on weld HAZs of low temperature steels.*

Steel Material Tested	Plate Thickness, mm	Welding Condition: Method and Heat Input	Welding Condition: Repair Welding	Specimen	Test Temperature, °C	Monotonically Loaded CTOD Test: No. of Specimens	Coefficient in Weibull Distribution: α[a]	Coefficient in Weibull Distribution: β[a]	δ_c of 2% Probability, mm[b]	δ_c by Fatigue CTOD Test, mm
Low-carbon fine-grain low-temperature steel[c]	25	SMAW (3G) 40kJ/cm	3 times	face cracked	−50	32	0.4464	1.636	0.035	0.040
Same as above but Ti-treated[c]	25	same	no	side cracked	−50	86	2.485	1.833	0.290	. . .
			no	face cracked	−50	76	1.719	1.227	0.071	. . .
			3 times	face cracked	−50	122	1.930	0.852	0.019	0.023
3½% Ni[c]	30	same	3 times	face cracked	−50	21	1.762	2.861	0.45	>0.7

[a] $P(\delta_c) = 1 - \exp[-(\delta_c/\alpha)^\beta]$; $P(\delta_c)$ = cumulative probability.
[b] Critical CTOD value, δ_c, for 2% cumulative probability calculated by α and β.
[c] Chemical compositions (wt%):
Low-carbon fine-grain steel: 0.10C, 1.41Mn, 0.22Ni, 0.0068N, Ceq = C + Mn/6 + Si/24 + Ni/40 = 0.35.
Low-carbon fine-grain steel (Ti treated): 0.07C, 1.07Mn, 0.24Ni, 0.011Ti, 0.0063N, Ceq = 0.26.
3½% Ni steel: 0.10C, 0.55Mn, 3.55Ni, 0.0047N, Ceq = 0.29.

found in the monotonic or fatigue CTOD test, therefore, surface-cracked wide plate tests were carried out on one of the welded joints that showed low δ_c values. Two specimens were prepared from the welded joint described in the fourth line of Table 1. A full breadth surface crack was produced by cyclic bending so that the fatigue crack tip lay in the same brittle zone as in the CTOD specimens. One of the two specimens, when tested at $-50°C$ by a 20 MN capacity tension test rig, showed brittle fracture with the stress lower than the yield stresses of the base and weld metals; the other specimen showed brittle fracture with a high stress. Analysis on the crack tip location of the latter specimen showed that the initial fatigue crack was too long and had its tip in the grain-refined HAZ. The fracture surface of the former specimen and its data are shown in Fig. 3. There was no indication of brittle fracture arrest. Subsequent study revealed that the fracture initiation point of this specimen was the same grain-coarsened region as in the CTOD specimens with the low δ_c result. The linear elastic fracture mechanics (LEFM) calculation for this wide plate specimen led to a δ_c value of 0.032 mm, which is almost consistent with the values from the monotonic and fatigue CTOD tests.

Agreement of δ_c values from CTOD tests and the wide plate test suggests that the results from the small-scale fracture toughness tests can be applied to prediction of the fracture initiation stresses of fatigue-cracked wide plate test specimens that simulate actual structures with fatigue crack defects. However, an infinite surface crack such as that used in the wide plate test is not expected in actual structures. It is important to investigate whether or not brittle fracture can propagate in the unnotched weld HAZs.

Development of the Short Crack Arrest (SCA) Test

The ESSO test has been used widely for the evaluation of brittle fracture arrest capability of steel materials including base plates and welded joints. The present authors, however, had quite often unexpected results from tests on welded joints in the as-welded condition. Figure 4 illustrates examples of fracture paths of as-welded joints. It is considered at the present time that the compressive weld residual stress, in the direction transverse to the weld line, at the specimen edges must be the cause of this deviation, since it is not observed in welded and center notched wide plate tests. This compressive weld residual stress at the edge is considerably high (Fig. 4). Since this situation is not expected in actual structures, modification of the test method is necessary.

The basic idea for the approach taken here is that the brittle fracture has to be initiated at the center of the specimen width to eliminate the effect of the compressive residual stress. The present authors recognize the importance of the concept proposed by an ASTM committee [*8*] in 1960. In this concept the arrest of a brittle fracture becomes possible, when it propagates just beyond the initial surface crack, due to the higher toughness associated with plane

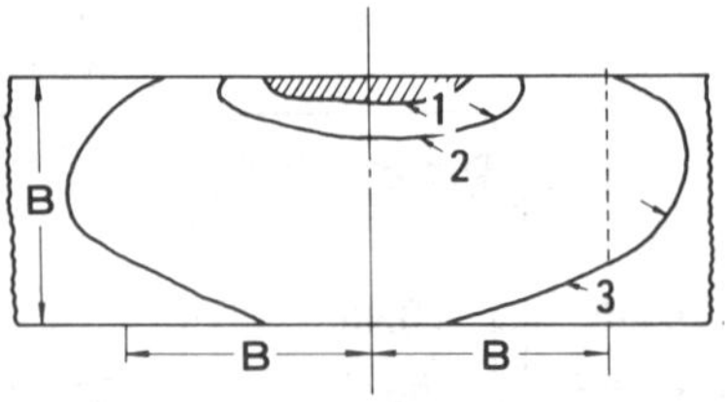

a. Sketch of fracture origin (1), slow growth to (2), fast fracture to (3) with arrest due to toughness associated with plane-stress situation, [8]

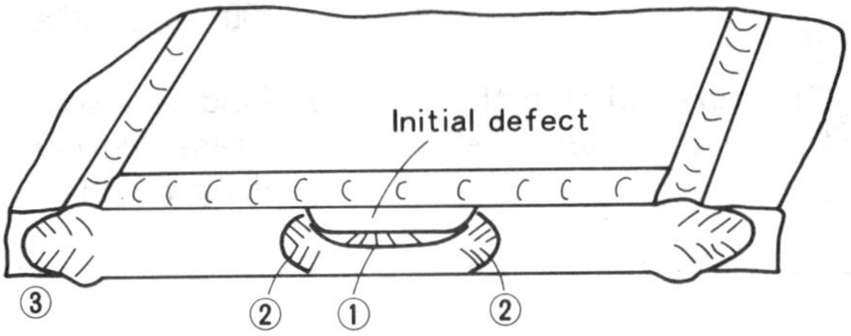

① Immediate arrest (pop-in)
② Short Crack Arrest (SCA)
③ Long Crack Arrest (LCA)

b. Chances of brittle fracture arrest for welded joint

FIG. 5—*Arrest of brittle fracture initiating in a weld joint.*

formation of a through-the-thickness crack with the same length as that of the surface crack (Fig. 6*c*). In the test a brittle fracture is initiated at the notch tip in the starter by loading it with a small jack having the capacity of 100 kN. The brittle crack travels through the brittle beads and penetrates into the specimen. Test results are divided into "Go" and "Arrest". Unless the brittle crack propagated to a total crack length more than the sum of the surface notch length and twice the plate thickness, the result is judged as "Arrest".

HAZ-SCA Test Results

The new test method has been applied to a wide range of steel materials for low-temperature and cryogenic application. Several specimens were prepared for each series of welded joints, and the test temperatures and the applied stresses were selected so as to obtain a critical temperature versus stress transition curve for each series. Examples of the HAZ-SCA test results and the transition curves are shown in Fig. 7. Fracture appearances of the test data indicated by (*a*), (*b*), and (*c*) in Fig. 7 are displayed in Fig. 8. Photograph (*a*)

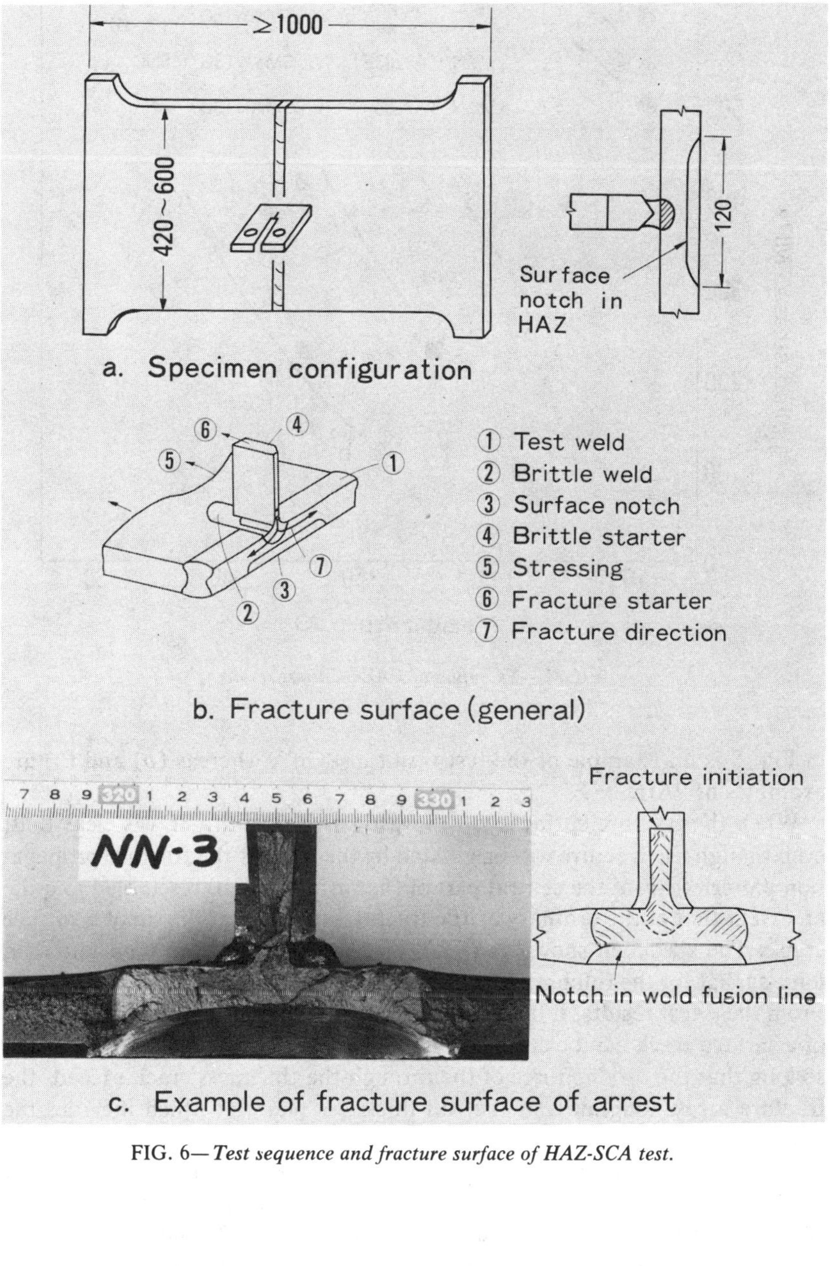

FIG. 6—*Test sequence and fracture surface of HAZ-SCA test.*

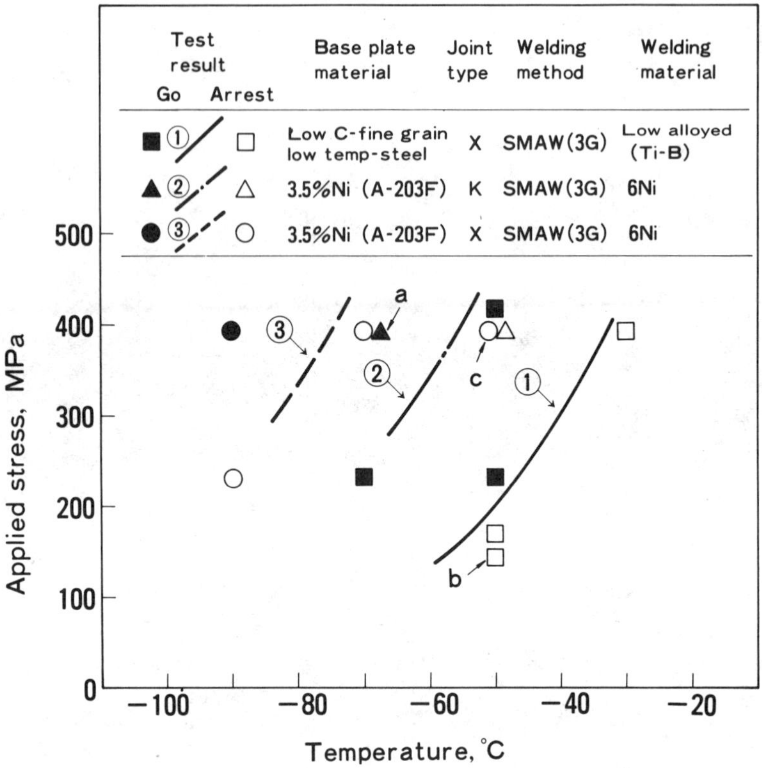

FIG. 7—*Examples of HAZ-SCA test results.*

in Fig. 8 is an example of the test result of "Go", whereas (*b*) and (*c*) are examples of "Arrest".

When the fracture surfaces were investigated carefully, it was clear that, even though the fracture was originated by the starter, the fracture propagation pattern (except the central part of the surface notch) resembled to quite an extent that of the ordinary brittle fracture initiating at the tip of a surface crack. The specimen shown as (*b*) in Fig. 8 was taken from the same weld joint as that for the full breadth surface crack wide plate test shown in Fig. 3. From these test results, it became clear that brittle fracture initiating at a tip of a surface crack can be arrested when the length of the surface crack is not so long that the driving force of the through-the-thickness crack exceeds the fracture arrest toughness of the full thickness material which includes the plane stress regions near the plate surfaces.

In Fig. 7 the significant effect of the weld joint type is observed. The K-joint of 3½% nickel steel shown as No. 2 has a critical temperature 20°C higher than that for the X-joint of the same steel shown as No. 3. The reason for this is that the weld HAZ of the X-joint is inclined against a line perpendicular to

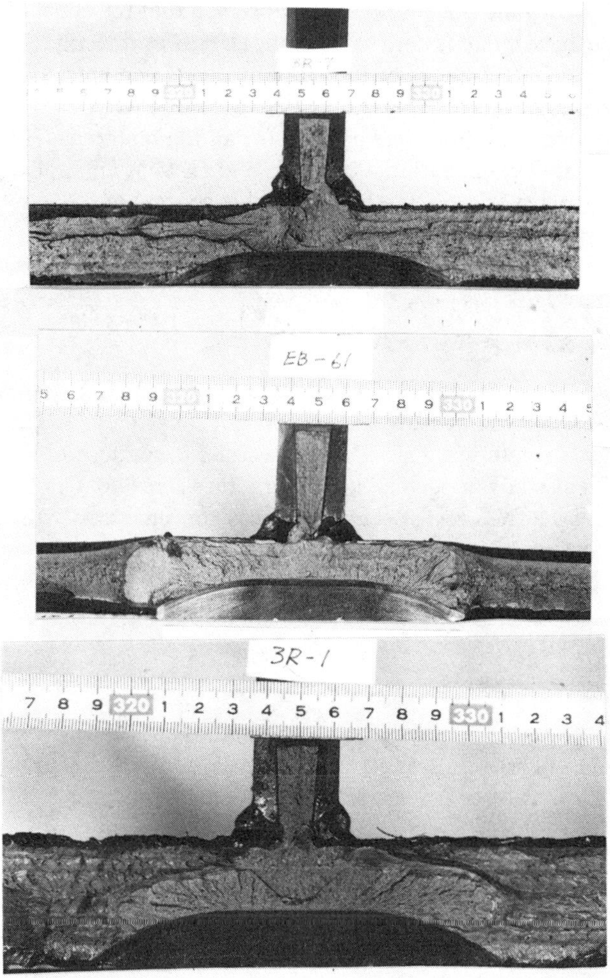

a. (Upper). 3½% Ni steel, K-joint, 6Ni-WM, tested at −70°C, 392MPa.
b. (Middle). LowC-fine grain low temperature steel, X-joint, Ti-B-WM, tested at −50°C, 147MPa.
c. (Lower) 3½%Ni steel, X-joint, 6Ni-WM, tested at −50°C, 392MPa.

FIG. 8—*Examples of fracture appearances of HAZ-SCA tests.*

the plate surface, whereas brittle fracture tends to propagate in a plane perpendicular to the plate surface. It is also observed that the effect of the nickel content in the base metal is noticeable. This effect was investigated further by testing a series of steel plates having various nickel contents. Figure 9 illustrates the results of this investigation where the same K-type joint shape and the same welding condition were applied for all the materials. It is seen that the brittle fracture propagation along the weld HAZ at an applied stress below yield took place at −50°C when the nickel content of the base metal was 2½% or less.

Compact Crack Arrest (CCA) Test

Since the SCA test requires a large-scale testing machine and a sophisticated apparatus for the fracture starter, it is not convenient for wider applications at many laboratories. The CCA test is easier to conduct and requires less material. It is also advantageous that the fracture arrest toughness value, K_a, is obtained by the CCA test. Because of these conveniences of the CCA test, the present authors also carried out the tests on the same welded joints as for the SCA tests.

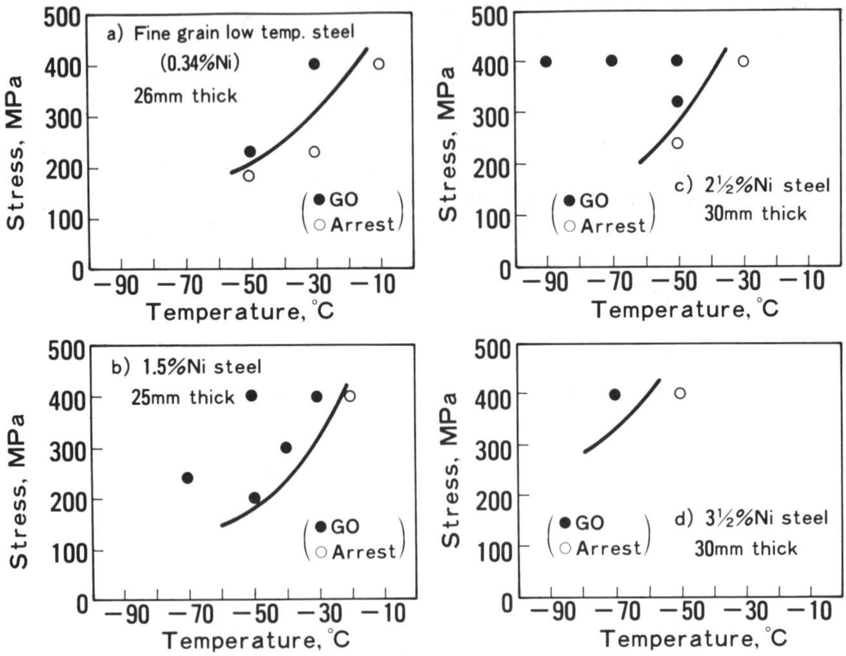

FIG. 9—*HAZ-SCA test results for weldments of nickel alloyed steel plates (SMAW, 3G, K-joint, 35 to 40 kJ/cm).*

During the investigation by means of the CCA test the authors made some modifications in the test and the calculation methods. These were as follows:

1. *Side Groove*—Specimens without side grooves were employed because fracture toughness values of the full thickness were of interest.

2. *Fracture Initiation Beads*—In order to control the fracture initiation stress intensity factor, K_Q, the following welding materials were applied:

- For the test at a temperature down to −30°C: Murex-Hardex.
- For a temperature between −30 and −100°C: Hardfacing electrode for $H_v = 300$.
- For a temperature between −100 and −150°C: electrode for 800 MPa class strength.
- For a temperature between −150 and −196°C: 6% nickel electrode.

3. *Application of Pre-Loading*—In order to avoid a premature fracture at a too low K_Q value the pre-loading method at a warm temperature was conducted. The same K value as an aimed K_Q value was applied at a warm temperature that would not cause fracture initiation.

4. *Calculation of* K_a *Values*—Even though the tests were conducted with careful design and preparation of specimens, many specimens showed higher K_Q and K_a values which exceeded the limitations set by Ripling et al [*10*] as a function of the specimen size and the materials' yield strength. In order to utilize as many data as possible, the following modifications in the K_a calculation method were made.

- Calculate the K_L value corresponding to the limit load for the specimen; this is obtained by means of the equation by Merkle et al [*11*].
- When the experimental K_Q value is larger than K_L, substitute the following equation for the original K_a formula which uses displacement for calculation of K_a. The new formula was obtained by Crosley et al [*12*] and applies the relationship between initiation K values, crack jump length, and K_a:

$$K_a = K_L\,[1 - 0.92\,(\Delta a/W) + 0.33\,(\Delta a/W)^2]$$

where Δa is the crack jump length and W is the specimen width.

The data obtained in this way were compared with the wide plate test results and were used for the analysis reported in the following section.

Discussion

Effect of Nickel Content of Base Plate on HAZ-SCA Test Results

As was observed in Figs. 7 and 9 the increase of the nickel content of the base metal improves the arrest capability of the weld HAZs. In order to show more clearly the effect of nickel content Fig. 10 was drawn. In this figure, the temperature at which the result of "Arrest" is obtained in the HAZ-SCA test under the applied stress of 400 MPa is taken as the critical temperature (i.e., SCA temperature) and is shown on the ordinate. This SCA temperature becomes lower when the nickel content is raised, K-joint is changed to X-joint, or welding material is changed from the ferritic type to the austenitic sort which is free from brittle fracture.

It was also noticeable that the SCA temperatures were very high compared with the ordinary transition temperatures, such as the 48 J (35 ft-lb) transition temperature in the Charpy test, on the welded joints used. When one has to select a material for an LPG storage tank operated at −45°C, and when one takes a combination of such severe conditions as 400 MPa of applied stress and a K-joint configuration caused by some unexpected coincidence of various consequences, then one may have to choose the 3½% nickel steel.

Relationship Between HAZ-SCA and Charpy Test Results

Although it is well known that the Charpy test is not a fracture toughness test, it has been used as an industrial small-scale test for a long time. It may therefore be useful to correlate Charpy test results with HAZ-SCA test results. For the correlation, *K* values applied to the HAZ-SCA tests were calculated by means of Irwin's tangent formula for a center cracked specimen together

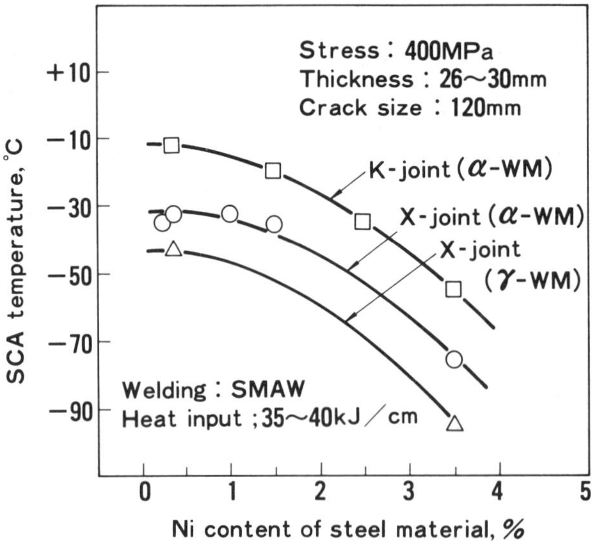

FIG. 10—*Effects of nickel content in steel, joint types, and types of weld metal on SCA temperature of weld HAZ.*

with the half crack length of 60 mm (i.e., the half of the length of the surface notch in the tests). Figure 11 shows the relationship between the HAZ-SCA and Charpy test results. The abscissa is the difference between the test temperature for the HAZ-SCA test and the fracture appearance transition temperature (FATT) in the Charpy test. In some areas in this figure, the results of "Go" and "Arrest" are mixed. The line shown in the figure, however, separates the "Arrest" data from the "Go" data. This line therefore may show a lower bound relationship between the K_a value and the Charpy test results for weld HAZs.

Modification of the SCA Test

Because the starter plate for the HAZ-SCA test can inject a brittle fracture into any part of a steel plate, it can be applied for other objectives. The base metal SCA (BM-SCA) test shown in Fig. 12 is one example of such a modified specimen. With this test method, fracture behavior of a welded plate having a brittle fracture initiation at the weld joint can be studied. The main points of this test method are (1) that the weld residual stress is incorporated in the same way as in the actual structures, and (2) that the method uses a center cracked and hence symmetrical specimen which eliminates the effect of bending moment found in single edge cracked specimens used in existing wide plate arrest tests.

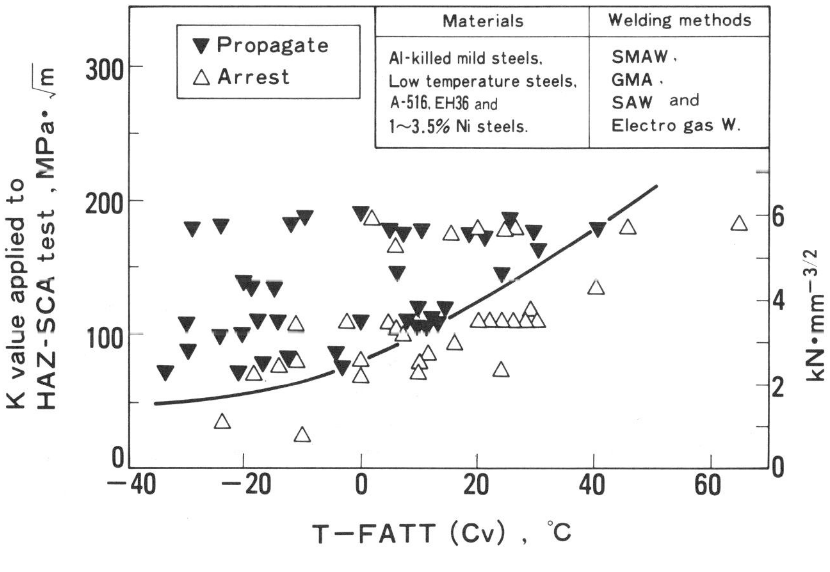

FIG. 11—*Correlation between HAZ-SCA and V-Charpy test results for low temperature steels with thickness range of 25 to 30 mm.*

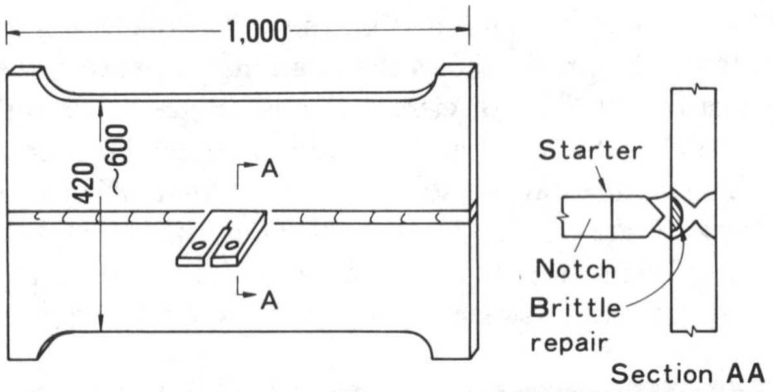

FIG. 12—*Base metal SCA (BM-SCA) test specimen.*

One of the present authors together with co-authors reported results of the BM-SCA test on low-temperature steels [*13*] and those on 9% nickel steels as well as the *K* analysis method by LEFM [*14*]. The data obtained by the BM-SCA test are also included in the analysis shown in the next section.

Correlation of CCA Test to SCA and Wide Plate Tests

The compact crack arrest test and the wide plate test results were brought together and analyzed by means of LEFM. Besides the two types of SCA tests, the duplex ESSO test has been conducted on some materials. The data of the duplex ESSO test were also included in the analysis. The applied *K* values are calculated without considering the dynamic situation of the run-and-arrest phenomenon of the brittle fracture. The results are summarized in Fig. 13. The ordinate is the applied *K* value in the wide plate arrest tests; the abscissa is the K_a value obtained by the CCA tests. Because all the wide plate tests are the "Go" and "Arrest" type of test, it can be said that a good correlation exists between CCA and other tests when solid and blank marks are separated by a line. Figure 13 seems to show a considerably good relationship between the small-scale and large-scale tests. This is very encouraging, since all the calculations were made by static methods. We may conclude that the static calculation can be applied to the brittle fracture arrest phenomenon when the crack length is not large. The designed half crack lengths were 60 mm in the HAZ-SCA and 150 mm in the duplex ESSO test. The half crack sizes observed in the BM-SCA tests with the results of "Arrest" were approximately 50 to 65 mm. Therefore the above conclusion may be valid up to a half crack length of 150 mm.

Figure 13 suggests that the CCA test when applied with the modified calculation is applicable for the evaluation of fracture arrest toughness of various steel plates and their weld joints.

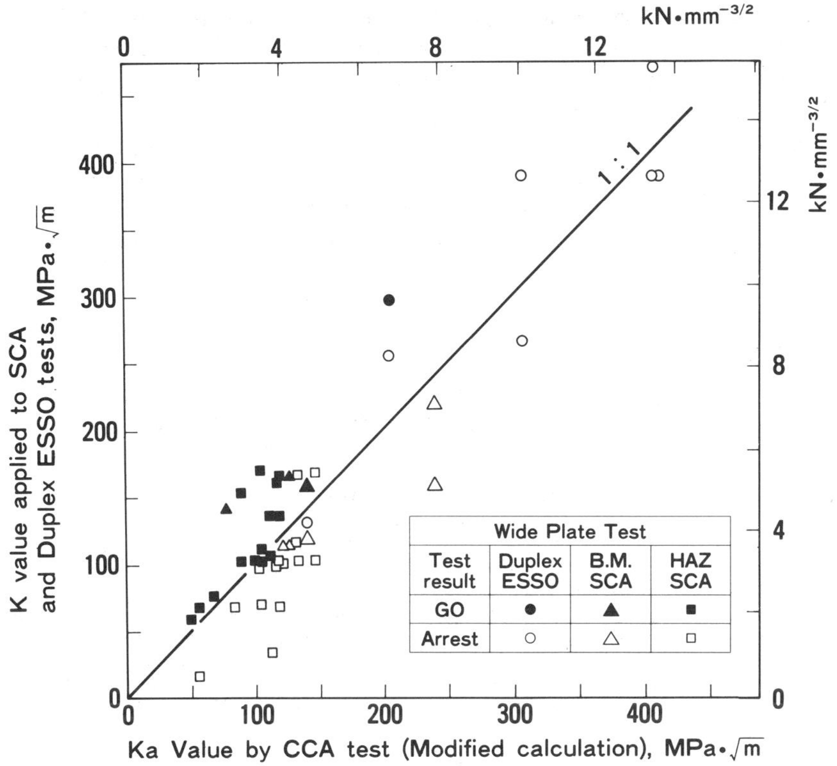

FIG. 13—*Relationship between CCA and various wide plate arrest test results.*

Conclusions

1. It was found that welded joints of steel materials which have been applied to actual important applications could show considerably low fracture initiation toughness when investigated with numerous CTOD or K_{Ic} specimens or examined by the fatigue CTOD test.
2. The material that showed a low fracture initiation toughness in the small-scale test will show also a low stress brittle fracture in the notched wide plate test when a crack tip is located at the same microstructure found to be brittle in the small-scale tests.
3. The SCA concept, which aims at the immediate arrest of brittle fracture when it propagates either beyond the length of the pre-existing surface defect or into a tougher material, seems to be the most realistic criterion when application of an arrest concept is necessary.
4. The HAZ-SCA test is suitable for the investigation on the practical arrest capability of steel plates and their weld joints.

5. The nickel content of the base plate, the joint geometry of the weld, and the type of the welding material applied affect the HAZ-SCA test results. When severe conditions are expected in service a considerably high nickel content of the base plate is required.

6. From K_a values obtained by the CCA test, results in the HAZ-SCA or BM-SCA test can be estimated. The CCA test seems to be a promising test method for fracture arrest toughness evaluation.

7. There was comparatively good correlation between the results of the HAZ-SCA and Charpy tests on the same weld joints.

References

[*1*] High Pressure Gas Safety Association, "Report of Investigation on Accidents Related with Fracture of Spherical Tanks," *High Pressure*, Vol. 7, No. 5, 1969, pp. 1-24 (in Japanese).

[*2*] Authors do not have direct information; see for general information: "Overseas Loss Information," No. 31, June 1977, Taisho Marine and Fire Insurance Co., Ltd. (in Japanese).

[*3*] Cuperus, N. J., "Cryogenic Storage Facilities for LNG and NGL," in *Proceedings*, 10th World Petroleum Congress, Panel Discussion 17, Paper 3, Heyden & Son, London, 1979.

[*4*] Tanaka, K., Sato, M., and Ishikawa, T., "Fatigue COD and Short Crack Arrest Tests," in *Proceedings*, International Conference on Fracture Toughness Testing: Method, Interpretation, and Application, Paper 18, London, March 1982.

[*5*] Ishikawa, T. and Tanaka, K., "Development of Fatigue CTOD Test for Investigation on Brittle Regions in Welded Joints," in *Proceedings*, Sixth International Conference on Fracture, New Delhi, Dec. 1984.

[*6*] Nibbering, J. J. W. and Lalleman, A. W., "Strength of EG Welded 34 mm Plate of Nb St. 52," SSL Report 143, Dec. 1969 (in Dutch) or IIW Doc. X-593-70, May 1970.

[*7*] Yokobori, T. and Aizawa, T., "A Proposal for the Concept of Fatigue Fracture Toughness," *Journal of the Japanese Society for Strength and Fracture of Materials*, Vol. 5, No. 2, 1970, pp. 54-58 (in Japanese).

[*8*] ASTM Committee on Fracture Testing of High-Strength Sheet Materials, *ASTM Bulletin*, No. 243, Jan. 1960, pp. 29-40.

[*9*] Irwin, G. R., *Welding Journal (Research Supplement)*, Vol. 41, Nov. 1962, pp. 519s-528s.

[*10*] Ripling, E. J. and Crosley, P. B., *Welding Journal (Research Supplement)*, Vol. 61, No. 3, 1982, pp. 65-74.

[*11*] Merkle, J. G. and Corten, H. T., *Journal of Pressure Vessel Technology, Transactions of ASME*, Vol. 96, No. 4, 1974, pp. 286-298.

[*12*] Crosley, P. B. and Ripling, E. J. in *Crack Arrest Methodology and Applications, ASTM STP 711*, American Society for Testing and Materials, Philadelphia, 1980, pp. 211-227.

[*13*] Tanaka, K., Ohno, Y., and Yamada, N., "Selection of Steel Materials for LPG Storage Tanks," in *Proceedings*, International Conference on Transport and Storage of LPG & LNG, Bruges, Vol. 2, 1984, pp. 9-16.

[*14*] Ishikura, N., Kohno, T., Maeda, H., Arimochi, K., and Tanaka, K., "Recent Development in Research on 9% Ni Steel and Its Weldments," in *Proceedings*, International Conference on Transport and Storage of LPG & LNG, Bruges, Vol. 1, 1984, pp. 1-11.

C. Y. Yang[1] *and W. H. Bamford*[1]

Variable Flaw Shape Analysis for a Reactor Vessel under Pressurized Thermal Shock Loading

REFERENCE: Yang, C. Y. and Bamford, W. H., **"Variable Flaw Shape Analysis for a Reactor Vessel under Pressurized Thermal Shock Loading,"** *Fracture Mechanics: Seventeenth Volume, ASTM STP 905*, J. H. Underwood, R. Chait, C. W. Smith, D. P. Wilhem, W. A. Andrews, and J. C. Newman, Eds., American Society for Testing and Materials, Philadelphia, 1986, pp. 41–58.

ABSTRACT: A study has been conducted to characterize the response of semi-elliptic surface flaws to thermal shock conditions which can result from safety injection actuation in nuclear reactor vessels. A methodology was developed to predict the behavior of a flaw during sample pressurized thermal shock events. The effects of a number of key variables on the flaw propagation were studied, including (1) fracture toughness of the material and its gradient through the thickness, (2) irradiation effects, (3) effects of warm prestressing, and (4) effects of the stainless steel cladding.

The results of these studies show that under thermal shock loading conditions the flaw always tends to elongate along the vessel inside surface from the initial aspect ratio. However, the flaw shape always remains finite rather than becoming continuously long, as has often been assumed in earlier analyses. The final shape and size of the flaws were found to be rather strongly dependent on the effects of warm prestressing and the distribution of neutron flux.

The improved methodology results in a more accurate and more realistic treatment of flaw shape changes during thermal shock events and provides the potential for quantifying additional margins for reactor vessel integrity analyses.

KEY WORDS: flaw shape, stress intensity factor, fracture toughness, crack initiation, crack arrest, thermal shock loading, warm prestressing, virtual crack extension (VCE) method

The most serious challenge to the integrity of a reactor pressure vessel comes from thermal shock loadings that can result when the safety injection system injects cold water into the normally hot (290°C) reactor vessel. This injection system is designed to cool the reactor core to prevent its overheating

[1]Westinghouse Electric Corporation, Nuclear Energy Systems, Pittsburgh, PA 15230.

in the event of a loss of water pressure in the system. This thermal shock can produce high stresses as well as lowering the temperature and therefore the material toughness of the vessel itself.

During the thermal shock, a flaw, if it exists, will be driven to elongate, since the highest stresses and lowest temperature occur near the vessel surface. The elongation of the flaw is resisted by the combined effect of the decreasing stress intensity factor at the surface as the crack lengthens and better material properties due to lower constraint and often more effective heat treatment of the plate or forging in that area. Elongation is also resisted by the presence of the stainless steel cladding.

The behavior of postulated flaws in the reactor vessel during a thermal shock has long been a subject of active investigation. Early procedures involved the use of simplified flaw shapes and approximations for calculating stress intensity factors. Analytical procedures have continually advanced to better deal with the complex interactions which occur between a flaw and the thermal shock loading. Several approaches are presently recommended for this assessment.

Section XI of the ASME Boiler and Pressure Vessel Code [*1*] provides a detailed procedure for fracture assessment of nuclear vessels. It is suggested that the flaw shape be assumed to retain its original aspect ratio after crack propagation has started. The Section XI guidelines state that crack initiation occurs when the maximum Mode I stress intensity factor, $(K_I)_{max}$, along the crack front exceeds the "K_{Ic}" value, which is the lower bound of the static crack initiation toughness, with a reference toughness curve for reactor vessel steels provided. Moreover, the Section XI guidelines also state that crack arrest will occur if $(K_I)_{max}$ falls below the "K_{Ia}" value, which is the lower bound of the dynamic crack initiation and arrest toughness, with another reference toughness curve also provided.

Another treatment of the flaw shape is the NRC recommendation [*13*] in which the initial flaw shape is assumed to be semi-elliptical with an aspect ratio of 6:1; the flaw is then assumed to become continuous (i.e., infinitely long) after crack initiation has started. Apparently, both the ASME Section XI guidelines and the NRC assumption are not realistic, because the final flaw shape in the reactor vessel beltline region is dependent on the type of load, the material properties, and other factors. The NRC assumption will normally lead to much more conservative results than those which are based on the ASME guidelines. A more realistic flaw shape would fall between these two bounds. This has been demonstrated in a series of thermal shock tests conducted at Oak Ridge National Laboratory (ORNL) on scale-model vessels, in which a thick-walled cylinder with a 19 mm (0.75 in.) deep semi-circular axial flaw on the inside surface was exposed to a severe thermal shock load. The results of the experiment showed that the crack propagated only axially and extended only to a finite length [*2*]. It should be mentioned that this experimental cylinder was not clad, and so the results do not reflect the complete set of interactions involved.

Many experimental programs were conducted in laboratories such as Framatome [14], Knolls Atomic Power Laboratory [20], and Oak Ridge Laboratories [21-23] to study the effect of stainless steel cladding on failure load of carbon steel specimens. All test results showed that the unirradiated cladding did significantly improve the load carrying capacity of the specimens.

The purpose of this paper is to study the flaw behavior as a function of the toughness related parameters while the flaw shape is allowed to vary during the crack propagation.

Method of Flaw Shape Change Analysis

In evaluating fracture behavior of a structural component both the material properties and the crack driving forces should be considered. Assume that very little plastic deformation is involved in the fracture process. Based on linear elastic fracture mechanics (LEFM), only the fracture toughness and the stress intensity need to be considered for the analysis. Methods of determination of material properties and the stress intensity factors are briefly described below.

Determination of Material Properties

Based on LEFM, K_{Ic} and K_{Ia} are the material properties that can be used to determine, respectively, whether crack initiation or arrest will or will not occur. For conservatism, the K_{Ic} and K_{Ia} data to be used in the analysis are taken from the lower bounds of all tests.

Since the fracture toughness varies with temperature, for convenience, a temperature scale is defined relative to the reference nil-ductility transition temperature, RT_{NDT}. The RT_{NDT} is a nonphysical constant related to the brittle-to-ductile fracture transition temperature.

The reactor vessel steels (A-533 and A-508) and the weld metal are susceptible to fast neutron fluence or irradiation damage. As a result of the irradiation damage the fracture toughness is decreased with the time of exposure. The reduction in toughness due to irradiation damage is enhanced with the increasing content of copper and nickel. The degree of irradiation damage can be assessed by measuring or determining the shift to higher temperatures of the reference transition temperature, ΔRT_{NDT}.

ΔRT_{NDT} is generally determined by so-called "trend curves", which correlate the fluence and copper content to ΔRT_{NDT}. Based on the initial RT_{NDT} value of the material at a specific location, the RT_{NDT} values at the end of the power plant design life can be determined from the trend curves. These final RT_{NDT} values are used to calculate K_{Ic} and K_{Ia}.

Determination of K_I *for Surface Flaws*

The stress intensity factors for most of the two-dimensional problems under simple loading conditions can be found in handbooks [e.g., 3]. In the case where the body is subjected to generally distributed symmetric loads, K_I may be computed by Bueckner's weight function method [*4,5*]. For surface flaws, three-dimensional calculations are involved in determination of the K_I-distribution along a crack front. A brief review of the works on this subject is given in Refs *6* and *7*. In general, numerical analysis is required to deal with this type of problem. Recently, Parks [*10*] developed the virtual crack extension (VCE) method to determine the J_I-values for three-dimensional problems. The VCE method is used in the present paper to determine the K_I-distribution for semi-elliptical cracks in a reactor vessel.

In practice, direct fracture mechanics analysis for the entire transient history experienced by a reactor vessel is prohibitively expensive because the load is time varying. More economic approaches have been developed by, for example, McGowan and Raymund [*6*], Heliot et al [*7*], and Raju and Newman [*8*], using the superposition technique. In the superposition technique, the stress intensity factors for a given crack are first computed for the uniform, linear, quadratic, and cubic (or higher order) loads which act on the crack surface. Then, in the actual evaluations, the stress (for example, the thermal stress due to a thermal transient) is given approximately by a polynomial function. The coefficients of the polynomial represent the weight of each load component that comprises the total load. Therefore the total stress intensity factor is determined by combining the contributions from each component load.

Assume that the normal stress distribution across a vessel section can be represented by a polynomial equation

$$\sigma = \sum_j A_j x^j \qquad (j = 0, 1, 2, 3) \tag{1}$$

where x is measured from the inside surface of the cylinder (Fig. 1) and A_j's are the coefficients of the stress distribution that can be determined by the least-square curve fitting technique.

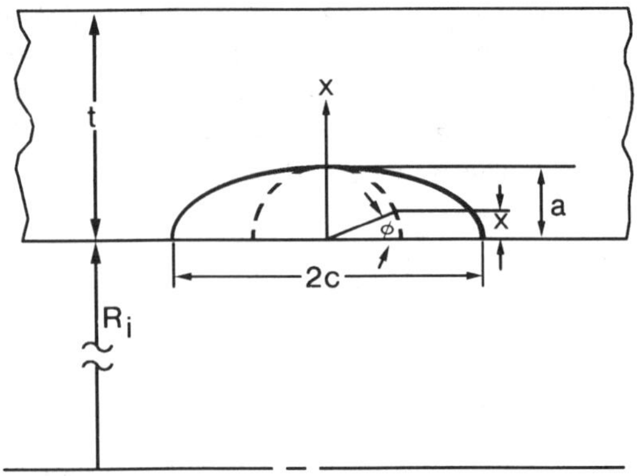

FIG. 1—*Schematic of a longitudinal semi-elliptical flaw on the inside surface of a cylinder.*

The corresponding K_I is given by [*8*]

$$K_I = \sqrt{\frac{\pi a}{Q}} \sum_j A_j G_j a^j \qquad (j = 0, 1, 2, 3) \tag{2}$$

where Q is the shape factor or the complete elliptical integral of the second kind and can be approximated by $Q = 1 + 1.464\ (a/c)^{1.65}$, and a and c are defined as in Fig. 1. G_j is the influence coefficient corresponding to load component σ_j. It should be noted that G_j is a function of the shape of crack, shape of the body containing the crack, and the position of the crack front considered.

In principle, if the G-data for a sufficient number of crack shapes are available, the K_I-value at any point along the crack front of a semi-elliptical crack of any shape and size within the range considered can be computed using Eqs 1 and 2 with the aid of an interpolation technique. The flaw shape analysis presented in this paper is mainly based on the results of Raju and Newman [*8*] for semi-elliptical cracks with a/c ratio ranging from 0.2 to 1.0 and a/t ratio ranging from 0.2 to 0.8 in a cylinder with $t/R = 0.1$. For the shape with $a/c = 0.1$, the results of Heliot [*9*] are used.

Owing to the fact that during a thermal transient the stress near the inside surface of the vessel is rather high and the temperature there relatively low, the crack would likely grow, if it initiates, primarily only in length to result in a very small a/c ratio. Therefore additional G-data for a longitudinal semi-elliptical surface crack with a very small aspect ratio ($a/c = 0.047$, $a/t = 0.2$, and $t/R = 0.1$) were generated using the VCE method [*10*] in conjunction with the finite element computer program ADINA [*11*]. This geometry was selected as a benchmark problem under a Westinghouse-ORNL exchange program [*24*].

The finite element model of this geometry consists of 216 isoparametric brick elements and 1282 nodes which results in a system approximately 3500 degrees of freedom. The model is shown in Fig. 2. In the VCE computation, nodes at the crack tip are perturbed to simulate a virtual crack extension [*12*]. The change in potential energy corresponding to such a perturbation, δP, can be evaluated in the finite element analysis. The J_I value is then calculated by

$$J_I = -\frac{\delta P}{\delta A} \tag{3}$$

where δA is the virtual crack area created by the nodal perturbation. The stress intensity factor can be determined using the well known relation $K_I = \sqrt{J_I E'}$, where $E' = E$ for plane stress condition and $E' = E/(1 - \nu^2)$ for plane strain condition, E = Young's modulus, and ν = Poisson's ratio.

In general, the J_I-value varies along the crack front, so the nodal perturbation has to be performed individually for all nodes along the crack front. For

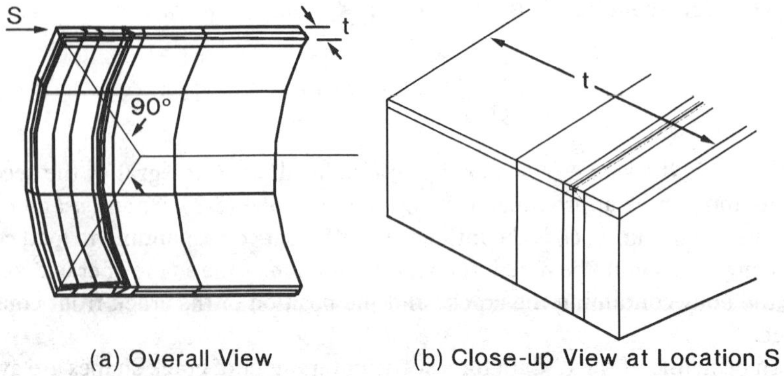

FIG. 2—*Finite element model of a cylinder with a semi-elliptical crack,* a/c = *0.047,* a/t = *0.2,* t/R = *0.1.*

each local nodal perturbation, the associated energy perturbation can be expressed by

$$\delta P = -\int_R J(s)\, \delta\ell\, (s) ds \tag{4}$$

where s is the arc length along the crack front, $\delta\ell(s)$ is the crack front displacement at s resulting from the nodal perturbation, and R is the region in which a local virtual crack extension is covered. Both $J(s)$ and $\delta\ell(s)$ can be expressed in terms of nodal values using interpolation functions. The J-distribution, or the J-value at the nodes along the crack front, is obtained by solving a system of linear equations transformed from a system of the integral equations (Eq 4).

The G-data corresponding to this crack shape are shown in Table 1. It was found that the Westinghouse and ORNL results are in very good agreement [*24*]. These data, together with those given by Raju and Newman [*8*] and Heliot [*9*], form the basis for the variable flaw shape analysis. Some of the data are plotted in G versus c/a frames (Figs. 3 and 4). It can be seen from these

TABLE 1—*Influence coefficients,* G, *for semi-elliptical surface flaw in cylinder* t/R = *0.1, longitudinal, inside surface,* a/t = *0.2,* a/c = *0.0472.*

$2\phi/\pi$	G_0	G_1	G_2	G_3
0	0.195	0.062	0.026	0.010
0.25	0.697	0.218	0.090	0.043
0.50	0.975	0.453	0.239	0.147
0.75	1.164	0.655	0.453	0.341
1.00	1.221	0.744	0.561	0.456

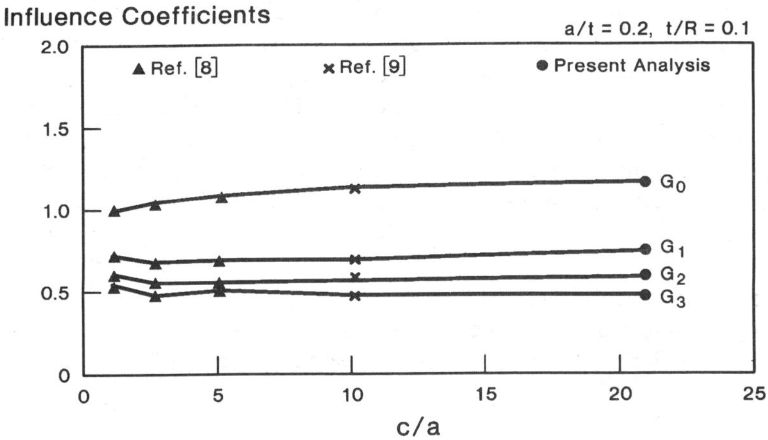

FIG. 3—*Influence coefficients as a function of aspect ratio,* c/a, *semi-elliptical crack on the inside surface of a cylinder, at* $\phi = 90°$ *(deepest point).*

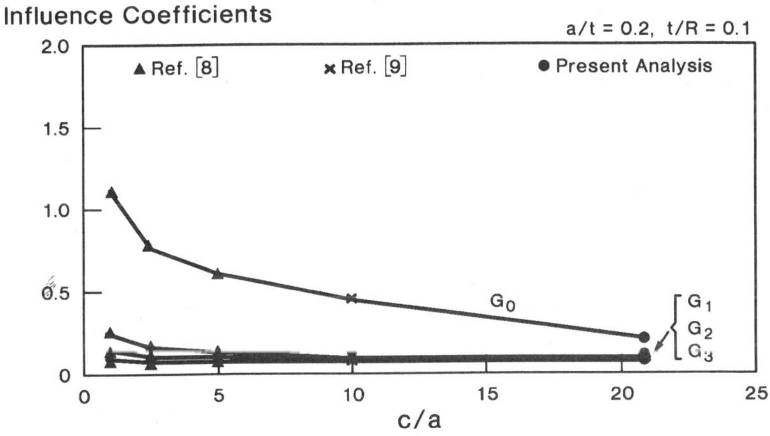

FIG. 4—*Influence coefficients as a function of aspect ratio,* c/a, *semi-elliptical crack on the inside surface of a cylinder, at* $\phi = 0°$ *(surface).*

figures that the *G*-data obtained in the present study are a reasonable extension of those given in Refs *8* and *9*. The *G*-data for another semi-elliptical surface flaw with $a/t = 0.2$, $a/c = 0.4$, and $t/R = 0.1$ were also computed using the VCE method. The results are in excellent agreement with those shown in Ref *8* and therefore are not shown in this paper.

For cracks with a/t ratios other than 0.2, the *G*-data corresponding to a/c ratios smaller than 0.1 (i.e., very long cracks) were extrapolated directly from the G versus c/a curves. However, the exact G-data for very long cracks should be determined for more accurate analyses.

The second step of variable flaw shape analysis is to determine the local values of K_{Ic} and K_{Ia} at points on the crack front. Crack initiation or arrest occurs when the local K_I/K_{Ic} or K_I/K_{Ia} ratios are, respectively, equal or greater than unity. Using these criteria the flaw shape can be predicted. It should be noted that the crack may propagate either radially (into the wall) or lengthwise (along the surface) depending upon the actual local K_I/K_{Ic} and K_I/K_{Ia} ratios. When K_I/K_{Ic} near the surface is greater than unity, a small increment is given to elongate the crack shape. Then the K_I/K_{Ic} and the K_I/K_{Ia} are re-evaluated for the "new crack". This process of crack propagation continues until the K_I/K_{Ia} ratios are less than unity all along the crack front.

Similar procedures are used for the radial propagation when K_I/K_{Ic} and K_I/K_{Ia} ratios in the region near the deepest point of the flaw are greater than unity. It should be noted that although the crack may independently propagate axially (or azimuthally in the case of circumferential flaws) and radially, the shape of the flaws remain semi-elliptical in the analysis. However, the geometry of the flaw (i.e., the aspect ratio) and its size are changed.

These procedures can be applied to any initially assumed semi-elliptical surface flaws for the entire history of a transient. The final shape and size of the postulated flaws are determined based on the type of transient and the material properties.

Control Parameters for Flaw Shape Studies

Parameters Evaluated

In order to study the flaw shape change phenomenon during a thermal transient, a systematic investigation was performed to determine the sensitivity of the factors that affect the crack propagation. The results of these investigations provide information regarding the history of the crack propagation from which more appropriate methods of fracture analysis may be developed. Basically, the variables that are used in the present flaw shape change study are RT_{NDT} at the inside surface, the added toughness contributed by the stainless cladding on the inside surface, the warm prestressing effect, the fracture toughness gradient on the surface, and one-dimensional versus two-dimensional flux (1D/2D) effects. The significance of these parameters is described below.

(1) RT_{NDT}—As stated earlier, RT_{NDT} is a parameter indexing the fracture toughness of the material. RT_{NDT} is the sum of the initial RT_{NDT} (RT_{NDTi}) and the ΔRT_{NDT} which is an additional shift due to irradiation damage. RT_{NDT} is a function of material residual elements and neutron fluence. Therefore, by specifying the RT_{NDT} value at the inside surface, along with a given copper,

nickel, and phosphorus content, the toughness at any point in the vessel wall can be determined. Curves correlating these data are called "trend curves".

The Guthrie trend curve [*13*], which represents the current NRC preferred method of determining the shift in RT_{NDT} in the reactor vessel beltline region, was selected for the present study. This trend curve is represented by

$$\Delta RT_{\mathrm{NDT}} = \alpha[-10 + 470\ \mathrm{Cu} + 350\ (\mathrm{Cu} \times \mathrm{Ni})] \left(\frac{f}{10^{19}}\right)^{0.27} \tag{5}$$

where Cu and Ni are the weight percent of copper and nickel, respectively; f is the fluence, $\mathrm{n/cm^2}$; and α is a unit conversion factor with $\alpha = 1$ for British units and $\alpha = 0.56$ for SI units.

The study was carried out based on a series of specified values of surface RT_{NDT}. Once this is specified, the value of RT_{NDT} (and thus toughness) as a function of distance into the vessel wall is a function of the fluence gradient and slope of the trend curve.

(2) *Added Toughness from Cladding*—The reactor vessel stainless steel cladding is believed to have superior toughness to the ferritic steel at the inside surface of the reactor vessel, particularly after irradiation has lowered the toughness of the ferritic steel. Although this benefit has not been quantified as yet, an analytical simulation of the effect was carried out as part of the parametric study. To accomplish the simulation, the innermost portion of the reactor vessel wall (4 mm) was assumed to have a constant fracture toughness equal to 165 MPa $\sqrt{\mathrm{m}}$ (150 ksi $\sqrt{\mathrm{in.}}$), and the remainder of the vessel wall had a toughness based on the ASME Code reference toughness, which is a function of the irradiation-induced change in RT_{NDT}. The clad toughness was based on results from a series of sandwich specimen tests conducted in France [*14*]. The stainless steel cladding does not have a transition to lower toughness values in the temperature range experienced during a thermal shock.

(3) *Warm Prestressing Effect*—Warm prestressing (WPS) [*15*] is an empirically observed phenomenon that prevents crack initiation during reactor vessel thermal shock transients as long as the stress intensity of the flaw is decreasing with time. A detailed review of the WPS phenomenon is given by Pickles and Cowan [*16*]. Basically the WPS effect is believed to be caused by two principal factors: increase of the yield strength due to temperature decrease, and formation of a compressive residual stress field at the crack tip due to unloading.

(4) *Fracture Toughness Gradient in the Base Metal*—It is well known that near the inside and outside surfaces of a unirradiated reactor vessel, the fracture toughness is significantly greater than the interior of the vessel material. This fact is indicated by the Charpy data [*17*] shown in Fig. 5. The data indicate that the surface layer has an approximately 28°C (50°F) lower value of RT_{NDT} than at the quarter thickness position. This toughness gradient can affect the shape of the flaw.

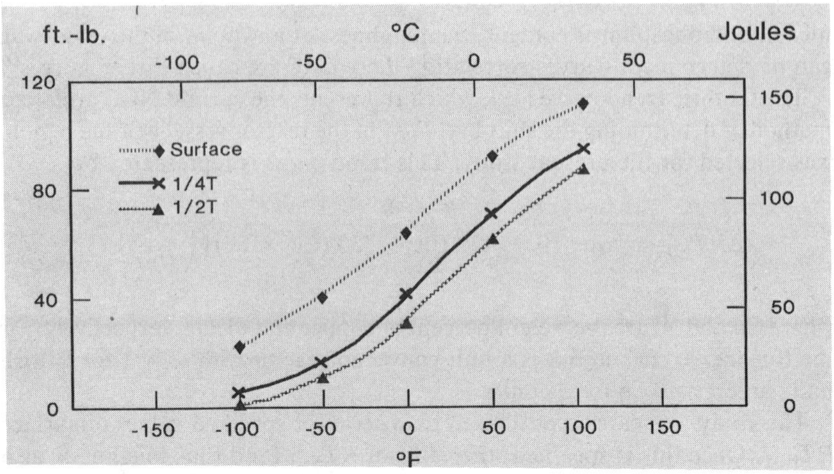

FIG. 5—*Variation of Charpy values with location where the specimens are cut from a plate.*

(5) *1D/2D Flux Effect*—It is well known that the neutron fluence (flux) varies axially, radially, and azimuthally. This means that a postulated crack started at the most severely damaged location will grow, if the local K_I/K_{Ic} ratio value is greater than 1, to a less damaged location during the thermal transient, and increases the possibility of arrest.

Parameters Not Evaluated

Although the parameters discussed in the previous section are the key variables involved in a pressurized thermal shock analysis, there are several other phases of the analytical procedure which have not been treated here. The interaction between the base metal and cladding was not dealt with completely, although a first attempt was made to account for the beneficial effect through increased toughness. This is a very complicated situation in reality, because in addition to the presence of the two materials, there is a complicated residual stress field which will be redistributed due to the presence of a flaw. This redistribution has not been characterized either analytically or experimentally, so it was not possible to deal with it here.

Another aspect of a pressurized thermal shock analysis that was not treated directly here is the temperature variation which can occur when safety injection flow does not mix completely with the water already in the system. This has no impact on the fracture analysis for most locations, because the flow which does not mix tends to form a layer whose effect on heat transfer is rather easily evaluated. The only case where this aspect could be important is where a cold water streaming effect occurs in the inlet nozzle. The presence of the hot-cold interface in these cases will affect the propagation of a flaw, but this interaction was not considered in the work reported here.

Sample Calculations

A small-break loss-of-coolant accident (LOCA) transient was selected for the analysis. This is a typical pressurized thermal shock (PTS) transient. The temperature history of the transient analyzed here is approximately represented by

$$T = 282 - 0.306t + 0.133 \times 10^{-3}t^2 - 0.187 \times 10^{-7}t^3 \tag{6}$$

where t is the time (seconds) and T is the temperature (°C). The pressure remains constant at 6.895 MPa (1000 psi) during the first 2000 s of the transient and then decreases nearly linearly to 2.19 MPa (318 psi) when t = 4000 s. Both the temperature and pressure remain constant thereafter.

The thickness-to-radius ratio is 0.1 and the thickness is 219 mm (8.625 in.). The crack is semi-elliptical on the inside surface in the longitudinal direction. The upper shelf toughness is 220 MPa$\sqrt{\text{m}}$ (200 ksi$\sqrt{\text{in.}}$); the initial RT_{NDT} is −18°C. Using these data, 16 cases were investigated to study the effect of the control variables described earlier. Descriptions of these cases are given in Table 2. For each case, five initial crack depths, ranging from 10 to 30% of the wall thickness, were evaluated. The shape of these cracks are all semi-elliptical with the initial aspect ratio $2c/a = 6.0$.

Results of Flaw Shape Analysis

The crack propagation history for each of these postulated flaws for all cases was calculated. For some cases there was no initiation at all, while for other cases the flaws initiated but were arrested. There were some cases where the flaw initiated and arrested but then re-initiated and eventually penetrated the wall. The crack propagation histories of all the cases studied are summarized in Table 3. This table provides a qualitative description of the manner of flaw shape change and its sensitivity to the various parameters studied. In general, the WPS effect provides significant benefits in resisting crack propagation. The benefits given by the other parameters are moderate.

TABLE 2—*Description of the cases analyzed.*

Case[a]	WPS?	Cladding Effect?	Toughness Gradient?[b]
1 to 4	no	no	no
5 to 8	no	no	yes
9 to 12	yes	no	no
13 to 16	no	yes	no

[a]Four RT_{NDT} in each case group 111, 139, 167, 194°C (200, 250, 300, 350°F).

[b]28°C (50°F) lower RT_{NDT} at surface than at $1/4T$ position.

TABLE 3—*Results of flaw shape change analysis.*

Case	RT_{NDT} (°F)	Flaw Shape Evolution Summary[a]	Remarks
1	200	E,D/P	No beneficial factors are considered in this group.
2	250	E,D/E,D/P	
3	300	E,D/E,D/P	
4	350	E,D/E,D/P	
5	200	E,D/P	Penetration only occurs on those with initial a/t = 0.2 flaws. Surface toughness gradient effect is moderate.
6	250	E,D/P	
7	300	E,D/P	
8	350	E,D/P	
9	200	E/	Strong benefit due to WPS is obvious.
10	250	E/	
11	300	E,D/E,D/	
12	350	E,D/E,D/P	
13	200	E,D/	Added toughness from cladding only benefits the lowest RT_{NDT} case.
14	250	E,D/P	
15	300	E,D/P	
16	350	E,D/P	

[a]Abbreviations: E = elongated, D = deepened, P = penetrated, / = arrest event.

The results of the evolution of the flaw shape during the postulated transient are presented in Figs. 6 and 7 to show the change in the aspect ratio (a/c) and the flaw depth (a/t). The ASME Code Section XI guideline would allow the initial flaw shape, $a/c = 0.333$, to remain constant, while the NRC approach described in Ref *13* would require the flaw to be treated as a continuous flaw, with infinite length. These approaches are shown graphically in the figures.

Four different surface RT_{NDT} values were investigated, namely 93, 121, 150, and 177°C (200, 250, 300, and 350°F), in association with other parameters. Point I of Figs. 6 and 7 represents the initial shape of the crack and is shared by four different cases in each diagram, as indicated by four different symbols in a large circle. When the transient occurs, the tough material, represented by low RT_{NDT} values, may be able to resist the load induced in the transient and no crack will initiate. In this case no curve can be developed in the flaw shape evolution diagram (i.e., Point I represents both the initial and the final geometric conditions). In other cases, the flaws will propagate and generally will be elongated and deepened. This is represented by the decreasing a/c values and the increasing a/t values.

There are cases where multiple arrest-reinitiation events occur. Each time when an arrest occurs, a symbol corresponding to the specified RT_{NDT} is marked in the diagram to show the new flaw geometry. This procedure may

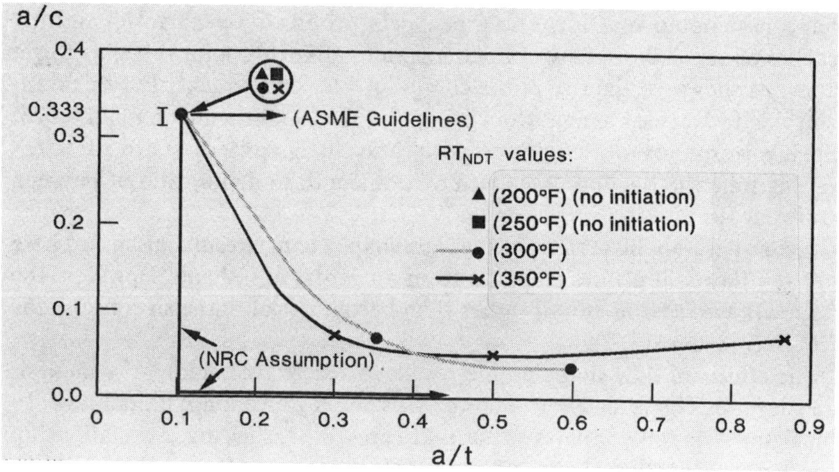

FIG. 6—*Flaw shape evolution based on various specified* RT_{NDT} *values (1D flux, WPS effect considered).*

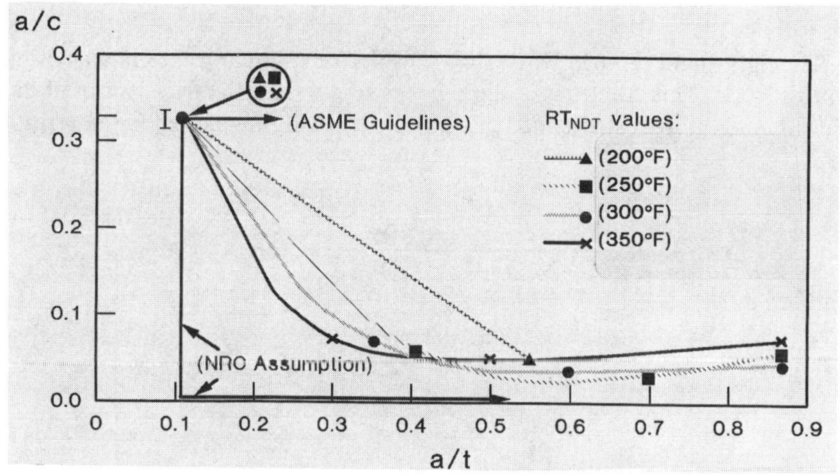

FIG. 7—*Flaw shape evolution based on various specified* RT_{NDT} *values (no beneficial factors are considered).*

continue depending upon the load and the material properties. It can be seen from these diagrams that for some cases the flaws are permanently arrested at a finite shape and depth, while for other cases the flaws will penetrate the wall after several reinitiations and re-arrests.

Figure 6 shows the flaw shape evolution with the presence of the benefit of the WPS effect, while Fig. 7 shows the results without WPS. In all cases, the postulated crack grows longer as well as deeper, if it initiates. The initial shape is denoted by "I". The crack gets progressively narrower until it reaches 50 to 60% of the wall thickness, and after this point it begins to get fatter, as shown by the rise in the curves for very deep cracks. For this transient the flaw shape changes from $a/c = 0.333$ to a minimum of about 0.025, which corresponds to a length-to-depth ratio changing from 6:1 to 80:1. After this time the flaw becomes fatter, to a length-to-depth ratio of between 30:1 and 50:1.

It can therefore be seen from the flaw shape change results discussed here that the flaw will always elongate from an initial 6:1 shape. However, the shape always remains finite, rather than becoming continuously long, as the NRC has assumed [*13*].

The effects of flaw shape on the results of reactor vessel integrity analysis are shown in Fig. 8 for a series of analyses employing a longitudinal flaw. In this figure, two sets of analysis results are presented, showing the relationship between the final flaw shape and the acceptable surface RT_{NDT}. The curve on the left was taken from results previously presented by the NRC [Fig. D-19 of Ref *13*]. Here the flaw shape was assumed to be constant as the flaw propagated. The results show that the acceptable surface RT_{NDT} of the $a/c = 0.333$ flaw (length-to-depth ratio 6:1) is considerably higher than that of the continuous flaw ($a/c = 0$).

The right-hand curve in Fig. 8 shows results for a similar transient analyzed in this work. This transient is slightly less severe than the transient used in Ref *13* but is more realistic. For this transient, results are plotted for an arrest

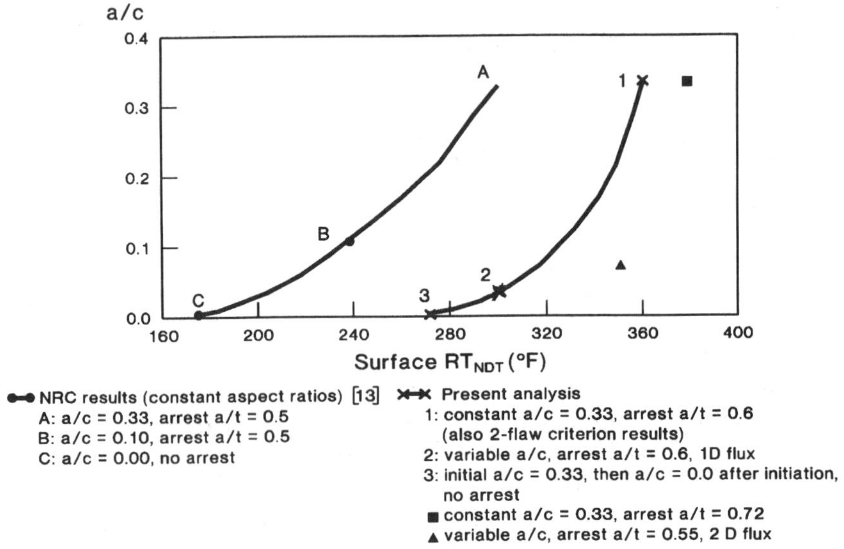

FIG. 8—*Effect of flaw shape treatment on reactor vessel integrity acceptability.*

depth of 60% as compared with 50% for the left-hand curve. The use of the 60% arrest depth resulted from the availability of results and should not imply any criterion for arrest. The use of the ASME Code suggested arrest depth of 75% of the vessel wall would result in another curve to the right of this curve, as shown by the square symbol plotted to the right of the curve.

Again it is seen that the shape affects the acceptable surface RT_{NDT}, with the trend similar to that observed in the left-hand curve. The end points of this curve were determined based on the ASME guidelines and the NRC assumptions, while Point 2 was a result of variable flaw shape treatment (Fig. 8).

An additional analysis using 2D flux variations was performed, where the flaw was allowed to take a variable shape. The use of this variable shape results in significant further improvement in results compared with the assumption that the flaw becomes continuous. Furthermore, use of the more accurate representation of the flux gradient (2D variation versus 1D variation) improves the result considerably, by nearly 28°C (50°F). It also results in a final flaw shape which is not as long, having a length-to-depth ratio of about 27:1.

For the transient considered in Fig. 8, the resulting acceptable RT_{NDT} for the 6:1 constant aspect ratio flaw is precisely the same as the result obtained using the two-flaw criterion to approximate the shape change. This criterion employs a 6:1 aspect ratio flaw, which is assumed to become a continuous flaw when the maximum value of K_I along the crack front equals or exceeds that for a continuous flaw of the same depth. This criterion has been used in many previously published analyses [e.g., *18*] and the variable shape change results obtained in Fig. 8 confirm its applicability for this transient. The results are the same for both the constant aspect ratio and the two-flaw criterion in this case because the stress intensity factor along the crack front never exceeds that for a continuous flaw.

Perhaps the key conclusion from this work is that the use of a variable shape flaw improves the results for this transient by approximately 44°C (80°F), in terms of RT_{NDT}, compared with the results obtained from the NRC-suggested methodology (where the flaw is assumed to become continuous as soon as it initiates). The arrest depth obtained from this more accurate treatment is shallower, and clearly the final flaw shape is less elongated than would have been expected. It is also important to realize that the transient analyzed is the small loss-of-coolant accident, which is a dominating transient with regard to risk to reactor vessel integrity.

Summary and Conclusions

One of the key elements in a fracture assessment of the integrity of the reactor vessel following a pressurized thermal shock is the size and shape of the postulated flaw. If the thermal shock event is sufficiently severe, the flaw will initiate, and its impact on the integrity of the reactor vessel is strongly dependent on the shape which the flaw takes as it extends. This report has been prepared to provide a state-of-the-art assessment of flaw shape change effects resulting from a range of material variables including the presence of the cladding.

The variable flaw shape analysis was performed to study the flaw behavior as a function of toughness-related parameters and to assess the conservatism of presently used analytical methods for thermal shock analyses. The results are shown in Table 3 and Figs. 6 to 8. The effects of the control parameters were evaluated and are summarized below:

1. Warm prestressing was found to have a significant impact on the analysis results. The major effect given by warm prestressing is that it can reduce the depth of flaw penetration considerably.
2. The cladding effect evaluated in the analysis is not as strong as the experimental results. The difference is primarily due to the approximations incorporated in the analysis.
3. The effect of the surface toughness gradient in the base metal (i.e., a 28°C lower value of RT_{NDT} at the surface than at the quarter thickness position) was found to be mild. The improved surface toughness would retard the crack initiation from the surface, but since this effect does not exist beyond the quarter thickness point, there is no impact on propagation in depth direction.
4. Two-dimensional fluence consideration was found to have a significant impact on the analysis results. The final flaw length is shorter and the depth is shallower.

The parametric study showed that the use of a variable flaw shape for the analysis of a small LOCA transient improved the results in terms of acceptable RT_{NDT} by about 44°C (80°F) compared with the results obtained from currently used methodology, where a continuous flaw is used for arrest, as shown in Fig. 8. Thus, accurately accounting for the shape change of a postulated flaw during a thermal shock can have a significant beneficial effect on the results of the analysis.

Finally, it should be noted that the present analysis is based on the LEFM theory and that plasticity, which may be significant in some cases, was not included. This generally leads to a conservative result. Moreover, the irradiation effects on the fracture of the cladding were not taken into account. It has been reported [e.g., *25,26*] that irradiation could seriously affect the toughness of austenitic stainless steels. The purpose of the present paper, however, was to develop a methodology of analysis that allows the crack to grow in a more realistic manner than those adopted currently. Any other significant factors that can affect fracture behavior may be included in the analysis. The accuracy and quality of the analysis can only be judged by comparing it with carefully performed experimental results.

Acknowledgments

The authors wish to acknowledge the financial support of the Westinghouse Owner's Group for this research program. The contributions of Dr. J. J. McGowan of Oak Ridge National Laboratories in beginning the development of the flaw shape change computer code are greatly appreciated. Appreciation is also due R. D. Cheverton, D. G. Ball, and R. Nanstead of Oak Ridge National Laboratories for their cooperation in comparing stress intensity factor solutions and in discussions of their cladding effects tests.

References

[*1*] "Rules for In-Service Inspection of Nuclear Power Components," Section XI, ASME Boiler and Pressure Vessel Code, New York, 1980.

[*2*] Cheverton, R. D., "Pressure Vessel Fracture Studies Pertaining to a PWR LOCA-ECC Thermal Shock: Experiments TSI-1 and TSI-2," ORNL/NUREG/IN-31, Oak Ridge National Laboratories, Oak Ridge, Tenn., Sept. 1976.

[*3*] Tada, H., Paris, P. C., and Irwin, G. R., *The Stress Analysis of Cracks Handbook*, Del Research Corp., Hellertown, Pa., 1973.

[*4*] Bueckner, H. F., "A Novel Principle for the Computation of Stress Intensity Factors," *Zeitschrift für Angewandte Mathematik und Mechanik*, Vol. 50, 1970, pp. 529–546.

[*5*] Rice, J. R., "Some Remarks on Elastic Crack-tip Stress Fields," *International Journal of Solids and Structures*, Vol. 8, 1972, pp. 751–758.

[*6*] McGowan, J. J. and Raymund, M., "Stress Intensity Factor Solutions for Internal Longitudinal Semi-Elliptical Surface Flaws in a Cylinder under Arbitrary Loadings," in *Fracture Mechanics (Eleventh Conference), ASTM STP 677*, C. W. Smith, Ed., American Society for Testing and Materials, Philadelphia, 1979, pp. 365–380.

[*7*] Heliot, J., Labbens, R. C., and Pellissier-Tanon, A., "Semi-Elliptical Cracks in a Cylinder Subjected to Stress Gradients," in *Fracture Mechanics (11th Conference), ASTM STP 677*, C. W. Smith, Ed., American Society for Testing and Materials, Philadelphia, 1979, pp. 341–364.

[*8*] Raju, I. S. and Newman, J. C., Jr., "Stress Intensity Factor Influence Coefficients for Internal and External Surface Cracks in Cylindrical Vessels," *Aspects of Fracture Mechanics in Pressure Vessel and Piping*, ASME, Vol. 58, 1982, pp. 37–48.

[*9*] Heliot, J., "Fissures Semi-elliptiques axiales de arande largeur ($a/c = 1/10$) debouchant a linterieur dun cylindre $t/R_i = 0.1$," Creusot-Loire, Sept. 1979.

[*10*] Parks, D. M., "The Virtual Crack Extension Method for Nonlinear Material Behavior," *Computer Methods in Applied Mechanics and Engineering*, Vol. 12, 1977, pp. 253–364.

[*11*] Bathe, K. J., "ADINA: A Finite Element Program for Automatic Dynamic Incremental Nonlinear Analysis," MIT Report 83448-1, Massachusetts Institute of Technology, Cambridge, Mass., 1978.

[*12*] Yang, C. Y. and Palusamy, S. S., "VCE Method of *J* Determination for a Pressurized Pipe under Bending," *Journal of Pressure Vessel Technology, Transactions of ASME*, Feb. 1983, pp. 16–22.

[*13*] NRC Policy Issue, Enclosure A, "NRC Staff Evaluation of Pressurized Thermal Shock," SECY-82-465, Nuclear Regulatory Commission, Washington, D.C., Nov. 23, 1982.

[*14*] Soulat, P., Almundheri, M., and Peault, B., "Etude de la Tenacite D'un Revetment Inoxydable D'acieu de Duve de Reacteur a Eau Sous Pression" Commissariat a L'Energie Atomique, Saclay, Report D. Tech SRMA (82) 1186, Oct. 1982 (in French).

[*15*] McGowan, J. J., "Application of Warm Prestressing Effects to Fracture Mechanics Analyses of Nuclear Reactor Vessels During Severe Thermal Shock," *Nuclear Engineering and Design*, Vol. 51, No. 3, Feb. 1979, pp. 431–444.

[*16*] Pickles, B. W. and Cowan, A., "A Review of Warm Prestressing Studies," *International Journal of Pressure Vessels and Piping*, Vol. 14, 1983, pp. 95–131.

[17] Hodge, J. M., "Properties of Heavy Section Nuclear Reactor Steels," WRC Bulletin 217, Welding Research Council, New York.
[18] "Summary Report on Reactor Vessel Integrity for Westinghouse Operating Plants," Westinghouse Report WCAP-10019, Dec. 1981.
[19] Bamford, W. H. and Yang, C. Y., "Flaw Shape Change and Clad Effects on Fracture Analyses of Reactor Vessels for Pressurized Thermal Shock," Westinghouse Report WCAP 10407, Sept. 1983.
[20] Report of KAPL/Bettis Technical Meeting with ORNL/NRC Personnel, Oak Ridge, Tenn. 27-28 Jan. 1983, Memorandum MDO/SE-83-18, Knolls Atomic Power Laboratory, Schenectady, N.Y., 8 Feb. 1983.
[21] "The HSST Program Vessel Integrity Review Group meeting, June 14, 1983," Oak Ridge National Laboratories, Oak Ridge, Tenn. (summary of Presentations).
[22] Robinson, G. C. and Corwin, W. R., "HSST Clad Plate Task Quick-Look Report for Specimens CP-1A, CP-3, CP-5, and CP-7," Report ORNL/CPT/001, Oak Ridge National Laboratories, Oak Ridge, Tenn., 21 June 1982.
[23] Robinson, G. C., "Small Scale Clad Effects Study," in *Proceedings*, Tenth Water Reactor Safety Information Meeting, Gaithersburg, Md., Oct. 1982.
[24] Yang, C. Y., letter to D. G. Ball, Westinghouse Report MT-SME 3170, 14 Nov. 1983.
[25] Mills, W. J., "Irradiation Effects on the Fracture Toughness of Austenitic Fe-Cr-Ni Alloys," Report HEDL-TME-82-17, UC-797, Tb, Th, Hanford Engineering Development Laboratory, Richland, Wash., July 1982.
[26] Dufresne, J., Henry, B., and Larsson, H., "Fracture Toughness of Irradiated AISI304 and 316L Stainless Steels," in *Effects of Radiation on Structural Materials, ASTM STP 683*, American Society for Testing and Materials, Philadelphia, 1979, pp. 511-528.

R. G. Forman[1] *and V. Shivakumar*[2]

Growth Behavior of Surface Cracks in the Circumferential Plane of Solid and Hollow Cylinders

REFERENCE: Forman, R. G. and Shivakumar, V., "**Growth Behavior of Surface Cracks in the Circumferential Plane of Solid and Hollow Cylinders,**" *Fracture Mechanics: Seventeenth Volume, ASTM STP 905*, J. H. Underwood, R. Chait, C. W. Smith, D. P. Wilhem, W. A. Andrews, and J. C. Newman, Eds., American Society for Testing and Materials, Philadelphia, 1986, pp. 59–74.

ABSTRACT: Experiments were conducted to study the growth behavior of surface fatigue cracks in the circumferential plane of solid and hollow cylinders. In the solid cylinders, the fatigue cracks were found to have a circular arc crack front with specific upper and lower limits to the arc radius. In the hollow cylinders, the fatigue cracks were found to agree accurately with the shape of a transformed semiellipse. A modification to the usual nondimensionalization expression used for surface flaws in flat plates was found to give correct trends for the hollow cylinder problem.

KEY WORDS: fatigue (materials), stress cycling, crack propagation, stress-intensity factor

One of the most important, but still little researched, problems in fracture mechanics is the growth behavior of surface cracks in the circumferential plane of solid and hollow cylinders. The problem occurs in almost all fields of fracture control, such as aircraft, aerospace, off-shore oil drilling, piping systems, or any other structural application containing cylindrical parts strained in either tension or bending. Not only is there a lack of applicable stress-intensity factor solutions to the problem, but even the geometrical modeling of the crack has not been sufficiently defined.

Most experimental results for the solid cylinder problem have shown that the crack front has the geometry of a circular arc. One of the first numerical solutions to the problem for tension loading was given by Johnson [*1*] using

[1]NASA, Lyndon B. Johnson Space Center, Houston, TX 77058.
[2]Lockheed-EMSCO, Houston, TX 77258.

the boundary integral method. Athanassiadis et al [*2*] also used the boundary integral method to obtain solutions for tension and bending problems, and showed experimental results that fitted a near circular crack front. Nezu et al [*3*] conducted finite element analysis of a solid cylinder in tension and confirmed the circular shape of the crack front with experimental results on aluminum and steel bars. Bush [*4*] obtained dimensionless stress-intensity factor results for single-edge-cracked (i.e., straight crack front) round bars loaded in tension by using experimental compliance measurements, and the results are applicable to the limiting condition of crack front radius. Daoud and Cartwright [*5*] also analyzed the single-edge-cracked round bar with the finite element method and calculated the strain energy release rate for the cases of both uniform tension and pure bending.

Finally, Wilhem et al [*6*] used fatigue crack growth test data on surface cracked round bars loaded in tension to obtain an empirical expression for the stress-intensity factor, K_I, that is applicable to the circular arc crack front geometry.

The research conducted on hollow cylinders with part-circumferential surface cracks is even more limited than that for solid cylinders. Delale and Erdogan [*7*] derived numerical results for exterior cracks of semielliptical shape by using the line-spring model and by approximating the cylinder as a shallow shell. No experimental work has been published which explicitly defines the geometry of the crack front or compares it with the semielliptical shape normally found for flat plates.

In this paper, experimental results are presented which define more precisely the natural shape of part-circumferential surface cracks grown in fatigue in cylindrical bodies. The circular shape of the crack front for the solid cylinder is confirmed, and equations are presented for the upper and lower bounds for the radius of curvature. Assuming the most severe case, a stress-intensity factor expression is presented for the solid cylinder with applied tension and bending stresses.

The hollow cylinder results for internal and external cracks are compared with a transformed shape of a semielliptical crack developed by Raju and Newman [*8*]. The results provide a more accurate and valid model for numerical analysis by finite element, boundary integral, or other methods.

Experimental Details

The crack growth studies were conducted using machine-notched specimens of 6061-T6 aluminum pipe for the hollow cylinder experiments and 6Al-4V annealed titanium and mild steel round bar for the solid cylinder experiments. The hollow cylinder dimensions were approximately 30 mm outer diameter and 22 mm inner diameter for the externally flawed specimens and 38 mm outer diameter and 25 mm inner diameter for the internally flawed specimens. The titanium rod diameter was 16 mm and the steel rod diameter was 25 mm.

The cyclic tension loads were applied using a hydraulic servo-operated fatigue machine. The bending loads were applied using a rotary cam machine with cantilever-type specimen support and the notch located approximately 2 mm from the support. All fatigue stress levels were less than the material yield stress levels, and the cyclic stress ratios were approximately zero for all experiments. The total number of fatigue cycles for each specimen was usually between 100 to 500 kilocycles, most of which occurred during the crack propagation stage. Since fatigue crack growth properties for the specific bar stock materials were not known, no attempt was made to obtain data for deriving or correlating ΔK versus da/dN type results.

The steel and titanium specimens were heat tinted at different increments of crack growth to obtain more data from each specimen. After fatigue cracking and breaking the specimens open to expose the fracture faces, the fatigue crack surfaces were photographed at approximately ×6.5 magnification. All the geometrical fitting for the solid cylinder results was performed directly on the photographs. Because the hollow cylinder analysis involved comparing the crack geometry with a modified elliptical equation, a digitizer connected to a desktop computer and plotter was used to digitize the crack shape and plot the fitted equation.

Results and Discussion

Solid Cylinder Results

Typical examples of the crack geometry obtained in the experiments are shown in Figs. 1 and 2 for the tension fatigue specimens and in Figs. 3 and 4 for the bending fatigue specimens. The circular arcs drawn in the figures for the tension specimens show the good fit and the accurate nature of the circular crack front. The bending fatigue results also show the circular crack shape but with a greater radius which is approximated by the equation

$$r = \frac{a(2R - a)}{2(R - a)} \tag{1}$$

In Eq 1, the radius, r, is obtained by using the assumption as shown in Fig. 5 that the crack front of depth a intersects the cylinder surface of radius R at perpendicular angles. The corresponding arc length $2b$ of the crack on the cylinder surface is given by

$$2b = 2R \tan^{-1}(r/R) \tag{2}$$

Also shown in Fig. 5 is the crack front for the case in which the radius is equal to the depth a. The arc length for this geometry is given by

$$2b' = 4R \sin^{-1}(a/2R) \tag{3}$$

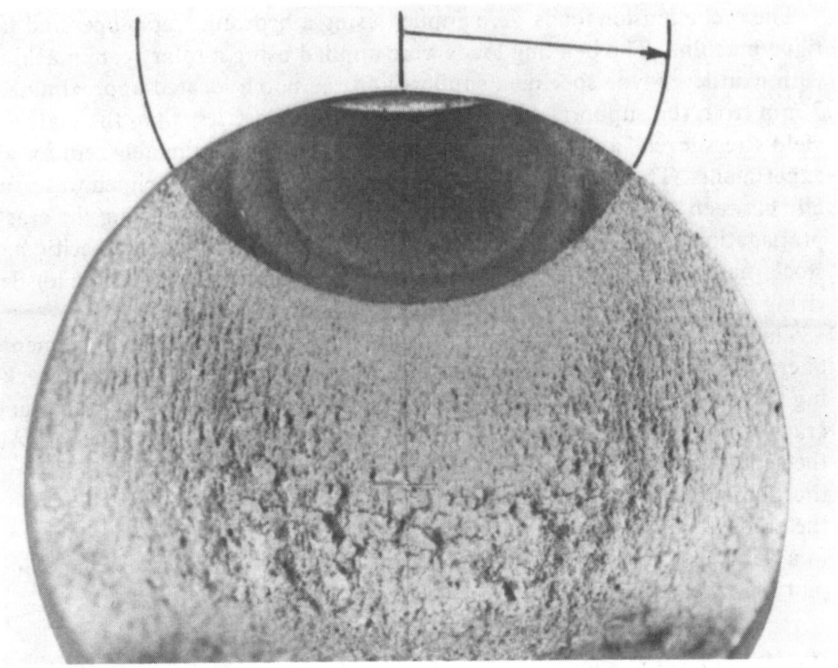

FIG. 1—*Fatigue crack in steel rod specimen cyclic loaded in tension.*

FIG. 2—*Fatigue crack in titanium rod specimen cyclic loaded in tension.*

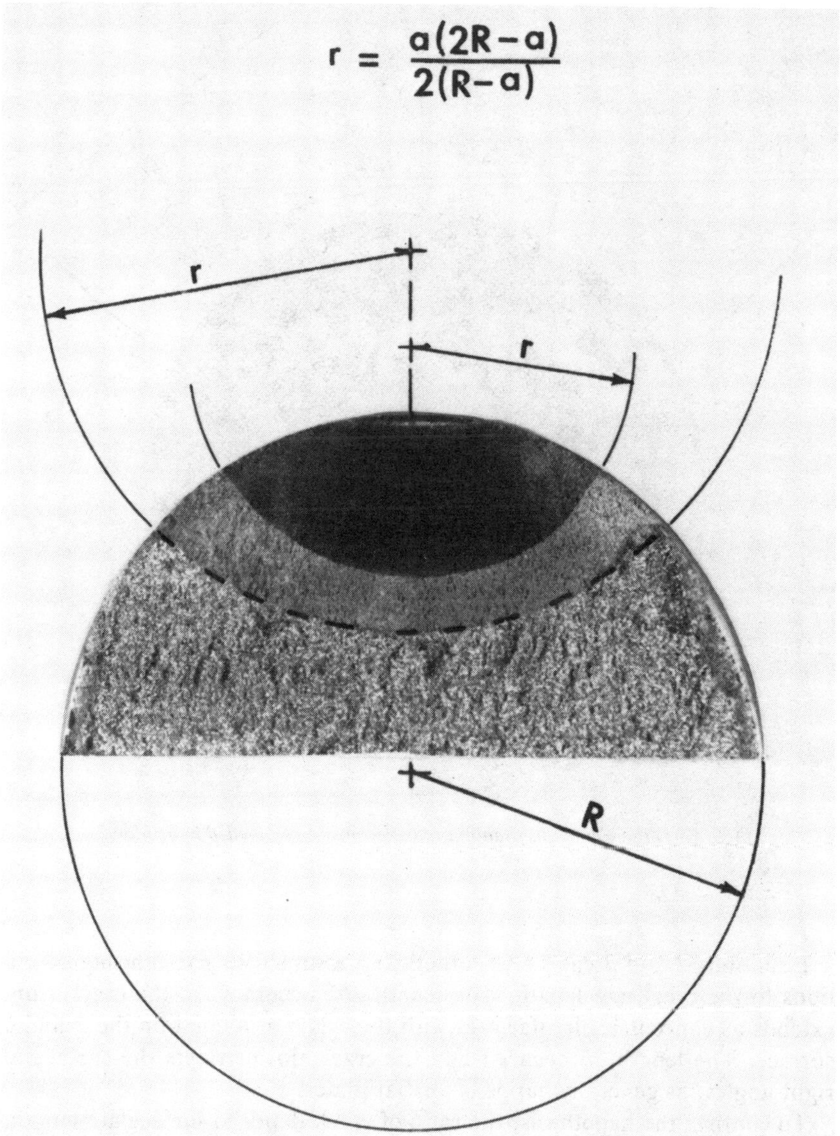

FIG. 3—*Fatigue crack in steel rod specimen cyclic loaded in bending.*

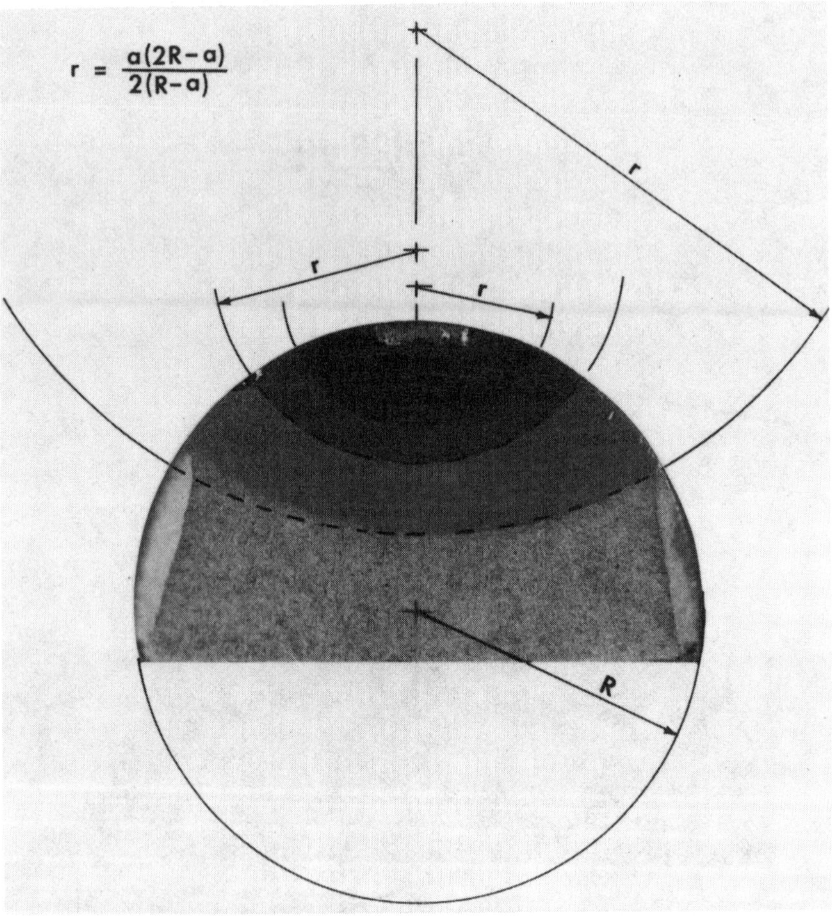

FIG. 4—*Fatigue crack in titanium rod specimen cyclic loaded in bending.*

Equations 2 and 3 can be hypothetically assumed to give limiting conditions to the crack arc length. The length $2b'$ occurs when the crack front extends as concentric circular arcs with the origin at a point on the cylinder surface. The length $2b$ occurs when the crack tip intersects the surface at right angles, as generally happens for flat plates.

To confirm the hypothesis, the ratio of crack depth to surface arc length, a/b, and depth to diameter ratio, a/D, were calculated from measurements on each specimen photograph; the results are plotted in Figs. 6 and 7. The lower curve in each figure is based on Eq 2 and the upper curve is based on Eq 3. The plotted data show that the ratio a/b is essentially bounded by the

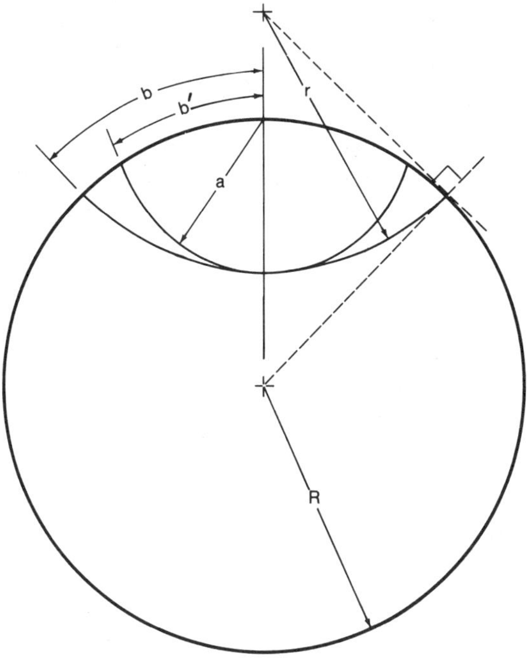

$$r = \frac{a(2R-a)}{2(R-a)}$$

$$b = R \tan^{-1}\frac{r}{R},\ b' = 2R \sin^{-1}\frac{a}{2R}$$

FIG. 5—*Geometry of surface crack in a solid cylinder.*

two curves. The limiting conditions thus reduce the a/b values of practical interest to between 0.64 and 1.0 for $a/D \leq 1/2$.

Published numerical results, both from the boundary integral analysis by Athanassiadis and the finite element analysis by Nazu, indicate that the stress-intensity factor, K_I, varies along the crack front for the solid cylinder problem. The boundary integral results also show that for both tension and bending, K_I is almost constant along the crack front for an a/b ratio equal to 0.7. Since the crack shape should stabilize at a geometry producing a constant K_I, this behavior appears to be confirmed by the bending test results shown in Fig. 7. The tension test results do not agree as closely with the crack front shape shown by the bending results, probably because of the high restraint against specimen rotation inherent in the tension fatigue load system and friction grips. In many practical applications, however, the perpendicular intersection criteria may give the best correlation for the stabilized flaw shape for tension fatigue loading.

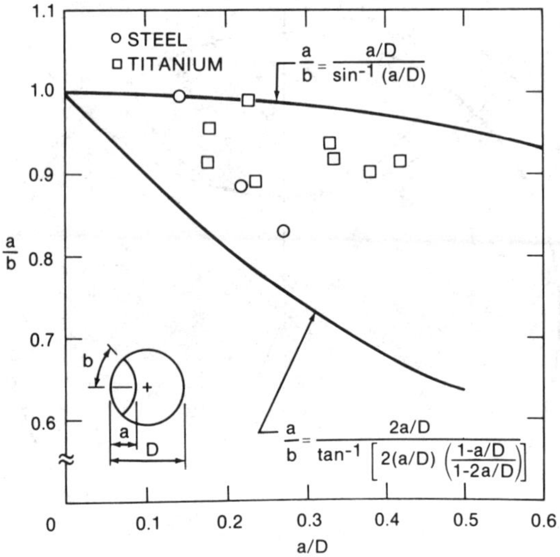

FIG. 6—*Comparison of crack shape with relative crack depth for solid cylinders cyclic loaded in tension.*

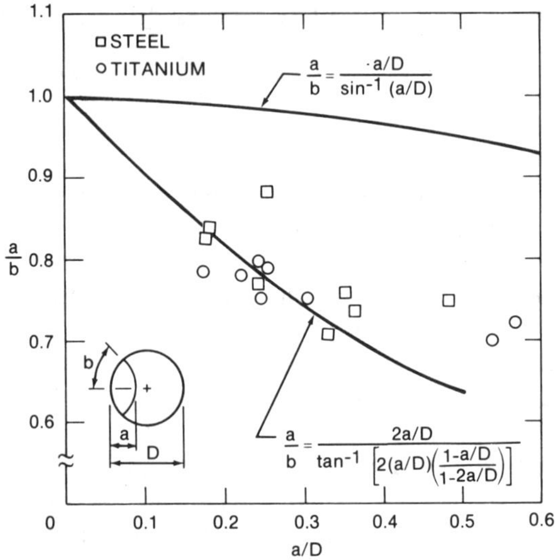

FIG. 7—*Comparison of crack shape with relative crack depth for solid cylinders cyclic loaded in bending.*

A common approach used for fatigue flaw growth analysis of semielliptical surface flaws is to predict the growth rate at the major and minor axes based on the stress-intensity factor values at these respective points. This approach requires a significantly more comprehensive set of stress-intensity factor solutions than for one-dimensional growth problems, such as through-type cracks in plates. Even though the crack geometry for a solid cylinder has been shown to be a circular arc, the variation in K_I makes it a two-dimensional growth problem (i.e., the increase in both depth and surface length must be calculated separately).

A simpler growth analysis method for the circular crack in a solid cylinder is to assume that the crack extends in the worst configuration. Since the ex-

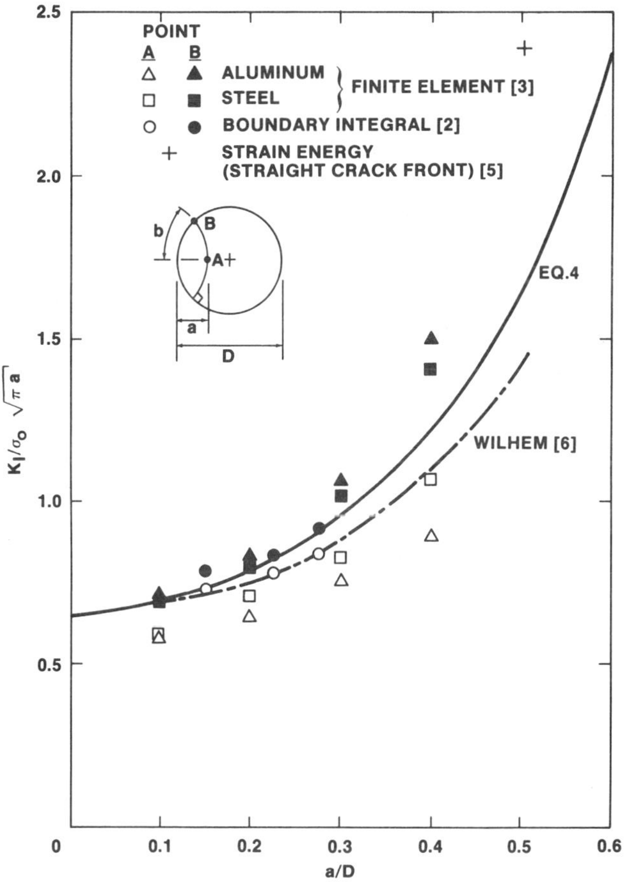

FIG. 8—*Variation of* K_I *as a function of relative crack depth for a solid cylinder loaded in tension.*

perimental measurements show significant scatter within the upper and lower limits, a reasonable assumption is to let the crack have the configuration producing the highest K_I value, or that prescribed by Eq 2. With the crack shape given explicitly, the growth rate analysis reduces to a one-dimensional problem.

Using the earlier referenced results, an approximate equation for K_I was developed for applied tension and bending stresses. Comparisons of the equation with published data are shown in Figs. 8 and 9. The equation should be applicable to fatigue crack growth analysis and is expressed as

$$K_I = [\sigma_0 F_0(\lambda) + \sigma_B F_B(\lambda)]\sqrt{\pi a} \tag{4}$$

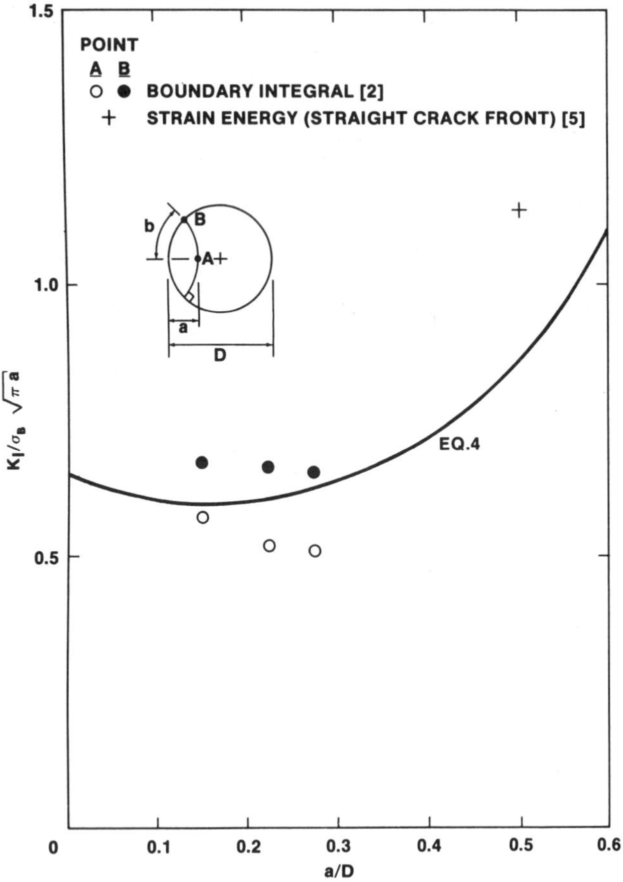

FIG. 9—*Variation of* K_I *as a function of relative crack depth for a solid cylinder loaded in bending.*

where σ_0 and σ_B are the applied uniform tension and bending stresses, and $F_0(\lambda)$ and $F_B(\lambda)$ are magnification factors given by

$$F_0(\lambda) = g(\lambda)[0.752 + 2.02\lambda + 0.37(1 - \sin \pi\lambda/2)^3]$$

$$F_B(\lambda) = g(\lambda)[0.923 + 0.199(1 - \sin \pi\lambda/2)^4]$$

where $g(\lambda) = 0.92(2/\pi)[(\tan \pi\lambda/2)/(\pi\lambda/2)]^{1/2}/\cos \pi\lambda/2$ and $\lambda = a/D$.

The magnification factors are essentially the rectangular bar solutions of Tada [9] which have been multiplied by the factors $(1.03/1.12)(2/\pi)$ to agree with the results of Smith [10] for a circular arc front. Equation 4 should have

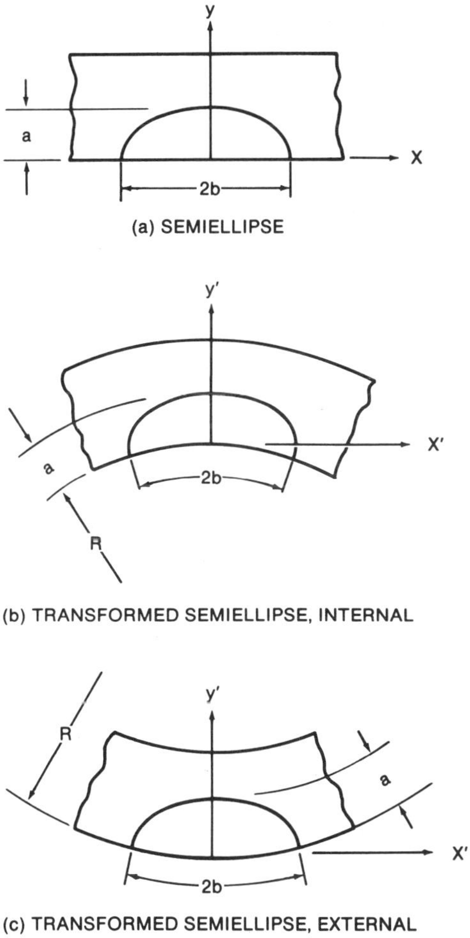

FIG. 10—*Transformed geometry for a semielliptical crack in a hollow cylinder.*

good accuracy for $a \ll D$, reasonable accuracy for $a < D/2$, and the proper limiting condition for $a \rightarrow D$.

Hollow Cylinder Results

The assumed transformation of a semielliptical surface crack in a flat plate to either an internal or external surface crack in a hollow cylinder is illustrated in Fig. 10 and given mathematically by the following equations:

(*a*) For the flat plate with a semielliptical crack:

$$\frac{x^2}{b^2} + \frac{y^2}{a^2} = 1 \tag{5}$$

(*b*) For the cylinder with an internal (+) or external (−) crack defined by radial and circumferential coordinates α and β:

$$\left[\frac{\beta}{b(1 \pm \alpha/R)}\right]^2 + \left[\frac{\ell n(1 \pm \alpha/R)}{\ell n(1 \pm \alpha/R)}\right]^2 = 1 \tag{6}$$

where

$$x' = (R \pm \alpha)\sin\left(\frac{\beta}{R \pm \alpha}\right)$$

$$y' = \pm\left[(R \pm \alpha)\cos\left(\frac{\beta}{R \pm \alpha}\right) - R\right]$$

$$0 \leq \alpha \leq a, \qquad -b \leq \beta \leq b$$

Comparisons of Eq 6 with digitized experimental results are shown in Figs. 11 and 12 for tension and bending fatigue cracks. The results verify that the transformation gives an accurate representation of the flaw geometry. Also, the results show that the transformed semielliptical geometry is attained relatively early with respect to the amount of fatigue crack growth for both the rectangular and circular arc starter notches.

In addition to the experimental studies of transformed semielliptical cracks, studies are being made concerning the stress-intensity factor representation. A method often used when expressing K_I for surface flaws in flat plates is to nondimensionalize the results with the embedded crack solution given by

$$K_I = \sigma\sqrt{\pi a}\ f(\phi)/\Phi \tag{7}$$

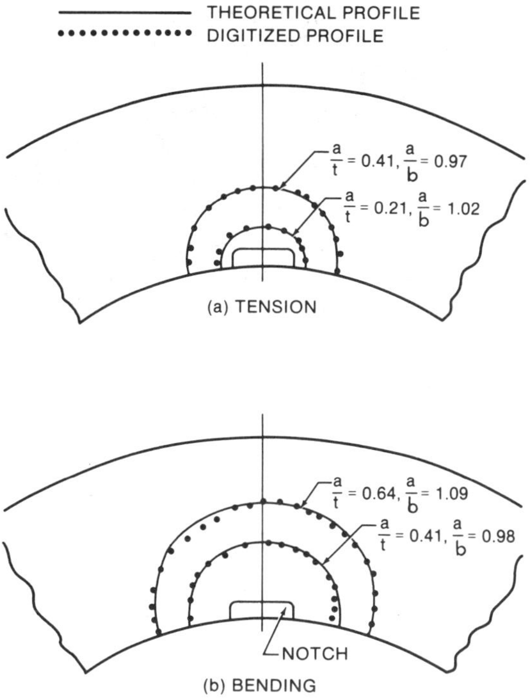

FIG. 11—*Profiles of internal cracks in hollow cylinder specimens.*

where Φ is the elliptic integral expressed as

$$\Phi = \int_0^{\pi/2} \sqrt{(\sin^2\phi + (a/b)^2 \cos^2\phi)} d\phi \tag{8}$$

and

$$f(\phi) = [\sin^2\phi + (a/b)^2 \cos^2\phi]^{1/4} \tag{9}$$

If a similar nondimensionalization procedure is used for the crack in a cylinder, a modified form of Φ and $f(\phi)$ is required. The authors are investigating this problem and have derived a preliminary expression for a transformed integral. Since Φ gives the ratio of the arc length of a semiellipse to the length of the major axis, a similar representation can be derived for the transformed ellipse. By performing this derivation, the modified integral, Φ', has the form

$$\Phi' = \int_0^{\pi/2} (1 \pm a/R)^{\sin\phi} \sqrt{\sin^2\phi + (A/B)^2 \cos^2\phi}\, d\phi \tag{10}$$

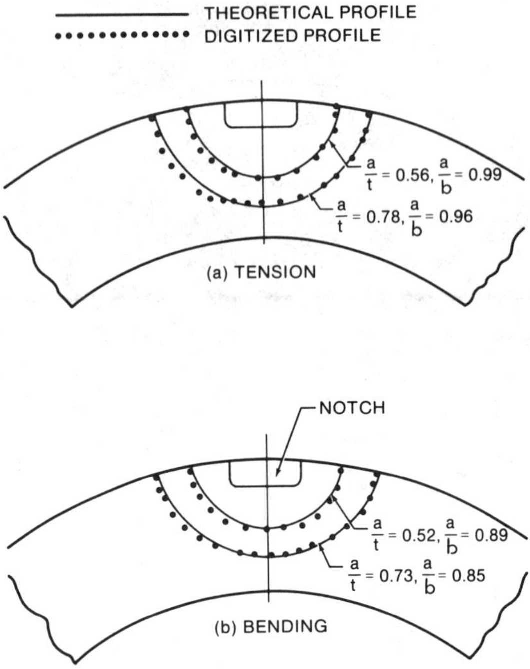

FIG. 12—*Profiles of external cracks in hollow cylinder specimens.*

where

$A = \ell n(1 \pm a/R)$, and
$B = b/R$.

Values for the transformed integral are plotted in Fig. 13 where Φ'/Φ is compared with a/b for different values of b/R. The results appear to have the proper trend. Specifically, the external crack which has a more flattened shape shows a decrease in Φ'/Φ (or increase in K_I at maximum crack depth). The internal crack which transforms into a more circularized shape has an increased Φ'/Φ or lower K_I than the flat plate crack. The modified integral also reduces to $\Phi' = \Phi = 1$ for $a/b = 0$. This agrees with the Nied and Erdogan [*11*] solution for the limiting case of an axisymmetric circumferentially cracked cylinder.

Conclusions

1. Experimental results for surface fatigue cracks in solid cylinders indicate that the geometry is accurately represented by a circular arc. The as-

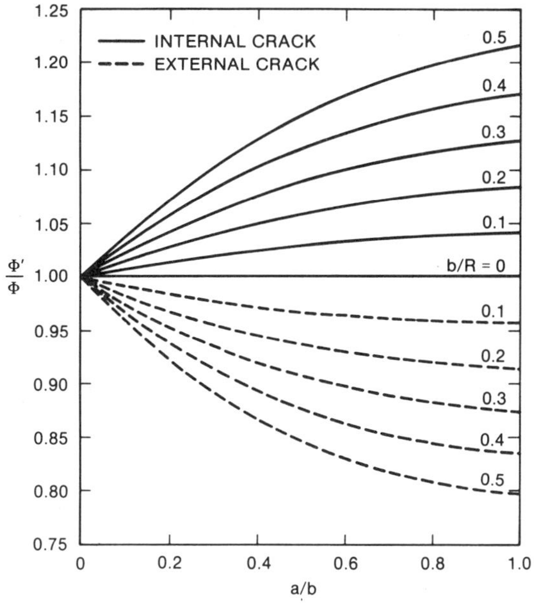

FIG. 13—*Variation of the transformed elliptic integral as a function of the shape factor,* a/b.

sumption of perpendicular intersection of the crack tip with the cylinder surface gives the maximum value for the arc radius of the crack.

2. A transformed semiellipse gives an accurate representation of the geometry for interior and exterior surface fatigue cracks in the circumferential plane of hollow cylinders.

3. A modified expression for nondimensionalizing stress-intensity factors for surface flaws in hollow cylinders is proposed and is shown to have the correct trends.

References

[1] Johnson, R. N., "Fracture of a Cracked Solid Circular Cylinder," Ph.D. dissertation, University of Wisconsin, Madison, 1972.

[2] Athanassiadis, A., Boissenot, J. M., Brevet, P., François, D., and Raharinaivo, A., *International Journal of Fracture*, Vol. 17, 1981, pp. 553-566.

[3] Nezu, K., Machida, S., and Nakamura, H. in *The 25th Japan Congress on Materials Research - Metallic Metals*, March 1982, pp. 87-92.

[4] Bush, A., *Journal of Testing and Evaluation*, Vol. 9, No. 4, July 1981, pp. 216-223.

[5] Daoud, O. E. K. and Cartwright, D. J., *Engineering Fracture Mechanics*, Vol. 19, No. 4, 1984, pp. 701-707.

[6] Wilhem, D., Fitzgerald, J., Carter, J., and Dittmer, D. in *Advances in Fracture Research*, D. François et al, Eds., Pergamon Press, Elmsford, N.Y., Vol. 1, 1980, pp. 11-21.
[7] Delale, F. and Erdogan, F., *Journal of Applied Mechanics, Transactions of ASME*, Vol. 49, March 1982, pp. 97-102.
[8] Raju, I. S. and Newman, J. C., this publication, pp. 789-805.
[9] Tada, H., Paris, P., and Irwin, G., *The Stress Analysis of Cracks Handbook*, Del Research Corp., Hellertown, Pa., 1973.
[10] Smith, F. and Alvi, M. in *Proceedings*, First International Conference of Pressure Vessel Technology, Delft, The Netherlands, 1969, pp. 793-800.
[11] Nied, H. F. and Erdogan, F., *International Journal of Fracture*, Vol. 22, 1983, pp. 277-301.

Walter L. Bradley,[1] *Keith E. McKinney,*[2] *and*
Paul C. Gerhardt, Jr.[3]

Fracture Toughness of Ductile Iron and Cast Steel

REFERENCE: Bradley, W. L., McKinney, K. E., and Gerhardt, P. C., Jr., **"Fracture Toughness of Ductile Iron and Cast Steel,"** *Fracture Mechanics: Seventeenth Volume, ASTM STP 905*, J. H. Underwood, R. Chait, C. W. Smith, D. P. Wilhem, W. A. Andrews, and J. C. Newman, Eds., American Society for Testing and Materials, Philadelphia, 1986, pp. 75–94.

ABSTRACT: The dynamic fracture behavior of ductile cast iron and cast steel has been studied to better explain why blunt notched Charpy bars indicate ductile iron has a quite inferior fracture toughness compared with cast steel with similar tensile properties (15 versus 75 J) while fracture mechanics tests indicate a much less significant difference in fracture toughness between the two materials. Using standard and fatigue precracked side-grooved Charpy specimens, it has been shown that the different predictions are due to a difference in constraint, the standard Charpy bar experiencing loss of constraint for the cast steel but remaining fully constrained for the ductile iron. For fatigue precracked and side-grooved Charpy specimens, initial crack extension occurred in both materials under conditions of full constraint, and the cast steel upper shelf fracture toughness was found to be only modestly better than the upper shelf fracture toughness of the ductile iron. Because of its lower ductile-to-brittle transition temperature, ductile iron has been found to have a superior fracture toughness to cast steel with similar tensile properties at temperatures below ambient.

KEY WORDS: ductile iron, cast steel, dynamic fracture, fracture toughness, K_{Id}

Ductile iron was introduced in the 1950s as an alternative to cast steel that was both more easily cast and machined. Unlike gray iron, it had good ductility which was achieved through the controlled addition of magnesium to the melt. The effect of magnesium additions was to cause the graphite to form as

[1]Professor of Mechanical Engineering, Texas A&M University, College Station, TX 77843.
[2]Formerly, Graduate Research Assistant, Mechanical Engineering Department, Texas A&M University, College Station, TX 77843; presently, Research Engineer, Shell Oil Company, New Orleans, LA 70068.
[3]Foundry Metallurgist, QIT-Fer et Titane, Chicago, IL 60603.

spherical nodules rather than as flakes as it does in gray cast iron. The spherical graphite nodules were much less detrimental to the ductility of the cast iron than were the graphite flakes, allowing the elongation in a 5 cm gage section to be increased from the 1 to 2% in gray iron to more than 20% in ductile iron.

Unfortunately, the timing of the application of magnesium to the melt to achieve spherical graphite nodules in the so-called ductile iron is rather critical, and occasionally gray iron with graphite flakes and poor ductility was accidentally made and marketed as ductile iron. The result was that ductile iron achieved a rather poor reputation for toughness in some industries in the 1950s. This poor reputation was further reinforced by its relatively low Charpy impact energy compared with cast steel. Thus ductile iron has been restricted to applications where toughness was not thought to be a significant design consideration.

In recent years, the evaluation of ductile iron using fracture mechanics testing has suggested that the toughness is much better than one would expect based on the low Charpy impact energies [*1-4*]. For example, recently measured values of critical stress intensity for ferritic ductile iron of 80 to 90 MPa $\sqrt{m}$ indicate that this material could tolerate a through-crack of 3.92 cm length in static applications approaching the yield strength (see Results and Discussion section). In fact, it has been certified to be suitable for construction of nuclear waste transport containers in Germany and is being considered for the same application in this country.

Reductions of 30% or more in component cost typically result when ductile iron is substituted for cast carbon steel due to a lower melting point and better machinability. Despite this impressive economic advantage and the new fracture mechanics indications of an acceptable level of toughness, ductile iron is still excluded from many applications because of concern over its presumed low fracture toughness, as indicated by Charpy impact values which seldom exceed 20 J [*5,6*].

The purpose of this project has been to carefully study the fracture behavior of ductile iron using Charpy specimens, modified to more nearly resemble the kinds of specimens used in fracture mechanics testing, to try to better understand this apparent discrepancy in the fracture toughness predicted by the two different methods. A cast steel with similar tensile properties has also been studied to further clarify the reason for the very different Charpy impact energies but modestly different K_{1c} values when one compares cast steel with ductile iron. Finally, two different quality ductile irons have been studied to determine the degree of sensitivity of the fracture toughness to processing.

In the next section, the rationale for the approach used in this work will be explained along with the details of the experimental approach that has been taken. The data analysis using fracture mechanics will also be outlined. In the following section, the results will be presented and their significance discussed. Finally, the conclusions of this work will be summarized.

Experimental Design, Procedures, and Analysis

The Charpy impact test differs from fracture mechanics tests in several important ways. Firstly, the fracture mechanics test coupon is always fatigue precracked before testing to give a very sharp crack (crack tip radius less than 0.0025 mm). The Charpy specimen, by comparison, is relatively blunt notched, having a V-notch with a root radius of 0.25 mm. Secondly, the fracture mechanics specimen is sufficiently large to give a triaxial tensile state of stress which approaches plane strain. Generally, fracture mechanics specimens are approximately 6.25 by 6.25 cm square and 2.5 cm thick. The standard Charpy specimen is only 1.0 cm thick, which is insufficient to give plane strain constraint in more ductile alloys. Thirdly, the Charpy specimen is tested under impact loading conditions at between 335 and 518 cm/s, whereas the standard fracture mechanics test for fracture toughness is conducted quasi-statically (ASTM Test for Plane-Strain Fracture Toughness of Metallic Materials [E 399]). Fourthly, fracture mechanics tests measure the energy required to initiate crack extension from a pre-existing flaw, but the Charpy test measures crack initiation energy plus crack propagation energy from a blunt notch.

The very different indications of relative toughness between ductile iron and cast steel previously noted for Charpy testing and fracture mechanics testing must somehow be related to these differences in test coupon geometry, rate of loading, and differentiation between fracture initiation energy and fracture propagation energy. These differences can have a significant effect on the ductile-to-brittle transition temperature as well as on the magnitude of the upper shelf fracture energy. For example, as one goes from a thin to a thick specimen, the state of stress may change from plane stress to plane strain, giving a significant increase in the transition temperature.

The presence of shear lips on the fractured surface of a Charpy specimen is clear evidence that plane stress conditions were present over a significant portion of the specimen adjacent to the surface. It is worth noting that Charpy specimens of cast ferritic/pearlitic steel always exhibit shear lips when tested at temperatures corresponding to the upper shelf. In contrast, ductile iron specimens do not develop shear lips at any temperature. Thus any comparison of upper shelf Charpy values for ductile iron and cast steel is an "apples to oranges" comparison that is bound to be misleading, particularly when the application envisioned involves components sufficiently thick to give plane strain constraint for both materials.

Materials also vary in their notch sensitivity (i.e, the reduction in strength as one goes from a blunt to a sharp notch). This variation might be due to differences in the respective fracture processes and the scale on which critical events must occur. Since manufacturing flaws or fatigue cracks introduced in service are sharp, the fracture behavior in the presence of sharp notches or cracks is far more important than the blunt notch behavior measured in a Charpy test.

To study the effect of notch acuity and constraint in Charpy tests, standard ductile iron and cast steel Charpy specimens were modified by side grooving and fatigue precracking. Half of the Charpy specimens were side grooved to a depth of 0.75 mm on each side from the Charpy V-notch to the back face of the specimen. Half of the regular and half of the side grooved Charpy specimens were then fatigue precracked to a notch plus crack depth of approximately 5.0 mm ($a/w = 0.5$). A total of 48 specimens of ductile iron and 48 specimens of cast steel were tested in this portion of the investigation. An additional 20 specimens of a better quality ductile iron were tested in a subsequent portion of the study to evaluate the sensitivity of the fracture toughness to materials processing. Since it was determined that side grooving had a minimal effect on the fracture toughness measured using fatigue precracked ductile iron specimens, these additional specimens were all tested without side grooving.

Material Selection and Sample Preparation

Ductile iron grade 60-40-18 (ASTM Specification for Ductile Iron Castings [A 536]) has a minimum yield strength of 276 MPa. The selection of a comparable grade of steel was based on yield strength. Valve grade WCB (ASTM Specification for Steel Castings, Carbon Suitable for Fusion Welding for High-Temperature Service [A 216]) was selected because it is a widely used steel casting alloy with a minimum yield strength of 248 MPa. Typical microstructures of the materials are shown in Figs. 1 and 2.

It was necessary to reduce the yield strength of the ductile iron to match that of the cast steel. This was accomplished by producing the iron with a silicon content of 2.20%, which is lower than "normal". A typical range for silicon concentration for this grade of ductile iron is from 2.50 to 2.80%.

The ductile iron used in this study was produced in two separate heats in grade 60-40-18 (as-cast) as specified by ASTM A 536. The metallic charge, consisting of 19% high purity pig iron, 38% steel scrap (#1 bushelings), and 43% ductile iron returns, was melted in an acid-lined channel induction furnace. The nodulizing treatment employed a 5% magnesium ferrosilicon in a covered ladle, with subsequent "post" inoculation using 75% ferrosilicon (inoculating grade). The molds used for the bars (8.75 by 8.75 by 45 cm) and keel blocks were made with an air setting resin/catalyst binder. Chemical composition and mechanical properties are summarized in Tables 1 and 2, respectively.

The first heat of ductile iron met the ASTM A 536 specification, but the toughness data were below expectation due to minor quantities of dross, microshrinkage, and degenerate graphite (Fig. 2). Appropriate changes were made in the processing so that the second heat of ductile iron was generally free of the imperfections of the first heat (Fig. 3). This was done by changing the gating and risering system and by changing the geometry from a 10 cm

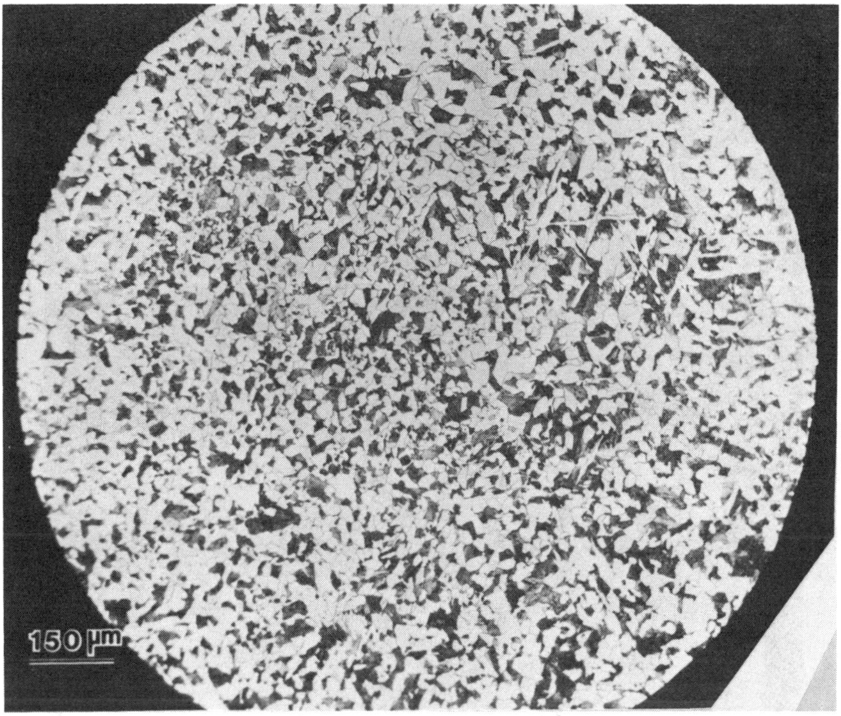

FIG. 1—*Microstructure of cast steel.*

square bar to a 2.5 cm thick plate. The first heat will be called *average quality* ductile iron throughout the remainder of this paper, while the second heat will be called *good quality* ductile iron.

The cast steel used in this study was produced in accordance with ASTM A 216 to give cast steel grade WCB (normalized at 940°C) in three bars measuring 7.5 by 7.5 by 50 cm. The metallic charge consisting of 40% steel scrap (cut structural and plate steel) and 60% steel returns was melted in a basic lined arc furnace. The molds used for the bars and keel blocks were made with an air setting resin/catalyst binder.

Experimental Procedures and Analysis

All Charpy specimens were tested using an instrumented Charpy impact tester to give a dynamic load-time record which could be analyzed to give the initiation energy, propagation energy, total fracture energy, and dynamic critical stress intensity factor for fracture. Instrumentation for the impact test consisted of an instrumented tup, dynamic response module, digital storage oscilloscope, *X-Y* plotter for hard copy output, and a microcomputer for data

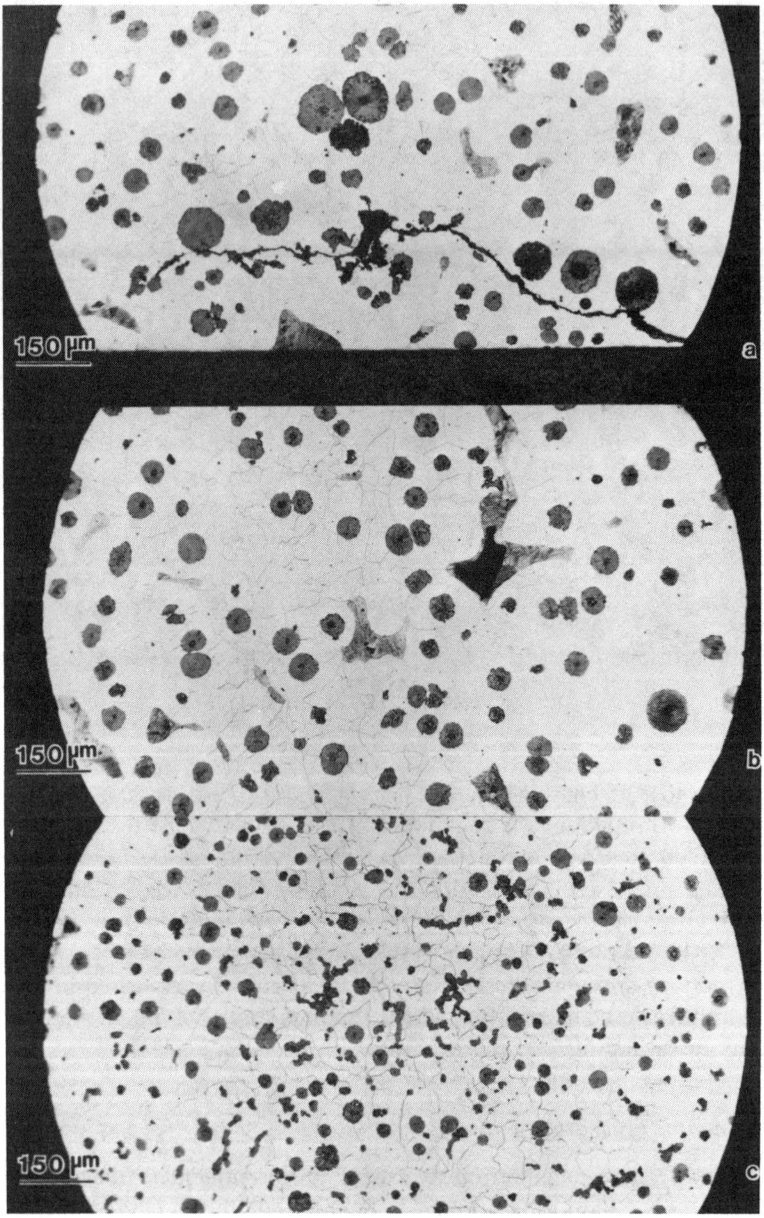

FIG. 2—*Microstructure of average-quality ductile iron showing* (a) *some slag*, (b) *shrinkage*, *and* (c) *vermicular graphite*.

TABLE 1—*Chemical composition (wt%) of materials studied.*

Material	C	Mn	Si	P	S	Cr	Ni	Mo	Cu	Mg
Average-quality ductile iron	3.77	0.294	2.15	0.015	0.020	0.042	0.030	0.017	0.063	0.05
Better-quality ductile iron	3.84	0.280	2.20	0.019	0.011	0.030	0.030	<0.01	0.110	0.047
Cast steel	0.25	0.82	0.38	0.014	0.012	0.190	0.180	0.080	0.280	...

TABLE 2—*Mechanical properties of materials studied.*

Material	Yield Strength, ksi (MPa)	Tensile Strength, ksi (MPa)	Elongation, %	Reduction of Area, %
Average-quality ductile iron	43.0 (297)	60.4 (417)	24	24
Better-quality ductile iron	46.4 (320)	65.0 (448)	23	23
Cast steel	46.0 (317)	80.3 (554)	28	50

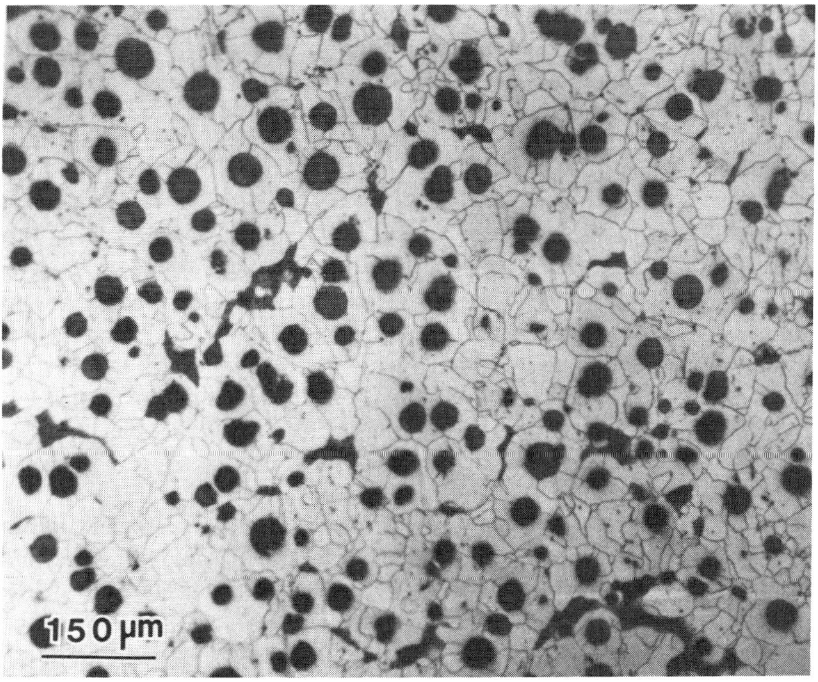

FIG. 3—*Microstructure of better-quality ductile iron.*

reduction. Standard momentum transfer equations were used to calculate the various energies of initiation and propagation from the measured load-time records as follows:

$$E = V_0 \int P\mathrm{dt} \cdot (1 - V_0 \int P\mathrm{dt}/4E_0) \tag{1}$$

where V_0 is the velocity of the tup on impact, E_0 is the initial potential energy, and P is the measured load as a function of time.

Crack initiation was assumed to occur at maximum load for dynamic loading. Such behavior has been verified for quasi-static fracture of ductile iron [*4*] and was also found to be the case in this study. On the other hand, the quasi-static fracture testing of the cast steel in this program indicated that crack extension began somewhat before maximum load. This would make the initiation energies and the fracture toughnesses calculated assuming crack initiation began at maximum load somewhat nonconservative upper bounds for the cast steel, overestimating the actual value of K_{Id} by 10 to 15%.

The dynamic critical stress intensity K_{Id} was estimated by first calculating the value of J_{Id} using the standard relationship $J = 2A/Bb$, where A is the area under the load-displacement curve to maximum load, B is the specimen thickness, and b is the uncracked ligament width, $W - a$ [*8*]. The equivalent dynamic stress intensity is then calculated using the relationship $K_{\mathrm{Id}} = \sqrt{JE}$, where E is the Young's modulus for the particular material. It is worth noting that ASTM E 813 requires the measurement of a *J-R* curve to obtain a valid J_{Ic}. Thus the J_{Id} values measured in this work by assuming crack growth begins at maximum load are not valid results as per ASTM E 813. It is believed by the authors that the upper shelf values of J_{Id} and K_{Id} calculated with the approach described herein are accurate. The most significant potential error introduced by this approach is that the actual transition temperature measured on larger specimens might be higher than the results obtained in this work.

Several fatigue precracked Charpy specimens and several compact tension specimens were tested quasi-statically to determine values for K_{Ic} on the upper shelf for the cast steel and ductile iron. The K_{Ic} values were calculated from J_{Ic} values as measured per ASTM E 813 using a single specimen approach with unloading compliance measurements made for the compact tension specimens. Crack growth was assumed to begin at maximum load for the fatigue precracked Charpy specimens tested quasi-statically.

Several of the fractured Charpy specimens were sectioned perpendicular to the fracture surface and polished so that a hardness map could be made using a Tukon hardness tester. A similar map was made after polishing an outside surface of fractured Charpy specimens. The hardness mapping allowed the extent of the plastic deformation zone to be determined for both initiation and propagation, at the center and on the surface of ductile iron and cast steel specimens. These results were subsequently correlated with the measured initiation and propagation energies.

Results and Discussion

Typical load-time traces measured during the dynamic fracture process are presented in Fig. 4. Note the different load scale used for the blunt notch specimen and different time scale used for the typical brittle fracture results. The initiation energy for the blunt notch specimens (assumed to be related to the area under the load-time trace to maximum load) is seen to be quite large compared with the initiation energy for fatigue precracked specimens. The propagation energies are much less sensitive to original notch geometry. The precipitous drop in load after crack initiation seen in Fig. 4*a* is a result of the fact that the load the specimen can support with an initially blunt notch is much greater than the load it can subsequently support once a sharp crack has grown from the blunt notch.

Figure 4*b* shows a typical load-time trace for ductile fracture of a fatigue precracked specimen. A best-fit curve was approximated through the measured curve, which includes some inertial effects. A maximum load in the absence of inertial effects was estimated, and is indicated by "*b*" in Fig. 4*b*, as distinct from the measured maximum load labeled "*a*" in Fig. 4*b*.

The first peak prior to maximum load in Figs. 4*a* and 4*b* is thought to be the result of displacement of the specimen to a more firm position against the anvil than it had when initially placed in position [*8*]. Thus the first portion of the load-time trace represents displacement of the specimen rather than bending of the specimen and therefore should be ignored in the calculation of the *J*-integral.

The total energy absorbed in fracture for the various ductile iron and cast steel specimens tested over a range of temperatures is presented in Fig. 5. Because some specimens were side-grooved, some fatigue precracked, and some both side-grooved and fatigue precracked, the total fracture energy was normalized to the area of material fractured to make the comparison from these various geometries more meaningful.

The decrease in fracture energy with fatigue precracking and side grooving is greater in the cast steel than in the ductile iron. It should be noted that the average quality ductile iron specimens with fatigue precracking had essentially the same results with and without side grooving. Thus the better quality ductile iron was tested with fatigue precracking only, without side grooving. When compared with the average quality ductile iron, the better quality ductile iron showed only a slightly improved Charpy impact energy for fatigue precracked specimens, whereas the dynamic fracture toughness as measured by K_{Id} will be seen presently to be quite improved. This suggests that the

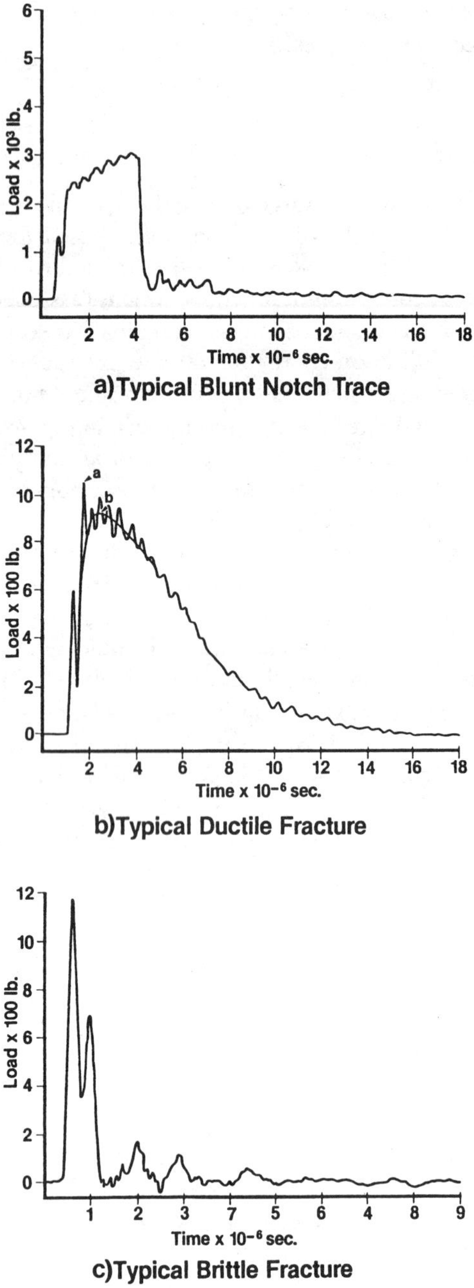

FIG. 4—*Typical load-time traces for* (a) *ductile fracture of a blunt notched specimen*, (b) *ductile fracture of a fatigue precracked specimen and* (c) *brittle fracture of a fatigue precracked specimen.*

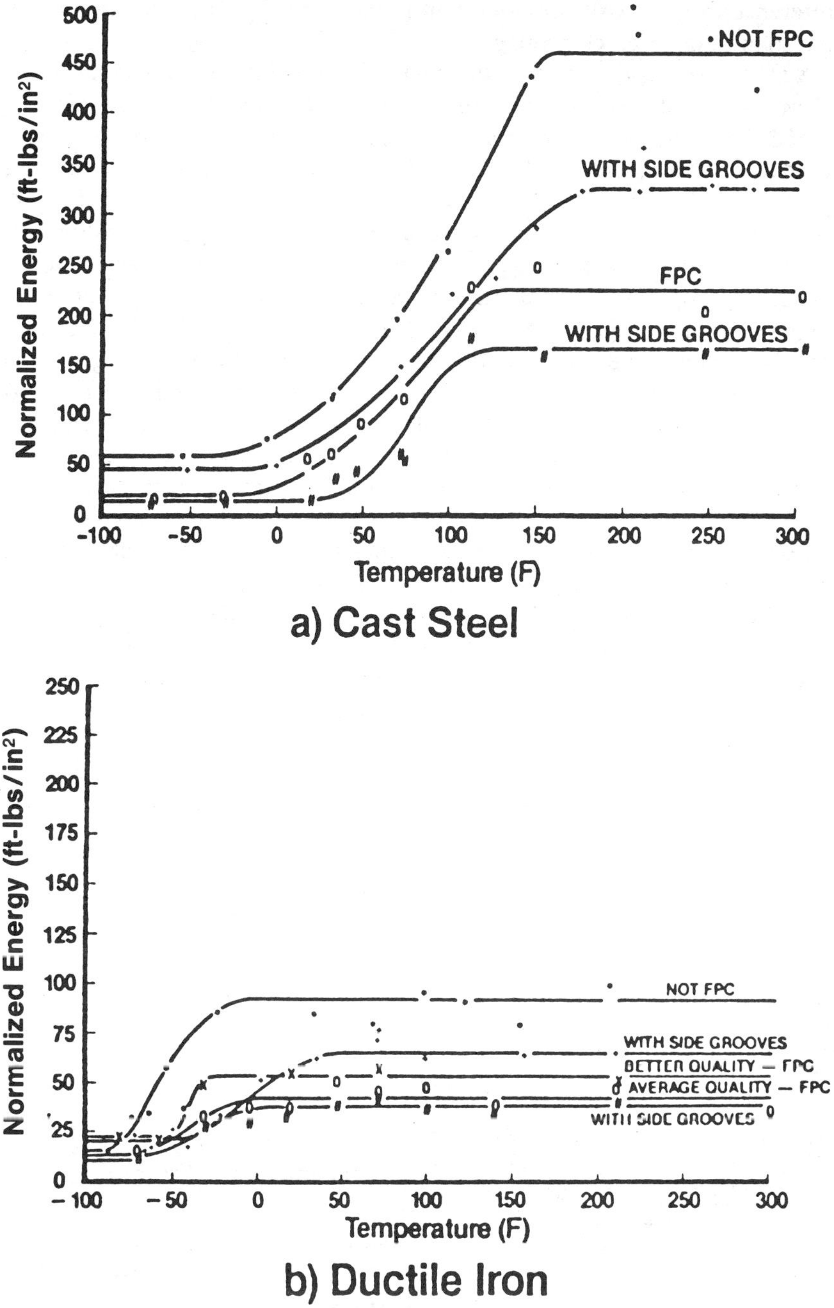

FIG. 5—*Total energy absorbed in dynamically fracturing standard Charpy specimens, Charpy specimens with side grooving, fatigue precracking, and side grooving/fatigue precracking for* (a) *cast steel and* (b) *ductile iron (1 ft-lb/in.*^{2} *= 2100 J/m*^{2}*).*

difference in quality of the ductile iron had a much larger effect on crack initiation than on crack propagation.

A more direct comparison of the ductile iron to cast steel is seen in Fig. 6, where total fracture energy for fatigue precracked Charpy specimens is presented. The cast steel is seen to be quite superior on the upper shelf; however, the ductile iron appears to be better below ambient temperatures due to a much lower ductile to brittle transition temperature.

The calculated values of K_{1d} for the average quality and good quality ductile iron are presented in Fig. 7. Though the standard Charpy values for the two ductile irons were similar, the critical stress intensity for crack initiation is seen to be much higher in the better quality ductile iron. The K_{1d} values for the cast steel and the better quality ductile iron are presented in Fig. 8. The upper shelf K_{1d} values for the cast steel should be considered upper bound values because crack growth probably began somewhat before maximum load (which was assumed to be the moment of crack growth) and because the ASTM E 813 values for B were not quite satisfied, meaning plane strain constraint was probably not achieved, even at the center of the test specimens. Subsequent quasi-static fracture mechanics tests on 2.5 cm thick compact tension specimens gave J_{1c} values with corresponding K_{1c} values of 150 to 160 MPa$\sqrt{m}$ compared with 180 MPa$\sqrt{m}$ for K_{1d} measured on side-grooved Charpy specimens and 200 + MPa$\sqrt{m}$ on Charpy specimens without side grooving.

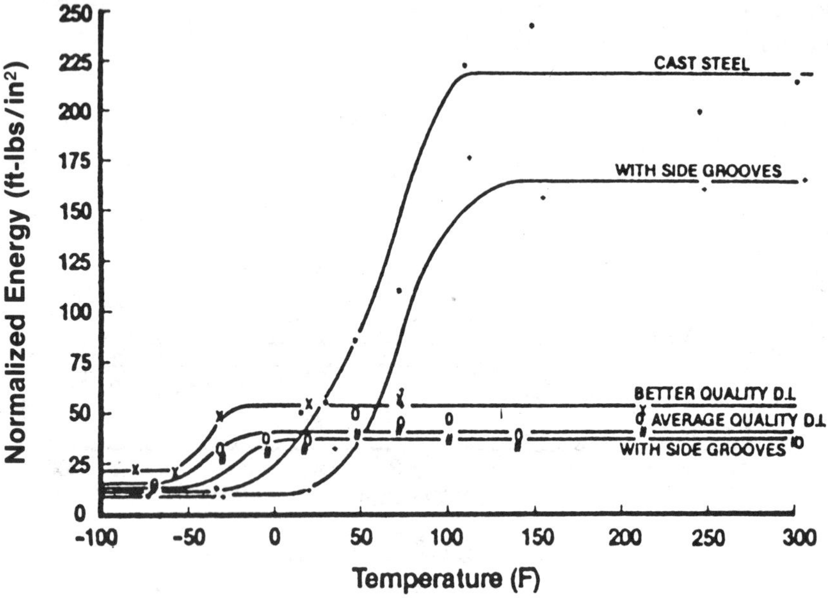

FIG. 6—*Comparison of total energy absorbed during dynamic fracture of fatigue precrack Charpy bars of cast steel and ductile iron (1 ft-lb/in.² = 2100 J/m²).*

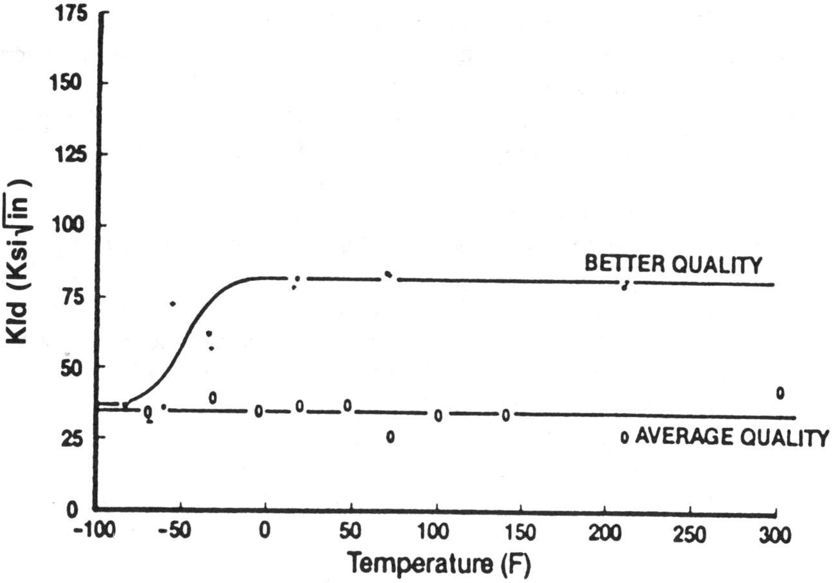

FIG. 7—*Comparison of dynamic critical stress intensity* K_{Id} *of average-quality ductile iron to better-quality ductile iron, calculated from* J_{Id} *values* (1.09 MPa$\sqrt{m}$ = 1 ksi$\sqrt{in.}$).

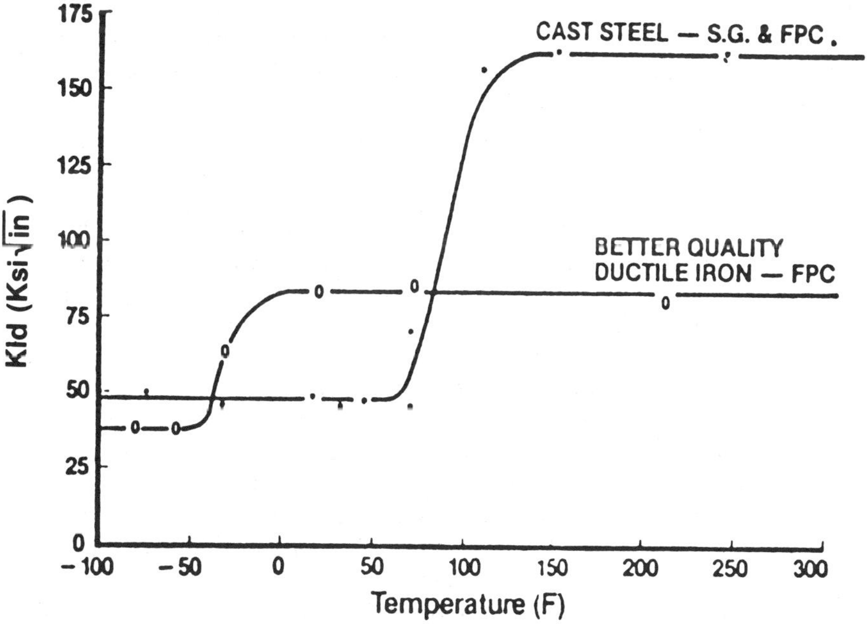

FIG. 8—*Comparison of dynamic critical stress intensity* K_{Id} *for ductile iron to cast steel, calculated from* J_{Id} *values* (1.09 MPa$\sqrt{m}$ = 1 ksi$\sqrt{in.}$).

The lower shelf values for K_{Id} are seen to be somewhat high than one would expect for cleavage in ductile iron or cast steel. The most likely explanation for this apparent overestimate of K_{Id} for brittle fracture is the effect of specimen inertia. There is still significant inertial loading to accelerate the specimen at the moment of crack extension for brittle fracture. Thus the load measured by the tup at this time is the sum of the load-producing acceleration of the Charpy specimen and the load that is bending the same specimen. Only the bending load should be included in the fracture mechanics calculations of K_{Id}. If one assumes that the average velocity of the specimen at the moment of crack growth is one half the tup velocity (ends of specimen being stationary and center moving at tup velocity), one can calculate the amount of kinetic energy imparted to the specimen during the fracture initiation process. If this kinetic energy is subtracted from the total energy used in a J-integral calculation, then a J_{Id} for specimen bending apart from specimen acceleration can be calculated. The lower shelf K_{Id} values calculated from the J_{Id} values are found to be approximately 30 and 25 MPa$\sqrt{m}$ for the cast steel and ductile iron, respectively, which are typical toughness values for cleavage in ferritic material.

The most interesting result from this work is the observation from Fig. 8 that the ductile iron has a superior fracture toughness as measured by K_{Id} when compared with cast steel from 25°C (77°F) down to −40°C (−40°F). Furthermore, the upper shelf K_{Id} values are not nearly as different as one might have expected based on a comparison of ductile iron Charpy values to cast steel Charpy values (15 versus 75 J). The superior behavior of the ductile iron below ambient temperature is due to a 60°C (140°F) lower ductile-to-brittle transition temperature.

The reason why the fracture toughness of ductile iron is superior to a cast steel with similar tensile properties when tested at room temperature or below is clearly seen in the fractographic results presented in Fig. 9. The cast steel fails by cleavage when loaded dynamically at temperatures below ambient, whereas the ductile iron fails by void coalescence, or ductile, fracture. For comparison purposes, quasi-static fracture of both materials at room temperature is also shown; both are seen to fracture in a ductile manner. The reason for this difference in transition temperature will be discussed later.

Returning to the question posed at the beginning of this paper, the discussion concludes by attempting to explain why the Charpy results indicate a relatively poor toughness for ductile iron while fracture mechanics results suggest a quite good toughness for this material. Since fracture initiation energy is related to critical stress intensity squared, one might expect the differences in K_{Id} determined in this work to give a 3× difference in Charpy energy rather than the 5× difference we actually measured. The upper shelf standard Charpy energy for the cast steel was approximately 83 J, for the average quality ductile iron 16 J, and for the good quality ductile iron 17 J.

The upper shelf Charpy results for the cast steel and the two different qual-

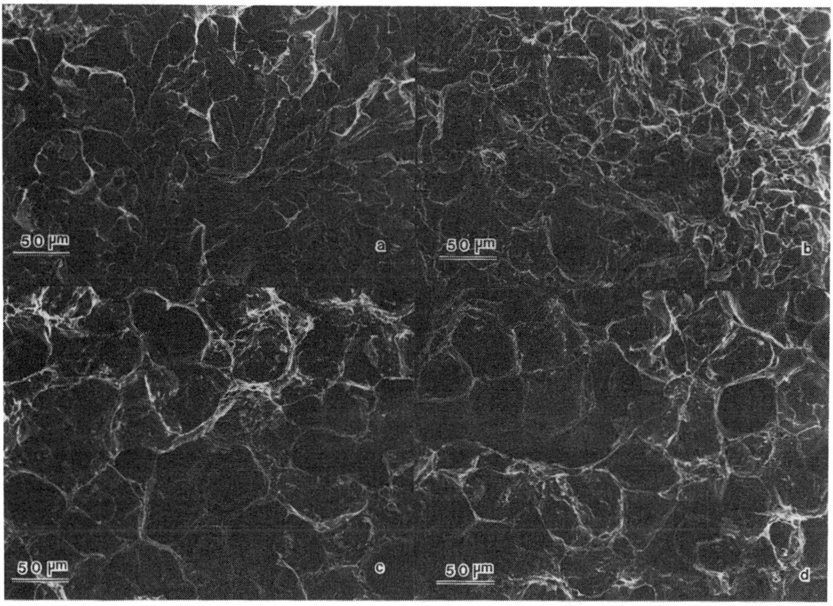

FIG. 9—*Fractographic results from SEM taken on fractured fatigue precracked Charpy bars of* (a) *cast steel, fractured dynamically at room temperature;* (b) *cast steel, fractured quasi-statically at room temperature;* (c) *ductile iron fractured dynamically at room temperature; and* (d) *ductile iron fractured quasi-statically at room temperature.*

ity ferritic ductile irons in the four types of specimens tested are presented in Table 3. The initiation and propagation energies are shown so that the effect of notch acuity and constraint on each of these can be determined. It should be noted that the standard Charpy test measures the total energy to fracture the specimen. Thus initiation and propagation energies are combined. By contrast, fracture mechanics considers only the initiation energy in calculating J_{1c} and J_{1d} or K_{1c} and K_{1d}.

Comparing the effect of notch acuity on initiation energy, the cast steel showed a 68% decrease in going from a blunt notched to a fatigue precracked specimen, while the good quality ductile iron showed a 53% decrease (Table 3). A more significant difference in the behavior of the two materials is seen by the effect of side grooving, where the initiation energy of the cast steel is seen to drop by an additional 35%. Side grooving the ductile iron produced essentially no difference in the fracture behavior for the fatigue precracked specimens. The combined effect of notch acuity (blunt notch versus fatigue precracking) and constraint (side grooved versus no side grooving) reduces the fracture initiation energy for the standard Charpy bar test of cast steel by 79%, while the same effects reduced the initiation energy of the good quality ductile iron by only 53%. Thus only 21% of the fracture initiation energy on

TABLE 3—*Normalized initiation and propagation energies* (J/mm^2) *for various Charpy impact tests on the upper shelf.*

	Cast Steel		Average-Quality Ductile Iron		Good-Quality Ductile Iron	
	Initiation Energy	Propagation Energy	Initiation Energy	Propagation Energy	Initiation Energy	Propagation Energy
Standard Charpy V-notch (CVN)	0.383	0.622	0.053	0.146	0.055	0.155
CVN with Side Grooving (SG)	0.292	0.374	0.042	0.083	...	...
CVN with Fatigue Precracking (FPC)	0.122	0.311	0.005	0.090	0.025	0.089
CVN, SG, FPC	0.079	0.253	0.005	0.068	[a]	...

[a]Assumed to be the same as CVN and FPC, based on results for average-quality ductile iron; namely, 0.025.

the upper shelf will be realized as initiation energy in a fatigue precracked, fully constrained specimen of cast ferritic/pearlitic steel, while 47% of the upper shelf fracture initiation energy measured in a good quality ductile iron Charpy test will be required to initiate crack extension in a fracture mechanics test or, more importantly, in a real component with a sharp flaw. This accounts for the very different indications of toughness found between a Charpy test of the ductile iron and a fracture mechanics test of the same material.

Microhardness mapping was used to define the region of plastic deformation which accompanied crack initiation and crack growth in the cast steel and ductile iron (Fig. 10). This information helped to further explain the different indications of relative toughness given by Charpy and fracture mechanics results for ductile iron and cast steel. The ductile iron was found to have a plastic zone that is fully constrained (surrounded by elastic material) at the moment of crack extension for both the blunt notch and fatigue precracked specimens. By contrast, the cast steel formed a fully plastic hinge for the blunt notched specimen prior to crack extension, whereas the fatigue precracked specimen was nearly fully constrained at the moment of crack exten-

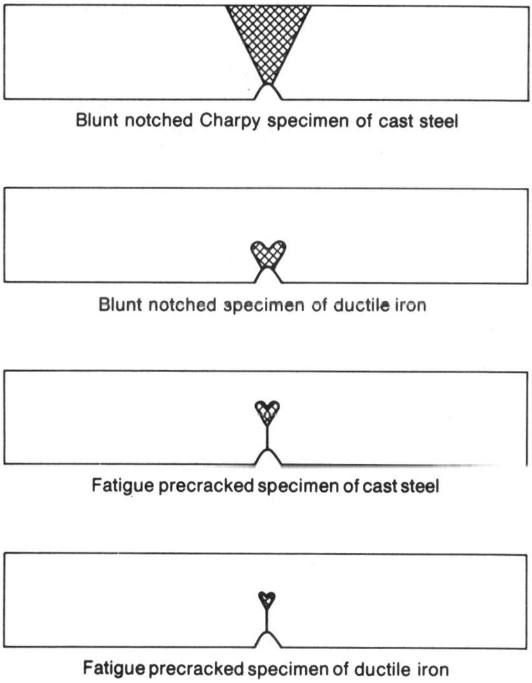

FIG. 10—*Plastic zone size at initiation of cracking for blunt notched and fatigue precracked Charpy specimens of ductile iron and cast steel (as inferred by microhardness testing).*

sion, since the plastic zone was fully surrounded by elastic material. Thus the comparison of blunt notch Charpy values for cast steel, which forms both shear lips and a plastic hinge before crack initiation, with blunt notch Charpy values for ductile iron, which forms no shear lip and is constrained by surrounding elastic material, is most inappropriate.

Since the fatigue precracked, side-grooved cast steel specimen has neither shear lips nor loss of constraint during Charpy impact testing, it gives a much more meaningful comparison to the ductile iron, which neither forms shear lips nor has loss of constraint. It is this significant difference in constraint in the standard Charpy test that gives such erroneous indications of relative toughness when comparing cast steel to ductile iron.

It is important to emphasize that the absolute toughness of the cast steel is still seen to be quite superior to ductile iron on the upper shelf. However, the degree of superiority is greatly overestimated by a Charpy test. Also, the superiority of the cast steel over the ductile iron used in this study is only observed at temperatures above room temperature, where one is generally less concerned about brittle fracture.

A comparison of the average quality and good quality ductile iron used in this study is given in Table 3. The data emphasize the importance of casting and metallurgical quality in achieving optimum toughness properties. Avoiding defects caused by shrinkage, slag, and other inclusions, as well as achieving a high degree of nodularity, are very important. It is also worth noting that the Charpy tests did not seem to be very sensitive to the very different fracture toughness behavior noted for the two ductile iron heats.

A few final comments are in order regarding the reasons why ductile iron has a lower value for upper shelf fracture toughness and a lower ductile to brittle transition. It is believed that the lower value for the upper shelf fracture toughness energy for the ductile iron is a result of graphite nodule debonding in the region of the crack tip, which leads to local softening of the matrix, strain localization, and final fracture. The cast steel having no graphite nodules will require a considerably greater degree of straining to achieve strain localization via void nucleation and growth at second-phase particles.

The lower ductile-to-brittle transition temperature of the ductile iron is thought to be the result of the different strengthening mechanisms employed in the two materials. While the yield strengths of the two alloys are similar, the matrix of the ductile iron is strengthened by silicon in solid solution, whereas the matrix of the ferritic cast steel is strengthened by a dispersion of pearlite. Cleavage cracks are more easily nucleated in pearlite than in ferrite, which may explain the higher observed ductile-to-brittle transition temperature of the cast steel.

Finally, the upper shelf fracture toughness of ferritic ductile iron needs to be converted into an equivalent critical flaw size to give a better physical feel for the material's flaw tolerance. Using the originally measured values of J_{1c} previously reported as equivalent values of K_{1c} (via $J_{1c}E = K_{1c}^2$), one may calculate the critical crack tip opening displacement for plane strain fracture toughness using the standard relationship

$$J_{1c} = 1.8 S_y \delta_c \tag{2}$$

Assuming a typical upper shelf value of 90 MPa$\sqrt{m}$ for K_{1c}, a value of 173 GPa for modulus, and a yield strength of 320 MPa, the critical crack opening displacement may be calculated to be 0.00817 cm. Then, using British Standard PD 6493:1980, "Guidance on Some Methods for the Derivation of Acceptance Levels for Defects in Fusion Welded Joints," one may determine the critical flaw size to be 1.96 cm for a specimen with a through-crack loaded to its yield strength. Finally, recognizing that the PD 6493 has a built in safety factor of 2×, the actual critical flaw size for a specimen with a through crack loaded to its yield strength would be 3.92 cm. This relatively large value indicates that ferritic ductile iron is much more flaw tolerant than is widely recognized.

Conclusions

The conclusions of this study may be summarized as follows:

1. Charpy impact tests give an erroneous indication of relative toughness when comparing ferritic ductile iron to cast steel due to the loss of constraint in the blunt notched Charpy bars of cast steel.
2. The fracture toughness of ductile iron as measured by the dynamic critical stress intensity factor K_{1d} is actually superior to that of cast steel at temperatures below ambient because of its much lower ductile-to-brittle transition temperature.
3. The upper shelf dynamic fracture toughness of cast steel is superior to that of ductile iron, but by a much smaller amount than one would predict based on the 5× larger blunt notched Charpy values.
4. Good quality ferritic ductile iron can tolerate a through crack up to 3.92 cm long at stresses up to its yield strength.
5. The fracture toughness of the ferritic ductile iron is sensitive to processing.
6. The blunt notch Charpy test may not be very sensitive to variations in processing in ductile iron which significantly affect the fracture toughness as measured by a fracture mechanics approach.

Acknowledgments

The authors wish to thank QIT-Fer et Titane Inc., Montreal, Quebec, for their financial support and encouragement for this project.

References

[*1*] Mead, H. E., Jr. and Bradley, W. L., "Fracture Toughness Studies of Ductile Cast Iron Using a J-Integral Approach," AFS Transactions, 1980, pp. 265-276.

[*2*] Bradley, W. L., "Fracture Toughness Studies of Gray, Malleable and Ductile Cast Iron," AFS Transactions, 1981, pp. 837-848.

[*3*] Bradley, W. L. and Andler, M. R., "Dynamic Fracture Behavior of Nodular Cast Iron," presented to International Conference on Fracture Prevention in Energy and Transportation Systems, Rio de Janiero, Dec. 1983.

[*4*] Salzbrenner, R. J., Van Den Avyle, J. A., Lutz, T. J., and Bradley, W. L., "Fracture Toughness Testing of Ductile Cast Irons," in *Fracture Mechanics: Sixteenth Symposium, ASTM STP 868*, American Society for Testing and Materials, Philadelphia, 1985, pp. 328-345.

[*5*] Andler, M. R., "The Dynamic Fracture Toughness of Nodular Cast Iron," S. M. thesis, Texas A&M University, College Station, Tex., 1983.

[*6*] *The Iron Castings Handbook*, C. F. Walton and T. J. Opar, Eds., Iron Casting Society, Chicago, Ill., 1981.

[*7*] Wullaert, R. A. in *Impact Testing of Metals, ASTM STP 466*, American Society for Testing and Materials, Philadelphia, 1970, pp. 148-164.

[*8*] Rice, J. R., Paris, P. C., and Merkle, J. G., "Some Further Results of J-Integral Analysis and Estimates," in *Progress in Flaw Growth and Fracture Toughness Testing, ASTM STP 536*, American Society for Testing and Materials, Philadelphia, 1973, pp. 231-245.

[*9*] Luyendisjk, L. and Nieswaag, H. in *Proceedings*, 49th International Foundry Congress, 1982, pp. 1-13.

[*10*] Dawes, M. G., "Elastic-Plastic Fracture Toughness Based on the COD and *J*-Contour Integral Concepts," in *Elastic-Plastic Fracture, ASTM STP 668*, J. D. Landes, J. A. Begley, and G. A. Clarke, Eds., American Society for Testing and Materials, Philadelphia, 1979, pp. 307-333.

B. M. Liaw,[1] *A. S. Kobayashi,*[1] *and A. F. Emery*[1]

Effect of Loading Rate on Dynamic Fracture of Reaction Bonded Silicon Nitride

REFERENCE: Liaw, B. M., Kobayashi, A. S., and Emery, A. F., **"Effect of Loading Rate on Dynamic Fracture of Reaction Bonded Silicon Nitride,"** *Fracture Mechanics: Seventeenth Volume, ASTM STP 905*, J. H. Underwood, R. Chait, C. W. Smith, D. P. Wilhem, W. A. Andrews, and J. C. Newman, Eds., American Society for Testing and Materials, Philadelphia, 1986, pp. 95–107.

ABSTRACT: Wedge-loaded, modified tapered double cantilever beam (WL-MTDCB) specimens under impact loading were used to determine the room temperature dynamic fracture response of reaction bonded silicon nitride (RBSN). The crack extension history, with the exception of the terminal phase, was similar to that obtained under static loading. Like its static counterpart, a distinct crack acceleration phase, which was not observed in dynamic fracture of steel and brittle polymers, was noted. Unlike its static counterpart, the crack continued to propagate at nearly its terminal velocity under a low dynamic stress intensity factor during the terminal phase of crack propagation. These and previously obtained results for glass and RBSN show that dynamic crack arrest under a positive dynamic stress intensity factor is unlikely in static and impact loaded structural ceramics.

KEY WORDS: dynamic fracture, dynamic stress intensity factor, dynamic crack arrest, ceramics, reaction bonded silicon nitride, impact loading

Nomenclature

a	Crack velocity
Δa	Crack extension
C_1	Dilatational stress wave velocity
K_{Id}	Dynamic initiation fracture toughness
K_{ID}	Dynamic fracture toughness
K_I^{dyn}	Dynamic stress intensity factor
t	Time

[1]Department of Mechanical Engineering, University of Washington, Seattle, WA 98195.

Introduction

Despite the inherent brittleness of structural ceramics, linear elastic fracture mechanics (LEFM) has been used to analyze ceramic failures with limited success. The difficulty lies in the quoted fracture toughness, which is an order of magnitude lower than those of common structural metals, resulting in critical flaw dimensions that approach the grain size. The stress intensity factor formulas, which are used to quantify ceramic fracture, thus violate the basic LEFM postulates of isotropy and homogeneity and lead to inconclusive or incorrect results. The smallness of the critical flaw size also taxes the sensitivity limit, which is barely adequate for detecting critical flaws in metals, of nondestructive inspection. On the positive side, LEFM has been used somewhat successfully for proof testing and for predicting the life of a ceramic structure that is subjected to stable crack growth. The life-time prediction procedure for ceramics is to relate the residual strength after proof testing to the uniaxial tensile strength without involving any fracture toughness or crack geometry. These and other unconventional applications of LEFM in ceramic fracture are the precursors to a mature state-of-science that is emerging from this field.

Despite extensive research efforts in dynamic fracture of metals, a comparable effort in structural ceramics is virtually nonexistent. The dynamic fracture initiation toughness, K_{Id}, which was found to be essentially constant throughout the range of room temperature to 1400°C, for silicon carbide has been reported [*1*]. The dynamic stress intensity factors of rapidly propagating cracks in plate glass [*2*] and reaction bonded silicon nitride [*3*], which were subjected to static overloading at room temperature, have also been determined. Unlike their metal counterpart, the terminal crack velocity in ceramics is preceded by a crack acceleration phase which has been observed in glass ceramics [*4*]. Also unlike its metal counterpart, the crack continues to propagate in ceramics under a dynamic stress intensity factor approaching zero. The unique gamma-shaped K_{ID} versus $\dot{a}$ relations of polymers [*5*] and metals [*6*] apparently do not exist and the propagating crack will not arrest in ceramics. The purpose of this paper is to further explore this unusual dynamic fracture response of structural ceramics under impact loading.

Method of Approach

A hybrid experimental-numerical procedure [*7,8*], where the experimentally determined crack extension history drives a finite element code in its generation mode, was used in this study. The dynamic finite element analysis in turn yields numerically the dynamic stress intensity factor [*9*] and energy partitions [*10*] during crack propagation.

Material and Specimen

The material used was a reaction bonded silicon nitride (RBSN) with 3% iron in its green state that was nitrided to 86% of its theoretical density of

2700 kg/mm^3. The fracture specimen, which is a wedge-loaded, modified-tapered double-cantilever beam (WL-MTDCB) specimen, was then finished ground to the dimensions shown in Fig. 1. The long, tapered beam section of the MTDCB specimen was designed to reduce the friction at the contact point between the loading wedge and the specimen. Extensive numerical experiments showed a 1% difference in the static stress intensity factors for the two extremes of friction effects (i.e., complete adhesion and complete slippage) for this specimen [*3*]. The specimen was loaded through a 60° wedge which was impacted by a 2.1-kg tup which in turn was dropped from a height of 140 mm. Figure 1 also shows the arrangement for static loading that was reported in Ref *3*.

Instrumentation

Crack extension was measured by a 25-mm long KRAK-GAGE[2] that was mounted on the flat side of the WL-MTDCB specimen. The transient voltage due to the increased electrical resistance of the tearing KRAK-GAGE was routed through a FRACTOMAT,[2] which was custom-built for full-scale output in 2 μs, to a waveform recorder.[3] This system for dynamic crack length

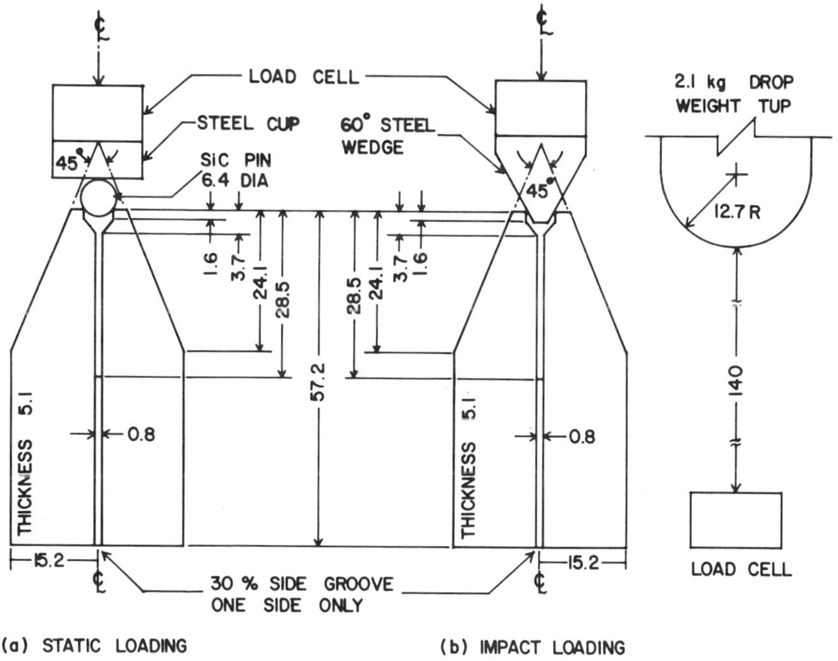

FIG. 1— *WL-MTDCB RBSN specimen (dimensions in millimetres).*

[2]TTI Division, Hartrum Corp., Chaska, Minn.
[3]Model 2805M Biomation, Gould, Inc., Santa Clara, Calif.

measurement was calibrated with a fracturing 7075-T6 single-edge-notch specimen, where the crack length recorded by an image converter camera was within 5% of that measured by the KRAK-GAGE [*11*]. The impact load, which was measured by a load transducer,[4] was also recorded on the waveform recorder.

Dynamic Finite Element Analysis

An implicit dynamic finite element code with the Newmark beta integration algorithm was executed in its generation mode by successively releasing crack-tip nodes according to the experimentally determined crack length versus time history. The experimentally determined wedge load versus time history was also prescribed. A linearly varying crack-tip nodal force release mechanism was used to model the smooth motion of the crack tip. The work done at the released crack-tip node was used to compute the dynamic stress intensity factor of the moving crack tip [*12*]. Crack opening displacements were used to compute the stress intensity factor prior to crack extension.

Figure 2 shows the finite element breakdown for the WL-MTDCB specimen. This finite element model was used to compute the static and dynamic stress intensity factors, K_I^{stat} and K_I^{dyn}, respectively. Past experiences [*8,9*] show that despite the coarseness in the finite element breakdown of Fig. 2, the static and dynamic stress intensity factors computed by the work done at the released crack tip are generally within 5% of the correct values.

Results

Experimental Results

Six RBSN WL-MTDCB specimens were impacted and two specimens were statically loaded to failure. Figures 3*a* and 3*b* show typical crack extension

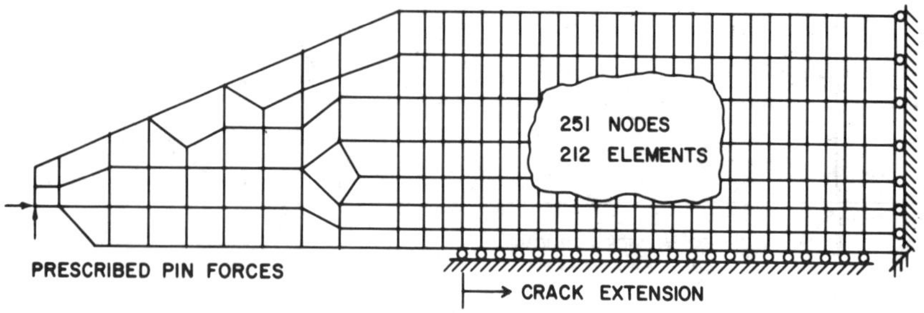

FIG. 2—*Finite element breakdown of WL-MTDCB specimen.*

[4]PCB Model 482A, Piezotronics, Inc., Depew, N.Y.

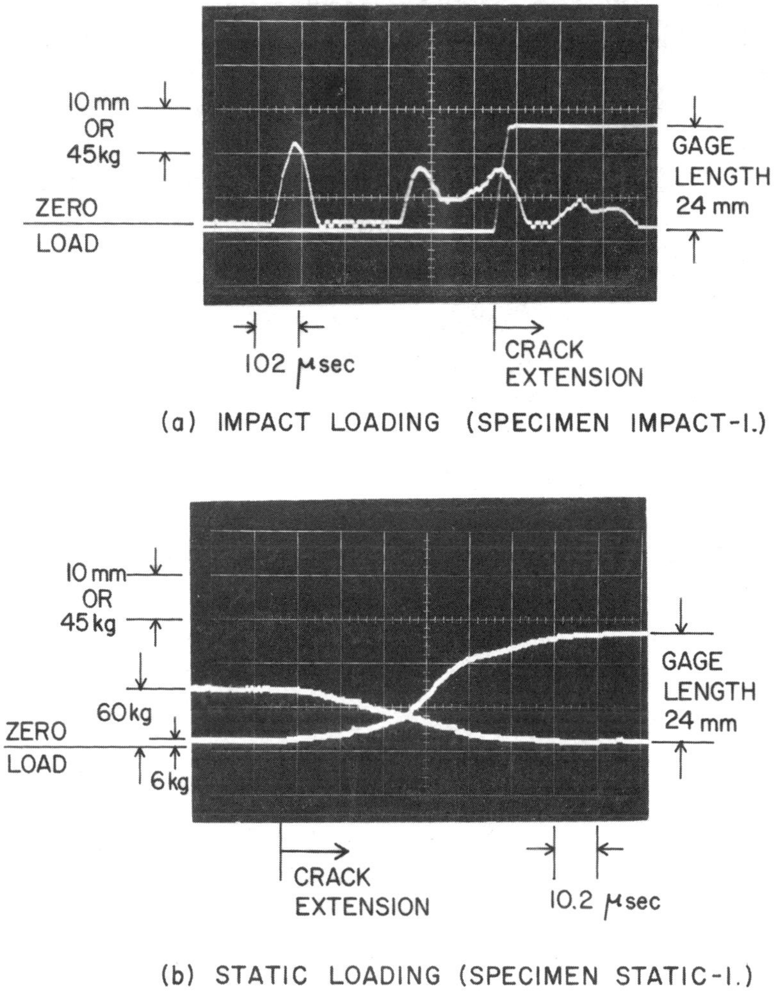

FIG. 3—*Crack extension and loads of fracturing RBSN specimens subjected to impact and static loadings.*

and load transducer records for impact and static loadings, respectively. Figures 4*a* and 4*b* show the crack extension, Δa, versus time histories, which were all within a narrow scatter band, of six impacted and two static specimens. The crack acceleration phase, which is prototypical of ceramic fracture specimens, at the onset of rapid crack propagation is noted. The maximum crack velocity of approximately 925 m/s or $\dot{a}/C_1 = 0.11$, where $\dot{a}$ and C_1 are the crack and dilatational wave velocities, respectively, is in agreement with that reported in Refs *2* to *4*.

Figure 5 shows the impact load variations prior to crack propagation for

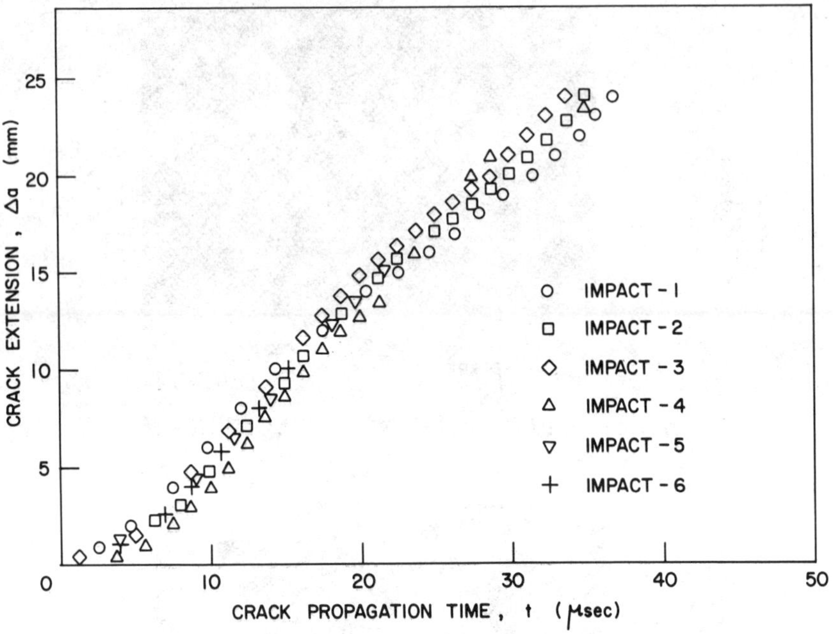

FIG. 4*a*—*Crack extensions in fracturing RBSN specimens subjected to impact loading.*

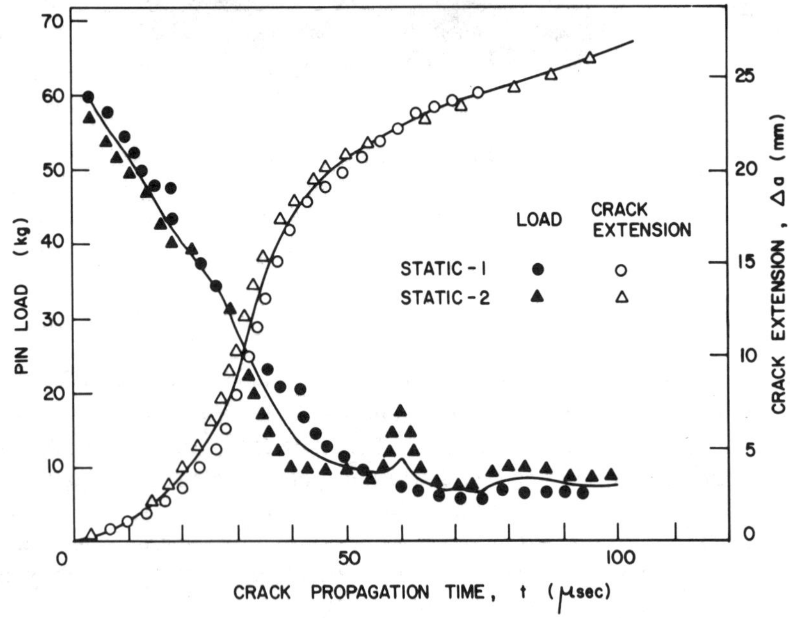

FIG. 4*b*—*Crack extensions and pin loads of fracturing RBSN specimens subjected to static loading.*

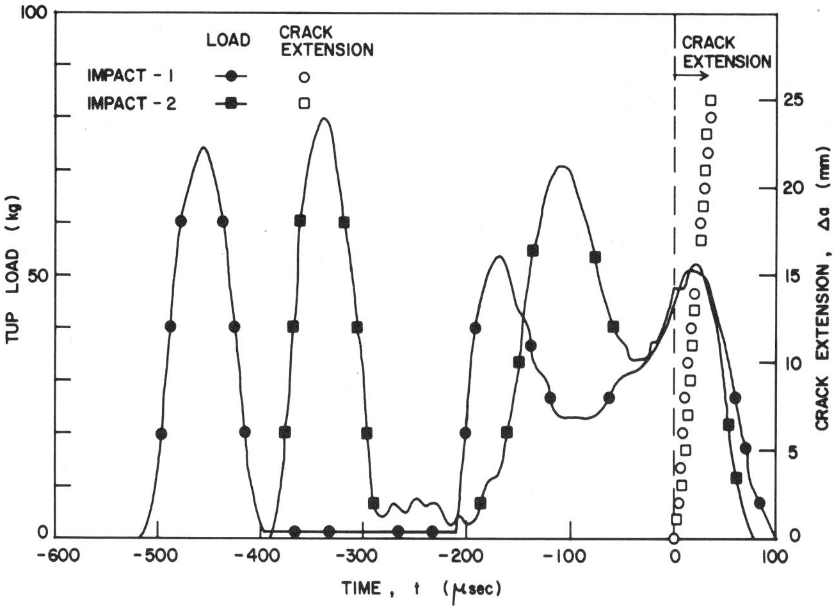

FIG. 5—*Tup loads and crack extensions of two RBSN specimens subjected to impact loading.*

two specimens. The 500 μs delay between the impact and the onset of rapid crack propagation at the blunt starter crack is consistent with other experimental observations [*13,14*]. Despite the 120 μs difference in the first peak loads, the subsequent load histories are in qualitative agreement with each other. Such reproducible load histories contributed to the reproducible crack extension histories shown in Fig. 4*a*.

Numerical Results

The precursor, which preceded the onset of dynamic fracture, requires a dynamic finite element analysis of the WL-MTDCB specimen corresponding to this 500 μs in addition to the 30 μs of rapid crack propagation. In an attempt to reduce, if not eliminate, this costly additional computation, the effect of this precursor on the computed dynamic stress intensity factors was assessed through a numerical experiment involving the entire precursor and its truncated version (Fig. 6). The former and latter for Impact test No. 1 (i.e., IMPACT-1) are designated as Cases 1 and 2, respectively. Figure 7 shows the differences in dynamic stress intensity factors when the entire precursor (Case 1) and its abbreviated version (Case 2) are used in the dynamic finite element analysis. Despite significant differences in the K_I^{dyn} variations with Δa for Cases 1 and 2, the $\dot{a}/C_1$ versus K_I^{dyn} relation in Fig. 8 exhibits little differences.

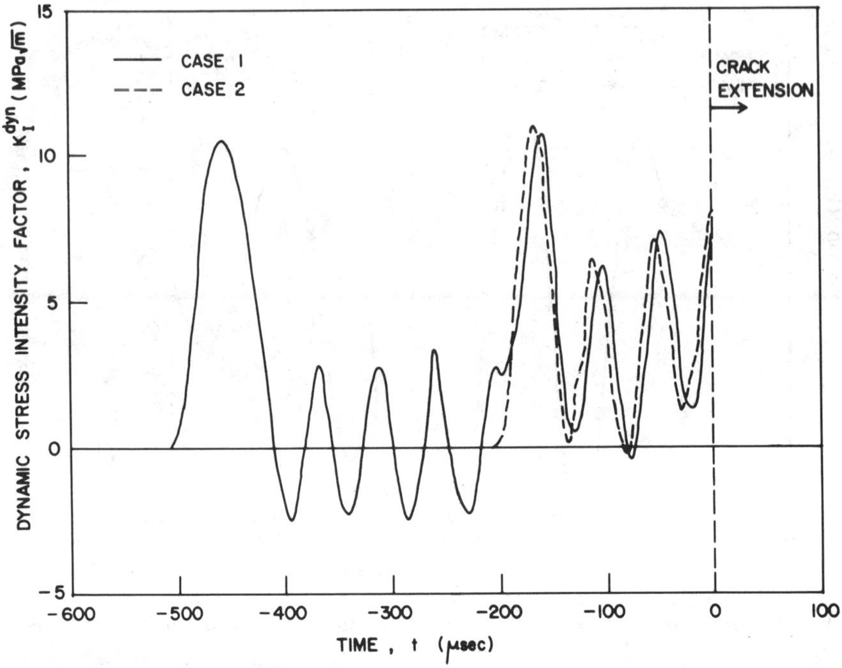

FIG. 6—*Transient dynamic stress intensity factors prior to fracture in a RBSN specimen subjected to impact loading (Specimen IMPACT-1).*

Figure 7 also shows the K_I^{dyn} versus crack Δa relation for specimen IMPACT-2 with the precursor shown in Fig. 5. Although the K_I^{dyn} versus Δa relations of IMPACT-1 and IMPACT-2 vary, these differences are obscured when replotted as $\dot{a}$ versus K_I^{dyn} in Fig. 8. For comparison purposes, Figs. 7 and 8 also show the corresponding K_I^{dyn} [*3*] for the RBSN WL-MTDCB specimens fractured under the static loading shown in Fig. 4*b*.

Discussion

The inertia effect and the necessity of dynamic analysis are shown in Fig. 6, where the onset of rapid crack propagation was not triggered at the maximum load peak. While such effects could be anticipated in massive notch bend steel specimens [*15*], Fig. 6 shows that inertia and stress wave effects cannot be ignored in the smaller and light, impacted ceramic specimens.

Despite the qualitative agreements between the K_I^{dyn} for static and impact loaded RBSN WL-MTDCB specimens in Fig. 7, the $\dot{a}$ versus K_I^{dyn} relations for the two loadings are different (Fig. 8). Under impact loading, the terminal crack velocity is observed under the high K_I^{dyn} as well as the low K_I^{dyn} at the terminal phase of crack propagation. The kinetic energy, which remained

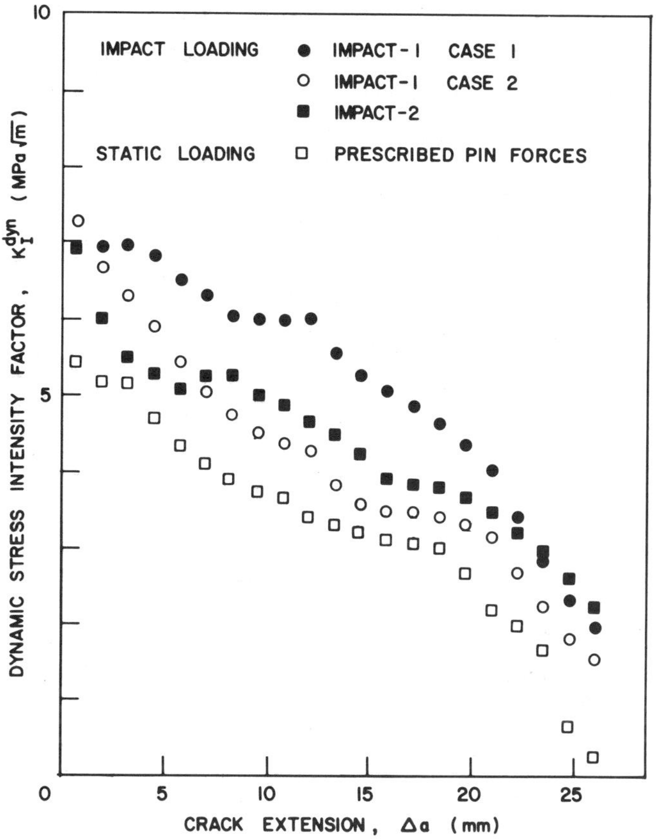

FIG. 7—*Dynamic stress intensity factors in a fracturing RBSN specimen subjected to static and impact loadings.*

negligible prior to crack propagation (Fig. 9), continually increased during crack propagation (Fig. 10). In contrast, the kinetic energy in the WL-MTDCB specimen loaded statically hovered about the zero line toward the terminal phase of crack propagation (Fig. 11). The strain energy also reached its minimum value in Fig. 10, while the strain energy in the impact loaded WL-MTDCB specimen continued to increase in Fig. 10.

The above results suggest that the rapidly propagating crack is not only driven by K_I^{dyn} but also by the surplus kinetic energy that rapidly separates the two open halves of the specimen as fast as the crack tip region fractures under the low K_I^{dyn}. Again in contrast, the low kinetic energy in the statically loaded specimen inhibits the separation of the two open halves and thus retards the crack propagation. It is interesting to note that limited experimental and numerical results also suggest that the terminal crack velocity in ductile fracture

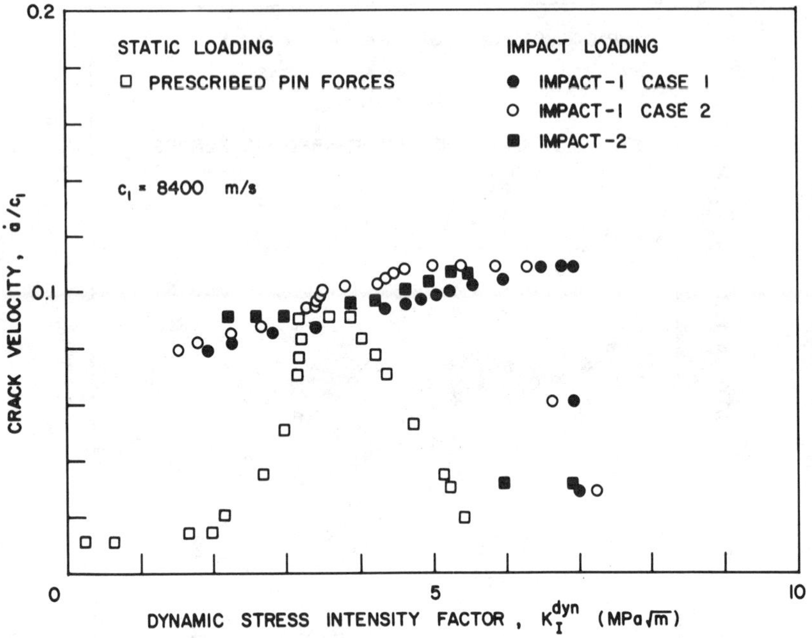

FIG. 8—*Dynamic stress intensity factor versus crack velocity relations of RBSN subjected to static and impact loadings.*

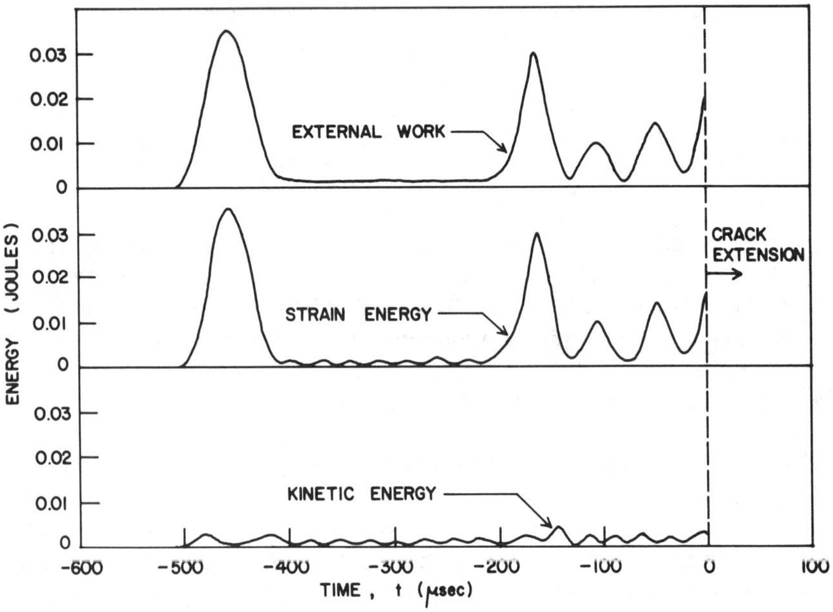

FIG. 9—*Energies prior to fracture in a RBSN specimen subjected to impact loading (Specimen IMPACT-1, Case 1).*

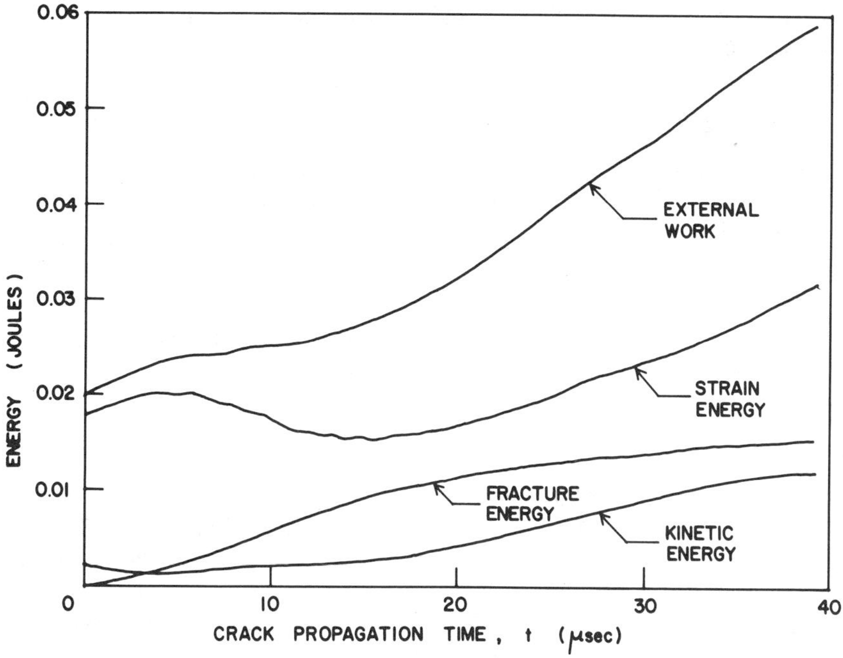

FIG. 10—*Energies in a fracturing RBSN specimen subjected to impact loading (Specimen IMPACT-1, Case 1).*

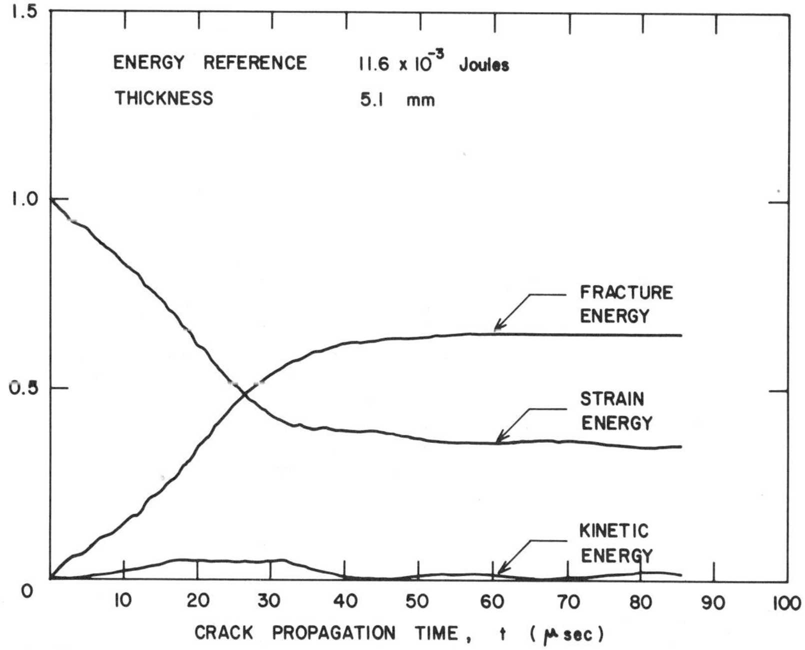

FIG. 11—*Energies in a fracturing RBSN specimen subjected to static loading.*

in low strength steel is likewise governed by the inertia of separating two halves [16,17].

Conclusions

1. The terminal phases of the dynamic stress intensity factor, K_I^{dyn}, versus crack velocity, $\dot{a}$, relations of impact and statically loaded RBSN WL-MTDCB specimens are different.
2. A crack can continue to propagate at its terminal velocity under a low dynamic stress intensity factor.
3. Crack velocity can be influenced by the inertia of the separating halves of the fracturing specimen.

Acknowledgments

This research was supported by the National Aeronautical and Space Administration under Grant NAGW 199 to the Ceramic Engineering Division, University of Washington. The authors wish to thank J. I. Mueller, principal investigator of the Ceramic Material Research Program, for his continuing support over the past six years, and C. W. Newquist, for his valuable assistance throughout the entire investigation.

References

[1] Gonczy, S. T. and Johnson, D. L. "Impact Fracture of Ceramics at High Temperature," in *Fracture Mechanics of Ceramics*, Vol. 3, R. C. Bradt, D. P. H. Hasselman, and F. F. Lange, Eds., Plenum Press, New York, 1978, pp. 495–506.
[2] Kobayashi, A. S., Emery, A. F., and Liaw, B. M., "Dynamic Fracture Toughness of Glass," in *Fracture Mechanics of Ceramics*, Vol. 6, R. C. Bradt, A. G. Evans, D. P. H. Hasselman, and F. F. Lange, Eds., Plenum Press, New York, 1983, pp. 47–61.
[3] Kobayashi, A. S., Emery, A. F., and Liaw, B. M., "Dynamic Fracture Toughness of Reaction-Bonded Silicon Nitride," *Journal of the American Ceramic Society*, Vol. 66, No. 2, Feb. 1983, pp. 151–155.
[4] Huang, J. L., Virkar, A. V., and Huber, R. J., "Fast Crack Propagation Across Interfaces," in *Fracture Mechanics of Ceramics*, Vol. 6, R. C. Bradt, A. G. Evans, D. P. H. Hasselman, and F. F. Lange, Eds., Plenum Press, New York, 1983, pp. 121–135.
[5] Kalthoff, J. F., Beinert, J., Winkler, S., and Klemm, W., "Experimental Analysis of Dynamic Effects in Different Crack Arrest Test Specimens," in *Crack Arrest Methodology and Applications, ASTM STP 711*, G. T. Hahn and M. F. Kanninen, Eds., American Society for Testing and Materials, Philadelphia, 1980, pp. 109–128.
[6] Hahn, G. T., Hoagland, R. G., and Cheverton, R. D., "Crack Arrest Methodology for Thick Sections," in *Advances in Fracture Research (Fracture '81)*, Vol. 2, R. François, Ed., Pergamon Press, New York, 1981, pp. 1043–1050.
[7] Kanninen, M. F., Gehlen, P. C., Barnes, C. R., Hoagland, R. G., Hahn, G. T., and Popelar, C. H., "Dynamic Crack Propagation under Impact Loading," in *Nonlinear and Dynamic Fracture Mechanics*, AMD-35, N. Perrone and S. N. Atluri, Eds., American Society of Mechanical Engineers, New York, 1979, pp. 195–200.
[8] Kobayashi, A. S., "Dynamic Fracture Analysis by Dynamic Finite Element Method: Generation and Propagation Analysis," in *Nonlinear and Dynamic Fracture Mechanics*, AMD-35, N. Perrone and S. N. Atluri, Eds., American Society of Mechanical Engineers, New York, 1979, pp. 19–36.

[9] Kobayashi, A. S., Seo, K., Jou, J. Y., and Urabe, Y., "A Dynamic Analysis of Modified Compact Tension Specimen Using Homalite-100 and Polycarbonate Plates," *Experimental Mechanics*, Vol. 20, 1980, pp. 73-79.

[10] Kobayashi, A. S., Emery, A. F., and Liaw, B. M., "Dynamic Fracture of Three Specimens," in *Fracture Mechanics: Fourteenth Symposium, Vol. II: Testing and Application, ASTM STP 791*, J. C. Lewis and G. Sines, Eds., American Society for Testing and Materials, Philadelphia, 1983, pp. II-251—II-265.

[11] Lee, O. S., Kobayashi, A. S., and Komine, A., "Further Studies on Crack Tip Plasticity of Rapid Tearing," *Experimental Mechanics*, Vol. 25, March 1985, pp. 66-74.

[12] Freund, L. B., "Crack Propagation in an Elastic Solid Subjected to General Loading: II—Non-Uniform Rate of Extension," *Journal of the Mechanics and Physics of Solids*, Vol. 20, 1972, pp. 141-152.

[13] Kalthoff, J. F., "Time Effects and Their Influences on Test Procedures for Measuring Dynamic Material Strength Values," in *Application of Fracture Mechanics to Materials and Structures*, G. C. Sih, E. Sommer, and W. Dahl, Eds., Martins Nijhoff Publishers, Leyden, The Netherlands, 1984, pp. 107-136.

[14] Mall, S., Kobayashi, A. S., and Urabe, Y., "Dynamic Photoelasticity and Dynamic Finite-Element Analysis of Dynamic-Tear-Test Specimens," *Experimental Mechanics*, Vol. 18, No. 12, Dec. 1978, pp. 449-456.

[15] Kobayashi, A. S., Ramulu, M., and Mall, S., "Impacted Notch Bend Specimens," *ASME Journal of Pressure Vessel Technology*, Vol. 104, No. 1, Feb. 1982, pp. 25-30.

[16] Kobayashi, A. S., Emery, A. F., Love, W. J., and Jain, A., "A Sensitivity Study on Numerical Analyses of Dynamic Girth Crack Propagation," *ASME Journal of Pressure Vessel Technology*, Vol. 103, No. 2, May 1981, pp. 169-174.

[17] Emery, A. F., Perl, M., Kobayashi, A. S., and Love, W. J., "The Use of the Split Ring in Modelling Ductile Axial Crack Extension in Pipes," *ASME Journal of Pressure Vessel Technology*, Vol. 103, No. 2, May 1981, pp. 151-154.

M. M. Ratwani[1] *and R. B. Deo*[1]

Resistance Curve Approach to Composite Materials Characterization

REFERENCE: Ratwani, M. M. and Deo, R. B., **"Resistance Curve Approach to Composite Materials Characterization,"** *Fracture Mechanics: Seventeenth Volume, ASTM STP 905*, J. H. Underwood, R. Chait, C. W. Smith, D. P. Wilhem, W. A. Andrews, and J. C. Newman, Eds., American Society for Testing and Materials, Philadelphia, 1986, pp. 108–123.

ABSTRACT: The delamination growth resistance of composites under Mode I (opening mode) has been investigated. It is shown that composites exhibit increasing resistance with increasing growth. The double cantilever beam specimen has been used to obtain delamination growth resistance curve over a wide range of delamination extension. The delamination growth resistance curve is shown to be path independent.

Delamination growth resistance curves have been obtained for current resin materials and high toughness materials. It is shown that the delamination growth resistance depends not only on the resin material but also on the fiber used. AS4/3501-6 material exhibits higher resistance than AS1/3501-6. The influences of temperature and porosity on delamination growth resistance have been investigated. It is found that AS1/3501-6 has lower resistance at −54°C (−65°F) than at room temperature ambient conditions. Porous laminates exhibit slightly lower resistance to delamination growth than nonporous laminates.

KEY WORDS: composites, delamination, fiber, matrix, Mode I, resistance curve, tough resins

Composite materials are finding increasing use in aerospace structures. Present day composite resins are susceptible to impact damage and hygrothermal environments. The limitations of the current resins and the mission requirements for future aircraft have led manufacturers to look into producing resins with improved properties. The desired improvements in properties are increased resistance to impact damage and to delamination growth, and higher glass transition temperature. In order to characterize different resins

[1]Manager, Structures Research Department, and Engineering Specialist, Strength & Life Assurance Research Department, respectively, Northrop Corporation, Hawthorne, CA 90250.

and compare their resistance to impact damage and delamination growth, the property measure commonly used is fracture toughness. For composite materials the fracture toughness measure is taken in terms of strain-energy release rate (G). The experimental evaluation of critical strain-energy-release rate (G_{Ic}) under Mode I has been investigated in Refs *1* to *9*. Three different specimen configurations, namely, double cantilever beam, delamination initiation, and free-edge delamination, have been discussed in these references. The relative merits and demerits of the specimens have been discussed in Ref *4*. It is pointed out in Ref *4* that the double cantilever beam specimen is the most promising in obtaining strain-energy-release rates.

The experimental results in Ref *1* showed fatigue crack growth under Mode I at strain-energy-release rates above G_{Ic}. The experimental results in Ref *9* showed increasing G_{Ic} values with increasing extensions. These results indicate that composite materials exhibit stable crack growth resistance above G_{Ic} and any failure criterion based on G_{Ic} will be conservative. G_{Ic} alone cannot define failure in a composite laminate.

A more realistic approach to describing delamination growth resistance of a composite laminate is by resistance curves. This concept has been used to describe plane-stress failure behavior of metallic materials [*10–12*]. A similar concept is developed for composites here.

Experimental Program

Specimen Geometry

Center crack tension and crackline wedge loaded specimens are commonly used to obtain resistance curves for metallic materials. The use of these specimen configurations to obtain resistance curves to characterize composite resins is not feasible, since the growth of delaminations in composites is an in-plane phenomenon and these specimen configurations would give resistance of the composites to fiber breakage which is not representative of resistance to delamination growth. The tapered double cantilever beam configuration has been used in Ref *13* to obtain resistance curves for adhesives. A specimen configuration similar to the double cantilever beam is considered suitable to obtain resistance (R) curves for composites under Mode I.

The untapered double cantilever beam (DCB) specimen was selected for the Mode I delamination growth tests conducted in the present program. The DCB specimen was 16 plies thick with a doubly symmetric $(0/\pm45/90/90/\mp45/0)_s$ layup and contained a simulated delamination located at the laminate midplane. Overall specimen geometry and delamination location are shown in Fig. 1. A similar specimen geometry with different laminate stacking sequences has been successfully used in several other Mode I delamination growth studies for composites [*1–4*] to determine G_{Ic} values and delamination growth rates in fatigue.

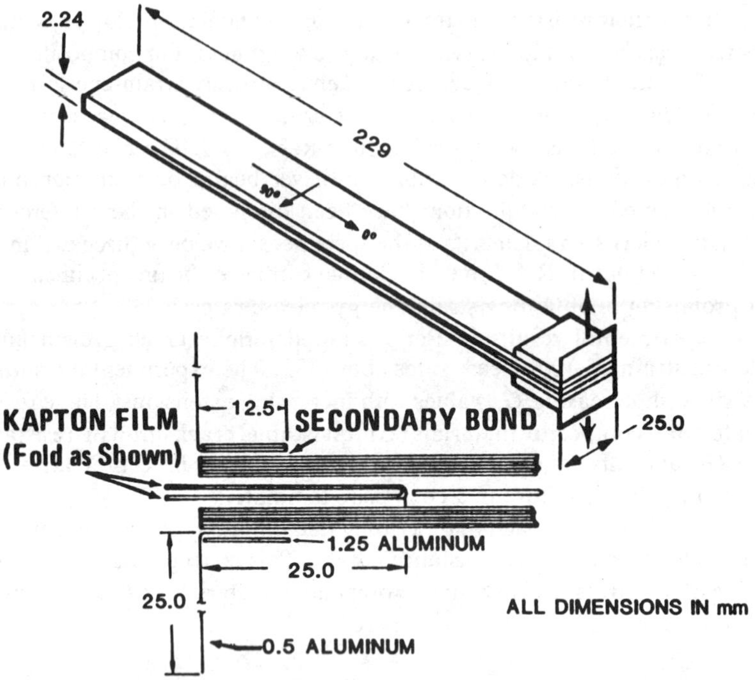

FIG. 1—*Double cantilever beam specimen (laminate layup: $(0/\pm 45/90_2/\pm 45/0)_s$.*

Test Procedure

Mode I tests were conducted in a servohydraulic machine under displacement control. A displacement-controlled loading was applied at the aluminum tabs to break the Kapton film simulating the delamination and then to cause slow delamination growth in predetermined length increments. After initiation of the delamination by breaking the Kapton film, the applied displacement was returned to zero and then increased at a slow constant rate to cause delamination growth. The applied displacement was then reduced to zero and this loading-unloading cycle started again to cause further delamination growth. The process was continued until the delamination had grown by approximately 102 mm (4 in.). During the loading and unloading tests, the load was continuously plotted as a function of the applied displacement, and the critical load value (P_{cr}) required to cause a fixed increase in delamination length was also noted for each case. A typical set of loading and unloading plots obtained under room temperature dry (RTD) environmental conditions for AS1/3501-6 material is shown in Fig. 2. These data were used to determine material resistance to delamination growth.

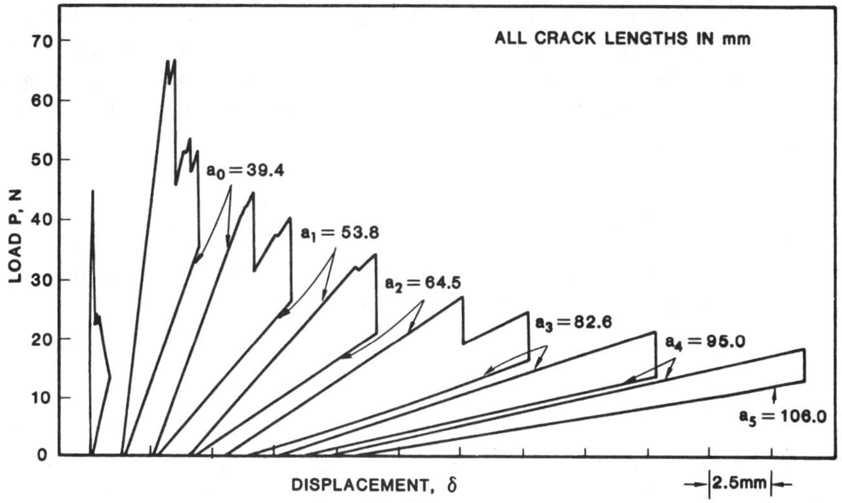

FIG. 2—*Loading/unloading data for AS1/3501-6 DCB specimen under RTD conditions (constant delamination growth).*

Calculation of Strain-Energy-Release Rate

The strain-energy-release rate (G_I) was calculated using a strength-of-materials analysis of the double cantilever beam which results in the following expression for G_{IR} [*1*]:

$$G_{IR} = P_{cr}^2 \left(\frac{dC}{da}\right)/2w \tag{1}$$

where C is the compliance and w is the specimen width.

Since P_{cr} is proportional to the inverse of the delamination length and the compliance is proportional to the cube of the delamination length, Eq 1 was written in the form

$$G_{IR} = 3A_1 \, A_2^2/2w \tag{2}$$

where

$$C = A_1 \, a^3 \tag{3}$$

and

$$P_{cr} = A_2/a \tag{4}$$

The constants A_1 and A_2 were determined from the load displacement plots of Fig. 2 by fitting Eqs 3 and 4 to the compliance versus delamination length and critical load for delamination growth versus delamination length data, respectively. Typical curves for compliance and critical load for the test data of Fig. 2 are shown in Fig. 3.

Development of Resistance Curves

Two different approaches, namely, incremental extension and cumulative extension, were used to obtain resistance curves.

Incremental Extension

In the incremental extension method, the delamination extension Δa_i was gradually increased with increasing number of load/unload cycles. For example, the delamination size was increased by 3.3 mm (0.13 in.) in the first cycle, 5.1 mm (0.20 in.) in the second cycle, 8.9 mm (0.35 in.) in the third cycle and so on, to a delamination length increment of 15.7 mm (0.62 in.) in the final cycle (Fig. 4). G_{IR} values were calculated for each cycle using Eq 1 and

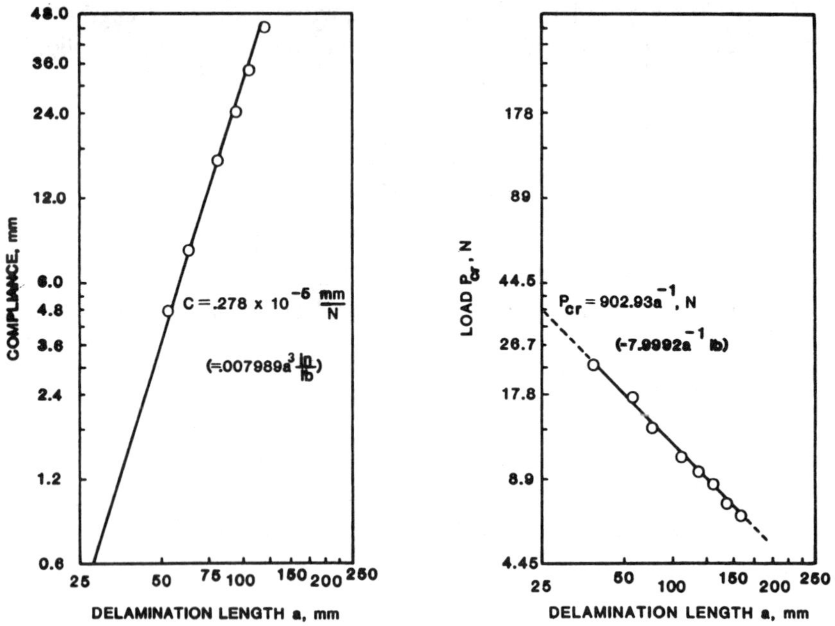

FIG. 3—*Determination of coefficients* A_1 *and* A_2 *(Eqs 3 and 4) from curve fit to* C *versus* a^3 *and* P_{cr} *versus* a^{-1} *data.*

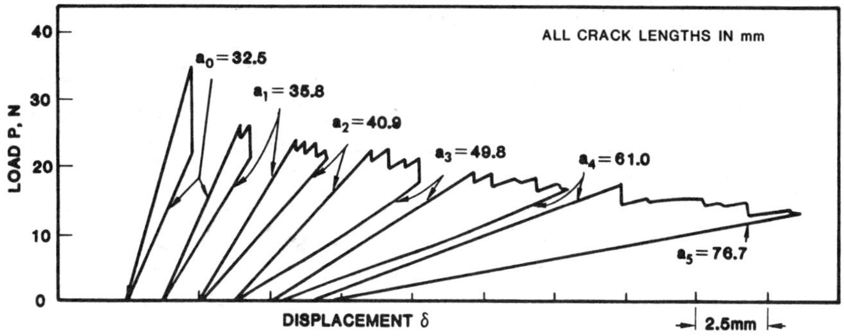

FIG. 4—*Loading/unloading data for AS4/3501-6 DCB specimen under RTD conditions (incremental delamination extension).*

the resistance curve obtained by plotting $\sqrt{G_I}$ versus the corresponding delamination length increase (Δa). This procedure, therefore, provides a direct measure of the resistance curve. A typical set of data for AS4/3501-6 composites is shown in Fig. 5.

Cumulative Extension

In the cumulative extension method, approximately constant extension Δa was used in the test procedure. The extension Δa_i is computed as the differ-

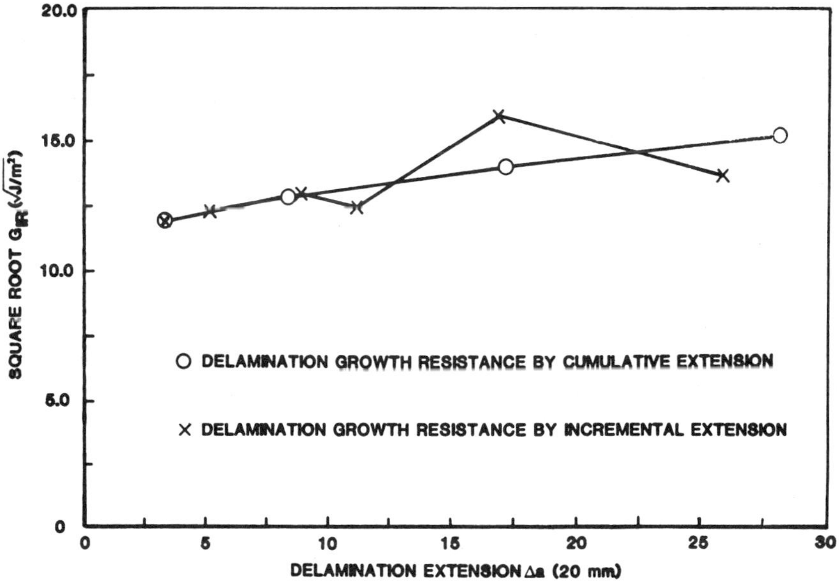

FIG. 5—*Comparison of resistance curves by cumulative extension and incremental extension methods.*

ence between the delamination length a_i corresponding to the ith loading-unloading cycle and the initial delamination length a_0. It should be noted that the crack length a_0 corresponds to the delamination length reached after the Kapton film has been broken through; thus the loading-unloading cycle to break through the Kapton is ignored. In order to calculate $\sqrt{G}$ the expression in Eq 1 is used where P_{cr} corresponds to the load level required to make the delamination grow from its initial length a_0. The rate of change in compliance dC/da for a delamination length increment Δa_i is computed using

$$\frac{dC}{da} \approx \frac{\Delta C}{\Delta a} = \frac{C_i - C_0}{a_i - a_0} \tag{5}$$

where C_i is the compliance determined from the loading/unloading curve for delamination length a_i. In this approach, it is assumed that the resistance offered by the material to a delamination length increase from an initial length a_0 to a final length a_{i+1} does not depend on whether or not a loading/unloading cycle was carried out at an intermediate delamination length a_i. The validity of this assumption is demonstrated by test data from DCB specimens.

The experimental resistance curve for AS4/3501-6, obtained using the cumulative extension procedure, is shown in Fig. 5. The figure also shows the resistance curve from the same specimen using the incremental extension procedure. It is seen that the resistance curves obtained by the two methods are very similar. The slight difference in the two curves is within experimental data scatter, and the cumulative extension procedure gives a smoother resistance curve. Figure 6 compares the two resistance curves in Fig. 5 with that derived from another specimen using the cumulative extension method. It is seen that scatter in the data from the two tests is well within the expected range for composites.

Experimental Results

Using the specimen configuration and method discussed earlier, resistance curves have been obtained for composite laminates, and the influences of various parameters such as matrix-fiber combination, porosity, and environment on the resistance curve have been investigated. The results obtained are discussed in the following subsections.

Resistance for AS1/3501-6 Laminate

Resistance curve data ($\sqrt{G}$ versus Δa) from three specimens are shown in Fig. 7. The test data show very little scatter, except for a data point from Specimen 3. The figure also shows a fitted resistance curve which has been extrapolated to the region of crack extensions less than 12.7 mm (0.5 in.) The

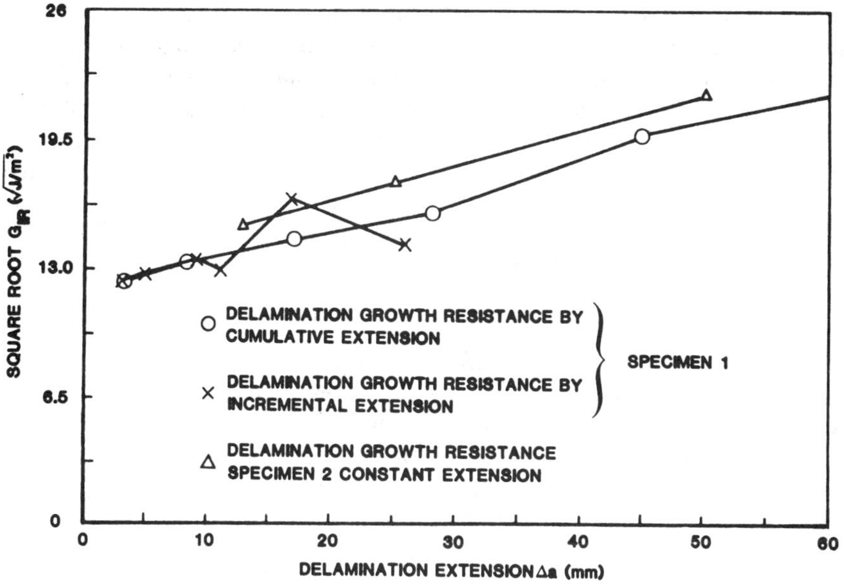

FIG. 6—*Resistance curves for AS4/3501-6 composites.*

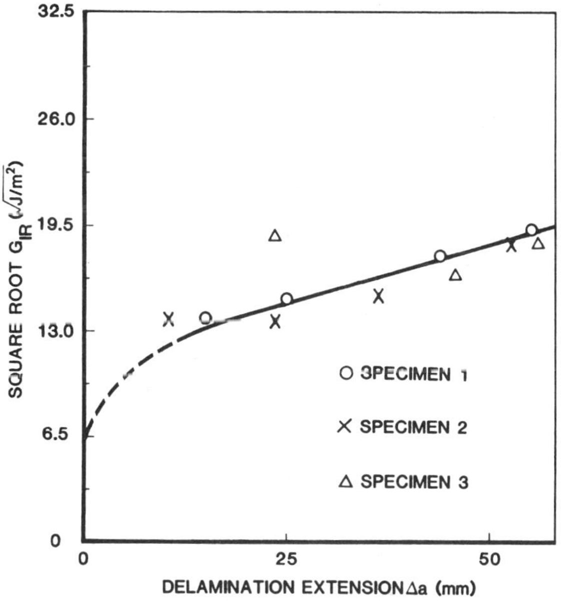

FIG. 7—*Mode I resistance curve for AS1/3501-6 composite.*

extrapolation of the $\sqrt{G_{IR}}$ curve to values of Δa below the experimental data is shown by dashed lines in Fig. 7 and all subsequent R-curve figures. This extrapolation was accomplished by analogy to the shape of R-curves obtained for metals and as such illustrates only trend information. It is recognized that additional data for delamination extensions less than approximately 12.7 mm (0.5 in.) are needed to verify the nature of the resistance curve for small extensions.

Influence of Fiber-Resin Combination on Resistance Curve

The resistance curve for AS4/3501-6 is shown in Fig. 8*a*. The figure also shows the resistance curve for AS1/3501-6 laminate. It is seen that the AS4/3501-6 resistance curve is higher than that for AS1/3501-6. Thus AS4/3501-6 material offers higher resistance to delamination growth than AS1/3501-6 even though the resin is the same in both laminates. Figure 8*b* compares SEM photographs of the AS1 and AS4 fibers. As can be seen from this figure, the AS1 fiber has a larger diameter and a considerably rougher surface than the AS4 fiber. These results indicate that the type of fiber and the fiber matrix interface have a definite influence on the resistance to delamination growth.

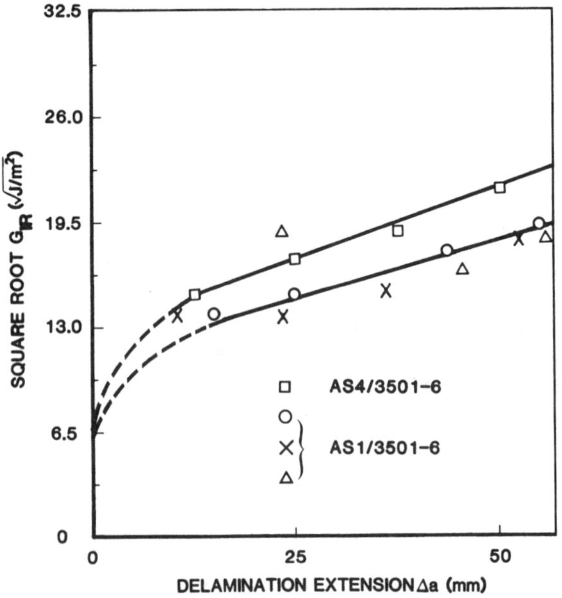

FIG. 8*a*—*Comparison of resistance curves for AS4/3501-6 and AS1/3501-6 composites.*

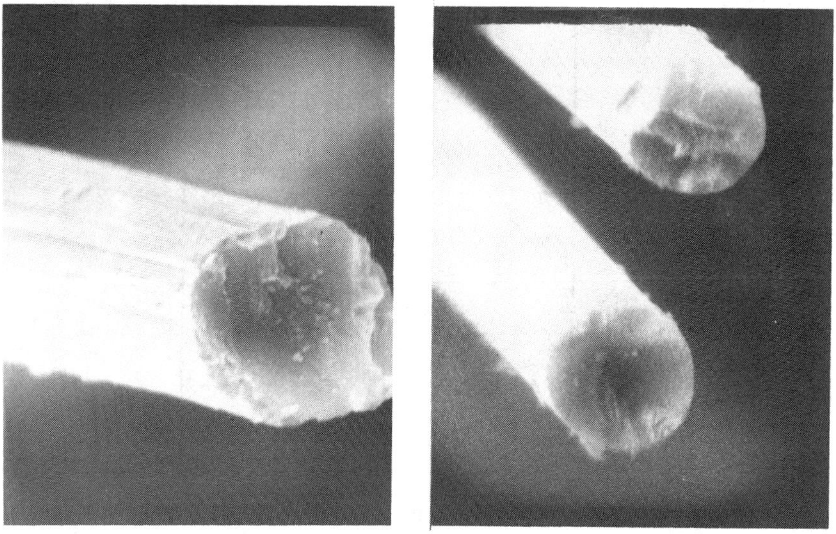

AS1 FIBER AS4 FIBER

FIG. 8*b*—*Comparison of AS1 and AS4 fiber microstructure (SEM photos originally at 5000×, reduced here 38%).*

Influence of Environment on Resistance

The resistance curves for AS1/3501-6 material, discussed earlier, were obtained under room temperature dry environment. The growth of delaminations is influenced by environment and hence the resistance curve should depend on the test environment. The resistance curves, obtained from two specimens, for AS1/3501-6 at −54°C are shown in Fig. 9. The test data from Specimen 1 are not considered reliable because the tests indicated ice formation at the edges of the delamination, resulting in compression loads when the specimen is unloaded. This resulted in a higher G value and thus a higher resistance curve. No such phenomenon was observed in Specimen 2, since adequate precautions were taken to minimize moisture in the environmental test chamber. Comparison of the resistance curves obtained under room temperature dry and −54°C dry environments is shown in Fig. 10. In this figure, only the central tendencies of the data are shown, and it is seen that the resistance curve at −54°C is lower than the resistance curve under room temperature dry environment, indicating that the material resistance is reduced at −54°C.

Influence of Porosity on Resistance Curves

The resistance curves obtained from three test specimens with about 2% porosity are shown in Fig. 11. It is seen that the test data have more scatter

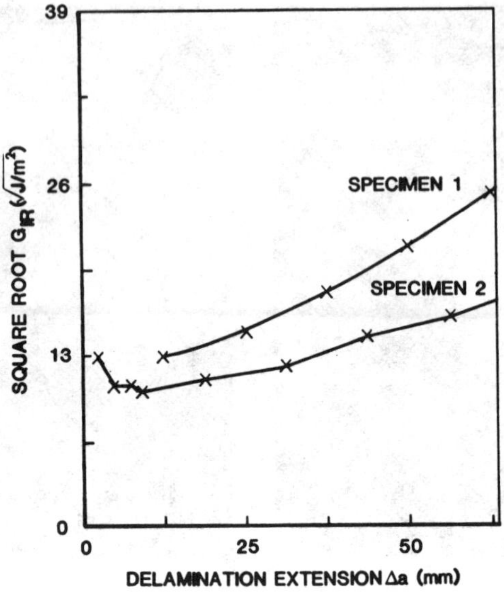

FIG. 9—*Resistance curve for AS4/3501-6 at −54°C.*

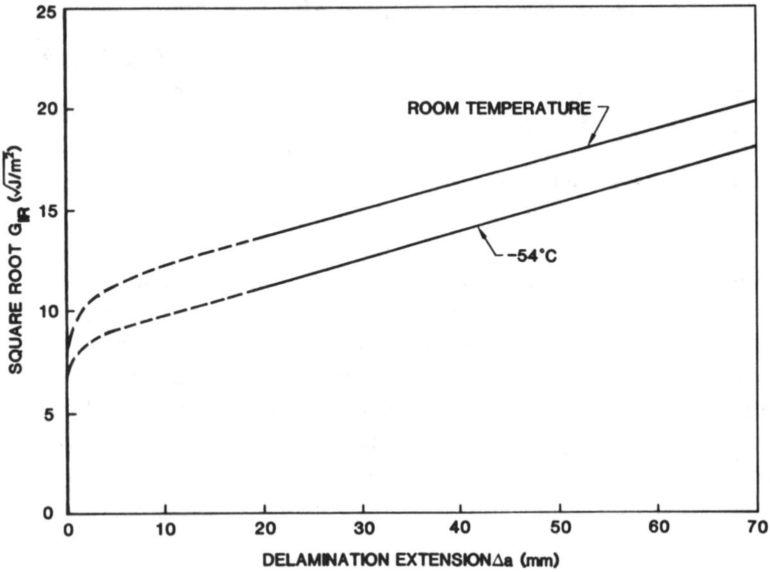

FIG. 10—*Comparison of resistance curves under room temperature and −54°C environments (AS1/3501-6).*

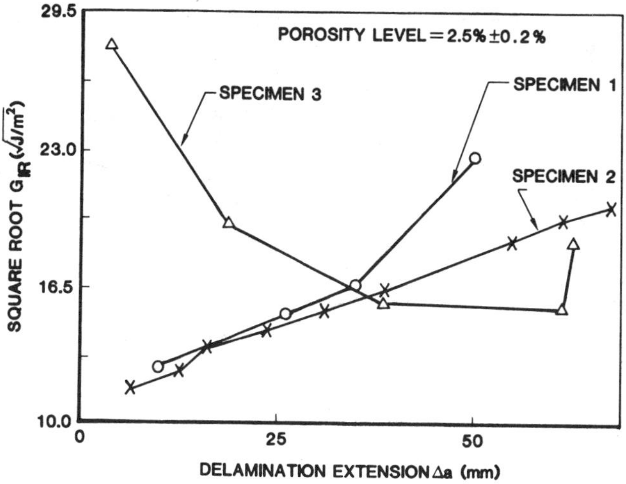

FIG. 11—*Resistance curves for porous laminates.*

than nonporous laminates and do not lie on a smooth curve for any specimen. This is caused by the delamination front progressing through a porous interface and resulting in larger growth than would normally occur. If the first two data points from Specimen 3 are ignored, the remaining data show scatter typical of porous laminates. The first two test points in Specimen 3 show higher values of $\sqrt{G}$, perhaps due to a strong bond between the Kapton insert simulating the delamination and the resin.

A comparison of resistance curves for AS1/3501-6 porous and nonporous laminates is shown in Fig. 12. The curves shown in this figure represent the central tendency of the test data. It is seen that the resistance curve for porous laminates is not significantly different from that for nonporous laminates and is within the scatter commonly encountered in composites. This does not imply, however, that delamination growth resistance is unaffected by porosity. An explanation for the absence of porosity effects was obtained from a microstructural examination of the specimen. The photomicrograph showed that porosity in the specimen was distributed across the cross section and was not uniform or concentrated at the delamination plane.

Resistance Curves for T300/5208

The G_{Ic} test data for T300/5208 obtained in Ref *2* were used to obtain the resistance curve for the material which has properties in the same class as AS1/3501-6 material. The central tendencies of the *R*-curve data for the two materials are shown in Fig. 13. It is evident that AS1/3501-6 has a higher resistance to delamination growth than T300/5208 material.

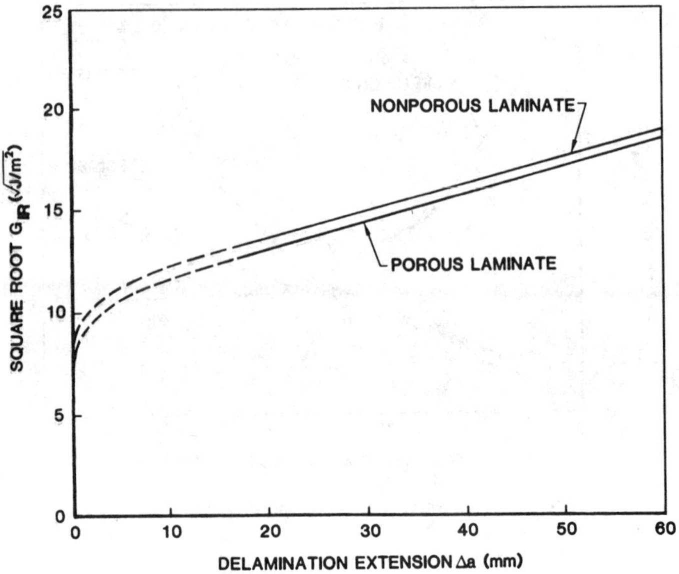

FIG. 12—*Comparison of resistance curves for porous and nonporous laminates.*

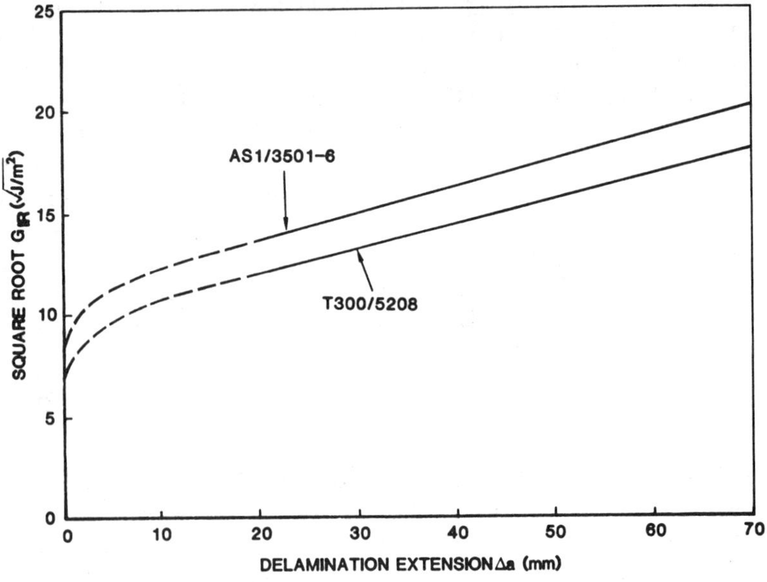

FIG. 13—*Comparison of resistance curves for AS1/3501-6 and T300/5208 materials.*

Resistance Curves for Tough Materials

The resistance curve approach was used to characterize materials with enhanced toughness such as AS4/5245C and AS4/X2220-1. The resistance curves for the two materials are shown in Fig. 14. It is evident that the resistance curve for AS4/5245C is much higher than the curves for the other materials, indicating that AS4/5245C has a significantly larger resistance to delamination growth (i.e., a much higher toughness). AS1/3501-6 has the lowest resistance curve or the lowest resistance to delamination growth. The surprising result is that the resistance for AS4/3501-6 is higher than the resistance curve for AS4/X2220-1. These results indicate that AS4/X2220-1 material does not have the higher toughness property expected.

Conclusions

The studies undertaken in the present research program have resulted in the following conclusions:

1. The resistance curve approach can be used to characterize composite laminates and to compare the toughness of new resin materials.
2. The delamination growth resistance of composite materials under Mode I depends not only on the resin but also on the fiber used in the material system. Laminates with AS4 fibers in 3501-6 resin show higher resistance to growth than AS1 fibers in the same resin.

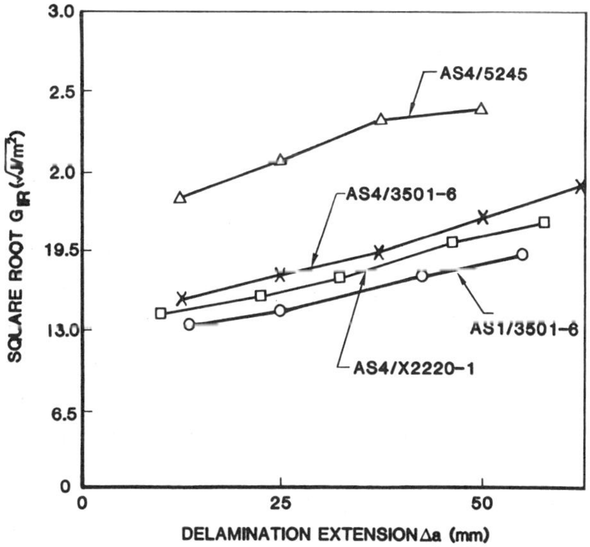

FIG. 14—*Resistance curve for new tough resins.*

3. The delamination growth resistance curve for a given material system can be obtained by either the incremental or the cumulative extension methods.

4. Porous laminates have lower resistance to delamination growth than nonporous laminates.

5. The AS1/3501-6 material system has lower resistance to delamination growth at −54°C (−65°F) than at room temperature ambient environmental condition.

Acknowledgments

The work reported here was performed under Northrop's Independent Research and Development Program. Special thanks to Mr. Dave Wilhem for useful discussions during the course of investigation.

References

[1] Wilkins, D. J. et al, "Characterizing Delamination Growth in Graphite/Epoxy," in *Damage in Composite Materials, ASTM STP 775*, American Society for Testing and Materials, Philadelphia, 1982, pp. 168–183.

[2] Ramkumar, R. L., "Performance of a Quantitative Study of Instability-Related Delamination Growth," NASA Contract Report 166045, National Aeronautics and Space Administration, Washington, D.C., March 1983.

[3] Ashizawa, M., "Improving Damage Tolerance of Laminated Composites Through the Use of New Tough Resins," presented to Sixth DoD/NASA Conference on Fibrous Composites in Structural Design, New Orleans, La.

[4] Whitney, J. M., Browning, L. E., and Hougsteden, W., "A Double Cantilever Beam Test for Characterizing Mode I Delamination of Composite Materials," *Journal of Reinforced Plastics and Composites*, Oct. 1982.

[5] Nicholls, D. J. and Gallagher, J. P., "Determination of G_{IC} in Angle-Ply Composites Using a Double Cantilever Beam Test," *Journal of Reinforced Plastics and Composites*, Vol. 2, Jan. 1983, pp. 2–17.

[6] Whitney, J. M. and Browning, C. E., "Materials Characterization for Matrix Dominated Failure Modes," in *Effects of Defects in Composite Materials, ASTM STP 836*, American Society for Testing and Materials, Philadelphia, 1984, pp. 104–124.

[7] Wang, S. S., "Delamination Crack Growth in Unidirectional Fiber-Reinforced Composites under Static and Cyclic Loading," in *Composite Materials: Testing and Design (Fifth Conference), ASTM STP 674*, American Society for Testing and Materials, Philadelphia, 1979, pp. 642–663.

[8] O'Brien, T. K., "Characterization of Delamination Onset and Growth in a Composite Laminate," in *Damage in Composite Materials, ASTM STP 775*, American Society for Testing and Materials, Philadelphia, 1982, pp. 140–167.

[9] Russell, A. J. and Street, K. N., "Moisture and Temperature Effects on the Mixed-Mode Delamination Fracture of Unidirectional Graphite/Epoxy," in *Delamination and Debonding of Materials, ASTM STP 876*, American Society for Testing and Materials, Philadelphia, 1985, pp. 349–370.

[10] Kraft, J. M. et al., "Effects of Dimensions on Fast Fracture Instability of Notched Sheets," in *Proceedings*, Crack Propagation Symposium, College of Aeronautics, Vol. 1, Cranfield, England, 1981.

[11] Wilhem, D. P. and Ratwani, M. M., "Crack Growth Resistance at Low Stress Intensity," *Journal of Engineering Materials and Technology*, Vol. 100, April 1978, pp. 134–137.

[12] Wilhem, D. P., Ratwani, M.M., and Zielsdorff, G. F., " A *J*-Integral Approach to Crack Resistance for Aluminum, Steel and Titanium Alloys," *Journal of Engineering Materials and Technology*, Vol. 99, April 1977, pp. 97–104.

[13] Hartman, A., "Crack Resistance of Some Structural Adhesives Determined on Adhesive Bonded TDCB Specimen," Report No. NLR-TR 75010U, National Aerospace Laboratory, Amsterdam, The Netherlands, Dec. 1974.

Charles E. Harris[1] *and D. H. Morris*[2]

A Comparison of the Fracture Behavior of Thick Laminated Composites Utilizing Compact Tension, Three-Point Bend, and Center-Cracked Tension Specimens

REFERENCE: Harris, C. E. and Morris, D. H., **"A Comparison of the Fracture Behavior of Thick Laminated Composites Utilizing Compact Tension, Three-Point Bend, and Center-Cracked Tension Specimens,"** *Fracture Mechanics: Seventeenth Volume, ASTM STP 905*, J. H. Underwood, R. Chait, C. W. Smith, D. P. Wilhem, W. A. Andrews, and J. C. Newman, Eds., American Society for Testing and Materials, Philadelphia, 1986, pp. 124-135.

ABSTRACT: An experimental study of the effects of specimen configuration and laminate thickness on the fracture behavior of notched laminated composites is presented. The behavior of $[0/\pm45/90]_{ns}$, $[0/90]_{ns}$ and $[0/\pm45]_{ns}$ graphite/epoxy T300/5208 laminates was studied. When fracture toughness was computed at the failing load of the center-cracked tension specimen, laminate thickness significantly influenced fracture toughness. If fracture toughness was computed at the interception of the 5% secant line with the load-COD record, fracture toughness was found to be independent of both laminate thickness and specimen configuration. Defined in this manner, fracture toughness can be physically interpreted as an indicator of the onset of significant notch-tip damage.

KEY WORDS: fracture toughness, composite, graphite/epoxy, center-cracked tension, compact tension, three-point bend, thick laminates

The effect of laminate thickness on the fracture behavior of laminated graphite/epoxy composites is the subject of an extensive, ongoing research program at Virginia Polytechnic Institute and State University. The primary objective is to experimentally study the influence of thickness on the fracture toughness of notched laminates and on the development of crack-tip damage prior to fracture. Associated with this objective is the comparison of fracture toughness and damage development for the center-cracked tension, compact

[1]Assistant Professor, Aerospace Engineering Department, Texas A&M University, College Station, TX 77843.
[2]Professor, Department of Engineering Science and Mechanics, Virginia Polytechnic Institute and State University, Blacksburg, VA 24061.

tension, and three-point bend specimen configurations. Discussion of test results from these three standard specimen configurations is the subject of this paper.

The first specimen configurations utilized for fracture toughness testing of metals and composites were the center-cracked and double-edge notched tension specimens. Problems associated with testing these configurations began to surface when test results identified the dependence of fracture toughness on specimen thickness. The thick specimens required higher loads that exceeded the load capacity of many testing machines. Also, problems developed with regard to gripping the specimens. (The gripping problem is especially acute for laminated composites.) Three-point bend and compact tension specimens were proposed as alternative configurations and were verified as being acceptable for plane strain fracture toughness testing of metals by ASTM [*1,2,* and ASTM Test Method E 399]. No such verification of the similarity of test results from the three specimen configurations has been established for composite materials.

The preponderance of experimental investigations of notched laminated composites have been conducted utilizing the center-cracked tension specimen geometry. This is because most laminates that have been investigated were thin sheets and, for thin specimens, the center-cracked tension specimen is a more stable geometry. However, there have been several investigators that utilized the three-point bend and compact tension specimens in addition to the double-edge or center-cracked tension specimen. Cruse and Osias [*3*] reported test results from ten angle-ply graphite/epoxy laminates utilizing the three-point bend and center-cracked tension specimens. They generally concluded that fracture toughness was independent of specimen geometry. However, other investigators who utilized several specimen types, such as Prewo [*4*], Reedy [*5*], and Shih and Logsdon [*6*], concluded that the modes of fracture and the fracture toughness varied considerably with specimen configuration.

In order to address this issue of specimen configuration, the authors studied the fracture behavior of center-cracked tension M(T), compact tension C(T), and three-point bend (SE(B)) specimens. Tests were conducted utilizing specimens obtained from laminates with three basic stacking sequences and thicknesses ranging from 6 to 120 plies. The specimen dimensions were selected in accordance with ASTM E 399 where practical, and fracture tests were generally conducted in accordance with this standard. The experimental program is described in more detail below, followed by a discussion of the program results.

Experimental Program

The laminate panels were prepared from graphite/epoxy, T300/5208, using a standard tape layup and autoclave curing process. The basic lamina tensile data were determined experimentally from $[0]_8$, $[90]_8$ and $[\pm 45]_{2s}$ coupons. Using the average of four or five replicate tests, the lamina properties were: $E_{11} = 138.7$ GPa (20.11×10^6 psi), $E_{22} = 10.77$ GPa (1.56×10^6 psi), $G_{12} = 5.98$ GPa (0.867×10^6 psi), and $\nu_{12} = 0.318$, where subscript "1" refers to the direction parallel to the fibers and "2" refers to the transverse direction. These property values are typical of those reported in the literature for T300/5208.

The laminate stacking sequences selected for the study were $[0/\pm 45/90]_{ns}$, $[0/90]_{ns}$ and $[0/\pm 45]_{ns}$ where the subscript "ns" means multiple layers of the repeated stacking sequence symmetric about the laminate midplane. The $[0/\pm 45/90]_{ns}$ and $[0/90]_{ns}$ M(T) laminates were tested at thicknesses of 8, 32, 64, 96 and 120 plies, where the individual ply thickness was approximately 0.127 mm (0.005 in.). Only the last three thicknesses were used for the C(T) and SE(B) specimens. The thicknesses for the $[0/\pm 45]_{ns}$ M(T) laminates were 6, 30, 60, 90 and 120 plies. Again, the last three thicknesses were used for the C(T) and SE(B) specimens. In all cases the 0° fiber orientation was perpendicular to the machined notch.

The M(T) specimen was 50.8 mm (2 in.) by 203 mm (8 in.) with a center notch cut by an ultrasonic cutting tool. The notch tip radius was 0.203 mm (0.008 in.); no attempt was made to sharpen the notch tips. The effect of the notch root radius on test results is summarized in Ref *7*. Based on reported results, the effect of radius (for all those types of specimens) on the fracture toughness values is negligible. For the M(T) specimens the crack length-to-width ratio ($2a/W$) was 0.50. Also, 50.8 mm (2 in.) C(T) specimens and 25.4 mm (1 in.) by 127 mm (5 in.) SE(B) specimens with a/W at 0.50 were tested at the three highest laminate thicknesses. In most cases four or more replicate tests were conducted. Further details may be found in Ref *7*.

All fracture tests were conducted at a constant crosshead displacement rate of 0.02 mm/s (0.05 in./min). The M(T) specimens were held in 51 mm (2 in.) wide wedge-action friction grips such that the specimen length between grip ends was approximately 127 mm (5 in.). The thin M(T) specimens, six or eight plies, were tested with an antibuckling support to prevent out-of-plane motion. The C(T) specimens were loaded through 9.5 mm (3/8 in.) loading pins. The SE(B) specimens were supported on 25.4 mm (1 in.) diameter cylindrical rods positioned 102 mm (4 in.) apart and loaded via a 19 mm (3/4 in.) radius ram. The supports were spring loaded to allow proper initial positioning and free movement once the test commenced. X-ray examinations of the loading regions of the C(T) and SE(B) specimens did not reveal any measurable amount of local bearing or cracking damage. Also, because of the thickness of the C(T) and SE(B) specimens (≥ 60 plies), no out-of-plane buckling occurred.

Crack-tip damage that formed prior to fracture was investigated using X-ray radiography and the laminate de-ply technique [*8*]. Zinc iodide enhanced X-ray examinations were used to establish the load levels where significant damage developed. Also, the progression of damage as the load approached the failure load was documented using X-ray radiography. Using a gold chloride agent to mark delamination regions, the through-the-thickness damage was studied using the destructive laminate de-ply technique. This technique allowed the study and documentation of the damage in each individual ply. The de-ply technique was especially useful for studying the details of broken fibers and for establishing the extent of delaminations as well as their exact interface location [*7*].

Determination of Fracture Toughness

Values of stress intensity factor were computed from a finite element analysis for each of the three specimen configurations [*7*]. The finite element code [*9*] computed the Mode I stress intensity factors and strain energy release rates based on linear elastic fracture mechanics (LEFM). The laminate material behavior was considered to be homogeneous and orthotropic.

Values of fracture toughness were computed at the maximum test load (P_m) and at the 5% secant intercept load (P_Q). The fracture toughness (K_m) at the maximum load was defined as the value of the LEFM orthotropic stress intensity factor at the maximum test load. Fracture toughness (K_Q) at the 5% secant load was defined as the value of the LEFM orthotropic stress intensity factor at the load P_Q which was defined in accordance with ASTM E 399. The load P_Q was defined as the intercept of the 5% secant line with the load-COD test record (P_5) or a higher load if one preceded P_5. The 5% secant line was drawn through the origin with a slope that was 0.95 times the slope of the initial linear portion of the load-COD test record.

Results and Discussion

The three specimen configurations exhibited fundamentally different load-COD test records. In the constant cross-head displacement mode of testing, the M(T) specimen failed catastrophically at the maximum test load. As illustrated in Fig. 1*a*, the typical M(T) specimen exhibited discontinuities in the load-COD record prior to failure. These discontinuities were associated with the formulation of discrete amounts of damage at the notch tip [*7*]. After the formation of significant damage, the specimen compliance changed and the specimen behavior was no longer linear elastic. On the other hand, the SE(B) specimen (Fig. 1*b*) and the C(T) specimen (Fig. 1*c*) did not fail catastrophically at the maximum test load. (Note that the scales in Figs. 1*a*, 1*b*, and 1*c* are different). This was because of the compression zone at the back face of the specimen opposite the advancing crack front. In addition, the SE(B) and C(T) specimens exhibited several load reductions at or near the maximum test load. These load reductions corresponded to the formation of discrete amounts of notch-tip damage [*7*]. As an example, Fig. 2 provides a compari-

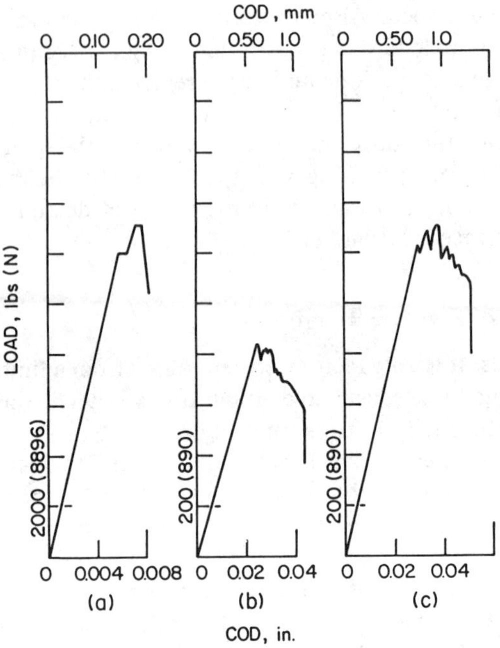

FIG. 1—*Typical load-COD test records for* (a) *center-cracked tension specimen,* (b) *three-point bend specimen, and* (c) *compact tension specimen.*

son of radiographs of the damage for the $[0/90]_{16s}$ laminate. In each case the radiograph was taken just after the first discontinuity in the load-COD record. As can be seen, the damage is similar in content and magnitude.

The effect of laminate thickness is best illustrated by computing fracture toughness at the maximum test load. This corresponded to catastrophic failure of the M(T) specimen and is consistent with the usual definition of notched laminate strength. Fracture toughness versus laminate thickness is shown in Figs. 3 to 5 for the $[0/90]_{ns}$, $[0/\pm45/90]_{ns}$ and $[0/\pm45]_{ns}$ laminates, respectively. The plotted values are the average of the four replicate tests. For the thin laminates the maximum deviation from the mean value was 8 to 12%, whereas for the thicker laminates the maximum deviation from the mean was 2 to 5%.

The change in fracture toughness was primarily attributed to the influence of laminate thickness on the notch-tip damage that formed prior to fracture. A detailed discussion of how the damage zone varied with laminate thickness is provided in Ref *7*.

There was very little difference in the values of fracture toughness for the $[0/90]_{ns}$ and $[0/\pm45/90]_{ns}$ laminates obtained from the three specimen configurations. The difference between the high and the low value for the $[0/90]_{ns}$ laminate was 952 MPa$\sqrt{\text{mm}}$ (27.4 ksi$\sqrt{\text{in.}}$) (C(T)) and 841 MPa$\sqrt{\text{mm}}$ (24.2

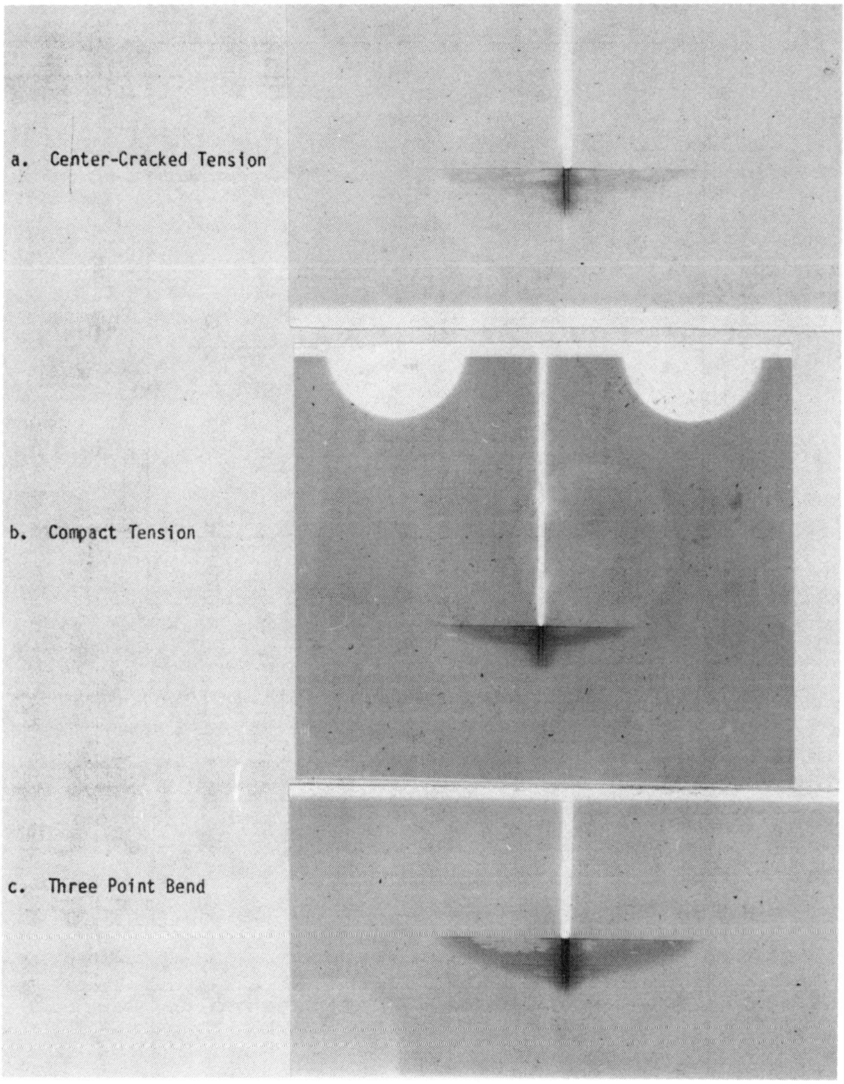

FIG. 2—*Comparison of damage at the first COD discontinuity for the* $[0/90]_{16s}$ *laminate.*

ksi$\sqrt{\text{in.}}$) (M(T)). For the $[0/\pm45/90]_{ns}$ laminate the high value was 1081 MPa$\sqrt{\text{mm}}$ (31.1 ksi$\sqrt{\text{in.}}$) (M(T)) and the low value was 973 MPa$\sqrt{\text{mm}}$ (28.0 ksi$\sqrt{\text{in.}}$) (SE(B)). However, there was considerable difference in the values of fracture toughness for the $[0/\pm45]_{ns}$ laminate, where the high value was 1387 MPa$\sqrt{\text{mm}}$ (39.9 ksi$\sqrt{\text{in.}}$) (M(T)) and the low value was 1112 MPa$\sqrt{\text{mm}}$ (32.0 ksi$\sqrt{\text{in.}}$) (C(T)). Previously, differences in the load-COD records of the M(T)

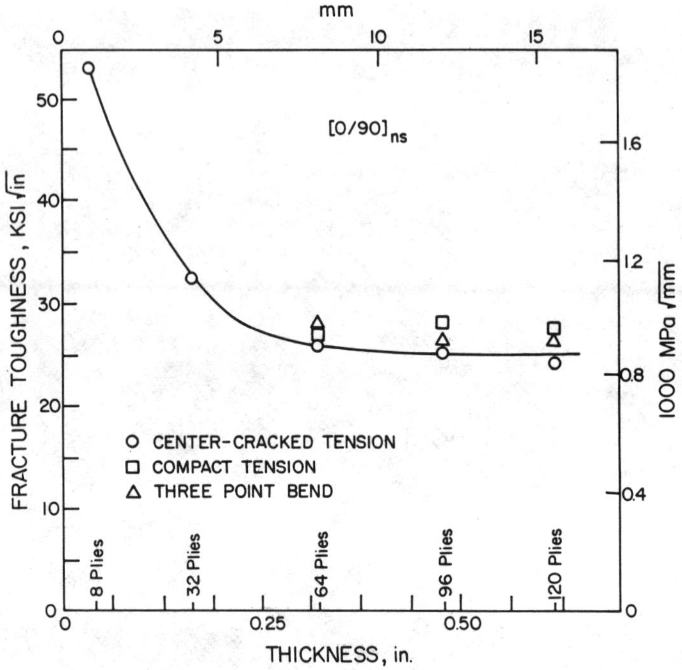

FIG. 3—*Fracture toughness at maximum load versus thickness for a [0/90]$_{ns}$ laminate.*

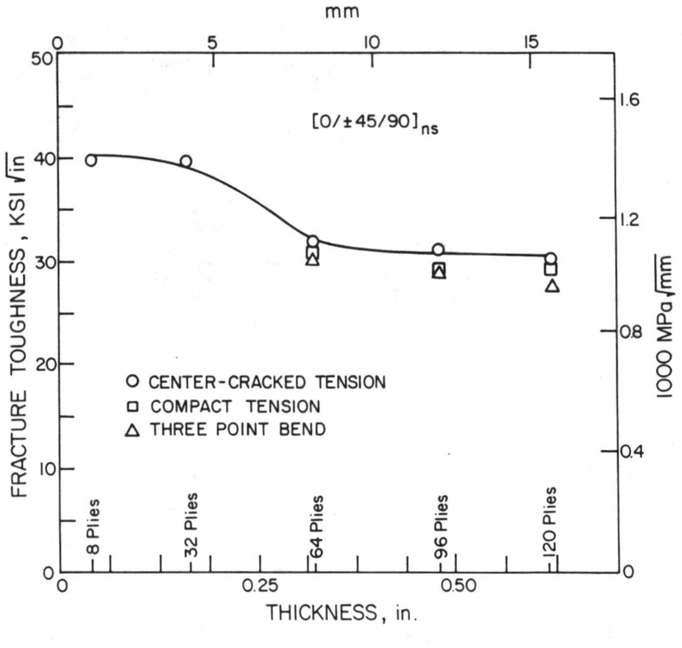

FIG. 4—*Fracture toughness at maximum load versus thickness for a [0/± 45/90]$_{ns}$ laminate.*

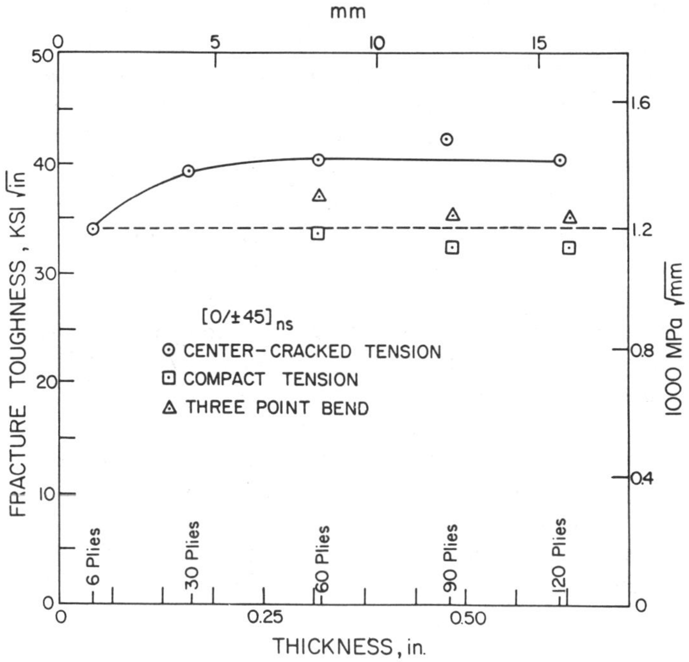

FIG. 5—*Fracture toughness at maximum load versus thickness for a [0/$\pm$45]$_{ns}$ laminate.*

specimen and the other two specimen configurations were discussed. Perhaps a more consistent comparison of fracture toughness values would occur if fracture toughness were computed at the first COD discontinuity for the M(T) specimen and at the first load reduction for the other two configurations. As the radiographs in Fig. 2 show, these load values were for a similar notch-tip damage zone, and the load records were nearly linear prior to the first load-COD discontinuity.

There were other differences in some load-COD records that further suggested the utilization of an alternative load to the maximum load. For example, sometimes one specimen from a set of four replicates would not have any discontinuities in the load-COD record, while the other specimens exhibited several discontinuities. The failure load of the specimen without discontinuities was very close to the loads where major discontinuities occurred in the load-COD records of the other specimens.

In the case of metals, ASTM E 399 provides a method for reconciling the types of differences in load-COD records discussed above. (While the basis for the approach outlined in ASTM E 399 may not be valid for a heterogeneous composite material, the approach may be applicable to the reduction of composite materials test data.) A secant line is constructed through the origin of the test record with a slope that is 0.95 times the slope of the initial portion of the load-COD record. The intersection of the 5% secant line with the test record defines a candidate load, P_5, for computing fracture toughness. The basis for using a 5% slope offset is due to the requirement to minimize crack-tip plasticity and stable crack growth (for metals) so that linear elastic fracture mechanics applies. (A detailed discussion of the basis for the 5% offset is given in Broek [*10*].) For the M(T) specimens of this study, the 5% secant line typically intercepted the load record at a discontinuity. Similarly, the load at the first load reduction was usually the P_5 value for the C(T) and SE(B) specimens. Therefore the load defined in accordance with ASTM E 399 at the 5% slope offset intercept was selected for computing the fracture toughness (K_5) of the laminates discussed herein. The fracture toughness versus laminate thickness is shown in Figs. 6 to 8 for the $[0/90]_{ns}$, $[0/\pm45/90]_{ns}$ and $[0/\pm45]_{ns}$ laminates, respectively. The K_5 fracture toughness can be physically interpreted as a measure of the onset of significant crack-tip damage.

As seen in Figs. 3 to 5 and Figs. 6 to 8 the K_m and K_5 values for the C(T) and SE(B) specimens are practically identical. This was because the load at the first load reduction was always quite close to the maximum value. By defining fracture toughness at the onset of damage, the influence of the notch-tip damage zone on the final fracture was minimized. Therefore the K_5 frac-

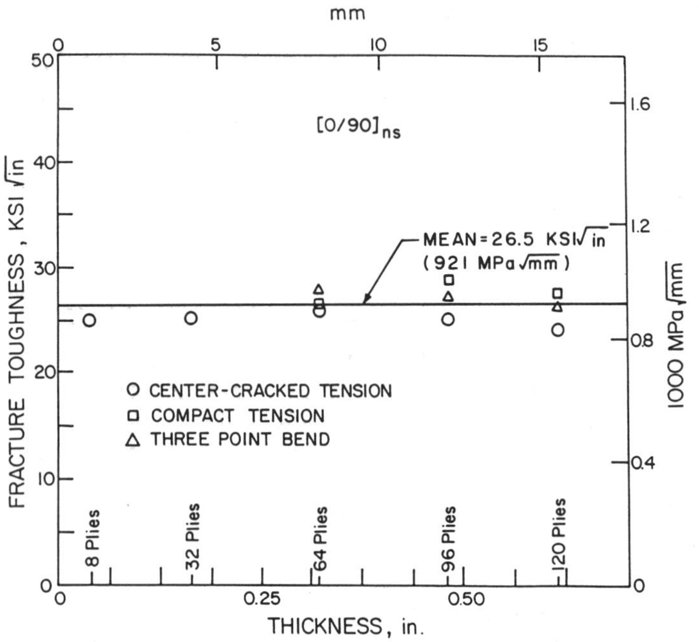

FIG. 6—*K_5 fracture toughness versus thickness for a $[0/90]_{ns}$ laminate.*

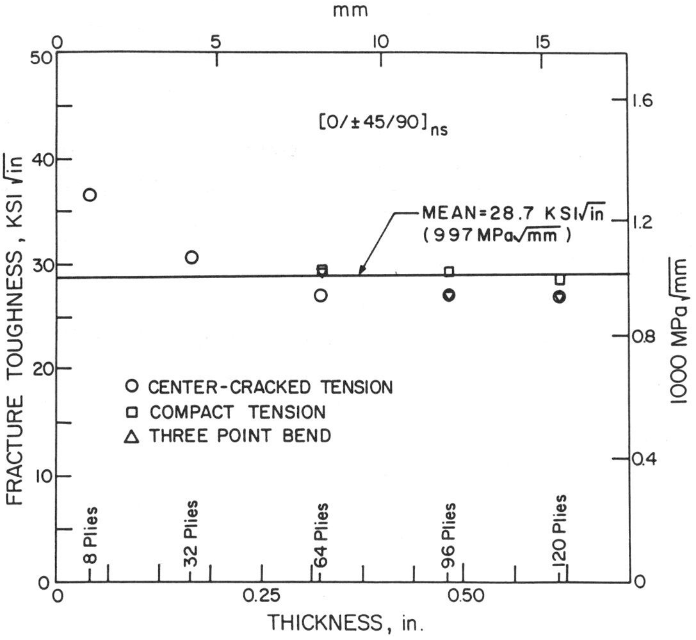

FIG. 7—*K_5 fracture toughness versus thickness for a $[0/\pm 45/90]_{ns}$ laminate.*

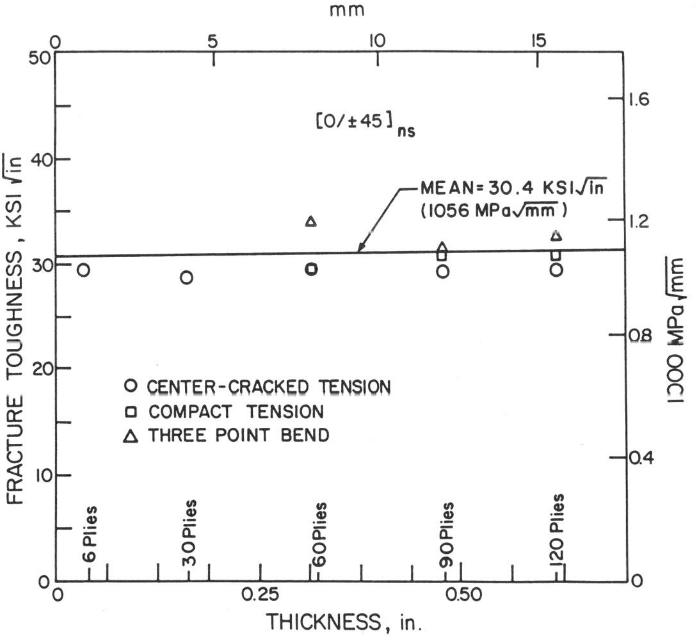

FIG. 8—*K_5 fracture toughness versus thickness for a $[0/\pm 45]_{ns}$ laminate.*

ture toughness is seen to be independent of laminate thickness. However, there is some thickness dependence for the $[0/\pm45/90]_{ns}$ laminate (Fig. 7).

The implication of this result is potentially quite significant. On the one hand, a design value of fracture toughness may be obtained by testing C(T) or SE(B) specimens. Alternatively, very thin center-cracked tension specimens may also yield a conservative design value when fracture toughness is defined at the onset of damage rather than at the failing load.

In conclusion, in a separate but related study the authors [*11*] have shown that fracture toughness computed at the onset of damage (K_5) was independent of laminate stacking sequence and thickness for M(T) specimens. Twenty-seven stacking sequences of graphite/epoxy T300/5208 laminates were included in the study. Including from three to six replicate tests in all cases, the mean value of fracture toughness was 990 MPa$\sqrt{\text{mm}}$ (28.5 ksi$\sqrt{\text{in.}}$) with a standard deviation of 104 MPa$\sqrt{\text{mm}}$ (2.99 ksi$\sqrt{\text{in.}}$). This compares favorably with the mean values shown in Figs. 6 to 8.

Conclusions

The results from this study showed that fracture toughness computed at the maximum test load was significantly influenced by laminate thickness. This was due to the effects of laminate thickness on the size and type of damage zone that formed at the notch-tip prior to fracture. However, fracture toughness was independent of laminate thickness when computed at the onset of damage (utilizing the 5% slope offset intercept similar to ASTM E 399). The values of fracture toughness obtained from thick center-cracked tension, compact tension, and three-point bend specimens were nearly identical when computed at the onset of damage. Furthermore, the type and magnitude of the notch-tip damage zone of the three specimen configurations were similar at the first discontinuity in the load-COD test record.

Acknowledgments

The authors wish to acknowledge the financial support and specimen preparation provided by the NASA-Langley Research Center through Grant NAG-264. Also, the authors wish to thank the contract monitor, C. C. Poe, Jr., for his helpful discussions.

References

[*1*] Brown, W. F., Jr., and Srawley, J. E., Eds., *Plane Strain Crack Toughness Testing of High Strength Metallic Materials, ASTM STP 410*, American Society for Testing and Materials, Philadelphia, 1966.

[*2*] Brown, W. F., Jr., Ed., *Review of Developments in Plane Strain Fracture Toughness Testing, ASTM STP 463*, American Society for Testing and Materials, Philadelphia, 1970.

[3] Cruse, T. A. and Osias, J. R., "Exploratory Development on Fracture Mechanics of Composite Materials," Air Force Materials Laboratory Report AFML-TR-74-111, Wright-Patterson AFB, Ohio, 1974.
[4] Prewo, K. M., "The Effect of Ply Lay-up Sequence on the Fracture Toughness of Boron Aluminum," *Journal of Composite Materials*, Vol. 12, Jan. 1978, pp. 40-52.
[5] Reedy, E. D., Jr., "On the Specimen Dependence of Unidirectional Boron/Aluminum Fracture Toughness," *Journal of Composite Materials Supplement*, Vol. 14, 1980, pp. 118-131.
[6] Shih, T. T. and Logsdon, W. A., "Fracture Behavior of a Thick-Section Graphite/Epoxy Composite," in *Fracture Mechanics: Thirteenth Conference, ASTM STP 743*, R. Roberts, Ed., American Society for Testing and Materials, Philadelphia, 1981, pp. 316-337.
[7] Harris, C. E. and Morris, D. H., "Fracture Behavior of Thick, Laminated Graphite/Epoxy Composites," NASA CR-3784, National Aeronautics and Space Administration, Washington, D.C., 1984.
[8] Freeman, S. M., "Characterization of Lamina and Interlaminar Damage in Graphite-Epoxy Composites by the Deply Technique," in *Composite Materials: Testing and Design (Sixth Conference), ASTM STP 787*, I. M. Daniel, Ed., American Society for Testing and Materials, Philadelphia, 1982, pp. 50-62.
[9] Chu, C. S., Anderson, J. M., Batdorf, W. J., and Aberson, J. A., "Finite Element Computer Program to Analyze Cracked Orthotropic Sheets," NASA CR-2698, National Aeronautics and Space Administration, Washington, D.C., 1976.
[10] Broek, D., *Elementary Engineering Fracture Mechanics*, 3rd rev. ed., Martinus Nijhoff Publishers, The Hague, The Netherlands, 1982.
[11] Harris, C. E. and Morris, D. H., "A Damage Tolerant Design Parameter for Graphite/Epoxy Laminated Composites," *Journal of Composites Technology and Research*, Vol. 7, No. 3, Fall 1985, pp. 77-81.

Robert A. Simonds[1]

Residual Strength of Five Boron/Aluminum Laminates with Crack-Like Notches After Fatigue Loading

REFERENCE: Simonds, R. A., "**Residual Strength of Five Boron/Aluminum Laminates with Crack-Like Notches After Fatigue Loading,**" *Fracture Mechanics: Seventeenth Volume, ASTM STP 905,* J. H. Underwood, R. Chait, C. W. Smith, D. P. Wilhem, W. A. Andrews, and J. C. Newman, Eds., American Society for Testing and Materials, Philadelphia, 1986, pp. 136–152.

ABSTRACT: Boron/aluminum specimens were made with crack-like slits in the center and with various proportions of 0° and ±45° plies. They were fatigue loaded and then fractured to determine their residual strengths. The fatigue loads were generally in the range of 60 to 80% of the static tensile strength of the specimen as determined from a previous study, and the stress ratio was 0.05. For virtually all of the specimens the fatigue loading was continued for 100 000 cycles. The spcimens were radiographed after the fatigue loading to determine the nature of the fatigue damage. A few specimens were sectioned and examined in a scanning electron microscope after being radiographed in order to verify the interpretation of the radiographs and also to get a better insight into the nature of the fatigue damage.

The results indicate that the fatiguing does not significantly affect the strength of the specimens tested. The results of the radiography and of the scanning electron microscopy indicate that the 45° plies suffer extensive damage in the form of split and broken fibers and matrix cracking in the vicinity of the ends of the slit. By contrast, the only significant damage to the 0° plies was a single 0° matrix crack growing from the ends of the slit.

KEY WORDS: boron/aluminum, fatigue, fracture

A study by Poe and Sova [*1*] establishes a relationship between "damage," in the form of a crack-like slit machined perpendicular to the load directions and in the center of a boron/aluminum tensile specimen, and the specimen's static fracture strength. The series of tests included specimens of five differ-

[1]Instructor, Engineering Science and Mechanics Department, Virginia Polytechnic Institute and State University, Blacksburg, VA 24060.

ent B/Al laminate configurations, several slit lengths, and several specimen widths. Thus the strength of "damaged" specimens could be compared with the strength of "undamaged" specimens determined by prior experiments [*2*]. The present study seeks to amplify that knowledge by adding fatigue damage to the ends of the slits. The specimens are fatigued at a level high enough that significant fatigue damage develops, yet not at such a high level that the specimen fails in fatigue before 100 000 cycles. The present study seeks to document the fatigue damage and to see what the effect of the fatigue damage is on the residual strength of the specimen.

Material and Specimens

The B/Al laminates were made by diffusion bonding 0.152-mm-diameter boron fibers and 6061 aluminum foil. The laminates were tested without any subsequent heat treatment. The laminate orientations were $[0]_{6T}$, $[0_2/\pm45]_s$, $[\pm45/0_2]_s$, $[0/\pm45]_s$, and $[\pm45]_{2s}$. The volume fractions were 50% for the $[0]_{6T}$ laminate and 45% for the cross-plied laminates. Fatigue/fracture, fracture, and tensile specimens were cut from each sheet of material. Results of the tensile tests were reported in Ref *2*; some of the average tensile results are given in Table 1. The results are for specimens 19.1 mm wide except for the $[\pm45]_{2s}$ laminate where the results are for 102-mm-wide specimens. There is a significant width effect for the $[\pm45]_{2s}$ laminate, so the $[\pm45]_{2s}$ data included in Table 1 are for the same width specimens as the specimens included in the present study. No appreciable width effect was found for the other laminate orientations.

The fatigue/fracture specimens were all 102 mm wide. Crack-like slits were machined into the center of each specimen with an electrical-discharge process. The slits were perpendicular to the loading direction and had lengths of 10.2, 30.5, and 50.8 mm. Figure 1 shows a sketch of the fatigue/fracture specimens that identifies fiber angles and shows the location of the grips.

The static fracture strengths of the various types of specimens were known prior to the present study from Ref *1*.

TABLE 1—*Tensile strengths of laminates.*

Laminate	Laminate Thickness, mm	Specimen Width, mm	Strength, MPa
$[0]_{6T}$	1.07	19.1	1670
$[0_2\pm45]_s$	1.49	19.1	800
$[\pm45/0_2]_s$	1.51	19.1	910
$[0/\pm45]_s$	1.11	19.1	581
$[\pm45]_{2s}$	1.52	102	221

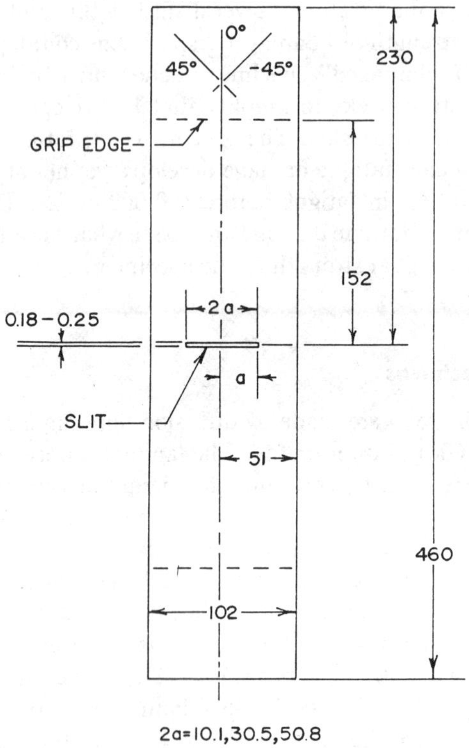

FIG. 1—*Specimen geometry (dimensions in millimeters).*

Test Procedures and Equipment

The specimens were tested in a hydraulically actuated, closed-loop, servo-controlled testing machine. The load, which was measured by a conventional load cell, was used for the feedback signal. For the fatigue portion of the tests, a function generator produced a sinusoidal command signal of 10.0 Hz. The testing machine was set so that the maximum cyclic load was a certain percentage of the average static failure load for each slit length and laminate orientation as determined in Ref *1*. The ratio of minimum to maximum stress was generally equal to 0.05. Most of the specimens were fatigued for 100 000 load cycles. The residual strength fracture tests were accomplished on the same testing machine by resetting the function generator to output a slow linear ramp command signal so that the specimen failed in about 2 mins.

Many of the specimens were radiographed after, and sometimes during, the fatigue portion of the testing in order to determine the damage caused by the fatigue loading. The radiographs were made with an industrial-type "soft" X-ray machine with a 0.254-mm beryllium window and a tungsten target. The voltage and current were set for 50 kV and 20 mA, respectively. The window of the X-ray tube was 607 mm from the specimen, and a high-resolution photographic plate was mounted directly on the opposite side of the specimen. The exposure time for each radiograph was 12 min.

A few specimens were destructively examined after the fatigue portion of the testing instead of being fracture tested. Typically, pieces were cut out of the specimen from the vicinity of the ends of the slits, and then leached in a sodium-hydroxide solution to enhance cracks in the aluminum matrix, to remove the matrix and expose the fibers, or to remove entire plies and expose the fibers of the underlying plies. The amount of material removed was dependent on the amount of time the piece was submerged in the solution. The leached pieces could then be examined in a scanning electron microscope.

Results and Discussion

Figure 2 compares the fracture strength of the fatigued specimens with the average strength of the specimens from Ref *1* that were fractured without any prior fatigue loading. The percentage of the static failure load that was used for the fatigue loading is indicated for each data point. More than one percentage indicates two or more specimens whose post-fatigue strength was the same. The graphs indicate that the difference in strength between the unfatigued specimens of Ref *2* and the fatigued specimens of the present study are relatively insignificant. This happened even though substantial damage often developed in the fatigued specimens as discussed below.

Fiber Damage

With the exception of one of the laminates, the preponderance of fiber damage occurred in the $\pm 45°$ plies rather than in the 0° plies. Figure 3 shows radiographs taken of the slit ends of Specimen 04022, which was a $[0]_{6T}$ laminate, after the specimen was fatigued for 100 000 cycles at 35% of the static fracture strength. The few fibers that are broken, as indicated by the gaps in the white line images of the tungsten fiber cores, all occur in the immediate vicinity of the ends of the slit. Specimen 04022 experienced a very slight increase in fracture strength over the average unfatigued strength.

Figure 4 is a radiograph of the ends or the slit of Specimen 43032, a $[0_2/\pm 45]_s$ laminate. Negligible damage to the fibers can be seen in the radiographs. Figure 5 shows the first and second plies, both 0° of the same specimen after matrix removal, as viewed in the scanning electron microscope. Zero degree fibers, except those actually cut by the slit, are intact. However, the 45° fibers, as shown in Fig. 6, show extensive damage that was not apparent in the radiographs. The 45° fibers appear to have split and to have washed away during matrix removal. The fibers in the $[\pm 45/0_2]_s$ laminates suffer similar damage as those in the $[0_2/\pm 45]_s$ laminate even though the stacking sequence

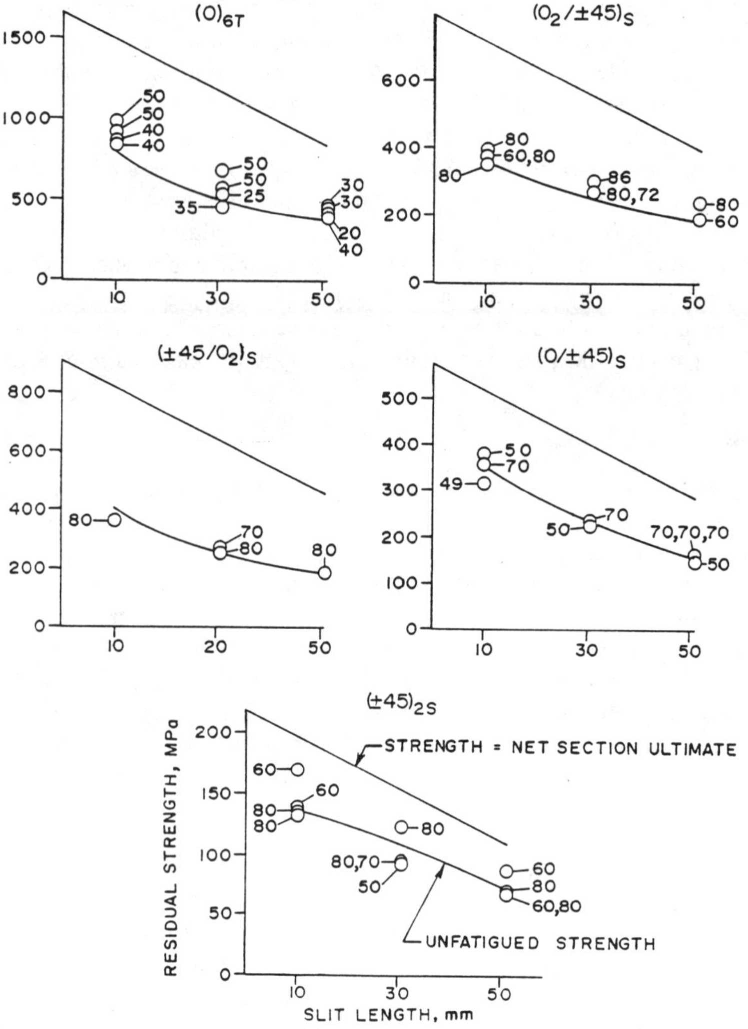

FIG. 2—*Strengths of fatigued specimens compared with strengths of unfatigued specimens.*

is different: negligible damage to 0° fibers and splitting of the 45° fibers and their subsequent washing away when the matrix is removed.

Laminate $[0/\pm45]_s$ was the only one of the five laminates to show significant damage to the 0° fibers as a result of the fatigue loading. Some of the radiographs of specimens of this laminate show broken 0° fibers as far as eight fibers away from the ends of the slit. Figure 7 is a radiograph of Specimen 31022 showing broken 0° fibers as well as broken 45° fibers. Figure 8 shows the results of fatiguing Specimen 31012 for 100 000 cycles at 70% of

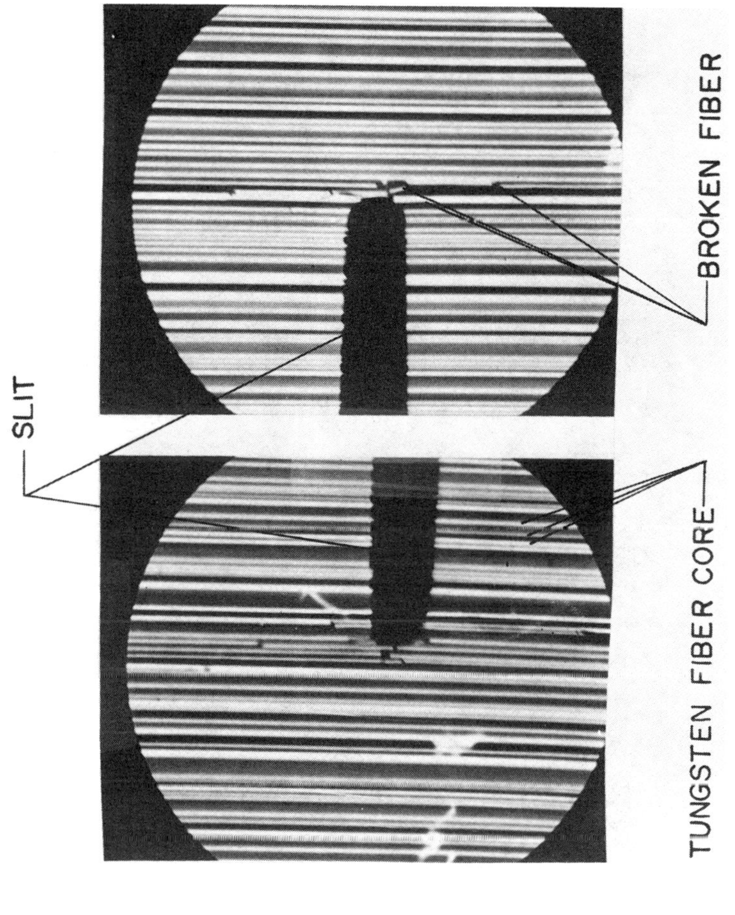

FIG. 3—*Radiographs of the ends of the slit of Specimen 04022 after fatiguing.*

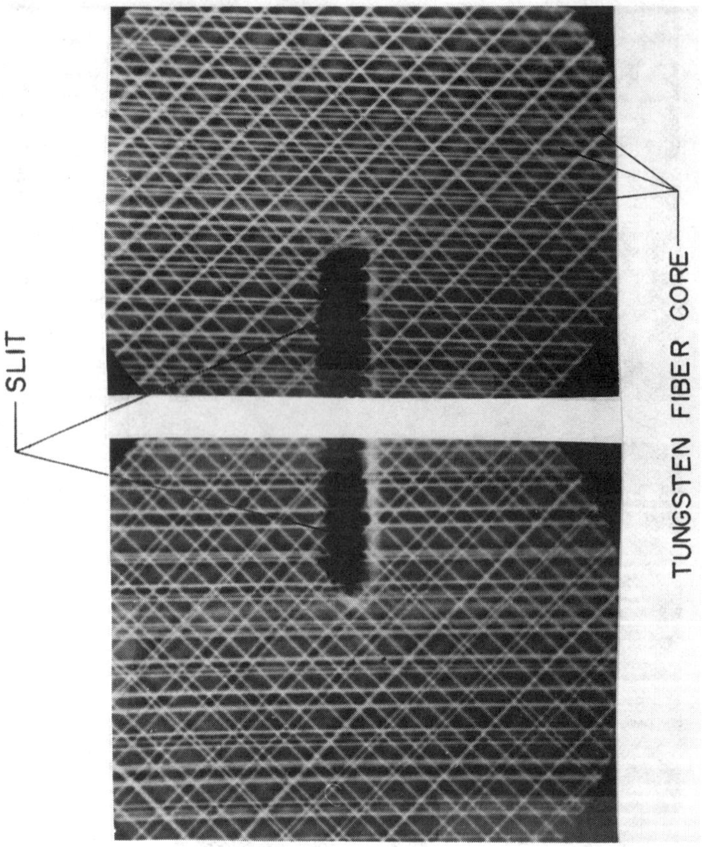

FIG. 4—*Radiographs of the ends of the slit of Specimen 43032 after fatiguing.*

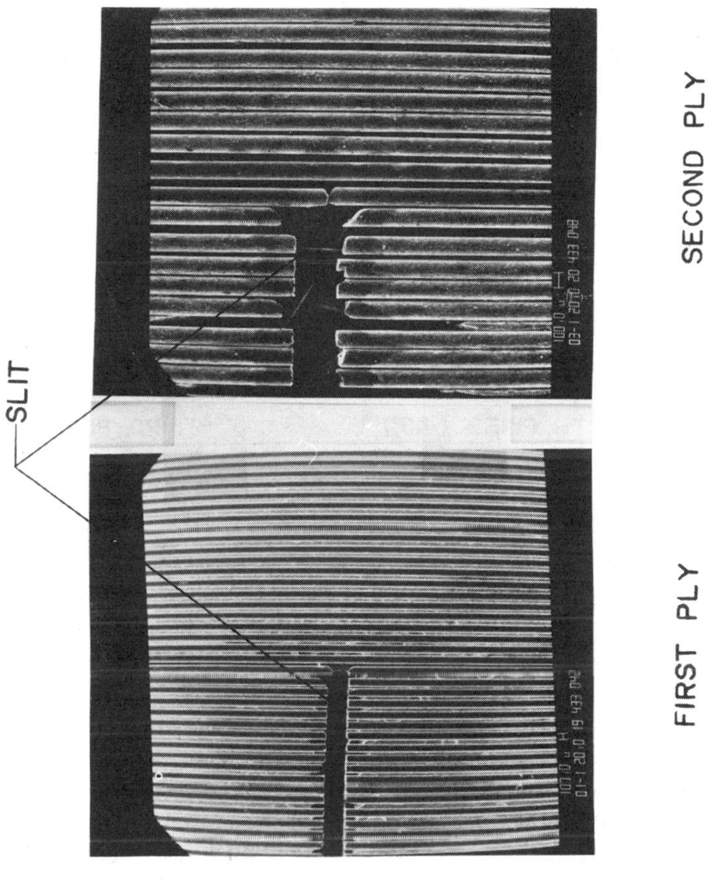

FIG. 5—*Scanning electron micrograph of the first and second (0°) plies of Specimen 43032 in the vicinity of one end of the slit with the matrix etched away after fatiguing.*

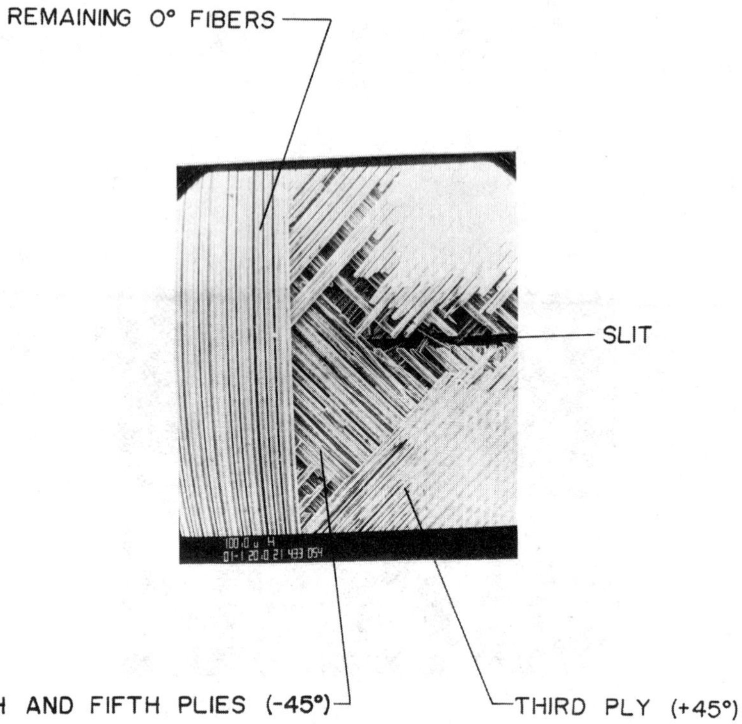

FIG. 6—*Scanning electron micrograph of 45° fibers of Specimen 43032.*

the estimated failure load, sectioning the specimen, then leaching the matrix away with sodium hydroxide to expose broken 0° fibers. Further leaching reveals extensively damaged 45° fibers as seen in the other laminates containing 45° fibers.

Damage to fibers in the $[\pm 45]_{2s}$ laminates appears to be similar to the damage to 45° fibers in the other laminates even though there are no accompanying 0° fibers. Figure 9 shows the results of fatiguing Specimen 13013 for 100 000 cycles to 70% of its average unfatigued static strength. A portion of the specimen taken from the area of one end of the slit was leached in sodium hydroxide to remove the top three plies to reveal the middle two plies, both parallel to one another, and portions of the underlying sixth ply. The fibers are extensively damaged and many of the broken pieces simply fell away from the section once the matrix was dissolved away.

Unpublished work by C. C. Poe, Jr., of NASA Langley Research Center indicates that the 45° fibers are subjected to large tensile stresses transverse to the fibers. Thus the 45° fibers could have broken because their transverse fatigue strength is low. The asymmetrical pattern of damage in Figs. 6 and 9 is

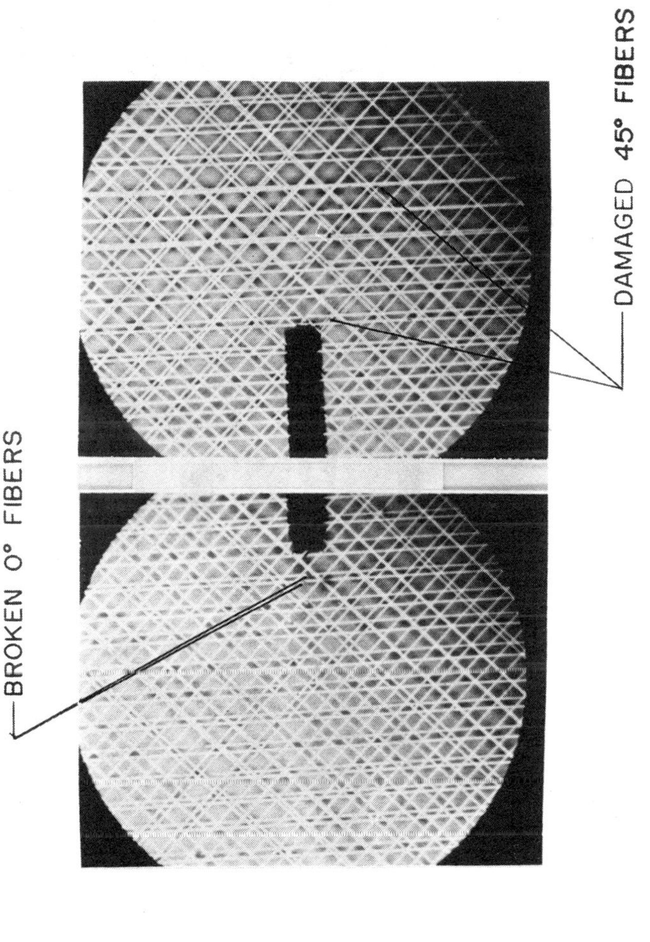

FIG. 7—*Radiograph of the ends of the slit of Specimen 31022 after fatiguing.*

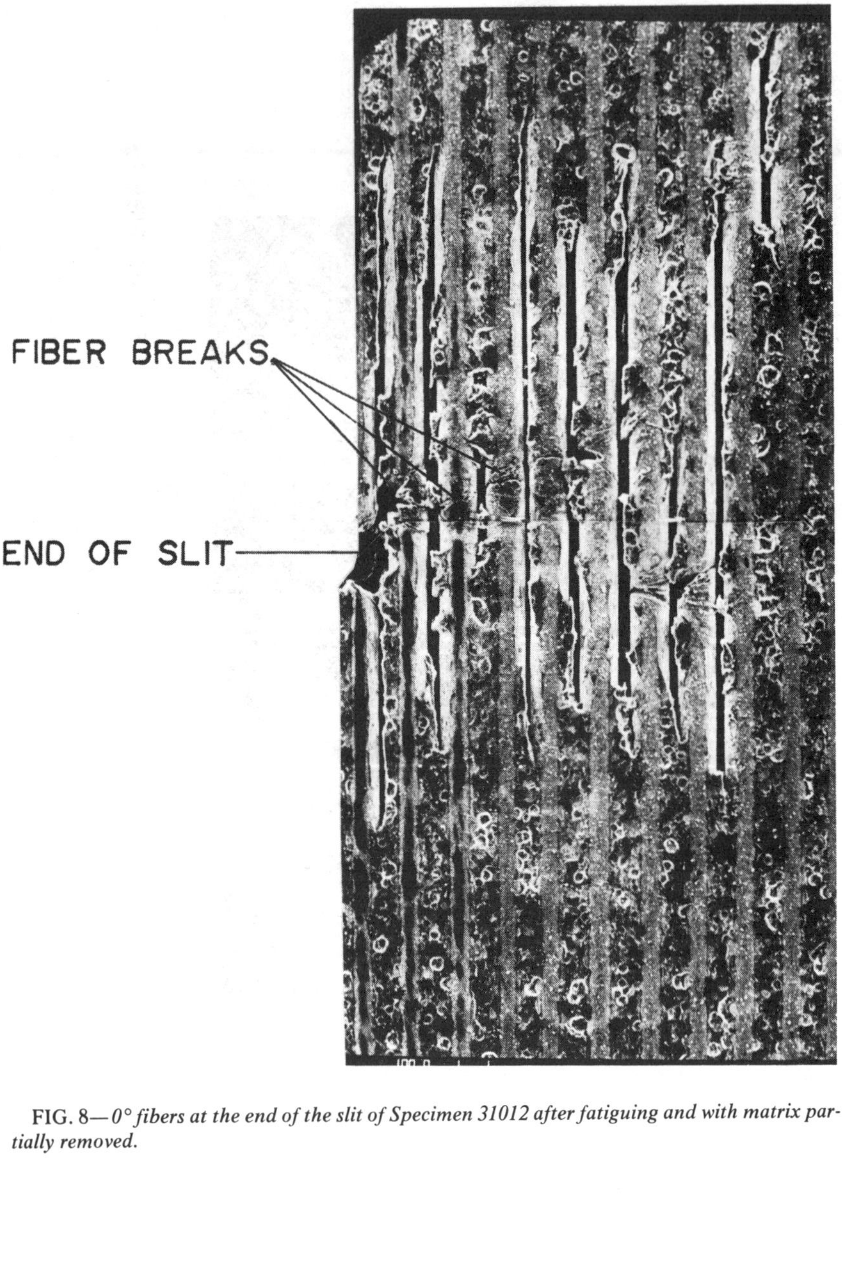

FIG. 8—*0° fibers at the end of the slit of Specimen 31012 after fatiguing and with matrix partially removed.*

FIG. 9—*Scanning electron micrograph of Specimen 13013 showing fourth and fifth (+45°) plies and sixth (−45°) ply at the end of the slit with matrix removed after fatiguing*

consistent with such a failure mode. Similar fiber damage was noted in Refs *3* and *4*.

Matrix Damage

Matrix damage in the boron/aluminum laminates subjected to fatigue loading appeared to take two forms: (1) cracks in the 0° plies which started at the ends of the slits, grew parallel to the 0° fibers, through the specimen thickness, and towards the grips with increasing numbers of cycles, and (2) cracks in the 45° plies originating at split 45° fibers. In addition, some cracks were observed between 0° plies and adjacent 45° plies. In the $[0]_{6T}$ laminates, cracks developed very quickly at the ends of the slits and grew very quickly

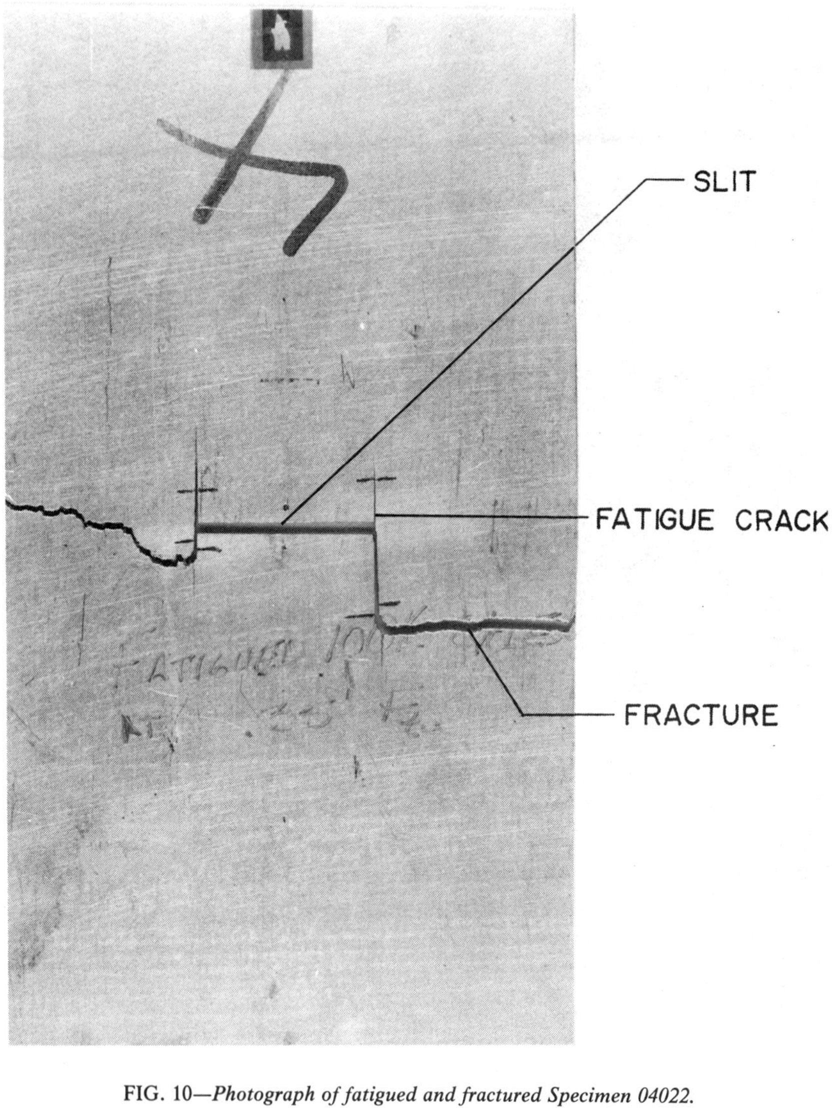

FIG. 10—*Photograph of fatigued and fractured Specimen 04022.*

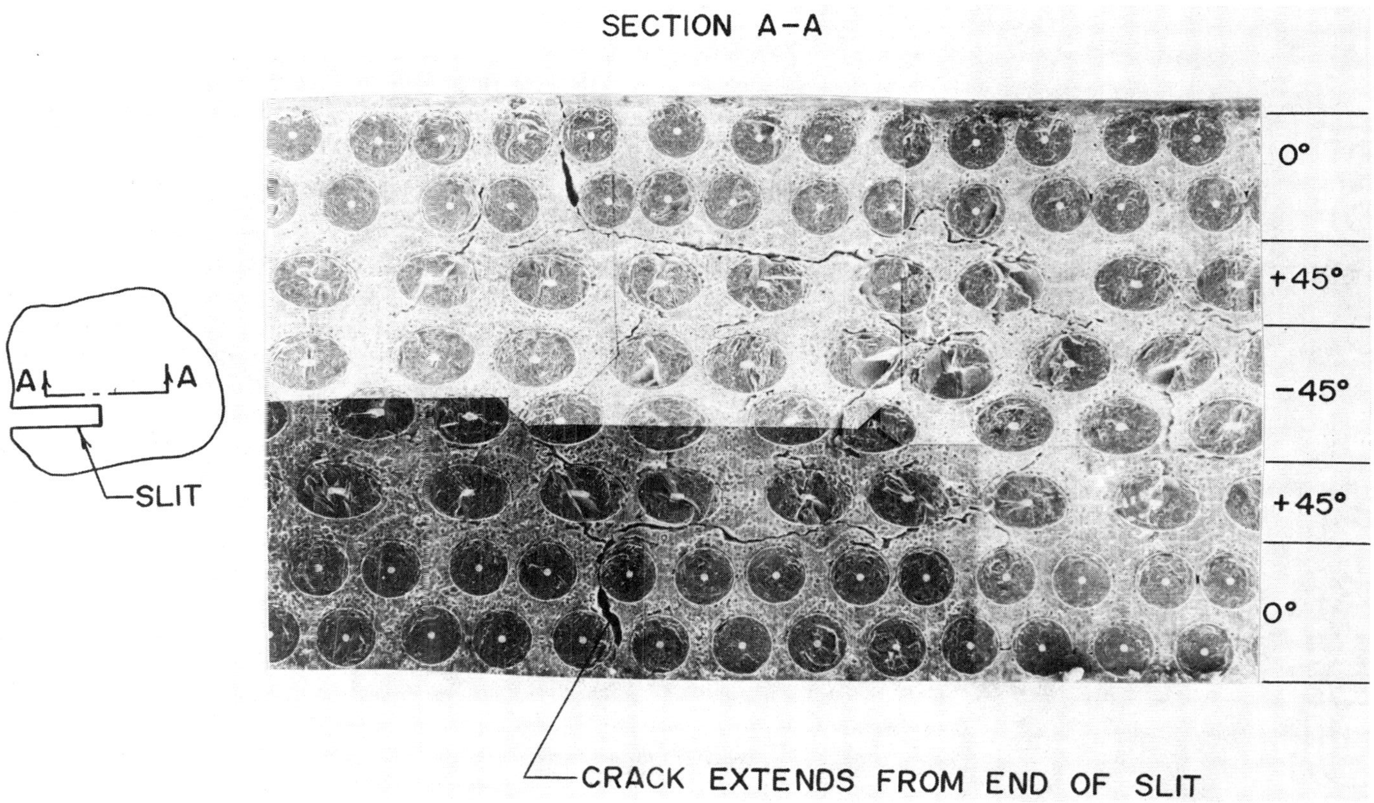

FIG. 11—*Scanning electron micrograph of a cross section of Specimen 43032 near one end of the slit after fatiguing. Section was etched to enhance cracks in matrix.*

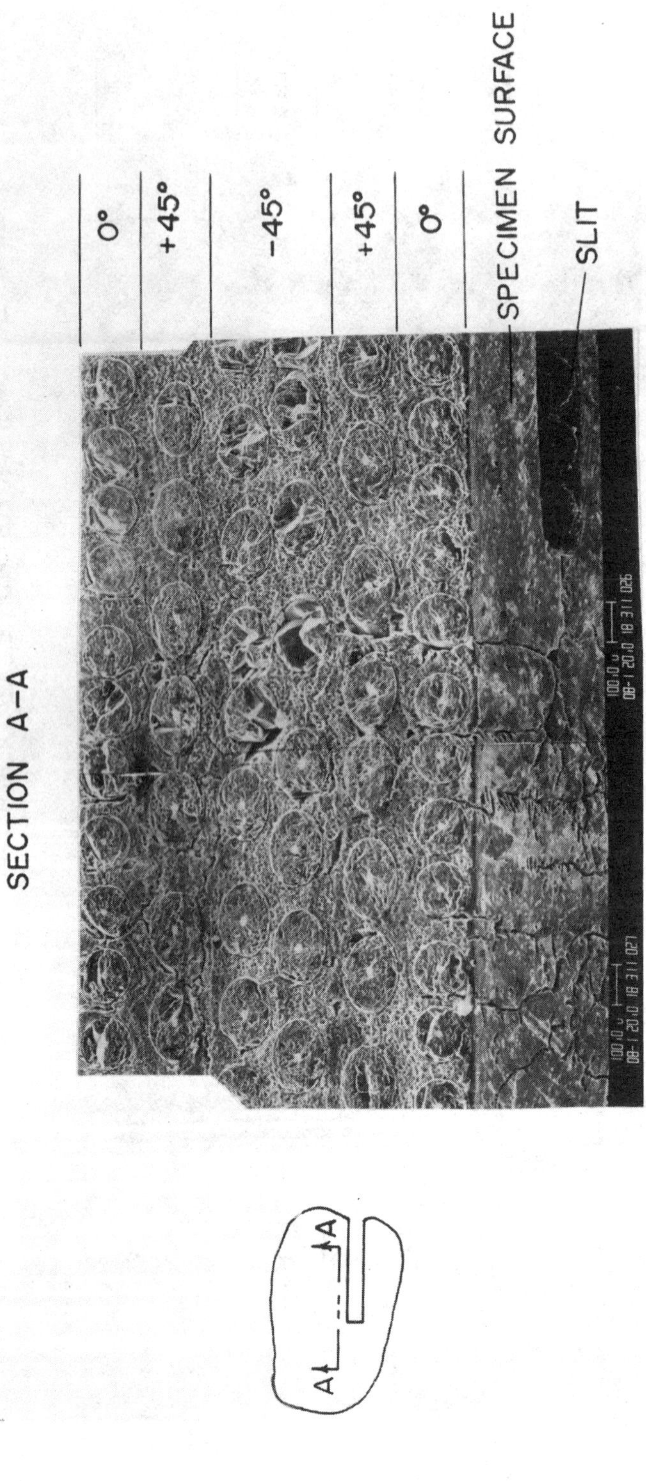

FIG. 12—Scanning electron micrograph of a cross section of Specimen 31032 in the vicinity of one end of the slit after fatiguing. Section was etched to enhance matrix cracks.

along the 0° fibers towards the grips. The matrix cracks in the $[0]_{6T}$ specimens grew so quickly that the $[0]_{6T}$ specimens were fatigued at lower percentages of their static fracture strengths. Figure 10 is a photograph of Specimen 04022 showing that the fatigue cracks grew from the ends of the slit and that the specimen then fractured at the ends of the fatigue cracks rather than at the ends of the slit. Pencil marks were made on the surface of the specimen at the visible ends of the fatigue cracks after the fatigue portion of the testing. During the subsequent fracture testing, the specimen fractured from the ends of the fatigue cracks rather than from the ends of the original slit. This behavior is in agreement with the predictions of shear lag analysis (Ref *5*). At fracture, there was some additional cracking in the 0° directions as there was in the fracture testing of unfatigued $[0]_{6T}$ specimens with slits (Ref *1*).

Figure 11 is a cross section of Specimen 43032, $[0_2/\pm 45]_s$, slightly above the slit. The section has been etched with sodium-hydroxide to enhance the matrix cracks. It shows cracks at the end of the slit in the 0° plies, between the 0° and +45° plies, and in the 45° plies. The $[\pm 45/0_2]_s$ specimens show virtually identical matrix cracking even though the stacking sequence is different: cracks between the 0° fibers growing from the ends of the slit and cracks in the 45° plies growing from the split fibers.

Fatiguing of $[0/\pm 45]_s$ laminates with slits does not appear to create a single dominating matrix crack in the 0° plies as it does in the other laminates containing 0° plies. Instead, several cracks appear between adjacent 0° fibers as can be seen in Fig. 12. Note that the matrix cracks in the 0° plies are between the fibers as opposed to connected with split fibers as in the 45° plies.

Matrix damage to $[\pm 45]_{2s}$ laminates took the form of matrix damage in the 45° plies as in other laminates: coincident with splits in the 45° fibers.

Summary

Boron/aluminum specimens containing crack-like slits in their centers perpendicular to the loading direction were fatigued. The fatigue loading was sufficient to cause damage to occur at the ends of the slits but not so high as to cause the specimens to fail in fatigue. The specimens were radiographed in an effort to determine the resulting damage. Broken fibers are fairly easy to detect in the radiographs; however, matrix damage is virtually indiscernible radiographically. In a few cases, specimens were sectioned and examined microscopically in order to verify the existence of the broken fibers as seen in the radiographs and in order to determine the matrix damage.

Fatigue damage at the ends of the slits takes the following forms:

1. Extensively broken 45° fibers in all the laminates containing 45° fibers.
2. In specimens in which at least 50% of the plies are 0° plies, there are cracks in the matrix in the 0° direction in the 0° plies starting at the ends of the slits and growing towards the grips.

3. Matrix cracking in the 45° plies associated with split and/or broken 45° fibers.

In addition, there are some broken 0° fibers in the $[0/\pm45]_s$ laminate specimens, and in those specimens containing both 0° plies and 45° plies there is a tendency of the matrix to crack between a 0° ply and an adjacent 45° ply.

In spite of the aforementioned damage, there is virtually no change in the strength of a fatigued boron/aluminum specimen containing a slit in the center compared with a similar specimen without fatigue.

Acknowledgments

The author would like to acknowledge the support of the NASA Langley Research Center through NASA Contract NAS1-17100. Sincere appreciation is extended to C. C. Poe, Jr., of NASA Langley Research Center for his encouragement and helpful discussions. Sincere appreciation is also extended to B. I. Wengert of VPI & SU for typing the report.

References

[*1*] Poe, C. C., Jr., and Sova, J. A., "Fracture Toughness of Boron/Aluminum Laminates with Various Proportions of 0° and ±45° Plies," NASA TP-1707, National Aeronautics and Space Administration, Washington, D.C., 1980.

[*2*] Sova, J. A. and Poe, C. C. Jr., "Tensile Stress Strain Behavior of Boron/Aluminum Laminates," NASA TP-1117, National Aeronautics and Space Administration, Washington, D.C., 1978.

[*3*] Johnson, W. S., "Mechanisms of Fatigue Damage in Boron/Aluminum Composites," in *Damage in Composite Materials, ASTM STP 775*, American Society for Testing and Materials, Philadelphia, 1982, pp. 83–102.

[*4*] Menke, G. D. and Toth, I. J., "The Time Dependent Mechanical Behavior of Metal Matrix Composites," AFML-TR-71-102, Air Force Materials Laboratory, Wright-Patterson AFB, Dayton, Ohio, Sept. 1971.

[*5*] Dharani, L. R., Jones, W. F., and Goree, J. G., "Mathematical Modeling of Damage in Unidirectional Composites," *Engineering Fracture Mechanics*, Vol. 17, No. 6, 1983.

Subcritical Crack Growth

T. Nicholas[1] *and T. Weerasooriya*[2]

Hold-Time Effects in Elevated Temperature Fatigue Crack Propagation

REFERENCE: Nicholas, T. and Weerasooriya, T., **"Hold-Time Effects in Elevated Temperature Fatigue Crack Propagation,"** *Fracture Mechanics: Seventeenth Volume, ASTM STP 905*, J. H. Underwood, R. Chait, C. W. Smith, D. P. Wilhem, W. A. Andrews, and J. C. Newman, Eds., American Society for Testing and Materials, Philadelphia, 1986, pp. 155–168.

ABSTRACT: An experimental investigation was conducted to evaluate the effects of hold times on the fatigue crack growth rate of Inconel 718 at 649°C using compact tension specimens. Tests were run under computer-controlled constant K conditions using compliance to determine crack length. Hold times ranging from 5 to 50 s were applied at maximum, minimum, and intermediate load levels. The data show that the hold times at maximum load were the most damaging in terms of crack growth rate. Hold times greater than 5 s led to purely time-dependent crack growth behavior which was predictable from sustained load data using K as a correlating parameter. Hold times at minimum or intermediate load levels had little or no effect on crack growth rate. A linear cumulative damage model based solely on fatigue and sustained load data was found to be adequate for spectrum loading so long as the hold times were at maximum load.

KEY WORDS: fatigue crack growth, time-dependent crack growth, creep/fatigue interactions, spectrum loading, nickel-base superalloy, cumulative damage model

Crack growth calculations for critical structural components in U.S. Air Force gas turbine engines have become an important part of the design and life management procedures in recent years. Accurate crack growth predictions are required as part of a damage tolerant design approach required by the Air Force for all new engines under the recently adopted Engine Structural Integrity Program (ENSIP) specifications. Further, implementation of a retirement-for-cause or on-condition lifting policy for existing engines requires the capability to predict crack growth rates in engine components un-

[1]Senior Scientist, Air Force Wright Aeronautical Laboratories, Wright-Patterson Air Force Base, OH 45433.

[2]Research Engineer, University of Dayton Research Institute, Dayton, OH 45469.

der actual operating conditions. These conditions include variations in temperature, cyclic frequency, stress ratio, and sustained load hold time, as well as interactive effects in a typical load spectrum. One of the major concerns is to address the effects of hold times at the highest operating temperatures, where turbine disk alloys demonstrate a varying degree of time-dependent material behavior. In particular, the effects of hold times between single or multiple cyclic loads are not fully understood.

Hold times in a typical load spectrum for a turbine engine disk represent a condition of constant engine speed and constant temperature. For high-performance military aircraft engines, hold times ranging from a few seconds to a few minutes are common throughout the load spectrum, which is comprised primarily of low frequency fatigue cycles of varying frequencies, amplitudes, and stress ratios. The hold times can occur at maximum load or at some intermediate load level between the maximum and minimum of the fatigue loading. These hold times can contribute to crack growth at high temperatures through a combination of creep and environmental degradation, primarily oxidation, near the crack tip. In nickel-base superalloys, sustained load crack growth is primarily an environmentally enhanced phenomenon with little or no creep present. In the absence of oxygen, sustained load crack growth rates have been observed to decrease by more than an order of magnitude [*1*].

Prior work has shown that sustained load crack growth in nickel-base superalloys can be characterized using the linear elastic fracture mechanics (LEFM) stress intensity factor, K, as the correlating parameter [*2,3*]. It has also been shown that for very low frequency cycles, fatigue crack growth rates can be deduced from sustained load crack growth data using simple integrations of the fatigue loading cycle [*4*]. Hold-time effects at maximum load between fatigue cycles were also well predicted from sustained load data. What has not been evaluated, however, is the contribution of sustained loads at stress levels below the peaks of the fatigue cycles and the interactions between cyclic and sustained loads in complex mission spectra. This paper presents the results of a systematic investigation to evaluate hold-time effects in elevated temperature fatigue and guidelines for the prediction of engine spectrum crack growth rates at high temperatures.

Experiments

A series of tests was performed on compact-type specimens of Inconel 718 involving combinations of cyclic loading and hold times. All test specimens were of identical dimensions having $H/W = 0.6$, $B = 10$ mm, and $W = 40$ mm. The Inconel 718 was heat treated to the standard treatment detailed in Table 1. All tests were conducted at a temperature of 649°C in an MTS servohydraulic test machine using a microcomputer in the feedback loop to control the load on the machine. In addition, a microcomputer was used to acquire and process the raw data in real time. All tests were conducted under condi-

TABLE 1—*Heat treatment for Alloy 718.*

Step 1:	Anneal at 968°C (1775°F) for 1 h, then air cool to temperature
Step 2:	Age harden at 718°C (1325°F) for 8 h, then furnace cool at 56°C/h (100°F/h) to 621°C (1150°F)
Step 3:	Age harden at 621°C (1150°F) for a total aging time in Step 2 and Step 3 of 18 h
Step 4:	Air cool to room temperature

tions of constant maximum stress intensity, K, until steady-state crack growth had been achieved. Crack length was determined indirectly from compliance measurements using digital displacement data from an MTS clip gage with quartz extension rods. The specimen was kept in a resistance-heated furance, while the air- or water-cooled extensometer was kept outside the furnace through the use of the extension rods. Temperature in the furnace was controlled to less than 1°C and the total temperature gradient on the crack path was kept to less than 10°C.

Digital load-displacement data were fit to a straight line using a least-squares minimization procedure incorporated in the microcomputer to determine compliance. Compliance was then converted to crack length using formulae detailed in previous works [*2,5*]. Crack lengths obtained from compliance data were verified by comparing them with the optical measurement data obtained from markings on the fracture surfaces. Some of the crack growth rate tests were repeated under identical conditions. The scatter of the growth rate data was found to be less than 10%.

Several types of loading spectra were used to evaluate the effects of hold times on the spectrum crack growth rates of Inconel 718 at 649°C. The first test involved a 1 Hz fatigue cycle with a hold time at maximum load between fatigue cycles of $R = 0.1$. The hold times ranged from 0 to 500 s, and tests were conducted for values of K_{max} of 40 and 27 MPa $m^{1/2}$. The second series of tests involved the application of a number of fatigue cycles with a hold time at maximum load interspersed between the blocks of cycles. The number of fatigue cycles ranged from 1 to 50 at $K_{max} = 40$ MPa $m^{1/2}$ using hold times of either 5 or 50 s. The third type of test involved a single fatigue cycle of 1.0 Hz with $R = 0.1$ with a hold time of either 5 or 50 s at a value less than K_{max}. In one group of tests, the hold was applied at the minimum load of the $R - 0.1$ fatigue cycles. Finally, a spectrum consisting of single fatigue cycle of 1 Hz at $R = 0.1$ followed by twelve cycles of 2 Hz at $R = 0.5$ and then a hold of 90 s at maximum load was applied. This spectrum was based on a design spectrum on the Air Force TF-34 engine.

Analysis

A linear cumulative damage model was used to predict crack growth rates for loading spectra consisting of combinations of fatigue cycles and hold times. The model involved the simple summation of the individual contributions of the cycles and the hold times. The fatigue crack growth rates were obtained from constant K tests at the appropriate frequency and R ratio. The hold time predictions were obtained from sustained load crack growth tests. Letting da/dN be the cyclic crack growth rate and da/dt the sustained load growth rate, the crack growth rate per cycle block consisting of N cycles and a hold time of t_H is

$$\frac{da}{dN_b} = N \cdot \frac{da}{dN} + t_H \frac{da}{dt} \tag{1}$$

where da/dN_b is the growth rate per cycle block. There are no interaction effects considered in this simple model.

Results and Discussion

The first series of tests involved the application of a single fatigue cycle of $R = 0.1$ at 1 Hz with a hold time at maximum load between cycles. Two values of maximum K were used, 40 and 27.8 MPa $m^{1/2}$ with hold times ranging from 0 (pure fatigue) to 500 s. The data are plotted in Fig. 1 as crack growth rate against total cycle time. Note that a total cycle time of 1 s corresponds to no hold time, since the fatigue cycle is at a frequency of 1 Hz. Shown also in Fig. 1 (dashed lines) is the analytical prediction from the linear cumulative damage model. Two things are evident from the results. First, the model predicts the growth rate extremely accurately for the entire range of hold times. Secondly, the behavior of this material is very time dependent. For hold times in excess of approximately 10 s, the behavior is essentially time dependent as evidenced by the 45° slope on the log-log plot of growth rate versus cycle time in Fig. 1.

The second series of tests involved the application of a block of fatigue cycles with a single hold time at maximum load interspersed between the cycles. The cycles were again applied at $R = 0.1$ and 1.0 Hz. The tests were all conducted using a maximum K of 40.0 MPa $m^{1/2}$. The data are plotted as the average growth rate, da/dt, against the number of fatigue cycles per block, N, in Fig. 2. N represents the number of fatigue cycles in a block or spectrum which consists of the fatigue cycles and the hold time. Thus $N = 0$ represents the condition of the hold time without any fatigue cycles. Values of N ranged from 1 to 100 using hold times of 5 and 50 s. The solid lines in Fig. 2 represent the predictions of the linear cumulative damage model for the two hold times. The arrow on the left, representing $N = 0$, represents da/dt for pure sustained load. The arrow on the right represents $N =$ infinite, which is the growth rate in mm/s for pure fatigue. The dashed lines represent the analytical predictions considering contributions for only sustained loading or only fatigue loading for the spectrum of N cycles and a single hold time at maxi-

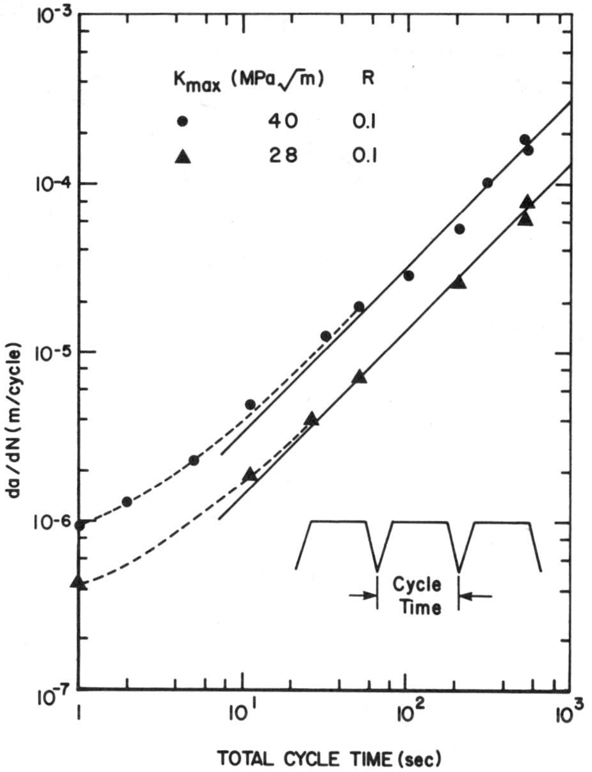

FIG. 1—*Crack growth rate for 1.0 Hz fatigue cycle with hold time at maximum load. Dashed lines show linear cumulative damage model predictions.*

mum load. It can be readily seen that for anything more than a few cycles up to 100 cycles per block, both sustained load and fatigue contribute to the overall crack growth in this material. Additionally, the linear cumulative damage model works very well in predicting the crack growth rate over the entire range of conditions. For the hold time of 50 s and larger numbers of cycles, the prediction appears to be slightly high. These data are replotted in Fig. 3 using N as the horizontal axis again. The vertical axis represents the crack extension that occurred during the application of the above block of loading. Three curves in the figure represent three different hold times: 0, 5, and 50 s. The hold time of zero represents the case of pure fatigue cycling. The model predictions are the straight lines parallel to the pure fatigue curve. The points along the vertical axis are determined from sustained load crack growth data.

The data points which represent the hold time of 5 s are parallel to the pure fatigue curve ($T_H = 0$ s) and shifted by a constant amount which represents the crack growth due to the sustained load of 5 s. In this case, Eq 1 (linear

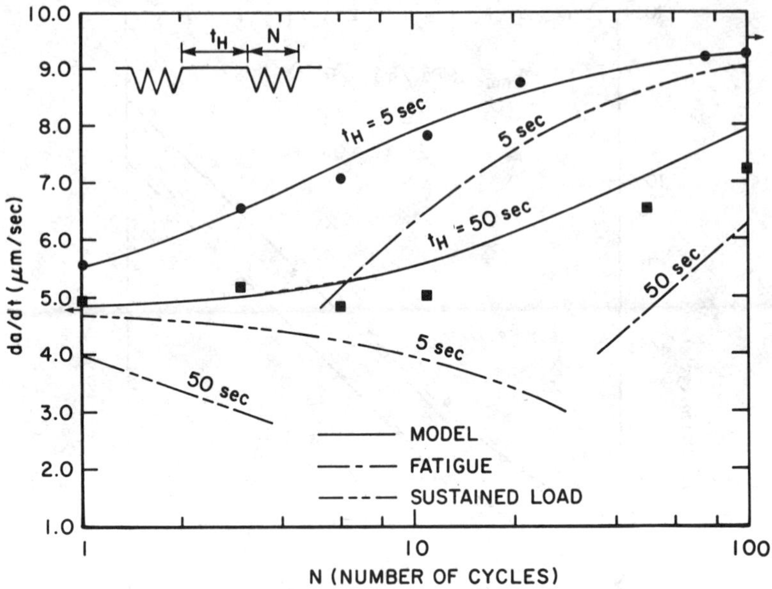

FIG. 2—*Crack growth rate for* N *fatigue cycles at 1.0 Hz with hold time at maximum load.*

summation model) accurately represents the crack growth per block for constant hold time. In the case of a hold time of 50 s, the crack growth per cycle became stabilized to the value of the steady-state fatigue crack growth rate only after a transition zone ahead of the crack tip was traversed. The size of this long hold-time affected zone is approximately equal to 0.015 mm. Since the growth rate within this zone is smaller than the steady-state value of the fatigue crack growth rate, the observed retardation is considered to be due to the blunting of the crack tip. However, for those cases where the maximum stress intensity of both hold time and cycling are equal, the crack growth retardation which occurs in the very small transition zone is insignificant in the total life prediction. Hence, in the life prediction methodology developed in this paper, the linear summation model given as Eq 1 is adequate and interactions, which are very small, are neglected.

The linear cumulative damage model has also been applied to similar data on IN100 [*6*]. At 649°C the behavior is almost exclusively cycle dependent, but at 732°C both cycle- and time-dependent behavior were observed as in Inconel 718. In both cases, the model accurately predicts the growth rate for blocks of fatigue cycles with interspersed hold time at maximum load.

The next series of tests was to evaluate the contribution of a hold time at minimum load when applied between fatigue cycles of $R = 0.1$. In this case, with maximum K values of 27.8 and 40 MPa $m^{1/2}$, the sustained load at minimum load ($K = 4$ and $K = 2.78$ MPa $m^{1/2}$) should have no contribution

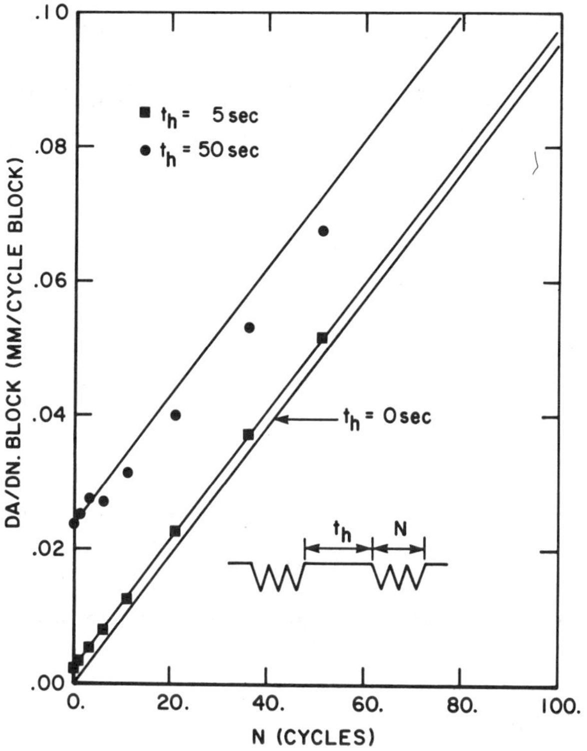

FIG. 3—*Crack growth per cycle block consisting of* N *cycles and a hold time at maximum load (same data as Fig. 2).*

since it is below the threshold values. The data, presented in Fig. 4, confirm this supposition. Hold times at minimum load of up to 500 s did not appear to influence the crack growth rate, although there was some variability in the data at the higher maximum K level of 40 MPa $m^{1/2}$. It should also be noted that these data were obtained under conditions of constant K only after steady-state (constant growth rate) conditions were achieved. In many of the tests, initially higher growth rates were observed in changing from other test conditions to conditions with hold times at minimum level. Sadananda and Shahinian [*7*] observed higher growth rates in constant load range tests using hold times at minimum load over those observed in pure fatigue tests. They attributed this to an environmentally enhanced degradation of properties even though the hold-time levels were below threshold.

To further evaluate the effects of sustained loads on fatigue crack growth rates, a series of tests was performed using single fatigue cycles with interspersed hold times at intermediate load levels. The fatigue cycles were applied under constant K conditions using maximum K = 40 MPa $m^{1/2}$, R = 0.1,

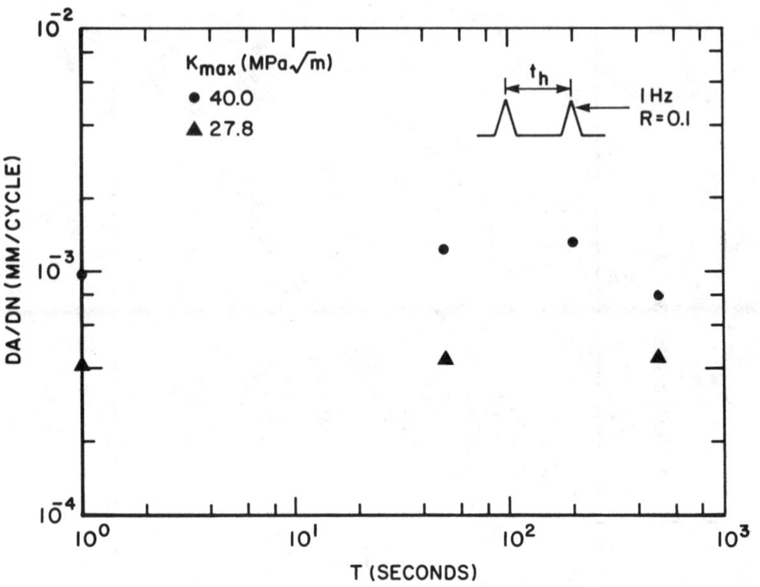

FIG. 4—*Crack growth rate for 1.0 Hz fatigue cycle of* R = *0.1 with hold time at minimum load.*

and a frequency of 1.0 Hz. Sustained load hold times of 5 and 50 s were applied at 75% (K = 30 MPa $m^{1/2}$) and 55% (K = 22 MPa $m^{1/2}$) of maximum load. In the latter case, the fatigue cycles were applied in two slightly different ways as depicted schematically in Fig. 5. The hold times were applied either after maximum load or after minimum load to evaluate the effect of the sequence of the fatigue loading on the crack growth rates. The main question which we sought to answer was whether the application of the entire range of ΔK on loading or unloading made any difference when there was a hold time between cycles. The results are shown in bar graph format in Fig. 5. For hold times of 5 s, the hold after minimum or maximum load made no difference. The analytical prediction of the growth rate due solely to the fatigue cycle matched the experimental data very closely. If the effect of the hold time were included in the prediction, the model overpredicted the crack growth rate. In this case, it appears that the hold time can be neglected in predicting the crack growth rate. When the hold time was increased by a factor of 10 to 50 s, the analytical contribution of the sustained load portion of the load spectrum becomes dominant. The experimental results, however, as seen in Fig. 5, show that the growth rate is close to that predicted by the fatigue cycle alone. In the case where the hold time was applied after maximum load (shown schematically in the figure), the growth rate was higher than when applied after minimum load and certainly higher than that due to fatigue only. There are two possible explanations. The first, and most plausible, is that steady-state

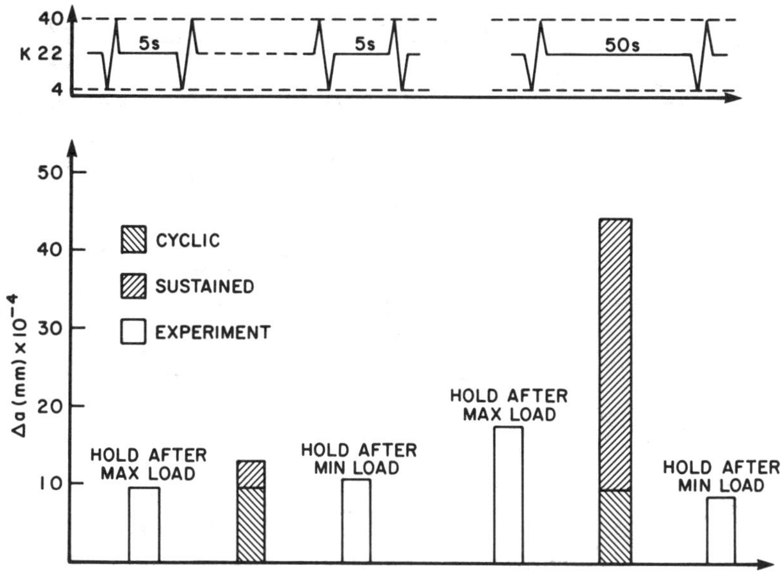

FIG. 5—*Crack growth rate for 1.0 Hz fatigue cycle with hold time at mean load.*

crack growth had not been achieved under constant K conditions. The total amount of crack extension was approximately 0.3 mm, which has been found to be adequate in most cases. We have observed in some data, however, that transient behavior can occur even after crack extensions of this magnitude. This was observed, in particular, in tests with hold times at minimum load between fatigue cycles. Only continued crack growth under identical conditions, which is very time consuming, could answer these questions. The second possible explanation would be an environmental degradation during the hold time which would accelerate the growth during the subsequent fatigue cycle where the ΔK range is applied during rising or increasing load. This area appears to warrant further study.

For the cases where the sustained load was applied at 75% of maximum load, the contribution of the hold time in the analytical prediction is significant for hold times of either 5 or 50 s. The experimental results show, however, that the actual growth rates are very close to those of pure fatigue cycling as shown in Fig. 6. Note that the analytical contribution of the hold time, particularly for the 50 s hold time, is very substantial. In these cases, with sustained loads at 75% of maximum load, it appears that the hold time does not contribute to the overall crack growth.

To further evaluate this hypothesis, the fatigue cycle was changed as shown schematically in Fig. 6. The maximum load of the fatigue cycle, the magnitude of the hold time, and the amplitude of the sustained load were the same as in the previous case, but the R-ratio was changed from 0.1 to 0.75. By

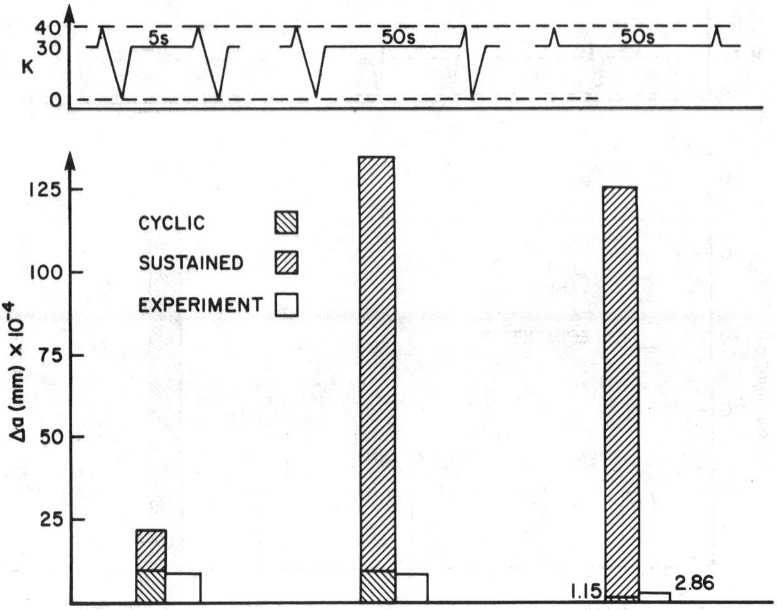

FIG. 6—*Crack growth rate for 1.0 Hz fatigue cycle with hold time at 75% maximum load.*

doing this, the growth rate of the fatigue cycle is significantly reduced because of the smaller range of ΔK. The experimental results are compared with the analytical prediction in Fig. 6 and show two things. First, the experimental growth rate is significantly less than that predicted due to the sustained load hold time alone. Second, the experimental growth rate is approximately 2.5 times that predicted due to fatigue cycling alone at $R = 0.75$. In the last case, neither fatigue nor sustained load predictions agree with the data, while the linear cumulative damage model grossly overpredicts the growth rate. This case can be thought of as a situation where sustained load crack growth is severely retarded by the application of periodic overloads at 50 s intervals. Alternatively, it can be viewed as fatigue cycling which is supplemented by a contribution due to sustained load at minimum load. The magnitude of this contribution is certainly non-zero but much less than that which is observed in pure sustained load tests with no fatigue cycles or periodic overloads.

Finally, to verify the applicability of the simple linear cumulative damage model to engine spectra, an experimental crack growth rate was obtained for the modified TF-34 load spectrum at high temperature. The spectrum, shown in Fig. 7, consists of one cycle of 1 Hz at $R = 0.1$, twelve cycles of 2 Hz at $R = 0.5$, and a 90 s hold time at the maximum stress intensity of 40 MPa $m^{1/2}$. Experimental growth for one total block of the spectrum and the analytical linear cumulative damage model predictions are also given in Fig. 7. For this modified spectrum, the analytically predicted cumulative sum of the growth

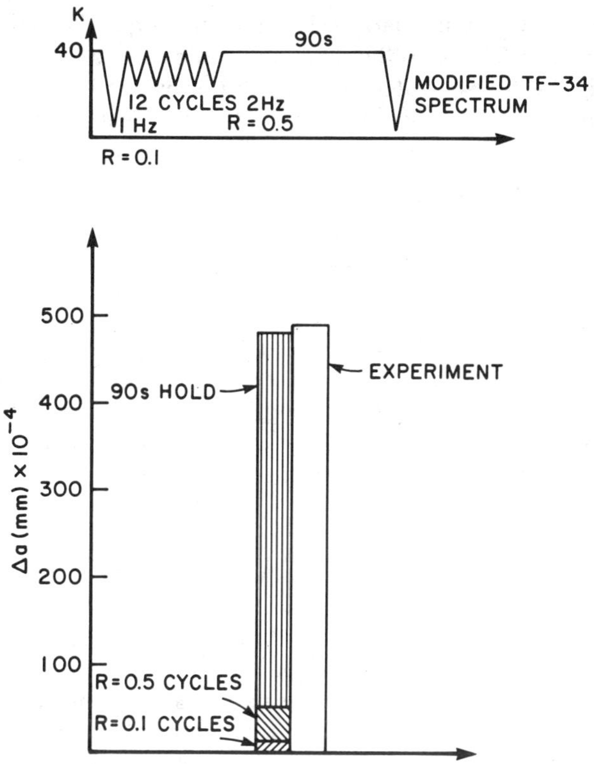

FIG. 7—*Crack growth rate for modified TF-34 spectrum.*

due to three sub-blocks of the spectrum is approximately equal to the experimentally predicted crack growth for the total spectrum. Thus the applicability of the linear cumulative damage model is verified for a typical engine spectrum, where the maximum stress intensity in each sub-block of the spectrum is equal. Further studies are required in cases when the maximum stress intensities at each sub spectrum block are not identical.

In all evaluations of crack growth rate behavior, the stress intensity, K, or stress intensity range, ΔK, was used for sustained load or cyclic behavior, respectively. For a creeping solid, the energy rate line integral C^*, derived by Landes and Begley [*8*], has been found to be more effective in correlating sustained load crack growth rates for some materials. Riedel and Rice [*9*] have defined a characteristic time t_1 as

$$t_1 = \frac{K_I^2(1 - \nu^2)}{E(n + 1)C^*} \tag{2}$$

where K is the stress intensity, ν is Poisson's ratio, E is Young's modulus, and C^* is the energy rate line integral. If the test time or hold time is sufficiently small compared with the characteristic time t_1 in Eq 2, then small-scale yielding prevails and K governs the stress and strain fields ahead of the crack tip. Conversely, for times greater than t_1, C^* is shown to be the correlating parameter for the stress and strain fields and, consequently, crack growth rates. The constant n is the exponent of the constitutive law for an assumed creeping solid given by

$$\frac{\dot{\epsilon}}{\epsilon_0} = \alpha \left(\frac{\sigma}{\sigma_0} \right)^n \quad (3)$$

where α is a proportionality constant and ϵ_0 and σ_0 are reference strain rates and stresses, respectively. The line integral C^* can be determined from simplified methods developed from analogy to elastic-plastic solutions for the J-integral as applied by Saxena [*10*], for example. In those methods, J-integral solutions, which are based on a constitutive law of the form

$$\frac{\epsilon}{\epsilon_0} = \alpha \left(\frac{\sigma}{\sigma_0} \right)^n \quad (4)$$

are easily converted to C^* solutions through the replacement of strain and displacement by strain rates and displacement rates, respectively, in the definition of J. The solutions for J, in this particular problem of a CT specimen, were taken from results obtained by Kumar et al [*11*] using an estimation scheme and computer program described in detail by Weerasooriya and Gallagher [*12*]. The value for n in Eq 3, obtained from a plot of secondary creep rates as a function of stress from an extensive series of tests, was found to be 18.4. The transition times are plotted in Fig. 8 for the cases of tests under constant K and constant P for a CT specimen of Inconel 718 at 649°C. It is clearly seen that the transition times, t_1, are orders of magnitude larger than the hold times or test times used in this investigation. It is also to be noted that, for each loading spectrum investigated, constant crack growth rates were obtained when tests were conducted under computer-controlled constant K conditions. Thus K has been established as a valid correlating parameter in this investigation.

Conclusions

Hold times at maximum load which occur between single or multiple fatigue cycles contribute to overall crack growth rate in Inconel 718 at 649°C. The magnitude of this contribution is determinable from sustained load crack growth data. A linear cumulative damage model predicts crack growth rates quite accurately over the range of conditions covered in this investigation. Application of hold times at less than maximum loads between fatigue cycles has less influence than that predicted from sustained load crack growth data.

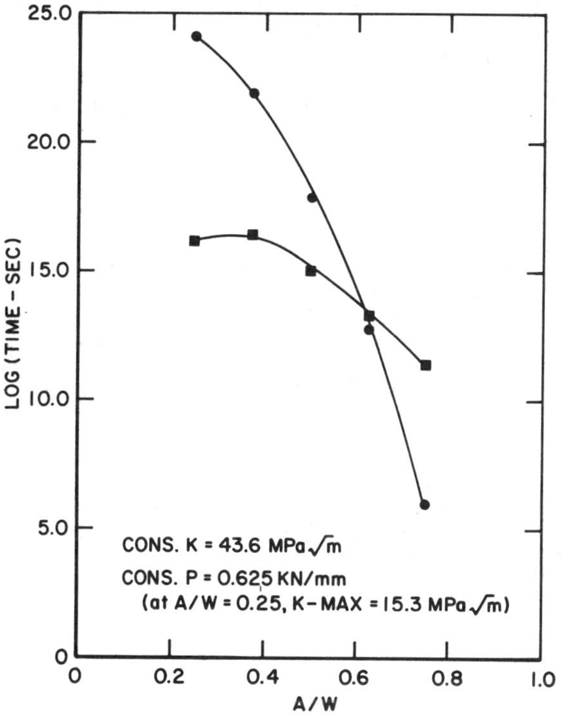

FIG. 8—*Characteristic time for constant* K *and constant load experiments in CT specimens of Inconel 718 at 649°C.*

Thus a linear cumulative damage model overpredicts crack growth rates. In most of the cases investigated, the contribution of the hold time could be neglected entirely. There was one case, however, where the hold time could not be neglected but was found to contribute only a small fraction of its effect as determined from sustained load test data. In general, sustained loads appear to be important only when applied at maximum load in a fatigue spectrum. When sustained loads are applied at less than maximum load, their contribution is greatly diminished but they create a situation which involves complex fatigue-sustained load interactions involving possible overload retardation effects.

References

[1] Stucke, M. et al, "Environmental Aspects in Creep Crack Growth in a Nickel Base Superalloy," to be published in *Proceedings*, Sixth International Conference on Fracture, Dec. 1984.

[2] Donath, R. C., Nicholas, T., and Fu, L. S., in *Fracture Mechanics: Thirteenth Conference, ASTM STP 743*, R. Roberts, Ed., American Society for Testing and Materials, Philadelphia, 1981, pp. 186-206.

[*3*] Ashbaugh, N. E., "Creep Crack Growth Behavior in IN718 in Laboratory Air," presented to ASTM 17th National Symposium on Fracture Mechanics, Albany, N.Y., Aug. 1984.
[*4*] Nicholas, T., Weerasooriya, T., and Ashbaugh, N. E., "A Model for Creep/Fatigue Interactions in Alloy 718," in *Fracture Mechanics: Sixteenth Symposium, ASTM STP 868*, American Society for Testing and Materials, Philadelphia, 1985, pp. 167-180.
[*5*] Saxena, A. and Hudak, S. J., Jr., *International Journal of Fracture*, Vol. 14, No. 5, Oct. 1978, pp. 453-468.
[*6*] Larsen, J. M. and Nicholas, T., *Engineering Fracture Mechanics*, Vol. 22, No. 4, 1985, pp. 713-730.
[*7*] Sadananda, K. and Shahinian, P., *Journal of Engineering Materials and Technology, Transactions of ASME*, Vol. 100, 1978, pp. 381-387.
[*8*] Landes, J. D. and Begley, J. A. in *Mechanics of Crack Growth, ASTM STP 590*, American Society for Testing and Materials, Philadelphia, 1976, pp. 128-148.
[*9*] Riedel, H. and Rice, J. R. in *Fracture Mechanics (Twelfth Conference), ASTM STP 700*, American Society for Testing and Materials, Philadelphia, 1980, pp. 112-130.
[*10*] Saxena, A. in *Fracture Mechanics (Twelfth Conference), ASTM STP 700*, American Society for Testing and Materials, Philadelphia, 1980, pp. 131-151.
[*11*] Kumar, V., German, M. D., and Shih, C. F., "Estimation Technique for the Prediction of Elastic-Plastic Fracture of Structural Components of Nuclear Systems," Combined Second and Third Semiannual Report, Feb. 1979 to Jan. 1980 for EPRI, SRD-80-094, General Electric Company, Schenectady, N.Y., 1980.
[*12*] Weerasooriya, T. and Gallagher, J., "Determining Crack Tip Field Parameters for Elastic-Plastic Materials via an Estimation Scheme," Technical Report AFWAL-TR-81-4044, Wright-Patterson Air Force Base, Ohio, 1981.

A. Petrovich,[1] *W. Bessler,*[1] *and W. Ziegler*[1]

Interactive Effects of High and Low Frequency Loading on the Fatigue Crack Growth of Inconel 718

REFERENCE: Petrovich, A., Bessler, W., and Ziegler, W., **"Interactive Effects of High and Low Frequency Loading on the Fatigue Crack Growth of Inconel 718,"** *Fracture Mechanics: Seventeenth Volume, ASTM STP 905*, J. H. Underwood, R. Chait, C. W. Smith, D. P. Wilhem, W. A. Andrews, and J. C. Newman, Eds., American Society for Testing and Materials, Philadelphia, 1986, pp. 169-184.

ABSTRACT: The effect of a loading profile containing a high frequency (200 Hz) loading component superimposed on a low frequency (~0.05 to 0.5 Hz) loading component on fatigue crack growth of Inconel 718 at 650°C was investigated. The high frequency sinusoidal loading was added during the upper level hold period of the low frequency trapezoidal waveform.

Fatigue crack growth tests were carried out on Inconel 718 with a constant low frequency ΔK and an increasing high frequency ΔK in such a manner that the regimes in which crack growth is low frequency cycle and high frequency cycle dominated were investigated. Particular attention was devoted to the interaction of the two loading components in the low frequency dominated crack growth regime and the dependence of the transition to high cycle domination on factors such as low frequency cycle ΔK and hold time. The effects of hold times in the range of 2 to 180 s and low frequency ΔK in the range of 15 to 40 MPa$\sqrt{m}$ were investigated.

In the high cycle dominated regime, the relationship between crack growth rate and high frequency ΔK had the characteristics of threshold crack growth data. In the low cycle dominated regime the high frequency cycle retarded crack growth under certain circumstances.

KEY WORDS: fatigue, crack propagation, high frequency loading, nickel-base alloy

The need for more accurate life prediction modelling of aircraft engine components motivates the experimental investigation of fatigue crack growth in turbine engine alloys under combined high and low frequency loading. The loading profile experienced by the engine disk, for example, includes a low frequency component associated with thermal gradients and centrifugal

[1]Mechanical Technology Incorporated, Latham, NY 12110.

forces with a superposition of high frequency loading associated with blade passage. The cycle period associated with the low frequency cycle (low cycle) loading is on the order of seconds to several hundred seconds. A wide range of loading rates and load levels may also be involved in the low cycle loading. The high frequency cycle (high cycle) loading would typically involve frequencies on the order of hundreds to several thousand hertz.

Previous studies of the high/low frequency load interaction effect reported by Powell et al [*1,2*] and Goodman and Brown [*3*] show that there is a transition value associated with the high frequency loading. High cycle loads above this transition value result in crack rates dependent primarily on the number of high frequency cycles, while below this load level the low cycle loading dominates crack growth. It is this transition value that is important to structural design engineering and the Air Force retirement-for-cause program. Engine components must be designed such that the high frequency stresses are always below this level, since crack growth rate would increase rapidly with increasing stress level above the transition value and there would be a risk of in-service failure.

Another important question associated with combined high/low cycle loading is the possibility of synergistic effects in the low cycle dominated regime. It was the purpose of the present investigation, therefore, to determine the variation of high cycle transition ΔK (ΔK_{tr}) and the interaction of high/low cycle loading in the low cycle dominated regime over a range of parameters characterizing the combined loading profile.

The present investigation provides the evaluation of the interactive effect of high and low frequency loading on a typical high temperature engine disk alloy (Inconel 718) over a wide range of low frequency cycle hold times and low frequency cycle ΔK. The study focused on the change in the relationship between crack growth rate and high frequency ΔK in the low cycle and high cycle dominated regime as the low cycle ΔK and cycle period change. The dependence of the value of high cycle ΔK_{tr} on low frequency ΔK and hold time was also established. All testing was performed at 649°C, which for Inconel 718 permits creep crack growth (time-dependent crack growth) as well as fatigue crack growth (cycle-dependent crack growth).

Experimental Procedures for High Frequency Fatigue Crack Growth Testing

A test system was constructed specifically for the present study. The system was designed to provide adequate load levels to 2000 Hz and to minimize the frequency ranges over which dynamic complications in load application are present. A purely servohydraulic system based on an Akashi voice-coil servovalve was used for all of the testing. The load frame and specimen were designed to minimize the opportunity for system resonances that complicate or invalidate measurement of specimen stress and provide undesirable loading patterns.

The specimen type used for this study was a center crack panel shown in Figs. 1*a* and 1*b*. A clevis arrangement with provisions to clamp the specimen ends was used to grip the specimen. By selecting the appropriate end clamps and thereby establishing the appropriate lateral stiffness, specimen resonances could be avoided at the selected test frequency. Both the high frequency and low frequency were sensed by a load cell. It was recognized that resonances in the load frame can disturb the correlation between the load cell measurement and stresses in the specimen and provide significant bending stresses associated with resonant lateral vibration. A modal analysis of the specimen was performed to determine its natural frequencies and mode shapes. The frequency of 200 Hz chosen for this study is far from the first natural frequency of 560 Hz. The absence of excessive bending stresses and a

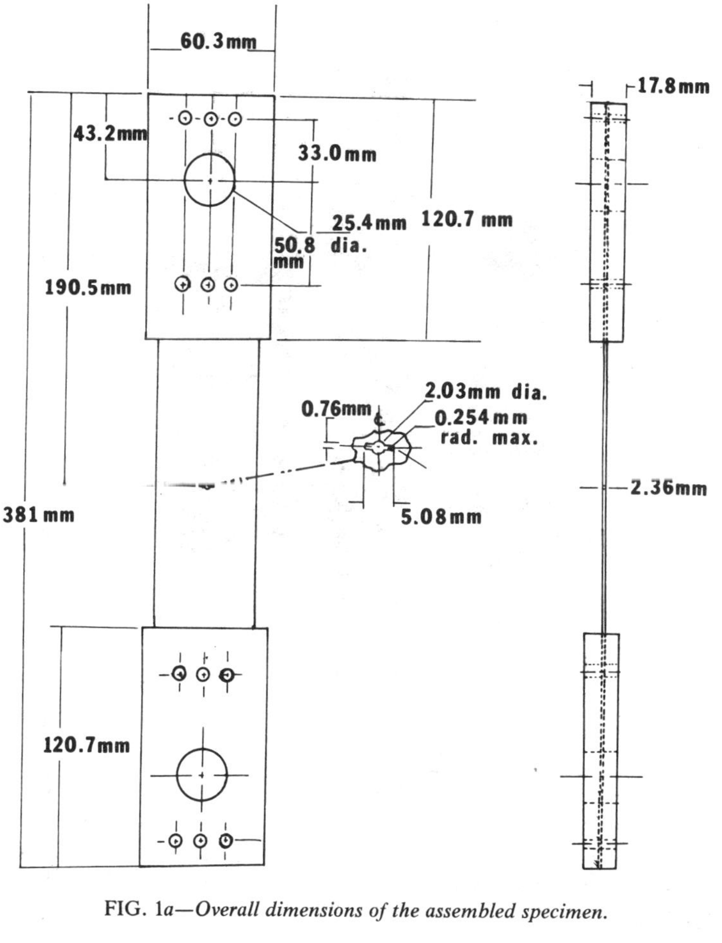

FIG. 1*a*—*Overall dimensions of the assembled specimen.*

FIG. 1*b*—*Specimen with end clamps disassembled.*

proper correlation between load cell measurement and specimen stresses were verified with strain gage measurement on the specimen. The precision in the high frequency ΔK measurement required for this study made these specimen dynamic evaluations essential.

The high/low frequency loading profile is shown in Figs. 2*a* and 2*b*. The low frequency component is a trapezoidal waveform with a rise (T_1) and fall

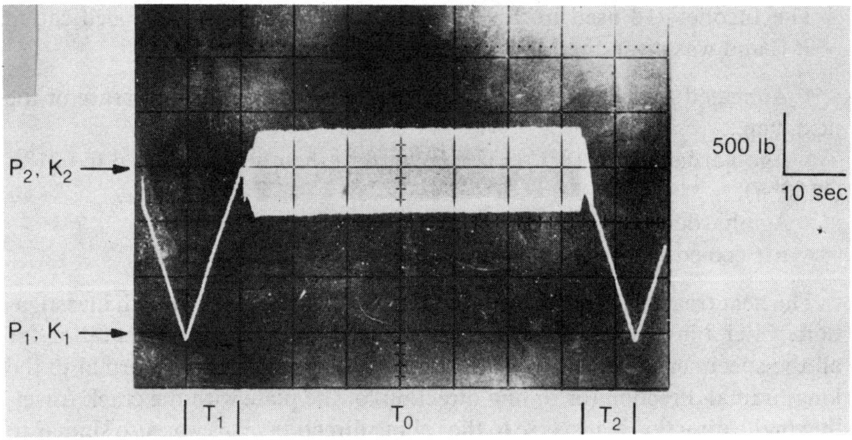

FIG. 2*a*—*Combined high/low frequency loading profile.*

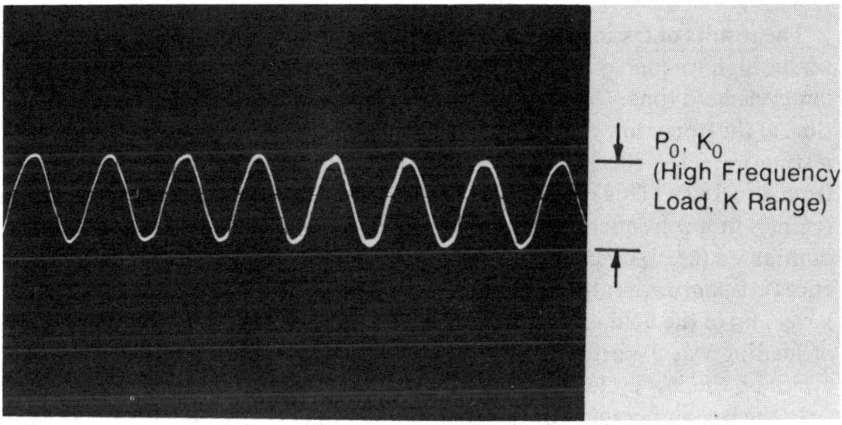

FIG. 2*b*—*High frequency component with expanded time scale.*

time (T_2) of 0.5 s and a hold time (T_0) of between 2 and 180 s. The interactive effect of high and low frequency loading was evaluated over this range of hold times. The high frequency loading was applied during the low frequency cycle hold period and typically ranged between 220 and 4450 N (50 to 1000 lb). The low frequency load levels P_1 and P_2 in Fig. 2*a* were varied during the tests such that the low frequency stress intensity factors K_1 and K_2 were maintained constant. The high frequency load range (P_0) was either increased during the test or maintained constant, which resulted in an increasing high frequency stress intensity factor range (K_0). The low cycle R ratio (P_1/P_2) was 0.1 for all of the testing. All testing was performed at 649°C.

The Inconel 718 used in this program conformed to AMS Specification 5596C and was given the following heat treatment:

- Annealed at 968°C (1775°F) for 1 h, then cooled to temperature of the next step.
- Age-hardened at 718°C (1325°F) for 8 h, then furnace-cooled to 621°C (1150°F).
- Age-hardened at 621°C (1150°F) for 10 h.
- Air-cooled to room temperature.

The heat treatment was the same as that used in the crack growth investigation of Ref *3* in order to allow comparison of data. The crack orientation for all the specimens tested was L-T; that is, the crack plane was normal to the longitudinal direction (or rolling direction) of the plate with the crack travelling in the direction transverse to the rolling direction. This was also similar to the investigations of Ref *3*.

Results and Discussion

The results of testing were a series of curves representing crack growth rate versus high frequency K range for constant low frequency cycle ΔK range and low cycle hold time. Crack growth rate is reported in terms of growth per unit time at the upper low cycle load level. The low cycle ΔK ranges included in the testing were 15, 20, 30, and 40 MPa$\sqrt{m}$. The low cycle hold times ranged from 2 to 180 s. This range of low cycle hold times was expected to cover the regimes in low frequency loading in which the low cycle crack growth is time dominated (creep crack growth) and in which the number of load cycle influences crack growth rate (combination of creep and fatigue crack growth). The lower end of the hold period range (i.e., 2 and 5 s) for the low cycle component of loading was expected to show the effect of the number of low frequency cycles on the low cycle crack growth rate.

In the curve of crack growth versus high frequency ΔK distinct regimes can be seen. As shown in Fig. 3, three types of behavior were observed over the range of low frequency ΔK and hold times investigated. In Type 1, the crack growth rate versus high cycle ΔK remained relatively constant in the low cycle dominated regime. Type 2 behavior was characterized by retardation of crack growth by the high frequency cycle in the low cycle dominated regime. Type 3 behavior was typical of the lowest low cycle ΔK studied, in which the low cycle ΔK was below the crack growth threshold and no crack growth could be measured in the low cycle dominated regime. In all these cases distinct low-cycle/high-cycle dominated regimes could be observed. However, the transition between these two regimes is not always distinct.

In many cases in the low cycle dominated regime the crack growth rate was fairly constant over a wide range of high frequency ΔK and increased abruptly at a distinct transition to the high cycle dominated regime. In other cases

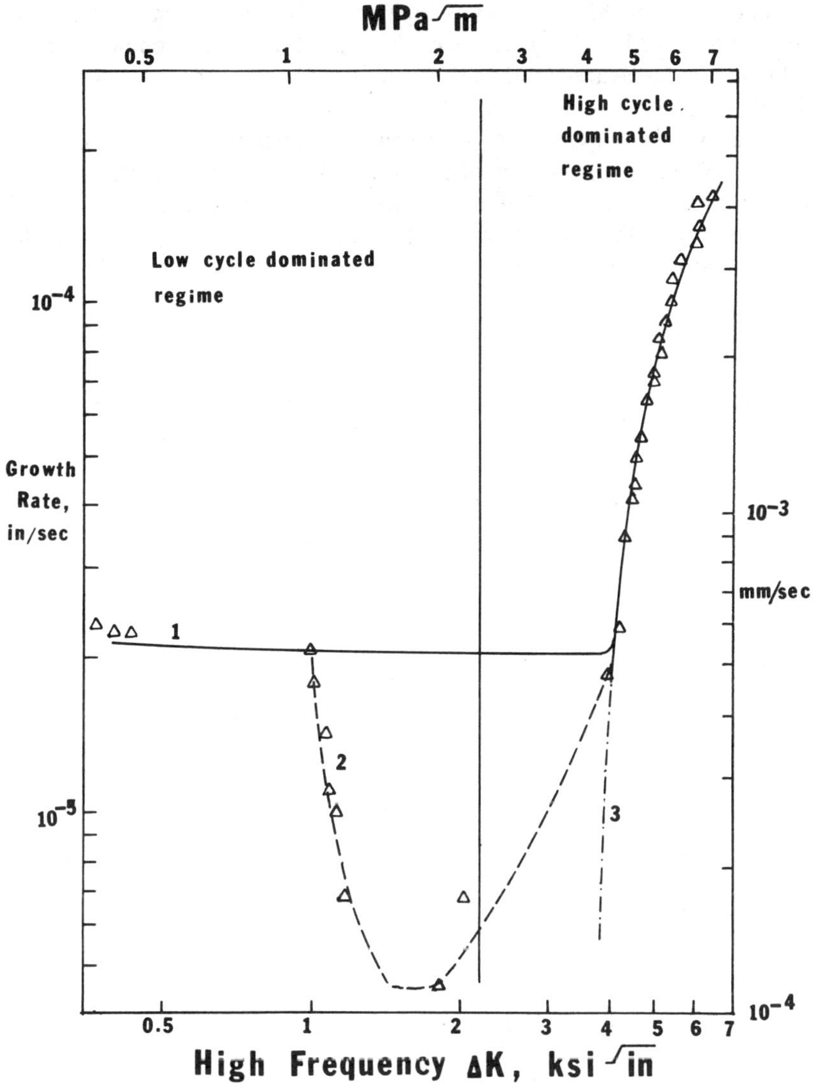

FIG. 3—*Characteristics of the high/low frequency interaction with the three types of behavior observed in this study. The points correspond to testing with a low frequency* ΔK *of 20 MPa*$\sqrt{m}$, *a low cycle time of 10 s, and a high cycle frequency of 200 Hz.*

there was a decreasing crack growth rate with increasing high frequency ΔK in the low cycle dominated regime before the rapid increase in growth rate at a less clearly defined high cycle transition. This retardation effect due to the high cycle component was especially pronounced at a low frequency ΔK level of 20 MPa$\sqrt{m}$ and existed to varying degrees of severity over the range of low frequency ΔK and hold times investigated. At the lowest value of low frequency ΔK (15 MPa$\sqrt{m}$) crack growth in the low cycle dominated regime could not be measured with either an increasing or decreasing loading scheme.

Figure 4 shows an example of such a curve corresponding to a low cycle ΔK of 15 MPa$\sqrt{m}$ and a hold time of 5 s. The data in this representation correspond to a test with increasing high frequency ΔK. Before data acquisition in the increasing high cycle ΔK mode, the crack was allowed to grow with a systematically decreasing ΔK until a crack growth rate on the order of 5 ×

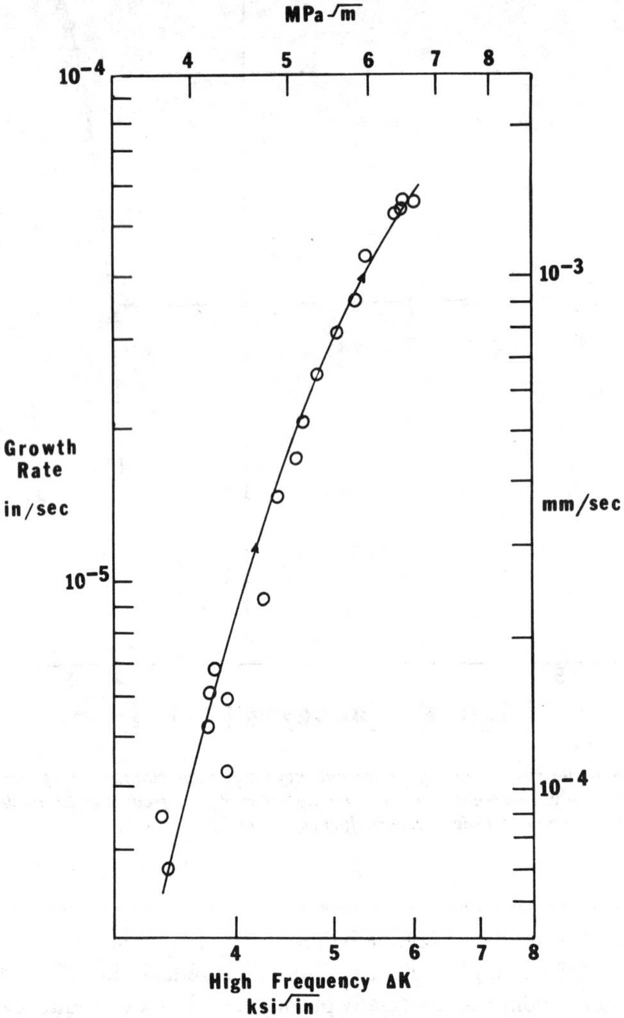

FIG. 4—*Results of a combined high/low frequency test with a low cycle* ΔK *of 15 MPa$\sqrt{m}$ and a low cycle hold time of 5 s.*

10^{-4} mm/s (2×10^{-6} in./s) was achieved. This precaution was taken to eliminate the effects on crack growth of the prior precycling. The data presented in Fig. 4 are characteristic of threshold fatigue crack growth data which generally exhibit increasing growth rate with increasing ΔK. The crack growth curve has a diminishing slope as is typically observed in the crack growth threshold regime when crack growth versus ΔK is plotted on log-log axes. The lower level of this curve corresponds to a growth rate of 3.30×10^{-7} mm per high frequency cycle (1.3×10^{-8} in./cycle) which would definitely be in the threshold regime for Inconel 718.

Figure 5 shows the results of a test conducted with a low frequency ΔK of 20 MPa$\sqrt{m}$ and a hold time of 5 s. It was carried out with a sequence of loads

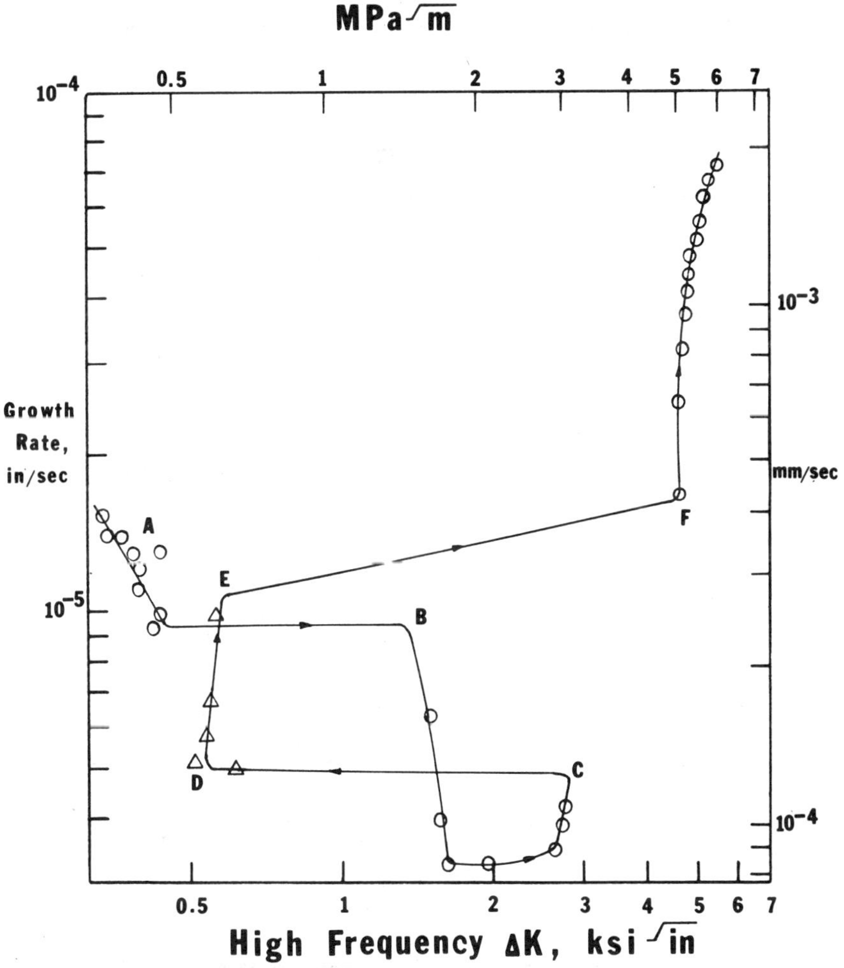

FIG. 5—*Results of combined cycle test with a low frequency* ΔK *of 20 MPa*$\sqrt{m}$ *and a hold time of 5 s. The line indicates the sequence of points.*

intended to illustrate an important aspect of the retardation effect that is very pronounced at a low frequency ΔK of 20 MPa$\sqrt{m}$. The line drawn through the experimental points has arrows that show the sequence of points as they occurred during the test. The initial loading up to Point A seems to give rise to a measurable retardation, and changing the high frequency load range to that at Point B rapidly accelerates the retardation. This results in a more severe retardation in the 0.762 mm (0.030 in.) of growth beyond Point B than was accomplished in the 2.79 mm (0.110 in.) of growth in the high cycle ΔK range around Point A. (Each point represents 0.254 mm [0.010 in.] of crack growth). Beyond Point B the crack growth rate decreases rapidly, reaches a minimum value, and then starts to increase. At Point C just beyond the minimum value of crack growth, a lower high frequency load range was applied (Point D). The crack growth rate increased from Point D to E, showing a gradual elimination of the retardation effect. At Point E the load range was

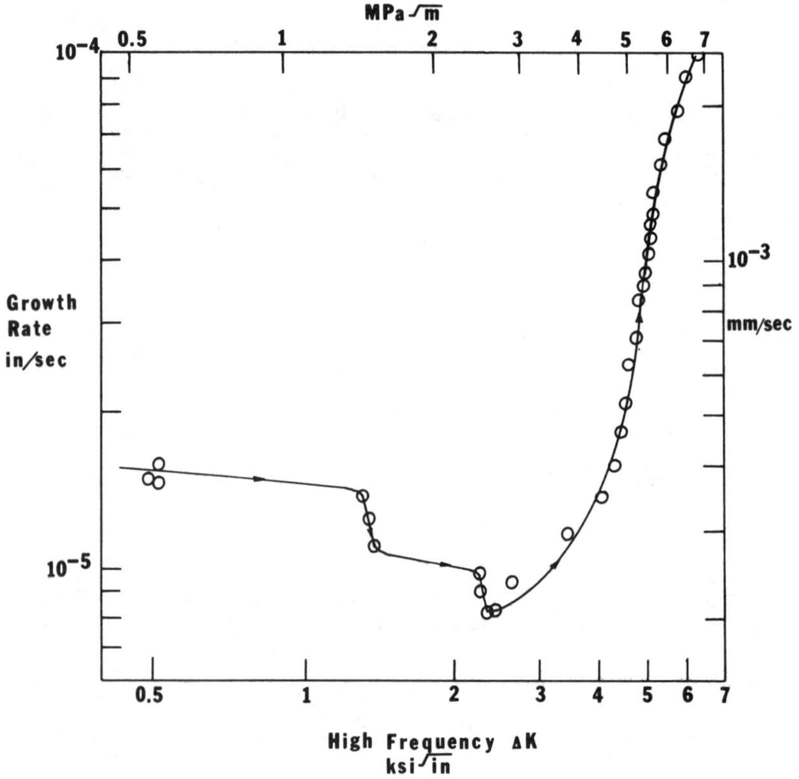

FIG. 6—*Results of a combined cycle test with a low frequency* ΔK *of 30 MPa*$\sqrt{m}$ *and a hold time of 5 s.*

again increased to Point F and crack growth continued in the high cycle dominated regime.

As the low frequency ΔK increases, the retardation effect generally becomes less pronounced. The data for a low cycle ΔK of 30 MPa$\sqrt{m}$ and a low cycle hold time of 5 s appear in Fig. 6. While the high frequency load results in a factor of 4 reduction in crack growth rate for a low cycle ΔK of 20 MPa$\sqrt{m}$, at a low cycle ΔK of 30 MPa$\sqrt{m}$ the reduction in crack growth

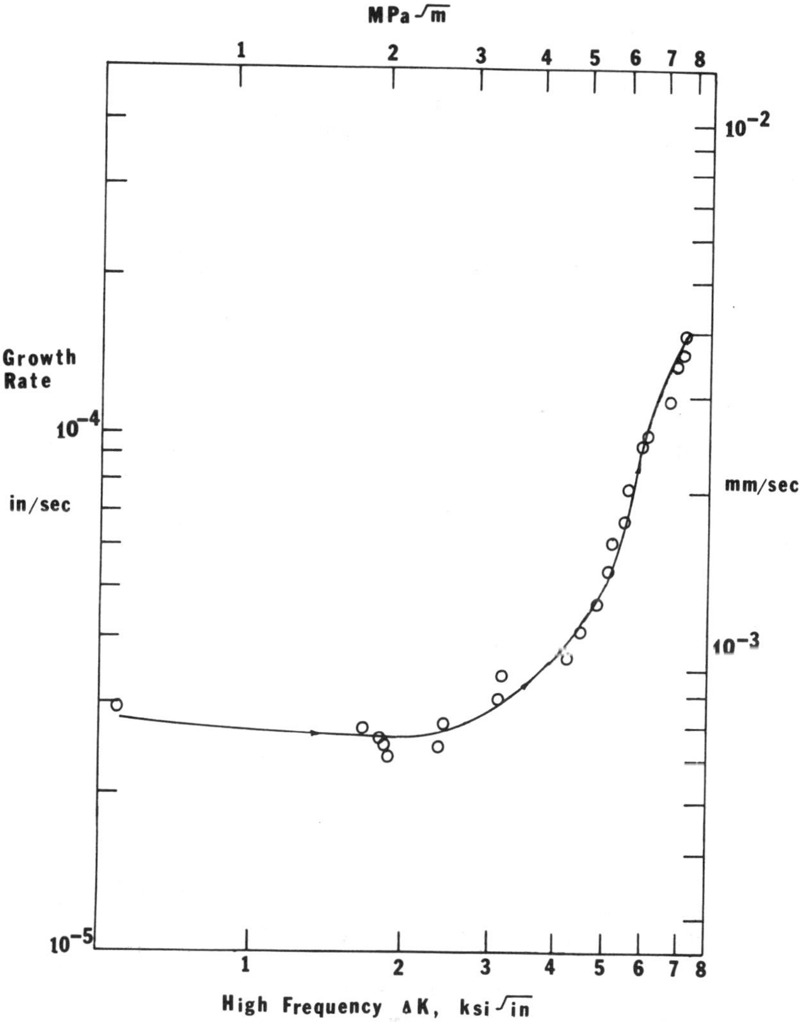

FIG. 7—*Results of a combined cycle test with a low cycle* ΔK *of 40 MPa$\sqrt{m}$ and a hold time of 5 s.*

rate is only a factor of 2. As shown in Fig. 7, a low cycle ΔK of 40 MPa$\sqrt{m}$ shows no measurable retardation associated with high frequency loading.

Figure 8 shows the effect of varying cycle time on the crack growth behavior with a low cycle ΔK of 15 MPa$\sqrt{m}$. No distinct trend is apparent and there is little deviation between these curves. Figure 9 shows the effect of cycle time ranging from 2 to 180 s on the crack growth behavior with a low cycle ΔK of 20

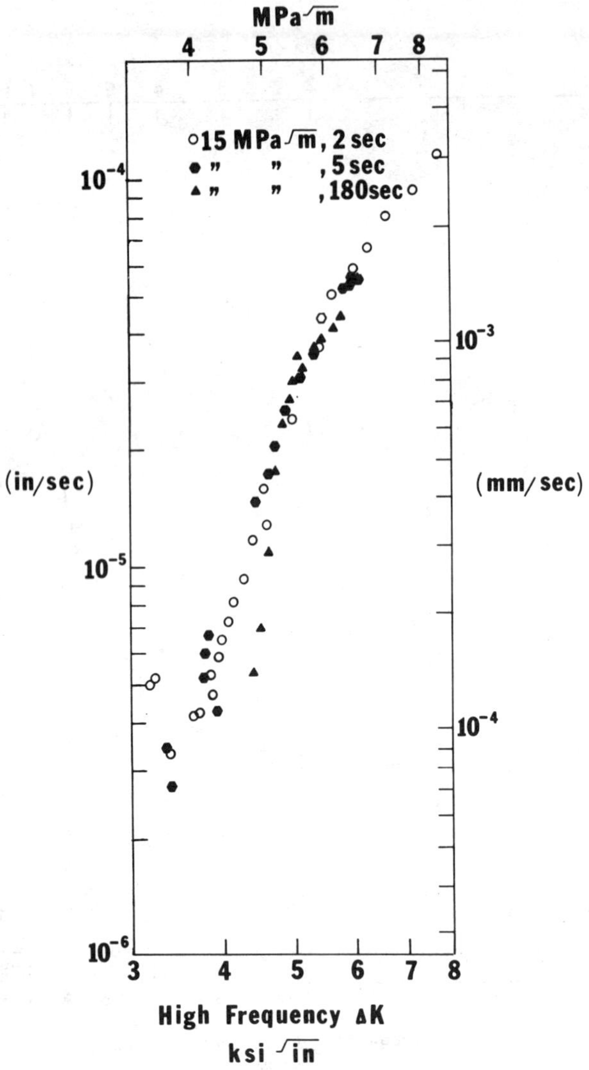

FIG. 8—*Comparison of crack growth rate versus high cycle* ΔK *for several hold times with a low cycle* ΔK *of 15 MPa*$\sqrt{m}$.

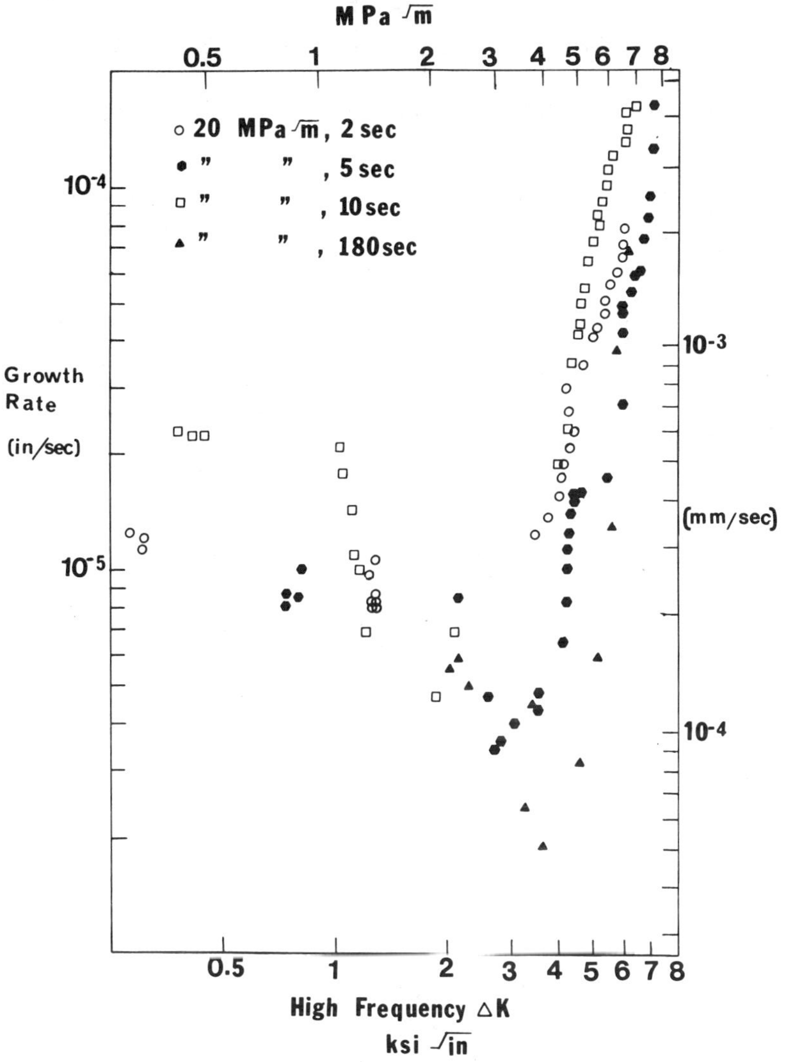

FIG. 9—*Comparison of crack growth rate versus high cycle* ΔK *for several hold times and a low cycle* ΔK *of 20* $MPa\sqrt{m}$.

$MPa\sqrt{m}$. The only significant feature in this group of tests is that with a 180 s hold time there appears to be a more severe retardation.

A comparison of crack growth rate versus high cycle ΔK for a hold time of 5 s and several values of low cycle ΔK appears in Fig. 10. As expected, the crack growth rate in the low cycle dominated regime increases as ΔK increases. In the high cycle regime, however, the curves seem to converge.

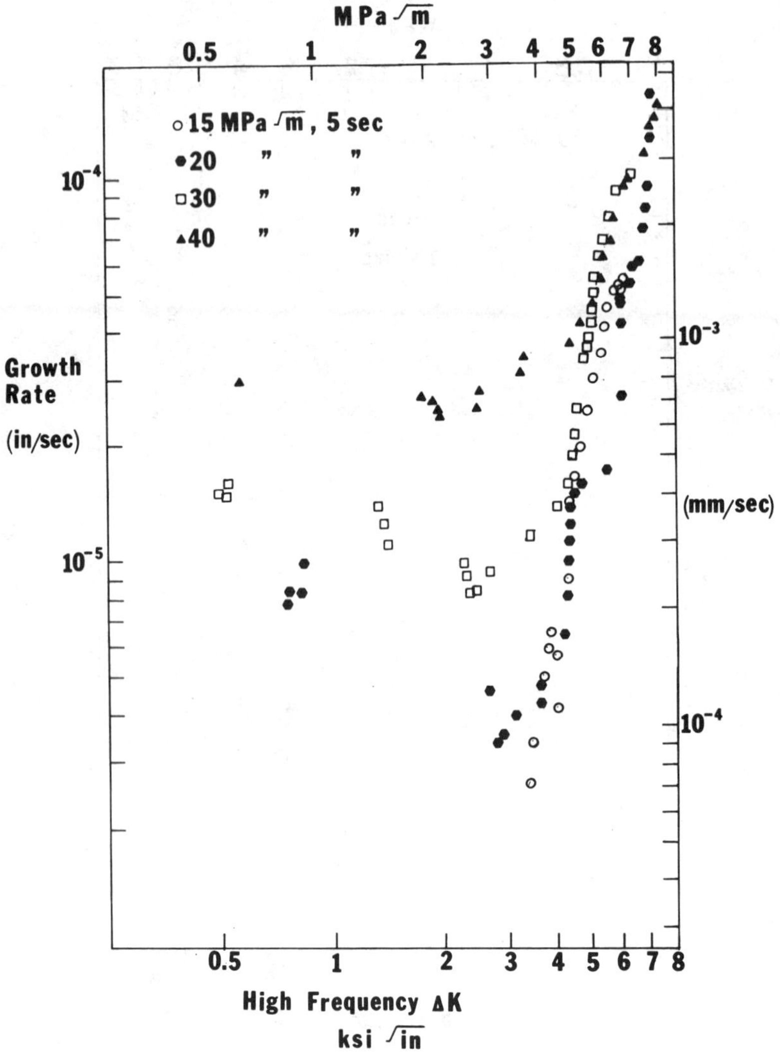

FIG. 10—*Comparison of crack growth rate versus high cycle* ΔK *tests for several low cycle* ΔK *ranging from 15 to 40 MPa*$\sqrt{m}$ *with a low cycle hold time of 5 s.*

A paper by Venkiteswaran et al [*4*] reports the results of a study involving the superimposition of a small vibratory stress on the axial creep behavior of a high temperature nickel-base alloy, Inconel X-750. This work demonstrated that the creep rate was lower and rupture life higher by an order of magnitude when a 500 to 900 Hz vibratory stress was applied transverse to the axial creep load. This effect was attributed to the formation of complex dislocation tangles, vacancy condensation along dislocation lines and crack tips, and a change in fracture mode from purely intergranular fracture to a mixture of intergranular, fatigue, and cleavage modes. It was suggested that the application of the high frequency loading, therefore, made creep crack propagation more difficult along the matrix containing γ' precipitates. Since the heat-treated Inconel 718 used in this study likewise contains γ' (Ni_3Al-Ti) as well as γ'' (Ni_3Cb) precipitates, this mechanism could apply in the present study.

The experiment summarized in Fig. 5 shows not only the rate (with respect to change in crack length) at which the retardation effect develops but also the rate at which it relaxes. There seems to be a crack growth interval of about 1.0 mm (0.0394 in.) required for the retardation effect to subside. The size of the plastically deformed region (R) ahead of the crack as given by the Dugdale model [*5*] neglecting the effects of creep is

$$R = \{\sec[{}^1\!/\!_2\pi(\sigma/\sigma_y)] - 1\}\, a \tag{1}$$

where σ is the applied stress, σ_y is the yield strength, and a is the half crack length.

As a point of reference the plastic zone size as calculated with this expression in the vicinity of the retardation relaxation in Fig. 5 assuming a yield strength of 980 MN/m^2 (140 ksi) is 0.20 mm (0.008 in.). A possible explanation of the fact that the affected region is considerably larger than the characteristic plastic zone size is that creep stress relaxation results in a larger characteristic zone. Reference *6* demonstrates that crack tip stresses can be modified significantly by creep. The most significant influence of creep relaxation as shown by Ref *6* is the reduction of the stress gradient beyond the crack tip (i.e., the development of a more uniform distribution of stress in a region that includes the above calculated "plastic zone" and an area further from the crack tip). An alternative explanation of the long relaxation interval is that the high frequency loading affects the creep rate versus stress constitutive properties; this was suggested in Ref *5*. This results in residual stresses in the crack tip plastic zone and residual plastic deformation remaining in the wake of the advancing crack (crack closure) [*7*]. Such a concept would allow the possibility of the effect persisting well beyond the above calculated plastic zone without postulating a significant modification in the crack tip stresses due to creep relaxation effects [*8*].

Reference *3* describes a study in which a similar loading profile was applied to Inconel 718 but with a high cycle frequency of 10 Hz rather than 200 Hz. The effect of the high cycle loading is qualitatively similar when a 10 Hz high cycle frequency is applied. However, when the retardation effect occurs with this lower high cycle frequency, a longer time period is required to reach the minimum crack growth rate. This would be expected if the retardation effect is related to the accumulated number of high frequency cycles.

Summary and Conclusions

From experiments conducted with a combined low amplitude high frequency cycle and a large amplitude low frequency cycle the following conclusions were drawn:

1. Two distinct regimes exist as the high cycle ΔK is varied with constant low frequency ΔK: one in which the low frequency cycle dominates crack growth behavior and one in which the high cycle frequency dominates crack growth behavior.
2. In the low cycle dominated regime a retardation effect is apparent.
3. This retardation effect is strongly dependent on the low cycle ΔK and generally decreases with increasing low cycle ΔK.
4. The effect of increasing frequency on this retardation effect is to bring about the minimum crack growth rate in a shorter period of time. The degree of retardation is therefore related to the accumulated high frequency cycles.

Acknowledgments

The authors wish to thank Dr. D. Jones, Dr. T. Nicholas, and T. Lagnese of the Materials Laboratory of the Air Force Wright Aeronautical Laboratories for their support and guidance in this project. The work presented in this paper was performed under Air Force Contract F33615-82-C-5056.

References

[*1*] Powell, B. E., Duggan, T. V., and Jeal, R. H., "The Influence of Minor Cycles on Low Cycle Fatigue Crack Propagation," *International Journal of Fatigue*, Vol. 4, No. 1, 1982.
[*2*] Powell, B. E. and Henderson, I., "Conjoint Action of High and Low Cycle Fatigue," Report on Grant AFOSR-82-0077, June 1982.
[*3*] Goodman, R. C. and Brown, A. M., "High-Frequency Fatigue of Turbine Blade Material," Report AFWAL-TR-82-4151 (Materials Laboratory, Air Force Wright Aeronautical Laboratories), Contract F33615-79-C-5108, Oct. 1982.
[*4*] Venkiteswaran, P. K., Ferguson, D. C., and Taplin, D. M. R., "Combined Creep-Fatigue Behavior of Inconel Alloy X-750," in *Fatigue at Elevated Temperatures, ASTM STP 520*, American Society for Testing and Materials, Philadelphia, 1973, pp. 462–472.
[*5*] Dugdale, D. S., "Yielding of Steel Sheets Containing Slits," *Journal of the Mechanics and Physics of Solids*, Vol. 8, 1960, p. 100.
[*6*] To, K. C., "A Phenomenological Theory of Subcritical Creep Crack Growth under Constant Loading in an Inert Environment," *International Journal of Fracture*, Vol. 11, No. 4, Aug. 1975, p. 641.
[*7*] Elber, W., "Fatigue Crack Closure under Cyclic Tension," *Engineering Fracture Mechanics*, Vol. 2, 1970, pp. 37–45.
[*8*] Bell, P. D. and Wolfman, A., "Mathematical Modeling of Crack Growth Interaction Effects," in *Fatigue Crack Growth under Spectrum Loads, ASTM STP 595*, American Society for Testing and Materials, Philadelphia, 1976, pp. 157–171.

Ashok Saxena[1]

Creep Crack Growth under Non-Steady-State Conditions

REFERENCE: Saxena, A., **"Creep Crack Growth under Non-Steady-State Conditions,"** *Fracture Mechanics: Seventeenth Volume, ASTM STP 905*, J. H. Underwood, R. Chait, C. W. Smith, D. P. Wilhem, W. A. Andrews, and J. C. Newman, Eds., American Society for Testing and Materials, Philadelphia, 1986, pp. 185–201.

ABSTRACT: The problem of characterizing crack growth under small scale creep and transition conditions is of great practical significance. This paper defines a crack tip parameter C_t for characterizing creep crack growth behavior under wide range creep conditions from small scale creep to steady-state creep. Under steady-state creep conditions, C_t is shown to reduce to the familiar C^*-integral. The physical basis for the C_t parameter is also provided. Wide range creep crack growth data on A470 Class 8 steel using two specimen geometries were obtained and correlated with the C_t parameter in the temperature regime of 482 to 538°C. The levels of creep deformation in the various specimens tested to obtain the crack growth rate data ranged from small scale creep to essentially steady-state creep conditions.

KEY WORDS: creep, fracture, cracks, C^*-integral, growth, transience

Creep crack growth is an important design concern for elevated temperature components and is also important in predicting the residual life of these components which are in service. Thus there is considerable interest in developing time-dependent fracture mechanics (TDFM) concepts for the creep regime.

Several crack tip parameters have been proposed for characterizing crack growth at elevated temperature when the cracked body is under the steady-state condition. These parameters include K for characterizing crack growth in nickel-base alloys [*1,2*] where environment dominates the crack growth process, J for when instantaneous plasticity dominated conditions exist [*3*], and C^* for creeping materials such as 304 stainless steel [*4,5*].

This paper focuses on characterizing the creep crack growth phenomenon

[1]Professor, School of Materials Engineering, Georgia Institute of Technology, Atlanta, GA 30332.

which is defined as subcritical cracking in the presence of creep deformation in the crack tip region. Both non-steady-state (including small scale creep and the transition regime) and steady-state creep conditions are considered. Specifically, a crack tip parameter, C_t, is proposed for characterizing creep crack growth behavior under non-steady-state creep conditions. Also, experimental evaluation of this parameter is conducted and expressions for estimating its magnitude are derived.

Figure 1 schematically shows the levels of creep deformation under which creep crack growth can occur. These consist of the small scale creep (SSC) region, the transition creep (TC) region, and the steady-state (SS) region. Under SSC, the creep zone size is small in comparison to the crack length and the pertinent dimensions of the body. Under SS conditions, the creep zone will have penetrated through the cracked body. The TC condition represents the intermediate region. The SSC and TC regions are under non-steady-state conditions and are characterized by a time-dependent load versus load-line deflection rate behavior. The crack tip stress distribution under non-steady-

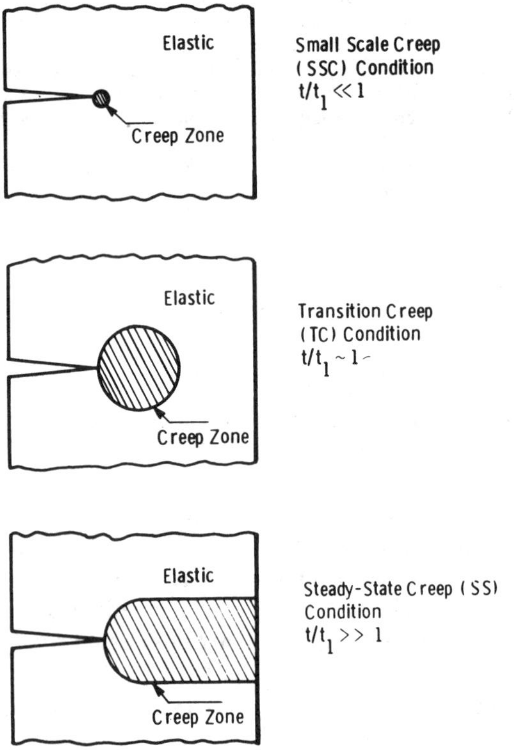

FIG. 1—*Schematic representation of the levels of creep deformation under which creep crack growth can occur* (t = *time,* t_1 = *transition time*).

state conditions is a function of time even when the crack is considered stationary. The steady-state region is characterized by a unique load versus load-line deflection rate behavior. Mathematically, steady-state and non-steady-state conditions may be defined as follows.

Consider a stationary Cartesian coordinate system in a cracked body in which the x axis lies along the crack plane in a direction normal to the crack front (Fig. 2). The crack itself is moving with a velocity $\dot{a}$, where the dot denotes derivative with respect to time and a is the instantaneous crack length. The normal stress in the y-direction is given by $\sigma(x,t)$, where t = time. The rate of change of stress at any point x is given by

$$\frac{d\sigma}{dt} = \dot{a}\frac{\partial \sigma}{\partial x} + \frac{\partial \sigma}{\partial t} \quad (1)$$

Ideal steady-state conditions exist when $\partial\sigma/\partial t$ is zero; therefore $(d\sigma/dt) = \dot{a}(d\sigma/dx)$. If the coordinate system moves with the crack tip, then $d\sigma/dt = 0$ or the stress at any fixed distance ahead of the moving crack tip is independent of time. Non-steady-state (or transient) conditions exist when $\partial\sigma/\partial t$ is not zero.

The problem of creep crack growth under transient conditions is of considerable practical significance to large components which are subjected to stress and temperature gradients. Because these components are designed to resist creep deformation, significant creep deformation is likely to occur during service only in localized regions near crack tips. The material surrounding the crack tip will most likely be under dominantly linear-elastic conditions. This is an ideal situation for crack growth under non-steady-state conditions to

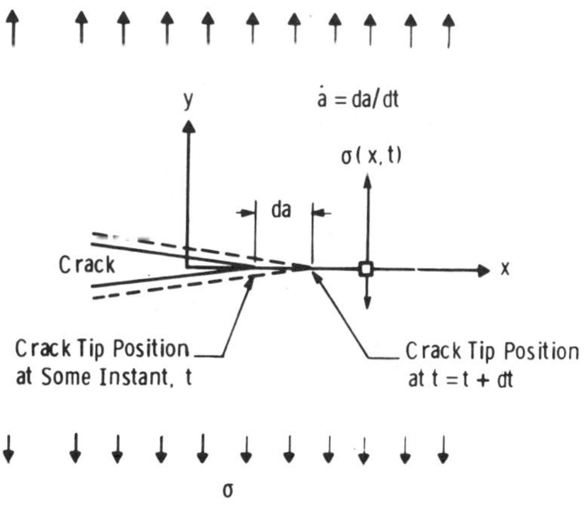

FIG. 2—*Crack tip coordinate system.*

occur. The primary focus of this paper is on developing a crack tip parameter for characterizing crack growth under such conditions.

The Crack Tip Parameter (C_t)

The transient field ahead of a stationary crack for SSC conditions has been derived by Riedel and Rice [*6*]. The results of this analysis (referred to as the RR analysis in the remainder of this paper) will be frequently used here; therefore it is appropriate to mention them briefly. A generalized version of the following equation was used in this analysis for describing the material deformation behavior:

$$\dot{\epsilon} = \frac{\dot{\sigma}}{E} + A\sigma^n \tag{2}$$

where $\dot{\epsilon}$ and $\dot{\sigma}$ are derivatives of stress and strain with time, respectively, E is Young's modulus, and A and n are material constants describing the material's secondary stage creep behavior. The stress, strain rate, and the creep zone size, r_c, ahead of the crack tip along the x-axis are given by Eqs 3 to 5. These equations are accurate within short times ($t \rightarrow 0$) following the application of the load.

$$\sigma \propto \left(\frac{K^2}{AErt}\right)^{1/n+1} \tag{3}$$

$$\dot{\epsilon} \propto \left(\frac{K^2}{AErt}\right)^{n/n+1} \tag{4}$$

$$r_c = \alpha\, K^2(EAt)^{2/n-1} \tag{5}$$

where r is distance from the crack tip, t is time, and K is instantaneously applied K level. A complete description of the equations is given in Ref *6*. For the purposes here, these simplified forms are sufficient. The constant α in Eq 5 depends on n, stress state, etc., and is defined later in Eq 17. The creep zone boundary is defined as the locus of points where the creep strain is equal to the elastic strain.

In the other limit, when the steady-state conditions are reached, the stress and strain rate are characterized by the C*-integral [*4, 7*]:

$$\sigma \propto \left(\frac{C^*}{r}\right)^{1/n+1} \tag{6}$$

$$\dot{\epsilon} \propto \left(\frac{C^*}{r}\right)^{n/n+1} \tag{7}$$

For a detailed description of the C^*-integral, readers are referred to earlier papers [*4, 7*].

The transition time, t_1, for K-dominated conditions to turn to C^*-dominated conditions under plane strain was also given in the RR analysis:

$$t_1 = \frac{K^2(1 - \nu^2)}{E(n + 1)C^*} \tag{8}$$

where ν is Poisson's ratio and the other constants are as defined earlier. These results have been verified by numerical analysis by other investigators [*8, 9*]. Next, we define a parameter C_t which is a promising candidate for unifying the SSC, and TC, and the SS creep cases in a single parameter. It will be obvious later that this parameter becomes identical to C^* in the steady-state regime, $t/t_1 \rightarrow \infty$.[2] The relevance of C_t as a crack tip parameter in the non-steady-state regime is explored in the subsequent discussion.

Definition of the C_t *Parameter*

Consider several identical pairs of cracked specimens. Within each pair, one specimen has a crack length, a, and the other has an incrementally differing crack length, $a + \Delta a$. The specimens of each pair are loaded to various load levels P_1, P_2, P_3 — P_i —, etc. at elevated temperature, and the load-line deflection as a function of time is recorded (Fig. 3*a*). The load-line deflection due to creep is V_c. It is assumed that no crack extension occurs in any of the specimens and the instantaneous response is linear-elastic. We first limit our consideration to the SSC conditions characterized by $t/t_1 \ll 1$. At a fixed time t, the load versus creep deflection rate, $\dot{V}_c$, behavior is plotted for all these specimens. A schematic of the expected behavior is shown in Fig. 3*b*. Several such plots can be generated from the above tests by varying time.

The area between the P-$\dot{V}_c$ curves for specimens of crack length a and $a + \Delta a$ is called ΔU_t^*. Physically, ΔU_t^* (the subscript denotes that this value is at a fixed time t) represents the difference in the energy rates (or power) supplied to the two cracked bodies with identical creep deformation histories as they are loaded to different load or deflection-rate levels. The C_t parameter is given by the equation

$$C_t = -\frac{1}{B}\frac{\partial U_t^*}{\partial a} \tag{9}$$

where B is the specimen thickness. As $t/t_1 \rightarrow \infty$, $C_t = C^*$ by definition of C^*.

[2]The time is scaled with respect to t_1 to provide a measure of the overall level of creep deformation. Thus $t/t_1 \ll 1$ implies SSC conditions, $t/t_1 \gg 1$ implies SS conditions, and the TC condition occurs when $t/t_1 \sim 1$ (Fig. 1).

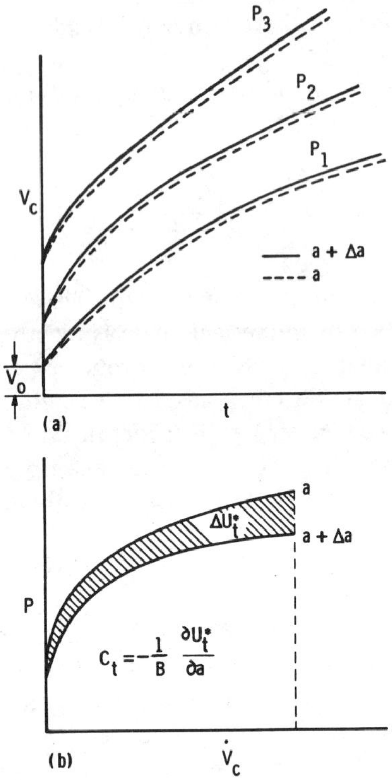

FIG. 3—(a) *Load-line deflection,* V_c, *as a function of time for bodies of crack lengths* a *and* a + Δa *at various load levels.* (b) *Definition of* C_t *parameter.*

Estimation of C_t *for Small Scale Creep*

In deriving the equation to calculate C_t for SSC conditions, analogy is made with the procedure for calculating the nonlinear contributions to the J-integral under small scale yielding conditions in the elastic-plastic regime [*10*]. Specifically, the Irwin concept of effective crack length, a_{eff}, is invoked for calculating load-line deflection rate as a function of applied K and time. Thus

$$a_{\text{eff}} = a_0 + r_c \tag{10}$$

where r_c is the creep zone size defined in Eq 5 and a_0 is the physical crack size. The additional load-line deflection due to creep, V_c, at any time, t, is given by the equation

$$V_c = V - V_0 = P\frac{dC}{da} r_c \tag{11}$$

where P is applied load, V_0 is instantaneous deflection, V is total deflection, and C is elastic compliance of the cracked body. From the relationship between the crack extension force, G, and K, the following equation can be derived [11]:

$$\frac{dC}{da} = \frac{2BK^2}{EP^2}(1 - \nu^2) \tag{12}$$

Substituting Eqs 5 and 12 into Eq 11 and differentiating with time, we obtain

$$\dot{V}_c = \frac{4\alpha(1 - \nu^2)}{E(n - 1)} \left(\frac{P}{B}\right)^3 \frac{F^4}{W^2} t^{3-n/n-1} (EA)^{2/(n-1)} \tag{13}$$

where $F = F(a/W) = (K/P)BW^{1/2}$ is the K-calibration factor. Equation 13 can also be written as

$$\dot{V}_c = \phi(a/W, t)P^3 \tag{14}$$

The C_t parameter can then be estimated as follows:

$$C_t = -\frac{1}{B}\frac{\partial U_t^*}{\partial a} = \frac{1}{Bda}\left[\int_0^{\dot{V}_c}\left(\frac{\dot{V}_c}{\phi}\right)^{1/3} d\dot{V}_c - \int_0^{\dot{V}_c}\left(\frac{\dot{V}_c}{\left(\phi + \frac{\partial\phi}{\partial a}\right)da}\right)^{1/3} d\dot{V}_c\right] \tag{15}$$

Equation 15 can be simplified to

$$C_t = \frac{P\dot{V}_c}{BW}\frac{F'}{F} \tag{16}$$

where $F' = dF/d(a/W)$.

Equation 16 provides a convenient method for calculating the value of C_t under SSC conditions. Experimental and numerical verification is needed to determine the maximum value of t/t_1 for which Eq 16 provides an estimate of C_t. By substituting Eq 13 into Eq 16, the value of C_t for SSC can be determined from the knowledge of the load, P, the K-calibration factors listed in handbooks [12], and the creep constants, A and n. The value of α in Eqs 5 and 13 is given by the equation

$$\alpha = \frac{1}{2\pi}\left(\frac{(n + 1)^2}{2n\,\alpha_n^{n+1}}\right)^{2/(n-1)} \tag{17}$$

where α_n^{n+1} is a factor with a value of approximately 0.69 for n values between 3 and 10.

Estimation of C_t *for Steady-State Conditions*

As discussed earlier, under SS conditions (or $t/t_1 \rightarrow \infty$), $C_t \rightarrow C^*$. Therefore the methods for estimating C_t are the same as those discussed in earlier papers for obtaining C^* [*4,5,13*]. One of the expressions for obtaining C^* in test specimens is [*5*]

$$C^* = \frac{P\dot{V}_c}{BW}\,\eta(a/W) \tag{18}$$

For center crack specimens, $\eta(a/W)$ is given by

$$\eta(a/W) = \frac{1}{2(1 - a/W)}\left[\frac{n-1}{n+1}\right] \tag{19}$$

where a and W are respectively the half crack length and the half width of the specimen. For compact type (CT) specimens:

$$\eta(a/W) = \frac{1}{(1 - a/W)}\,\frac{n}{n+1}\,(\gamma - \beta/n) \tag{20}$$

where the values of γ and β are as given in earlier papers [*3,5*].

Table 1 compares the values of $\eta(a/W)$ with those of F'/F for CT and CCT specimens for various n values including $n = 10.5$ which applies to ASTM grade A470 Class 8 steel at 538°C tested in this study. For CT specimens, η

TABLE 1—*Comparison of the values of* F'/F *with* $\eta(a/W,n)$ *for CT and CCT specimens.*[a]

	CT Specimen				CCT Specimen			
		$\eta(a/W,n)$				$\eta(a/W,n)$		
a/W	F'/F	$n = 3$	$n = 6$	$n = 10.5$	F'/F	$n = 3$	$n = 6$	$n = 10.5$
0.2	2.976	1.966	2.278	2.434	2.755	0.312	0.446	0.516
0.3	2.590	2.307	2.661	2.843	2.0668	0.357	0.510	0.590
0.4	2.644	2.725	3.133	3.346	1.821	0.417	0.595	0.688
0.5	3.078	3.274	3.757	4.008	1.785	0.50	0.714	0.826
0.6	3.920	4.063	4.654	4.962	1.914	0.625	0.893	1.033
0.70	5.330	5.340	6.110	6.510	2.255	0.833	1.190	1.377
0.80	7.949	7.856	8.983	9.570	3.042	1.25	1.785	2.065

[a] a and W are half crack length and half width for CCT specimens and full crack length and full width for CT specimens.

and F'/F values are comparable. This result is in agreement with the elastic-plastic case for determination of the plastic contribution to J-integral. For this latter case the η values for small scale yielding (SSY) (analogous to SSC) and the fully plastic conditions are comparable. On the other hand, the values of η and F'/F are quite different for the CCT specimen. Therefore in a test performed on a CCT specimen where the scale of creep deformation varies during the test, an interpolation scheme is needed to compute C_t values. However, if the entire test lies in either the SSC or SS regime, C_t values can be accurately obtained from Eq 16 or 18, respectively.

Next, we discuss creep crack growth data obtained for evaluating the C_t parameter. Further discussion of the C_t parameter is deferred until after the data are presented in the Results and Discussion section.

Experimental Procedure

Standard 25.4 mm (1 in.) thick compact specimens and 50.8 mm (2 in.) wide (half width W = 25.4 mm) and 25.4 mm thick center crack tension (CCT) specimens were obtained from a large cylindrical forging manufactured in accordance with ASTM Specification for Vacuum-Treated Carbon and Alloy Steel Forgings for Turbine Rotors and Shafts (A 470, Class 8). The notches in these specimens were oriented along the radial direction of the forging. The tensile properties at various temperatures are given in Table 2. Secondary creep rate as a function of applied stress is shown in Fig. 4 at 538°C (1000°F) and at 482°C (900°F), the two test temperatures during creep crack growth testing.

Creep Crack Growth Testing

Creep crack growth tests were conducted at 538°C (1000°F) and at 482°C (900°F) using the constant load method and the constant-deflection rate method. The latter has been discussed in detail in earlier publications [*13*]. The constant load tests were conducted on dead weight type creep machines. The load-line deflection and crack length as a function of time were recorded periodically during the test. The crack length was monitored using the electric potential method [*14,15*].

TABLE 2—*Tensile properties of the material (average of three tests).*

Temperature		0.2% Yield Strength		Ultimate Strength		% Elongation	Reduction in
°C	(°F)	MPa	(ksi)	MPa	(ksi)	(5 cm gage length)	Area, %
24	(75)	623	(90.4)	775.6	(112.5)	14.2	39
427	(800)	515.7	(74.8)	624.6	(90.6)	14.2	53
538	(1000)	464	(67.3)	522.6	(75.8)	17.5	75

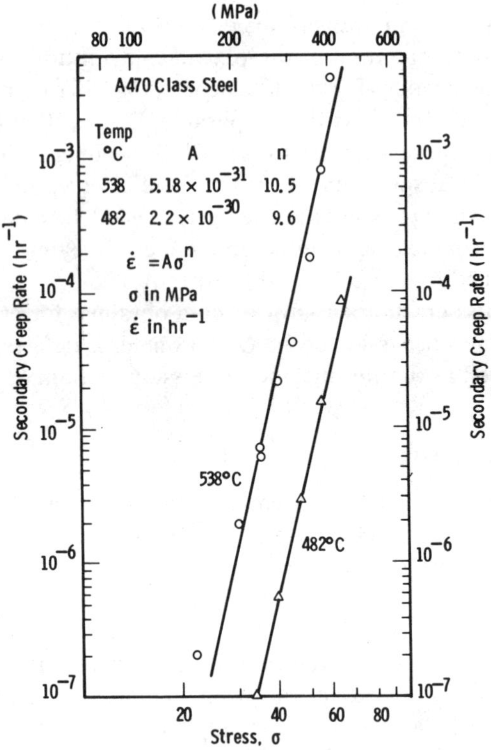

FIG. 4—*Steady-state creep deformation behavior A470 Class 8 steel.*

The creep crack growth rate measured during the various tests was plotted against the C_t parameter. In tests in which the instantaneous response is dominantly elastic, the creep-induced deflection rate $\dot{V}_c$ can be calculated from the equation [5]

$$\dot{V} = \frac{\dot{a}B}{P}\left[\frac{2K^2}{E}\right] + \dot{P}C + \dot{V}_c \tag{21}$$

where $\dot{V}$ is the measured deflection rate, $\dot{a}$ is the crack growth rate, C is the elastic compliance, and $\dot{P}$ is the rate of load change (for constant load tests $\dot{P} = 0$). Thus all quantities needed for calculating C_t are known. If the time-independent plasticity becomes significant, it can be accounted for as discussed in an earlier paper [5].

Results and Discussion

Experimental Evaluation of the C_t *Parameter*

Figure 5 shows the creep crack growth rate, da/dt, at 538°C (1000°F) plotted as a function of the C_t parameter. These data include results from several

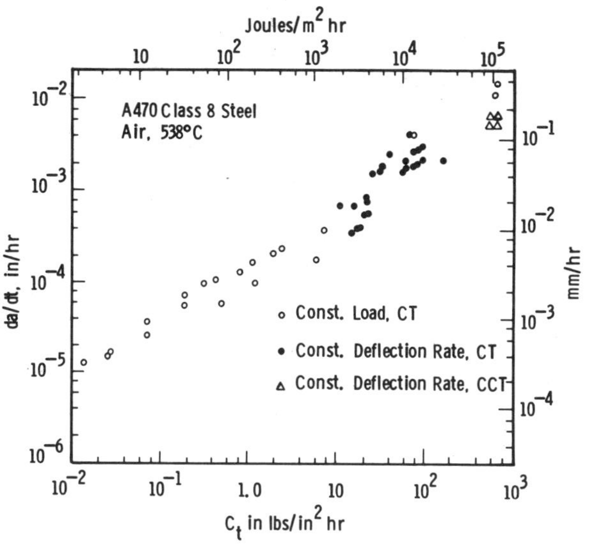

FIG. 5—*Creep crack growth rate behavior of A470 Class 8 steel at 538°C in air.*

CT specimens loaded to various load and deflection rate levels and from one CCT specimen loaded under deflection rate control. The data also range over five orders of magnitude in C_t values and three orders of magnitude in crack growth rates. Excellent correlation between da/dt and C_t values were obtained between the various tests, suggesting that C_t is a good parameter for characterizing creep crack growth behavior.

To determine the extent of creep deformation in the specimens at various times during the test, a parameter τ is defined as

$$\tau = \frac{C^*}{C_t - C^*} \tag{22}$$

where C^* is the steady-state value of C_t and can be determined by analogy to fully plastic J-solutions listed for various cracked specimens in handbooks [*16*]. A value of $\tau \rightarrow 0$ indicates highly non-steady-state conditions and $\tau \rightarrow \infty$ indicates steady-state conditions characterized by $C_t = C^*$. The concept of τ is similar to t/t_1, except t/t_1 applies to stationary cracks only. The values of τ for the data reported in Fig. 4 range from on the order of 10^{-3} to essentially ∞ (Table 3). Therefore a range of creep conditions from SSC to essentially SS existed in the various test specimens. Also, during a given test the τ value changed substantially. Despite the varying conditions, the uniqueness of the relationship between an average value of $\dot{a}$ computed over a time interval Δt and the average value of C_t computed over the same time interval was maintained. This aspect deserves a more in-depth experimental and analytical

TABLE 3—*Values of τ during various tests.*

Specimen Geometry and Number	a/W	$\tau = C^*/(C_t - C^*)$
CT-36	0.518	4.88×10^{-3}
CT-37	0.520	1.96×10^{-3}
CT-74	0.565	0.12
CT-74	0.60	0.65
CT-74	0.648	∞
CT-74	0.663	1.0
CCT-1	0.544	2.67×10^{-2}
CCT-1	0.594	2.3×10^{-2}
CCT-1	0.653	7.3×10^{-3}
CCT-1	0.714	5.1×10^{-3}

evaluation because it is of considerable practical significance as explained below.

One of the primary motivations for characterizing creep crack growth behavior is to use the data generated on small laboratory specimens to predict crack growth in large components. The scale of creep deformation in a small specimen, however, is likely to be much different from the scale of creep deformation in a large component. If the uniqueness in the da/dt versus C_t relationship can be experimentally demonstrated, such as in the results of this study, and analytically justified, it can be argued that the difference in the scales of creep deformation between the specimen and component is not a factor in determining the creep crack growth behavior, provided that C_t can be accurately estimated.

One of the limitations of the C_t parameter is that its use should be restricted to cavitating materials in which creep deformation (or damage) is necessary in the crack tip region for crack growth to occur. In such materials, even if creep crack growth occurs under dominantly elastic conditions, K (or J) will not be the correct parameter for characterizing the growth rate. This is clearly shown in Fig. 6, where creep crack growth rates are plotted as a function of K or $\sqrt{EJ}$. The same data correlated with C_t as shown earlier in Fig. 5. Under these conditions, no unique relationship between crack tip stress and K exists because of stress relaxation; therefore no unique correlation between $\dot{a}$ and K should be expected.

Another restriction (or limitation) that applies to the use of C_t for characterizing creep crack growth is the condition that the damage (in the form of cavities) be localized. In other words, the process zone in which the cavities nucleate and grow should be small in size in comparison to the region in which the crack tip equations, such as Eqs 3 to 7, apply. A detailed discussion of this issue in the context of creep crack growth can be found in papers by Riedel [*17*] and by Hutchinson [*18*].

Besides the two limitations of C_t discussed above, there are several impor-

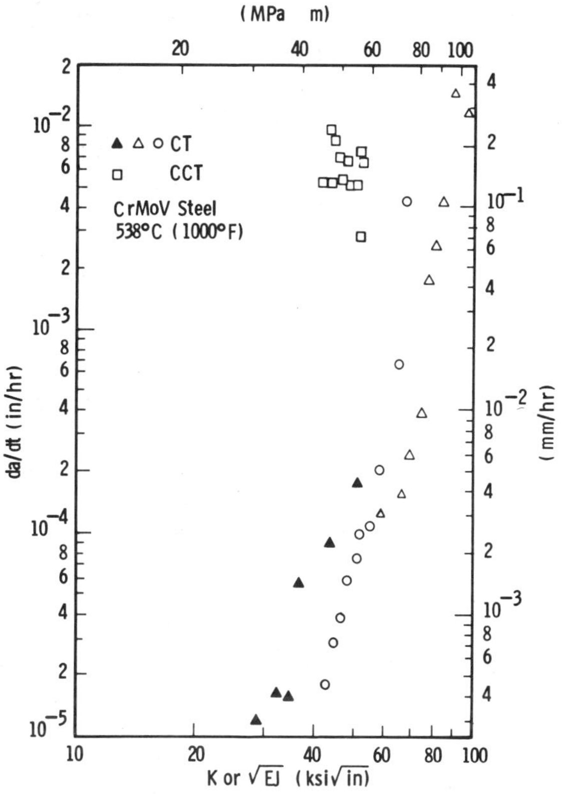

FIG. 6—*Creep crack growth rate plotted as a function of the stress intensity parameter. (A different symbol is used for each test.)*

tant questions that should be answered before C_t can be used in life prediction of elevated temperature components. The first deals with the method for estimating the magnitude of C_t in the transition region. The transition region is the time domain between SSC and SS and is characterized by t/t_1 or $\tau \sim 1$. Some preliminary ideas may be developed on the basis of the limited finite element analysis of Ehlers and Riedel [*9*]. They have suggested a crack tip parameter $C(t)$:

$$C(t) = C^*(1 + t_1/t) \tag{23}$$

It is likely that $C(t)$ in Eq 23 is an estimate of C_t as defined in this paper.

The above method is analogous to the method of determining J for elastic-plastic situations by combining the elastic and the fully plastic values of J [*16*]. Numerically, it was shown by Ehlers and Riedel that C_t calculated from Eq 23 approximates the amplitude of the Hutchinson [*19*] and Rice and Rosengren [20] (HRR) field over t/t_1 ranging from essentially 0 to ∞. Therefore Eq 23 appears to at least be a good starting point for developing an interpolation scheme for determining C_t values. More numerical analysis of various cracked configurations should be performed to develop and verify an interpolation scheme. The second limitation deals with computing the value of C_t in a component containing a crack which is growing. The value of C_t is affected by crack growth in a manner somewhat different from the other field parameters such as K and J. The magnitude of C_t is time dependent even at a fixed crack length and it depends on the extent of creep in the specimen. The extent of creep is characterized by the value of t/t_1 for stationary cracks. For a growing crack, two things happen. First, the transition time continuously decreases due to shrinking of the remaining ligament if the load is held constant. This can perhaps be accounted for without much difficulty. More importantly, the reference from which t is measured is lost for growing cracks. This makes it difficult to determine the value of C_t without actually measuring $\dot{V}_c$. In order to make any approximations, it is necessary to have numerical and experimental results available on cracked configurations.

Another question that remains about C_t is its relationship to the HRR fields as time t asymptotically approaches zero and also under SSC conditions. Under SS conditions, the readers are reminded that $C_t = C^*$; thus the one-to-one relationship between C_t and the HRR field has already been demonstrated [21]. Under conditions of t asymptotically approaching zero, C_t should be consistent with the RR analysis described earlier. No expression for estimating C_t under these conditions is available; therefore the question whether C_t characterizes the HRR field under these conditions is still open. Under SSC conditions, the expression for calculating C_t is derived as described earlier in this paper. A comparison of the amplitudes of the stress fields in the HRR format from C_t and from the limited finite element results of Ehlers and Riedel [9] is made for $t/t_1 < 0.1$ in Fig. 7. In order to compute C_t, Eq 13 was substituted into Eq 16 to eliminate $\dot{V}_c$ from Eq 16. There appears to be a reasonable correlation between the amplitudes of stress fields from the two methods. This further substantiates the validity of the C_t parameter; however, it is recommended that more finite element work be done on different geometries for a detailed evaluation.

Influence of Temperature on the Creep Crack Growth Behavior

Figure 8 shows the influence of temperature on the creep crack growth behavior of A470 Class 8 steel. Comparing the data at 538°C (1000°F) and 482°C (900°F) shows only a marginal difference between the da/dt versus C_t behavior even when the creep deformation rates are vastly different for the two temperatures (Fig. 4). It appears that the C_t parameter provides a first-order normalization of the influence of temperature on the creep crack growth behavior. The arguments for rationalizing this trend were developed previously [4] when similar behavior was observed for 304 stainless steel.

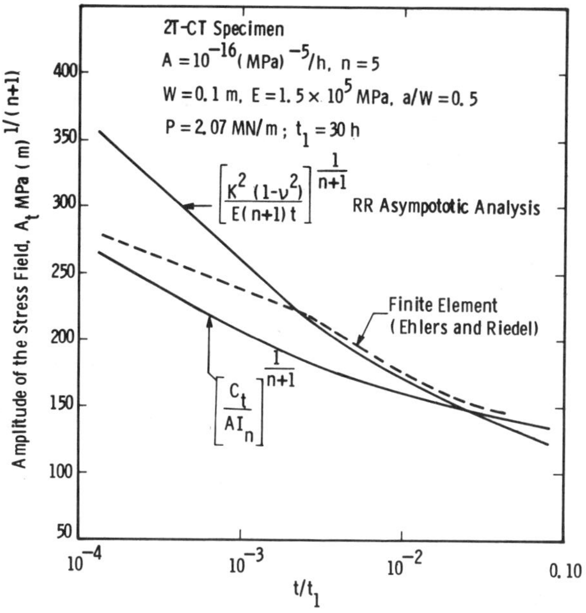

FIG. 7—*Comparison between the amplitude of the HRR stress field obtained from various analyses for small scale creep conditions.*

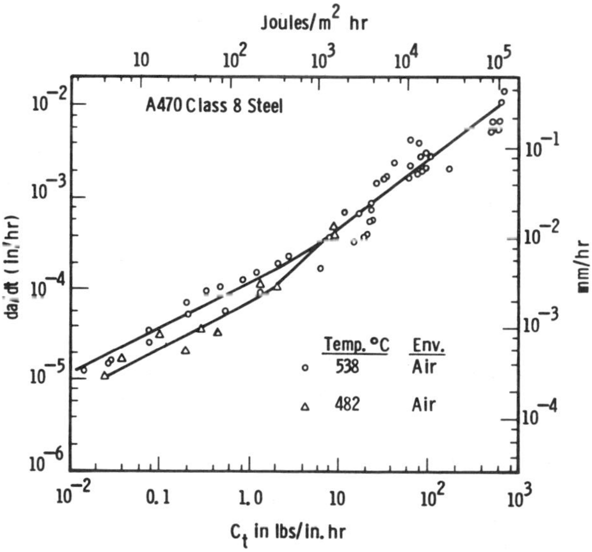

FIG. 8—*Creep crack growth behavior of A470 Class 8 steel as a function of temperature.*

Summary and Conclusions

The following conclusions can be drawn from this study:

1. A crack tip parameter C_t is proposed which characterizes the creep crack growth behavior in A470 Class 8 steel under a variety of creep conditions ranging from small scale creep to steady-state creep behavior. The correlation is independent of specimen geometry and has been demonstrated over a wide range of da/dt and C_t values.
2. Under steady-state creep conditions, C_t reduces to the path-independent integral C^*. Also, C_t can be interpreted as the *instantaneous* value of the difference between the energy rates supplied to two creeping cracked bodies which are identical in all respects except incrementally differing crack lengths.
3. To a first-order approximation, da/dt versus C_t behavior is independent of temperature in the range of 482 to 538°C.
4. A formula for estimating C_t in the small scale creep regime is derived in this paper. Also, the limitations of the C_t parameter are discussed in detail.

Acknowledgments

The author wishes to acknowledge the financial support of the Westinghouse Steam Turbine-Generator Division for conducting this study and the encouragement received from Dr. V. P. Swaminathan of that division. He also wishes to acknowledge several helpful discussions with Drs. J. D. Landes and H. A. Ernst of Westinghouse R&D Center and Prof. J. L. Bassani of University of Pennsylvania, Prof. J. W. Hutchinson of Harvard University, and Dr. H. Riedel from the Max Planck Institute in W. Germany. C. M. Fox of Westinghouse R&D Center helped in conducting the creep crack growth experiments.

References

[1] Sadananda, K. and Shahinian, P., *Metallurgical Transactions*, Vol. 8A, 1977, pp. 439-449.
[2] Pelloux, R. M. and Huang, J. S. in *Creep-Fatigue-Environment Interactions*, R. Pelloux and N. Stoloff, Eds., TMS-AIME, 1980, pp. 151-164.
[3] Saxena, A., Ernst, H. A., and Landes, J. D., *International Journal of Fracture*, Vol. 23, 1983, pp. 245-257.
[4] Saxena, A. in *Fracture Mechanics: Twelfth Conference, ASTM STP 700*, American Society for Testing and Materials, Philadelphia, 1980, pp. 131-151.
[5] Saxena, A. and Landes, J. D., "Characterization of Creep Crack Growth in Metals," presented at Sixth International Conference on Fracture, ICF-6, New Delhi, Dec. 1984.
[6] Riedel, H. and Rice, J. R. in *Fracture Mechanics: Twelfth Conference, ASTM STP 700*, American Society for Testing and Materials, Philadelphia, 1980, pp. 112-130.
[7] Landes, J. D. and Begley, J. A. in *Mechanics of Crack Growth, ASTM STP 590*, American Society for Testing and Materials, Philadelphia, 1976, pp. 128-148.
[8] Bassani, J. L. and McClintock, F. A., *International Journal of Solids and Structures*, Vol. 7, 1981, pp. 479-492.

[9] Ehlers, R. and Riedel, H. in *Advances in Fracture Research*, ICF-5, D. Francois, Ed., Vol. 2, 1981, pp. 691-698.
[10] Shih, C. F. in *Mechanics of Crack Growth, ASTM STP 590*, American Society for Testing and Materials, Philadelphia, 1976, pp. 3-22.
[11] Irwin, G. R. and Kies, J. A., *Welding Journal, Research Supplement*, Vol. 33, 1954, pp. 193-198.
[12] Tada, H., Paris, P., and Irwin, G. R., *The Stress Analysis of Cracks Handbook*, Del Research Corp., Hellertown, Pa., 1973.
[13] Saxena, A., Shih, T. T., and Ernst, H. A. in *Fracture Mechanics: Fifteenth Symposium, ASTM STP 833*, R. J. Sanford, Ed., American Society for Testing and Materials, Philadelphia, 1984, pp. 516-531.
[14] Saxena, A., *Engineering Fracture Mechanics*, Vol. 13, 1980, pp. 741-750.
[15] Schwalbe, K. H. and Hellmann, D., *Journal of Testing and Evaluation*, Vol. 9, May 1981, pp. 218-221.
[16] Kumar, V., German, M. D., and Shih, C. F., "An Engineering Approach to Elastic-Plastic Fracture Analysis," EPRI NP 1931, Electric Power Research Institute, Palo Alto, Calif., July 1981.
[17] Riedel, H. in *Fundamentals of Deformation and Fracture*, Proceedings of the Eshalby Memorial Symposium, Sheffield, April 1984, K. J. Miller, Ed., Cambridge University Press, England.
[18] Hutchinson, J. W., *Acta Metallurgica*, Vol. 31, 1983, pp. 1079-1088.
[19] Hutchinson, J. W., *Journal of the Mechanics and Physics of Solids*, Vol. 16, 1968, pp. 13-31.
[20] Rice, J. R. and Rosengren, G. F., *Journal of the Mechanics and Physics of Solids*, Vol. 16, 1968, pp. 1-12.
[21] Goldman, N. L. and Hutchinson, J. W., *International Journal of Solids and Structures*, Vol. 11, 1975, pp. 575-591.

Isao Soya,[1] *Hironori Takashima,*[1] *and Yoichi Tanaka*[1]

An Application of Stress Intensity Factor to Fatigue Strength Analysis of Welded Invar Sheet for Cryogenic Use

REFERENCE: Soya, I., Takashima, H., and Tanaka, Y., **"An Application of Stress Intensity Factor to Fatigue Strength Analysis of Welded Invar Sheet for Cryogenic Use,"** *Fracture Mechanics: Seventeenth Volume, ASTM STP 905*, J. H. Underwood, R. Chait, C. W. Smith, D. P. Wilhem, W. A. Andrews, and J. C. Newman, Eds., American Society for Testing and Materials, Philadelphia, 1986, pp. 202–225.

ABSTRACT: Fatigue strength of several types of welded Invar joints were investigated at room temperature and 111 K. Lap fillet welded joints, in which a 1.5 mm thick Invar sheet was welded to a 1.5 mm or 3.0 mm thick Invar sheet or 12 mm thick stainless steel, were made by tungsten-inert-gas (TIG) welding. A joint, in which a 0.5 mm thick Invar sheet was inserted between two overlapped 1.5 mm thick Invar sheets, was made by resistance seam welding. In these joints, the root of the weld is the region most susceptible to fatigue crack initiation.

Since *S-N* relationship depends upon the joint type, specimen size, and loading method in a fatigue test, a fatigue strength parameter independent of these conditions is necessary for the fatigue evaluation of structures. In this study, equivalent stress intensity factor range ΔK_{eq} was adopted and investigated as such a parameter. By simulating the measured cyclic strain behaviors by elastic-plastic FEM analyses, the cyclic ranges of tensile, bending, and shear stresses that arise in the vicinity of the weld can be estimated. Using these ranges, ΔK_{I}, ΔK_{II}, and accordingly ΔK_{eq} were calculated for each specimen.

As a result, it was found that the relationships between ΔK_{eq} and cycles-to-failure in the range above 10^4 cycles were well within a narrow scatter band, regardless of the type of joints. It was concluded that ΔK_{eq} is a very effective parameter for prediction of fatigue strength when the sharp root of a weld is the problem.

KEY WORDS: Invar, lap welds, resistance seam welds, cryogenics, fatigue, notch sensitivity, stress intensity factor, cyclic loads, strain measurement, finite element method, elastic-plastic analysis

Stainless steel and 9Ni alloy steel are extensively used in cryogenic structures such as liquefied natural gas (LNG) plants. Recently, however, Invar

[1]Senior Research Engineer, Chief Research Engineer, and Research Engineer, respectively, R&D Laboratories II, Nippon Steel Corporation, Sagamihara, Japan.

(36Ni alloy steel) has also come to be used widely in cryogenic applications, because the coefficients of thermal expansion of Invar are as small as 1.5×10^{-6}/K.

Fatigue failures which result from cyclic variations in pressure, temperature, etc., are an important problem for the assessment of the integrity of the cryogenic structures. From this point of view, several studies have been made on the fatigue properties of stainless steel for cryogenic use [*1–3*]. When Invar sheet is used in structures, it should be considered that the welded joint is the region most susceptible to fatigue crack initiation.

In fatigue tests of welded sheets with complicated joint profiles which are asymmetric with respect to the loading axis, *S-N* relationships are greatly dependent on the configuration of the specimen, apparatus used, and loading method, because the bending and shear stresses arise in the vicinity of the weld even if the test is conducted under the tensile loading. In general, it is difficult to obtain *S-N* diagrams for all the combinations of joint profiles and loading conditions. To assess the integrity of structures, therefore, it is desirable to describe the fatigue strength in terms of a parameter which is not dependent upon joint profile and loading conditions.

The strain amplitude at the fatigue crack initiation point, which has been applied already to the fatigue design rules of the American Society of Mechanical Engineers, is one such parameter. For the welded joints considered in this study, it is very likely that a fatigue crack would initiate from the root of the weld during service and the fatigue test. As it is very difficult to predict exactly the strain amplitude at the sharp root by measurement or calculation, the results of the fatigue test cannot be arranged in terms of the strain amplitude.

Since the tip of the weld root is in the form of a very sharp notch, the stress intensity factor may be applicable. Some studies have been made on applicability of the stress intensity factor range to the fatigue strength of the weld root, for example, in the butt welded joints of pipes [*4*], the fillet welded joints of plates [*5*], and the spot or seam welded joints of sheets [*6*].

For fatigue failure that occurs by Modes I and II, the equivalent stress intensity factor range ΔK_{eq}, which is a combination of ΔK_I and ΔK_{II}, can be a measure that is not dependent upon joint profiles and loading conditions. Thus the method for calculation of ΔK_{eq} in the fatigue test of the welded Invar sheets and the effectiveness of ΔK_{eq} as a measure for fatigue strength were investigated in this study.

Materials and Test Method

A 1.5 mm thick Invar sheet was welded to 3.0 and 0.5 mm thick Invar sheets and 12 mm thick stainless steel plate (SUS304) to make various types of welded joints. The chemical composition and tensile properties of these Invar sheets are given in Tables 1 and 2, respectively. Combining these sheets,

TABLE 1—*Chemical composition of Invar sheets.*

Thickness, mm	Composition, wt%								
	C	Si	Mn	P	S	Ni	Cr	Mo	Co
1.5	0.023	0.17	0.32	0.002	0.001	35.50	0.17	0.01	0.21
3.0	0.028	0.15	0.23	0.003	0.001	35.72	0.23	0.12	0.21
0.5	0.023	0.17	0.32	0.003	0.001	35.50	0.17	0.01	0.21

TABLE 2—*Tensile properties of Invar sheets.*

Thickness, mm	Temperature	Yield Stress, MPa	Tensile Strength, MPa	Elongation, %
1.5	room temperature	297	483	44.6
	77 K	649	926	53.9
3.0	room temperature	300	477	38.0
	77 K	627	919	41.0

five types of lap fillet welded joints and a resistance seam welded joint were made under the welding conditions given in Table 3. Hereafter these welded joints are referred to as Joint N, Joint A, etc., as shown in Table 3.

The profiles of the specimens cut from the lap fillet welded joints are shown in Fig. 1. Joints N, A, B, C, and D are gripped by the jig as shown in Fig. 2. For Joint A, 150 mm wide specimens were also tested to study the effect of the width of specimen. The profile of the specimen of Joint E and the jig used for its setting are shown in Fig. 3, and the details of the cross section are shown in Fig. 4.

Fatigue tests were conducted on these specimens at room temperature (RT) and at the temperature of LNG (111 K). Servohydraulic testing machines of 50 and 100 kN capacity were used. The specimens of lap fillet welded joints were tested under pulsating load control with a stress ratio R (minimum applied stress/maximum applied stress) in the range from 0.05 to 0.1, but the specimen of Joint E was tested under pulsating displacement control with R of 0.1. The test frequency ranged from 1 to 50 Hz. In the test at 111 K, a thermocouple was attached to each specimen to control the temperature within the deviation of ± 2 K using the liquefied nitrogen or low-temperature nitrogen gas.

Apart from these fatigue tests, the behavior of the specimen of each type of joint under cyclic loading at RT and 111 K was measured by attaching strain gages and displacement gages to each specimen at the positions shown in Figs. 1 and 4. In this measurement, the increasing and decreasing load was applied statically up to two or three cycles.

TABLE 3—*Joint welding conditions.*

Type of Joint	Name of Joint	Combination of Thicknesses, mm	Welding Procedure	Welding Conditions				Throat Depth, mm
				Wire	Current, A	Voltage, V	Speed, cm/min	
Lap fillet welded joint of two sheets	N	1.5 by 1.5	automatic TIG	...	90	12	30	1.1
	A	1.5 by 1.5	automatic TIG	...	85	10 ~ 11	20	1.1
	B	1.5 by 1.5	manual TIG	...	65	9 ~ 10	10 ~ 11	1.2
	C	1.5 by 3.0	manual TIG	...	65	9 ~ 10	8 ~ 9	0.8
	D	1.5 by 12.0* (*:SUS304)	semi-automatic TIG	(ESAB OUTROD16.94G (1.6 mmϕ))	100	10 ~ 12	6 ~ 7	2.2
Seam of three sheets	E	1.5 by 0.5 by 1.5	resistance seam welding	...	2700	(electrode force 25 MPa)	93	(nugget width 2.3)

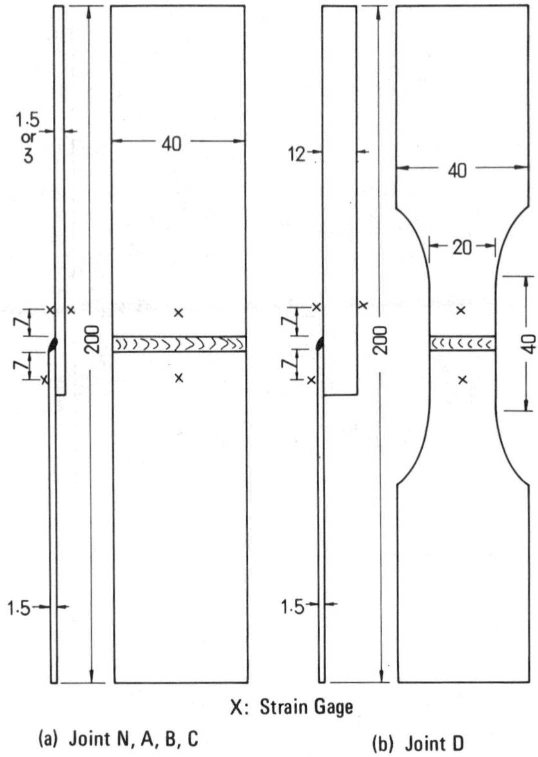

FIG. 1—*Fatigue test specimens of lap fillet welded joints (dimensions in millimetres).*

Method of Stress Analysis

In this study, elastic-plastic analysis and elastic analysis were conducted using the finite element method (FEM) in order to analyze the conditions of cyclic stresses in the vicinity of the weld and to calculate the stress intensity factors of the weld root. Using a program based on triangular elements of uniform strain, the analysis was carried out under the plane strain condition. An example of the finite element subdivision is shown for Joint A in Fig. 5. Similar profiles were also determined for other joints so that the cross section of each joint could be closely simulated.

In the elastic-plastic analysis under cyclic loading, it is necessary to consider the Bauschinger effect. Accordingly, the program used here is based on the mixed hardening theory [*7*]. In this theory, the magnitude of the Bauschinger effect is expressed by the mixed hardening parameter M. This theory agrees with the isotropic hardening theory when M is 1, but with the kinematic hardening theory when M is 0. In this study M was assumed to be 0.5, because it is unknown what value of M represents Invar.

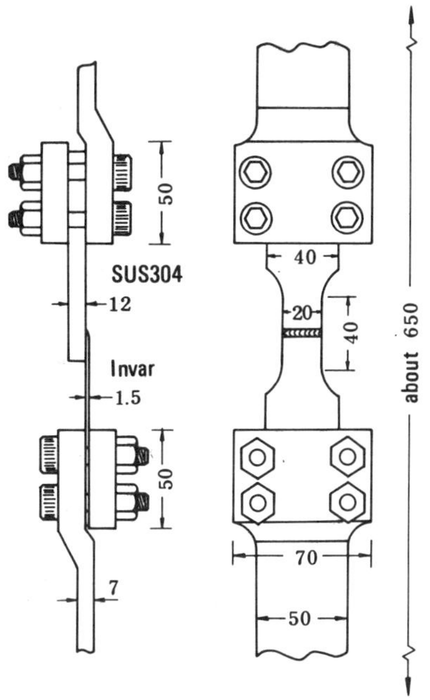

FIG. 2—*Fatigue test assembly for Joint D (dimensions in millimetres).*

The stress intensity factors were calculated by the method proposed in Ref *8*. This method makes it possible to calculate K_I and K_{II} separately using the stresses in the elements i and j and the displacements of the nodes m and n shown in Fig. 5*c*.

The material properties for the FEM analyses are shown in Fig. 6. Of these properties, the stress-strain curve, yield stress, and Young's modulus of Invar were obtained by experiments. Other properties are the estimated or assumed values. The stress-strain curve was input through multi-linear approximation as shown in Fig. 6.

Results of Fatigue Test

The results of fatigue test are shown in Figs. 7 to 9 in which the tensile stress range $\Delta\sigma$ implies the nominal value for the cross section of the 1.5 mm thick Invar sheet. For Joint E, the displacement range Δd was controlled to be constant during the test, and the nominal stress range $\Delta\sigma$ was calculated from the relation between Δd and the load measured by the load cell.

In most specimens, failure occurred from the weld root as shown in Fig. 10. In the few specimens marked with an asterisk in Figs. 7 and 8, however, final failure occurred from the toe of weld but a fatigue crack was also observed in

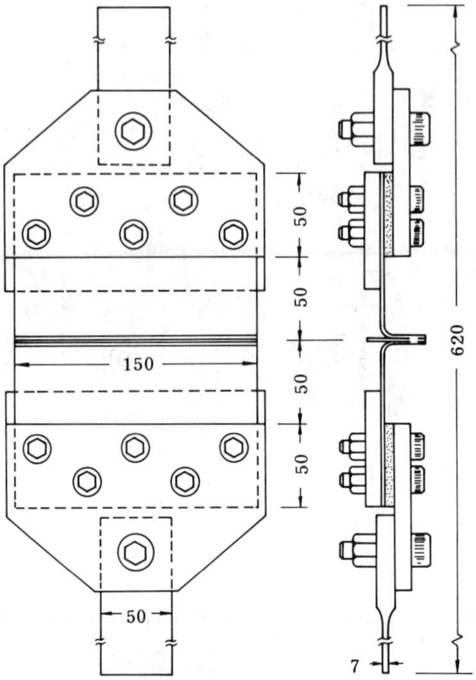

FIG. 3—*Fatigue test assembly for Joint E (dimensions in millimetres).*

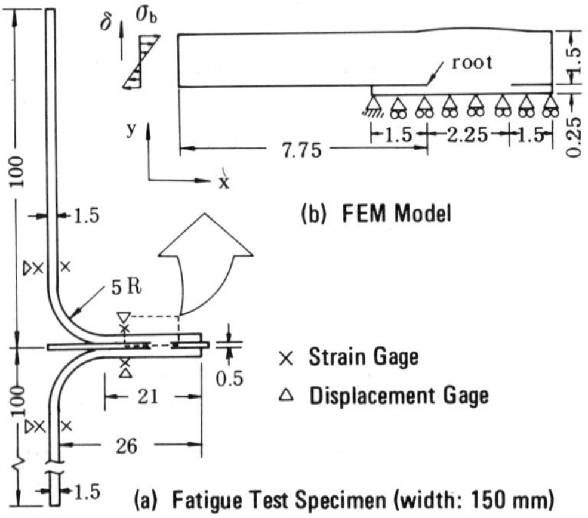

FIG. 4—*Details of Joint E specimen (dimensions in millimetres).*

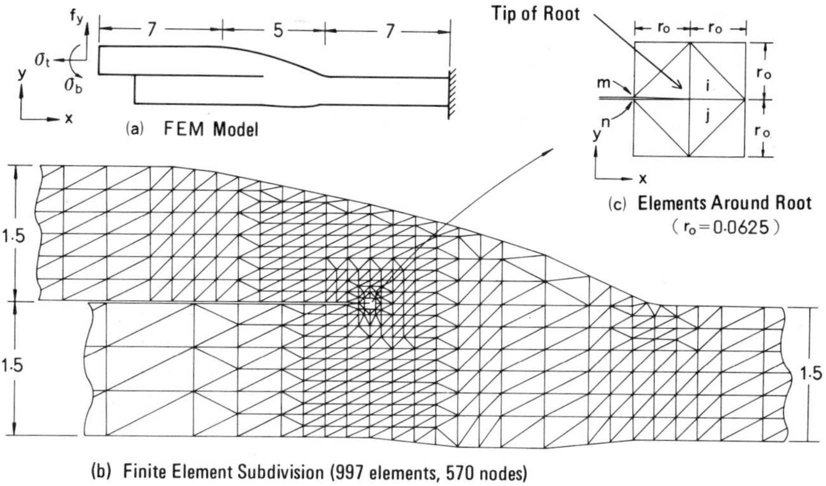

FIG. 5—*Model of Joint A for FEM analysis (dimensions in millimetres).*

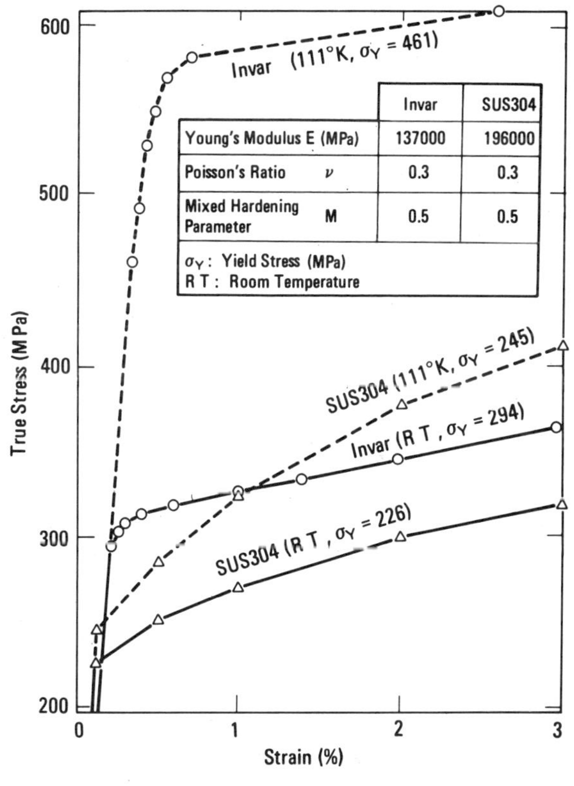

FIG. 6—*Material properties for FEM analysis.*

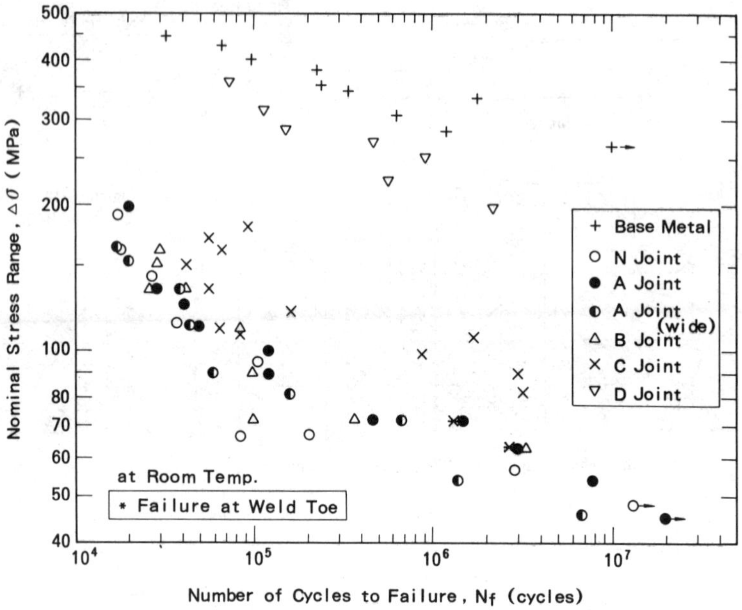

FIG. 7—S-N *relationships of lap fillet welded joints at room temperature.*

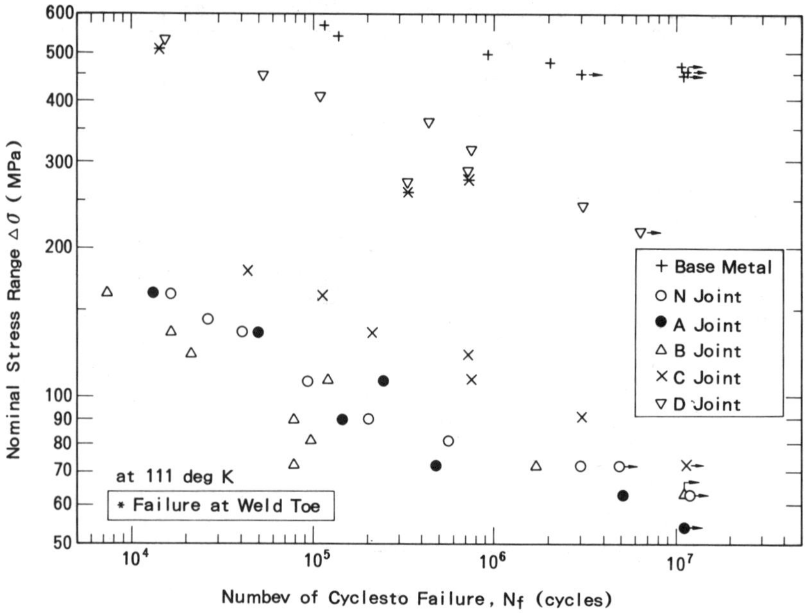

FIG. 8—S-N *relationships of lap fillet welded joints at 111 K.*

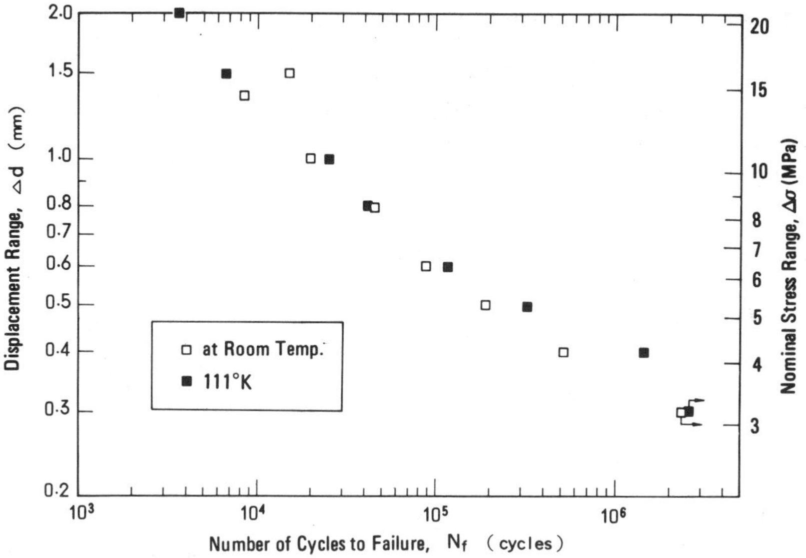

FIG. 9—*Relationships between displacement range and cycles-to-failure for Joint E.*

FIG. 10—*Cross section of failed Joint B specimen.*

the root. On the other hand, in some specimens of Joint D, the fatigue crack was also observed in the toe of the weld, even if the failure occurred from the root. In Joint E, the heat-affected zone bordering on the nugget was also forged by electrode force of resistance seam welding, and failure occurred from the root of the forged zone. For Joint A, the results for 150 mm wide specimens were nearly the same as those for 40 mm wide specimens at RT. Thus it seems that the effect of the width of specimen is negligibly small.

The fatigue strength varied widely depending on the type of joint, because their geometries were different (i.e., throat depth, leg length, etc., of the weld). Also, the effect of secondary stress induced by tensile load differs from specimen to specimen, since these specimens are of the eccentric type. Thus, in order to obtain the general fatigue properties of the welded joints, it is necessary to apply a fatigue strength parameter that is independent of the joint profiles and loading conditions.

Analyses of Test Results and Discussion

Behavior of Specimens under Cyclic Loading

Behavior of Joint A—An example of the strain behavior of Joint A under the cyclic nominal tensile stress σ_t in the range from 0 to 159 MPa is shown in Fig. 11. Strain gages were attached at the positions G1, G2, and G3 as shown in the figure. The strains at all positions exhibited a complicated behavior for the following reason: Because the specimens are of the eccentric type, the bending stress and shear stress would arise at each position of the strain gage, and the combination ratio of such stresses would change with increase or decrease in nominal tensile stress.

To clarify how the tensile stress is combined with the bending and shear stresses, the elastic-plastic FEM analysis was conducted on the model shown

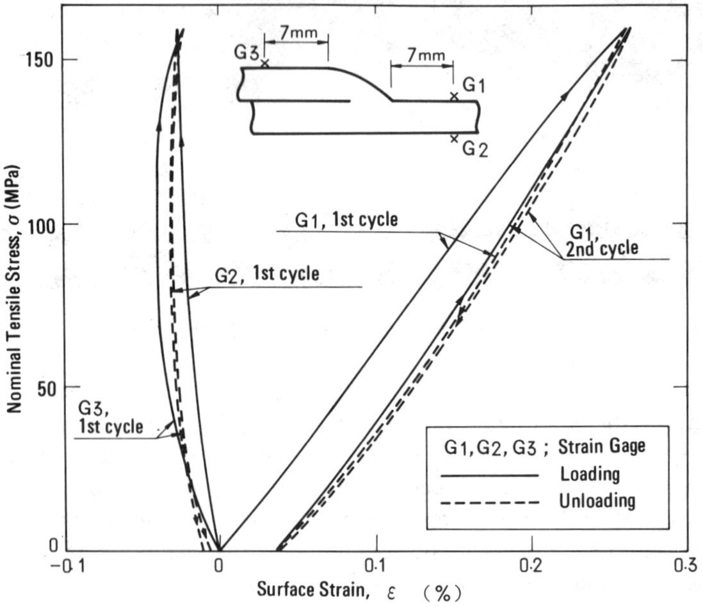

FIG. 11—*Strain behavior in Joint A specimen at room temperature.*

in Fig. 5. In this model, the loaded end and fixed end coincided with the positions of the strain gages attached to the specimen.

Thus elastic-plastic analysis was conducted by applying various combinations of tensile stress σ_t, bending stress σ_b, and shear force f_y in the y-direction to the left-hand end of the model, and the combination in which the strains at the loaded and fixed ends behaved as shown in Fig. 11 was obtained by trial and error. Here, the bending stress σ_b is expressed by the value at the surface of the specimen, and f_y implies force per unit width of the specimen.

A comparison between the measured strain values and the strain behavior determined by FEM is shown in Fig. 12. The results of the FEM analysis in the figure were obtained for the combination of the boundary stresses shown in Fig. 13. The combination in the first loading process differs from that in the unloading process. After the maximum load in the first cycle, however, the behavior in the loading process is just the reverse of that in the unloading process.

Figure 14 shows the behavior of stress and strain in the stress-concentrated areas (the root and the toe) in the case when the boundary stresses shown in Fig. 13 are applied. Shown also in Fig. 14 is the behavior at the root in the case where the strains measured at 111 K are simulated. This figure shows

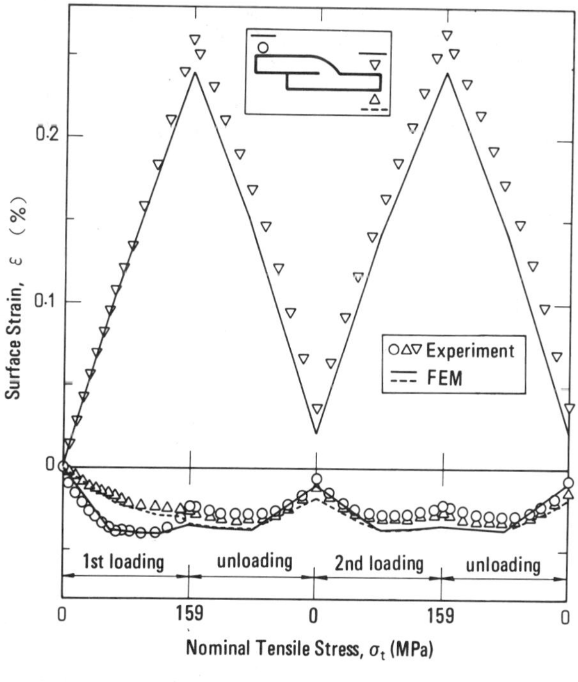

FIG. 12—*Measured and simulated strain behavior for Joint A specimen at room temperature.*

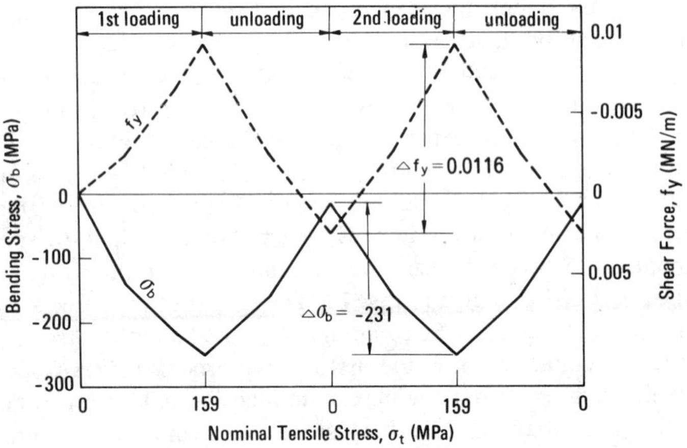

FIG. 13—*Behavior of bending stress and shear force in Joint A specimen at room temperature.*

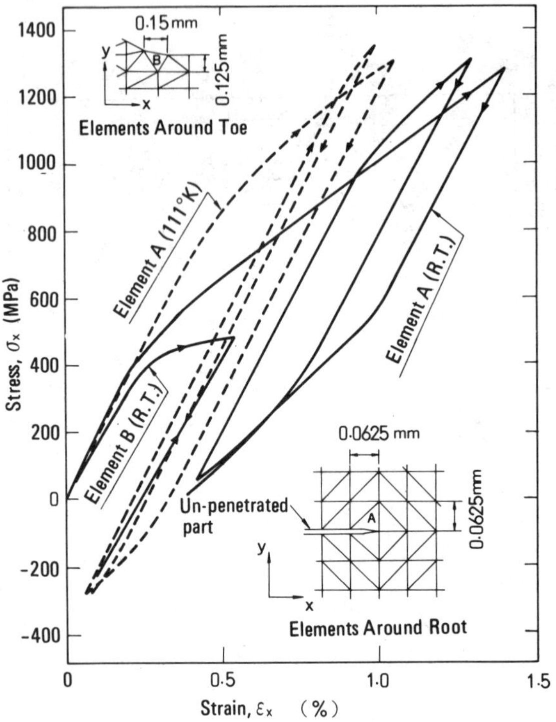

FIG. 14—*Stress-strain behavior at the stress-concentrated region in Joint A.*

that large plastic deformation occurs in the first loading process. In the second cycle, however, the plastic deformation due to the re-yield of the root is small and the hysteresis loop seems to be stabilized as the number of cycles increases. At the toe, shakedown takes place after the maximum load of the first cycle.

In the analysis for RT, about 200 elements around the root and toe yielded at the first maximum load, but only 8 elements around the root yielded at the maximum load in the second cycle. In other welded joints also, large plastic deformation occurred at the root in the first loading process but the plastic deformation caused after that was small. It is therefore reasonable to assume that the plastic deformation caused after the first cycle is of small-scale yielding and to apply the linear fracture mechanics concepts.

In the analysis of Joint A at RT, the stress combination ratios for the second cycle as shown in Fig. 13 were $\Delta\sigma_b/\Delta\sigma_t = 231/159 = -1.45$ and $\Delta f_y/\Delta\sigma_t = 0.0116/159 = 72.6 \times 10^{-6}$ m. Accordingly, ΔK_{eq} is calculated as described later, assuming that these ratios are the same for specimens having different nominal tensile stress ranges $\Delta\sigma_t$.

Behavior of Joint D—In the specimen of Joint D, 12 mm thick stainless steel is very stiff compared with 1.5 mm thick Invar sheet, and the amount of eccentricity of the specimen is large. Accordingly, the behavior under cyclic loading differs considerably from those of the specimens of Joints N, A, B, and C. The cyclic strain behavior under the nominal tensile stress σ_t in the range from 0 to 265 MPa is shown in Fig. 15.

In Joints N, A, B, and C, two overlapping Invar sheets are deformed in such a manner that they are separated, when a tensile load is applied. With Joint D, however, the Invar sheet deforms as if it moved into the stainless steel plate, when elastic-plastic analysis is conducted so as to simulate the strain behavior shown in Fig. 15 by combining σ_t, σ_b, and f_y only. As such a deformation is impossible in actual welded joints, it is necessary to take account of the contact reaction force R_y (force per unit width of specimen) that arises at the contact area between Invar sheet and stainless steel plate as shown in the model of Fig. 16.

The condition at which the strain behavior as shown in Fig. 15 is reproduced by combining σ_t, σ_b, f_y, and R_y was also determined by the trial-and-error method of elastic-plastic analysis, the results of which are shown in Fig. 17. The behavior in the loading process of the first cycle differed considerably from that in the unloading process. In the second cycle, however, the difference is not so significant. This difference in the second cycle is due to the fact that the difference between the strain behavior of the gage G3 in Fig. 15 in the loading process and that in the unloading process was taken into consideration. In this case, the average value of the stress combination ratios in the loading and unloading processes of the second cycle was used for the calculation of ΔK_{eq} as described later.

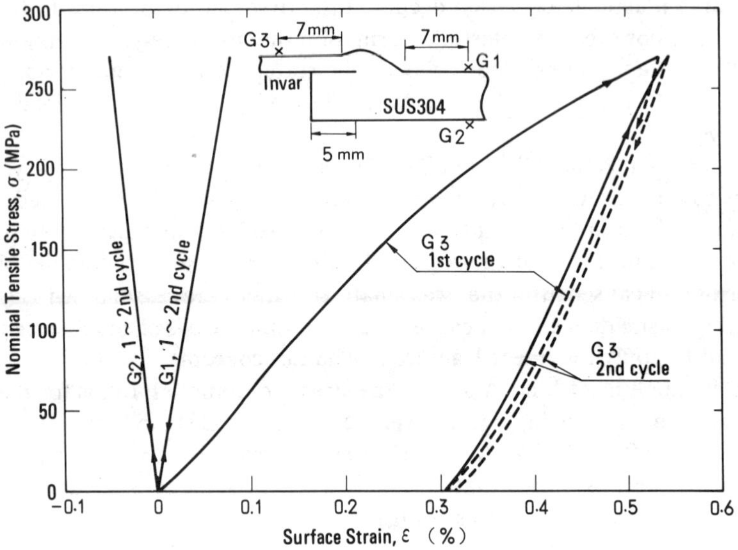

FIG. 15—*Strain behavior in Joint D specimen at room temperature.*

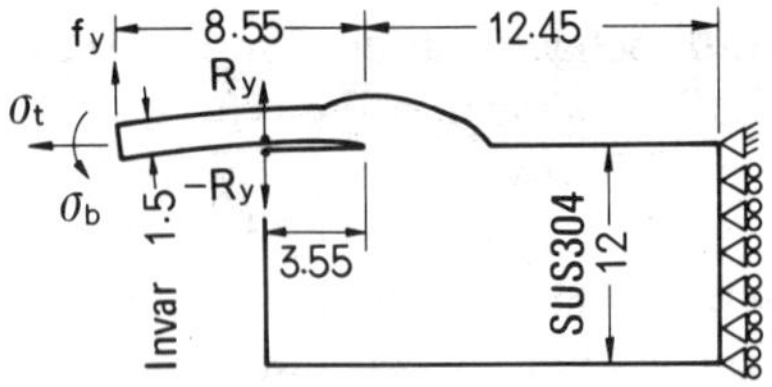

FIG. 16—*Model of Joint D for FEM analysis (dimensions in millimetres).*

Similar analyses were conducted for Joints B and C. When the stress combinations derived from these analyses are expressed by Eq 1, the values of α_1, α_2, and α_3 for each welded joint are as given in Table 4:

$$\Delta\sigma_b = \alpha_1 \Delta\sigma_t; \qquad \Delta f_y = \alpha_2 \Delta\sigma_t; \qquad \Delta R_y = \alpha_3 \Delta\sigma_t \tag{1}$$

Since the throat depth and cross-sectional profile of Joint N are nearly the same as those of Joint A, the results of the analysis for Joint A were applied to Joint N.

Behavior of Joint E—For Joint E, the behavior under cyclic loading was measured using displacement gages in addition to strain gages. Figure 18 shows the relationship of the displacement between the upper and lower chucks of the testing machine to those between *A* and *A′* and between *B* and

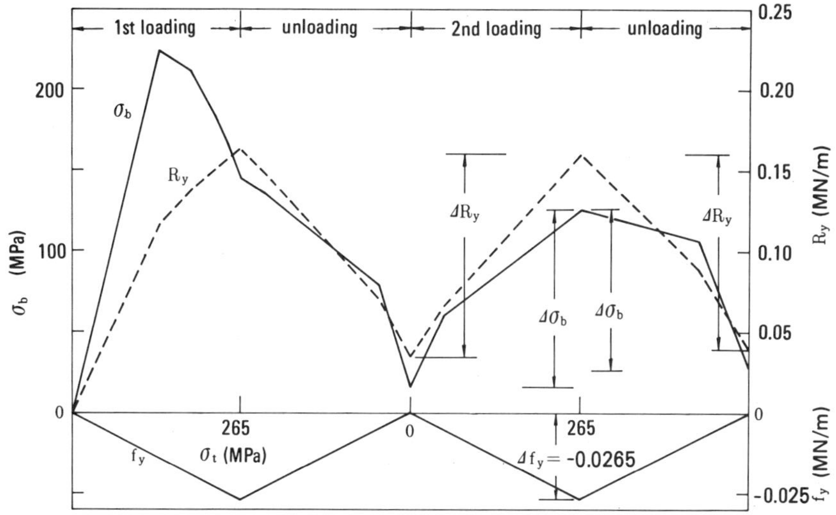

FIG. 17—*Behavior of bending stress, shear force, and reaction force in Joint D specimen at room temperature.*

B' measured by the displacement gages. Figure 19 also shows the relationship of the displacement 2δ between A and A' to the strains ϵ at the positions of A (average of A and A') and B (average of B and B'). In view of the control accuracy of the testing machine, the test for measurement of behavior was conducted statically under the load control. Accordingly, just a small residual deformation was observed after the unloading process of the first cycle.

After the first unloading process, however, the relationship between displacement and strain shows a straight line. On the other hand, the elastic behavior is exhibited at A and A'; therefore the following relationship exists between the bending stress (surface stress) range $\Delta\sigma_b$ and $\Delta\epsilon$:

$$\Delta\sigma_b = E\Delta\epsilon/(1 - \nu^2) \tag{2}$$

Using the linear relationship between $2\Delta\delta$ and $\Delta\epsilon$ after the first unloading shown in Fig. 19, Eq 2 becomes

$$\Delta\sigma_b = -2.25 \times 10^6 \Delta\delta \tag{3}$$

where stress is in megaPascals and displacement is in metres. Furthermore, since a linear relationship exists between Δd and $2\Delta\delta$ as shown in Fig. 18, the relationship between the displacement range between the chucks, Δd, and $\Delta\delta$ is written as

$$\Delta\delta = 0.128\Delta d \tag{4}$$

TABLE 4—*Coefficients for calculation of* K_{eq} *in lap fillet welded joints (stress in MPa, shear force in MN/m, and stress intensity factor in* $MPa\sqrt{m}$*).*

Name of Joint	Temperature	Coefficients of Eq 1			Coefficients for ΔK_I				Coefficients for ΔK_{II}				$\beta = \frac{\Delta K_{eq}}{\Delta \sigma_t}$
		α_1	α_2 (μm)	α_3 (μm)	A_1 ($m^{1/2}$)	B_1 ($m^{1/2}$)	C_1 ($m^{-1/2}$)	D_1 ($m^{-1/2}$)	A_2 ($m^{1/2}$)	B_2 ($m^{1/2}$)	C_2 ($m^{-1/2}$)	D_2 ($m^{-1/2}$)	($m^{1/2}$)
N, A	room temperature	−1.45	72.6	0	−0.0351	−0.0235	604	...	0.000686	0.0119	−289	...	0.0569
	111 K	−1.45	61.5	0									0.0496
B	room temperature	−0.950	51.3	0	−0.0181	−0.0184	459	...	−0.00278	0.0102	−241	...	0.0338
	111 K	−1.18	50.6	0									0.0379
C	room temperature	−0.760	60.0	0	−0.0503	−0.0298	702	...	−0.00392	0.0135	−295	...	0.0351
	111 K	−0.480	40.0	0									0.0235
D	room temperature	0.407	−100	230	−0.00503	−0.00699	200	113	−0.0181	0.00398	−89.8	−32.6	0.0151
	111 K	0.795	−80	255									0.0162

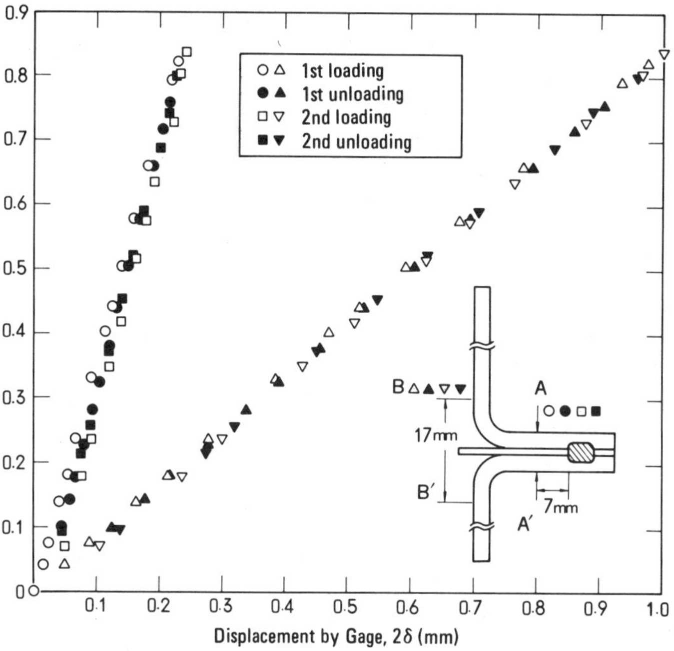

FIG. 18—*Displacement in Joint E specimen at room temperature.*

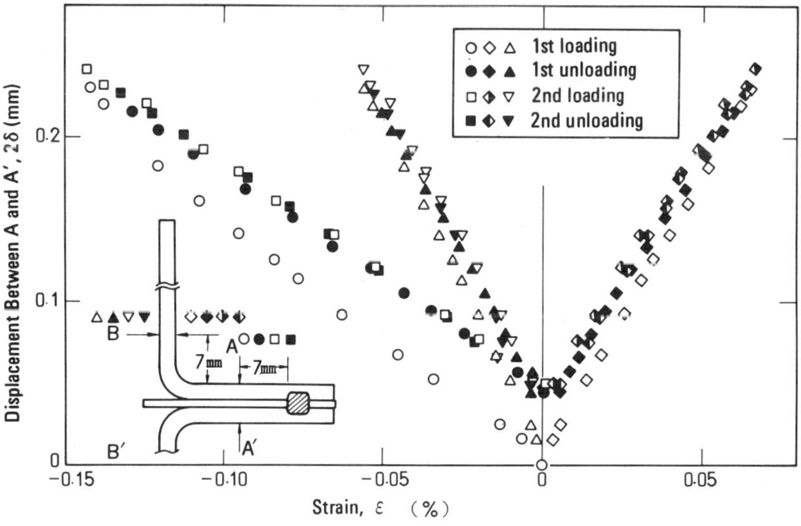

FIG. 19—*Displacement-strain behavior in Joint E specimen at room temperature.*

Using Eqs 3 and 4, the displacement and bending stress ranges at the positions A and A' can be calculated from the displacement range Δd which was controlled in the fatigue test.

Similar measurements were made at 111 K. The behavior at this temperature was nearly the same as that at RT. Equations 3 and 4, therefore, are well applicable to the behavior at 111 K.

Calculation of ΔK_{eq}

Conducting the elastic stress analyses of the models (Figs. 5, 16, etc.) of the lap fillet welded joints under each boundary condition of σ_t, σ_b, f_y, or R_y, stress intensity factors K_I and K_{II} for the weld root can be calculated for individual boundary conditions. Superimposing these results, stress intensity factor ranges for combined boundary condition can be written as

$$\begin{aligned} \Delta K_I &= A_1 \Delta\sigma_t + B_1 \Delta\sigma_b + C_1 \Delta f_y + D_1 \Delta R_y \\ \Delta K_{II} &= A_2 \Delta\sigma_t + B_2 \Delta\sigma_b + C_2 \Delta f_y + D_2 \Delta R_y \end{aligned} \tag{5}$$

The coefficients A_1, B_1, etc., are listed in Table 4.

If the stress combinations of Eq 1 are substituted into Eq 5, the stress intensity factor ranges can be expressed by $\Delta\sigma_t$ only as shown below:

$$\Delta K_I = \beta_1 \Delta\sigma_t; \qquad \Delta K_{II} = \beta_2 \Delta\sigma_t \tag{6}$$

The energy release rate of the crack in which Modes I and II coexist is proportional to $K_I^2 + K_{II}^2$ [9]. Since the equivalent stress intensity factor can be defined to be a stress intensity factor which results in the same energy release rate as that for two modes, K_{eq} is expressed as

$$K_{eq} = \sqrt{K_I^2 + K_{II}^2} \tag{7}$$

The equivalent stress intensity factor range ΔK_{eq}, therefore, can be expressed as follows by substituting Eq 6 into Eq 7:

$$\Delta K_{eq} = \sqrt{\beta_1^2 + \beta_2^2}\ \Delta\sigma_t = \beta \Delta\sigma_t \tag{8}$$

Using this equation, the data for lap fillet welded joints as shown in Figs. 7 and 8 can be converted into ΔK_{eq}.

The coefficients in these equations are summarized in Table 4. Since R_y does not exist in Joints N, A, B, and C, D_1 and D_2 were not obtained for these joints. In case of Joint D, β_1 was taken to be zero in application of Eq 8, because ΔK_I was negative value of which the absolute value is very small.

As for Joint E, arbitrary deformation near the welded part can be expressed by superimposing Figs. 20*a* and 20*b*. Therefore ΔK_I and ΔK_{II} can be derived by superimposing the results of individual calculations of the models shown in Figs. 20*a* and 20*b* and become

$$\begin{aligned} \Delta K_I &= 69.2 \times 10^3 \Delta\delta + 7.75 \times 10^{-3} \Delta\sigma_b \\ \Delta K_{II} &= -26.8 \times 10^3 \Delta\delta - 2.34 \times 10^{-3} \Delta\sigma_b \end{aligned} \tag{9}$$

Using Eqs 3, 4, and 9, ΔK_{eq} becomes

$$\Delta K_{eq} = \sqrt{\Delta K_I^2 + \Delta K_{II}^2} = 56.1 \times 10^3 \Delta\delta = 7.18 \times 10^3 \Delta d \tag{10}$$

Then the data shown in Fig. 9 can be converted into ΔK_{eq}.

Relationship between ΔK_{eq} *and Cycles-to-Failure*

Figures 21 and 22 show the relationship between ΔK_{eq}, which was calculated as described above, and cycles-to-failure N_f at RT and 111 K, respectively. The nominal stress range varies widely depending on the types of welded joints, but the scatter band of ΔK_{eq} for all types of joints is relatively narrow. Even for Joint E, for which the *S-N* diagram in terms of nominal stress range as shown in Fig. 9 is considered to be meaningless, ΔK_{eq} is well within the scatter band for lap fillet welded joints as shown in Figs. 21 and 22. This fact may imply that ΔK_{eq} is an effective parameter for fatigue strength in engineering sense.

It is considered that the resistance against fatigue failure from the weld root can be expressed by ΔK_{eq}. The propagation process of fatigue crack, however, is not taken into consideration in the analysis, so this parameter cannot be applied to the joints which have quite different ratios of crack initiation life to failure life from those of the joints in this study.

Figure 23 shows an appearance of the fracture surface in which the stria-

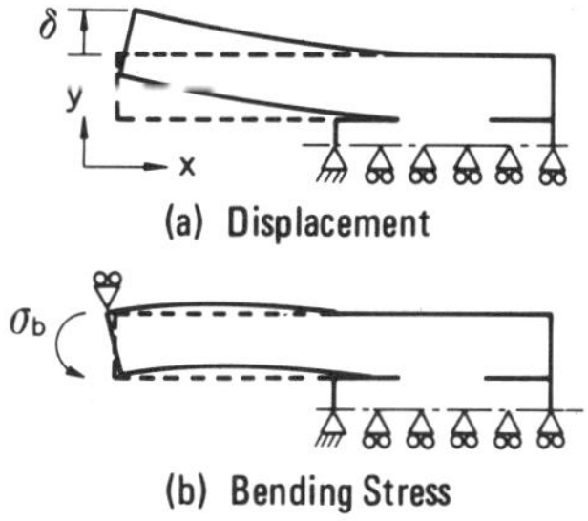

FIG. 20—*Boundary conditions of Joint E model for elastic analysis.*

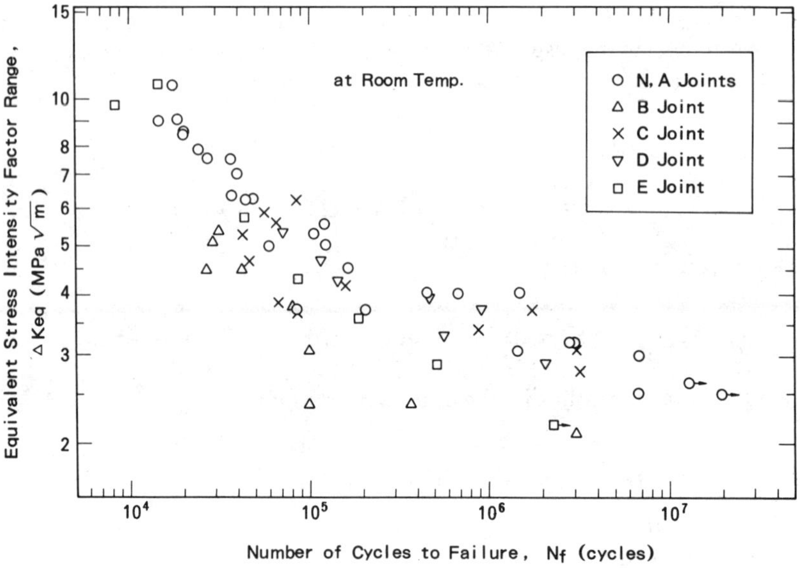

FIG. 21—*Relationships between* ΔK_{eq} *and cycles-to-failure at room temperature.*

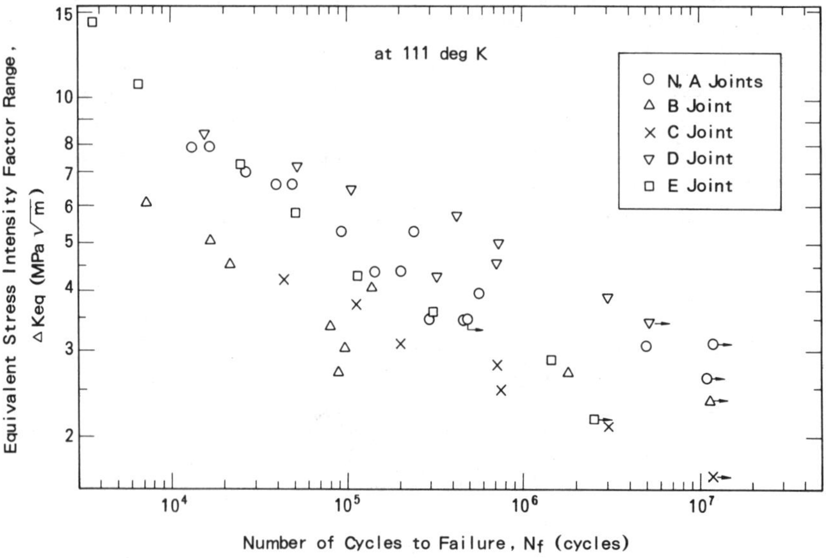

FIG. 22—*Relationships between* ΔK_{eq} *and cycles-to-failure at 111 K.*

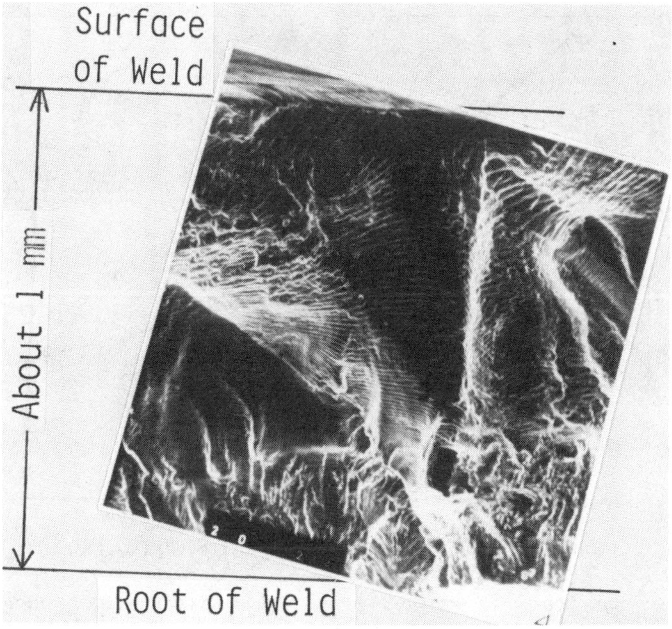

FIG. 23—*Fracture surface of Joint B specimen at room temperature.*

tions are observed through two thirds of the crack propagation distance. A fatigue crack started at the root of the weld and propagated to the surface of the weld. Estimating from the striations, it has taken about 97% of the number of cycles to failure for the crack to grow up to the first one third of the whole distance. This fact was also true for other joints in the case when the striations were observed.

The results obtained at RT and 111 K are summarized in Fig. 24. On the whole, a significant difference is not observed between the results at RT and those at 111 K. As the tensile strength of Invar at 111 K is considerably higher than that at RT, the notch susceptibility may be increased at low temperatures, because the weld root is produced in the form of a very sharp notch. It is considered that the results at both temperatures become nearly equal, because the effects of increased strength and increased notch susceptibility on the fatigue strength at 111 K were cancelled by each other. In Joint D, however, there is a clear difference between ΔK_{eq} at RT and at 111 K. The reason is that ΔK_I is nearly zero and the root does not clearly show its effect as a sharp notch because the overlapped portion of Invar and stainless steel is deformed as if they moved into each other as described previously.

The best-fit lines in Fig. 24 represent the average property of fatigue

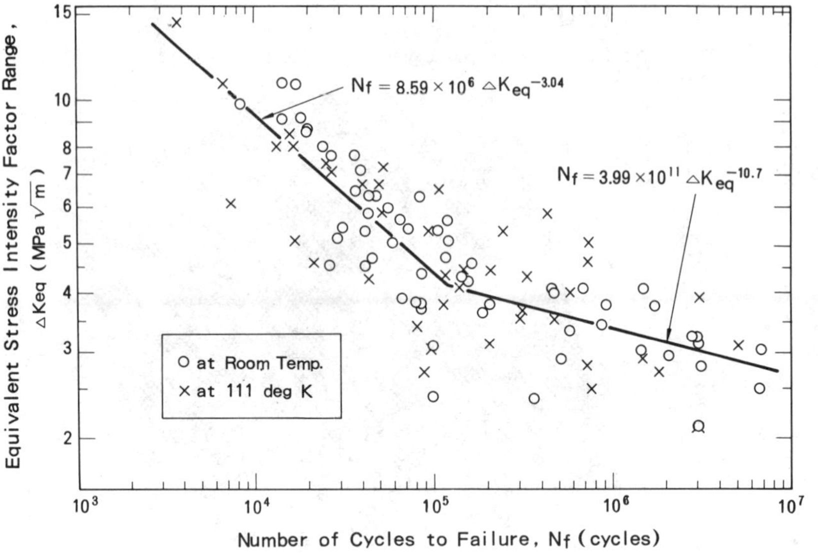

FIG. 24—*Comparison of* ΔK_{eq} *to cycles-to-failure relationships at room temperature and 111 K.*

strength for the weld roots of Invar sheets. Using these relationships, it is possible to evaluate the integrity of a structure against fatigue failure in the case where Invar sheets are applied to the cryogenic structures.

Conclusions

Invar sheet is a useful material for the cryogenic structures. But, in order to assess systematically the integrity of the structure, it is necessary to describe the fatigue strength of the weld root in terms of a parameter that is not dependent upon joint profile and loading conditions.

In this study, the equivalent stress intensity factor range ΔK_{eq} was adopted as such a parameter and its effectiveness was investigated. Fatigue tests were conducted on five types of lap fillet welded joints and one type of resistance seam welded joint of Invar sheets at RT and 111 K. Study was then made of a method for arranging the test results in terms of ΔK_{eq} and on the applicability of linear fracture mechanics concepts. The following conclusions were drawn:

1. In the fatigue test, the combinations of tensile stress, bending stress, and shear force that arise in the specimen in the cycles after the first unloading are stabilized even if large plastic deformation takes place in the first loading process. Moreover, new large plastic deformation does not occur around the weld root after the first unloading. It is therefore reasonable to apply the linear fracture mechanics to the fatigue test in this study, because

one can regard the plastic deformation under cyclic loading as small-scale yielding.

2. The combinations of tensile stress, bending stress, and shear force that arise in the vicinity of a weld in a specimen can be estimated by elastic-plastic FEM analysis to simulate the measured strain behavior. Using these combinations, ΔK_{eq} can be calculated for the weld root.

3. The results of fatigue test arranged in terms of the relationship between ΔK_{eq} and cycles-to-failure are well within a narrow scatter band, regardless of the type of joint. It is considered that this relationship represents the fatigue property of welded Invar sheets which is not dependent upon joint profile and loading conditions.

As described above, it has been verified that ΔK_{eq} is a very effective parameter of fatigue strength for cycles-to-failure above 10^4. For the fatigue life shorter than 10^4 cycles, however, further study is required on the applicability of this method because large cyclic plastic deformation would occur.

Acknowledgments

The authors wish to express grateful acknowledgment to Toyo Kanetsu K. K. which has made available most of the welded joints used in the study. Acknowledgment is also extended to Professor Toshie Okumura, Tokyo Denki University, Ex-Professor Takeyasu Inoue, Yokohama National University, and Professor Hideo Kitagawa, Yokohama National University, for their valuable comments and suggestions.

References

[1] Udoguchi, T., Kasahara, K., Kitagawa, M., Tsunenari, T., Ichihara, T., and Sakamoto, N., *Pressure Engineering*, High Pressure Institute of Japan, Vol. 19, No. 4, 1981, pp. 169-175 (in Japanese).

[2] Okazaki, S., Nose, H., and Tsunenari, T., Technical Review No. 75, Kawasaki Heavy Industries, Ltd., July 1980, pp. 1-9 (in Japanese).

[3] Uemura, T., Kimura, T., Nakamura, Y., and Sakai, K., *Pressure Engineering*, High Pressure Institute of Japan, Vol. 20, No. 5, 1982, pp. 16-23 (in Japanese).

[4] Hanzawa, M., Yokota, H., Ishiguro, T., and Okuma, T. in *Proceedings*, Symposium on Material Strength Study and Design, The Japan Society of Mechanical Engineers, No. 800-8, 1980, pp. 69-72 (in Japanese).

[5] Usami, S. and Shida, S., *Pressure Engineering*, High Pressure Institute of Japan, Vol. 20, No. 2, 1982, pp. 42-50 (in Japanese).

[6] Pook, L. P., NEL Report No. 588, National Engineering Laboratory, United Kingdom, 1975.

[7] Axelsson, K. and Samuelsson, A., *International Journal for Numerical Methods in Engineering*, Vol. 14, 1979, pp. 211-225.

[8] Miyata, H., Shida, S., and Kusumoto, S. in *Proceedings*, Symposium on Mechanical Behavior of Materials, Kyoto, Japan, Aug. 1974, pp. 63-80.

[9] Nuismer, R. J., *International Journal of Fracture*, Vol. 11, No. 2, 1975, pp. 245-250.

James M. Larsen[1]

An Automated Photomicroscopic System for Monitoring the Growth of Small Fatigue Cracks

REFERENCE: Larsen, J. M., **"An Automated Photomicroscopic System for Monitoring the Growth of Small Fatigue Cracks,"** *Fracture Mechanics: Seventeenth Volume, ASTM STP 905*, J. H. Underwood, R. Chait, C. W. Smith, D. P. Wilhem, W. A. Andrews, and J. C. Newman, Eds., American Society for Testing and Materials, Philadelphia, 1986, pp. 226–238.

ABSTRACT: An automated photomicroscopic system for monitoring the growth of small fatigue cracks was developed. Using a specimen that localized crack initiation within a small region, a 35 mm camera mounted on a reflected light microscope was employed to record the growth of small fatigue surface cracks of approximate lengths ranging from 25 μm to 2 mm. A microcomputer was used to control the testing, operate the camera, and record pertinent data. Upon completion of the test, photographs of the small cracks were projected onto a computer digitizing tablet in order to obtain digital crack length data. The precision of the resulting crack length measurements was approximately 1 μm for cracks on the order of 25 μm in length. In order to utilize the available crack growth data fully, a modified incremental polynomial method for reduction of data to the form of da/dN versus ΔK was developed. In general, the photomicroscopic system provided an efficient, cost effective method for monitoring the growth of small, naturally initiated fatigue cracks.

KEY WORDS: automated, crack propagation, fatigue (materials), fracture mechanics, mechanical properties, microcomputer, microcrack, microscope, photomicroscopy, short crack, small crack, test methods

Recently, a number of investigators have reported that small fatigue cracks may propagate at rates that are significantly faster than long cracks subjected to a nominally equivalent crack driving force. Additionally, small cracks have been observed to grow under loading conditions that were well below the threshold stress intensity factor range (ΔK_{th}) established for long cracks in conventional specimens. This anomalous behavior, which has been the sub-

[1]Materials Research Engineer, Materials Laboratory, Air Force Wright Aeronautical Laboratories, AFWAL/MLLN, Wright-Patterson Air Force Base, OH 45433.

ject of several review articles [*1–3*], may lead to anti-conservative damage tolerant life calculations based on linear elastic fracture mechanics (LEFM). The significance of such errors has prompted considerable research in this area; however, the difficulty of generating the appropriate experimental data has hampered progress. Although specialized optical methods have been used effectively to monitor the initiation and early growth of fatigue cracks [*4–7*], conventional optical techniques employed in testing of specimens containing large cracks have generally been inadequate to study small crack behavior, and a number of alternative methods have been used. Typically, length measurement of small cracks has been accomplished through the use of acetate replicas, by periodic examination with a scanning electron microscope (SEM), or by electric potential methods. Valuable data have been produced by each of these methods; however, there are disadvantages associated with each. Although replica and SEM methods offer excellent resolution, the use of these methods to track the propagation of very small cracks may be very time consuming, thereby limiting the number of data that can be realistically obtained. The use of electric potential monitoring generally requires an artificial crack starter to initiate the crack between the electric potential leads. The physical size of the crack starter defines a lower limit of crack size that can be investigated, and the crack starter may alter the normal mode of crack initiation. Clearly, the selection of the crack length measurement technique to be used in the study of small cracks depends on the goals of the research program.

The primary objective of the project discussed here was to study naturally initiated cracks in nominally smooth specimens fatigued under elastic loading conditions over a range of crack sizes on the order of 25 μm to 2 mm. The 25 μm size was chosen as a practical limit of crack size that might be realistically addressed in life prediction of actual structures, while cracks of length greater than approximately 2 mm can generally be monitored with conventional experimental methods. The approach taken utilized optical photomicroscopy for data acquisition, and a microcomputer was employed to perform the potentially tedious tasks of operating the fatigue machine and acquiring the photographic and printed data. This paper discusses the development of the photomicroscopic system and presents representative data of the growth of small surface cracks in a high strength titanium alloy. A modified data reduction procedure that is useful for both short and long cracks is presented.

Experimental Procedure

Crack growth testing was performed using the specimen shown in Fig. 1. This simple geometry, which resembled a design used earlier by Lankford [*8*], employed a mild notch to initiate surface cracks naturally in a localized field. The elastic stress concentration factor (K_t) due to the notches was originally estimated using tabulations from a handbook [*9*], and a subsequent two-di-

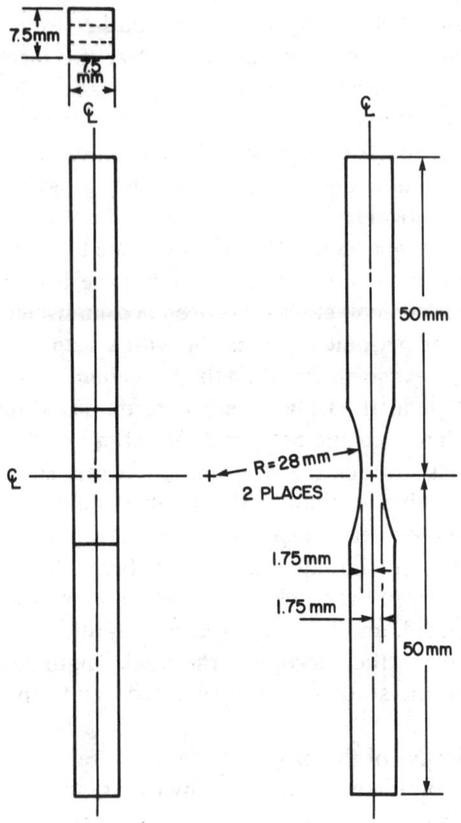

FIG. 1—*Small crack fatigue specimen.*

mensional elastic finite element stress analysis confirmed that $K_t = 1.027$. Thus the through-thickness stress was essentially uniform, while the reduced gage section effectively localized crack initiation within a small region that could be conveniently photographed. In order to eliminate surface residual stresses [*10*] and roughness produced during machining of the titanium alloy tested, the gage sections of all specimens were carefully electropolished to a depth of at least 0.20 mm. The electropolishing also produced a highly reflective surface that enhanced the detection and resolution of small surface cracks, while highlighting the material's microstructure. In order to limit crack initiation to only one of the two notches, one side of the specimen was mechanically polished lightly following the electropolishing. This produced a shallow residual compressive stress that, for the alloy tested, was sufficient to inhibit early crack initiation on the mechanically polished surface, and crack initiation occurred preferentially on the opposite surface.

Crack growth data were acquired photographically using a metallurgical microscope mounted with a 35 mm camera having a 250 frame film magazine and powered by a standard motor drive. Figure 2 shows this setup mounted on a custom-built precision three-dimensional translation stage. For a completely automated test, the film plane magnification was limited to less than 5× in order to obtain a view of the full specimen width. It was found, however, that crack initiation could be conveniently detected by periodic visual microscopic examination of a small region of the specimen notch. Thus, a small crack could be initiated, and the subsequent growth of the crack could be photographed at an increased magnification (usually 20 to 40×) which provided improved image resolution. For the material and test conditions used, surface cracks of half-length less than 25 μm were routinely located in a precracking period of approximately 1 h. Crack visibility was significantly enhanced through the use of reflected light illumination. Initially, microscope illumination was provided by a continuous lighting source; however, improved image resolution was achieved by lighting with an electronic flash. Fatigue cycling was periodically interrupted, and photographs were taken while the specimen was held at maximum load for a period of approximately 1 s. Using Kodak Panatomic-X film, photographs were taken with virtually no loss of the microscope's available image resolution [*11*].

FIG. 2—*Microscope and camera mounted on a three-dimensional translation stage.*

The testing was automated using the system shown schematically in Fig. 3. This system employed an IBM Personal Computer to control all aspects of the testing. By programing a Wavetek 175 function generator, the microcomputer controlled the servo-hydraulic fatigue machine. The camera was controlled by Tecmar PC-Mate Lab Master digital to analog (D/A) converter, and the associated analog to digital (A/D) converter was used to provide feedback of fatigue machine performance throughout the test. During each test the microcomputer recorded pertinent data including the photograph frame numbers and the corresponding values of fatigue cycle count. At the conclusion of the test the film was developed, and a standard photographic enlarger was used to project the negative images of the small crack onto a computer digitizing tablet. The projected image magnification (up to approximately 600×) was calibrated using photographs taken of a microscope stage micrometer slide. The digitizing tablet was used to convert the photographs into data of surface crack length (calculated as the length projected onto a plane normal to the axis of loading), and these data were merged with their associated cycle counts to produce a computer file of surface crack length ($2c$) versus cycles (N). The shapes of the surface cracks were determined by a heat tinting procedure to be discussed later. Using these shapes, the crack growth data were reduced to the form of dc/dN versus ΔK using the surface crack stress intensity solution of Newman and Raju [*12*] and a newly developed modified incremental polynomial method (see Appendix).

In order to assess the capability and accuracy of the photomicroscopic system and data reduction procedure, material and test conditions were selected that would minimize effects of microstructure and crack tip plasticity on the growth of small cracks. All testing was performed on the alloy Ti-6Al-2Sn-4Zr-6Mo in the cast-and-forged condition and heat treated to produce a fine duplex microstructure of equiaxed primary α phase (hexagonal close packed) in a matrix of Widmanstatten $\alpha + \beta$ (body centered cubic) phase as shown in Fig. 4. The forging and heat treatment produced disks of high strength mate-

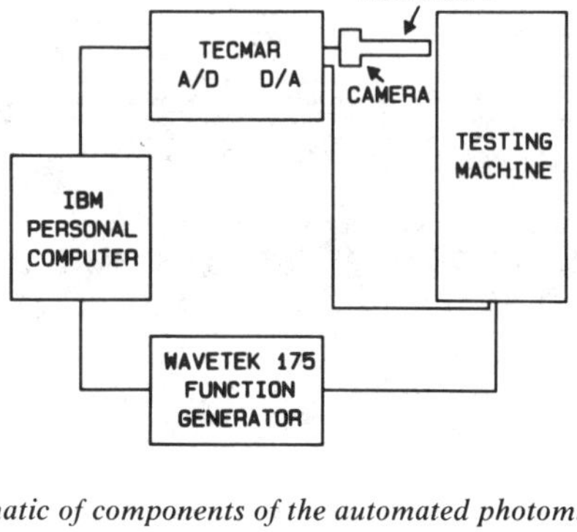

FIG. 3—*Schematic of components of the automated photomicroscopic system.*

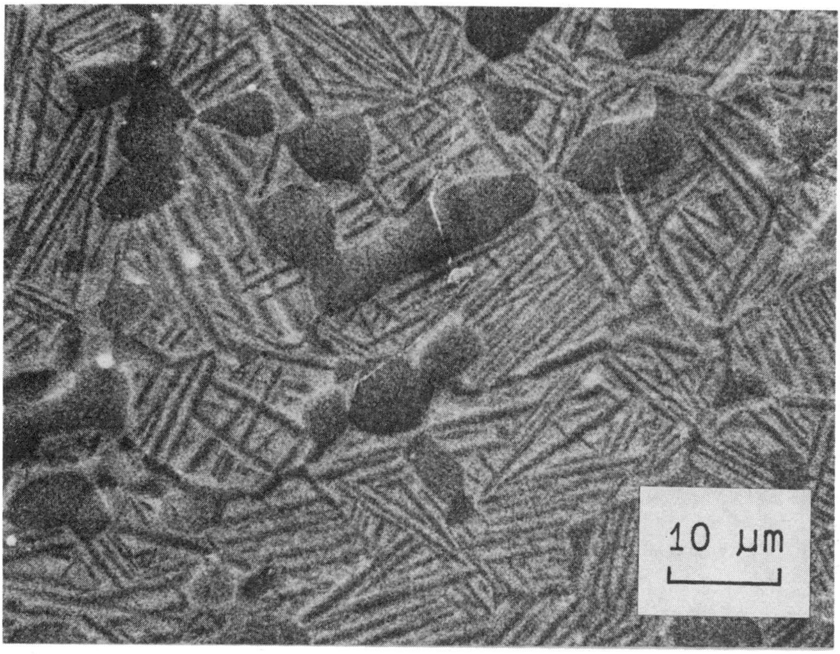

FIG. 4—*Microstructure of Ti-6Al-2Sn-4Zr-6Mo.*

rial (yield strength = 1160 MPa; ultimate strength = 1230 MPa) that was essentially isotropic, as determined by sonic and mechanical testing. Small crack test specimens (Fig. 1) were machined from the disks with the specimen axis oriented circumferentially and the notches oriented radially. This orientation is designated C-R in ASTM Test for Plane-Strain Fracture Toughness of Metallic Materials (E 399). Compact-type (CT) specimens used in the investigation were also of C-R orientation, having the load line oriented circumferentially and crack plane oriented parallel to the disk radius.

Results and Discussion

A typical photograph of a small fatigue crack under maximum load is presented in Fig. 5. For the titanium alloy tested, a single small surface crack generally formed and quickly took on an orientation that was normal to the loading axis. Thereafter, crack growth followed a relatively flat path having a low level of surface roughness. For more than 80% of the specimens, crack initiation occurred at a location well removed from a specimen corner, and the crack propagated as a surface crack until fracture occurred. In order to determine the shape of the surface cracks, a number of tests were interrupted while the cracks were still short, and the specimens were heated to a tempera-

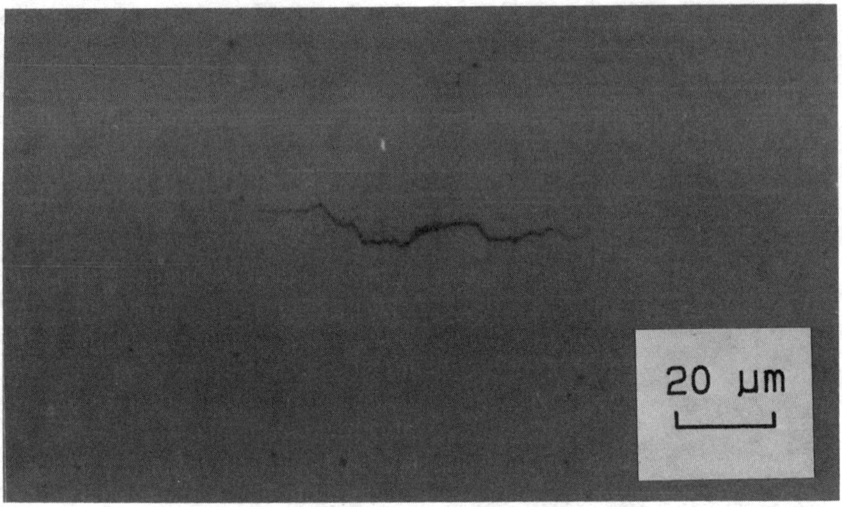

FIG. 5—*Typical photograph of a small fatigue crack in an unetched specimen under load.*

ture not exceeding 400°C to oxidize the crack surface. Upon failure of the specimens, the heat-tinted surface cracks were measured to determine their shapes as projected onto a plane normal to the axis of loading. The resulting data of crack depth (a) versus surface crack half-length (c) are presented in Fig. 6. The cracks were nearly semicircular over the full range of sizes examined and had an average aspect ratio of $a/c = 1.03$.

Figure 7 presents representative data (c versus N) for growth of a small crack subjected to fatigue loading with a maximum stress equal to 75% of the material's yield strength. This specimen was tested in room temperature air at a frequency of 20 Hz with a load ratio (minimum load/maximum load) of $R = 0.1$. The 25 μm initial crack length shown for this specimen was a typical detection size for the photomicroscopic system. For this crack size, the standard deviation of carefully repeated measurements of crack length (c) was approximately 1 μm. This precision was considered excellent, since the value approached the theoretical optical resolution of the microscope. As the crack extended, the increasing level of crack tip plasticity made locating the crack tip more difficult, and the variability in measurement of crack length increased. As shown in Fig. 7, photographs of the small cracks were taken frequently. This provided a measure of variation in crack growth rate as the crack extended, and the large number of data points could be treated statistically to determine mean behavior (see Appendix), thereby minimizing the influence of the random error associated with the crack length measurement.

The data of Fig. 7 were reduced to the form of dc/dN versus ΔK, and these data are presented in Fig. 8 along with data from a number of additional tests

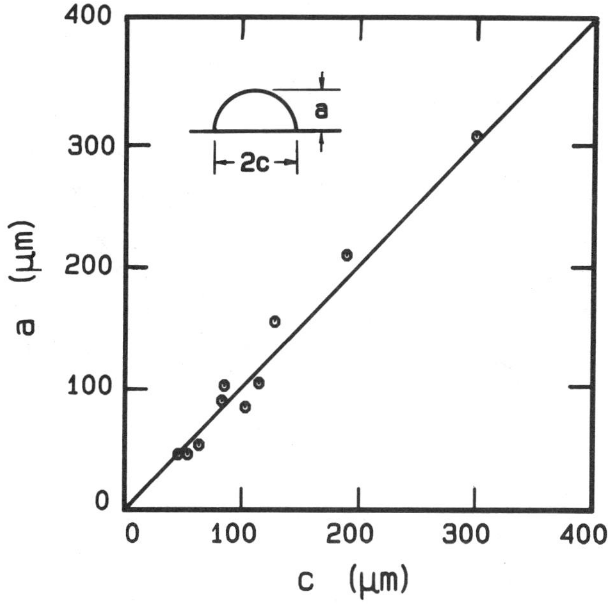

FIG. 6—*Measurements of crack depth* (a) *versus half-length* (c) *for surface cracks in Ti-6Al-2Sn-4Zr-6Mo. The solid line corresponds to* a/c = *1.*

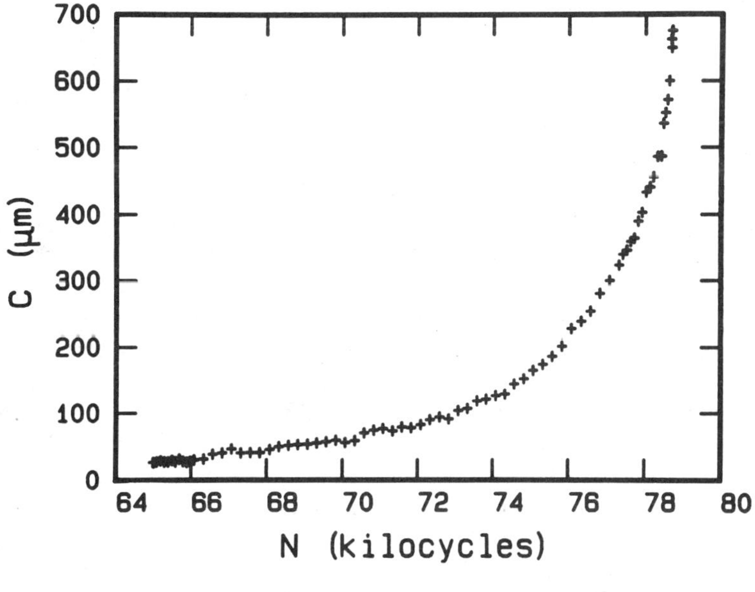

FIG. 7—*Typical data of surface crack half-length* (c) *versus cycles* (N).

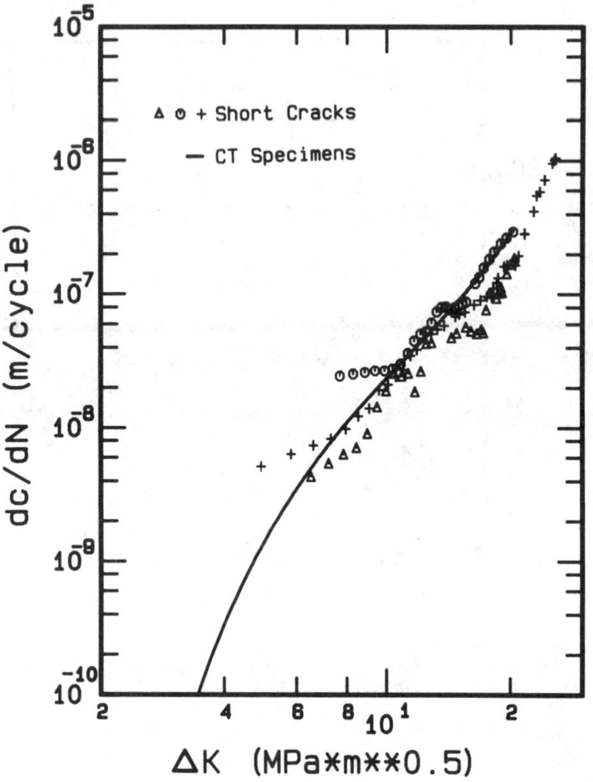

FIG. 8—*Typical crack growth rate data for fatigue with a load ratio of* R = *0.1.*

at the same conditions. Each symbol represents data from a different short crack test. The solid curve represents crack growth in conventional CT specimens that were 10 mm in thickness. Except for the shortest crack lengths, the data from the small surface cracks corresponded quite closely with the long crack behavior. The small cracks did, however, display a tendency to propagate discontinuously. This was an expected result, since the small cracks interacted with only a few microstructural features and were measured on the surface of the specimen, while the propagation of a through-thickness crack in a CT specimen monitored by elastic compliance represented average, or bulk, behavior.

For the data shown, the growth rates for short and long cracks became essentially equivalent above an approximate surface crack size of $c \geq 75$ μm. This transition crack size was in agreement with the independent findings of James and Morris [*13*] on the same material, who found that, above this size, LEFM effectively correlated data of the growth of surface cracks. The 75 μm crack size was slightly larger than the value of the short crack correction factor defined by El Haddad et al [*14*], which was calculated as $\ell_0 = (\Delta K_{th}/\Delta\sigma_e)^2/\pi = 23$ μm, where $\Delta K_{th} = 3.4$ MPa$\sqrt{\text{m}}$ and $\Delta\sigma_e = 400$ MPa are the long crack threshold stress intensity factor range and the fatigue limit respectively. The transition crack size was somewhat larger than the size of the two primary microstructural dimensions, the primary α phase grain size ($\sim 6\ \mu m$) and the prior β phase grain size ($\sim 15\ \mu m$). For all tests performed, the ratio of the size of the monotonic plastic zone (estimated as $(K/\sigma_y)^2/\pi$) [*15,16*] to the crack length was 1/4 or less, and the similar ratio involving the cyclic plastic zone (estimated as $(K/2\sigma_y)^2/\pi$) [*15*] was less than 1/16.

The fine scale of alloy microstructure, limited crack tip plasticity, and limited tendency for anomalous growth of small cracks under the test conditions investigated provided an ideal demonstration of the capabilities of the photomicroscopic system. The excellent agreement between the data for small surface cracks and long cracks in CT specimens illustrated the accuracy of the experimental and analytical procedures. In general, the photomicroscopic system was found to be an accurate, efficient, and cost effective method for monitoring the growth of small, naturally initiated fatigue cracks.

Summary

The development and application of an automated photomicroscopic system for monitoring the propagation of small surface cracks were discussed. A specimen was designed to localize crack initiation within a small region in order to facilitate microscopic examination and data acquisition by photography. A 35 mm camera mounted on a reflected light microscope was used to record the growth of small surface cracks of approximate sizes ranging from 25 μm to 2 mm. Crack visibility and resolution were enhanced by photographing under maximum load with lighting provided by an electronic flash. The photomicroscopic system employed an IBM Personal Computer to control fatigue crack growth testing, operate the camera, and record pertinent data. Upon completion of a test, photographs of the small crack were projected onto a computer digitizing tablet which was used to obtain numerical crack length data. The capabilities of the photomicroscopic system were demonstrated in testing of the high strength titanium alloy Ti-6Al-2Sn-4Zr-6Mo for which the data of small and large cracks were similar for cracks sizes greater than approximately 75 μm. The measurement precision of the complete system was found to be approximately 1 μm for cracks of length of approximately 25 μm. In order to fully utilize the large number of crack growth data generated by the system, a modified incremental polynomial method for reduction of data to the form of da/dN versus ΔK was developed. In general, the photomicroscopic system provided an efficient and cost effective method for monitoring the growth of small, naturally initiated fatigue cracks.

Acknowledgments

This research was performed under Materials Laboratory in-house research project number 2307P102, Air Force Wright Aeronautical Laboratories, Wright-Patterson Air Force Base, Ohio. The author wishes to acknowledge the helpful discussions of Dr. T. Nicholas of the Materials Laboratory and Dean J. C. Williams and Professor A. W. Thompson of Carnegie-Mellon University, Pittsburgh, Pennsylvania. Grateful acknowledgment is also extended to Dr. A. M. Rajendran of the University of Dayton Research Institute, Dayton, Ohio, who performed the finite element stress analysis of the small crack specimen. The research discussed in this paper was performed as partial fulfillment of the requirements for the degree Doctor of Philosophy in Metallurgical Engineering and Materials Science at Carnegie-Mellon University.

APPENDIX

Modified Incremental Polynomial Method of Data Reduction

The use of the automated photomicroscopic system to monitor the propagation of small surface cracks required a nonstandard method to reduce the crack growth data ($2c$ versus N) to the form of dc/dN versus ΔK. In order to study the discontinuous nature of propagation of small fatigue cracks and track any unusual events, photographs were taken frequently, and the resulting crack extension (Δc) between successive photos was often vanishingly small. Although this procedure provided valuable data on the instantaneous crack length, the precision of individual crack length measurements was large compared with Δc, possibly leading to significant errors in the determination of dc/dN. Following guidelines set forth in ASTM Test for Constant-Load-Amplitude Fatigue Crack Growth Rates Above 10^{-8} m/Cycle (E 647), the data should have been edited such that Δc was at least ten times the crack length measurement precision. Although this approach was relatively effective, an alternative method was developed in order to utilize all the available crack length data without introducing artificially large errors in the calculated values of dc/dN.

According to ASTM E 647, an incremental polynomial regression is to be performed such that each successive seven crack length points are fitted with a second-degree polynomial (parabola), and da/dN and ΔK are determined at the fourth of the seven points. Since the standard specifies that the crack length data be taken at approximately equal increments of Δa, and that Δa be ten times the measurement precision, the error in da/dN is maintained approximately uniform as the crack extends. A slight modification to the ASTM incremental polynomial method provided similar control of the error in da/dN, while allowing all the data to be used. Figure 9 illustrates the general approach schematically. The primary difference from the standard method lies in the choice of the data that are regressed incrementally. For the present approach, all crack length data falling within an interval (Δa_{reg}) were fitted with the second-degree polynomial, and ΔK and da/dN were calculated at the midpoint of the interval. This process was performed repeatedly with the sliding regression being successively incremented by an amount Δa_{inc}. For the small crack data presented in this paper, Δa_{inc} was set equal to ten times the measurement precision, and Δa_{reg} was taken as six times Δa_{inc}. The values of Δa_{inc} and Δa_{reg} were chosen to provide good definition of crack growth rate for the cracks at their smallest size, while limiting the variability in da/dN due to errors in the individual measurements.

The accuracy of the modified incremental polynomial method was evaluated in a manner similar to that used by Hudak et al [*17*]. A random error in crack length was

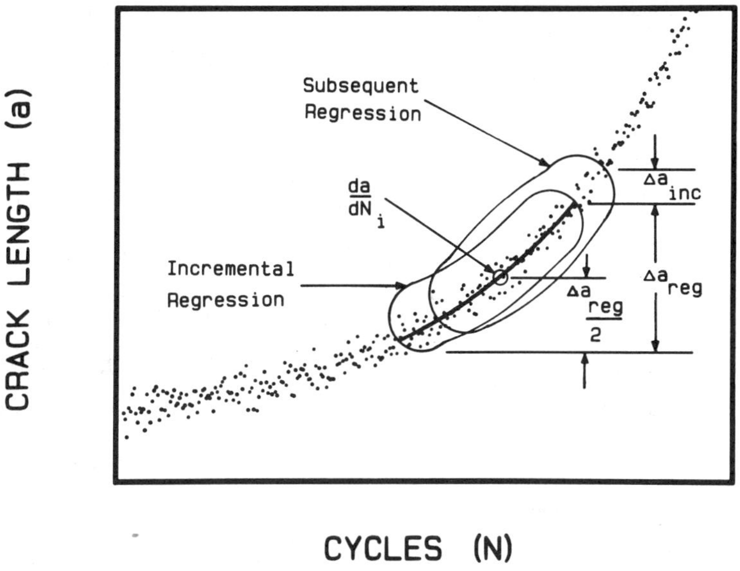

FIG. 9—*Schematic illustrating the approach used in the modified polynomial method for reduction of fatigue crack growth data.*

added to each of the points in an artificial crack growth data set (a versus N) that was generated from a realistic analytical expression giving da/dN as a function of ΔK. The modified incremental polynomial method was used to reduce the artificial data to the form of da/dN versus ΔK, and these data were compared with the original analytical expression. The reduced data and the analytical expression were found to agree extremely well. Based on this success, the modified incremental polynomial method was used to reduce crack growth data from CT specimens that were tested under computer control with crack length determined by compliance measurements. As many as 500 crack length measurements were taken at approximately equal intervals throughout the test, and the crack growth rate data were obtained from this large statistical sampling. In general, the method was found to be very effective in reducing data from a number of different types of crack growth tests. The method provided the added benefit of using a large number of data to define the crack growth rate behavior, thereby achieving a high degree of statistical confidence in the calculations.

References

[*1*] Hudak, S. J., Jr., "Small Crack Behavior and the Prediction of Fatigue Life," *Journal of Engineering Materials and Technology, Transactions of ASME*, Vol. 10, 1981, pp. 26–35.

[*2*] Suresh, S. and Ritchie, R. O., "Propagation of Short Fatigue Cracks," *International Metals Reviews*, Vol. 29, 1984, pp. 445–476.

[*3*] Leis, B. N., Kanninen, M. F., Hopper, A. T., Ahmad, J., and Broek, D., "A Critical Review of the Short Crack Problem in Fatigue," Air Force Wright Aeronautical Laboratories Report AFWAL-TR-83-4019, Wright-Patterson Air Force Base, Ohio, 1983.

[*4*] Papirno, R. and Parker, B. S., "An Automatic Flash Photomicroscopic System for Fatigue Crack Initiation Studies," in *Cyclic Stress-Strain Behavior: Analysis, Experimentation, and*

Failure Prediction, ASTM STP 519, American Society for Testing and Materials, Philadelphia, 1973, pp. 98–108.

[5] Lankford, J., Cook, T. S., and Sheldon, G. P., "Growth of Small Fatigue Cracks at High Temperatures," in *Advances in Fracture Research, ICF-5*, D. Francois et al, Eds., Pergamon Press, New York, 1980, pp. 2449–2456.

[6] Kim, Y. H., Mura, T., and Fine, M. E., "Fatigue Crack Initiation and Microcrack Growth in 4140 Steel," *Metallurgical Transactions*, Vol. 9A, 1978, pp. 1679–1683.

[7] Kung, C. Y. and Fine, M. E., "Fatigue Crack Initiation and Microcrack Growth in 2024-T4 and 2124-T4 Aluminum Alloys," *Metallurgical Transactions*, Vol. 10A, 1979, pp. 603–610.

[8] Lankford, J., "The Growth of Small Fatigue Cracks in 7075-T6 Aluminum," *Fatigue of Engineering Materials and Structures*, Vol. 5, 1982, pp. 233–248.

[9] Peterson, R. E., *Stress Concentration Factors*, Wiley, New York, 1974, p. 37.

[10] Pardee, W. J., Morris, W. F., and Addison, R. C., "Quantitative Nondestructive Evaluation (NDE) for Retirement-for-Cause," Annual Technical Report No. 1 on Defense Advanced Research Projects Agency Contract No. MDA903-80-C-0641, Rockwell International Science Center, Thousand Oaks, Calif., 1981, p. 9.

[11] *Kodak Professional Black-and-White Films*, 2nd ed., Eastman Kodak Co., Rochester, N.Y., 1976.

[12] Newman, J. C., Jr., and Raju, I. S., "Stress-Intensity Factor Equations for Cracks in Three-Dimensional Finite Bodies," in *Fracture Mechanics: Fourteenth Symposium—Volume I: Theory and Analysis, ASTM STP 791*, J. C. Lewis and G. Sines, Eds., American Society for Testing and Materials, Philadelphia, 1983, pp. I-238–I-265.

[13] James, M. R. and Morris, W. L., "Effect of Fracture Surface Roughness on Growth of Short Fatigue Cracks," *Metallurgical Transactions*, Vol. 14A, 1983, pp. 153–155.

[14] El Haddad, M. L., Smith, K. N., and Topper, T. H., "Fatigue Crack Propagation of Short Cracks," *Journal of Engineering Materials and Technology, Transactions of ASME*, Vol. 101, 1979, pp. 42–46.

[15] Rice, J. R., "Mechanics of Crack Tip Deformation and Extension by Fatigue," in *Fatigue Crack Propagation, ASTM STP 415*, American Society for Testing and Materials, Philadelphia, 1967, pp. 247–309.

[16] Broek, D., *Elementary Engineering Fracture Mechanics*, Martinus Nijhoff Publishers, Leyden, The Netherlands, 1984, pp. 91–93.

[17] Hudak, S. J., Jr., Saxena, A., Bucci, R. J., and Malcolm, R. C., "Development of Standard Methods of Testing and Analyzing Fatigue Crack Growth Rate Data," Air Force Materials Laboratory Report AFML-TR-78-40, Wright-Patterson Air Force Base, Ohio, 1978.

A. F. Grandt, Jr.,[1] *A. B. Thakker,*[2] *and D. E. Tritsch*[3]

An Experimental and Numerical Investigation of the Growth and Coalescence of Multiple Fatigue Cracks at Notches

REFERENCE: Grandt, A. F., Jr., Thakker, A. B., and Tritsch, D. E., **"An Experimental and Numerical Investigation of the Growth and Coalescence of Multiple Fatigue Cracks at Notches,"** *Fracture Mechanics: Seventeenth Volume, ASTM STP 905*, J. H. Underwood, R. Chait, C. W. Smith, D. P. Wilhem, W. A. Andrews, and J. C. Newman, Eds., American Society for Testing and Materials, Philadelphia, 1986, pp. 239-252.

ABSTRACT: This paper describes results of an experimental and numerical study concerned with the growth and coalescence of multiple fatigue cracks. A multi-degree-of-freedom algorithm is developed to predict the cyclic growth of separate cracks which develop along the bore of a circular notch. The crack shapes and sizes are allowed to develop naturally as they join into a single flaw which then propagates to failure. Experiments are described with multiflawed specimens of titanium and Waspaloy. The numerical model does an excellent job of predicting the crack coalescence life in all the specimens.

KEY WORDS: fatigue cracks, crack coalescence, life predictions, surface flaws, fracture mechanics

This paper describes results of an experimental and numerical study concerned with the growth and coalescence of multiple fatigue cracks located at notches. As shown schematically in Fig. 1, various combinations of surface and corner cracks may initiate along the bore of a hole in a plate. Upon remote cyclic loading, individual flaws will extend by fatigue, eventually linking up into a single dominant crack which grows to final failure. The right-hand path of Fig. 2, for example, shows two initial surface flaws which first coa-

[1]Professor and Head, School of Aeronautics and Astronautics, Purdue University, W. Lafayette, IN 47907.

[2]Project Manager, Rolls Royce, Inc., Atlanta, GA 30349; formerly employed by Pratt & Whitney Aircraft Group, West Palm Beach, FL 33402.

[3]Former Graduate Assistant, School of Aeronautics and Astronautics, Purdue University, W. Lafayette, IN 47907; currently with Sikorsky Aircraft, Stratford, CT 06601.

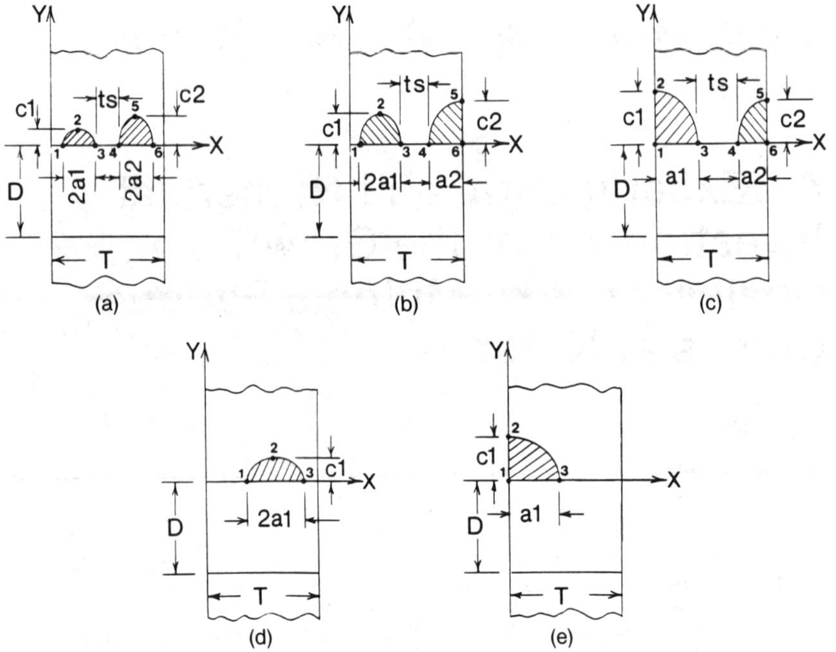

FIG. 1—*Schematic representation of cracks located along bore of a hole.* (a) *Two surface cracks.* (b) *Surface and corner crack.* (c) *Two corner cracks.* (d) *Single surface crack.* (e) *Single corner crack.*

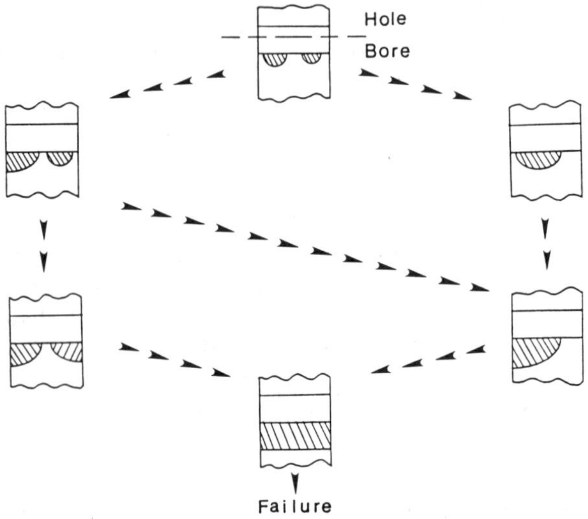

FIG. 2—*Schematic representation of various ways in which two surface cracks at a hole can transition into a single through-thickness crack.*

lesce into a single surface crack, which in turn grows into a corner flaw, and then finally transitions into a through-the-thickness crack. The left-hand route shows another possibility as the initial surface cracks grow into corner flaws before coalescing into a single through-crack.

The manner in which individual cracks grow, coalesce, and propagate to failure depends on initial crack sizes, shape, spacing, and applied load. Although techniques are well developed to predict the cyclic life of large, single cracks, the analysis and interaction of multiple flaws have received less attention. Stress intensity factors have been obtained for interacting cracks by several authors [*1-5*], but application of these results to fatigue life prediction is quite limited [*3,6-7*]. The current paper describes a predictive scheme for computing the total life of multiple cracked notches. A set of experiments is conducted with turbine engine alloy double-notch test specimens, and the numerical model is used to predict the measured fatigue crack growth shapes and lives.

Numerical Model

This section briefly reviews the predictive model developed to analyze multi-cracked notches. Since many of the details are reported elsewhere [*4,6-7*], the discussion here is intentionally brief.

Returning to Fig. 1, a computer program has been written to determine the cyclic life of a multi-cracked hole in a plate loaded in remote tension. Various combinations of semielliptical surface and quarter elliptical corner cracks are assumed to lie along the bore of the hole. The hole diameter is D, the plate thickness is T, and a remote tensile stress σ is applied perpendicular to the cracks, which are assumed to lie in the same radial plane. The semiaxes of one crack are given by dimensions a_1 and c_1, while those of the second flaw are a_2 and c_2. The crack separation (measured along the bore of the hole) is t_s.

Stress intensity factors are determined at the key crack tip locations (Numbers 1 to 6 in Fig. 1). The corresponding crack growth rates are then computed at these points from the fatigue crack growth model for the material of interest. A small crack extension is assumed at one location, and the cycles required for this growth increment are computed. Assuming the same cyclic interval, the appropriate crack extensions are computed at the other tips and added to the existing flaw dimensions. Repeating this procedure in an iterative manner allows the cracks to develop naturally as a function of elapsed cycles. Note that the multi-degree-of-freedom model does not require prior assumptions on crack shape as a function of flaw size.

When crack tips touch a free surface or another flaw, a new uniform surface, corner, or through-the-thickness flaw shape is assumed (as shown schematically in Fig. 2). If two surface cracks come in contact, for example, a single semielliptical surface flaw is assumed to encompass the original flaws. If adjacent corner flaws touch, or the bore dimension of a single crack penetrates completely through the specimen thickness, the model assumes a uniform through-the-thickness crack. This instant transition of coalescing flaws into a uniform surface or corner crack, as well as the rapid transformation into a uniform through-crack, is a conservative assumption employed to simplify the analysis, but is not believed to cause a significant error. Numerical and experimental results for single surface or corner flaws at holes [*8*], for example, indicate that transition to a uniform through-the-thickness crack can occur quite rapidly in the context of total specimen life. In addition, experimental results for coalescing corner cracks in transparent polymer test specimens [*9*] indicate that a uniform through-thickness flaw can develop quite rapidly once linkup occurs.

A key step in the numerical algorithm is the computation of stress intensity factors at the appropriate crack tip locations. The current model employs solutions presented by Newman and Raju [*10*] for *single* surface and corner flaws. Their solutions are readily coded for computer use and provide an effective means for computing K along the border of arbitrary crack shapes. These single crack results are then modified by an "interaction factor" which increases the single crack K value at Locations 3 and 4 to account for the influence of the adjacent crack.

The interaction factor is based on a three-dimensional stress intensity factor solution [*4*] for multiple corner cracks obtained by the finite element-alternating method (Fig. 3). Here the interaction factor γ is defined as the ratio of the hole bore stress intensity factor for a symmetric corner crack divided by the corresponding result for a single corner flaw, and is presented as a function of flaw shape a/c and dimensionless crack separation t_s/a. Polynomial expressions (represented by the solid lines in Fig. 3) were fit through the original finite element-alternating method results (open symbols) and incorporated into the life prediction program. The polynomials are given in the following form, and the coefficients A_i are summarized in Table 1.

$$\gamma = \sum_{i=0}^{9} A_i (t_s/a)^i \tag{1}$$

Although the crack interaction analysis [*4*] was only performed for symmetric corner cracks, the interaction factor is applied in the present scheme to nonsymmetric surface and corner flaw combinations as well. Only the adjacent hole bore locations (Points 3 and 4) have stress intensity factors modified in this manner. The other points employ the original Newman-Raju [*10*] single crack results for K. Once the multiple cracks join into a single surface or corner flaw, the original Newman and Raju [*10*] solutions are once more employed. If the crack continues to grow into a through-the-thickness flaw, then the appropriate through-crack K solution is used. Additional modifications to consider semicircular edge notches used for the current experiments are described in a later section.

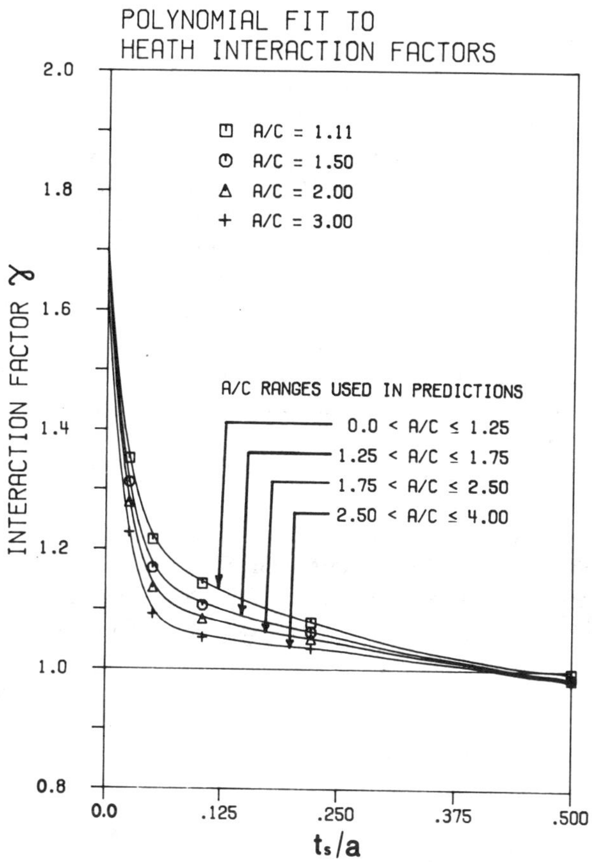

FIG. 3—*Crack interaction factor* γ *for corner cracked holes as a function of crack shape* a/c *and spacing* t_s/a.

TABLE 1—*Coefficients for the ninth-order polynomial representations of the Heath interaction factors.*

Polynomial Coefficients	a/c			
	1.11	1.50	2.00	3.00
A_0	1.6982016	1.7005499	1.6801349	1.6311495
A_1	−20.954392	−23.838719	−24.892797	−25.188174
A_2	375.84251	440.43727	468.38829	482.88140
A_3	−3966.9614	−4689.1275	−5018.3680	−5203.5340
A_4	25939.076	30780.019	33008.082	34290.504
A_5	−108168.84	−128555.52	−137864.00	−143223.29
A_6	287544.18	341837.22	366223.56	380109.65
A_7	−470866.72	−559568.72	−598611.06	−620466.37
A_8	432388.13	513486.98	548416.94	567576.27
A_9	−170229.96	−201990.33	−215374.49	−222555.28

Experimental Procedure

A set of experiments were conducted with double notched multiflawed specimens manufactured from Waspaloy and Ti-6-2-4-6. Both are widely used turbine engine materials in which multiple cracks have been observed. Waspaloy is a heat treatable nickel-based alloy which offers useful strengths at the elevated temperatures seen in the high pressure turbine section of turbine engines, while Ti-6-2-4-6 is an alpha-beta titanium alloy which responds to heat treatment and is used for the fan and lower stages of the compressor. Baseline fatigue crack growth data for the current test conditions were obtained for both materials in the form of the hyperbolic sine [*11*] model.

The experimental program employed the double-notched specimens shown in Fig. 4. This specimen design was selected because the stress gradient at the notch is representative of that typically seen at fastener holes in turbine disks. Small EDM slots were used to initiate fatigue cracks at various points along the bore of one of the edge notches. The slot dimensions were approximately 1 mm (0.04 in.) wide and 0.38 mm (0.015 in.) deep. The crack profiles were marked at periodic cyclic intervals by heat tinting techniques so that flaw dimensions could be determined after specimen fracture. The heat tinting temperatures were selected to mark the crack fronts at least three times during the specimen life without the temperatures being high enough to influence

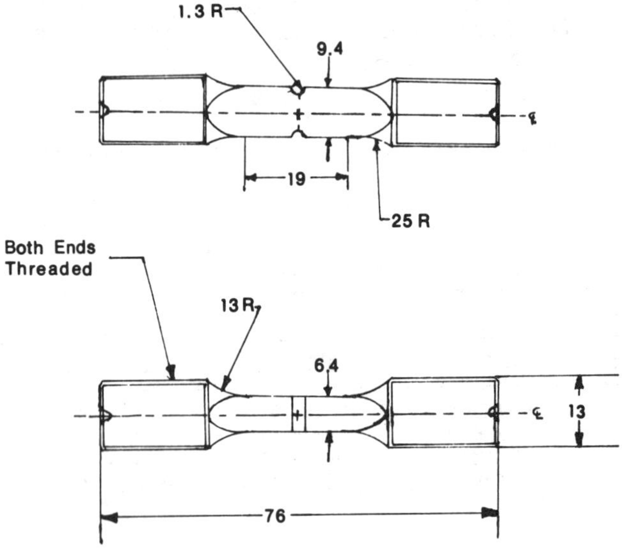

FIG. 4—*Drawing of test specimen showing specimen dimensions and location of two semicircular edge notches (dimensions in millimetres).*

subsequent crack growth at the test temperature. The specimens were heated in clam-shell, resistance-type furnaces, and thermocouples were used to monitor specimen temperature during testing.

Constant-amplitude loading (triangular waveform) was applied to the edge notched specimens by a closed-loop electrohydraulic servocontrolled fatigue machine. The titanium specimens were tested at room temperature, while the Waspaloy tests were conducted at 204°C (400°F) (a relatively low temperature for this material). The cyclic frequency was fixed at 10 cpm and the stress ratio was 0.05 in all cases. The maximum nominal stress per cycle was 345 MPa (50 ksi) for the titanium specimens, and 758 MPa (110 ksi) for all the Waspaloy tests, except for one Waspaloy experiment (Specimen 10A) conducted at a peak stress of 662 MPa (96 ksi).

Surface crack lengths were also measured from surface replicas to obtain crack lengths as a function of elapsed cycles. Composite crack profiles were obtained from the heat tint marker bands using an optical sliding microscope and the surface replica measurements. The initial flaw dimensions (measured after the first heat tinting cycle when fatigue cracks had formed at the EDM slots) are summarized in Table 2. The initial flaw configuration consisted of two surface cracks (Fig. 1*a*) for all tests, except for Specimen 14B which contained initial surface and corner flaws (Fig. 1*b*) and Specimen 16B which had three nearly symmetric surface flaws. In the latter case, the middle crack was located on the centerline, and the two outer flaws were assumed to be mirror images of each other for analysis purposes.

TABLE 2—*Summary of initial crack dimensions for semicircular edge notch specimens (all dimensions given in millimetres).*

Test	Crack 1 a_1	Crack 1 c_1	Crack 2 a_2	Crack 2 c_2	Separation (t_s)	Material
17	0.62	0.28	0.71	0.30	1.70	Waspaloy
18	0.72	0.28	0.79	0.33	1.57	Waspaloy
3A[a]	0.51	0.41	0.51	0.38	0.51	Waspaloy
7A[a]	0.51	0.41	0.51	0.38	0.46	Waspaloy
8A	0.76	0.46	0.94	0.51	1.57	Waspaloy
10A	0.89	0.41	0.97	0.48	1.42	Waspaloy
8B	0.53	0.38	0.53	0.36	1.94	titanium
10B	0.75	0.53	0.67	0.41	0.33	titanium
12B	0.83	0.48	0.57	0.43	1.85	titanium
14B[b]	0.495	0.36	0.61	0.51	0.66	titanium
16B[c]	0.28	0.30	0.29	0.28	0.85	titanium

[a]Dimensions based on EDM slot size. Initial flaws coalesced during precracking before first heat timing cycle.

[b]Crack 2 is a quarter-elliptical corner flaw.

[c]Test with three symmetric surface flaws. Cracks 1 and 3 are mirror images of each other.

Discussion

Six Waspaloy and five titanium specimens were tested to fracture. Fatigue crack growth curves for a typical Waspaloy test are given in Fig. 5; results for a titanium experiment are presented in Fig. 6. The open symbols represent the various crack dimensions (defined in the legend) measured from the fracture surfaces, while the solid and dotted lines are numerical predictions.

Since relatively few heat tinting cycles were used to record crack profiles for measurement after fracture, it was not possible to determine exactly when the cracks joined together in the experiments. Nevertheless, total cycles to failure is known precisely, and the coalescence life is bounded fairly well by the last measurement with separate cracks and the first measurement of a single flaw. In Fig. 5, for example, coalescence occurred sometime between 4500 cycles (when two distinct cracks were present) and final fracture at 5806 cycles. Coalescence in Fig. 6 is again bounded between 4000 cycles, the last time two cracks were observed, and 5091 cycles, when the resulting single flaw caused specimen fracture.

Recall that the predictive analysis was developed for flawed holes in tension plates, and employed the Newman and Raju stress intensity solutions for single surface or corner flaws at holes [*10*] in conjunction with an interaction factor developed for coalescing corner cracks at holes [*4*]. Prior results have indicated that the original algorithm gives a good estimate for crack growth shape and life for single and double cracked holes [*6, 7*].

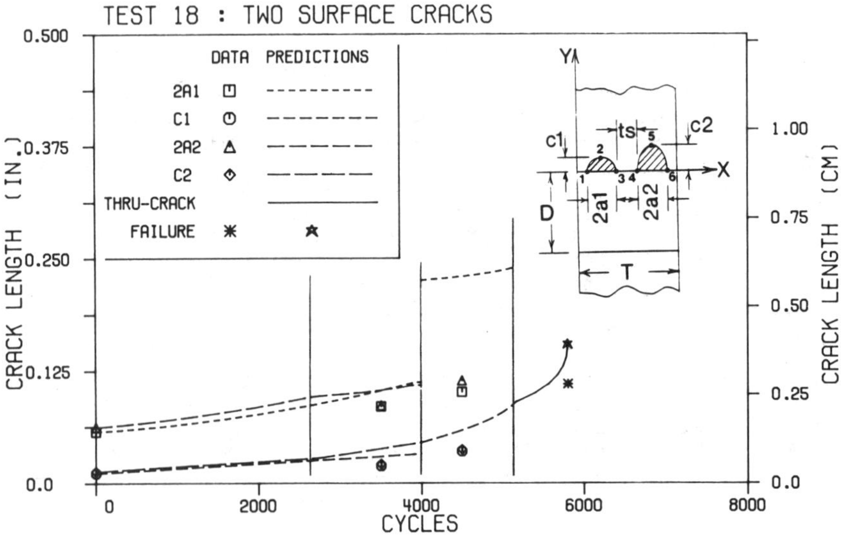

FIG. 5—*Comparison of predicted and measured fatigue crack growth curves for Waspaloy specimen which contained two initial surface cracks along bore of semicircular notch.*

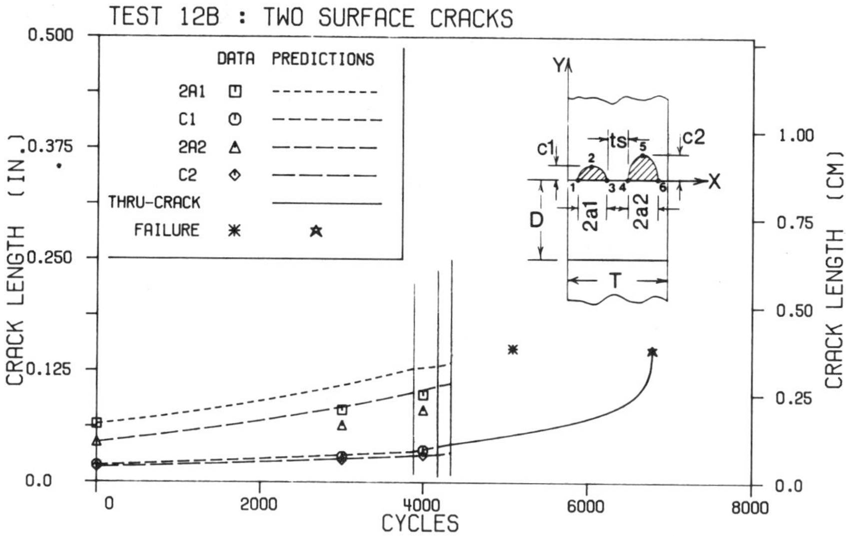

FIG. 6—*Comparison of predicted and measured fatigue crack growth curves for titanium specimen containing two initial surface cracks along bore of semicircular notch.*

To apply the predictive program to the present experiments, it was necessary to obtain stress intensity factors for the double-edge-notch specimen geometry. An edge notch "correction factor" was obtained by determining the ratio of stress intensity factors for through-cracks in the edge-notched specimen divided by K for a through-cracked hole in a plate loaded in remote tension. The through-cracked notch solution was obtained here by a weight function for the flawed hole geometry [*12*] which computed K from the unflawed elastic stress distribution for the double-edge-notch specimen (obtained here by a finite element analysis). The well known Bowie [*13*] analysis was used as K for the through-cracked hole geometry. The dimensionless correction factor F_n is given in Fig. 7 as a function of dimensionless distance c/D from the edge of the hole. Thus stress intensity factors for the edge notch specimen were obtained by multiplying the cracked hole K [*4,10*] by the notch factor F_n from Fig. 7. At the notch bore locations (Points 1, 3, 4, and 6 in Fig. 1) $c/D = 0$ and $F_n = 0.72$, while at Location 2, F_n was evaluated for $c = c_1$ and at Crack Tip 5 for $c = c_2$.

Returning to Figs. 5 and 6, compare the predicted growth of the edge notch cracks with the measured results. Waspaloy Test 18 (Fig. 5) consisted of two nearly symmetric initial surface cracks (Fig. 1*a*) that grew into a through-the-thickness flaw which caused specimen fracture at 5800 cycles. Note that at approximately 2650 cycles, the numerical model predicts that Crack 2 transitions into a corner crack (Fig. 1*b* configuration). The surface and corner cracks then coalesce at approximately 4000 cycles to form a single corner flaw

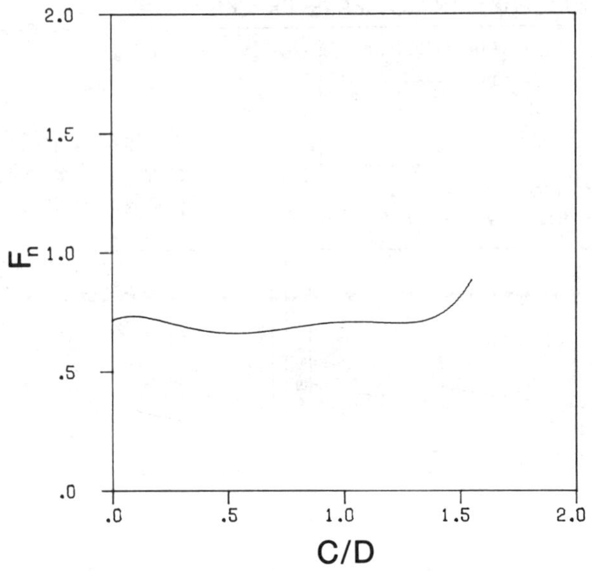

FIG. 7—*Dimensionless "correction factor"* F_n *used to modify cracked hole stress intensity factor solutions for application to semicircular edge-notched specimen.*

(Fig. 1*e*). The corner flaw grows and transitions into a through-the-thickness crack at 5168 cycles, and causes final fracture at 5806 cycles. The total predicted life is 5800 cycles, practically identical to the experimental life.

Although details of the manner in which the individual flaws actually coalesced and propagated to failure are unavailable from the heat tinting profiles, and the predicted crack shapes cannot be evaluated in this region, the predicted coalescence life agrees well with the experimental bounds, and the model does give an excellent estimate of total specimen life.

Typical results for one of the titanium tests are given in Fig. 6. In this experiment, the initial surface flaws (Fig. 1*a*) joined into a through-the-thickness crack which then caused fracture at 5091 cycles. Note that the analysis predicts that the two surface cracks transition into separate corner cracks (Fig. 1*c* geometry), one shortly after the other. The first corner crack transition occurs at 3921 cycles, the second at 4168 cycles. The two corner cracks then join into a single through-crack at 4337 cycles, which finally fractures at 6784 cycles. Again, although it is not possible to determine exactly when coalescence actually occurred, final fracture is predicted fairly well.

Figures 8 and 9 summarize the predictions for the eleven test specimens. Figure 8 plots the computed crack coalescence life versus the actual time required for the cracks to join in each test. Since the exact coalescence period could not be found experimentally, the measured lives are again expressed in terms of the limits discussed previously. With the exceptions of Specimens 3A

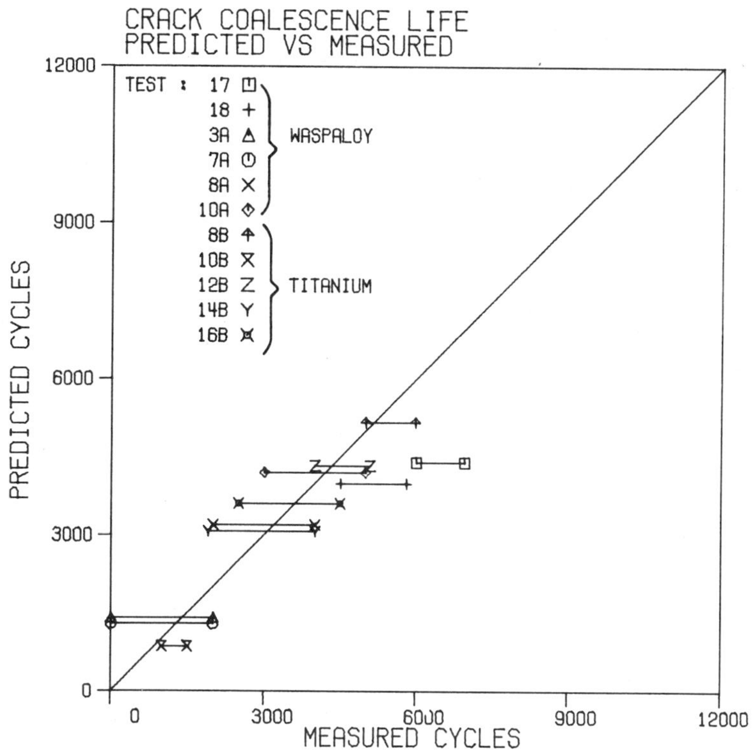

FIG. 8—*Comparison of predicted crack coalescence life with measured cycles for flaw linkup in Waspaloy and titanium test specimens.*

and 7A, the experimental coalescence life is bounded by the last heat tint cycle before coalescence and the first tint following linkup into a single flaw. The cracks formed at the EDM slots in Waspaloy Tests 3A and 7A coalesced before the first heat tinting cycle. Thus the initial crack dimensions used for those two specimens are based on the EDM slots, not on the actual fatigue cracks employed for all the other tests. Total predicted time to fracture is plotted versus actual specimen life (which was precisely measured) in Fig. 9.

Note that the numerical model gives excellent predictions for the coalescence period and for the total cycles to failure. Although the total life calculations usually slightly exceed the experimental results, all predictions are well within a scatter factor of 1.6. Since the predicted and measured coalescence lives agree more closely than the total life results, there may be some error in the prediction of the final single flaw. Although the predicted fracture lives were based on the final crack reaching K_{Ic} and no consideration was given to the possibility of plastic yielding, a more likely source for the difference in predicted and measured total lives is the fact that the stress intensity factors

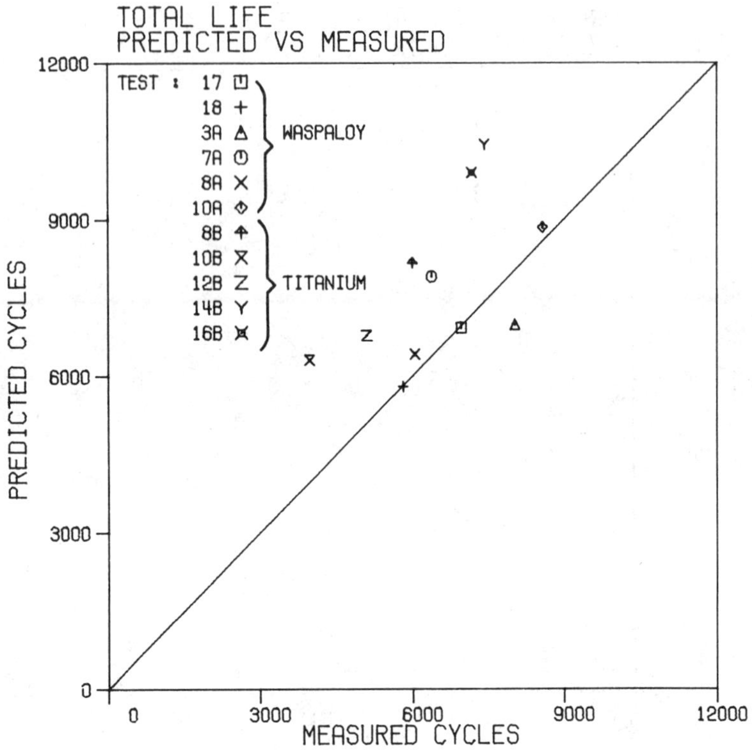

FIG. 9—*Comparison of total predicted life versus measured cycles to failure for Waspaloy and titanium test specimens.*

for the relatively large through-cracks may be in error. This observation is consistent with the fact that the "correction factor" given in Fig. 7 would be expected to be less appropriate for larger flaw sizes. If desired, this portion of the analysis scheme could be improved with a more detailed stress intensity factor solution for large cracks in the edge-notched specimen. Nevertheless, the numerical scheme and approximations employed here give excellent estimates of crack coalescence life and generally good predictions for total specimen life.

Concluding Remarks

This paper has described results of an experimental and numerical study of the growth and coalescence of multiple cracks at notches. An algorithm is presented to predict the growth and coalescence of individual surface and corner cracks along the bore of a notch, their coalescence into a single crack, and the propagation to fracture of the final flaw. The analysis employs conventional fracture mechanics techniques to predict the size and shape of the individual cracks. The Newman and Raju [*10*] stress intensity factor solutions for single surface or corner cracks are modified by an interaction factor [*4*] to account for the presence of an adjacent flaw. A further modification is described to convert the cracked hole results to the semicircular edge-notched specimen geometry employed for the experimental program.

Experiments were conducted with Waspaloy and titanium specimens which contained initial surface and/or corner flaws along the bore of a notch. Heat tinting techniques were used to mark the crack profile as the specimens were cycled to failure. Post-fracture measurements of the specimen dimensions are compared with results from the predictive model and, in general, agree quite well with the analysis. The initial life required for the individual flaws to grow together is predicted very well for all the experiments. The remaining life of the single crack is slightly overpredicted in most cases, although the total life estimates are generally quite good.

The results of these experiments, along with those described in another effort with multicracked holes in transparent polymer specimens [*6*], indicate that the analysis procedure is capable of giving good estimates of the period required for cracks to coalesce into single flaws. The Newman and Raju [*10*] stress intensity factors, when modified by the crack interaction analysis described in Ref *4*, give reasonable K approximations along the border of multiple hole flaws. An engineering factor developed to apply these flawed hole results to semicircular edge notches performed well for small flaws, although the procedure may be less accurate for large cracks.

Acknowledgments

The experimental program was performed by A. Thakker while employed by the Government Products Division of the Pratt & Whitney Aircraft Group. The authors gratefully acknowledge the financial support provided by the Air Force Office of Scientific Research through Grant AFOSR-82-0041 and by the AFWAL Materials Laboratory through Contract F33615-80-C-5160. Major D. A. Glasgow served as the AFOSR technical monitor, and Dr. W. Reimann as the AFWAL project engineer.

References

[*1*] Kamei, A. and Yokoburi, T., "Two Colinear Asymmetrical Elastic Cracks," *Report of the Research Institute of Strength of Materials*, Tohoku University, Japan, Vol. 10, Dec. 1984.

[*2*] Murakami, Y. and Nemat-Nasser, S., "Interacting Dissimilar Semi-Elliptical Surface Flaws under Torsion and Bending," *Engineering Fracture Mechanics*, Vol. 16, No. 3, 1982, pp. 373–386.

[*3*] Chang, R., "On Crack-Crack Interaction and Coalescence in Fatigue," *Engineering Fracture Mechanics*, Vol. 16, No. 5, 1982, pp. 683–693.

[*4*] Heath, B. J. and Grandt, A. F., Jr., "Stress Intensity Factors for Coalescing and Single Corner Flaws Along a Hole Bore in a Plate," *Engineering Fracture Mechanics*, Vol. 19, No. 4, 1984, pp. 665–673.

[5] Baek, T. H., Burger, C. P., and Voloshin, A. S., "Stress-Intensity Factors for Interacting Semicircular Surface Cracks," in *Proceedings*, Fifth International Congress on Experimental Mechanics, Society for Experimental Stress Analysis, Montreal, 10-15 June 1984.

[6] Grandt, A. F., Jr., Perez, R., and Tritsch, D. E., "Cyclic Growth and Coalescence of Multiple Fatigue Cracks," in *Advances in Fracture Research*, Proceedings of the Sixth International Conference on Fracture, New Delhi, Dec. 1984, Vol. 3, pp. 1571-1578.

[7] Tritsch, D. E., "Prediction of Fatigue Crack Lives and Shapes," M.S. thesis, School of Aeronautics and Astronautics, Purdue University, W. Lafayette, Ind., Aug. 1983.

[8] Grandt, A. F., Jr., Harter, J. A., and Heath, B. J., "The Transition of Part-Through Cracks at Holes into Through-the-Thickness Flaws," in *Fracture Mechanics: Fifteenth Symposium, ASTM STP 833*, American Society for Testing and Materials, Philadelphia, 1984, pp. 7-23.

[9] Perez, R., "Initiation, Growth, and Coalescence of Fatigue Cracks," M.S. thesis, School of Aeronautics and Astronautics, Purdue University, W. Lafayette, Ind., Aug. 1983.

[10] Newman, J. C. and Raju, I. S., "Stress-Intensity Factor Equations for Cracks in Three-Dimensional Finite Bodies," NASA Technical Memorandum 83200, NASA Langley Research Center, Hampton, Va., Aug. 1981.

[11] Annis, C. G., Jr., Wallace, R. M., and Sims, D. L., "An Interpolative Model for Elevated Temperature Fatigue Crack Propagation," Technical Report AFML-TR-76-76, Air Force Materials Laboratory, Wright-Patterson Air Force Base, Ohio, Nov. 1976.

[12] Grandt, A. F., Jr., "Stress Intensity Factors for Some Through-Cracked Fastener Holes," *International Journal of Fracture*, Vol. 11, No. 2, April 1975.

[13] Bowie, O. L., "Analysis of an Infinite Plate Containing Radial Cracks Originating from the Boundary of an Internal Circular Hole," *Journal of Mathematics and Physics*, Vol. 35, 1956, pp. 60-71.

W. N. Sharpe, Jr.,[1] *and J. J. Lee*[1]

Near-Tip Crack Displacement Measurements During High-Temperature Fatigue

REFERENCE: Sharpe, W. N., Jr., and Lee, J. J., **"Near-Tip Crack Displacement Measurements During High-Temperature Fatigue,"** *Fracture Mechanics: Seventeenth Volume, ASTM STP 905,* J. H. Underwood, R. Chait, C. W. Smith, D. P. Wilhem, W. A. Andrews, and J. C. Newman, Eds., American Society for Testing and Materials, Philadelphia, 1986, pp. 253-264.

ABSTRACT: A laser-based interferometric technique was used to measure crack opening displacements near the tips of fatigue cracks in superalloy compact tension specimens at 23 and 650°C. Measured compliances showed reasonable agreement with predictions of linear elastic fracture mechanics at both temperatures. The opening load ratios were easily determined by the minicomputer-controlled measurement system and where found to be independent of temperature, crack length, and precracking level when they were measured at positions more than half the specimen thickness away from the tip.

KEY WORDS: fatigue, compliance, closure, superalloy, crack opening displacement, interferometry, high temperature

The details of deformation and displacement around the tip of a fatigue crack arc obviously of considerable interest, but the small size of the region makes measurement difficult. Ideally one would like to study the deformation at various points along the crack front, but this is possible only in transparent specimens. Pitoniak et al [*1*] and Barker and Fourney [*2*] have examined the displacement behind the tip of a crack by optical techniques in plastic specimens.

For metals, one is restricted to surface strain or displacement measurements, and again various optical techniques are used. Examples are to be found in the works of Evans and Luxmoore [*3*] and Macha et al [*4*]. The latter work measured both in-plane and out-of-plane displacements and used the

[1]Professor and Chairman, and Graduate Assistant, respectively, Department of Mechanical Engineering, The Johns Hopkins University, Baltimore, MD 21218.

resulting data in a study of the closure phenomenon. Out-of-plane displacements were obtained by optical interference between the polished specimen surface and an optical flat. In-plane displacements were measured by optical interference from closely spaced indentations illuminated by a laser. This latter technique was an improvement on an earlier study by Sharpe et al [5] based on the same optical principle but using reflective lines scribed parallel to the crack.

There are other methods of measuring displacements near the tip of a crack, ranging from the scanning electron microscope technique of Nisitani and Kage [6] to the small mechanical clip gage of Ohta et al [7]. These techniques have their advantages but are somewhat cumbersome. The electron microscopy method requires the preparation of replicas, while the clip gage must be removed from a specimen while the crack is being grown.

The study of creep and fatigue at high temperature has gained importance with the initiation of a "retirement for cause" approach to turbine disks in aircraft jet engines. A more complete understanding of the material's and the component's response to loading in the presence of a crack is demanded, and this has placed an additional burden on the experimentalist wishing to examine in detail the crack tip behavior at high temperature.

This paper is a report of some crack opening displacement measurements at positions within a few millimetres of the tip of a fatigue crack in a superalloy specimen at 650°C. The technique is laser-based interferometry from two indentations astride the fatigue crack, the same technique as used earlier by Macha et al [4]. However, the system is controlled by a minicomputer that permits real-time measurements and storage of the data for subsequent analysis. Emphasis is placed on measuring the displacements to permit observations of the behavior on the surface as the crack is loaded. The data acquired can be used to establish an opening load in the closure sense, and the effects of measurement location, temperature, and crack length are reported. Compliances are also easily determined, and the feasibility of using these near-tip displacements as an indicator of crack growth is discussed. The measurement system has been adequately described elsewhere, so the emphasis here is on the results rather than the techniques.

Measurement System

The principle underlying the laser-based interferometric technique for measuring in-plane displacements is quite simple. Two very small indentations are placed approximately 100 μm apart on the specimen surface with a Vicker's microhardness tester. When they are illuminated using a laser, interference fringe patterns are produced because of the path differences between light rays reflecting (diffracting) from the sides of the indentations. Motion of these fringes is proportional to the relative displacement between the two indentations. Fringe motion is monitored on a realtime basis by servocontrolled scanning mirrors and by photomultiplier tubes which are controlled by a minicomputer. The minicomputer does the appropriate calculations and outputs a voltage signal equal to the relative displacement between the indentations. It also controls the electrohydraulic test machine and stores the load-displacement data for later analysis.

The minicomputer-controlled Interferometric Strain/Displacement Gage (ISDG) has been used to measure the cyclic plastic strains at the roots of notches at high temperature [*8*]. The two reflecting indentations were placed 100 μm apart at the root of the notch, and strain was recorded at 60 load levels during the tension-compression cycle. Essentially the same system has been used in a recent study [*9*] of the threshold values for creep and fatigue. In this work, the indentations were placed across the fatigue crack to measure the crack opening displacement near its tip. In both cases, the material was Inconel 718 and the tests were conducted at 650°C. Figure 1 is a photomicrograph of a pair of indentations used for the measurements reported herein.

References *8* and *9* describe the ISDG in some detail, so only the specifications established during those studies are given here:

Gage length: 50 to 200 μm
Range: 100 μm
Resolution: 0.02 μm
Sampling rate: 10/s
Relative uncertainty: ±3%
Temperature range: to 730°C on certain superalloys

This unique measurement technique offers the advantages of a short gage length and the capability of operation in a hostile environment such as high temperature. The specimen is heated by an electric resistance furnace mounted on the test machine. This furnace has three quartz windows for the incident laser beam and the two exiting fringe patterns.

The ISDG is not fast enough to record load-displacement for every cycle of a reasonable fatigue test, so tests are periodically interrupted for a recording cycle. This cycle takes 24 s, recording 120 data points on loading and 120 during unloading. The data reported in this work were recorded for cracks that were grown to a particular length. The load-displacement cycles were then run at values much less than the maximum fatigue load. This permits measurement of the load-displacement behavior at several positions behind the crack tip and on both sides of the specimen.

Material and Specimens

The specimen material is Inconel 718, a nickel-based superalloy widely used in the turbine disks of aircraft jet engines. It received a standard heat treatment of 968°C for 1 h, was quenched at 718°C for 8 h, and then air cooled. In another study [*10*], the elastic moduli for a similarly treated material were determined to be 203EXP3 MPa and 161EXP3 MPa at room tem-

FIG. 1—*Photomicrograph of a pair of indentations across a fatigue crack after testing at 650° C. The two indentations are 100 μm apart.*

perature and 650°C respectively. Yield stresses were found to be 997 MPa at 23°C and 844 MPa at 650°C.

The specimens were the standard compact tension geometry with overall dimensions of 48 by 50 mm and a dimension of $W = 40$ mm from the load line to the furthest edge. They were machined to a thickness of 10.0 mm from 12.5-mm-thick plate with the notch oriented in the T-L configuration. All precracking of the specimens was done at room temperature; the loads are given in the Results section. Before precracking, the specimens were mechanically polished on various grades of silicon carbide paper with the final polish using 0.5 μm alumina paste on a felt wheel. After polishing, the specimens were preoxidized at 650°C for 1 h; this process gives a protective coating to the specimen surface that improves the reflectivity of the indentations at temperature.

One specimen, number 7-035, was used for all the tests reported herein. After testing was completed, the specimen was broken open and the crack lengths measured. Figure 2 is a sketch of the crack shapes.

Data Analysis

A typical set of load-displacement records at five positions behind the crack tip at 650°C is shown in Fig. 3. The increased compliance as one moves back

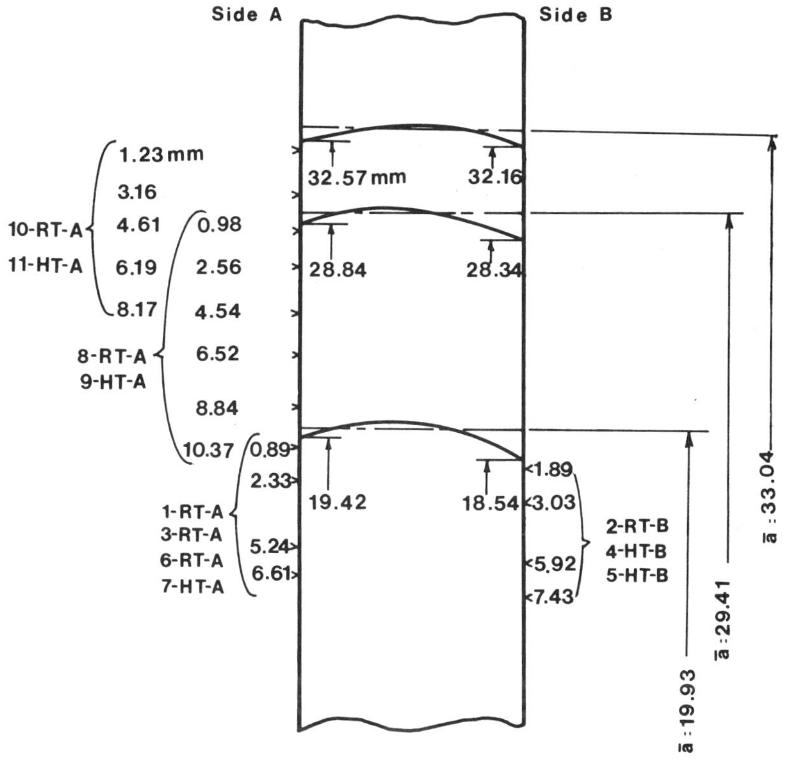

FIG. 2—*Sketch of the crack shapes measured after testing was completed. The larger labels refer to test numbers, the smaller ones are dimensions in millimetres. The three average crack lengths are measured from the load line of the compact specimen. Other dimensions are distances behind the average crack position.*

from the tip is quite obvious as is the decrease in the opening load. Computer programs, described below, were written to establish the compliance and the opening load in an unbiased manner.

The compliance was established by a least-squares straight line fit to the upper portion of the load-displacement curve. As many data points were used as would yield a fit with an "R" value of 0.9900 or higher (the number of data points used would vary among records). Then, the uppermost data point was taken as zero displacement and the fitted straight line added to the displacement values for lower loads. Figure 4 is a plot of the original and the reduced data. Note that there is a slight difference between the fitted straight line for loading and unloading.

The opening load was arbitrarily established as that load at which the reduced displacement data deviated from zero by more than 10% of the maximum reduced displacement. A deviation on the order of 10% is necessary in order to avoid spurious results from noisy data. This automated approach is less subjective than visual determination of the onset of linearity.

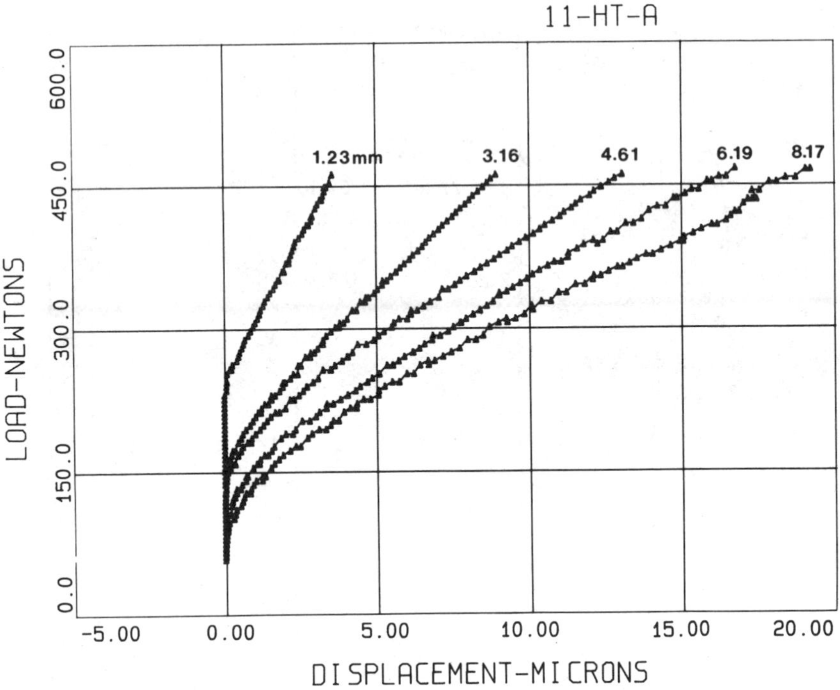

FIG. 3—*Load-displacement records at various positions behind the tip of the longest crack.*

Compliance data are presented in the following section, and comparisons are made with predictions by linear elastic fracture mechanics. Those theoretical compliances are calculated by the procedures developed by Saxena and Hudak [*11*]. The crack lengths used in the calculations were the surface crack lengths measured before the tests and the average crack length determined after the specimen was broken open (the average was computed from the length at the midpoint and the two quarter-points as prescribed by ASTM E 399).

Results

The test numbers are listed in Fig. 2 and are selected to aid the reader in interpreting the following plots. The test numbers are in chronological order, and the letters "RT" and "HT" represent tests at room temperature and 650°C respectively. The last letter refers to either Side A or Side B. Test 1 was at 23°C on Side A at four positions behind the crack tip. The next test, Test 2, was at 23°C on Side B at four positions. Test 11 was the last test at 650°C on Side A at five positions for a longer crack. The specimen was turned around

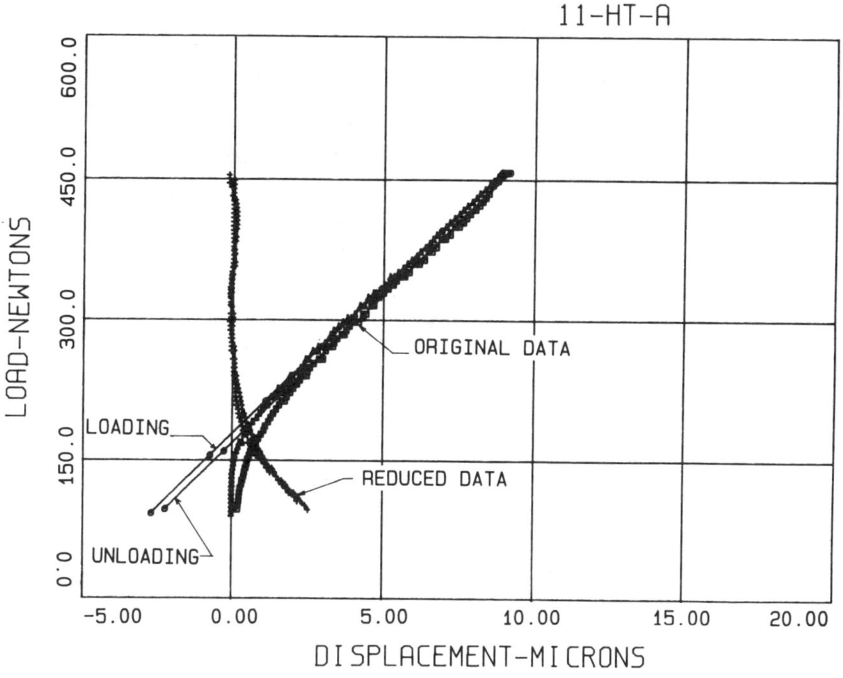

FIG. 4—*Original load-displacement data and the reduced data used for determining the opening load ratio.*

in the grips to measure on the two sides. In the following plots of opening load ratios, the maximum load corresponds to the peak load when the specimen was precracked at room temperature.

Compliances

Figure 5 presents the results of compliance measurements for the first crack with an average length of 19.93 mm corresponding to an a/W ratio of 0.498. The precracking was done with an R-ratio of 0.1 and a ΔK of 14.8 MPa m$^{1/2}$. Tests were run on Sides A and B at room and high temperature. Three tests were run on Side A at 23°C and two tests on Side B at 650°C to evaluate the reproducibility of the measurements. The reproducibility was better than ±3% except for the measurements closest to the crack tip where it was slightly larger. The specimen was removed from the test machine between experiments.

Theoretical compliances as a function of position behind the crack tip are plotted in Fig. 5. First, the compliances for Sides A and B are calculated at 23°C under the assumption of plane stress; note that the surface cracks differ

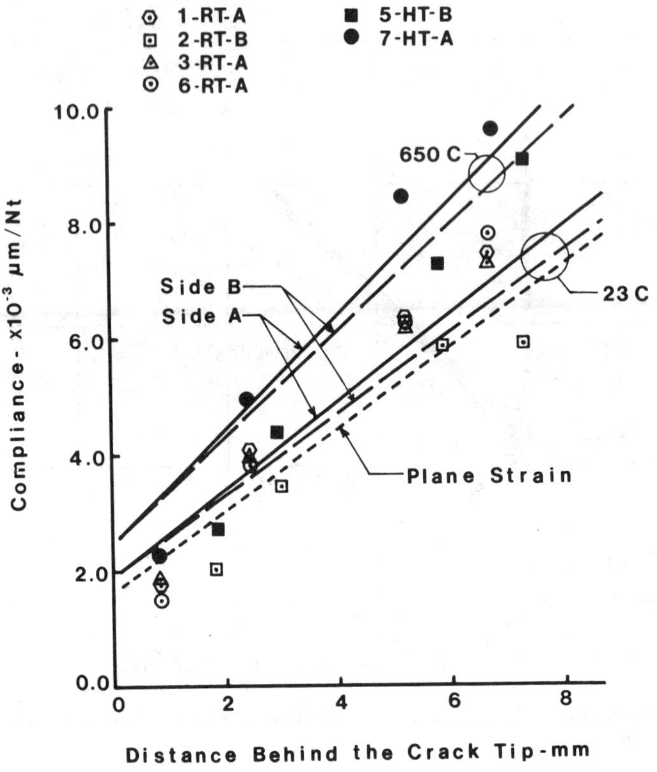

FIG. 5—*Compliances measured at various positions behind the crack tip at 23° C* (open symbols) *and 650° C* (filled symbols). *The lines are calculated compliances: for Sides A and B at both temperatures using plane stress, and for the average crack length at 23° C using plane strain.*

in length by almost 1 mm. Then, the compliance was calculated assuming plane strain and using the average crack length. The measured values for Sides A and B agree reasonably well except for the data point furthermost from the tip on Side B. That measurement was repeated and verified.

The measured compliances at high temperature show a similar behavior and reflect the reduced elastic modulus at elevated temperature. Note that Test 6 at 23°C was interspersed between Tests 5 and 7 at 650°C. The reproducibility of the data at 23°C confirms that the loads applied during the single load-displacement cycle were well within the elastic range and that the thermal cycling did not change the behavior of the cracked specimen.

The measurement positions of Fig. 5 are not even a full specimen thickness behind the crack tip; the furthermost is 7.43 mm back, and the specimen is 10 mm thick. Given that the crack lengths on the two sides are almost 1 mm different, it is not too surprising that the measured compliances do not agree better with the theoretical ones.

Compliance measurements for the other two cracks shown in Fig. 2 yielded similar results. In fact, the measured compliances for the longest crack agreed quite well with the predicted surface values, but note that it is a very straight crack.

Opening Load Ratios

Figure 6 presents the opening load ratios for the tests in Fig. 5. Again, the replicated data on Side A at 23°C are quite reproducible before and after Test 5. If closure is due to residual stresses, that fact indicates there has been no stress relief during Test 5-HT-B. This was confirmed in Test 7-HT-A by making measurements 45 and 90 min after the first set of measurements; the results were essentially the same. The behavior at 650°C is similar to that at room temperature. In both cases, the ratios are different for Sides A and B when determined near the crack tip, but ratios further from the tip turn out to be the same.

The opening load ratio drops off quite rapidly with distance from the crack tip as reported earlier by Macha et al [*4*]. They discovered that opening load ratios at positions more than 5 mm behind the crack tip in IN-100 compact

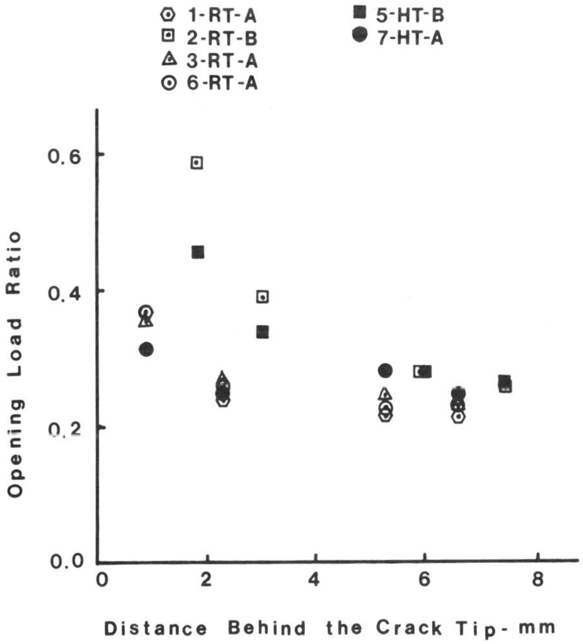

FIG. 6—*Opening load ratios at various positions behind the crack tip at 23 and 650° C for the tests in Fig. 5.*

tension specimens at room temperature were the same as that measured at the crack mouth with a clip gage. And, their result was independent of crack length.

Figure 7 presents the opening load ratios for three crack lengths at 23 and 650°C. The second crack (Tests 7 and 8) was precracked at a higher value of ΔK (22.0 MPa m$^{1/2}$) than the other two (14.8 MPa m$^{1/2}$). The data show that the opening load ratio, once one moves back roughly half the specimen thickness form the crack tip, is independent of crack length, precracking level, and temperature.

Discussion and Summary

The measured compliances are in reasonable agreement with linear elastic predictions both at room and elevated temperature. In view of the fact that the measurement positions are less than one specimen thickness behind the crack tip, the lack of agreement is not too surprising.

Whereas one cannot use the compliances measured so close to the tip to determine the crack length with great precision, one can use them to determine changes in crack length. For example, assume a crack in these specimens initially 20.0 mm long with indentations placed 1.0 mm behind the tip. If that crack grows 0.2 mm, the compliance will change by 11%. Therefore, if

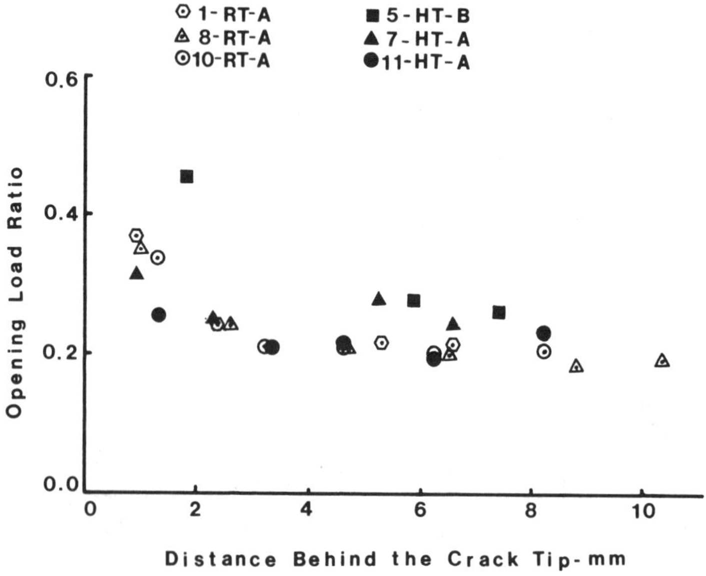

FIG. 7—*Opening load ratios at various positions behind the crack tip for different crack lengths, precracking, and temperature.*

the ISDG is used at one location, it can effectively monitor small changes in crack length.

Calculated compliances were based on surface crack lengths and plane stress. Measurements far removed from the crack tip should be based on an average crack length and plane strain. There is an advantage in surface measurements in that one can examine the behavior near the tip, but bulk measurements such as the crack mouth displacement have the advantage of smoothing out some of the details. The calculated plastic zone size for these cracks was less than 0.1 mm, so most of the crack front experienced plane strain. A compliance curve based on plane strain and the average crack length is plotted in Fig. 5, but it shows no better agreement with the measurements.

The opening load ratios decay to a value independent of position behind the crack tip as observed by Macha et al [*4*]. As one moves back from the crack tip, the displacement becomes more dependent on the bulk behavior of the specimen rather than on the surface behavior. Opening loads measured close to the crack tip reflect only the surface behavior there. The appropriate place to measure the opening load is therefore somewhat removed from the crack tip.

The opening load ratio is shown here to be independent of temperature and precracking load as well as crack length. Chana and Beevers [*12*] observe from electron microscopy studies that crack closure in nickel-based superalloys at 600°C is caused by an irregular crack trajectory and the presence of an oxide film. The results obtained here tend to support that view. Otherwise, some difference in the ratio would have been observed for the different precracking levels.

Other investigators have measured crack opening displacements near crack tips; the significance of the measurements reported here is that they were conducted at high temperature. The behavior at 650°C for this particular superalloy is quite the same as it is at 23°C. The ISDG has been shown to be a useful technique for measurements under the hostile conditions of high temperature and can be used for further studies; it is especially useful for opening load ratio studies.

Acknowledgments

The authors acknowledge the support of the Air Force Materials Laboratory and the encouragement of Dr. T. Nicholas. The support of The Johns Hopkins University is also greatly appreciated.

References

[*1*] Pitoniak, F. J., Grandt, A. F., Jr., Montulli, L. T., and Packman, P. F., *Engineering Fracture Mechanics*, Vol. 6, No. 4, 1984, pp. 663-670.

[*2*] Barker, D. B. and Fourney, M. E., *Experimental Mechanics*, Vol. 17, No. 7, 1977, pp. 241-247.

[*3*] Evans, W. T. and Luxmoore, A., *Engineering Fracture Mechanics*, Vol. 6, No. 5, 1974, pp. 735–743.
[*4*] Macha, D. E., Corbly, D. M., and Jones, J. W., *Experimental Mechanics*, Vol. 19, No. 6, 1979, pp. 207–213.
[*5*] Sharpe, W. N., Jr., and Grandt, A. F., Jr., in *Mechanics of Crack Growth, ASTM STP 590*, American Society for Testing and Materials, Philadelphia, 1976, pp. 302–320.
[*6*] Nisitani, H. P. and Kage, M. N. in *Fracture 1977*, ICF4, Vol. 2, Waterloo, Canada, 1977, pp. 1099–1108.
[*7*] Ohta, A., Kosuge, M., and Sasaki, E., *International Journal of Fracture*, Vol. 14, 1978, pp. 251–264.
[*8*] Sharpe, W. N., Jr., and Ward, M., *Journal of Engineering Materials and Technology*, Vol. 105, 1983, pp. 235–241.
[*9*] Sharpe, W. N., Jr., "A New Optical Technique for Rapid Determination of Creep and Fatigue Thresholds and High Temperature", AFWAL-TR-84-4028, Air Force Wright Aeronautical Laboratories, Wright-Patterson AFB, Ohio, April 1984, 105 pp.
[*10*] Sharpe, W. N., Jr., Domas, P. A., Ward, M., and Yau, J., "Benchmark Notch Test for Life Prediction, NASA CR-165571, NASA-Lewis, Cleveland, Ohio, June 1982.
[*11*] Saxena, A. and Hudak, S. J., *International Journal of Fracture*, Vol. 14, 1978, pp. 453–468.
[*12*] Chana, M. S. and Beevers, C. J. in *Advances in Crack Length Measurement*, Chameleon Press, London, 1982, pp. 375–394.

Roy Wilson[1] *and Anthony Palazotto*[2]

Viscoplastic Fatigue in a Superalloy at Elevated Temperatures

REFERENCE: Wilson, R. and Palazotto, A., **"Viscoplastic Fatigue in a Superalloy at Elevated Temperatures,"** *Fracture Mechanics: Seventeenth Volume, ASTM STP 905*, J. H. Underwood, R. Chait, C. W. Smith, D. P. Wilhem, W. A. Andrews, and J. C. Newman, Eds., American Society for Testing and Materials, Philadelphia, 1986, pp. 265-275.

ABSTRACT: This study involved extending existing analyses of the stress field and plastic zone ahead of a crack tip in a IN-100 compact tension specimen through a larger number of load cycles considering a frequency of 2.5 Hz and an *R* ratio of 0.1 at 732°C. The Bodner-Partom viscoplastic constitutive equations are utilized in describing the material behavior.

KEY WORDS: fracture mechanics, fatigue, elevated temperature, nickel-base superalloy, finite element

This study focuses on the stress/strain fields around a crack tip and how a cyclic load affects them. The authors used an in-house finite element computer program named VISCO incorporating constant strain triangles with the Bodner-Partom viscoplastic constitutive equations [*1–4*]. The modeling was accomplished using a compact tension specimen made from Gatorized® IN-100 (a superalloy used in the manufacturing of turbine disks) at 732°C. The program was modified to incorporate load cycling considering a frequency of 2.5 Hz and a *R* ratio of 0.1 with a maximum stress intensity of $K_1 = 38.5$ MPa $m^{1/2}$ (*R* is defined as the ratio of applied minimum load to maximum load). Thirteen load cycles were carried out.

[1]Engineer, Air Force Weapons Laboratory, Kirtland Air Force Base, Albuquerque, NM 87140. Former Graduate Student.

[2]Professor, Aeronautics and Astronautics Department, Air Force Institute of Technology, Wright-Patterson AFB, OH 45433.

Bodner-Partom Constitutive Law

The Bodner-Partom equations have been extensively discussed elsewhere [1-4]. For completeness the main characteristics of the time-dependent stress/strain relations will be shown.

The elastic shear-strain rate relation can be derived through Hooke's Law, while the plastic strain rate $\dot{\epsilon}_{ij}^{P}$ can be expressed as

$$\dot{\epsilon}_{ij}^{P} = \lambda\, S_{ij} \tag{1}$$

where S_{ij} = deviatoric stress components and

$$\lambda = [D_2^P/J_2]^{1/2} \tag{2}$$

where J_2 = second invariant of the deviatoric stress and $D_2^P = 1/2\, \dot{\epsilon}_{ij}^{P}\dot{\epsilon}_{ij}^{P}$ is the second invariant of the inelastic strain rate tensor. Bodner-Partom expressed D_2^P as

$$D_2^P = D_0^2 \exp\left[-\frac{(Z^2)^n}{3J_2} \frac{n+1}{n} \right] \tag{3}$$

This expression is based on extensive experimental data and has been modified to fit results found by several researchers. D_0 is the limiting value of the plastic strain rate in shear. The parameter n controls strain rate sensitivity. Z is the measure of material hardening and is a function of plastic work.

Analysis of the variables in the equation for the second invariant of the plastic strain rate shows that D_0 is the upper limit of plastic strain rate and therefore can be arbitrarily chosen as any high value. For convenience, it is generally chosen as $D_0 = 10^4\ s^{-1}$ unless very high rates of straining are present.

Bodner's elastic-viscoplastic theory is based on an internal material state variable Z. Bodner's Z hardness parameter is a basic material property and is deformation history dependent. The following relationship has been used for the Z parameter:

$$\dot{Z} = m\,(Z_1 - Z)\,\dot{W}_p - AZ_1 \left(\frac{Z - Z_2}{Z_1} \right)^r \tag{4}$$

where m, Z_1, A, r, and Z_2 are material constants. The first term in Eq 4 is the increase in hardening due to inelastic work, W^P, while the second term is the thermal recovery or softening. The rate of plastic work is

$$\dot{W}_p = Z(D_2^P J_2)^{1/2} \tag{5}$$

The Bodner constants for IN-100 at 732°C are given in Table 1.

TABLE 1—*Bodner coefficient for IN-100 at 732°C.*

Material Parameter	Description	Value
E	Elastic modulus	18.133×10^4 MPa (26.3×10^3 ksi)
n	Strain rate exponent	0.7
D_0	Limiting value of strain rate	10^4 s^{-1}
Z_0	Limiting value of hardness	6304 MPa (915.0 ksi)
Z_1	Maximum value of hardness	6993 MPa (1015.0 ksi)
Z_2	Minimum value of hardness	4134 MPa (600.0 ksi)
m	Hardening rate exponent	0.37273 MPa^{-1} (2.57 ksi^{-1})
A	Hardening recovery coefficient	1.9×10^{-3} s^{-1}
r	Hardening recovery exponent	2.66
	(1 KBAR = 100 MPa = 14.504 ksi)	

Methods of Analysis

The Computer Program

A two-dimensional constant strain finite element program named VISCO was used throughout this study. VISCO accounts for nonlinear viscoplastic material behavior. The accuracy of the program has been verified by Smail and Palazotto [*3*] and Keck [*5*].

The Bodner-Partom viscoplastic constitutive equations in VISCO are solved using the Gauss-Seidel iterative equation solver with overrelaxation, eliminating costly stiffness matrix factorization. Time integration of the Bodner equations is accomplished for each element using an Euler extrapolation scheme. A variable time step algorithm is included that maximizes the time step size during the analysis while maintaining good accuracy. During each time step, equilibrium tolerances are checked. If the tolerances are exceeded, the time step is reduced until equilibrium is obtained.

VISCO's loading function was modified for the sawtooth pattern cycle response incorporated in this analysis. Using a linear load equation solver to model the load spectra, a percentage of total load was calculated at each time step. An additional characteristic of the program is the incorporation of a crack algorithm which can be used, if required, for varying boundary conditions behind the crack tip brought about by crack closure considering, in particular, negative R ratios. In all of this work, an assumption of isotropic behavior is made even under the reversal of loading. A more complete study would in general have to consider the problem of isotropic versus anisotropic or kinematic hardening. The constants shown in Table 1 for IN-100 have been found to represent simple cyclic behavior quite well [*6*]. Node displacement and element strain, stress, and the Bodner Z-hardness were output at requested time intervals.

Finite Element Modeling

The two-dimensional fatigue modeling was accomplished using a standard compact tension specimen geometry (Fig. 1). Owing to symmetry, only half of the compact tension specimen was modeled using constant strain triangular elements. The finite element mesh shown in Fig. 2, consisting of 382 elements, was selected to model the top half of the compact tension specimen. The pattern of elements within the mesh arrangement allows for unlimited size reduction and ensures that no two adjacent elements differ in size by more than a factor of 2. Except for elements near the loading pin holes, element aspect ratios varied from 0.5 to 1.0. Elements near the crack tip have an area of 3.1494×10^{-5} cm^2. Since crack growth was not specifically studied, the length of the crack was held constant. The specimen thickness was established at a constant 5.46 mm.

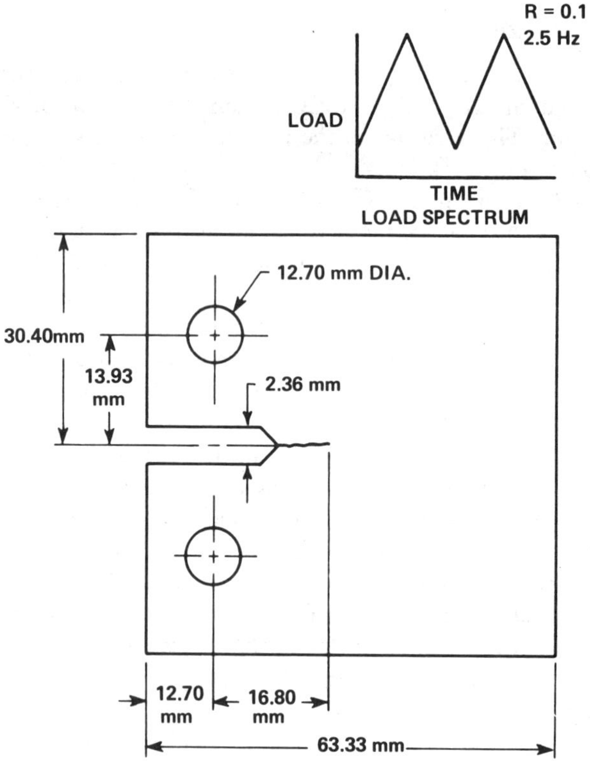

FIG. 1—*Compact tension specimen geometry.*

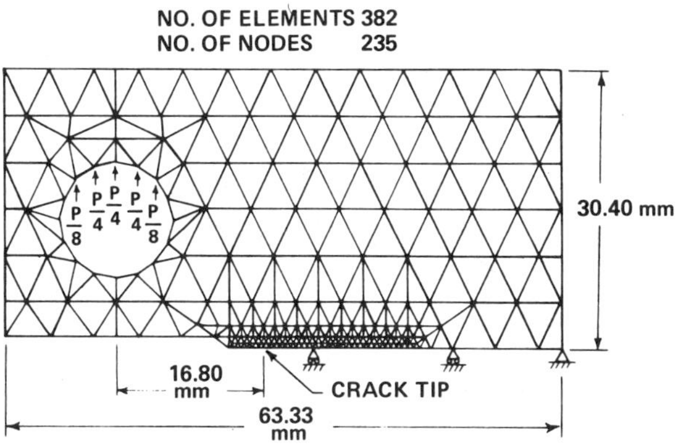

FIG. 2—*Finite element model.*

Results and Discussion

The mesh presented previously has been compared against a more refined mesh especially in the vicinity of the crack tip. The comparison was made with a model consisting of 543 elements (327 nodes). It has been shown [*7*] that the stress field, plastic zone, and far-edge displacements were within 2% of the more refined model. It is thus assumed that the finite element model used in this study is adequate.

There is definite benefit derived from this 382 element model. Current research has been able to achieve only a moderate number of load cycles before computer operating time becomes prohibitive [*6*]. For example, the 543 element mesh allows somewhat less than four complete load cycles in 2000 central processing unit (CPU) seconds. With equal accuracy, the present model is able to complete 13.5 cycles in the same amount of computer time.

Keck et al [*6*] carried out work considering load cycles. It should be pointed out that this reference considered the effective stress-strain plot. In addition, there appeared to be a decreasing accumulation of plastic strain at each cycle in the material ahead of the crack tip. The work reported upon in this paper duplicated the finding mentioned above, but also investigated the stress and strain functions in the y direction. Since more cycles were accomplished, a better approximation of the material interaction could be traced for the $R = 0.1$ loading. The reader is reminded that the possible developments put forth subsequently are predicated upon zero crack extension, and thus may only be considered a first step in attempting to incorporate an elastic-viscoplastic constitutive relation into a cycle loading analysis.

The characteristics of crack closure in the consideration of cyclic loading are important phenomena. It is observed in Fig. 3 that the vertical (y) dis-

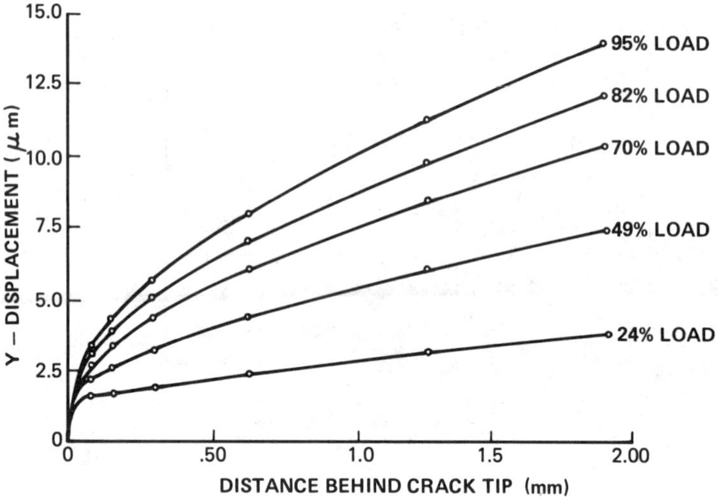

FIG. 3—*Displacement versus distance behind the crack tip.*

placements for any of the 13 load cycles examined, behind the crack tip, do not exhibit any closure. This is due to the nature of the loading ($R = 0.1$, $K = 38.5$ MPa m$^{1/2}$, 2.5 Hz) and specimen geometry as well as the assumption of no crack growth.

The displacement of the corner of the crack mouth for the first load cycle is shown in Fig. 4. When this figure is superimposed over the crack mouth displacement for the second load cycle, it is found that the slopes of the curves are identical. Throughout the 13 load cycles examined in this study, there was no change in the plot of crack mouth displacement versus load. The results show that the compact tension specimen behaves elastically at points on the specimen boundary and does not see the localized inelastic behavior near the crack tip.

Figure 5 illustrates the y stress/strain behavior for the element above the crack tip. Shown are several complete cycles. One may observe that reversed yielding occurs as well as a slight reduction in maximum stress. These phenomena were expected. Further examination of Fig. 5 reveals that a rapid increase in plastic deformation occurs during the first load cycle. Each successive cycle has less plastic deformation.

In order to pursue this last finding further, a listing in Table 2 of the total y strain after each load cycle along with the change in strain from the previous load cycle is given for the first element ahead of the crack tip. Figure 6 shows a plot of the tabular values. The curve is then extrapolated to approximate what appears to be a constant or stable amount of plastic strain in terms of numbers of cycles. After approximately 23 load cycles, it appears that no more plastic straining will take place.

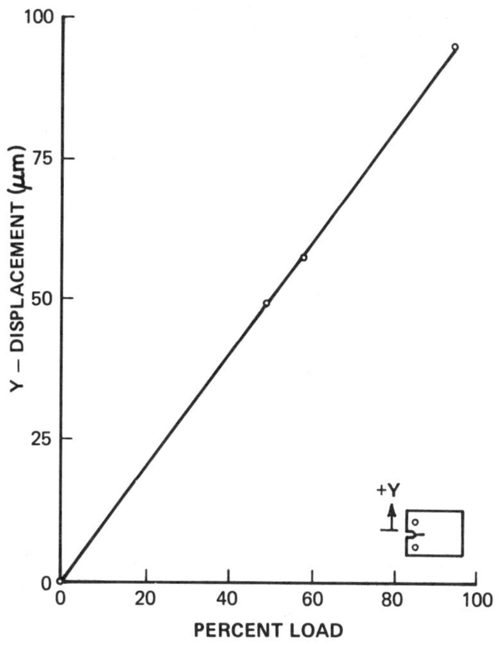

FIG. 4—Y *displacement versus percent of load.*

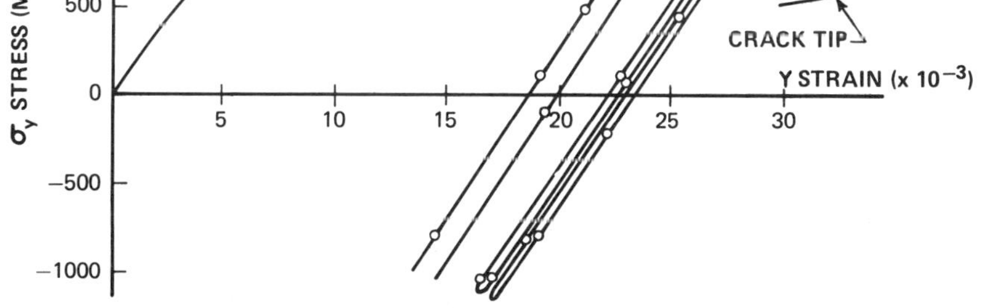

FIG. 5—*Compact tension specimen stress/strain behavior.*

The behavior of the Bodner Z hardness material parameter is illustrated in Fig. 7. The elements immediately ahead of the crack tip along the line of symmetry reach a saturation value within the first load cycle. Elements which are farther out than 2.4% of the crack length ahead of the crack tip see no

TABLE 2—*Total strain values after each load cycle.*

Cycle	Strain 10^{-3}	Strain Increase 10^{-3}
0	0.0	0.0
1	14.03	14.03
2	15.25	1.22
3	16.07	0.827
4	16.58	0.51
5	17.0	0.42
6	17.33	0.33
7	17.60	0.27
8	17.84	0.24
9	18.09	0.25
10	18.29	0.20
11	18.49	0.20
12	18.67	0.18
13	18.81	0.14

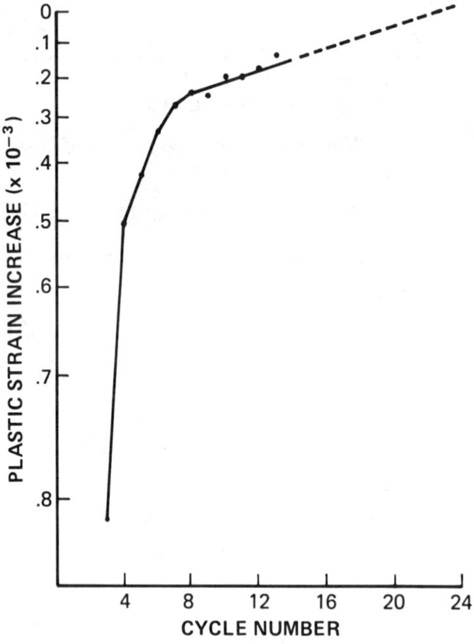

FIG. 6—*Projected number of cycles required for stabilization.*

change from the initial hardness value specified. The elements in between show a slight increase in their hardness value through approximately three load cycles where no more increase is indicated.

The stress and strain fields ahead of the crack tip, at full positive load, are shown in Figs. 8 and 9. The stress magnitude is decreasing over the 13 load

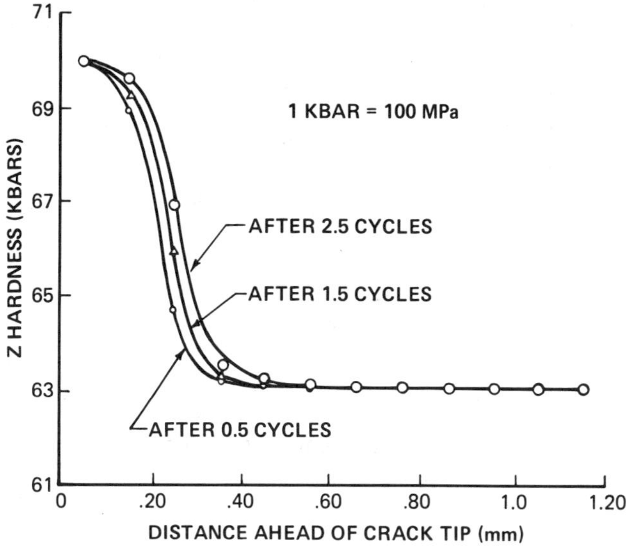

FIG. 7—Z *hardness versus distance ahead of crack tip.*

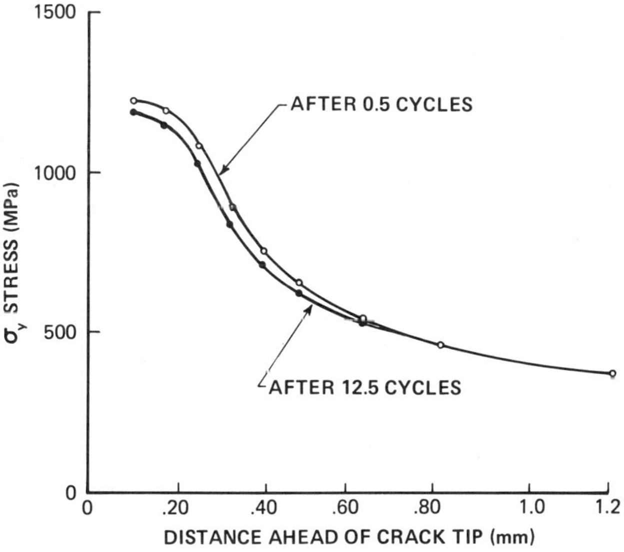

FIG. 8—Y *stress versus distance ahead of crack tip.*

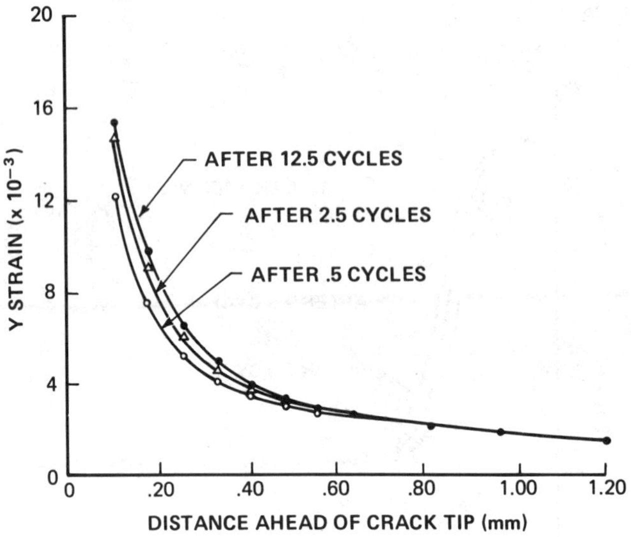

FIG. 9—Y *strain versus distance ahead of crack tip.*

cycles. The strain field shows a slight increase over the 13 cycles with the majority occurring early in the load history. This is again showing the trend towards a stable strain-time history after approximately 23 load cycles.

Conclusions

The results of the computations carried out using the isotropic Bodner viscoplastic constitutive equations with no crack extension allowed to model the behavior of IN-100 at elevated temperatures yield the following conclusions for an R ratio of 0.1:

1. The increase in strain realized after each load cycle under an R ratio of 0.1 decreases in a manner which indicates that after approximately 23 load cycles the material will no longer undergo plastic strain increase.
2. The large majority of plastic straining occurs within the first three load cycles.
3. The stress field ahead of the crack tip remains relatively constant after one to three load cycles.
4. The cyclic behavior of the compact tension specimen near the crack tip is neither stress nor strain controlled.
5. The far-field overall behavior of the compact tension specimen for the conditions considered is elastic and independent of load level.

Acknowledgments

The authors would like to express their appreciation to Dr. Ted Nicholas of Metals Behavior Branch, Air Force Wright Aeronautical Material Laboratory, Wright-Patterson Air Force Base, Ohio, for his assistance throughout this project.

References

[*1*] Hinnerichs, T., "Viscoplastic and Creep Growth Analysis by the Finite Element Methods," Ph.D. dissertation, Department of Aeronautics and Astronautics, Air Force Institute of Technology, Wright-Patterson AFB, Ohio, 1980.

[*2*] Hinnerichs, T., Palazotto, A., and Nicholas, T., "Application of a Numerical Procedure of Creep Crack Growth from Displacement Measurements in IN-100," in *Fracture Mechanics: Fourteenth Symposium—Volume II: Testing and Applications, ASTM STP 791*, American Society for Testing and Materials, Philadelphia, 1983, pp. II-166–II-181.

[*3*] Smail, J. and Palazotto, A., "The Viscoplastic Crack Growth Behavior of a Compact Tension Specimen Using the Bodner-Partom Flow Law," *Engineering Fracture Mechanics*, Vol. 19, No. 1, 1984, pp. 137–158.

[*4*] Bodner, S. and Partom, Y., "Constitutive Equations for Elastic-Viscoplastic Strain Hardening Materials," *Journal of Applied Mechanics, Transactions of ASME*, Vol. 42, 1975, pp. 385–389.

[*5*] Keck, J. "The High Temperature Viscoplastic Fatigue Behavior of a Compact Tension Specimen Using the Bodner-Partom Flow Law," M.S. thesis, Department of Aeronautics and Astronautics, Air Force Institute of Technology, Wright Patterson AFB, Ohio, 1982.

[*6*] Keck, J., Nicholas, T., and Palazotto, A., "High Temperature Viscoplastic Fatigue Behavior of a Compact Tension Specimen," *Engineering Fracture Mechanics*, Vol. 22, No. 1, 1985, pp. 77–91.

[*7*] Wilson R., "The High Temperature Viscoplastic Fatigue Behavior of IN-100 Using the Bodner-Partom Flow Law," M.S. thesis, Department of Aeronautics and Astronautics, Air Force Institute of Technology, Wright-Patterson AFB, Ohio, 1983.

Fracture Testing

J. H. Underwood,[1] J. A. Kapp,[1] and M. D. Witherell[1]

Fracture Testing with Arc Bend Specimens

REFERENCE: Underwood, J. H., Kapp, J. A., and Witherell, M. D., "**Fracture Testing with Arc Bend Specimens,**" *Fracture Mechanics: Seventeenth Volume, ASTM STP 905*, J. H. Underwood, R. Chait, C. W. Smith, D. P. Wilhem, W. A. Andrews, and J. C. Newman, Eds., American Society for Testing and Materials, Philadelphia, 1986, pp. 279–296.

ABSTRACT: A limited review of existing stress, stress intensity factor, and displacement analyses was compared with new work in order to select arc bend geometries appropriate for fracture testing. Results from the literature for rectangular and arc bend specimens were compared with finite element and boundary collocation results from the present work.

Two series of comparative tests were performed, one with arc specimens cut from a steel forging with outer-to-inner radius ratio of 2.5, the other from an aluminum cylinder with outer-to-inner radius ratio of 1.3. Fracture toughness tests, K_{Ic}, and J_{Ic}, when appropriate, were performed with standard arc tension specimens and with three-point arc bend specimens with both arc and chord support.

Conclusions were drawn regarding the appropriate stress intensity factor, crack mouth displacement, and load-line displacement solutions for arc bend fracture specimens. Recommendations were offered for practical ranges of specimen geometry and for reliable test procedures.

KEY WORDS: arc bend, bend specimen, fracture testing, *K* solution, displacement solution, hollow cylinder

The arc tension specimen, essentially a section of a hollow disk with tension loading (Fig. 1*a*), is now used routinely. It has been part of ASTM Test for Plane-Strain Fracture Toughness of Metallic Materials (E 399) since 1978. For some applications the same type of section loaded in bending would be more convenient. The overall objectives of this paper are to review the available analytical results related to arc bend specimens, perform additional analyses and tests, and propose some standardized procedures for fracture testing with arc bend specimens. Specific objectives are: (1) to review pub-

[1]U.S. Army Armament R&D Center, Watervliet, NY 12189.

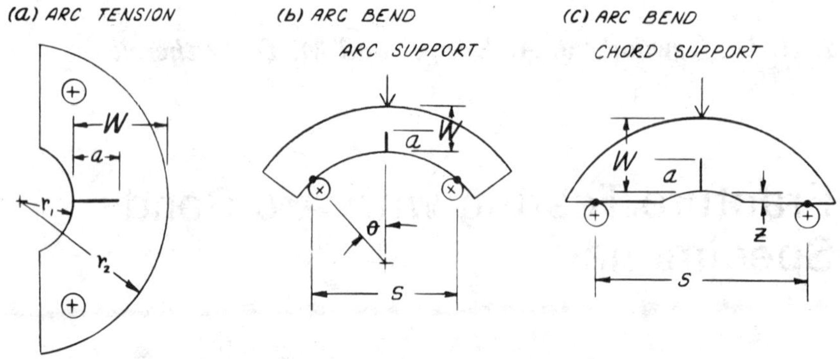

FIG. 1—*Arc shaped fracture specimen geometries.*

lished analytical results of stress intensity factor, K, crack mouth opening displacement, v, and load-line displacement, δ, for rectangular and arc bend specimens; (2) to perform additional analytical calculations of K, v, and δ, using boundary value collocation and finite element methods; (3) to perform fracture toughness tests of example geometries of arc bend specimens for experimental verification of K, v, and δ; and (4) to propose arc bend test procedures, including a range of specimen geometries and associated K, v, and δ information suitable for accurate, wide-range expressions in a standard test method.

Some prior work has been done with arc bend geometries as fracture specimens. Most work included three-point bending in which the outer load points are a free rolling support on the inner radius, termed *arc support* here (Fig. 1*b*). This type of testing has the advantage of loading directly on the existing inner and outer radius surfaces, if they are smooth and regular enough. Other work investigated arc bend specimens with support on a flat chordal surface, termed *chord support* here (Fig. 1*c*). This specimen, although requiring a machined surface, is more similar to the rectangular bend specimen for which standard test procedures are already available. Jones [*1*] investigated arc support geometries with outer-to-inner radius ratio, r_2/r_1, between 1.05 and 1.25 and various support angles, θ. Tracy [*2*] analyzed arc support specimens with r_2/r_1 between 1.25 and 2.00 with one value of θ, 45°. Ritter and Rea [*3*] considered a chord support geometry with $r_2/r_1 = 1.31$ with various support spans, S. This work gave useful guidance for analysis and testing of arc bend specimens, but it did not provide the repeatability and accuracy of analytical and experimental results over a wide range of test conditions which are needed for a standard test method.

The recent boundary collocation results of Gross and Srawley [*4*] gave additional guidance for arc bend testing as well as accurate analytical results for a wide range of arc support geometries. The authors provided tabular results of stress intensity factor, K, and crack mouth opening displacement, v, for r_2/r_1 between 1.10 and 2.50 and θ between 11.5 and 90°. This includes much of the geometry range of interest in arc bend fracture testing.

The present work is described in two parts, analysis and testing. First, analysis results from the literature and the present investigation are compared on a common basis. By considering the basic geometry of arc specimens along with certain deep-crack limit solutions, K and v results from rectangular, arc bend-arc support and arc bend-chord support geometries can be compared using the same, nearly constant-valued parameter. Literature results are compared with boundary collocation and finite element results from the present work for various geometries. The second part of this paper describes two series of tests in which arc bend specimens were made from two hollow cylinders, one a high strength steel cylinder with outer-to-inner radius ratio, r_2/r_1, of about 2.5, the other a high strength aluminum alloy cylinder with r_2/r_1 about 1.3. K_{Ic} and some J_{Ic} test results from arc bend specimens with both arc and chord support were compared with results from arc tension tests of the same material.

Analysis

Common Comparison

The earlier K, v, and δ results for bend specimens can be compared with the current results using a parameter which takes account of most of the important mechanics of this type of specimen. By using such a common parameter, the earlier and new results can be compared directly for the purpose of mutual verification. In addition, since the parameter has a nearly constant value for all rectangular and arc geometries, the results in this form lead directly to simple and accurate interpolation for whatever specimen geometries are of interest.

The basis for a common comparison of K results from rectangular and arc bend geometries is the combination of two deep-crack K limit solutions [5]:

$$\underset{a/W\to 1}{K} = \frac{3.975M}{B(W-a)^{3/2}} + \frac{1.4635P^*}{B(W-a)^{1/2}} \tag{1}$$

where a is crack depth, B is specimen thickness, W is specimen depth, M is bending moment, and P^* is the horizontal component of force exerted at the specimen support (Fig. 2). Equation 1 is an expression for the deep-crack limit K (as $a/W \to 1$) for an arc bend-arc support specimen. The first term in Eq 1 is the K due to a pure bending moment, M, and the second term is the K due to the pure tension loading of force, P^*, applied in line with the center of

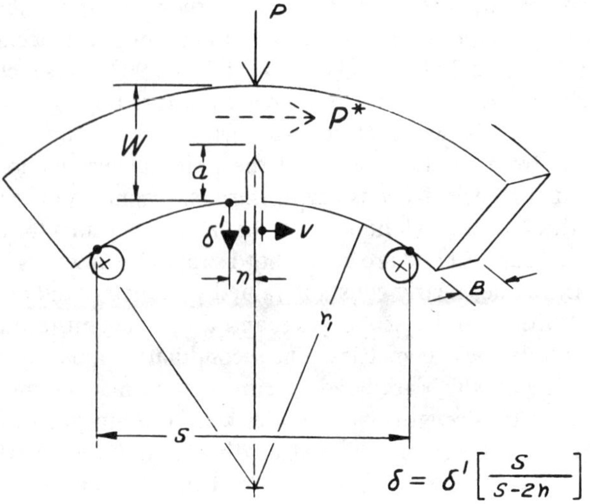

FIG. 2—*Arc bend-arc support specimen and nomenclature.*

the uncracked ligament. Expressions for M, P^*, and S can be obtained from plane geometry as

$$M = \frac{P}{2}\tan\theta\left(r_1 + \frac{W}{2} + \frac{a}{2}\right) \tag{2}$$

$$P^* = \frac{P}{2}\tan\theta \tag{3}$$

$$S = 2r_1\sin\theta \tag{4}$$

where P is the center load applied to the specimen and M is the moment about the center of the uncracked ligament.

Combining Eqs 1 to 4 gives

$$\frac{KBW^{1/2}(1 - a/W)^{3/2}}{P\left[\dfrac{S}{W\cos\theta} + \tan\theta(1 + a/W)\right]\left[1 + \dfrac{0.7366(1 - a/W)}{\dfrac{S}{W\sin\theta} + (1 + a/W)}\right]} = Y = \underset{a/W\to 1}{0.995} \tag{5}$$

a dimensionless K parameter for use in comparing K results from rectangular and arc bend geometries. Important features of Eq 5 are that it approaches the exact deep-crack limit solution and that it includes both bending and normal force loading. For rectangular bend specimens and chord support arc bend specimens, which have no normal force loading and $\theta = 0$, Eq 5 reduces to

$$\frac{KBW^{1/2}(1 - a/W)^{3/2}}{PS/W} = Y \underset{a/W \to 1}{=} 0.995 \tag{6}$$

A similar common comparison of v results from various bend geometries can be made, based on the deep-crack v limit solution [5]:

$$\underset{a/W \to 1}{v} = \frac{15.8\, MW}{BE(W - a)^2} \tag{7}$$

Using a similar approach to that described by Eqs 1 to 6 and related discussion, gives

$$\frac{EBv\left[\dfrac{(1 - a/W)^2}{(1 + a/W)^2}\right]}{P\left[\dfrac{S}{W\cos\theta} + \tan\theta(1 + a/W)\right]\left[1 + \dfrac{0.7366(1 - a/W)}{\dfrac{S}{W\sin\theta} + (1 + a/W)}\right]} = Y_v \underset{a/W \to 1}{=} 0.9875 \tag{8}$$

a dimensionless K parameter for use in comparing v results from rectangular and arc bend geometries.

Comparison of load-line displacement, δ, for different types of bend specimens is expected to be less straightforward, because δ is more affected by uncracked specimen geometry than are K and v. The approach taken here is to use a δ parameter which approaches the proper deep-crack δ limit solution for load-line displacements due only to the presence of the crack, and further due only to bending. Normal stress effects and uncracked specimen effects are ignored in this δ parameter. The deep-crack δ limit solution used is [5]

$$\underset{a/W \to 1}{\delta} = \frac{3.95\, MS}{BE(W - a)^2} \tag{9}$$

so that the dimensionless δ parameter for use in comparing rectangular and arc bend geometries becomes

$$\frac{EBW\delta(1 - a/W)^2}{PS\left[\dfrac{S}{W}\cos\theta + \tan\theta(1 + a/W)\right]} = Y_\delta \underset{a/W \to 1}{=} 0.9875 \tag{10}$$

Stress Analysis

Several arc bend geometries were modeled by boundary collocation and finite element methods. The K, v, and δ values obtained were compared with results from the literature using the basis of comparison described previously. The boundary collocation method was similar to that developed by Hussain et al [*6*] and used some of the modeling techniques of Gross and Srawley [*4,7*]. The finite element method was based on that used by Kapp and Pu [*8*] with important use of enriched finite elements [*9*]. The type of element array used for K determination of chord support geometries is shown in Fig. 3. The necessary configurational changes were made to model the arc support and rectangular geometries. Changes in the element density were made to properly model displacements, as will be described in the following discussion of stress analyses results. Nine categories of results were obtained including K, v, and δ for rectangular, arc support, and chord support geometries. Literature results extensive enough for comparison were available in five of these categories.

The comparison of collocation and finite element K results with appropriate data from the literature is shown in Fig. 4 using the parameter Y defined by Eq 5. The solid line in each plot is for the standard rectangular bend specimen [*5*]. The dashed lines in Fig. 4 are "eyeball" best fit lines considering all the data presented. In Fig. 4*a* the Gross and Srawley [*4*] arc support collocation K results are seen to be in close agreement with the present results. For $r_2/r_1 = 1.5$ and 2.0 the two independent sets of collocation results are often indistinguishable. For $r_2/r_1 = 1.1$ the arc support results compared well with the very similar rectangular bend geometry. In Fig. 4*b* arc support finite element K results are close to the rectangular bend, particularly where expected, for $r_2/r_1 = 1.1$. In Fig. 4*c* chord support finite element K results compare well with the rectangular bend results. Note the direct comparison of

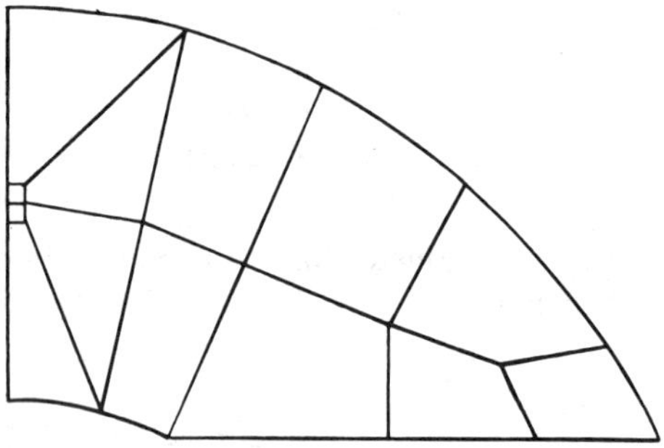

FIG. 3—*Finite element arrangement for arc bend-chord support specimen;* $a/W = 0.5$, $r_2/r_1 = 2.0$, $S/W = 3.347$, $Z/W = 0.1$; *crack tip element size* $= 0.05W$.

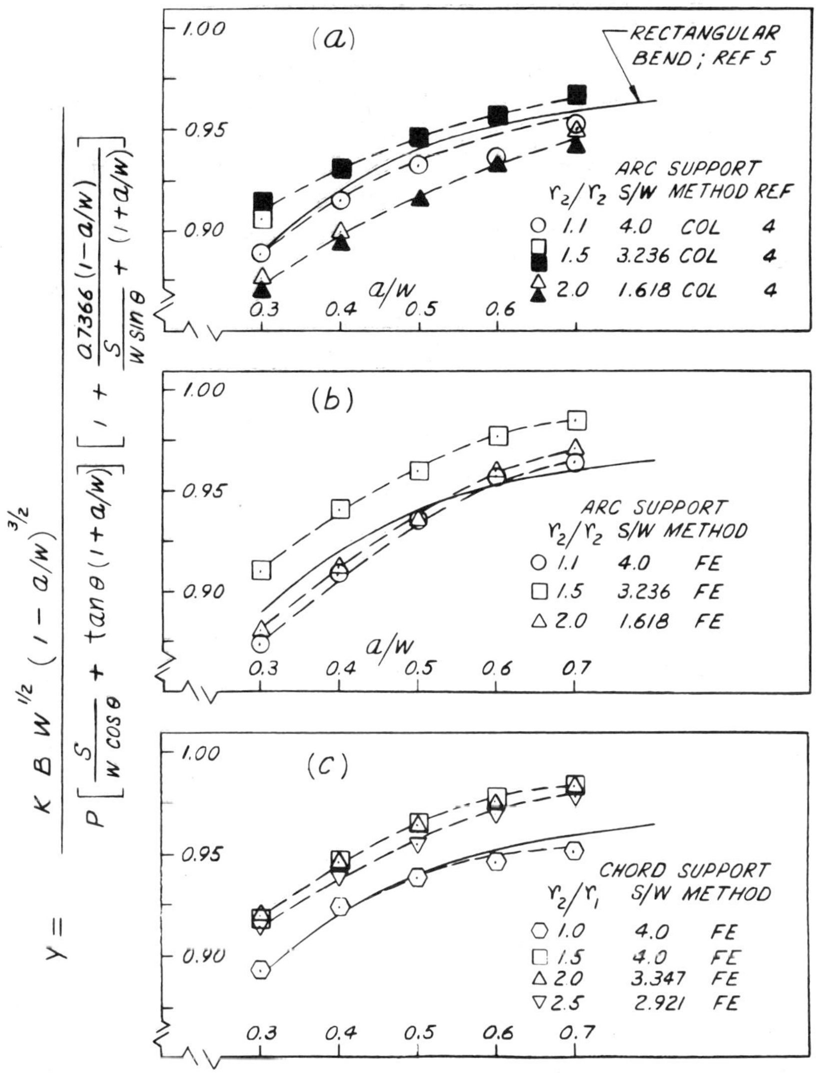

FIG. 4—*Stress intensity,* K, *versus* a/W *for bend specimens of various geometries.*

rectangular geometries and the good agreement between finite element results with $r_2/r_1 = 1.0$ and the rectangular bend geometry.

The general trend of all the results in Fig. 4 is that the K parameters for seven significantly different bend geometries agree within about 4%, and that the K parameter tends to increase slightly with increasing r_2/r_1 or S/W. These results indicate that the Eqs 1 to 4 input to the K parameter properly accounts for the important effects of arc-shaped geometry, method of support, and deep crack limit conditions of the arc bend specimen.

The comparison of v results is shown in Fig. 5 using the parameter Y_v defined by Eq 8. Again, the solid line in both plots is for the standard rectangular bend specimen [5]. The finite element v results here and the δ results in Fig. 6 were obtained using an element array of the same type as in Fig. 3 but denser. A total of 29 elements were used rather than the 13 shown in Fig. 3. The denser array, particularly along the load-line, produces a more faithful simulation of the actual displacements in the specimen. In Fig. 5a the collocation results [4] are in reasonable agreement with the rectangular bend results. In Fig. 5b the finite element results for the rectangular geometry, that

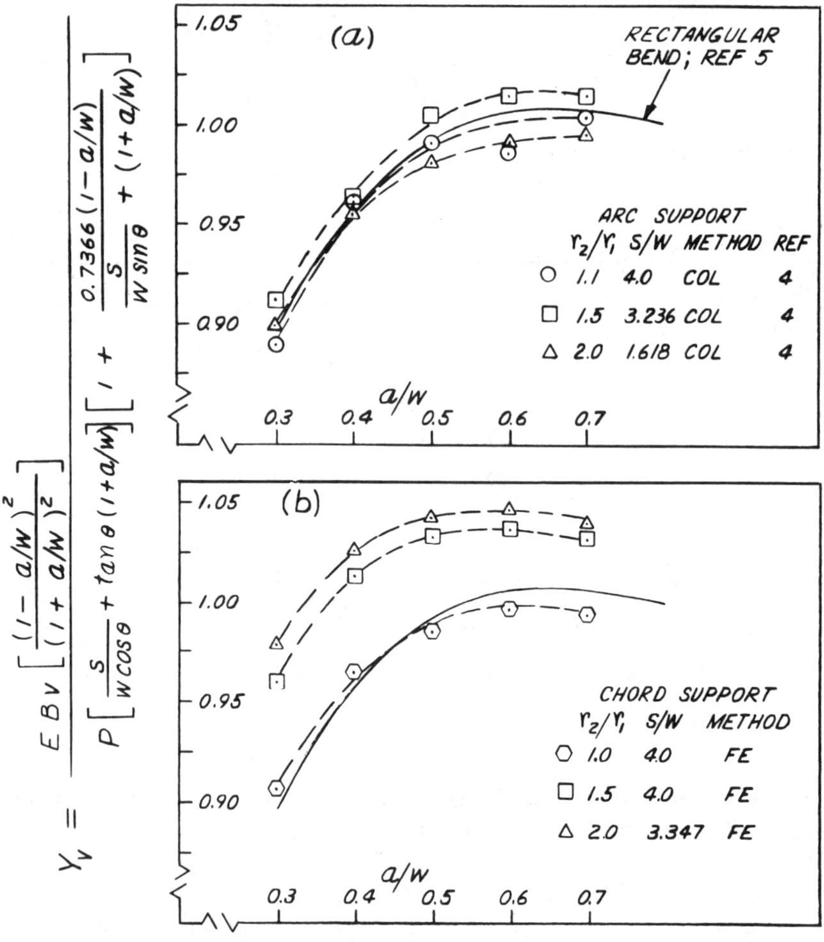

FIG. 5—*Crack mouth displacement,* v, *versus* a/W *for bend specimens of various geometries.*

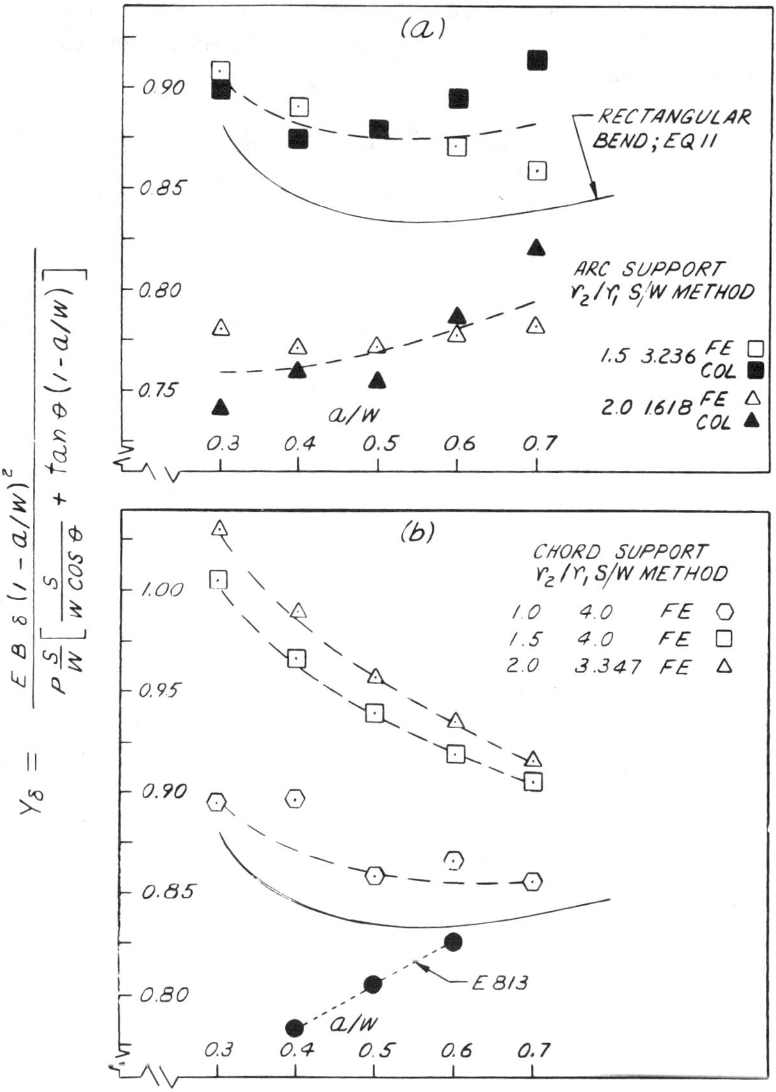

FIG. 6—*Load-line displacement, δ, versus a/W for bend specimens of various geometries*

is for $r_2/r_1 = 1.0$, agree well with the rectangular bend results from the literature. The general trend for all the v results is that separate calculations of Y_v for similar geometries agree within about 2% and v tends to increase with increasing r_2/r_1 or S/W. The increases in Y_v with these variables are larger than those seen with Y (Fig. 4).

The comparison of δ results is shown in Fig. 6 using the parameter Y_δ defined by Eq 10. The solid line in each plot is obtained from the total load-line displacement, δ, for the standard rectangular bend specimen, determined in the following manner:

$$\delta = (\delta_{bend} + \delta_{shear})_{no\ crack} + \delta_{crack} \tag{11}$$

$$\delta_{bend} = \frac{MS^2}{BW^3E} \tag{12}$$

$$\delta_{shear} = \frac{3.12M}{BWE} \tag{13}$$

$$\delta_{crack} = \frac{6MS}{BW^2E} f_\delta \tag{14}$$

$$f_\delta = fn(a/W)$$

where δ_{bend} and δ_{shear} are the components of δ due to the pure bending and shear of the uncracked specimen, and δ_{crack} is the component due to presence of the crack. The expressions for δ_{bend} and δ_{shear} are from mechanics [*10,11*] and δ_{crack} is from Tada et al [*5*].

A comparison of the results from the load-line displacement expression for a rectangular bend specimen (Eq 11), can be made with data from ASTM Test for J_{Ic}, A Measure of Fracture Toughness (E 813). A table of load-line displacements for the rectangular bend specimen is included in ASTM E 813 for use in checking the accuracy of some of the experimental measurements of the method. Three of these data are shown in Fig. 6*b* in the form of Y_δ. The agreement with Eq 11 is within about 7%, but we believe the agreement should be closer. We suggest that the component of displacement due to shear of the uncracked specimen, as described by Eq 13, has been omitted from the data in ASTM E 813. When this component is added to the three data points from ASTM E 813, the agreement with Eq 11 is within 0.5%.

As indicated by Eqs 11 to 14, there are three major contributions to δ, so separate calculations of δ might not be expected to agree as well with each other as do K and v calculations. This is apparent in Fig. 6. Separate calculations of δ for the same geometry agree within about 1 to 6%. The tendency towards increasing δ with increasing r_2/r_1 or S/W is more pronounced than with K and v.

Experiments

Fracture toughness tests were performed both as a direct physical check on the analyses and as a means of identifying unanticipated problems with arc bend testing. Table 1 outlines the test conditions. The arc tension tests of

TABLE 1—*Fracture toughness test conditions.*

Material	Yield Strength, MPa	Nominal Fracture Toughness, MPa $m^{1/2}$	r_2/r_1	Geometry — Number of Specimens: Type of Displacement Measure: Arc Tension	Arc Bend; Arc Support	Arc Bend; Chord Support
A732 steel	1180	137	2.50	$7:v, \delta$	...	$7:v, \delta$
7075-T6 aluminum	540	28	1.29	$5:v, \delta$	$4:v$ $3:\delta$	$4:v$ $3:\delta$

both aluminum and steel were the zero offset geometry of ASTM E 399, also shown sketched in Fig. 1*a*. For this geometry the displacement measured at the crack mouth is also load-line displacement, $v = \delta$. Steel arc bend-chord support specimens were tested so that v and δ could be measured simultaneously. Aluminum arc bend specimens were tested by measuring v from four specimens for each type of support and measuring δ from three specimens each.

Steel Tests

Fourteen specimens were made with the *C-R* orientation from a steel hollow cylinder forging and tested so that both K_{Ic} and J_{Ic} could be determined. Table 2 shows the nominal specimen dimensions and the results. Actual specimen dimensions varied by up to about 5% from nominal and were taken into account in the K_{Ic} and J_{Ic} calculations. The results show that, as suspected before testing, the fracture toughness of the steel was very close to the value which separates a valid from an invalid K_{Ic} for the specimen size used. Only three of the seven arc tension specimens yielded a valid K_{Ic}. The arc bend results are listed as K_Q, because the arc bend is not a standard geometry. One of the arc bend results, #4, passed two critical ASTM E 399 requirements, that $K_{max}/K_Q \leq 1.1$ and $(K_Q/\sigma_y)^2/B \leq 0.4$. Putting aside the validity concern, it is clear that the arc tension and arc bend fracture toughness measurements were in close agreement.

A comparison of J_{Ic} measurements from arc tension and arc bend specimens is shown in Fig. 7. Four of the seven combined K_{Ic} and J_{Ic} tests for each of the two groups were interrupted near maximum load, and the specimens were heat tinted. The resulting J versus heat tint Δa plots are shown. It must be emphasized that the calculation of J was approximate at best, since ASTM E 813 procedures for the compact and rectangular bend specimens were used for the arc tension and arc bend specimens, respectively. These J_{Ic} results should be considered only as some indication of how appropriate or inappro-

TABLE 2—*Steel fracture toughness results (nominal dimensions:* r_1 = *47.4 mm,* B = *35.6 mm,* W = *71.1 mm).*

Arc Tension, $a/W = 0.51$, $X/W = 0$					Arc Bend; Chord Support, $a/W = 0.51$, $S/W = 2.92$, $Z/W = 0.1$			
Specimen No.	$\frac{2.5\,(K_Q/\sigma_y)^2}{B}$	K_{max}, MPa $m^{1/2}$	K_{Ic}, MPa $m^{1/2}$	K_Q, MPa $m^{1/2}$	Specimen No.	$\frac{2.5\,(K_Q/\sigma_y)^2}{B}$	K_{max}, MPa $m^{1/2}$	K_Q, MPa $m^{1/2}$
1	0.99	150.4	139.3	...	1	0.93	154.6	135.6
2	0.99	153.4	139.4	...	2	0.84	...	128.9
3	1.01	155.8	...	141.6	3	0.96	...	137.7
4	1.07	161.1	...	145.1	4	1.00	155.1	141.0
5	0.91	145.6	133.6	...	5	0.97	...	138.3
6	1.06	157.2	...	144.2	6	0.96	159.8	137.8
7	1.07	162.8	...	145.3	7	1.01	158.4	141.4
	Mean:	155.2	137.4			Mean:	157.0	137.3
	Standard deviation:	6.0	3.3			Standard deviation:	2.5	4.2

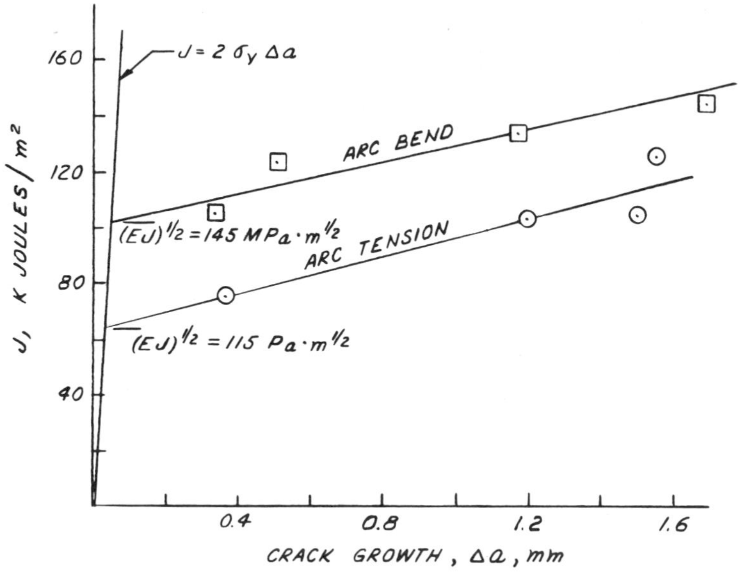

FIG. 7—J *versus* Δa *from arc tension and arc bend-chord support steel specimens described in Table 2.*

priate it is to use these existing procedures for the new arc specimen geometries. Comparing the results in Fig. 7 with the K_{Ic} results of Table 2, the compact procedure applied to the arc tension specimen gives a low measure of J_{Ic} and K_{Ic} and the rectangular bend procedure applied to the arc bend specimen gives a high measure of J_{Ic} and K_{Ic}. A possibly oversimplified analysis of these results is that the arc tension specimen is less compliant than the compact specimen and thus yields a low measure of J_{Ic} and K_{Ic}, and that the arc bend specimen is more compliant than the rectangular bend specimen and thus yields a high measure of J_{Ic} and K_{Ic}. A detailed analysis of J_{Ic} tests with arc tension specimens is described by Kapp and Bilinsky [*12*], work related to this paper.

Aluminum Tests

Twenty specimens were made with C-R orientation from an aluminum hollow cylinder extrusion and tested so that K_{Ic} and accurate crack mouth and load-line displacements could be determined. Roller bearings were used for the arc support tests with the intent that free rolling support on the inner radius was maintained during the test with no movement of the center of the rollers. Table 3 lists the results, which indicate that the fracture toughness measured by both the arc and chord support geometries was close to that from the arc tension tests. The three mean values differed by less than 4%.

TABLE 3—*Aluminum fracture toughness results (nominal dimensions:* r_1 = *87.6 mm,* B = *12.7 mm,* W = *25.4 mm).*

Arc Tension $X/W = 0$			Arc Bend; Arc Support $S/W = 4$			Arc Bend; Chord Support $S/W = 4$; $Z/W = 0.1$		
Specimen No.	K_{max}, MPa $m^{1/2}$	K_{Ic}, MPa $m^{1/2}$	Specimen No.	K_{max}, MPa $m^{1/2}$	K_Q, MPa $m^{1/2}$	Specimen No.	K_{max}, MPa $m^{1/2}$	K_Q, MPa $m^{1/2}$
2	31.3	28.6	1	27.6	27.4	1	28.7	...
3	29.0	28.0	2	28.2	28.1	2	27.4	26.6
4	29.0	29.0	3	28.7	27.9	3	28.5	...
5	28.4	28.2	4	29.2	...	4	27.2	27.2
6	29.0	27.6	5	27.6	...	5	29.4	...
			6	31.4	...	6	28.1	28.2
			7	26.2	25.7	7	27.8	26.7
Mean:	29.3	28.3	Mean:	28.4	27.3	Mean:	28.2	27.2
Standard deviation:	1.1	0.5	Standard deviation:	1.6	1.1	Standard deviation:	0.8	0.7

Also, unlike the steel tests, all the K_{Ic} and K_Q results were well within the maximum load and specimen size requirements of ASTM E 399.

The aluminum tests were planned to provide a comparison of measured and calculated displacements for arc bend specimens. The comparison is given in Table 4. The measured parameters of v and δ were from the one specimen of each of the four groups which had S/W and a/W closest to the nominal dimensions; the calculations were based on values taken from Figs. 5 and 6 for the nominal dimensions. The generally good agreement between measured and calculated displacements is an indication of the precision and accuracy of the analytical and experimental results in the work here and the literature cited. The apparent trend towards lower measured than calculated displacements for arc support could be explained by movement of the point of contact between roller and specimen so as to decrease θ and S/W and thus decrease the specimen compliance.

Conclusions

The arc support and chord support-arc bend specimens are suitable for fracture toughness testing, each in its own range of geometry. The arc support specimen is best suited for a constant value of relative span, $S/W = 4$, and for r_2/r_1 between 1.0 and 1.4, which corresponds to θ between 0 and 53°. The chord support specimen is best suited to r_2/r_1 between 1.1 and 2.0, with the relative offset of the chordal surface at a constant value, $Z/W = 0.1$. This will allow a constant relative span, $S/W = 4$, for r_2/r_1 between 1.1 and 1.6, and a S/W gradually decreasing to 3.35 for r_2/r_1 between 1.6 and 2.0. The largest r_2/r_1 geometry of each of these specimen geometry ranges is shown in Fig. 8.

The main reason for the upper limit on r_2/r_1 for each of the specimen types is as follows. For arc support specimens with θ much above 50°, the contact point between specimen and roller may move enough during the test to significantly change the K, v, and δ of the specimen. For chord support specimens with S/W much below 3.35, the amount of shear relative to bending is significant enough that some materials may not fracture in the pure opening mode which is intended in K_{Ic} and J_{Ic} tests.

TABLE 4—*Comparison of measured and calculated displacement,* v *and* δ*, for aluminum arc bend specimens;* $r_2/r_1 = 1.29$, $S/W = 4$, $a/W = 0.53$, $E = 68\,950$ *MPa.*

	EBv/P		*EBδ/P*	
	Measured	Calculated	Measured	Calculated
Chord support	43	43.2	70	66.0
Arc support	55	66.4	86	93.4

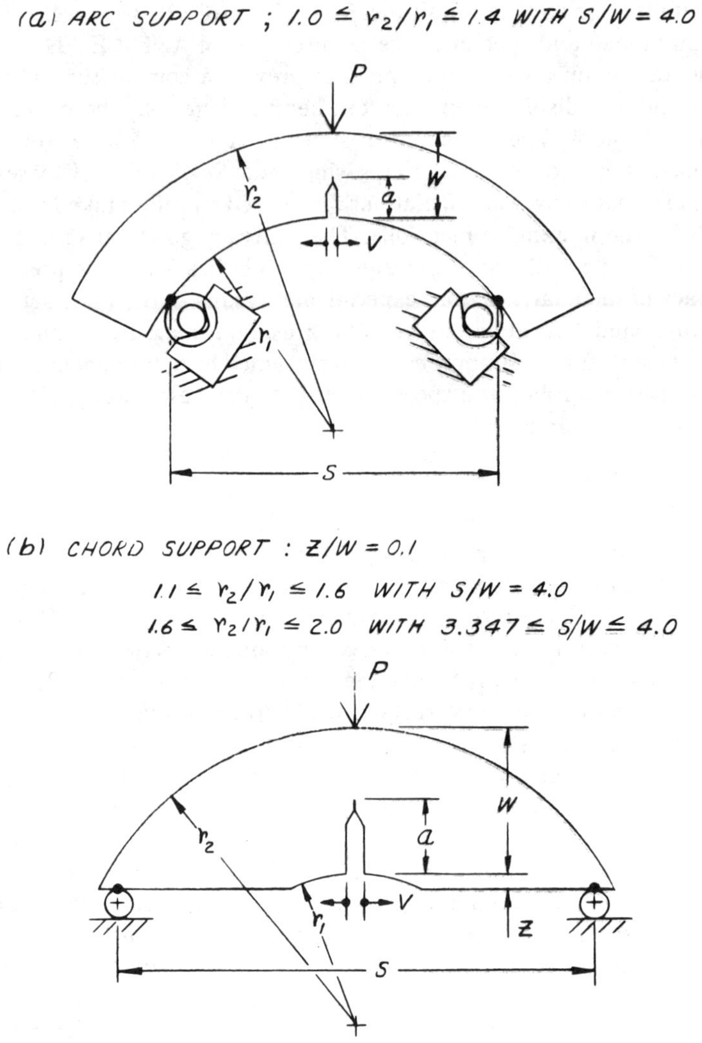

FIG. 8—*Recommended geometries for arc bend fracture testing; largest* r_2/r_1 *of recommended range is shown.*

We believe that K, v, and δ expressions can be determined from the results here with sufficient accuracy for K_{Ic} and J_{Ic} tests. The dashed lines in Figs. 4, 5, and 6 are believed to be accurate within about 1%, 2%, and 6%, respectively, for K, v, and δ over the range of a/W from 0.3 to 0.7. For the range of specimen geometries described above and for a/W between 0.45 and 0.55, the K, v, and δ results here are believed to be accurate within 0.5%, 1%, and 3%, respectively.

Regarding J_{Ic} tests with arc bend geometries, the chord support specimen is most suitable, with $S/W = 4.0$ and $Z/W = 0.1$. As r_2/r_1 approaches 1.0, the testing and analysis of chord support J_{Ic} specimens will become identical to that of the rectangular bend specimen in ASTM E 813.

APPENDIX

An expression which is useful for analyzing arc bend-chord support specimens can be obtained from plane geometry (Fig. 9). Using

$$S = 2r_u \sin \phi$$

a nondimensional form can be written

$$S/W = 2\left[r_1/W + U/W\right] \sin\left[\cos^{-1}\left(\frac{r_1/W - Z/W}{r_1/W + U/W}\right)\right] \tag{15}$$

An example of the use of Eq 15 is where, for $r_2/r_1 = 1.625$, $r_1/W = 1.6$, $U/W = 0.9$, and $Z/W = 0.1$, S/W is calculated to be 4.0. This demonstrates that, for $U/W = 0.9$ and $Z/W = 0.1$, $r_2/r_1 = 1.625$ is the upper limit of arc bend geometries for chord support with $S/W = 4.0$.

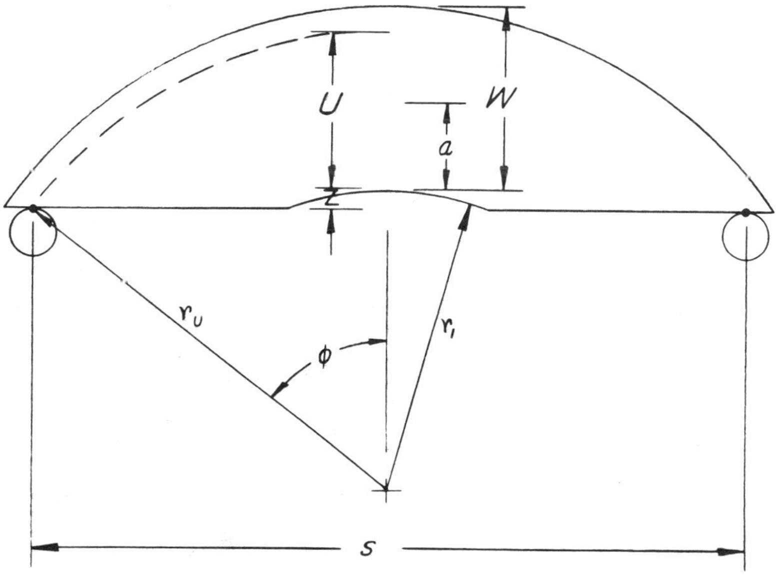

FIG. 9—*Arc bend-chord support specimen geometry.*

References

[1] Jones, A. T., "Fracture Toughness Testing with Sections of Cylinders," *Engineering Fracture Mechanics*, Vol. 6, 1974, pp. 653-662.

[2] Tracy, P. G., "Analysis of a Radial Crack in a Circular Ring Segment," *Engineering Fracture Mechanics*, Vol. 7, 1975, pp. 253-260.

[3] Ritter, J. C. and Rea, T. W., "A Curved Beam Fracture Toughness Specimen," *International Journal of Pressure Vessels and Piping*, Vol. 5, 1977, pp. 275-286.

[4] Gross, B. and Srawley, J. E., "Stress Intensity and Displacement Coefficients for Radially Cracked Ring Segments Subject to Three-Point Bending," NASA Technical Memo. 83059, National Aeronautics and Space Administration, Washington, D.C., March 1983.

[5] Tada, H., Paris, P. C., and Irwin, G. P., *The Stress Analysis of Crack Handbook*, Del Research Corp., Hellertown, Pa., 1973.

[6] Hussain, M. A., Lorenson, W. E., Kendall, D. P., and Pu, S. L., "A Modified Collocation Method for C-Shaped Specimens," Benet Weapons Laboratory Technical Report R-WV-T-X-6-73, Watervliet, N.Y., 1973.

[7] Gross, B. and Srawley, J. E., "Analysis of Radially Cracked Ring Segments Subject to Forces and Couples," in *Developments in Fracture Mechanics Test Methods Standardization, ASTM STP 633*, W. F. Brown, Jr., and J. G. Kaufman, Eds., American Society for Testing and Materials, Philadelphia, 1977, pp. 39-56.

[8] Kapp, J. A. and Pu, S. L., "Fatigue Design of Thick-Walled Cylinders Considering the OD as a Failure Initiation Site," in *Pressure Vessel Design*, G. E. O. Widera, Ed., American Society of Mechanical Engineers, New York, 1982, pp. 115-128.

[9] Gifford, I. N., Jr., and Hilton, P. D., "Stress Intensity Factors by Enriched Finite Elements," *Engineering Fracture Mechanics*, Vol. 10, 1978, pp. 485-496.

[10] Roark, R. J. and Young, W. C., *Formulas For Stress and Strain*, McGraw-Hill, New York, 1975.

[11] Haggag, F. M. and Underwood, J. H., "Compliance of a Three-Point Bend Specimen at Load Line," *International Journal of Fracture*, Vol. 26, 1984, pp. R63-R65.

[12] Kapp, J. A. and Bilinsky, W. J., "J_{Ic} Testing Using Arc Tension Specimens," this publication, pp. 297-306.

J. A. Kapp[1] *and W. J. Bilinsky*[2]

J_{Ic} Testing Using Arc-Tension Specimens

REFERENCE: Kapp, J. A. and Bilinsky, W. J., "**J_{Ic} Testing Using Arc-Tension Specimens,**" *Fracture Mechanics: Seventeenth Volume, ASTM STP 905*, J. H. Underwood, R. Chait, C. W. Smith, D. P. Wilhem, W. A. Andrews, and J. C. Newman, Eds., American Society for Testing and Materials, Philadelphia, 1986, pp. 297-306.

ABSTRACT: J_{Ic} was determined in two materials (6061-T651 aluminum and ASTM A-723 Grade 1, Class 4 pressure vessel steel) using arc-tension (A(T)) and compact tension (C(T)) specimens. The *J-R* curves were determined by using the multispecimen method and the compliance unloading method. *J* was determined for the A(T) specimen by the Merkle-Corten method of analysis as modified by Clarke and Landes. A correction factor was included to account for the tensile loading component, and both the plastic and elastic components of *J* were necessary when using the A(T) specimen. With the proper formulas for *J* in the A(T) specimen, the same *J-R* curves were determined in both materials using either A(T) or C(T) specimens. This preliminary study suggested that the A(T) specimen was totally adequate for J_{Ic} testing and should be included in subsequent versions of ASTM Test for J_{Ic}, A Measure of Fracture Toughness (E 813).

KEY WORDS: J_{Ic} testing, fracture testing, specimen design, fracture mechanics

Fracture toughness, whether measured as K_{Ic} or J_{Ic}, is a valuable measure of a material's tolerance of pre-existing defects and has many engineering applications. The present method for determining J_{Ic} (ASTM E 813 on J_{Ic}, A Measure of Fracture Toughness) allows J_{Ic} to be measured by either a compact tension (C(T)) specimen or a three-point bend specimen (SE(B)). Sometimes it is difficult to obtain specimens of these geometries from certain structural components such as cylindrical pressure vessels. The arc-tension (A(T)) specimen is easily obtained from these cylindrical components. This paper summarizes the results of an initial study into the feasibility of using A(T) specimens for J_{Ic} testing.

[1]Materials Research Engineer, U.S. Army Armament, Munitions, and Chemical Command, Armament Research and Development Center, Large Caliber Weapon Systems Laboratory, Benet Weapons Laboratory, Watervliet, NY 12189.

[2]Mechanical Engineer, General Electric Company, Selkirk, NY 12518.

Calculation of J for A(T) Specimens

The A(T) specimen (Fig. 1) encompasses a large range of possible geometries. This is due to the variability of R_2, R_1, and X. For this specimen to be of the most use, restrictions on R_2 and R_1 cannot be allowed, but restricting X in J testing should cause few difficulties. X is made variable in K testing mainly for load efficiency, but if substantial plasticity is allowed as in J testing, load efficiency is not as important. Thus we restrict our thinking to A(T) specimens with $X = 0$, which has the further experimental advantage in displacement measurement of the load line intersecting the crack mouth.

To determine J for the arc-tension specimen, we proceed in the manner outlined by Merkle and Corten [*1*] as modified by Clarke and Landes [*2*]. Under fully plastic conditions on the uncracked ligament, the load-displacement trace is idealized as Fig. 2, and J is (Eq 9, Ref *2*)

$$J = \frac{2}{bB}\left[\frac{1+\alpha}{1+\alpha^2} - \alpha\frac{(1-2\alpha-\alpha^2)}{(1+\alpha^2)^2}\right]A_T + \frac{2\alpha}{bB}\frac{(1-2\alpha-\alpha^2)}{(1+\alpha^2)^2}P^*\delta_T + G$$
$$- \frac{2}{bB}\left[\frac{1+\alpha}{1+\alpha^2} + \alpha\frac{(1-2\alpha-\alpha^2)}{(1+\alpha^2)^2}\right]A_e \quad (1)$$

where b and B are as shown in Fig. 1; A_T, P^*, δ_T and A_e are from Fig. 2; G is the elastic energy release rate; and

$$\alpha = 2((a/b)^2 + a/b + 1/2)^{1/2} - 2(a/b + 1/2) \quad (2)$$

where a is the crack length shown in Fig. 1.

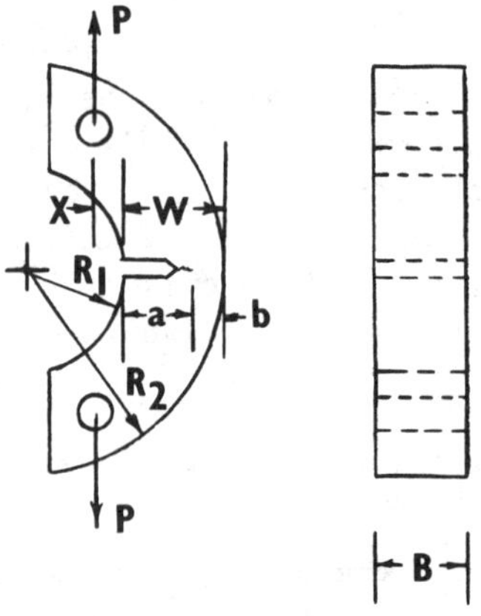

FIG. 1—*Arc-tension specimen.*

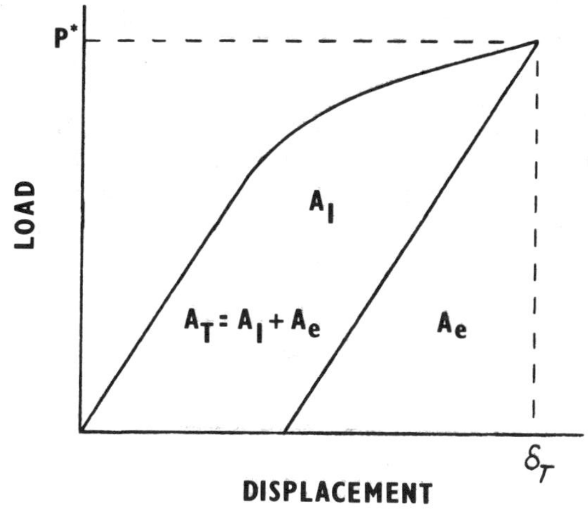

FIG. 2—*Idealized elastic-plastic load-displacement trace.*

Equation 1 is too cumbersome to be useful experimentally. The first step in simplifying is to assume that the higher order terms involving A_T and $P^*\delta_T$ cancel:

$$J = \frac{2A_T}{bB}\,\frac{1+\alpha}{1+\alpha^2} + G - \frac{2A_e}{bB}\left[\frac{1+\alpha}{1+\alpha^2} + \frac{\alpha(1-2\alpha-\alpha^2)}{(1+\alpha^2)^2}\right] \tag{3}$$

It is convenient to represent G in terms of the elastic area A_e. It can be shown that [1,2]

$$G = \left[\frac{k(1-a/w)Y^2}{EB}\right]\frac{2A_e}{Bb} \tag{4}$$

where

$$Y = \frac{KB\sqrt{w}}{P}$$

and k is the elastic stiffness of the specimen or the initial slope of the load-displacement record, K is the stress intensity factor, and E is the elastic modulus. Finally, we can write Eq 3 in the simple form

$$J = \lambda_1\,\frac{2A_T}{Bb} + \lambda_2\,\frac{2A_e}{Bb} \tag{5}$$

$$\lambda_1 = \frac{1+\alpha}{1+\alpha^2} \tag{6}$$

$$\lambda_2 = \frac{k(1 - a/w)Y^2}{EB} - \left(\lambda_1 + \frac{\alpha(1 - 2\alpha - \alpha^2)}{(1 - \alpha^2)\alpha^2}\right) \tag{7}$$

These equations are exactly the same equations that have been developed for the C(T) specimen [1,2]; but for that specimen when a/w is 0.5 or larger, λ_2 is negligible. Such is not the case with the arc-tension specimen. We evaluate λ_2 by using wide range expressions fit to numerical stress analysis results [3,4]:

$$Y = \frac{KB\sqrt{w}}{P} = [3x/w + 1.9 + 1.1a/w] \times [1 + 0.25(1 - a/w)^2(1 - R_1/R_2)]f(a/w) \tag{8}$$

$$f(a/w) = ((a/w)^{1/2}/(1 - a/w)^{3/2})(3.74 - 6.30a/w + 6.32(a/w)^2 - 2.43(a/w)^3)$$

$$k = \frac{P}{\delta} = \frac{EB(1 - a/w)^2}{(2x/w + 1 + a/w)F(a/w, r_1/r_2)} \tag{9}$$

$$F(a/w, R_1/R_2) = [0.34 + 13.75a/w - 12.67(a/w)^2 + 6.47(a/w)^2 + (1 - a/w)^{0.05}(1 - R_1/R_2)(0.8 - 0.5R_1/R_2)]$$

To compare the quantities necessary to determine J, it is convenient to use the notation

$$\lambda_1 = \frac{1 + \alpha}{1 + \alpha^2}$$

$$\lambda^* = \frac{\alpha(1 - 2\alpha - \alpha^2)}{(1 + \alpha^2)^2}$$

$$\lambda_J = \lambda_1 + \lambda^*$$

$$\lambda_e = \frac{k(1 - a/w)Y^2}{EB}$$

These quantities are computed for various R_1/R_2 and comparisons made in Table 1. The tabulation shows that in the crack length range important to J_{Ic} testing $a/w \geq 0.45$, λ_2 is not negligible. Indeed, should A_e be a significant portion of A_T, neglecting to account for this elastic portion of J would introduce errors approaching 20%. Therefore, unlike J_{Ic} testing with C(T) speci-

TABLE 1—*Comparison of the quantities used to determine* J *for the arc-tension,* X/W = *0 specimen.*

a/w	λ_1	λ^*	λ_J	λ_e	λ_e/λ_J	λ_2
			R_1R_2 = 0.91			
0.3000	1.1773	0.0963	1.2736	1.8404	1.4450	0.5668
0.3500	1.1674	0.1013	1.2687	1.7164	1.3528	0.4476
0.4000	1.1565	0.1034	1.2600	1.6092	1.2772	0.3492
0.4500	1.1448	0.1029	1.2477	1.5183	1.2169	0.2706
0.5000	1.1325	0.1000	1.2325	1.4423	1.1703	0.2099
0.5500	1.1196	0.0950	1.2146	1.3793	1.1356	0.1647
0.6000	1.1063	0.0882	1.1946	1.3271	1.1110	0.1326
0.6500	1.0929	0.0800	1.1728	1.2835	1.0944	0.1107
0.7000	1.0793	0.0705	1.1498	1.2459	1.0837	0.0962
0.7500	1.0657	0.0600	1.1257	1.2120	1.0766	0.0863
0.8000	1.0522	0.0488	1.1009	1.1790	1.0709	0.0780
			R_1R_2 = 0.67			
0.3000	1.1773	0.0963	1.2736	1.8850	1.4800	0.6114
0.3500	1.1674	0.1013	1.2687	1.7501	1.3794	0.4814
0.4000	1.1565	0.1034	1.2600	1.6332	1.2962	0.3732
0.4500	1.1448	0.1029	1.2477	1.5338	1.2293	0.2861
0.5000	1.1325	0.1000	1.2325	1.4505	1.1769	0.2182
0.5500	1.1196	0.0950	1.2146	1.3814	1.1373	0.1668
0.6000	1.1063	0.0882	1.1946	1.3241	1.1085	0.1296
0.6500	1.0929	0.0800	1.1728	1.2763	1.0883	0.1035
0.7000	1.0793	0.0705	1.1498	1.2355	1.0746	0.0857
0.7500	1.0657	0.0600	1.1257	1.1991	1.0652	0.0734
0.8000	1.0522	0.0488	1.1009	1.1644	1.0576	0.0634
			R_1R_2 = 0.5			
0.3000	1.1773	0.0963	1.2736	1.8987	1.4908	0.6251
0.3500	1.1674	0.1013	1.2687	1.7591	1.3865	0.4904
0.4000	1.1565	0.1034	1.2600	1.6375	1.2997	0.3776
0.4500	1.1448	0.1029	1.2477	1.5339	1.2294	0.2862
0.5000	1.1325	0.1000	1.2325	1.4469	1.1740	0.2145
0.5500	1.1196	0.0950	1.2146	1.3746	1.1318	0.1600
0.6000	1.1063	0.0882	1.1946	1.3147	1.1006	0.1201
0.6500	1.0929	0.0800	1.1728	1.2648	1.0784	0.0919
0.7000	1.0793	0.0705	1.1498	1.2223	1.0631	0.0725
0.7500	1.0657	0.0600	1.1257	1.1846	1.0524	0.0590
0.8000	1.0522	0.0488	1.1009	1.1493	1.0439	0.0484
			R_1R_2 = 0.4			
0.3000	1.1773	0.0963	1.2736	1.9007	1.4923	0.6270
0.3500	1.1674	0.1013	1.2687	1.7593	1.3867	0.4906
0.4000	1.1565	0.1034	1.2600	1.6357	1.2982	0.3757
0.4500	1.1448	0.1029	1.2477	1.5302	1.2263	0.2824
0.5000	1.1325	0.1000	1.2325	1.4414	1.1696	0.2090
0.5500	1.1196	0.0950	1.2146	1.3676	1.1260	0.1530
0.6000	1.1063	0.0882	1.1946	1.3064	1.0936	0.1118
0.6500	1.0929	0.0800	1.1728	1.2554	1.0704	0.0826
0.7000	1.0793	0.0705	1.1498	1.2121	1.0543	0.0624
0.7500	1.0657	0.0600	1.1257	1.1740	1.0429	0.0483
0.8000	1.0522	0.0488	1.1009	1.1385	1.0341	0.0375

mens, we must include the contribution of A_e to obtain an accurate measurement of J when using A(T) specimens.

Although λ_2 could be calculated from Eqs 8 and 9, significant computation is involved. Noting that λ_2 is virtually independent of R_1/R_2, a simple polynomial in a/w can be found from which λ_2 is more easily determined. Using least squares, the polynomial is

$$\lambda_2 = 1.919 - 6.235(a/w) + 6.935(a/w)^2 - 2.557(a/w)^3 \qquad (10)$$

In the range of $0.5 \leq a/w \leq 0.6$, Eq 10 agrees with the computed values of λ_2 within about 5% for R_1/R_2 between 0.4 and 0.9.

Experimental Results

The J_{Ic} tests were performed on a pressure vessel steel and an aluminum alloy using A(T) and C(T) specimens. Both methods for determining crack growth outlined in ASTM E 813 (compliance unloading and fracture surface measurement) were used. The aluminum alloy was 6061-T651 supplied in rolled sheet form 1.27 cm (0.5 in.) thick. Specimens were obtained such that the L-T direction was tested. The C(T) specimens were of standard geometry with through-thickness (B) of 1.27 cm (0.5 in.). The A(T) specimens were produced such that the outside radius R_2 was 5.08 cm (2.0 in.) and the inside radius R_1 was 2.54 cm (1.0 in.). The pressure vessel steel was ASTM A-723 Grade 1, Class 4. Specimens were obtained from a long hollow cylindrical forging that had an outside radius of 11.7 cm (4.6 in.) and an inside radius of 4.8 cm (1.9 in.). Specimens were obtained in the C-R orientation with a B value of 3.4 cm (1.35 in.). After testing the A(T) specimens, C(T) specimens were machined from one of the broken halves of the A(T) specimen. The largest possible specimen was obtained, one with a B value of 2.3 cm (0.8 in.).

The results of the testing of these samples appear in Figs. 3 to 6. The results from the pressure vessel steel (Figs. 3 and 4) suggest very little difference in J_{Ic} by using different specimens. With the multispecimen method (Fig. 3), the slope of the resistance curve is the same for either specimen, but the intersection with the blunting line is somewhat greater when using compact tension specimens. Although J_{Ic} appears to be affected by the specimen, the difference between the two test results can be easily attributed to scatter. With compliance unloading (Fig. 4), more variation is evident. The two resistance curves from the arc-tension specimens are approximately parallel but offset such that different values of J_{Ic} were obtained. Comparing these with the two curves generated using compact tension specimens, we find that the least squares fit to these data give a slope which is substantially steeper than those obtained with arc-tension specimens. The J_{Ic} values obtained from these tests are very close (76 kJ/m^2 and 81 kJ/m^2) and compare very favorably with one of the two arc-tension specimen J_{Ic} measurements. Again, one might consider

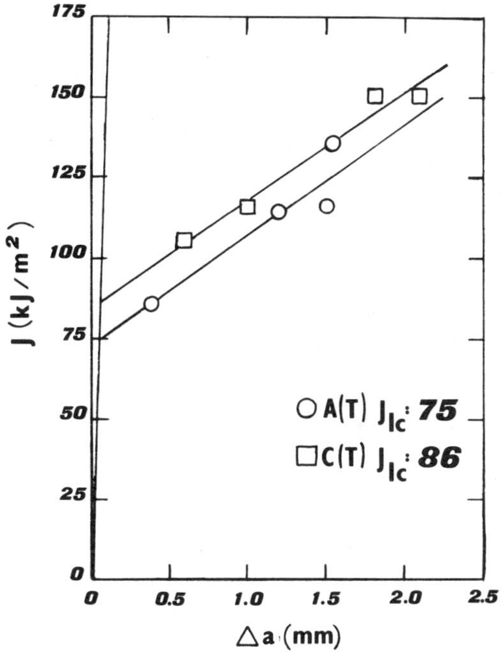

FIG. 3—*Multispecimen* J-R *curves for A-723 pressure vessel steel.*

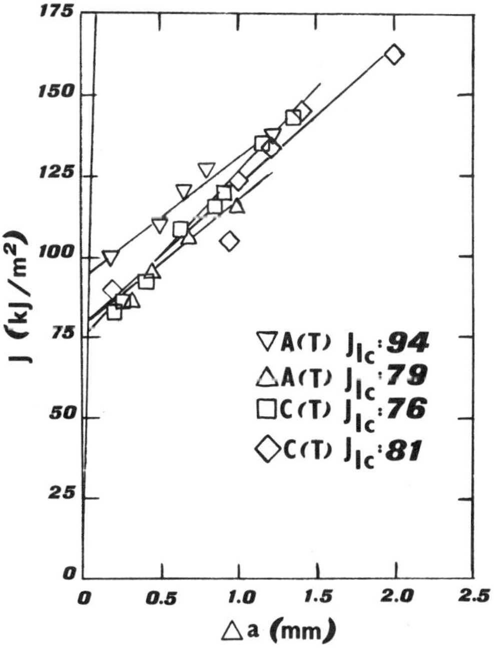

FIG. 4—*Compliance unloading* J-R *curves for A-723 pressure vessel steel.*

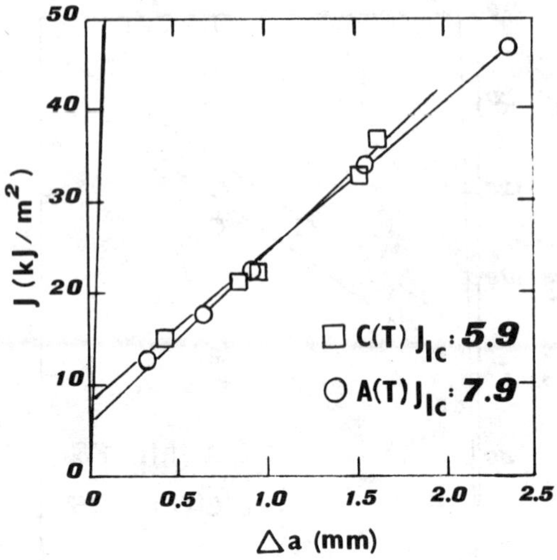

FIG. 5—*Multispecimen* J-R *curves for 6061-T651 aluminum.*

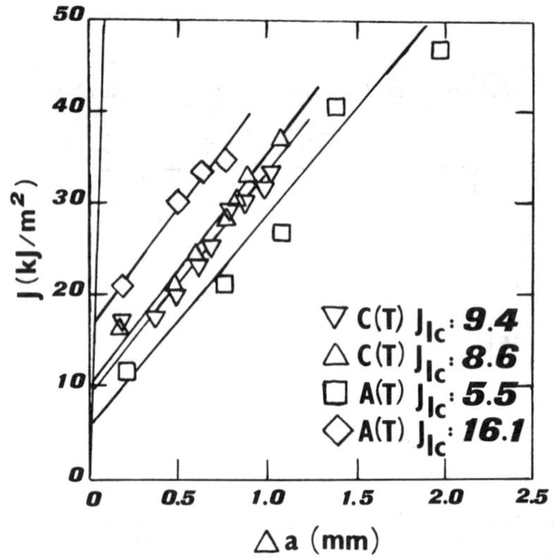

FIG. 6—*Compliance unloading* J-R *curves for 6061-T651 aluminum.*

the differences as the result of using different specimen geometries, but it is safe to account for these differences as material scatter.

Comparing the results of testing the 6061-T651 aluminum alloy also shows little or no difference between arc-tension and compact tension specimens. With multispecimen testing (Fig. 5), essentially the same resistance curve was obtained using either specimen geometry. The least squares lines fit to these data give a small difference in slope and somewhat different J_{Ic} value, but again these differences are material property scatter. With compliance unloading, similar to the steel results, more scatter was observed. All the resistance curves had essentially the same slopes, but J_{Ic} varied a great deal with these specimens. The two compact tension J_{Ic} values agreed very well, but the two arc-tension specimens differed substantially, although the average of the two compact tension values compared very well with the average of the two arc-tension values.

For further comparison, the J_{Ic} results are presented in Table 2. For each material and specimen type, the three J_{Ic} values generated (one by the multi-specimen method and two by the compliance unloading method) are treated as a single statistical population. The mean value and standard deviation are given for each population. It is clear that the mean value of J_{Ic} from the A(T) and C(T) specimens of pressure vessel steel is essentially the same, although there seems to be somewhat more scatter generated when the A(T) specimens are used. The reason for the scatter cannot be established at this time. Further testing would be required to determine if the difference is due to specimen geometry or natural material scatter. The same can be stated for the aluminum. Reasonable agreement between the mean values of J_{Ic} from either specimen type was observed, but as in the case of the aluminum tests, scatter is much greater when A(T) specimens are used than when C(T) specimens are used. Again whether the cause of the scatter is the specimen geometry or material related, it would have to be determined by testing a larger number of specimens.

Regardless of the reasons for increased scatter in the J_{Ic} results from A(T) versus C(T) specimens, the fact that the mean values of J_{Ic} for two different materials using both A(T) and C(T) specimens is encouraging. Further test-

TABLE 2—*Statistical comparison of* J_{Ic} *measured with A(T) and C(T) specimens.*

Material		J_{Ic}, kJ/m^2	
		A(T)	C(T)
Pressure vessel steel		75	86
		94	76
		79	81
	Mean	82.6	81.0
	Standard deviation	10.0	5.0
6061-T651 aluminum		7.9	9.4
		5.5	8.6
		16.1	5.9
	Mean	9.8	8.0
	Standard deviation	5.56	1.83

ing should be performed, perhaps in coordination with ASTM Task Group E24.08.04, as a laboratory round robin to establish if the A(T) specimen should be included in later versions of ASTM E 813.

Summary and Conclusions

Analyses and tests were performed to determine the applicability of using A(T) specimens for J_{Ic} testing. The analysis showed that to determine J for these specimens a more involved calculation is necessary, namely, J must be broken down into its elastic and plastic components. Two areas under the load displacement curve must be measured. This is more complex than the analysis involved with compact tension specimens, but the additional amount of analysis involved should not restrict the use of A(T) specimens. The J_{Ic} measurements were made using compact tension and arc-tension specimens. The A(T) specimen should be considered as an alternative specimen in future versions of ASTM E 813.

References

[*1*] Merkle, J. G. and Corten, H. T., *Journal of Pressure Vessel Technology, Transactions of ASME*, Vol. 96, Nov. 1974, pp. 286-292.

[*2*] Clarke, G. A. and Landes, J. D., *Journal of Testing and Evaluation*, Vol. 7, No. 5, Sept. 1979, pp. 264-269.

[*3*] Kapp, J. A., Newman, J. C., Jr., and Underwood, J. H., *Journal of Testing and Evaluation*, Vol. 8, No. 6, Nov. 1980, pp. 314-317.

[*4*] Kapp, J. A., Leger, G. S., and Gross, B., "Wide-Range Displacement Expressions for Standard Fracture Mechanics Specimens," in *Fracture Mechanics: Sixteenth Symposium, ASTM STP 868*, American Society for Testing and Materials, Philadelphia, 1985, pp. 27-44.

Jacques H. Giovanola[1]

Investigation and Application of the One-Point-Bend Impact Test

REFERENCE: Giovanola, J. H., **"Investigation and Application of the One-Point-Bend Impact Test,"** *Fracture Mechanics: Seventeenth Volume, ASTM STP 905*, J. H. Underwood, R. Chait, C. W. Smith, D. P. Wilhem, W. A. Andrews, and J. C. Newman, Eds., American Society for Testing and Materials, Philadelphia, 1986, pp. 307–328.

ABSTRACT: This paper investigates a dynamic fracture test that produces smoothly varying stress intensity histories of controlled amplitude and duration. A test configuration has been adopted in which a simple edge-cracked specimen is loaded in bending by impacting it at the midsection without supporting it at the outer edges. The stress intensity history is measured by a strain gage near the crack tip and has approximately a sinusoidal time dependence.

The dependence of stress intensity amplitude and duration on the impact velocity and on the specimen material and dimensions has been determined both analytically and experimentally. The limit on the specimen aspect ratio to obtain a smooth stress intensity history controlled only by the first mode of vibration of the specimen has also been established. Stress intensity pulses of durations varying between 90 and 500 μs can be readily achieved, and the maximum amplitude can be independently adjusted over an order of magnitude (20 to 200 MPa $m^{1/2}$).

The new experimental procedure has been used to measure the dynamic fracture toughness of 4340 steel (HRC 50) at three loading rates. Comparison of the dynamic results with the value of the static toughness indicates little strain rate sensitivity of the fracture toughness of 4340 steel in the range of loading rates investigated (quasi-static to 3×10^6 MPa $m^{1/2}$ s^{-1}).

KEY WORDS: dynamic fracture, stress intensity history, impact bend test, unsupported bend specimen, dynamic fracture toughness, 4340 steel, rate sensitivity

Nomenclature

- a Crack length
- $\dot{a}$ Crack velocity
- B Specimen thickness
- C_{CR} Elastic compliance of bend specimen due to crack

[1]Research Engineer, Department of Metallurgy and Fracture Mechanics, SRI International Menlo Park, CA 94025.

C_{UN} Elastic compliance of uncracked bend specimen
$d(x,t)$ Specimen deflection
d_{CR}^{CL} Specimen deflection along center line due to crack
d_{UN}^{CL} Uncracked specimen deflection along center line
E Young's modulus
E_{MAX}^{KIN} Maximum kinetic energy of specimen
E_{MAX}^{STR} Maximum strain energy of specimen
$K_I(t)$ Dynamic stress intensity factor
K_{Ic} Static fracture toughness
K_{Id} Dynamic initiation fracture toughness
K_I^{MAX} Maximum stress intensity amplitude
m Specimen mass
M Hammer mass
P_H Hammer load
P_H^{MAX} Maximum hammer load
P_{MAX}^{EFF} Maximum effective dynamic load on the specimen
S Specimen length
t Time
t^{MAX} Time at which maximum stress intensity amplitude is reached
T_0 Duration of stress intensity history
T_1 First natural period of oscillation of specimen
T_{UN} First natural period of oscillation of the uncracked specimen
V_H Hammer velocity at t^{MAX}
V_{imp} Hammer impact velocity
V_S Velocity of the specimen center of mass at t^{MAX}
W Specimen width
x Coordinate along specimen axis
ρ Mass density of specimen material
ν Poisson's ratio

Introduction

A dynamic fracture test having an adjustable and smoothly varying load history is desirable for unambiguous determination of dynamic fracture behavior. Unfortunately, conventional tests, such as the Charpy impact test or the dynamic tear test, do not produce a smoothly varying stress intensity history. Thus interpretation of the results is not straightforward. This paper discusses a modification of the conventional impact bend test, first proposed and investigated by Kalthoff and co-workers [1-3], which produces a smoothly varying stress intensity history. The amplitude and duration of the stress intensity history are easily varied by varying the impact velocity and the specimen dimensions.

We have investigated the influence of specimen dimensions and impact velocity on the stress intensity history both analytically and experimentally. We have then used the new test procedure to measure the dynamic fracture toughness of 4340 steel at three strain rates.

Background

Dynamic effects in impact-loaded, three-point-bend specimens (3PBS) have been discussed in several recent publications. In particular, Kalthoff and co-workers have shown that the stress intensity history results from the superposition of the specimen free oscillations on the fundamental oscillation of the hammer-specimen-support system [*1,2,4–6*]. Böhme [*6*] has proposed a semi-analytical description of the behavior of the hammer-specimen-support system to predict the influence of specimen dimensions, impact velocity, and material properties on the stress intensity history. He has further demonstrated that the dynamic effects due to the specimen oscillations can be estimated for specimens of arbitrary size and material from a set of normalized master curves. Each master curve depends only on the ratios of crack length to width, a/W, and specimen length to width, S/W.

Böhme and Kalthoff [*1*] have also observed that soon after impact the 3PB specimen momentarily loses contact with the supports. Therefore, during the initial phase of the test, the supports do not influence the loading, and the stress intensity history obtained for specimens impacted with or without supports is identical. The crack tip loading is then strictly achieved by inertia. Only at later times, when the specimen makes contact with the supports again, does its behavior differ from that of the unsupported specimen; the stress intensity history for the supported specimen continues to rise, whereas that for the unsupported specimen decreases to zero.

On the basis of these observations, Kalthoff and co-workers [*2,3*] have demonstrated that the bend specimen impacted without supports—called *one-point-bend test* (1PBT)—provides a convenient configuration to obtain controlled and smoothly varying stress intensity histories.

The test arrangement and loading mechanism for the 1PBT are illustrated in Fig. 1. Figure 2 shows the resulting stress intensity history in a specimen with a blunt notch, as measured with a strain gage placed near the notch tip. This stress intensity history is roughly sinusoidal and is characterized by the half period T_0 and the maximum amplitude K_I^{MAX}. Figure 2 also shows the corresponding hammer load. On impact, the center portion of the specimen is accelerated away from the hammer, whereas the end portions of the specimen lag behind because of inertia. This causes the specimen to bend and to load the crack tip. The specimen deflection also progressively unloads the hammer tip until the maximum deflection (and hence the maximum stress intensity) is reached. As the specimen bends back, the stress intensity decreases, whereas the hammer load increases again because the relative displacement of the hammer and of the specimen's center portion are now in opposite directions [*1*].

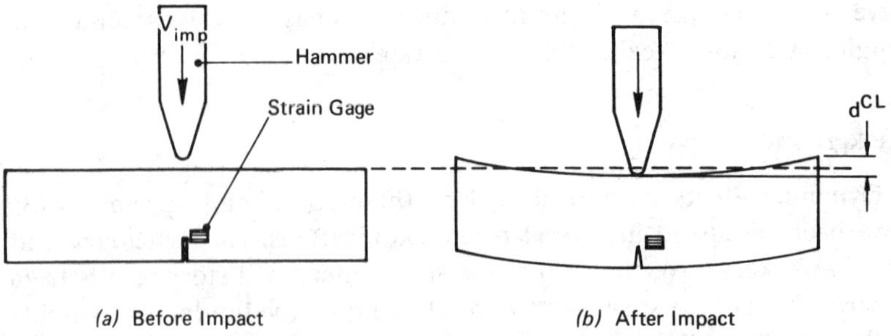

FIG. 1—*Test arrangement and loading mechanism for the 1PBT.*

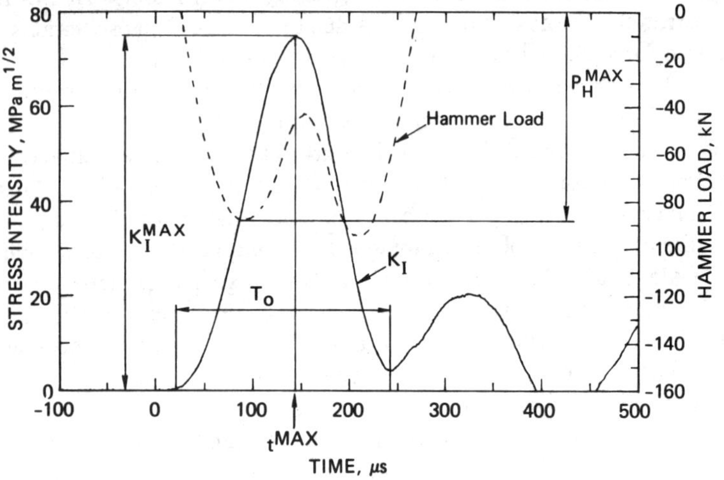

FIG. 2—*Typical stress intensity and load histories obtained in 1PBT.*

According to this interpretation of the stress intensity records, the first natural period of oscillation of the specimen controls the duration of the approximately sinusoidal stress intensity history. Thus the test duration can be controlled by changing the dimensions of the specimen to change the natural frequency. Furthermore, the results of Böhme [*6*] suggest that the maximum stress intensity amplitude is proportional to the impact velocity.

The purpose of the present investigation was to verify the foregoing predictions, to establish the range of load durations that could be achieved with the 1PBT, to determine more precisely the role of specimen dimensions and material properties on the stress intensity history, and to apply the 1PBT to dynamic fracture toughness testing.

Investigation of the Dynamic Behavior of the 1PBT

Analysis

Here we assume that the stress intensity history is sinusoidal with a period $2T_0$ equal to the first natural period of the specimen and with an amplitude K_I^{MAX}. We then determine how these two parameters depend on the specimen material and dimensions and on the impact velocity.

Details of this approximate analysis are presented in the Appendix together with references to earlier analyses. We find that T_0 and K_I^{MAX} are given by

$$T_0 = T_{UN}\, h_1\left(\frac{S}{W}, \frac{a}{W}, \nu\right) = W\sqrt{\frac{\rho}{E}}\, h_2\left(\frac{S}{W}, \frac{a}{W}, \nu\right) \tag{1}$$

and

$$K_I^{MAX} = V_{imp}\sqrt{W}\sqrt{\rho E}\, h_3\left(\frac{S}{W}, \frac{a}{W}, \nu\right) \tag{2}$$

where T_{UN} is the first natural period of oscillation of the uncracked beam with free ends; h_1, h_2, and h_3 are dimensionless functions of the specimen dimensions and of Poisson's ratio; and the other symbols are as defined in the Nomenclature section of this paper. In Eq 2, it has been assumed that the hammer mass, M, is much larger than the specimen mass, m.

Equations 1 and 2 show that for geometrically similar specimens (and neglecting a mild dependence on Poisson's ratio), T_0 scales with the specimen size and with the square root of the ratio of mass density to elastic modulus; K_I^{MAX} is proportional to the impact velocity and scales with the square root of specimen size, mass density, and elastic modulus. Both T_0 and K_I^{MAX} are independent of specimen thickness and increase with crack length for given values of S, W, and V_{imp}. Equations 9 and 18 in the Appendix can be used to obtain estimates of T_0 and K_I^{MAX}. The estimates of K_I^{MAX} will only yield approximate trends, however, because two coefficients are not known precisely in Eq 18.

Experimental Verification

To verify the predictions of Eqs 1 and 2, we have tested several steel and aluminum specimens. Table 1 gives a matrix of the specimens tested and their dimensions and aspect ratios. The specimens had a blunt notch of 0.5 mm (20 mil) tip radius to prevent fracture during the tests. We measured the stress intensity history with a single element strain gage placed a small distance—3 to 5 mm (0.1 to 0.2 in.)—away from and directly above the notch tip. The gage was oriented to measure strains perpendicular to the crack. The stress

TABLE 1—*Specimen dimensions and material, and impact conditions for experimental investigation of 1PBT.*

Specimen Size Number	Materials Tested	S		W		B		a		S/W	A/W	V_{imp}	
		mm	(in.)	mm	(in.)	mm	(in.)	mm	(in.)			m/s	(ft/s)
1	steel aluminum	177.8	(7)	25.4	(1)	12.7	(0.5)	7.6	(0.3)	7	0.3	5.3	(17.4)
2	steel aluminum	355.6	(14)	25.4	(1)	12.7	(0.5)	7.6	(0.3)	14	0.3	5.3	(17.4)
3a	steel aluminum	241.3	(9.5)	63.5	(2.5)	12.7	(0.5)	19.1	(0.75)	3.8	0.3	5.3	(17.4)
3b	steel	241.3	(9.5)	63.5	(2.5)	12.7	(0.5)	25.4	(1.0)	3.8	0.4	5.3	(17.4)
3c	steel	241.3	(9.5)	63.5	(2.5)	12.7	(0.5)	31.8	(1.25)	3.8	0.5	5.3	(17.4)
4	steel aluminum	355.6	(14)	63.5	(2.5)	12.7	(0.5)	19.1	(0.75)	5.6	0.3	3.5, 5.3	(11.5, 17.4)
5	steel aluminum	101.6	(4)	38.1	(1.5)	12.7	(0.5)	12.7	(0.5)	2.7	0.33	1.0 to 5.3	(3.3 to 17.4)
6	steel	228.6	(9)	88.9	(3.5)	9.5	(0.375)	26.9	(1.06)	2.6	0.3	3.5	(11.5)

intensity can be calculated from the strain measurement by two methods. In the first method, stress intensity and strain are related using a static calibration procedure developed by Loss [*7,8*]. In the second method, the elastic singularity solution [*9,10*] is used to relate near-crack-tip strain and stress intensity factor; a plane stress situation is assumed at the gage location. Böhme and Kalthoff have discussed the uncertainties associated with this approach [*11*]. By comparing *K* values measured simultaneously with either of the two methods, we have found that for the specimen geometries considered here, the two methods agree to within 5%. Because of its simplicity, we used the singularity solution method for the blunt notch experiments. The impact tests were performed on a standard pendulum impact machine modified to accommodate the 1PBT specimen and a redesigned instrumented hammer. Figure 3 shows a schematic of the test setup. In some experiments we varied the impact velocity between 1 and 5.3 m/s by changing the drop height of the hammer. The impact velocity was measured by two pairs of light-emitting diodes/photodiodes. For each specimen and test condition, we repeated the experiment several times to check reproducibility. Figures 4 to 7 summarize the experi-

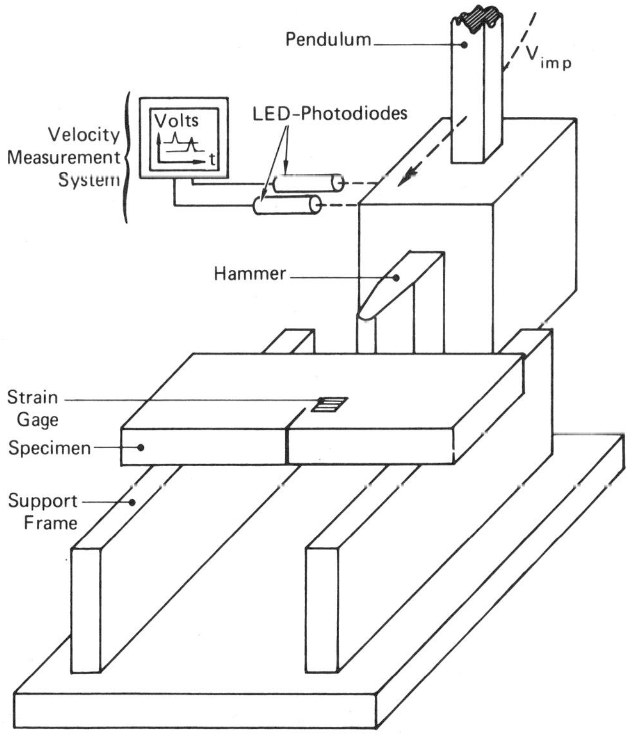

FIG. 3—*Experimental setup used to perform 1PBT.*

mental results. They agree well with the analytical predictions (Eqs 1 and 2) and demonstrate that the interpretation of the specimen behavior in the one-point-bend test and the assumptions made in the analysis are correct.

Figure 4 represents the stress intensity history for several steel and aluminum specimens impacted at 5.3 m/s. First we observe that for specimens with high S/W values, higher modes of oscillation also significantly affect the

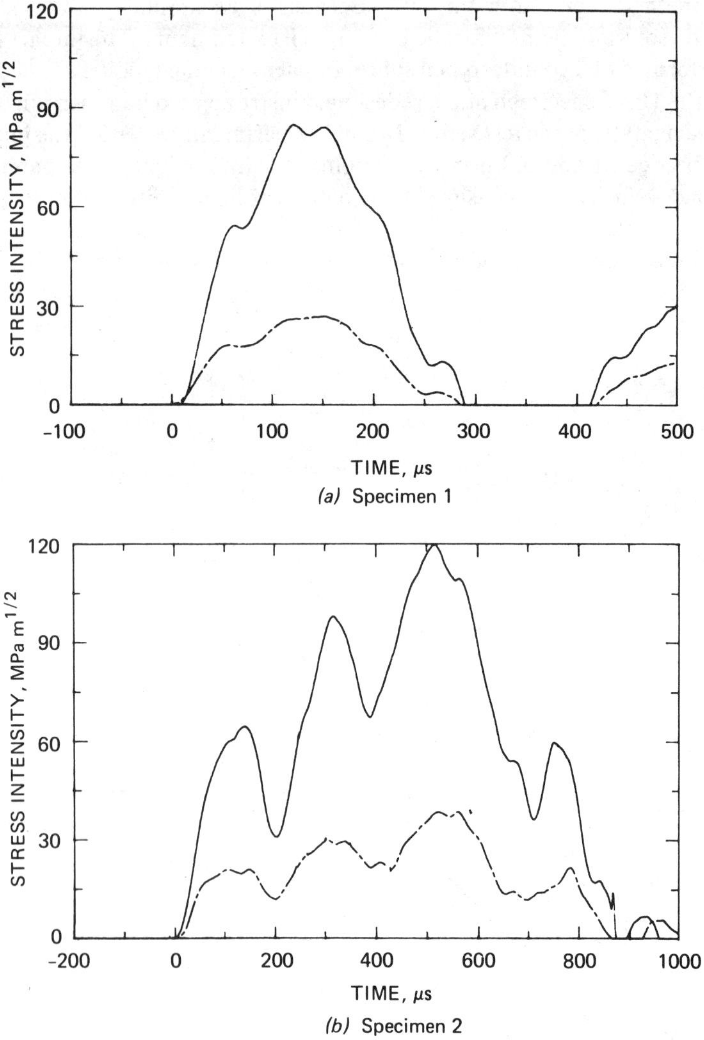

(a) Specimen 1

(b) Specimen 2

FIG. 4—*Stress intensity histories obtained for six different geometries. Legend—Solid line: steel specimen; dash line: aluminum specimen. Specimen numbers refer to numbers in Table 1.* V_{imp} *= 5.3 m/s.*

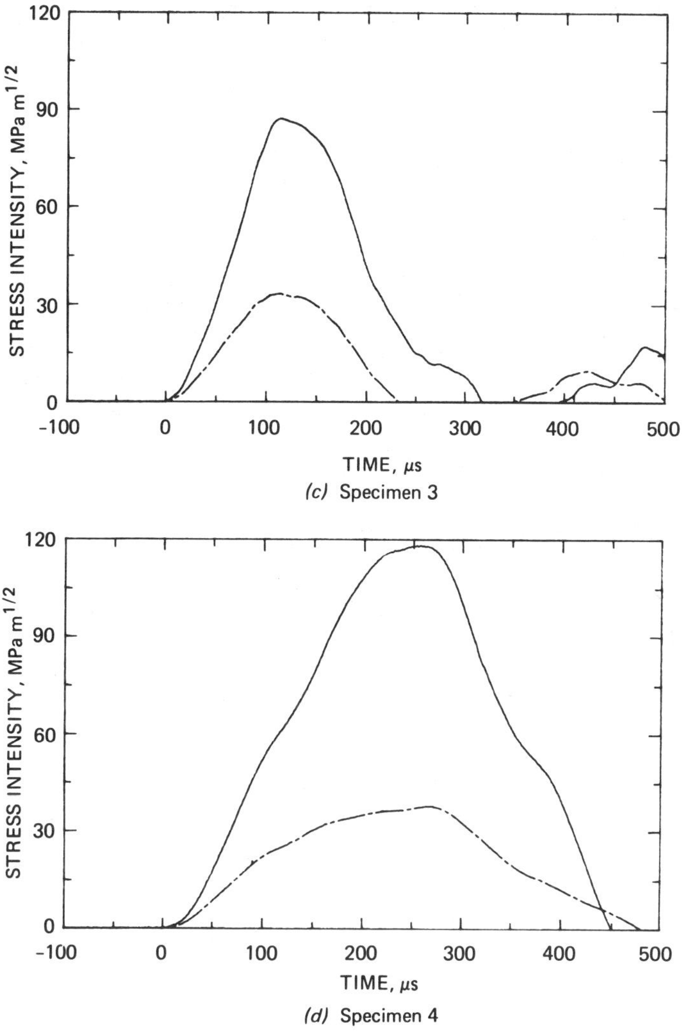

FIG. 4—*(continued).*

stress intensity history. Beyond a value of $S/W = 7$, the stress intensity history displays large oscillations superimposed on the fundamental oscillation (Fig. 4*b*). This behavior complicates the interpretation of fracture tests. Thus specimens with S/W values less than seven are preferable. (We have not investigated whether a/W ratios higher than 0.3 would suppress the higher oscillations in geometries with $S/W > 7$). By taking into account this limitation on S/W, test durations T_0 in the range of 90 to 500 μs can be readily achieved with 1PB specimens of practical size.

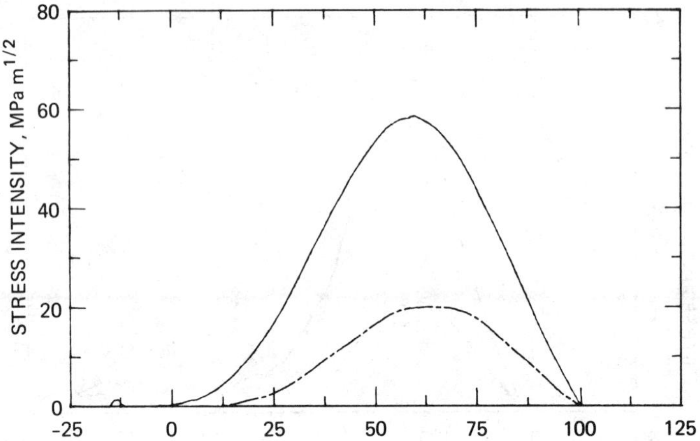

(e) Specimen 5

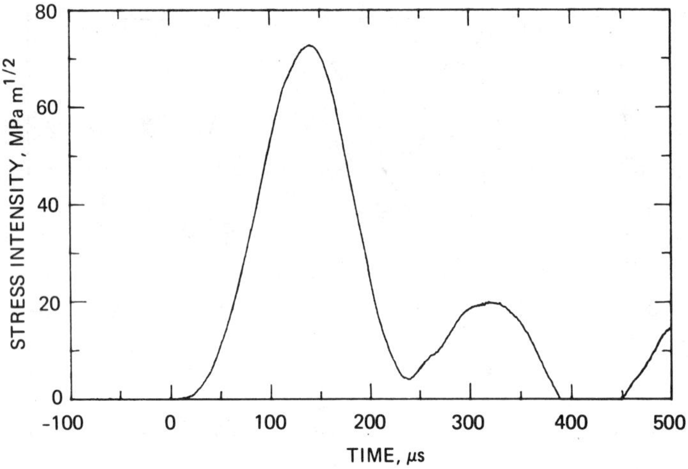

(f) Specimen 6

FIG. 4—*(continued)*.

Test durations measured in Fig. 4 and predictions from Eq 9 in the Appendix are compared in Table 2. Estimates are given only for specimens for which the compliance can be evaluated with reasonable confidence from handbook tabulations. The agreement between predictions and measurements is sufficiently good to confirm that the first mode of oscillation of the specimen largely controls the test duration for specimens with $S/W < 7$. The largest error is on the order of 30%. The greatest discrepancies arise for those stress intensity histories that show marked inflections, and they may be attributable to the influence of higher mode oscillations.

TABLE 2—*Comparison of predicted and measured 1PBT durations.*

Specimen Size Number	Predicted Half Period ($T_1/2$), μs	Measured Duration 1PBT (T_0), μs	$\frac{T_0 - 1/2\, T_1}{T_0} \times 100$	S/W
1	300	284	−6.0	7.0
2	1123	854	−31.5	14.0
3a	261	304	14.3	3.8
3b	294	300	2.0	3.8
3c	345	340	−1.5	3.8
4	508	434	−17.1	5.6

Figure 4 further demonstrates that the stress intensity histories for steel and aluminum specimens are remarkably similar except for the amplitude. Again, this finding is expected from Eqs 1 and 2. Because the ratio ρ/E is about the same for steel and aluminum, specimens with the same dimensions will have the same fundamental period. Because higher oscillations are multiples of the fundamental, we anticipate that steel and aluminum specimens will have a stress intensity history with the same global shape. According to Eq 2, K_I^{MAX} values for steel and aluminum specimens of identical geometry should be approximately in the ratio of 3 to 1. The measured ratios for specimen sizes 1 to 5 in Table 1 are 3.2, 3.1, 2.6, 3.1, and 2.9 to 1, respectively.

Tests on the same specimen at several impact velocities show that K_I^{MAX} is proportional to V_{imp}, as predicted by Eq 2. This is illustrated in Fig. 5, which shows the stress intensity histories for steel specimen 4 tested at 3.5 and 5.3 m/s, plotted after scaling with the impact velocity.

The effect of geometrical scaling is represented in Fig. 6. The stress inten

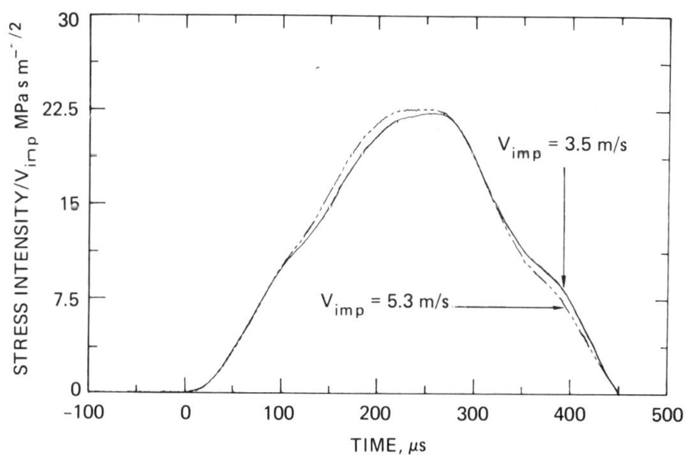

FIG. 5—*Normalized stress intensity histories for specimen 4 impacted at 3.5 and 5.3 m/s.*

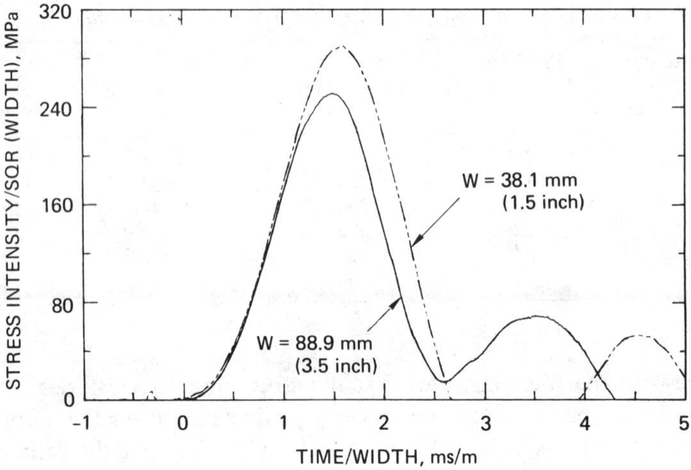

FIG. 6—*Normalized stress intensity histories for two geometrically similar specimens of different sizes.*

sity divided by the square root of the specimen width has been plotted as a function of time divided by the specimen width for two similar specimens (5 and 6). The agreement of the two curves is relatively good considering that the scaling of the in-plane dimensions, in particular of the crack length, was not exact (see Table 1).

Figure 7 shows the influence of increasing the crack length on the stress intensity history when all other dimensions are kept constant. Both $T_0/2$ and

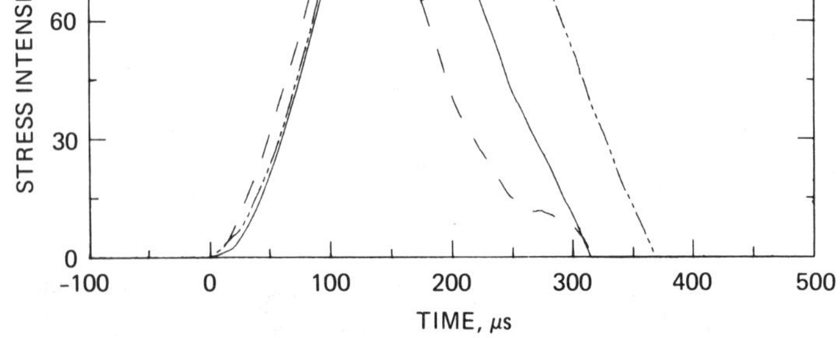

FIG. 7—*Stress intensity histories for specimen 3 with three different crack lengths.*

K_I^{MAX} increase with crack length as predicted by the analysis. Note, however, that the initial portion of the stress intensity histories is the same for all three crack lengths. This result occurs because in the early part of the loading, the dynamic stress intensity factor depends on time and not on crack length. Only after elastic waves have traversed the crack length several times does the stress intensity become proportional to the square root of crack length [*12,13*].

Finally, multiple tests on the same specimen at the same impact velocity have demonstrated that the stress intensity histories are quite reproducible for large specimens (less than 6% deviation on K_I^{MAX}). For small specimens, the scatter increases somewhat (15% deviation on K_I^{MAX}), presumably because small specimens are more sensitive to misalignment errors.

Dynamic Fracture Tests

As part of an investigation of dynamic crack instability criteria, we have used the 1PBT to determine the dynamic fracture toughness of commercial aircraft-quality 4340 steel at three different strain rates. Specimens corresponding to sizes 3 and 4 in Table 1 were cut from 63.5 by 12.7 mm (2.5 by 0.5 in.) bars, and quenched and tempered to a Rockwell C hardness of 50. The specimens were notched, fatigue precracked in accordance with ASTM Standard Test Method for Plane-Strain Fracture Toughness of Metallic Materials (E 399), and instrumented with a strain gage.

Six tests were performed at each of three impact velocities—3.5, 5.3, and 50 m/s (11.5, 17.4, and 164 ft/s)—to achieve stress intensity rates of 3.3×10^5, 5×10^5, and 3×10^6 MPa $m^{1/2}$ s^{-1} (3.0×10^5, 4.6×10^5, and 2.7×10^6 ksi in.$^{1/2}$ s^{-1}). The experiments at impact velocities of 3.5 and 5.3 m/s were performed on the modified pendulum tester with specimens of size 4 with $a/W = 0.3$ and 0.38, respectively. To achieve the 50 m/s impact velocity, we resorted to a compressed air launcher to accelerate a steel hammer plate. The test arrangement remained otherwise the same as for the experiments on the pendulum tester. The high-impact-velocity specimens were of size 3. For comparison, we also measured the static fracture toughness in triplicate following the standard procedure in ASTM E 399.

For the dynamic fracture tests, we obtained the stress intensity from the strain measurements using the static calibration method. The dynamic initiation fracture toughness, K_{Id}, was determined from the experimental stress intensity history as the stress intensity value at the point where crack extension is first noticeable on the record. Crack extension manifests itself by either a clear drop in the measured strain or by a gradual deviation from an essentially linearly rising stress intensity curve.

Figures 8 to 10 show typical stress intensity histories measured at the three loading rates, together with the point of crack initiation. The curves in Figs. 8 to 10 represent the true stress intensity history only up to the point of crack initiation, because once the crack propagates the stress intensity is no longer directly proportional to the recorded strain. Figure 8 shows that for the tests at 3.3×10^5 MPa $m^{1/2}$ s^{-1}, the crack started to propagate and then arrested after extending 2 to 5 mm (0.1 to 0.2 in.). At the higher rates, the specimens fractured completely.

At low impact velocities, crack initiation causes a clear drop in the measured strain, and the point at which K_{Id} is evaluated can therefore be easily determined (Fig. 8). At the highest impact velocity, the strain record has a smoother appearance; thus the point of crack initiation was chosen as the point of separation of the recorded strain history from a straight line fitted to the main portion of this curve (Fig. 9). For the intermediate impact velocity, some test records showed a significant departure from linearity before a sharp drop in strain; others only showed a rapid drop in strain. Therefore, for the intermediate impact velocity, either one or the other definition of the point of crack initiation was used, as appropriate for the particular test record (Fig. 10).

The results of the static and dynamic fracture experiments are summarized in Fig. 11. Although the scatter in data increases with loading rate, the average value of the dynamic fracture toughness K_{Id}, 58.5 MPa $m^{1/2}$ (53.5 ksi $in.^{1/2}$), is essentially the same for the three rates, and it is only slightly lower than the static K_{Ic} value, 63.7 MPa $m^{1/2}$ (58 ksi $in.^{1/2}$).

Discussion

Results of the Fracture Tests

The dynamic fracture toughness measurements reported here demonstrate that in the relatively narrow range of strain rates 3.3×10^5 to 3.0×10^6 MPa

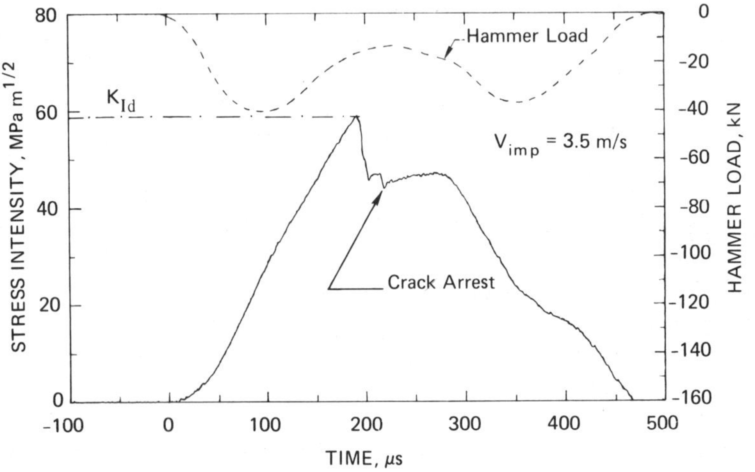

FIG. 8—*Stress intensity and hammer load histories for fracture test at low stress intensity rate.*

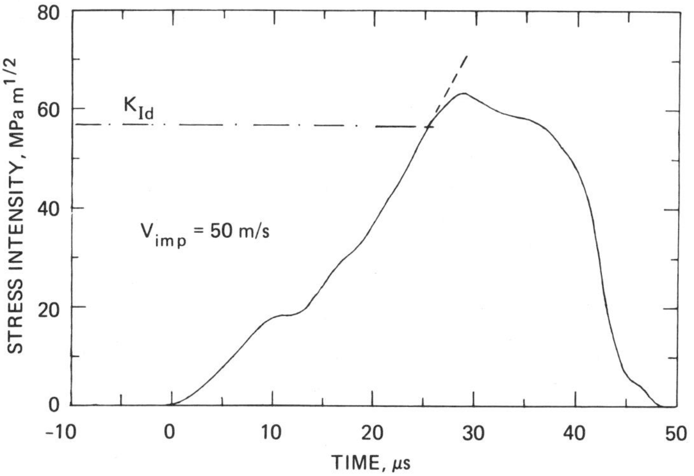

FIG. 9—*Stress intensity history for fracture test at high stress intensity rate.*

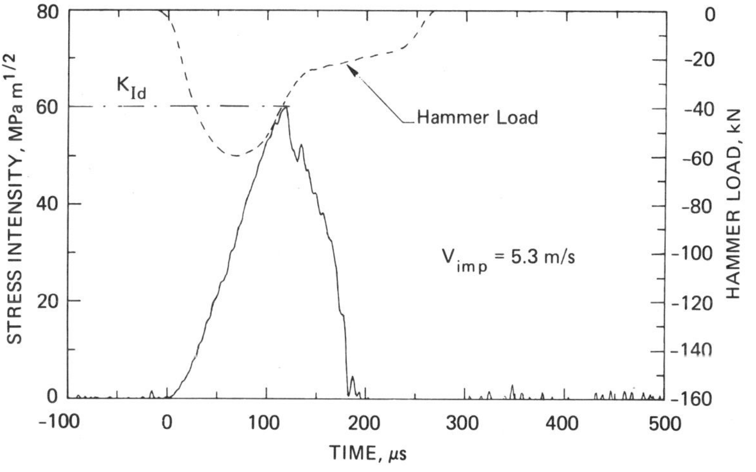

FIG. 10—*Stress intensity and hammer load histories for fracture test at intermediate stress intensity rate.*

$m^{1/2}$ s^{-1} (3×10^5 to 2.7×10^6 ksi in.$^{1/2}$ s^{-1}), K_{Id} for 4340 steel HRC 50 is not sensitive to changes in loading rate. Furthermore, comparison of the dynamic fracture toughness value 58.5 MPa $m^{1/2}$ (53.3 ksi in.$^{1/2}$) with the static fracture toughness value 63.7 MPa $m^{1/2}$ (58.0 ksi in.$^{1/2}$) shows that the overall strain-rate sensitivity of the fracture toughness is only mild. If it is assumed that the strain-rate sensitivity of the fracture toughness is a consequence of the strain-rate sensitivity of the flow stress, then our results are consistent

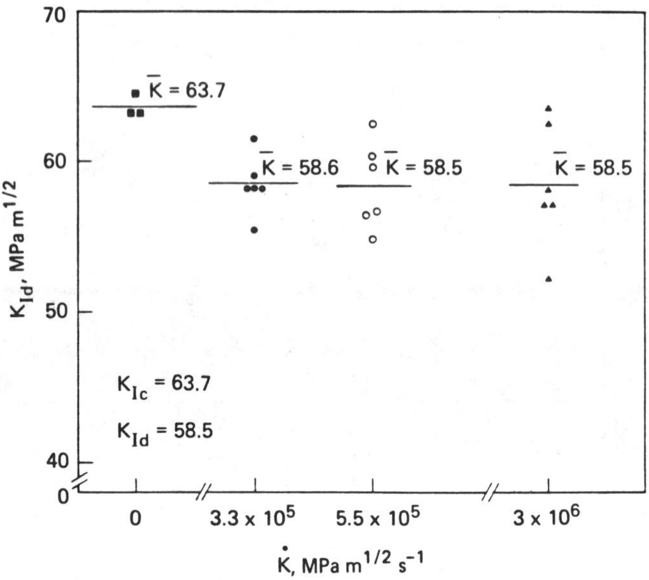

FIG. 11—*Summary of fracture test results for four stress intensity rates.*

with dynamic flow stress data for 4340 steel [*14,15*]. These data, obtained for lower hardnesses (HRC 40 and HRC 30), indicate little strain-rate sensitivity for the flow stress over a range of strain rates from 10^{-4} s^{-1} to 10^4 s^{-1}.

However, our results contradict the dynamic fracture data of Homma et al [*16,17*] for nominally the same material and for comparable loading rates. Homma et al [*16,17*] report a K_{Id} value of 31.7 MPa $m^{1/2}$, more than 30% lower than the static toughness. This discrepancy may reflect differences in microfailure modes. Homma et al [*16*] observed a transition from fully dimple fracture at low strain rates to a mixture of dimples and cleavage facets at high strain rates. Scanning electron microscope (SEM) observations did not reveal a change in microfracture mode in our experiments. Under both static and dynamic conditions, fracture occurred by what appears a mixture of dimple, intergranular, and transgranular failures.

The difference between the results of Homma et al and the present results may also be due to a difference in the definition of the point of crack initiation. Homma et al measure K_{Id} after 20 to 50 μm of crack extension, whereas the K_{Id} value we report probably corresponds to 200 to 500 μm of crack extension.

Significance of the 1PBT

The present investigation has confirmed the suggestion by Kalthoff et al [*2,3*] that the 1PBT provides an attractive test configuration for dynamic fracture experiments. It produces a smoothly varying stress intensity history that can be easily characterized in a calibration experiment. Furthermore, the loading rate and the pulse amplitude and duration can be readily adjusted over an order of magnitude range for each parameter. The analytical and experimental results given in this paper provide the information necessary to choose the appropriate combination of specimen dimensions and impact velocity to achieve desired test conditions.

The results for the 1PBT expressed in Eqs 1 and 2 also supplement the results obtained for the impacted 3PBT reported by Böhme [6]. They provide an independent analysis of the free specimen oscillations that were shown to play an important role in the impacted 3PBT, and they yield the same conclusion regarding the representation of these dynamic effects by master curves that depend only on specimen aspect ratios.

The maximum stress intensity amplitude that can be obtained in a 1PBT using a conventional pendulum impact tester is rather low, particularly in experiments with lighter alloys such as aluminum or titanium alloys. For these materials, devices capable of higher impact velocities are required. This problem may also be circumvented by attaching ballast plates to the specimen extremities. The high-strain-rate fracture experiments also showed that the point of crack initiation is not always clearly defined on the records of strain as a function of time. Further study of this problem is desirable to obtain an unambiguous definition of the point of dynamic crack initiation.

An interesting aspect of the 1PBT was brought to light by the fracture experiments in which crack arrest occurred. They indicated that the one-point-bend test could provide a convenient experimental procedure to study not only dynamic crack initiation but also crack propagation and arrest. The advantage of the one-point-bend test for crack propagation investigations would be that because of the inertial loading, the driving force on the running crack tip and the crack velocity could be easily controlled by changing the impact velocity. Further, because of its very simple boundary conditions, the one-point-bend test should be much easier to simulate in numerical calculations than the conventional wedge-loaded compact tension crack arrest test configuration.

Summary and Conclusions

The stress intensity history obtained by impacting an unsupported, cracked, bend specimen has been investigated. A combination of experiments and analysis was used to establish the influence of specimen material and dimensions and impact velocity on the maximum stress intensity amplitude and the test duration. Stress intensity pulses of durations varying between 90 and 500 μs can be readily achieved. For steel, maximum amplitudes exceeding 100 MPa $m^{1/2}$ have been produced with a pendulum impact tester.

The applicability of the test procedure has been illustrated by measuring the dynamic fracture toughness of 4340 steel (HRC 50) at three different strain rates. The tests indicated little strain-rate sensitivity for the fracture toughness of this material.

Finally, it is suggested that the one-point-bend test could also conveniently serve to investigate dynamic crack propagation and arrest.

Acknowledgments

This work was supported by the Air Force Office of Scientific Research under Contract AFOSR/F49620-81-K-0007 and was monitored by Dr. Alan H. Rosenstein. The author would like to thank Drs. J. F. Kalthoff, and W. Böhme, Fraunhofer Institut für Werkstoffmechanik, Freiburg, West Germany, for suggesting the use of the one-point-bend test and for supplying preliminary information about the test. Many helpful discussions during the course of this study with Dr. D. A. Shockey are also gratefully acknowledged.

APPENDIX

Over the years several authors have presented analyses of the impacted-bend specimen to help in evaluating dynamic fracture experiments [*6,18–22*]. Here we present simple derivations to obtain the test duration T_0 and the maximum stress intensity amplitude K_I^{MAX} for the 1PBT. The resulting expressions are to be considered more as a dimensional analysis than as precise quantitative predictions.

Estimation of the Duration T_0

To evaluate the test duration T_0, we assume that it is equal to one half the fundamental period of oscillation T_1 of the cracked beam. Ireland has previously proposed an empirical formula to estimate the period of inertial oscillation in the impacted 3PBT [*21*]. We estimate T_1 using Rayleigh's method. To do so, we assume for the deflection d of the beam the following function of position x along the axis:

$$d(x) = d_{UN}^{CL} \sin\left(\frac{\pi}{S}x\right) + 2\frac{x}{S} d_{CR}^{CL}; \quad 0 < x < \frac{S}{2} \tag{3}$$

The first term on the right of Eq 3 represents the center line deflection of the uncracked beam, and the second term represents an estimate of the deflection due to the crack.

Next we express d_{UN}^{CL} and d_{CR}^{CL} in terms of the respective elastic compliances C_{UN} and C_{CR}, and of an effective dynamic load amplitude P_{MAX}^{EFF}, where P_{MAX}^{EFF} can be regarded as the maximum hammer load corrected to account for the inertia forces in the sense of the analysis reported in Refs *19* and *20*. Thus

$$d_{UN}^{CL} = C_{UN} P_{MAX}^{EFF} \quad \text{and} \quad d_{CR}^{CL} = C_{CR} P_{MAX}^{EFF} \tag{4}$$

The beam deflection at any point x and time t is given by

$$d(x,t) = \left[C_{UN} \sin\left(\frac{\pi}{S}x\right) + C_{CR}\frac{2x}{S}\right] P_{MAX}^{EFF} \sin\left(\frac{2\pi}{T_1}t\right); \quad 0 < x < \frac{S}{2} \tag{5}$$

Using Eq 5, we calculate the maximum kinetic energy of the specimen

$$E_{\text{MAX}}^{\text{KIN}} = \rho BW(P_{\text{MAX}}^{\text{EFF}})^2\left(\frac{2\pi}{T_1}\right)^2\left[C_{\text{UN}}^2\frac{S}{4} + \frac{4S}{\pi^2}C_{\text{UN}}C_{\text{CR}} + \frac{S}{6}C_{\text{CR}}^2\right] \tag{6}$$

The maximum strain energy is simply

$$E_{\text{MAX}}^{\text{STR}} = \frac{1}{2}(P_{\text{MAX}}^{\text{EFF}})^2(C_{\text{UN}} + C_{\text{CR}}) \tag{7}$$

Equating kinetic energy and strain energy yields

$$T_1 = 2\pi\sqrt{2\rho BW\frac{\left(\frac{C_{\text{UN}}^2}{4} + \frac{4C_{\text{UN}}C_{\text{CR}}}{\pi^2} + \frac{C_{\text{CR}}^2}{6}\right)}{(C_{\text{CR}} + C_{\text{UN}})}} \tag{8}$$

Substituting

$$C_{\text{UN}} = \frac{S^3}{4EBW^3}\left[1 + 3(1+\nu)\left(\frac{W}{S}\right)^2\right]$$

and rearranging yields

$$T_1 = T_{\text{UN}}\sqrt{\frac{\pi^4}{24}}\sqrt{\frac{\frac{1}{4} + \frac{4}{\pi^2}\beta + \frac{1}{6}\beta^2}{(1+\beta)}}\sqrt{1 + 3(1+\nu)\left(\frac{W}{S}\right)^2} \tag{9}$$

where $\beta = C_{\text{CR}}/C_{\text{UN}}$ and $T_{\text{UN}} = (4\,S^2/\pi\,W)\sqrt{3\rho/E}$ is the fundamental period of the uncracked beam.

The dependence of T_1 on specimen dimensions and material properties given by Eq 9 is the same as the dependence anticipated from Ireland's formula [*21*].

Estimation of the Maximum Amplitude $K_{\text{I}}^{\text{MAX}}$

To estimate the maximum stress intensity amplitude, we express it as

$$K_{\text{I}}^{\text{MAX}} = \frac{3}{2}\frac{S\sqrt{\pi a}}{BW^2}F(a/W)P_{\text{MAX}}^{\text{EFF}} \tag{10}$$

where $F(a/W)$ is the handbook-tabulated stress intensity calibration appropriate for the given S/W ratio [*23*]. We then proceed to express $P_{\text{MAX}}^{\text{EFF}}$ using Newton's law and conservation of momentum and energy.

At every instant t, the force P_{H} on the hammer is given by

$$P_{\text{H}}(t) = P_{\text{H}}^{\text{MAX}}g\left(\frac{t}{T_1}\right) \tag{11}$$

where $P_{\text{H}}^{\text{MAX}}$ is defined in Fig. 2 and $g(t/T_1)$ is a nondimensional function that can be obtained experimentally.

Integration of Newton's law for the hammer from the initial time of impact $t = 0$ to the time t^{MAX} at which the specimen deflection, and hence the stress intensity, is maximum gives

$$\int_0^{t^{MAX}} P_H^{MAX} g\left(\frac{t}{T_1}\right) dt = T_1 P_H^{MAX}\left[G\left(\frac{t^{MAX}}{T_1}\right) - G(0)\right] \tag{12}$$

$$= T_1 P_H^{MAX} \alpha = M(V_{imp} - V_H)$$

where M is the hammer mass, V_{imp} is the initial impact velocity, V_H is the residual hammer velocity at time t^{MAX}, and α is a coefficient resulting from the integration of $g(t/T_1)$. Considering the approximately sinusoidal shape of the hammer load in Fig. 2 (period $\approx T_0 \approx T_1/2$), we see that α is on the order of $1/2\pi \approx 0.15$.

From Eq 12 we obtain

$$V_H = V_{imp} - \frac{\alpha T_1 P_H^{MAX}}{M} \tag{13}$$

Application of momentum conservation to the hammer-specimen system, with substitution of Eq 13 for V_H, yields

$$V_S = \frac{\alpha T_1 P_H^{MAX}}{m} \tag{14}$$

where m is the specimen mass and V_S is the velocity of the center of mass of the specimen, which at t^{MAX} is also the uniform velocity of the whole specimen.

Finally, the energy conservation equation can be written

$$\frac{1}{2} V_{imp}^2 M = \frac{1}{2} V_H^2 M + \frac{1}{2} m V_S^2 + \frac{1}{2} (P_{MAX}^{EFF})^2 (C_{UN} + C_{CR}) \tag{15}$$

where the last term of the right-hand side represents the strain energy, P_{MAX}^{EFF}, C_{UN}, and C_{CR} have been defined above.

We now write P_{MAX}^{EFF} as a multiple of P_H^{MAX}:

$$P_{MAX}^{EFF} = \frac{1}{\lambda} P_H^{MAX} \tag{16}$$

The results of Ref *20* indicate that $1/\lambda$ should be between 0 and 1.

By substitution of Eqs 13, 14, and 16 in Eq 15 we obtain

$$P_{MAX}^{EFF} = 2V_{imp} T_1 \alpha \left[\frac{1}{\lambda T_1^2 \alpha^2 \dfrac{1}{M} + \dfrac{1}{m} + \dfrac{1}{\lambda}(C_{UN} + C_{CR})}\right] \tag{17}$$

Finally, combining Eqs 9, 10, and 17, we obtain

$$K_{\mathrm{I}}^{\mathrm{MAX}} = V_{\mathrm{imp}} \sqrt{\pi a}\, \sqrt{12\rho E}\, \frac{3\alpha}{\pi^2} F(a/w)\left(\frac{S}{W}\right)^3 \sqrt{1 + 3(1+\nu)\left(\frac{W}{S}\right)^2}\, L(\beta)$$

$$\times \frac{1}{\dfrac{1}{M} EB(C_{\mathrm{UN}} + C_{\mathrm{CR}}) + \lambda\alpha^2\left[1 + \dfrac{m}{M}\right]\dfrac{12}{\pi^4} L^2(\beta)\left(\dfrac{S}{W}\right)^3\left[1 + 3(1+\nu)\left(\dfrac{W}{S}\right)^2\right]} \tag{18}$$

where

$$L(\beta) = \sqrt{\frac{\pi^4}{24}\left(\frac{1/4 + 4/\pi^2\beta + 1/6\,\beta^2}{1+\beta}\right)}$$

Equations 9 and 18 are the expanded versions of Eqs 1 and 2. The assumption $T_0 = T_1/2$ has been made in writing Eq 1.

References

[1] Böhme, W. and Kalthoff, J. F., *International Journal of Fracture*, Vol. 20, No. 1, 1982, pp. 139-143.

[2] Kalthoff, J. F., Winkler, S., Böhme, W., and Shockey, D. A. in *Proceedings*, International Conference on the Dynamical Properties and Fracture Dynamics of Engineering Materials, Brno, Czechoslovakia, June 1983.

[3] Kalthoff, J. F., Böhme, W., and Winkler, S., "Verfahren zum Bestimmen mechanischer Werkstoffkennwerte und Vorrichtung zur Durchführung des Verfahrens," Patent Disclosure Deutsches Patentamt, June 1983.

[4] Kalthoff, J. F., Winkler, S., and Beinert, J., *International Journal of Fracture*, Vol. 13, No. 4, Aug. 1977, pp. 528-531.

[5] Kalthoff, J. F., Winkler, S., Klemm, W., and Beinert, J. in *Instrumented Impact Tests*, Proceedings of 5th International Conference on Structural Mechanics in Reactor Technology, Berlin, Vol. G (4/5), North-Holland, Amsterdam, 1979.

[6] Böhme, W., "Eine Einfache Methode zur Bestimmung der dynamischen Risspitzenbeanspruchung bei Schlagbelasteten Dreipunktbiegeproben," FhG Report Z 2/84, Fraunhofer-Institut für Werkstoffmechanik, Freiburg, W. Germany, March 1984.

[7] Loss, J. F., Hawthorne, J. R., and Griffis, C. A., "Fracture Toughness of Light Water Reactor Pressure Vessel Materials," Naval Research Laboratory Memorandum Report 3036, Washington, D.C., April 1975.

[8] Loss, J. F., paper presented to Third Water Reactor Safety Research Information Meeting, Oct. 1975.

[9] Sneddon, I. N., *Proceedings of the Physical Society, London*, Vol. 187, 1946, pp. 229-260.

[10] Williams, M. L., *Journal of Applied Mechanics*, Vol. 24, 1957, pp. 109-114.

[11] Böhme, W. and Kalthoff, J. F., "Der Einfluss der Probengrösse auf dynamische Effekte by der K_{Id}-Bestimmung in Kerbschlagbiegetest," FhG Report W 3/83, Fraunhofer-Institut für Werkstoffmechanik, Freiburg, W. Germany, 1983.

[12] Freund, L. B., *Journal of the Mechanics and Physics of Solids*, Vol. 21, 1973, pp. 47-61.

[13] Shockey, D. A., Kalthoff, J. F., and Erlich, D. C., *International Journal of Fracture*, Vol. 22, 1983, pp. 217-229.

[14] Duffy, J., Brown University, Providence, R.I., private communication, 1983.

[15] Chartagnac, P. F., "Analysis of the SRI Symmetric Taylor Test Data on 4340 Steels," SRI International Internal Report, Menlo Park, Calif., 1983.

[16] Homma, H., Shockey, D. A., and Murayama, Y., *Journal of the Mechanics and Physics of Solids*, Vol. 31, No. 3, 1983, pp. 261-279.
[17] Homma, H., Shockey, D. A., and Hada, S., this publication, pp. 683-696.
[18] Nash, G. E., *International Journal of Fracture Mechanics*, Vol. 5, No. 4, 1969, pp. 269-286.
[19] Radon, J. C. and Turner, C. E., *Engineering Fracture Mechanics*, Vol. 1, No. 3, 1969, pp. 411-428.
[20] Turner, C. E. in *Impact Testing of Metals, ASTM STP 466*, American Society for Testing and Materials, Philadelphia, 1970, pp. 93-114.
[21] Ireland, D. R. in *Proceedings*, International Conference on Dynamic Fracture Toughness, London, England, July 1976.
[22] Gudmundson, P., *Journal of the Mechanics and Physics of Solids*, Vol. 31, No. 4, 1983, pp. 329-345.
[23] Tada, H., Paris, P. C., and Irwin, G. R., *The Stress Analysis of Cracks Handbook*, Del Research Corp., Hellertown, Pa., 1973.

Robert J. Buzzard,[1] *Bernard Gross,*[1] *and John E. Srawley*[2]

Mode II Fatigue Crack Growth Specimen Development

REFERENCE: Buzzard, R. J., Gross, B., and Srawley, J. E., **"Mode II Fatigue Crack Growth Specimen Development,"** *Fracture Mechanics: Seventeenth Volume, ASTM STP 905*, J. H. Underwood, R. Chait, C. W. Smith, D. P. Wilhem, W. A. Andrews, and J. C. Newman, Eds., American Society for Testing and Materials, Philadelphia, 1986, pp. 329–346.

ABSTRACT: A novel Mode II test specimen has been developed that has potential application in understanding phenomena associated with mixed-mode fatigue failures in high performance aircraft engine bearing races. The attributes of the specimen are as follows: it contains one single-ended notch, which simplifies data gathering and reduction; the fatigue crack grows in-line with the direction of load application; a single axis test machine is sufficient to perform testing; and the Mode I component is vanishingly small.

KEY WORDS: Mode II, fatigue, edge sliding mode, anti-plane shear

Nomenclature

a	Crack length
B	Specimen thickness
E'	E plane stress, $= E/(1 - \nu^2)$ plane strain
F	Pin shear reaction force
H	Specimen arm height
J	Strain energy release rate (path-independent integral)
ΔJ	Variation of J from P_{min} to P_{max}
K_I	Mode I stress intensity factor
K_{II}	Mode II stress intensity factor
P	Applied end load
R	Minimum fatigue load/maximum fatigue load

[1]Materials Engineer, Structural Materials, NASA Lewis Research Center, Cleveland, OH 44135.

[2]Scientific Consultant to Fatigue and Fracture Branch, NASA Lewis Research Center, Cleveland, OH 44135.

r	Polar coordinate referred to crack tip
S	Extent of end reaction parabolic shear distribution
U	Total relative crack mouth displacement in x direction
W	Specimen width
x, y	Cartesian coordinate referred to crack tip
Y_I	Dimensionless stress intensity coefficient for Mode I
Y_{II}	Dimensionless stress intensity coefficient for Mode II
θ	Polar coordinate referred to crack tip
ν	Poisson's ratio
σ	Applied end stress $= P/BH$
χ	Stress function

Introduction

Current development of high performance rolling element bearings for aircraft engines (up to 3 million *DN*, where *DN* is the product of shaft diameter in millimetres and speed in revolutions per minute) has aroused concern about fatigue crack growth in the inner bearing race that leads to catastrophic failure of the bearing and the engine.

The following basic model is suggested. During a period of steady engine speed, such as cruise operation, the hoop stress in the inner race (shrink plus centrifugal stresses) is substantial but steady. Consider now a microscopic region of the inner race close to the bearing surface where there is a defect. Each time the race rotates relative to the bearing cage, this region will be contacted once by every rolling element in the cage. The Hertzian contact pressure during a particular encounter will depend on the angular position of the defective region in relation to the direction of the overall radial bearing load. The rotational speed is so high that the frequency of significant contacts is on the order of kilohertz. Eventually a microcrack will develop at this site, so that a fully reversed cyclic stress intensity field may be postulated. The Mode II component changes abruptly in sign as the center of Hertzian pressure moves across the crack. The magnitude of the alternating Mode II field will follow a predictable spectrum that will be essentially repeated at intervals of a certain number of shaft rotations, as can readily be deduced from Section 24 of Ref *1*. The steady hoop stress component of Mode I stress intensity is much greater in magnitude than the cyclic Mode II component, and will tend to keep the crack faces apart, thus avoiding rubbing which would inhibit Mode II growth. The hertzian stress distribution gives a maximum shear stress at some small distance below the contact surface. The cyclic variation of this stress as the bearings move along the inner race is thought to cause crack initiation [*2*]. From a recent paper by Hahn et al [*3*], the "predominantly Mode II cyclic crack growth driving force" is discussed. The final failure is Mode I due to the interference and centrifugal stresses. While some investigators believe crack initiation occurs at the surface in Mode I, this belief is by no means universal.

To address this problem in the laboratory, initially a test specimen and loading method is required whereby fatigue crack growth under predominantly Mode II loading may be isolated and studied. The present work is directed towards developing such a specimen and method.

Background

A fatigue specimen has been developed at NASA-LeRC (Lewis Research Center) by which fatigue data relative to the aforementioned bearing cracking problem may be obtained. This specimen was developed after an extensive literature search revealed that a specimen meeting all the following requirements did not exist. The requirements of a suitable specimen are:

1. The specimen contains a single notch, which simplifies analysis and the monitoring of crack propagation.
2. The fatigue crack grows in-line with the centerline of the machined notch at or near a zero degree angle to the notch axis.
3. A uniaxial testing machine is sufficient for performing tests.
4. The Mode I loading stresses are insignificant compared with the Mode II stresses.

Prior to this developmental work, several of the various Mode II specimens described in the literature were examined using acrylic plastic and 7075-T6 aluminum as specimen materials. The results of these examinations are described briefly below.

A V-notched specimen type (Fig. 1*a*) designed by Iosipescu has been used by several investigators as a monotonically loaded shear fracture test specimen [*4*]. Specimens of this type fabricated from plastic and analyzed photoelastically using polarized lighting did not exhibit the classical Mode II stress pattern; that is, concentric semiellipses centered at the notch tip (e.g., see Fig. 2). Nor was this expected by Iosipescu, who believed this type of pattern signified compressive rather than shear forces. When loaded monotonically to failure, the V-notched specimens exhibited cracking at 45° to the notch tip. By revising the notches from the V-configuration to a slot configuration, a weak elliptical pattern was observed at very high loads; however, the centerline of the pattern was not in line with the specimen notch centerline. The most nearly successful test of this general specimen type was with a specimen that was fully restrained from bending by redesign of the test fixture. This specimen failed at the notch centerline but contained additional large cracks starting at the notch tips and propagating at about a 70° angle to the notch centerline towards the tensile loaded zones of the specimen.

Many variations of the specimen and method of loading met with the same

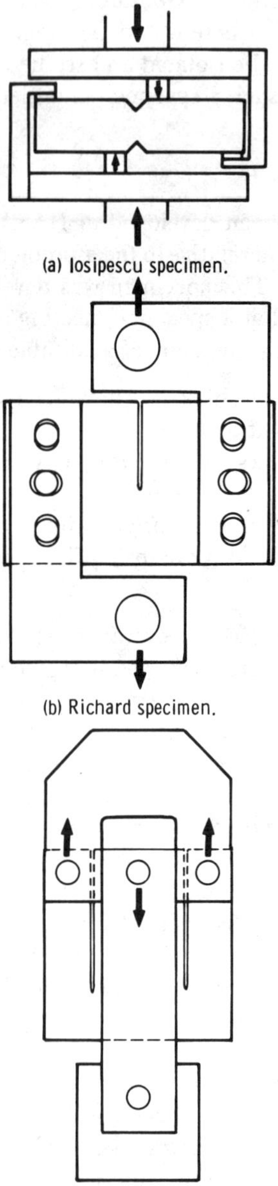

(a) Iosipescu specimen.

(b) Richard specimen.

(c) Chisholm specimen.

FIG. 1—*Three candidate Mode II test specimens.*

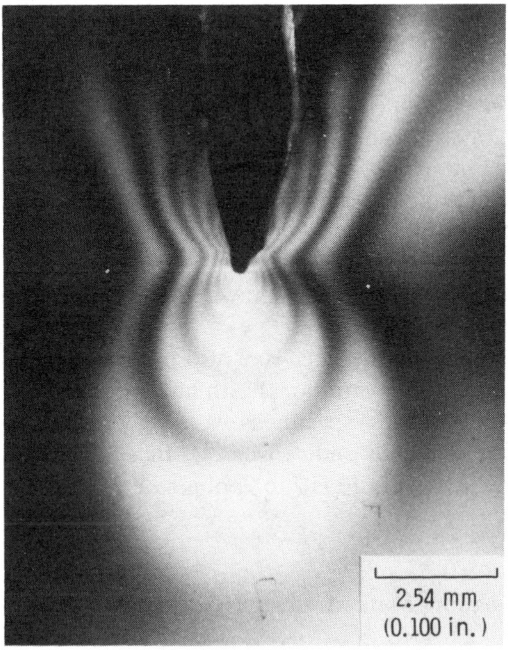

FIG. 2—*Mode II photoelastic pattern at notch tip of Columbia resin model of NASA LeRC Mode II specimen.*

generally unsatisfactory results. Consequently further efforts with this specimen type were discontinued. A decided advantage of this specimen type over other designs would have been its simplicity and economy of material. Even had the preliminary tests been successful, however, the monitoring of two cracks simultaneously and the resulting fatigue data reduction may have been problematic.

A specimen type designed by Richard [*5*] for Mode II fatigue crack propagation was also investigated. This specimen design and fixturing is shown in Fig. 1*b*. A very good Mode II photoelastic stress pattern was observed with this specimen design. However, at high loads this specimen tends to fail at the grip holes, while at moderate loads the fatigue crack propagates from the notch root towards the tensile-loaded leg of the specimen at an angle of about 70°.

Variations to this specimen included notching at both the top and bottom as well as the use of side grooves to reduce the web cross section. These modifications, however, did not improve the results.

A double-notched fatigue crack propagation specimen developed by Chisholm [*6*] is shown in Fig. 1*c*. This specimen exhibited a very good Mode II pattern at both notch tips, and the monotonic fracture paths were in line with the machined notches (that is, a crack angle of near zero degrees). At high loads, the fatigue cracks in 7075-T6 Al specimens proceeded also at near zero degree angles. At low fatigue loads, however, fatigue cracks ran approximately 70° to the machined notches. It should be noted that according to tangential stress or strain energy density criteria crack growth is predicted to occur at an angle of 70 to 80° from the crack line. However, by applying a large enough ΔJ (plastic deformation is observed) crack propagation is constrained along the crack line.

Although the double-notched specimen is an attractive candidate for fatigue work, a single-notched specimen with simpler fabrication and testing considerations is preferable.

Various other specimen designs were also investigated (about nine major designs, with about twenty variations) with unfavorable results. The primary failure mode of most of these specimens was a 70° crack moving into the tensile loaded side of the notch, indicative of a Mode I failure. These specimen designs were as described in Refs *7* to *9*, plus several original NASA designs.

Specimen and Loading Method

The final design of the LeRC Mode II test specimen is shown in Fig. 3*a*. Its overall dimensions were initially chosen as 76 mm (3 in.) by 102 mm (4 in.), because this was a convenient size for preliminary photoelastic examination of the specimen's stress field. Photoelastic models were fabricated from both 19.1 mm (3/4 in.) and 6.4 mm (1/4 in.) thick acrylic plastic. Crack growth specimens were made from 6.4 mm (1/4 in.) and 3.2 mm (1/8 in.) thick 7075-T6 aluminum sheet. This thickness was chosen so that loading would not exceed the maximum capacity of the testing machine (a 10-kip, hydraulically operated servocontrolled facility).

The testing fixture is shown in Fig. 3*b*. It was fabricated from Type 300 maraging steel and allowed clearance for a 19.1 mm (3/4 in.) thick specimen. V-notches were used to engage the top loading pins so that adjustments in horizontal pin spacing could be accomplished by use of steel shims.

The testing machine applies a compressive load to the load train. As the ram of the testing machine is raised, the lower part of the test fixture, along with the specimen, moves upward. The left upper pin experiences a downward reaction force exerted by the upper part of the testing fixture. Rotation of the specimen is prevented by the lower central pin. This pin passes through and is held in place by the specimen, and contacts the upright leg of the test fixture. This pin retains its position within the specimen, but can move in a vertical direction along the upright leg of the test fixture. The slot in the specimen at the lower pin hole allows the right half of the specimen to move upward, with no pin interference, as the specimen is deformed during testing.

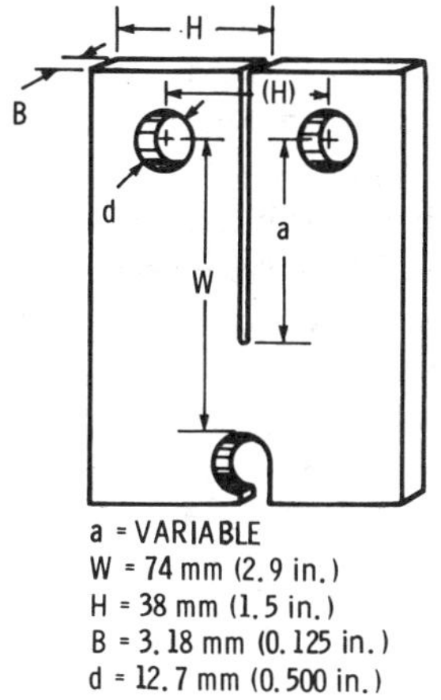

(a) Mode II test specimen.

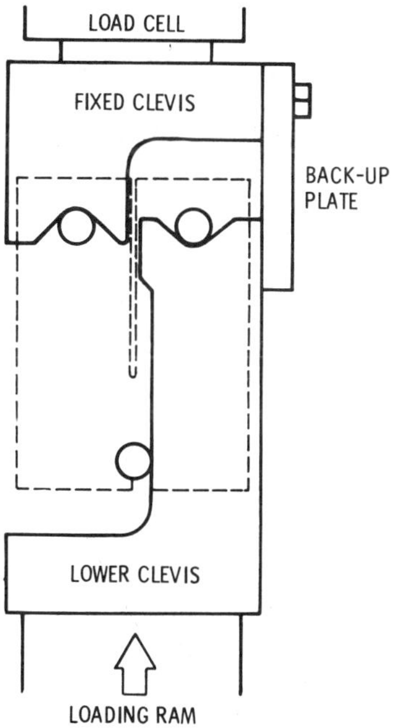

(b) Specimen pin-mounted in loading fixture.

FIG. 3—*NASA LeRC Mode II test specimen and loading arrangement.*

Specimen Analysis

The objectives of this analysis were to qualitatively obtain the magnitudes of K_{II} and K_I and their variation with respect to crack length, and to compare analytically obtained displacements with experimental results. The specimen stress intensity and displacements were analyzed by a method of analysis described in detail by Gross and Mendelson [*10*]. This method consists of finding a stress function χ that satisfies the biharmonic equation $\nabla^4\chi = 0$ and the specimen boundary conditions. The biharmonic equation and the boundary conditions along the crack line are satisfied by the Williams stress function [*11*]. Satisfying equilibrium of forces and moments (Fig. 4) results in $P = 2\sigma H^2B/W$. The shear stress distributions are taken as parabolic along boundary AB, $D'D$, and FG. The pin reaction loads $((P/4) - F)$ and $((P/4) + F)$ were adjusted to satisfy the observed condition of negligible crack mouth opening. Because of the load asymmetry (Fig. 4), the stress function consists of an infinite series of even and odd functions:

$$\chi(r, \theta) = \sum_{n=1}^{\infty} d_{4n-3}r^{n+1/2}\left[\cos(n - 3/2)\theta - \left(\frac{2n-3}{2n+1}\right)\cos(n + 1/2)\theta\right]$$

$$+ d_{4n-2}r^{n+1}[\cos(n - 1)\theta - \cos(n + 1)\theta]$$

$$+ d_{4n-1}r^{n+1/2}[\sin(n - 3/2)\theta - \sin(n + 1/2)\theta]$$

$$+ d_{4n}r^{n+1}\left[\sin(n - 1)\theta - \left(\frac{n-1}{n+1}\right)\sin(n + 1)\theta\right]$$

The boundary values of χ and its normal derivative to the boundary are obtained from the assumed model (Fig. 4):

Along AB:

$$\chi = -\frac{\sigma y^2}{2}$$

$$\frac{\partial\chi}{\partial x} = \left(\frac{3\sigma}{WH} - \frac{6F}{BH^3}\right)\left(\frac{y^3}{3} - \frac{Hy^2}{2}\right)$$

Along BC:

$$\chi = -\frac{\sigma H^2}{2}\left(1 + \frac{a}{W} + \frac{x}{W}\right) + \frac{F(x + a)}{B}$$

$$\frac{\partial\chi}{\partial y} = -\sigma H$$

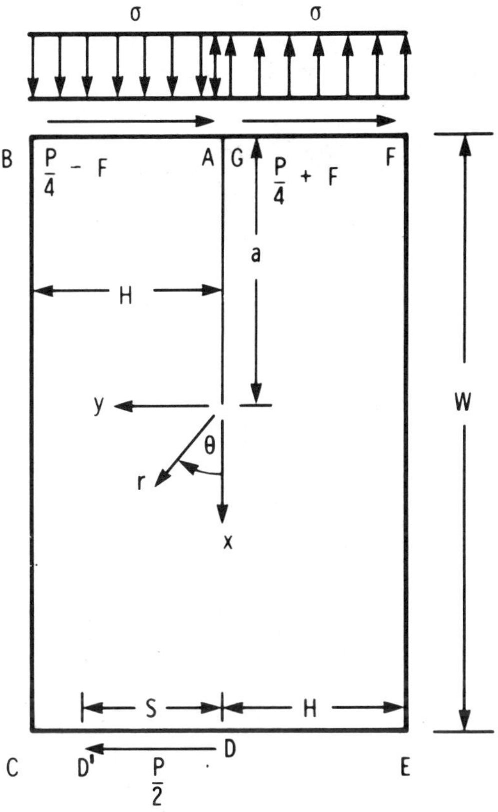

FIG. 4—*Analytical model of the NASA-LeRC Mode II specimen subject to asymmetric loading;* H/W = *0.5.*

Along CD':

$$\chi = -\sigma H y + \frac{FW}{B}$$

$$\frac{\partial \chi}{\partial x} = -\frac{\sigma H^2}{2W} + \frac{F}{B}$$

Along $D'D$:

$$\chi = -\sigma H y + \frac{FW}{B}$$

$$\frac{\partial \chi}{\partial x} = \frac{\sigma H^2}{2W}\left[1 + \frac{12}{S^3}\left(\frac{y^3}{3} - \frac{Sy^2}{2}\right)\right] + \frac{F}{B}$$

Along *DE*:

$$\chi = -\sigma H y + \frac{FW}{B}$$

$$\frac{\partial \chi}{\partial x} = \frac{\sigma H^2}{2W} + \frac{F}{B}$$

Along *EF*:

$$\chi = \frac{\sigma H^2}{2}\left(1 + \frac{x}{W} + \frac{a}{W}\right) + \frac{F}{B}(x + a)$$

$$\frac{\partial x}{\partial y} = -\sigma H$$

Along *FG*:

$$\chi = \frac{\sigma y^2}{2}$$

$$\frac{\partial \chi}{\partial x} = \left(\frac{3\sigma}{HW} + \frac{6F}{BH^3}\right)\left(\frac{y^3}{3} + \frac{Hy^2}{2}\right)$$

The pin reaction loads $((P/4) - F)$ and $((P/4) + F)$ were adjusted to satisfy the observed condition that the crack mouth opening was small.

Preliminary K_{II} and K_I calculations indicate that the specimen type presented here is subjected to a predominantly Mode II condition when under load. Values of K are expressed as $K_{(I,II)} = Y_{(I,II)}(P/Ba^{1/2})$; values for the stress intensity coefficient Y for both modes are presented in Table 1 for various crack length to specimen width (a/W) ratios. For example, a specimen used in preliminary fatigue testing had an a/W ratio of 0.679; the corresponding Y_{II} and Y_I values indicate a K_{II}/K_I ratio of about 65 to 1 at this a/W ratio. The solutions obtained for the Mode II specimen shown in Fig. 4 are given in Table 1 for $S = H$. When $S = 0.8H$, an overall reduction of less than 1% was obtained for both the stress intensity and displacement coefficients.

While the analytical model does not accurately simulate the actual complex load conditions, displacement coefficients have been computed and are reported in Table 1.

Results and Discussion

Photoelastic examination of 9.5 mm (3/8 in.) thick plastic specimens shows a symmetrical Mode II pattern at the notch tip. The axis of symmetry is inline with the notch centerline; that is, at an angle of zero degrees to the cen-

TABLE 1—*Stress intensity factor and crack mouth displacement coefficients for Mode II specimen of proportions* H/W = *0.5 and* S = H.

a/H	a/W	Y_{II} $\left(K_{II}\frac{Ba^{1/2}}{P}\right)$	Y_I $\left(K_I\frac{Ba^{1/2}}{P}\right)$	$\frac{E'U}{\sigma A}$ $\theta = 180°$	$\theta = -180°$
1	0.50	1.34	−0.035	2.176	−2.322
1.2	0.60	1.55	−0.020	2.271	−2.441
1.4	0.70	1.70	0.044	2.358	−2.522
1.5	0.75	1.73	0.104	2.411	−2.541

terline (Fig. 2). This symmetry and alignment is stable with increasing load and increases in size as load is increased. This suggests that the direction of applied load remains constant during loading. Supportive of this, the notch width measured by use of a feeler gage remains constant during loading. This was true also for tests on aluminum sheet specimens.

Aluminum specimens were initially fabricated from 6.4 mm (1/4 in.) thick sheet; however, the loads required to fracture them monotonically were near the limit of the testing facility. The test section of such specimens was therefore reduced to a thickness of 3.2 mm (1/8 in.) and later specimens were made from 3.2 mm (1/8 in.) sheet entirely. Even at this thickness, no buckling or out-of-plane movement of the specimen tangs was observed.

Initial test data were obtained as monotonic load versus time plots using a ram movement rate of 1.27 mm (0.05 in.) per minute. A typical test record is shown in Fig. 5*a*. The photograph of a failed specimen in Fig. 5*b* shows that the fracture path is at an angle near zero degrees to the machined notch. A partially failed specimen that had been blackened and scribed (Fig. 6*a*) shows the type of displacement which occurs under load. The photograph (Fig. 6*b*) shows the same specimen after partial cracking caused by monotonic loading.

Having established by several additional test runs that the fracture path of this type of specimen is in the desired direction, preliminary fatigue tests were performed to determine the crack propagation direction in Mode II fatigue loading. As predicted by the tangential stress and strain energy density criteria, at low loads the crack propagates at about 70° from the notch towards the tensile-loaded leg of the specimen. At increased loads (about 50% of the monotonic breaking load), however, the fatigue crack path was at or near zero degrees to the machined notch direction. The "threshold" load and mechanism responsible for one or the other types of crack propagation has yet to be investigated more thoroughly. It was observed, however, by use of scribed lines on a specimen as in Fig. 6, that if the specimen is fatigued at a maximum load which is great enough to cause one side of the specimen's entire web section to be displaced vertically relative to the other side, the crack grows in line with the machined notch. At lower loads, those at which the

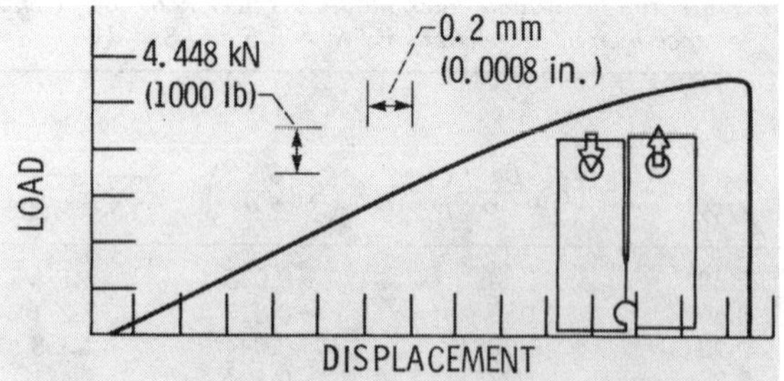

(a) Load versus crosshead displacement test record for NASA LeRC Mode II specimen No. 3-10-83.

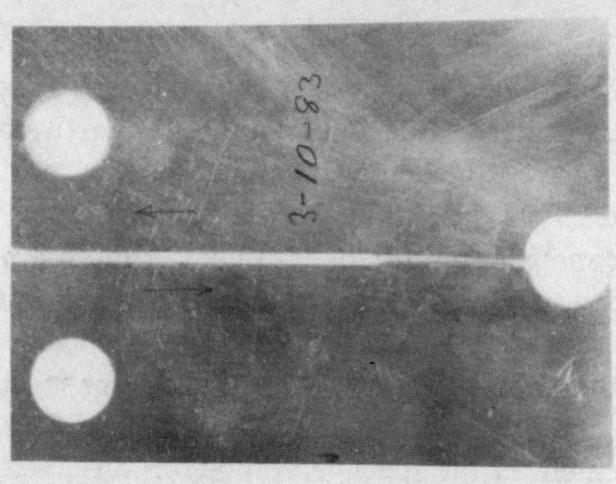

(b) Mode II specimen monotonically loaded to fracture. Material: 3.2 mm (1/8 in.) thick 7075-T6 aluminum.

FIG. 5—*Test record* (a) *and failed specimen* (b).

bottom horizontal scribe line indicates no relative displacement, the crack grows at approximately 70° to the machined notch.

From a continuum point of view, we have large amounts of plastic shear deformation at the cyclic loads favoring crack propagation along the crack-line. The original crack tip is no longer at the end of the deformed crack. The maximum tangential stress theory is no longer applicable in determining the direction of crack growth in this case. Some bending in the specimen legs produces a Mode I component, which is negligible compared with the Mode II value.

This would at first suggest that for in-line Mode II crack propagation to be operative the overall displacement resulting from the applied load must be great enough to cause one half of the specimen to shear past the other half (here in a vertical direction) in its entirety. If the load or displacement is too low, or if the material is very brittle, then there is no relative sliding motion between the two specimen halves. The specimen may now be considered as

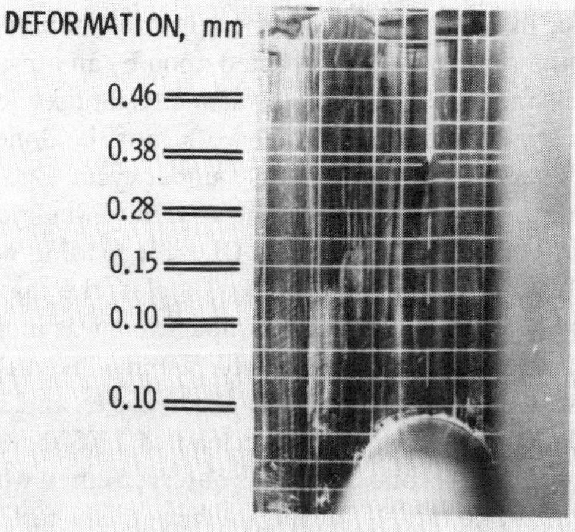

(a) Deformation at 13.344 kN (3000 lb) load (~57% of failure load). Vertical grid spacing = 2.5 mm (100 mils).

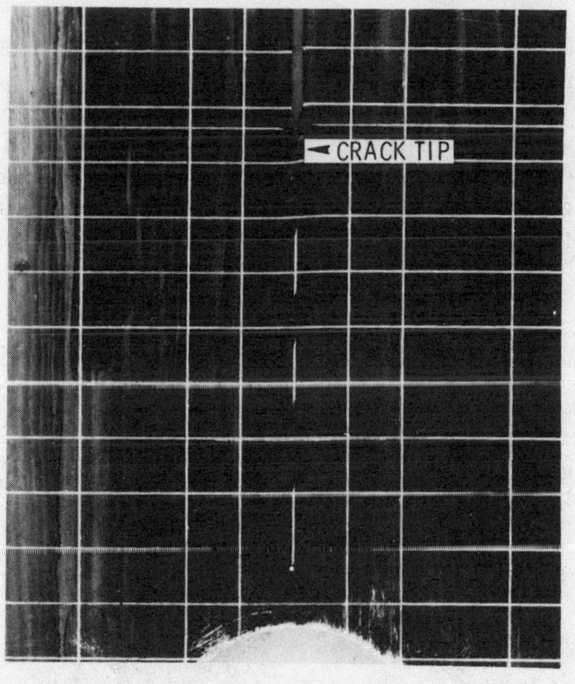

(b) Permanent deformation after unloading from 23.57 kN (5300 lb); crack length ~1.3 mm (50 mils).

FIG. 6—*Examples of Mode II specimen deformation while under load* (a) *and after unloading from 98% of failure load* (b) *for 3.18 mm (0.125 in.) thick 7075-T6 aluminum specimen.*

being held in place in the test fixture by the upper left-hand pin and the lower pin (Fig. 3), and the right-hand side is acted upon by an upward tensile force. The top of the V-shaped notch root acts as a stress-raiser, causing eventual tensile failure to originate there. Further work must be done in this area.

The specimen's satisfactory performance under cyclic loading is verified by the preliminary data shown in Fig. 7. The specimen was cycled at 2 Hz at a maximum load of 13.789 MN (3100 lb). All cyclic loading was performed at an R ratio of 0.1. After approximately 2000 cycles, the maximum load was reduced to 13.344 MN (3000 lb). Crack propagation was monitored with reference to scribe lines spaced at 1.27 mm (0.050 in.) intervals along the web centerline. The test was interrupted after 4000 cycles and subsequently restarted and loaded to a maximum cyclic load of 13.522 MN (3040 lb). At about 4800 total cycles, a second crack was observed emanating at an angle of about 70° from the top of the "V" at the notch root. The test was then discontinued, since the in-line (Mode II) crack growth stopped with the advent of this secondary crack growth. The exact time of secondary crack initiation is not known. However, it is possible that it initiated during initial stages of the restart procedure as the cyclic load was being increased gradually to its maximum level.

The fatigue data obtained from this initial test run (Fig. 7) indicate that for each load the crack growth as a function of the number of cycles is fairly linear, simplifying conversion to crack growth rates. K_{II} is a function of crack length. However, the experimental results over the applied load range 13.344 MN (3000 lb) $\leq P \leq$ 13.789 MN (3100 lb) indicate that da/dN is not a function of crack length. Over the load range tested it appears that da/dN is independent of K_{II} and strongly dependent upon the applied load.

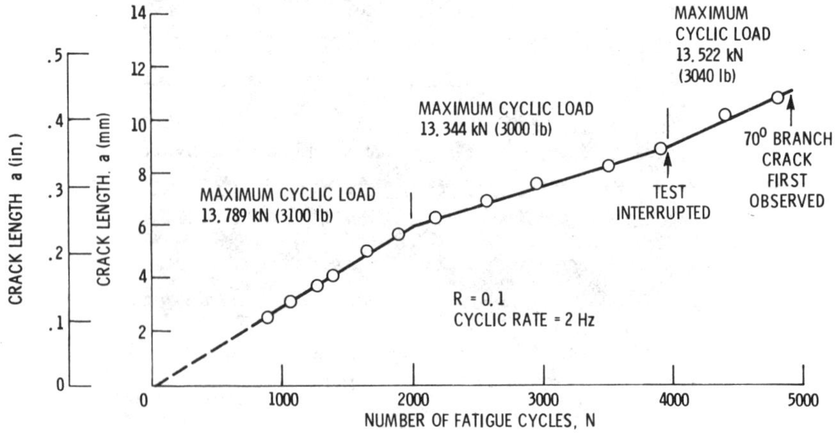

FIG. 7—*Results of Mode II cyclic loading of 7075-T6 aluminum NASA LeRC Mode II specimen.*

Photographs of the failed specimen are shown in Fig. 8. The initial direction of cracking was at 0 to 2° to the notch centerline, and gradually changed to about 10° after attaining a length of about 2.5 mm (0.1 in.) (Fig. 8*a*). This agrees with the observation of Chisholm for this material [*6*], and with data obtained for a Chisholm-type specimen used in this study. A photograph of the fracture surface (Fig. 8*b*) shows a dark coloration most likely resulting from the formation of oxide during rubbing of the mating surfaces. This would be expected, since opening mode forces are negligible for this specimen.

Scanning electron microscope photographs (Fig. 9) reveal the same evidence of oxided and rubbed surfaces. That the presence of some Mode I is desirable to prevent such rubbing is debatable. A truly Mode II situation would perhaps by its nature also include the effects of such rubbing in the

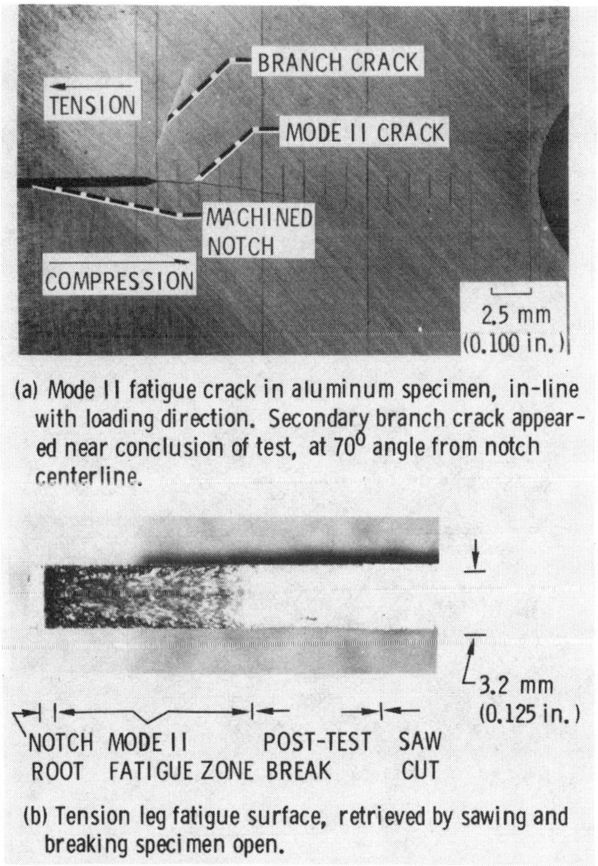

(a) Mode II fatigue crack in aluminum specimen, in-line with loading direction. Secondary branch crack appeared near conclusion of test, at 70° angle from notch centerline.

(b) Tension leg fatigue surface, retrieved by sawing and breaking specimen open.

FIG. 8—*Notch zone of 7075-T6 aluminum Mode II test specimen after 4800 cycles (total) at 13.788 kN (3100 lb), 13.344 kN (3000 lb), and 13.477 kN (3040 lb).*

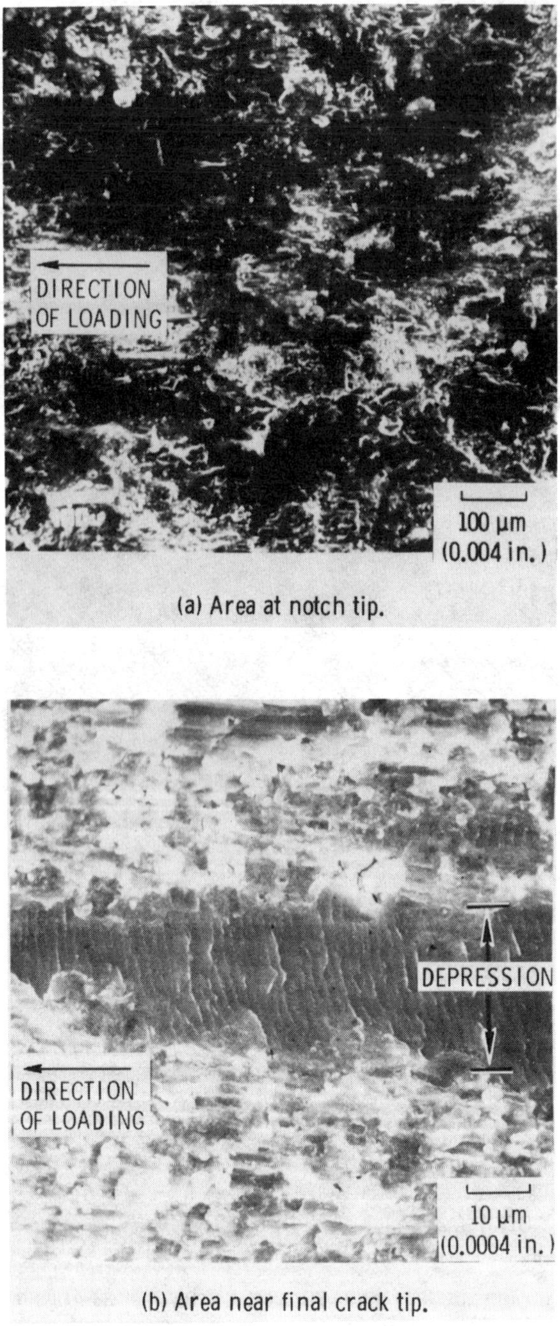

FIG. 9—*Scanning electron microscope photographs of Mode II crack surface. Note white particles believed to be abraded oxides.*

characterization of a material's properties. Should the intentional introduction of Mode I forces be desired, it could probably be accomplished by dimensional adjustment of the loading points of the specimen or by altering the length of the machined notch.

Further development of this type of specimen will be associated with determining a suitable compliance calibration method and in further analyses by use of finite element methods. An optimum design of both the specimen and the test fixture is also required, since the information presented herein is based upon an initial design of both.

References

[1] Palmgren, A., *Ball and Roller Bearing Engineering*, SKF Industries, Philadelphia, 1945, pp. 58-61.
[2] Keer, L. M. and Bryant, M. D., "A Pitting Model for Rolling Contact Fatigue," *Transactions of ASME*, Vol. 105, April 1983, pp. 198-205.
[3] Hahn, G. T., Bhargava, V., Yoshimura, H., and Rubin, C., "Analysis of Rolling Contact Fatigue and Fracture," presented at the Sixth International Conference on Fracture, New Delhi, India, Dec. 1984.
[4] Iosipescu, N., "New Accurate Procedure for Single Shear Testing of Metals," *Journal of Materials*, Vol. 2, No. 3, Sept. 1967, pp. 537-566.
[5] Richard, H. A., "A New Compact Shear Specimen," *International Journal of Fracture*, Vol. 17, 1981, pp. R105-R107.
[6] Jones, D. L. and Chisholm, D. B., "An Investigation of the Edge-Sliding Mode in Fracture Mechanics," *Engineering Fracture Mechanics*, 1975, Vol. 7, pp. 261-270.
[7] Arcan, M., Hashin, Z., and Voloshin, A., "A Method to Produce Uniform Plane-Stress States with Applications to Fiber-Reinforced Materials," *Experimental Mechanics*, Vol. 18, 1978, pp. 141-146.
[8] Raju, K. R., "On the Sliding Mode Stress Intensity Factors for a Three-Point Bend K_{II} Specimen and Mode II Fracture Toughness," *International Journal of Fracture*, Vol. 17, 1981, pp. R193-R197.
[9] Watkins, J., "Fracture Toughness Test for Soil-Cement Samples in Mode II," *International Journal of Fracture*, Vol. 23, 1983, pp. R135-R138.
[10] Gross, B. and Mendelson, A., "Plane Elastic Analysis of V-Notched Plates," *International Journal of Fracture Mechanics*, Vol. 8, No. 3, Sept. 1972, pp. 267-276.
[11] Williams, M. L., "Stress Singularities Resulting from Various Base Conditions in Angular Corners of Plates in Extension," *Journal of Applied Mechanics*, Vol. 19, No. 4, pp. 526-528.

DISCUSSION

L. Banks-Sills[1] (*written discussion*)—The authors pose an interesting physical problem that requires understanding of both mixed-mode and Mode II fracture mechanisms. One of the design requirements of the Mode II specimen presented here calls for crack propagation in a self-similar manner.

[1]Department of Mechanics, Materials, and Structures, Faculty of Engineering, Tel Aviv University, Ramat Aviv, Israel.

The direction of crack growth is governed primarily by the properties of the material being tested and the conditions at the crack tip. If there is "brittle" fracture as occurs with Perspex, a crack in a Mode II field propagates at angles between 60 and 70° with respect to the parent crack [*1,2*]. In elastic materials, this coincides with the direction of maximum tangential stress [*2*]. For elastic-plastic materials, finite element studies show that depending upon the strain hardening parameter, the maximum tensile stress direction is between 70 and 83° [*3*]. If, however, there is much plastic deformation ahead of the crack tip, resulting from high loads and/or plane stress conditions, it is observed in the laboratory that cracks grow in a self-similar manner. This experimental phenomenon would seem to indicate that there is considerable void growth ahead of the crack which causes propagation in that direction.

These two mechanisms, namely maximum tangential stress and void growth, compete in not only Mode II deformation but also Mode I for determining crack propagation direction [*4*]. This direction is controlled by the behavior of the material at the crack tip. For Mode II, when elastic behavior is dominant, the crack propagates at angles between 60 and 80° with respect to the parent crack as predicted by the maximum tensile stress criterion. When plasticity predominates, the crack propagates self-similarly by means of void growth.

R. J. Buzzard et al (*authors' closure*)—The authors wish to thank Dr. L. Banks-Sills for her discussion. We intend to perform a scanning electron microscope study for void growth in the crack tip region.

Discussion References

[*1*] Erdogan, F. and Sih, G. C., *Journal of Basic Engineering*, Vol. 85, 1963, pp. 519-527.
[*2*] Banks-Sills, L. and Arcan, M., this publication, pp. 347-363.
[*3*] Shih, C. F. in *Fracture Analysis, ASTM STP 560*, American Society for Testing and Materials, Philadelphia, 1974, pp. 187-210.
[*4*] Hutchinson, J. W., personal communication.

Leslie Banks-Sills[1] *and Mircea Arcan*[1]

A Compact Mode II Fracture Specimen

REFERENCE: Banks-Sills, L. and Arcan, M., **"A Compact Mode II Fracture Specimen,"** *Fracture Mechanics: Seventeenth Volume, ASTM STP 905*, J. H. Underwood, R. Chait, C. W. Smith, D. P. Wilhem, W. A. Andrews, and J. C. Newman, Eds., American Society for Testing and Materials, Philadelphia, 1986, pp. 347–363.

ABSTRACT: A compact Mode II fracture specimen together with its loading frame is analyzed and tested. This system is a further development of an earlier circular Mode II fracture specimen presented by the authors. For fatigue precracking in two different loading configurations, formulas relating the Mode I stress intensity factor, applied load, and specimen geometry are developed. In order to determine K_{IIc} from the fracture load, a relationship between K_{II}, the applied load, and specimen/loading frame geometry is obtained. All of these relations are calculated by means of the finite element method.

These formulas are then employed to perform a series of Mode II fracture toughness tests with Perspex (also called Plexiglas or polymethyl methacrylate [PMMA]). The average measured K_{IIc} value was 0.94 MPa$\sqrt{m}$. The K_{Ic} tests were performed with the compact tension specimen, yielding $K_{Ic} = 1.06$ MPa$\sqrt{m}$; thus K_{Ic}/K_{IIc} was found to be 1.127. In addition, crack propagation angles were found to be between 63 and 70°. These results are compared with the maximum tangential stress and maximum tangential principal stress criteria.

KEY WORDS: fracture, Mode II, fracture toughness, Perspex, polymethyl methacrylate (PMMA), fracture criteria, Mode II specimen

Interest in Mode II fracture testing has increased during the last few years. Several specimens designed with the aim of measuring K_{IIc} have appeared in the literature [*1–5*]; each of these has various deficiencies. Their main shortcoming is the absence of a uniform pure shear stress field before introduction of the crack; for a full discussion see Refs *6* and *7*.

The work described herein is a further development of earlier studies by the authors [*6–10*], leading to a practical compact specimen for K_{IIc} testing. In Refs *6* to *10* a circular Mode II specimen was analyzed and tested, demonstrating the presence of a nearly uniform pure shear field before introduction

[1]Senior Lecturer and Professor, respectively, Department of Mechanics, Materials, and Structures, Faculty of Engineering, Tel-Aviv University, Ramat-Aviv, Tel-Aviv, Israel.

of the crack. It was further observed that all isostatics pass through this region. These two characteristics define the significant region of the specimen. In addition, it was shown that for all cases considered, K_I/K_{II} was less than 0.025 [*9,10*], so that K_I may be neglected. Calibration formulas for fatigue precracking and fracture testing were determined. Tool steel and Perspex, both brittle materials, were tested [*9,10*] and K_{IIc} values were determined.

The new compact shear specimen in its loading frame is examined by means of photoelasticity. Calibration formulas required for fatigue precracking and K_{IIc} determination are obtained from finite element analysis of the specimen. The testing procedure is described, and test results for K_{Ic}, K_{IIc} and the crack propagation angle in Perspex are reported. These results are then compared with crack propagation theories.

Specimen

The compact shear specimen is illustrated in Fig. 1*a*. The lines marked *CD* are isostatic directors which function together with the isostatic directing contours of the loading frame. Forming an angle of 45° with the vertical specimen axis, these lines cause the isostatics to follow ±45° directions through the significant region of the specimen. This region is located between *DD* and

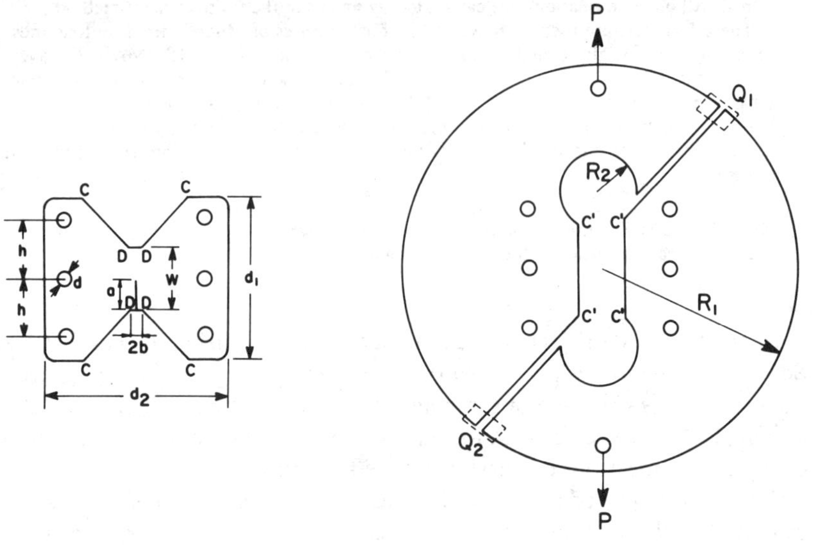

(a) (b)

FIG. 1—(a) *Compact shear specimen with crack of length* a *(*W $=$ *30 mm,* 2b $=$ *6 mm,* d_1 $=$ *82 mm,* d_2 $=$ *98 mm,* d $=$ *8 mm,* h $=$ *31 mm).* (b) *Loading frame (*R_1 $=$ *95 mm,* R_2 $=$ *20 mm). The 45° isostatic directors are located at* CD *in the specimen continuing at* C′ *in the frame.*

DD. All the isostatics pass through this region at ±45° with respect to the vertical axis, creating a nearly uniform pure shear field before crack introduction. In order to induce these conditions, the compact shear specimen is employed with the loading frame illustrated in Fig. 1*b*. The load is applied as shown so that its position relative to the central vertical specimen axis, together with the isostatic directing contours, creates the significant region.

This region is examined by means of photoelasticity. Figure 2*a* is a photograph of an uncracked aluminum specimen attached to the loading frame constructed from ASSAB 705 steel (a Swedish steel similar to SAE 4337). This loading frame is employed in the subsequent studies. A photoelastic coating (PS-1,a Vishay Intertechnology product designation) applied to the aluminum specimen illustrates the region of constant τ_{max}. The 45° isoclinic exhibited in Fig. 2*b* is produced in a Perspex specimen loaded in the loading frame. It may be noted that in the region of constant τ_{max}, the directions of the principal stresses are oriented at angles of ±45° with respect to the vertical specimen axis. Thus Figs. 2*a* and 2*b* demonstrate the existence of a pure shear region before introduction of a crack.

Finite Element Analysis

In order to carry out fracture toughness testing with the compact shear specimen, Modes I and II calibration formulas are required, the former for precracking and the latter for K_{IIc} determination. Because fatiguing the specimen in Mode II would, in many cases, cause the crack to turn out of its plane, the precracking procedure is performed under Mode I loading. In this way, the fatigue crack propagates in a self-similar manner, so that at failure the crack experiences only Mode II conditions.

For all the finite element calculations, the square root singular stresses at the crack tip are modeled by means of four rectangular quarter-point eight-noded elements surrounding the crack tip. It has been shown that this element has a bounded strain energy and is square root singular in a region adjacent to the crack tip [*11*]. Results accurate to at least 2% in Mode I and/or Mode II have been achieved by extrapolating the displacements on the crack faces to the crack tip [*11,12*]; that is, by employing the expressions

$$
\begin{aligned}
K_I^* &= \frac{\sqrt{2\pi}\mu}{\kappa + 1} \frac{v_2 - v_1}{\sqrt{r}} \\
K_{II}^* &= \frac{\sqrt{2\pi}\mu}{\kappa + 1} \frac{u_2 - u_1}{\sqrt{r}}
\end{aligned} \tag{1}
$$

where μ is the material shear modulus, $\kappa = 3 - 4\nu$ for plane strain and $(3 - \nu)/(1 + \nu)$ for generalized plane stress, ν is Poisson's ratio, r is the distance from the crack tip, u_2 and u_1 are the upper and lower crack face x-direction

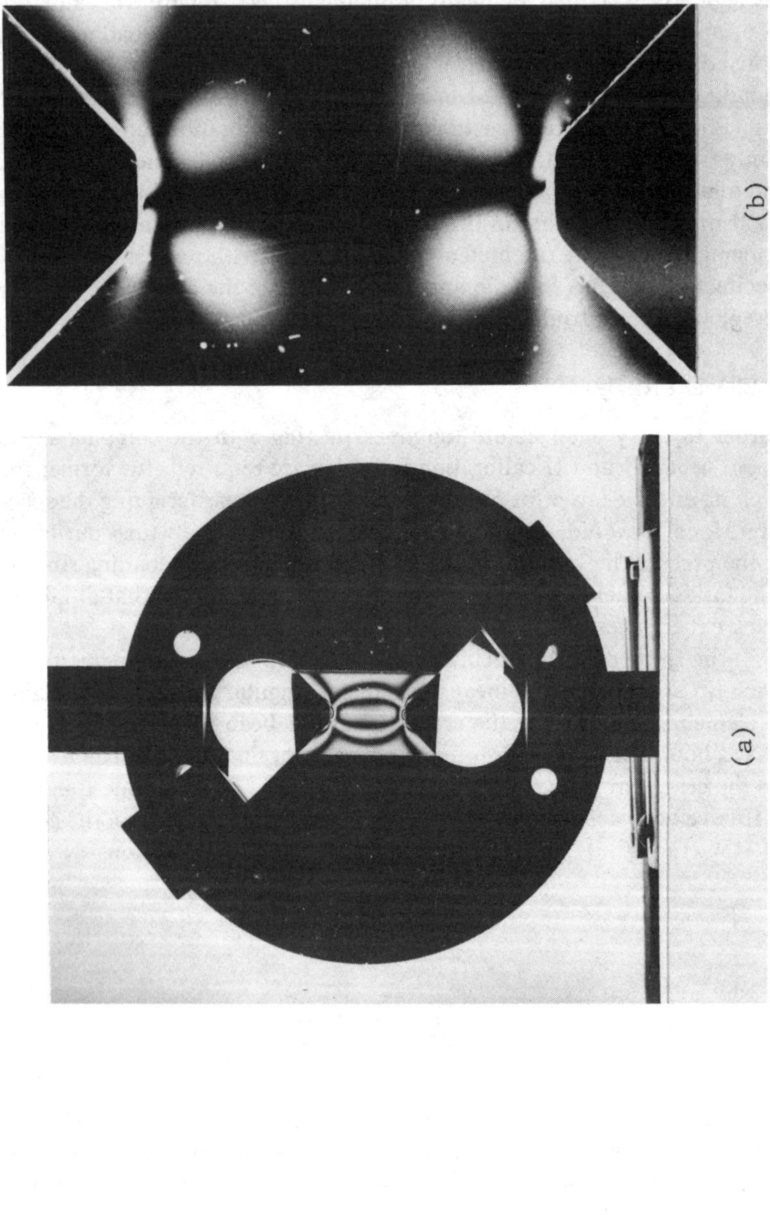

FIG. 2—*(a) Isochromatic pattern and (b) 45° isoclinic of the uncracked compact specimen loaded in pure shear.*

displacements, respectively, and v_2 and v_1 are the respective y-direction displacements (Fig. 3). Then

$$K_{\rm I} = \lim_{r \to 0} K_{\rm I}^{*}$$
$$K_{\rm II} = \lim_{r \to 0} K_{\rm II}^{*} \qquad (2)$$

The Mode I results are described first. Analysis was carried out for the two loading cases illustrated in Fig. 4. Because of the moment induced by the loading in Fig. 4*b*, a much larger $K_{\rm I}$ is produced for equivalent applied loads. Since Perspex has a rather low fracture toughness in tension, a calibration formula is determined for the loading configuration in Fig. 4*a*. For completeness, a formula associated with Fig. 4*b* is also produced. This latter relationship may be employed with materials such as aluminum and steel, for which this loading configuration is more convenient. The mesh employed is exhibited in Fig. 5. Because of symmetry, only half of the specimen is analyzed. Thus there are only two distorted elements surrounding the crack tip. All elements are eight-noded isoparametric elements. The loading is modeled as a point force located at the hole center, with the holes neglected. Finite element

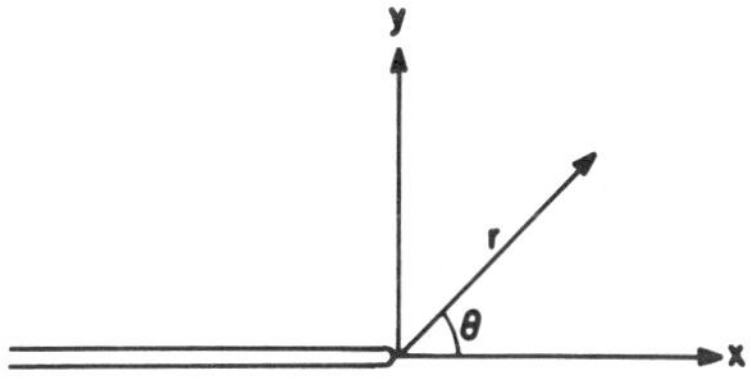

FIG. 3—*Crack tip coordinates.*

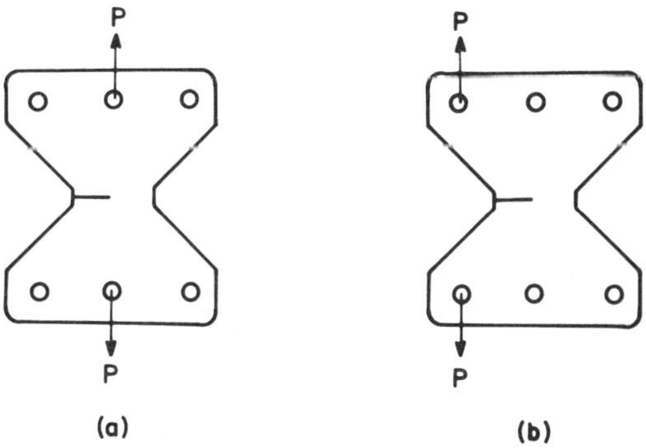

FIG. 4—*Loading configurations of compact shear specimen for fatigue precracking.*

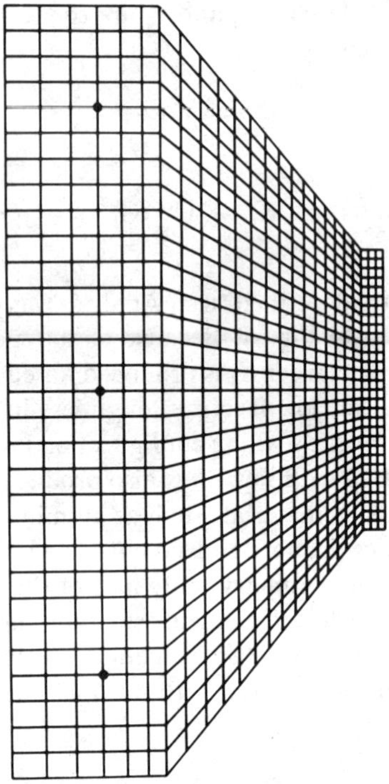

FIG. 5—*Finite element mesh of compact shear specimen for Mode I analysis. Darkened nodes indicate possible location for load application.*

calculations for $0.3 \leq a/W \leq 0.7$ at 0.1 intervals were carried out and a fourth-order polynomial fit through these points. For the loading in Fig. 4*a*:

$$K_{\rm I} = \frac{P\sqrt{\pi a}}{Wt}\{23.09 - 151.4(a/W) + 544.2(a/W)^2 - 845.0(a/W)^3 + 533.3(a/W)^4\} \quad (3a)$$

and for Fig. 4*b*:

$$K_{\rm I} = \frac{P\sqrt{\pi a}}{Wt}\{5.528 - 42.29(a/W) + 159.8(a/W)^2 - 254.1(a/W)^3 + 162.5(a/W)^4\} \quad (3b)$$

where a is the crack length, W is the specimen width, and t is the specimen thickness (see Fig. 1*a*). These formulas are used in determining the applied load required for fatigue precracking the compact shear specimen.

After the notch is extended with a fatigue crack, the specimen is inserted into the loading frame (Fig. 1*b*) and loaded to fracture. Then K_{IIc} may be calculated from the recorded load at fracture. Thus a relationship between applied load and the stress intensity factor K_{II} is required. To accomplish this, a finite element analysis of the loading frame and specimen is performed. In the calculation, the values of E and ν for both the steel loading frame (E = 206 GPa, ν = 0.3) and the Perspex specimen (E = 3.4 GPa, ν = 0.39) are employed. The finite element mesh of the loading frame/specimen assembly is shown in Fig. 6. In the finite element model, the two structures do not overlap, so that the analyzed structure is slightly less stiff than the actual one. Moreover, some unpublished studies show that changing the relative stiffness of loading frame and specimen affect K_I rather than K_{II}. It may be concluded that neglecting this extra material should not affect the relationship between K_{II} and the applied force P. The connection between the specimen and loading frame is through the six darkened nodes shown in Fig. 6; that is, the pins are modeled as point forces again with the holes neglected. Other studies will be carried out to examine the effect of this form of modeling. As with Eqs 3*a* and 3*b*, calculations for $0.3 \leq a/W \leq 0.7$ were performed at intervals of 0.1 and a fourth-order polynomial fit through the results, yielding

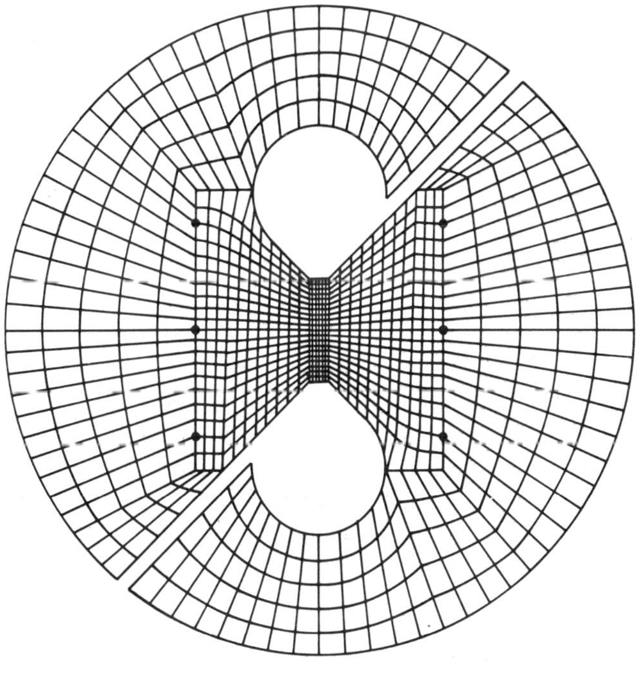

FIG. 6—*Finite element mesh of compact shear specimen and loading frame for Mode II analysis. Darkened nodes indicate connection between the two structures.*

$$K_{II} = \frac{P\sqrt{\pi a}}{Wt}\{0.882 - 0.795(a/W) - 0.131(a/W)^2 - 2.063(a/W)^3 + 3.421(a/W)^4\} \quad (4)$$

Values of K_I were also calculated at each interval and seen to be less than 0.3% of K_{II}. In previous results obtained for the circular specimen, K_I/K_{II} was less than 2.5% [*10*]. Note in that case, as well as in the present study, K_I was found to be negative, indicating a slight closing of the crack. These new results demonstrate that there is a negligibly small K_I for a loading frame/specimen assembly when the loading frame is much stiffer than the specimen. Hence in this case an even better Mode II fracture specimen is obtained.

To determine if Eq 4 is material dependent, similar calculations employing the mesh of Fig. 6 were performed for a loading frame/specimen assembly of the same material. The relative thickness of each was considered. The resulting relationship between K_{II} and P was determined to be

$$K_{II} = \frac{P\sqrt{\pi a}}{Wt}\{1.006 - 0.313(a/W) + 3.344(a/W)^2 - 6.691(a/W)^3 + 5.649(a/W)^4\} \quad (5)$$

Equations 4 and 5 are plotted on the same graph and exhibited in Fig. 7. The largest difference of 2% occurs for an a/W of 0.3, decreasing to less than 0.1% as a/W increases. Hence it may be concluded that either Eq 4 or 5 may be employed for all material combinations of the loading frame and specimen, and that there is no need for a special formula for each material tested.

Testing

Both K_{Ic} and K_{IIc} testing of Perspex was performed. Following ASTM Test for Plane-Strain Fracture Toughness Testing of Metallic Materials (E 399), four compact tension specimens were tested. The crack opening displacement (COD) gage that was employed exerts approximately 14.5 N on the specimen at fracture. Since this force is approximately 5% of the fracture load, the K_{Ic} values were adjusted. To this end, the stress intensity factor for the geometry in Fig. 8 was employed [*13*]. With a force of 14.5 N, K_I was calculated and superposed with the value of K_{Ic} calculated for a compact tension specimen (Table 1). The average corrected K_{Ic} value was determined as 1.062 MPa$\sqrt{m}$.

Tests to determine the yield stress (σ_y) were also performed. As previously observed [*10*], the yield stress and ultimate stress are indistinguishable. The average from three tests showed an order of magnitude value for the ultimate stress to be 50 MPa. In accordance with ASTM E 399, a minimum specimen thickness was determined as 1.2 mm. All specimens were of 10 mm nominal thickness.

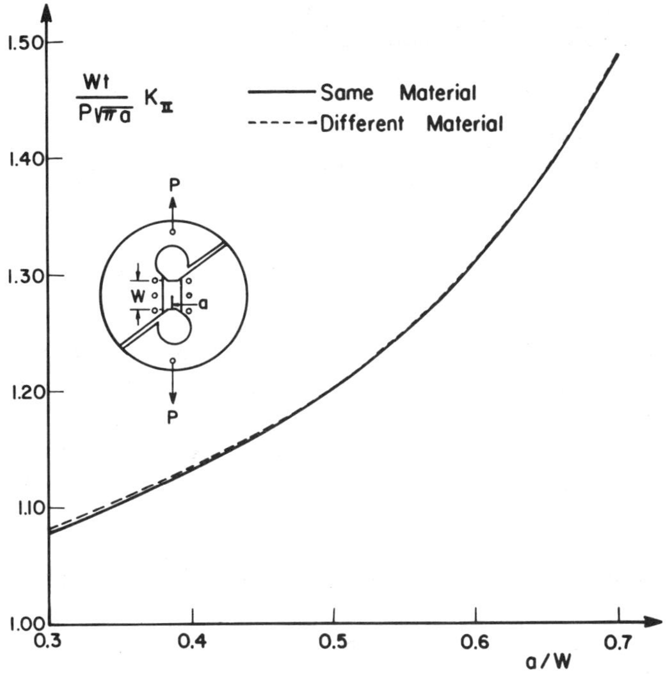

FIG. 7—*Comparison of* K_{II} *versus crack length for specimen and loading frame composed of identical and different materials.*

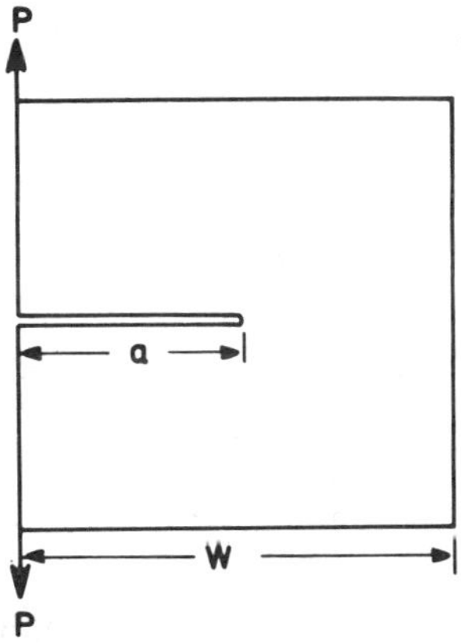

FIG. 8—*Geometry employed to evaluate effect of COD upon* K_{Ic}.

The procedure for K_{IIc} testing followed as closely as possible that for K_{Ic} testing. Specimen dimensions are shown in Fig. 1*a*. Chevron notches 1 mm thick were inserted into the Perspex specimens. Employing Eq 3*a*, appropriate for the loading configuration in Fig. 4*a*, fatigue precracking was performed in two stages. In the first stage, the stress intensity factor was less than

TABLE 1—*Mode I test results for Perspex.*

Specimen	W, mm	a, mm	t, mm	P_Q, kN	K_{Ic}, MPa$\sqrt{m}$	K_{Ic},* MPa$\sqrt{m}$
121 P	59.9	27.8	9.63	0.258	0.963	1.033
122 P	60.1	30.0	9.67	0.238	0.964	1.043
130 P	59.6	27.4	9.71	0.278	1.004	1.073
131 P	59.4	27.5	9.70	0.281	1.029	1.098

*K_{Ic} values corrected for load exerted by COD.

0.8 K_{Ic}, and in the second less than 0.6 K_{Ic}, until the final crack length a was reached (Table 2). The crack length was measured after fracture as prescribed in ASTM E 399 at three locations along the crack front (in the center and at one fourth the thickness from each end). Good fatigue precracks could be attained from the chevron starter notch. Difficulties were encountered when V-starter notches were employed, so that this type of notch was not pursued.

After completion of the precracking procedure, the specimen was inserted into the loading frame (as in Fig. 2*a*). Note the restraining devices at points Q_1 and Q_2 in Fig. 1*b*; these are employed to prevent the slight out-of-plane tendency. The crack opening displacement gage has been adapted here for measuring, in this case, the relative crack face sliding. The "knife edges" employed in this application are exhibited in Fig. 9*a*; Fig. 9*b* illustrates the placement of the knife edges relative to the crack. Note that the placement of this device is just behind the fatigue precrack so that it will not interfere with the fracture process.

TABLE 2—*Mode II test results for Perspex.*

Specimen	W, mm	a, mm	t, mm	P_Q, kN	K_{IIc}, MPa$\sqrt{m}$	K_{IIc},* MPa$\sqrt{m}$	α
115 P	30.4	15.4	9.57	1.00	0.910	0.923	...
116 P	30.4	16.5	9.59	...	...	...	−66.3°
117 P	30.3	15.5	9.58	1.08	0.991	1.005	−66.3°
118 P	30.3	15.1	9.59	0.98	0.888†	...	−69.7°
119 P	30.4	15.2	9.57	...	...	...	−67.7°
120 P	30.3	15.6	9.58	1.00	0.924†	...	−66.3°
123 P	30.3	16.4	9.61	0.85	0.820	0.835	...
124 P	30.3	15.2	9.65	0.90	0.810†	...	−64.3°
125 P	30.3	15.1	9.67	1.04	0.930	0.944	−62.7°
126 P	30.3	15.6	9.65	0.98	0.902	0.916	−63.0°
127 P	30.3	15.1	9.60	1.17	1.050	1.064	−66.0°
128 P	30.3	15.5	9.61	0.97	0.890	0.904	−64.7°

*K_{IIc} values corrected for load exerted by COD.
†Invalid test results; $P_{max}/P_Q > 1.1$.

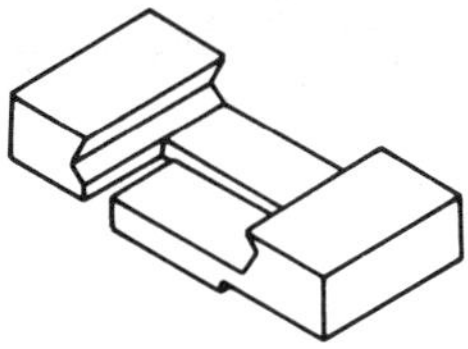

(a)

(b)

FIG. 9—(a) *Knife edges for compact shear specimen.* (b) *Location of these on the specimen indicated by cross-hatch.*

The specimens were fractured and a load-displacement graph recorded. All records were similar to the one shown in Fig. 10 which is a Type I record. As shown in Fig. 10 and following ASTM E 399, the load P_Q was determined at the point of intersection of the load-displacement curve and the line whose slope is 95% of that of the tangent to the linear portion of the load-displacement record. Perhaps reconsideration of the method of determining P_Q is required for K_{IIc} testing; after more experience is gained in testing metals, this should be possible. In the Mode II testing, the force exerted on the crack by the COD gage is between 1 and 2% of the fracture load. Since it is nearly on the load line, it was added to P_Q; the small couple resulting from the distance between the knife edges and the crack was neglected. Values of K_{IIc} corrected for this additional force (15 N) are exhibited in Table 2. For an experiment in which the crack length and load met all requirements for a valid Mode I test, K_{IIc} was determined by means of the calibration formula in Eq 4. Several invalid results appear in Table 2; in those cases, the ratio $P_{max}/P_Q > 1.1$. The average K_{IIc} value from the seven valid experiments is 0.94 MPa$\sqrt{m}$. Thus K_{Ic}/K_{IIc} is 1.127.

Also exhibited in Table 2 is the angle at which the crack propagated relative to the original crack. This angle was measured by a Nikon comparator of 20× magnification at three locations along the crack front: at the intersec-

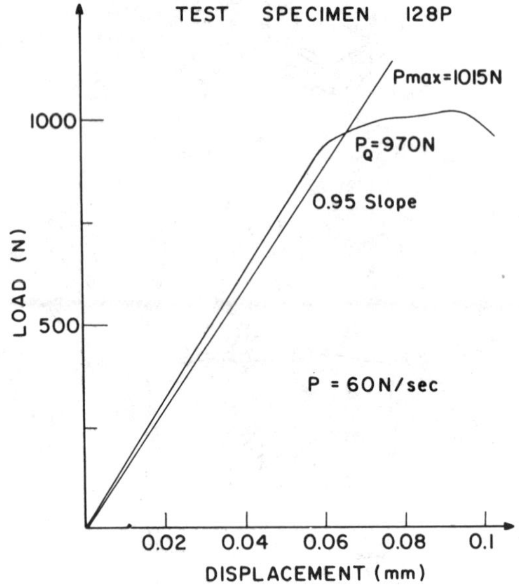

FIG. 10—*Load-displacement record of a Mode II fracture toughness test.*

tion of the specimen surfaces and the crack front and at the maximum crack length along the crack front. Those values appearing in Table 2 are the average of these three values.

Discussion

In this section theoretical models for the propagation of cracks under mixed-mode loading in brittle materials based on the maximum tangential stress [*14*] are discussed, and comparison of K_{IIc} measurements and crack propagation angles α made with these theories. Mode II is viewed as a special case of mixed-mode fracture.

In 1963, Erdogan and Sih [*14*] proposed that a crack propagates along a ray extending from the crack tip (Fig. 3) on which $\sigma_{\theta\theta}$ is maximum. Employing the first term in the Irwin-Williams expansion with $\partial\sigma_{\theta\theta}/\partial\theta = 0$, they found the angle of propagation α to satisfy

$$K_{I} \sin \alpha + K_{II}(3 \cos \alpha - 1) = 0 \tag{6}$$

This result was also established in Ref *15* from energy considerations. Equation 6 implies that not only $\partial\sigma_{\theta\theta}/\partial\theta = 0$ (i.e., $\sigma_{\theta\theta}$ is the maximum tangential stress), but also $\sigma_{r\theta} = 0$ (see Eqs 22 and 23 in the Appendix); thus $\sigma_{\theta\theta}$ is also the maximum principal tangential stress. Note that only the singular term of $0(r^{-1/2})$ is considered. Moreover, in Refs *14* and *15*, it was proposed that the crack propagates when $\sigma_{\theta\theta}(\theta = \alpha) = \sigma_{cr}$ (a critical value of the stress) or equivalently

$$K_{Ic} = \frac{1}{2} \cos(\alpha/2) \, [K_I(1 + \cos \alpha) - 3K_{II} \sin \alpha] \tag{7}$$

Since only Mode II deformation is being considered, $K_I = 0$; thus Eqs 6 and 7 imply

$$\alpha = -70.5° \tag{8a}$$

and

$$K_{Ic} = 1.15 \, K_{IIc} \tag{8b}$$

It has been pointed out [*16,17*] that it is insufficient to consider only the singular term in the stress expansion. If the next order non-zero term is included, it may be seen from Eq 24 that a maximum tangential stress occurs when

$$\left(\cos \frac{\alpha}{2} + 3 \cos \frac{3\alpha}{2}\right) - \frac{5}{4}\left(\frac{r}{a}\right)\left(\cos \frac{\alpha}{2} - 5 \cos \frac{5\alpha}{2}\right) = 0 \tag{9}$$

The angle of crack propagation α, as determined from Eq 9, depends on the distance r/a, at which it is measured from the crack tip. It may be further observed from Eq 22 that $\sigma_{r\theta}$ is not zero at this angle (i.e., $\sigma_{\theta\theta}$ is not a principal stress). Hence as suggested in Ref *18*, it is also possible to predict the crack propagation angle as the root of $\sigma_{r\theta} = 0$ or

$$\left(\cos \frac{\alpha}{2} + 3 \cos \frac{3\alpha}{2}\right) \quad \frac{3}{4}\left(\frac{r}{a}\right)\left(\cos \frac{\alpha}{2} - 5 \cos \frac{5\alpha}{2}\right) = 0 \tag{10}$$

The criterion of Eq 10 implies that crack propagation occurs along the direction of pure Mode I deformation [*19*]. For several values of r/a, crack propagation angles are calculated and exhibited in Table 3.

Employing these angles, a relationship between K_{Ic} and K_{IIc} may be determined. Rewriting Eq 21 as

$$\sigma_{\theta\theta} = \frac{3K_{II}}{4\sqrt{2\pi r}}\left\{-\left(\sin \frac{\theta}{2} + \sin \frac{3\theta}{2}\right) + \frac{5}{4}\left(\frac{r}{a}\right)\left(\sin \frac{\theta}{2} - \sin \frac{5\theta}{2}\right)\right\} \tag{11}$$

implies that

$$K_{Ic} = \beta \, K_{IIc} \tag{12a}$$

TABLE 3—*Crack propagation angle α and ratio of Mode I and Mode II fracture toughnesses β at several distances r/a from the crack tip.*

	$\sigma_{r\theta} = 0$		$\partial\sigma_{\theta\theta}/\partial\theta = 0$	
r/a	α	β	α	β
0.01	−70.0°	1.150	−69.6°	1.150
0.03	−68.9°	1.142	−67.9°	1.142
0.05	−67.9°	1.135	−66.1°	1.136
0.07	−66.8°	1.130	−64.6°	1.131
0.09	−65.8°	1.125	−63.0°	1.128

where

$$\beta = \frac{3}{4}\left\{-\left(\sin\frac{\alpha}{2} + \sin\frac{3\alpha}{2}\right) + \frac{5}{4}\left(\frac{r}{a}\right)\left(\sin\frac{\alpha}{2} - \sin\frac{5\alpha}{2}\right)\right\} \tag{12b}$$

Values of β are also exhibited in Table 3 corresponding to values of α calculated for the maximum tangential stress (Eq 9) and the principal tangential stress (Eq 10). Comparison of the two theories shows only slight differences between the angle of crack propagation α as well as the ratio β of K_{Ic}/K_{IIc}.

Consideration of the experimental results in Table 2 shows measured values of α to be compatible with either of the requirements $\partial\sigma_{\theta\theta}/\partial\theta = 0$ or $\sigma_{r\theta} = 0$ when measurement is made at some small distance from the crack tip. In addition, the ratio of the average K_{Ic} and K_{IIc} values from Tables 1 and 2 yields 1.127, a value somewhat lower but still consistent with those ratios shown in Table 3. Perhaps for a more precise correlation, the coefficients of each term of the asymptotic expansions in Eqs 9, 10, and 12*b* should be determined from the finite element analysis of the compact shear specimen, rather than from consideration of an infinite plate with a far field uniform applied shear stress as is carried out in the Appendix. Nonetheless, the experimental and analytical results tend to support each other. On the other hand, from these experimental results, it does not appear possible to discern between the two different mechanisms of crack growth, namely crack propagation perpendicular to the direction of maximum tangential stress ($\partial\sigma_{\theta\theta}/\partial\theta = 0$) or in the direction of Mode I conditions ($\sigma_{r\theta} = 0$). Indeed either one seems a logical candidate for a crack propagation theory of brittle materials from a physical as well as an experimental viewpoint.

Conclusions

A Mode II compact specimen and its loading frame have been presented. Photoelastic verification of a pure shear region in the specimen before crack introduction was carried out. Calibration formulas for Mode I fatigue precracking of the specimen, as well as a relationship between K_{II} and the applied fracture load, were determined. The relationship is universal in that it applies for all material combinations of loading frame and specimen. Mode I and Mode II fracture toughness experiments on Perspex were performed. Employing the compact shear specimen presented here, an average K_{IIc} value of 0.94 MPa$\sqrt{m}$ was determined. In addition, an average K_{Ic} value was determined as 1.06 MPa$\sqrt{m}$, so that K_{Ic}/K_{IIc} was found to be 1.127. The average measured crack propagation angle was found to be 65.7°. Both the experimentally determined K_{Ic}/K_{IIc} ratio and the crack propagation angle α conform to analytically obtained values of these quantities from assumptions that in brittle material crack propagation occurs either perpendicular to the direction of maximum tangential stress or in the direction of Mode I conditions.

Acknowledgments

This work was supported by the Israel National Council for Research and Development and by the Department of Solid Mechanics, Materials, and Structures, Tel Aviv University.

APPENDIX

An asymptotic expansion for the stresses surrounding a crack tip with a remotely applied pure shear field (Fig. 11) is developed in order to discuss crack propagation theories. Following Refs *17* and *20*, the solution may be developed from complex variable theory. The pair of holomorphic functions which represents a solution to the problem is given as

$$\Phi(z) = \phi'(z) = -\frac{i\tau}{2}\frac{z}{(z^2 - a^2)^{1/2}} + \frac{i\tau}{2} \tag{13}$$

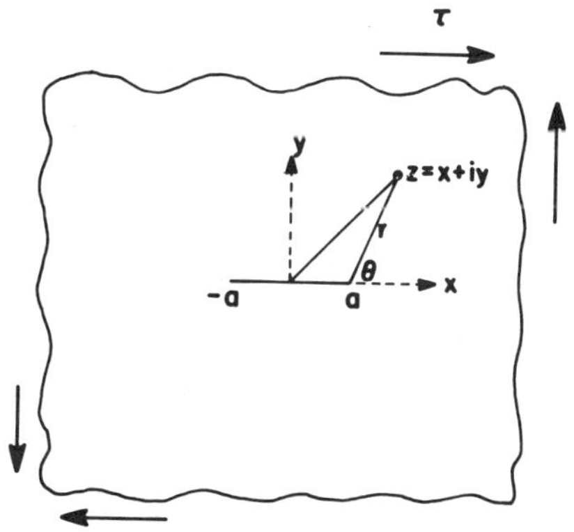

FIG. 11—*Central crack in an infinite plate with a pure shear stress applied far from the crack.*

$$\Omega(z) = -\frac{i\tau}{2}\frac{z}{(z^2 - a^2)^{1/2}} - \frac{i\tau}{2} \tag{14}$$

In terms of the functions $\Phi(z)$ and $\Omega(\mathbf{z})$, the stress components may be written as [20]

$$\sigma_{xx} + \sigma_{yy} = 2[\Phi(z) + \overline{\Phi(z)}] \tag{15}$$

$$(\sigma_{yy} - \sigma_{xx}) + 2i\sigma_{xy} = 2[(\bar{z} - z)\Phi'(z) + \bar{\Omega}(z) - \Phi(z)] \tag{16}$$

To expand these equations near the crack tip $x = a$, take

$$z - a = \zeta = re^{i\theta} \tag{17}$$

so that

$$\begin{Bmatrix} \Phi(\zeta) \\ \Omega(\zeta) \end{Bmatrix} = -\frac{i\tau}{2\sqrt{2}}\left[\left(\frac{\zeta}{a}\right)^{-1/2} + \frac{3}{4}\left(\frac{\zeta}{a}\right)^{1/2} + \cdots\right] \pm \frac{i\tau}{2} \tag{18}$$

Consideration of Eqs 15 to 18 yields

$$\sigma_{xx} + \sigma_{yy} = -\sqrt{2}\,\tau\left(\sqrt{\frac{a}{r}}\sin\frac{\theta}{2} - \frac{3}{4}\sqrt{\frac{r}{a}}\sin\frac{\theta}{2}\right) \tag{19}$$

and

$$\begin{aligned}\sigma_{yy} - \sigma_{xx} + 2i\sigma_{xy} = \sqrt{2}\,\tau\Bigg\{&\sqrt{\frac{a}{r}}\left[\sin\frac{\theta}{2}\left(1 + \cos\frac{\theta}{2}\cos\frac{3\theta}{2}\right) + i\cos\frac{\theta}{2}\left(1 - \sin\frac{\theta}{2}\sin\frac{3\theta}{2}\right)\right] \\ &- \frac{3}{4}\sqrt{\frac{r}{a}}\left[\sin\frac{\theta}{2}\left(1 + \cos^2\frac{\theta}{2}\right) - i\cos\frac{\theta}{2}\left(1 + \sin^2\frac{\theta}{2}\right)\right]\Bigg\}\end{aligned} \tag{20}$$

For our purposes, only terms up to $0(r^{1/2})$ are retained.

From stress transformation equations, stresses of interest $\sigma_{\theta\theta}$ and $\sigma_{r\theta}$ are determined as

$$\sigma_{\theta\theta} = \frac{3\sqrt{2}}{8}\tau\left\{-\sqrt{\frac{a}{r}}\left(\sin\frac{\theta}{2} + \sin\frac{3\theta}{2}\right) + \frac{5}{4}\sqrt{\frac{r}{a}}\left(\sin\frac{\theta}{2} - \sin\frac{5\theta}{2}\right)\right\} \tag{21}$$

and

$$\sigma_{r\theta} = \frac{\sqrt{2}}{8}\left\{\sqrt{\frac{a}{r}}\left(\cos\frac{\theta}{2} + 3\cos\frac{3\theta}{2}\right) - \frac{3}{4}\sqrt{\frac{r}{a}}\left(\cos\frac{\theta}{2} - 5\cos\frac{5\theta}{2}\right)\right\} \tag{22}$$

Note

$$\cos\frac{\theta}{2} + 3\cos\frac{3\theta}{2} = 2\cos\frac{\theta}{2}(3\cos\theta - 1) \tag{23}$$

Finally, the maximum value of $\sigma_{\theta\theta}$ may be found from

$$\frac{\partial \sigma_{\theta\theta}}{\partial \theta} = \frac{3\sqrt{2}}{16} \tau \left\{ -\sqrt{\frac{a}{r}} \left(\cos\frac{\theta}{2} + 3\cos\frac{3\theta}{2} \right) + \frac{5}{4}\sqrt{\frac{r}{a}} \left(\cos\frac{\theta}{2} - 5\cos\frac{5\theta}{2} \right) \right\} = 0 \quad (24)$$

References

[1] Liu, A. F., "Crack Growth and Failure of Aluminum Plate Under In-Plane Shear," Paper 73-252, American Institute of Aeronautics and Astronautics, 1973.
[2] Shah, R. C. in *Fracture Analysis, ASTM STP 560*, American Society for Testing and Materials, Philadelphia, 1974, pp. 29–52.
[3] Chisholm, D. B. and Jones, D. L., *Experimental Mechanics*, Vol. 17, 1977, pp. 7–13.
[4] Richard, H. A., *International Journal of Fracture*, Vol. 17, 1981, pp. R105–R107.
[5] Raju, K. R., *International Journal of Fracture*, Vol. 17, 1981, pp. R193–R197.
[6] Arcan, M. and Banks-Sills, L., "Mode II Fracture Specimen-Photoelastic Analysis and Results," in *Proceedings*, Seventh International Conference on Experimental Stress Analysis, Haifa, Israel, 1982, pp. 187–201.
[7] Banks-Sills, L. and Arcan, M., *Experimental Mechanics*, Vol. 23, 1983, pp. 257–261.
[8] Banks-Sills, L., Arcan, M., and Bui, H. D., *International Journal of Fracture*, Vol. 22, 1983, pp. R9–R14.
[9] Banks-Sills, L., Arcan, M., and Gabay, H., *Engineering Fracture Mechanics*, Vol. 19, 1984, pp. 739–750.
[10] Banks-Sills, L. and Arcan, M., "Mode II Fracture Testing with Application to PMMA," in *Application of Fracture Mechanics to Materials and Structures*, G. C. Sih, E. Sommer, and W. Dahl, Eds., Martinus Nijhoff Publishers, The Hague, 1984, pp. 337–344.
[11] Banks-Sills, L. and Bortman, Y., *International Journal of Fracture*, Vol. 25, 1984, pp. 169–180.
[12] Banks-Sills, L., Arcan, M., and Bortman, Y., *Engineering Fracture Mechanics*, Vol. 20, 1984, pp. 145–157.
[13] Rooke, D. P. and Cartwright, D. J., *Compendium of Stress Intensity Factors*, Her Majesty's Stationery Office, England, 1976, p. 101.
[14] Erdogan, F. and Sih, G. C., *Journal of Basic Engineering*, Vol. 85, 1963, pp. 519–527.
[15] Nuismer, R. J., *International Journal of Fracture*, Vol. 11, 1975, pp. 245–250.
[16] Williams, J. G. and Ewing, P. D., *International Journal of Fracture Mechanics*, Vol. 8, 1972, pp. 441–446.
[17] Eftis, J. and Subramonian, N., *Engineering Fracture Mechanics*, Vol. 10, 1978, pp. 43–67.
[18] Maiti, S. K. and Smith, R. A., *International Journal of Fracture*, Vol. 24, 1984, pp. 5–22.
[19] Cotterell, B. and Rice, J. R., *International Journal of Fracture*, Vol. 16, 1980, pp. 155–169.
[20] Muskhelishvili, N. I., *Some Basic Problems of the Mathematical Theory of Elasticity*, 4th ed., Noordhoff, Groningen, The Netherlands, 1963.

G. E. Sutton[1] *and M. G. Vassilaros*[1]

Influence of Partial Unloadings Range on the J_I-R Curves of ASTM A106 and 3-Ni Steels

REFERENCE: Sutton, G. E. and Vassilaros, M. G., **"Influence of Partial Unloadings Range on the J_I-R Curves of ASTM A106 and 3-Ni Steels,"** *Fracture Mechanics: Seventeenth Volume, ASTM STP 905,* J. H. Underwood, R. Chait, C. W. Smith, D. P. Wilhem, W. A. Andrews, and J. C. Newman, Eds., American Society for Testing and Materials, Philadelphia, 1986, pp. 364–378.

ABSTRACT: An investigation was performed to evaluate the effects of elastic unloadings on the *J*-integral–resistance curves of ASTM A106 Class C steel and 3-Ni steels. Compact specimens (1T) were tested using (1) the multispecimen technique, (2) the direct-current potential drop technique, and (3) an elastic unloading compliance technique with unloadings ranging from 10 to 90%. The two former techniques were 0% unloading procedures used to generate the reference *J-R* curves. These reference curves were compared with *J-R* curves obtained using elastic unloading for the two steels. Results indicated that use of the elastic unloading compliance technique caused no significant difference in the *J-R* curves.

KEY WORDS: *J*-integral, resistance curves, single-specimen tests, multispecimen test, direct-current potential drop, elastic unloading compliance, deep unloadings, *J-R* curves, testing techniques

Suitable procedures for testing fracture toughness play an integral part in evaluating and qualifying materials for nuclear pressure vessel piping. The test procedure for determining the plane-strain fracture toughness, K_{Ic}, is widely accepted and well documented (ASTM Test for Plane-Strain Fracture Toughness of Metallic Materials [E 399]). This fracture mechanics concept has also been used to characterize high-cycle fatigue crack growth (ASTM Test for Constant-Load-Amplitude Fatigue Crack Growth Rates Above 10^{-8} m Cycle [E 647]). However, in the design and construction of many modern structures the elastic assumptions are violated, which precludes the appli-

[1]Research Metallurgist, David Taylor Naval Ship R&D Center, Bethesda, MD 20084.

cation of linear elastic fracture mechanics in characterization of the fatigue and fracture response. Instead, the behavior of these ductile alloys must be analyzed in terms of elastic-plastic fracture mechanics concepts.

The J-integral test procedure (ASTM Test for J_{Ic}, A Measure of Fracture Toughness [E 813]) is a widely used and accepted method for measuring fracture toughness under elastic-plastic conditions. Rice [1] developed the J-integral and showed its relationship to crack initiation. Later Hutchinson and Paris [2] and Ernst et al [3] extended this concept to account for growing cracks.

These developments with regard to the J-integral were based on deformation theory plasticity. This is a nonlinear elastic theory; one restriction to its use is that unloading is not permitted. Unloading can be produced by two different sources.

The first source of an unloading effect results from partial unloading of the specimen during a test, thereby elastically unloading elements near the crack tip which have already experienced plastic (or assumed nonlinear elastic) loading. If the nominal elastic unloadings are large enough, the material near the crack tip might even experience some plastic unloading which would result in some significant hysteresis. This case was treated analytically by Clarke et al [4] and calculated to be negligible for 10% elastic unloadings. These calculations were performed using the argument that the plastic zone size resulting from the ΔK of a small elastic unload would be much smaller than the fracture process zone size in a J-integral test. This effect and possible fatigue crack growth interactions were investigated in this study.

The second source of unloading occurs only to the elements in the wake of a growing crack. This effect has been discussed by Hutchinson and Paris [2]. Although it may produce a significant cumulative effect on the J-R curve, we did not address it in this study.

As first introduced, the J-integral test procedure consisted of a minimum of four specimens that were individually loaded to produce varying amounts of crack extension, after which they were heat tinted to mark the crack extension and then pulled to failure so the crack length could be measured. The J-integral was estimated from the area under the load versus load-point-displacement diagram for each specimen. Finally, extrapolation of a least-squares fit of the data to the blunting line was used to determine the critical value of J for crack initiation under plane strain, J_{Ic}.

The multispecimen test is costly and time consuming. Clarke et al [4] proposed a method for determining J_{Ic} from a single specimen. They measured crack extension repeatedly during the test by partially unloading the specimen to determine the elastic compliance. Elastic compliance, the reciprocal of the unloading slope, was compared with a calibration curve expression to estimate the crack length. Thus the single specimen with many elastic unloadings replaced the more time-intensive multispecimen test. This is the procedure presented in the Annex of ASTM E 813.

The foregoing work pertained to the determination of *J* for crack initiation J_{Ic}. The *J*-integral has also been thought to have a role during crack growth. This led to the use of the *J*-Resistance (*J-R*) curve in evaluating a material's tearing resistance. Paris et al [*5*] defined a tearing (*T*) modulus in terms of the *J-R* curve and proposed an instability theory which stated that a flawed member will tear stably until the material tearing modulus is exceeded. At that point, tearing instability will occur. More recently, Paris [*6*] developed a crack stability criterion which uses a *J* versus *T* plot to diagrammatically present conditions that will lead to crack instability in a material. This method has been used to assess the safety of nuclear reactor pressure vessels. Tearing instability concepts are strongly dependent on the *J-R* curve characterization for a specific material.

The significance of approaching safety assessment with the *J-T* diagram underlines the importance of developing an accurate *J-R* curve for characterizing a material. Two concerns have been noted with respect to use of the single-specimen test procedure to establish the *J-R* curve. Firstly, the restriction from deformation theory plasticity states that unloading is not permitted. The concern arises as to whether the elastic unloading of the single specimen is a significant violation, and what its possible effects are on the *J-R* curve. As mentioned earlier, Clarke et al [*4*] treated the question analytically, not experimentally, for the material investigated in their study.

Secondly, Landes and McCabe [*7*] studied the effect of load history on the *J-R* curve of HY-130 and ASTM A508 Class 2 steels. They applied 13 to 75 individual cycles of plastic loading and plastic unloading to produce a *J-R* curve from 1T compact specimens of the materials. These curves were compared with *J-R* curves generated with 10% elastic unloading compliance techniques. Cyclic loading had no effect on the *J-R* curve for HY-130 steel, but significantly reduced the *J-R* curve of the ASTM 508 Class 2 steel. The first type of behavior was labeled "*R*-curve-dominated crack growth"; the second type was called "cyclic-dominated crack growth". The authors did not ascribe the observed behavior to some material property. They attributed the difference in the behavior of the two steels to results of the relationship between the size of the cyclic loops and the overall monotonic toughness of the two steels. In a similar study on a modified 4340 steel, Landes and Leax [*8*] also reported "*R*-curve-dominated crack growth". Their tests were performed by applying 7 to 45 complete cycles of unstated size to 1T compact specimens during a *J*-integral elastic unloading compliance test.

Our objective was to establish the effect of elastic unloadings on the *J-R* curve of ASTM A106 Class C steel and 3-Ni steel. To study the effects, *J-R* curves were generated in which the elastic unloading was the parameter varied, ranging from 0 to 90%. *J-R* curve tests using multispecimen and direct-current potential drop (DCPD) test techniques represented the 0% unloadings. These *J-R* curves were compared with results from elastic unloading compliance tests in which unloadings ranged from 10 to 90%.

Although alternative methods exist for *J-R* curve testing (namely DCPD, ACPD, and the dual-gage procedure of Andrews [*9*], the unloading compliance single specimen method is widely utilized. Its effect on the *J-R* curve must be known and understood if the *J-R* curve is to be properly used.

Materials

This investigation was conducted using two different steels, ASTM A106 Class C steel and a 3-Ni steel. The chemical analysis for these steels is shown in Table 1. The A106 Class C steel was received in the form of pipe with 98-mm (3⁷/₈-in.) thick wall,[2] and all specimens were tested at 135°C (275°F) to ensure full upper shelf fracture response. The 3-Ni steel was procured as 25.4-mm (1-in.)-thick plate. The mechanical properties for both materials are presented in Table 2. All specimens tested were of the modified compact 1T geometry with 20% side grooves (Fig. 1). The ASTM A106 specimens were fabricated in the L-C orientation (the longitudinal axis of the pipe is the direction normal to the crack plane and the circumferential direction is the direction of crack propagation). The 3-Ni specimens were of the T-L orientation (the transverse direction is normal to the crack plane and the crack is propagating in the longitudinal direction). All specimens were precracked as prescribed in ASTM E 399 so that the ratio of crack length to specimen depth (a/W) was 0.65.

Test Procedures

The test arrangement used in this study was detailed by Joyce and Gudas [*10*] and can be described as a computer-interactive procedure in which digitized load-displacement data are analyzed during testing. To evaluate the unloading effects, four distinct tests were conducted: (1) a series of multispecimen tests; (2) single-specimen tests in which the amount of elastic unloading was kept at 10 to 15%; (3) single-specimen tests in which the unloading ranged from 30 to 90% (called "deep elastic unloading"); and (4) direct-cur-

TABLE 1—*Chemical composition of materials tested.*

Material	Chemical Constituents, wt%							
	C	S	P	Si	Ni	Mn	Cr	Mo
ASTM A106 Class C steel	0.26	0.017	0.028	0.22	0.26	0.90	0.11	...
3-Ni steel	0.15	0.014	0.010	0.21	2.70	0.34	1.26	0.23

[2]Original measurements were made in inch-pound units.

TABLE 2—*Mechanical properties of materials tested.*

Material	0.2% Yield Strength, MPa (ksi)	Ultimate Tensile Strength, MPa (ksi)	Elongation in 50 mm, %	Reduction in Area, %
ASTM A106 Class C steel	317 (45.0)	503 (72)	28	72
3-Ni steel	582 (83.5)	708 (101.5)	27	68

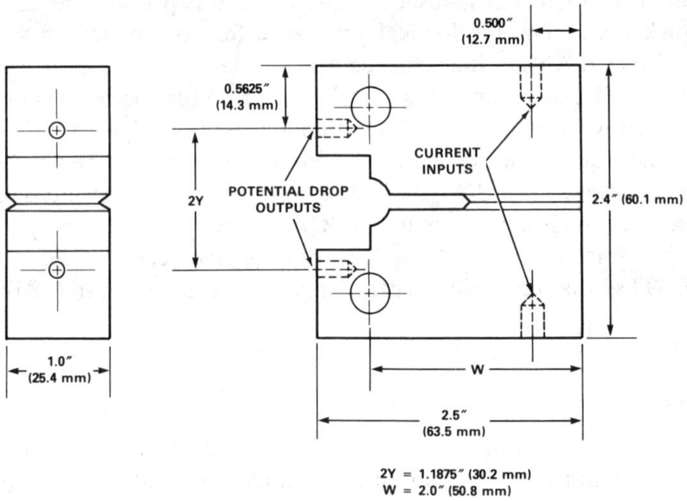

FIG. 1—*Modified 1T compact specimen.*

rent potential drop (DCPD) tests, which represented zero-unloading of a specimen. For the 10% unloading tests and the deep unloading tests, the compliance calibration expression of Saxena and Hudak [*11*] was used to estimate crack length at each unloading.

The formula for calculating J in all four test cases is the Merkle-Corten equation [*12*] as modified by Clarke and Landes [*13*]:

$$J = \frac{\beta A}{Bb}$$

where

$\beta = 2\,[(1+\alpha)/(1+\alpha^2)]$,
$\alpha = [(2\,a/b)^2 + 2(2\,a/b) + 2]^{1/2} - (2\,(a/b) + 1)$,
A = area under the load-displacement curve,

B = net specimen thickness,
b = remaining ligament, and
a = crack length.

This expression was selected as the basis for comparison because it consists of both bending and tension components and because it does not have the crack growth correction terms found in the more recent single-specimen formulation [3]. The crack-growth-corrected expression of Ernst et al [3] is difficult to apply to results of a multispecimen J test. Since all specimens tested in this task had the same crack length, we believe that the crack-growth-corrected J expression would not significantly change the differences in J-R curves.

The value of J at the initiation of crack growth, J_{Ic}, was determined for the multispecimen and single-specimen elastic unloading tests by computing the intersection point of the blunting line ($J = 2\,\sigma_y \cdot \Delta a$, where σ_y is flow stress and Δa is crack extension) and the least-squares fit of the data points that are at least 0.15 mm (0.006 in.) beyond the blunting line but do not exceed 1.5 mm (0.060 in.) of crack extension as measured from the blunting line. The tearing modulus, T, was calculated for the same range of crack extension as above by using the expression

$$T = \frac{dJ_I}{da} \cdot \frac{E}{\sigma_y}$$

where

dJ_I/da = least-squares linear regression slope of J_I versus crack extension curve,
E = elastic modulus, and
σ_y = flow stress = $(\sigma_{ys} + \sigma_{uts})/2$.

Multispecimen Tests

Ten specimens of the 3-Ni steel and nine specimens of the A106 steel were used in the multispecimen technique. The program in the on-line computer was modified to print the area and corresponding approximation of J on the monitor screen as it integrated the area under the load-displacement curve during the test. Instead of unloading the specimen at selected different levels of clip-gage displacement, these specimens were unloaded at various values of applied J. After testing, all specimens were first heat tinted at 370°C (700°F) to mark the maximum crack extension and then broken open at liquid nitrogen temperature. The initial and final crack lengths were determined using the eight-point technique of ASTM E 813. These measured crack lengths were used in the evaluation of the J-integral.

Single Specimen Tests: 10% Elastic Unloading

The single-specimen unloading compliance procedure of Joyce and Gudas [*10*] was followed, in which periodic partial unloadings were made during the test to determine crack size. The unloadings were on the order of 10 to 15% of the maximum load; a sample load-displacement record for such a test is shown in Fig. 2.

Single Specimen Tests: Deep Elastic Unloading

The single specimen unloading compliance procedure [*10*] was again used. Typical load-displacement traces are presented in Fig. 2 for 55 and 90% unloading. The desired amount of unloading was decided before the test and

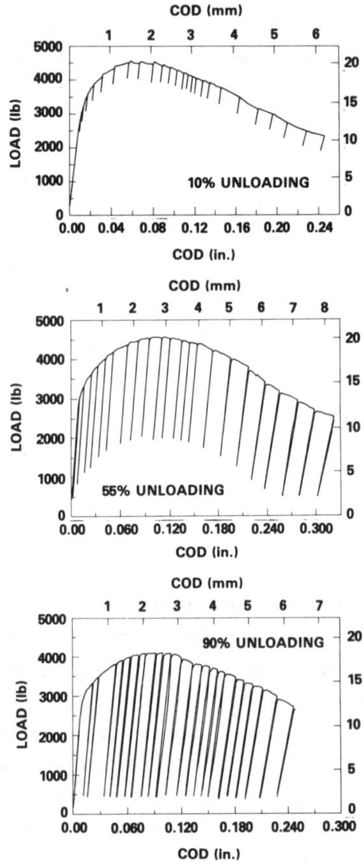

FIG. 2—*Sample load displacement record (load versus crack opening displacement (COD)) for single-specimen tests with 10, 55, and 90% unloading.*

then applied at each unloading, with the limitation that the load would not be allowed to fall below 2.22 kN (500 lb) so as to prevent possible movement of the loading pins. The actual percent unloading was computed after the test by dividing the unloading range by the maximum load achieved during the test. Figure 3 shows the hysteresis loop obtained in one deep unloading cycle. The data used in the compliance expression to determine crack length were from the high density data acquired in the first 10% of the unloading, so that this result could be compared with results of the single-specimen 10% unloading tests. Only data from the unload were used for crack length estimation.

Direct-Current Potential Drop

Specimens of each material were tested using the direct-current potential drop (DCPD) technique. This procedure is based on a correlation between changes in electrical potential and changes in ligament area. Johnson [*14*] showed that a theoretical expression can correlate potential drop with crack length, and that this expression is sensitive only to specimen depth (W) and the location on the specimen of current inputs and potential outputs. Lowes and Fearnehough [*15*] observed that the plot of crack opening displacement versus potential will be linear to the point of crack initiation, and that deviation from linearity can be used to predict crack initiation. For the placement of current and potential leads shown in Fig. 1, Vassilaros and Hackett [*16*] experimentally established the expression

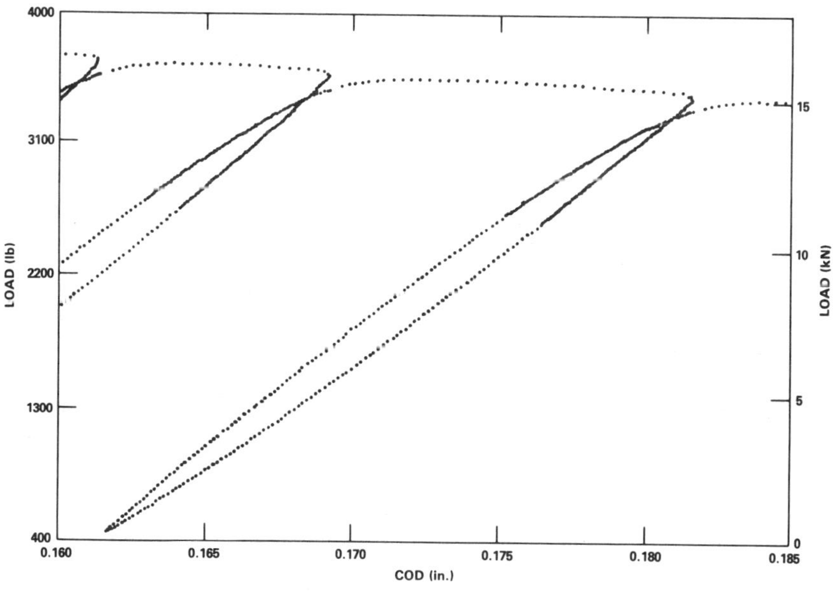

FIG. 3—*Hysteresis loop for a 90% unloading cycle.*

$$P = (1.5 \times a/W)^{2.9} + 0.605$$

over a range of crack lengths (a/W) from 0.6 to 0.8.

After each individual test, the specimen was heat tinted and then broken open at liquid nitrogen temperature. The initial and final crack lengths were measured on the fracture surface using the eight-point method.

Unloading Effect on the *J-R* Curve

Multispecimen Test Results

The results for the nine A106 steel specimens tested for the multispecimen *J*-resistance curve are shown as numbered filled triangles in Fig. 4. In every case there was measurable crack extension, which ranged from 0.229 to 3.531 mm (0.009 to 0.139 in.). Specimens 3 to 7 are in the ASTM-specified range of 0.15 to 1.50 mm (0.006 to 0.060 in.) from the blunting line (shown as offset lines in Fig. 4), and using the analysis of ASTM E 813 results in a J_{Ic} of 268 kJ/m^2 (1531 in.-1b/in.^{2}) and a tearing modulus of 233.

Figure 5 shows the multispecimen test results for the ten specimens of 3-Ni steel. For the first data point of Fig. 5 there was no measurable crack exten-

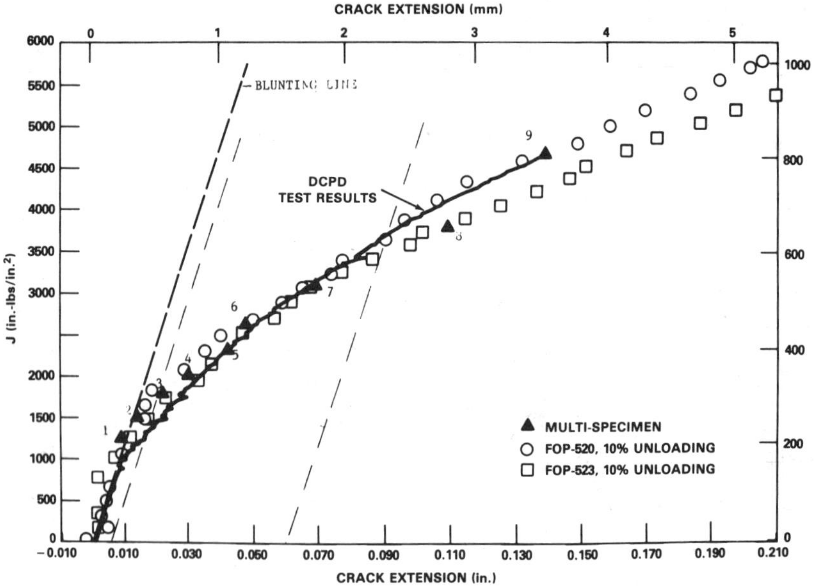

FIG. 4—*Comparison of* J-R *curves for ASTM A106 Class C steel: the multispecimen test; the two single-specimen, 10% elastic unloading compliance tests; and the direct-current potential drop test.*

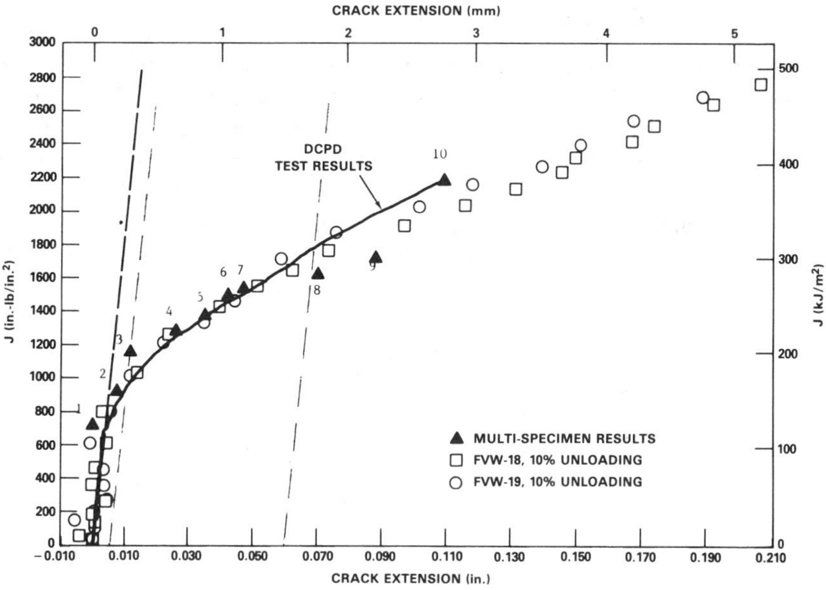

FIG. 5—*Comparison of* J-R *curves for 3-Ni steel: the multispecimen test; two single-specimen, 10% elastic unloading compliance tests; and the direct-current potential drop test.*

sion, while the remaining nine specimens had measurable crack extension ranging from 0.178 to 2.769 mm (0.007 to 0.109 in.). The J_{Ic} for the multispecimen results is 178 kJ/m^2 (1016 in.-1b/in.^{2}) and the tearing modulus is 42.5.

10% Compliance Unloading Results

Figure 4 presents the results of single-specimen tests of ASTM A106 steel with nominal 10% unloadings for crack length estimation by compliance, and compares these results with the results of the multispecimen tests. The single-specimen and multispecimen results are in good agreement. The calculated J_{Ic} values for the two single specimens presented in Fig. 4 are 309 kJ/m^2 and 236 kJ/m^2 (1766 and 1347 in.-lb/in.^{2}); tearing modulus values are 206 and 243. These values for J_{Ic} and T bound the multispecimen results, and the average J_{Ic} for the 10% unloading is 272 kJ/m^2 (1556 in.-lb/in.^{2}), which is in good agreement with the multispecimen data.

Figure 5 compares the multispecimen results for the 3-Ni steel with results from the 10% compliance unloading tests of single specimens. There is good agreement in the *J-R* curves, especially in the indicated ASTM range of 0.15 to 1.5 mm. (0.006 to 0.060 in.). The average J_{Ic} of 166 kJ/m^2 (950 in.-lb/in.^{2}) is slightly lower, while the tearing modulus average of 44.5 is slightly higher, than in the multispecimen results.

Deep Compliance Unloading Results

Results for single-specimen tests in which the unloading for A106 steel ranged from 35 to 90% of the maximum load are presented in Fig. 6. There appear to be two distinct sets of *J-R* curve results in this figure; however, there is no discernible pattern to the behavior. The lower group of data represents 35, 50, and two 90% unloading tests, while the upper group is from 70 and 90% unloading tests. Since no trend showed up in the results, this scatter can probably be attributed to material variability. This assumption is reasonable considering that these specimens were fabricated from pipe with a wall thickness of 98 mm (3⅞ in.)

The J_{Ic} and T values for the deep elastic unloading, single-specimen tests are given in Table 3. The J_{Ic} values range from 222 to 299 kJ/m^2 (1266 to 1708 in.-lb/in.^{2}), generally lower than the multispecimen results. Looking at the band of *J-R* curves in Fig. 6, the range of J at 2.5 mm (0.100 in.) of crack extension is 753 kJ/m^2 (4300 in.-lb/in.^{2})$\pm$10%, which is within the typical spread for *J-R* tests.

Figure 6 compares the 10% elastic unloading results with the deep unloading tests. Agreement is good among the curves, with the deep unloading curves being slightly higher. This is not interpreted as an effect of unloading. As in Landes and co-workers' work [*7,8*], an effect of unloading or cycling

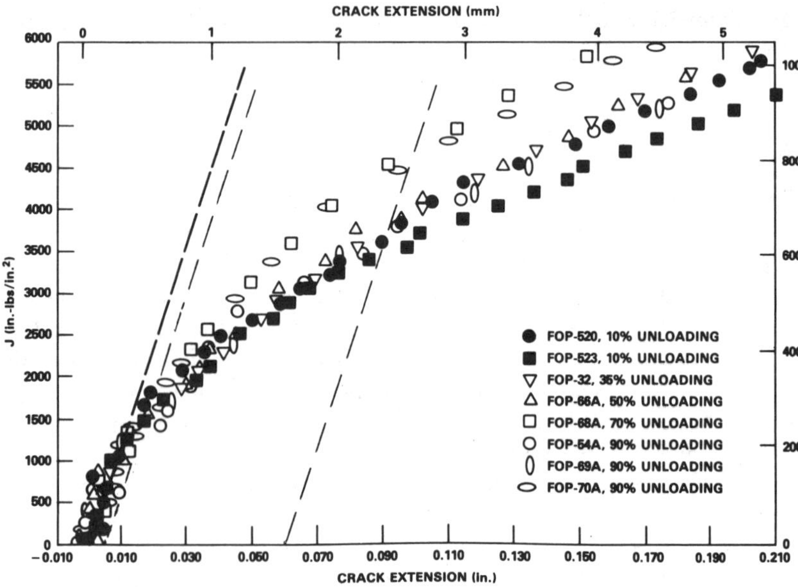

FIG. 6—*Comparison of* J-R *curves for ASTM A106 steel: results from single-specimen, 10% unloading and deep unloading compliance tests.*

TABLE 3—J *and* T *results for multi- and single-specimen tests of A106 Class C steel at 135°C (275°F).*

Specimen/Test	J_{Ic} kJ/m² (in.-lb/in.²)	T
Multispecimen	268 (1531)	233
10% Elastic Unloading		
FOP 520	309 (1766)	206
FOP 523	236 (1347)	243
35% Elastic Unloading		
FOP 32	233 (1333)	270
50% Elastic Unloading		
FOP 66A	229 (1306)	292
70% Elastic Unloading		
FOP 68A	299 (1708)	326
90% Elastic Unloading		
FOP 54A	224 (1281)	273
FOP 69A	222 (1266)	278
FOP 70A	232 (1328)	378

would manifest itself in a shift of the resistance curve to the right and down, which is not the case here.

Figure 7 presents the results of single-specimen J-integral tests for 3-Ni steel performed with the partial unloading compliance range of 10 to 90%. The 10% elastic unloading results are in the middle of the range of J-R curves, and over the first 1.5 mm (0.060 in.) of crack extension the J-R curves are within a tight band. The range of J at 2.54 mm (0.100 in.) of crack extension is 333 kJ/m² (1900 in.-lb/in.²) ± 12%. This represents a reasonable spread in the data, and there are no obvious trends in the J-R curve behavior. The J_{Ic} and T values given in Table 4 suggest a slight decrease in J_{Ic} and increase in T with increased unloading range, which is not evident in Fig. 7.

Because Landes and McCabe [7] observed a cyclic-dominated behavior, the fracture surfaces of specimens subjected to large unloadings were examined with a scanning electron microscope (SEM). There was no evidence of fatigue striations observed on either the A106 or the 3-Ni steel specimens.

Direct-Current Potential Drop (DCPD) Results

The DCPD results for A106 steel are presented in Fig. 4, with the data from the multispecimen test and the single-specimen, 10% elastic unloading tests. Good agreement is evident in these comparisons.

Figure 5 presents the DCPD results for 3-Ni steel, and it likewise shows good agreement with multispecimen and single specimen, 10% unloading compliance tests.

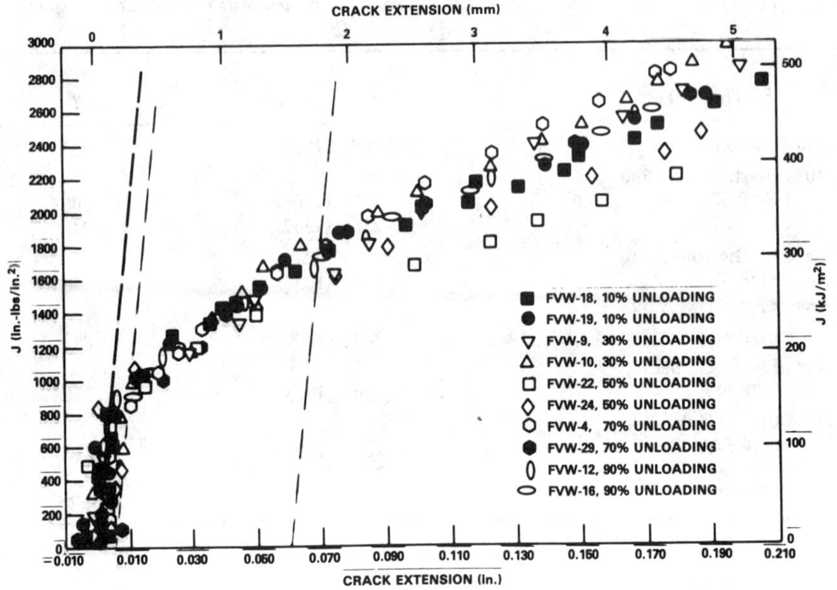

FIG. 7—*Comparison of* J-R *curves for 3-Ni steel: results from single-specimen, 10% percent unloading and deep unloading compliance tests.*

TABLE 4—J *and* T *results for multi- and single-specimen tests of 3-Ni steel.*

Specimen/Test	J_{Ic} kJ/m² (in.-lb/in.²)	T
Multispecimen	178 (1016)	42.5
10% Elastic Unloading		
FVW-18	171 (975)	41.3
FVW-19	162 (925)	47.7
30% Elastic Unloading		
FVW-9	154 (877)	41.2
FVW-10	133 (760)	59.9
50% Elastic Unloading		
FVW-22	. . .[a]	. . .
FVW-24	. . .[a]	. . .
70% Elastic Unloading		
FVW-4	135 (769)	56.8
FVW-29	120 (683)	60.4
90% Elastic Unloading		
FVW-12	173 (989)	36.9
FVW-16	148 (846)	51.2

[a]Insufficient data points in ASTM E 813 range.

Summary and Conclusions

The objective of this investigation was to experimentally evaluate the effects of elastic unloadings used in the single-specimen test on the *J*-resistance curve. The single variable in this study was the range of unloading, with the extreme being 90% unloading to magnify any unloading effect. The deep unloading test results were compared with the standard 10% unloading results, with multispecimen results, and with DCPD results, which represented 0% unloading.

There appeared to be a slight decrease in J_{Ic} with increasing unloading range as compared with the multispecimen results. This trend was evident in both materials. It could be that the larger unloading ranges affect J_{Ic} in the region of the transition from blunting behavior to tearing, but that the rest of the *J-R* curve does not see this effect. Perhaps a better explanation would be that this transition is sensitive to the Bauschinger effect which is important in cyclic plasticity studies. What is important is that elastic unloading had no apparent effect on the resulting *J-R* curve. This is consistent with the results reported by Landes and co-workers [*7,8*]. Landes cyclically loaded two steels and studied the corresponding behavior on the *J-R* curve. Where upward of 70 large cycles were applied, he observed little effect in the *J-R* curve for HY-130 steel and termed the crack growth "*R*-curve dominated," but in a similar cyclic loading of A508 Class 2 steel, he observed a significant effect on the *J-R* curve and called this crack growth "cyclic-dominated." However, when he subjected both materials to a total of 13 cycles per specimen, he observed no cyclic effect in either steel. In a usual single-specimen *J* test, the number of unloadings is on the order of 15 to 20. The apparent lack of fatigue crack growth interactions was also evident from the scanning electron microscope examination of the fracture surfaces of specimens subjected to large unloading. The SEM results of the A106 and 3-Ni steels indicated no evidence of fatigue striations.

From the various comparisons, the following conclusions are drawn for the two materials investigated:

1. The *J-R* curves from the single-specimen, 10% elastic unloading tests agree very well with the results of the multispecimen tests.

2. The *J-R* curves from the single-specimen, 10% elastic unloading tests are in excellent agreement with the zero-unloading results obtained from the DCPD test.

3. The *J-R* curves obtained from the single-specimen, deep unloading range study show no significant difference from the single-specimen, 10% unloading compliance results; however, a slight reduction in the calculated J_{Ic} was evident. This could reflect a slight lowering of the flow stress of the two materials due to a Bauschinger effect.

Acknowledgments

The authors wish to acknowledge the support and encouragement of Mr. Milton Vagins and Mr. Jack Strosnider of the U.S. Nuclear Regulatory Commission during the performance of this research task.

References

[*1*] Rice, J. R., "A Path Independent Integral of the Approximate Analysis of Strain Concentration by Notches and Cracks," *ASME Journal of Applied Mechanics,* Vol. 35, June 1968, pp. 379-386.

[*2*] Hutchinson, J. W. and Paris, P. C., "The Theory of Stability Analysis of *J*-Controlled Crack Growth," in *Elastic-Plastic Fracture, ASTM STP 668,* J. D. Landes, J. A. Begley, and G. A. Clarke, Eds., American Society for Testing and Materials, Philadelphia, 1979, pp. 37-64.

[*3*] Ernst, H. A., Paris, P. C., and Landes, J. D., "Estimation of *J*-Integral and Tearing Modulus *T* From a Single Specimen Test Record," in *Fracture Mechanics: Thirteenth Conference, ASTM STP 743,* R. Roberts, Ed., American Society for Testing and Materials, Philadelphia, 1981, pp. 476-502.

[*4*] Clarke, G. A., Andrews, W. R., Paris, P. C., and Schmidt, D. W. in *Mechanics of Crack Growth, ASTM STP 590,* American Society for Testing and Materials, Philadelphia, 1976, pp. 27-42.

[*5*] Paris, P. C., Tada, H., Zahoor, A., and Ernst, H. in *Elastic-Plastic Fracture, ASTM STP 668,* J. D. Landes, J. A. Begley, and G. A. Clarke, Eds., American Society for Testing and Materials, Philadelphia, 1979, pp. 5-36.

[*6*] Paris, P. C., "A Method of Application of Elastic-Plastic Fracture Mechanics to Nuclear Vessel Analysis," in *Resolution of the Task All Reactor Vessel Materials Toughness Safety Issue,* NUREG-0744, Vol 1, Rev. 1, Appendix B, Oct. 82.

[*7*] Landes, J. D. and McCabe, D. E., "Load History Effects on the J_R-Curve," *Elastic-Plastic Fracture: Second Symposium, Vol. II—Fracture Resistance Curves and Engineering Applications, ASTM STP 803,* C. F. Shih and J. P. Gudas, Eds., American Society for Testing and Materials, Philadelphia, 1983, pp. 723-738.

[*8*] Landes, J. D. and Leax, T. R., "Load History Effects on the Fracture Toughness of a Modified 4340 Steel," in *Fracture Mechanics: Fifteenth Symposium, ASTM STP 833,* R. J. Sanford, Ed., American Society for Testing and Materials, Philadelphia, 1984, pp. 436-448.

[*9*] Andrews, W. R., "An Investigation of the Relationships Between *J* and CTOD," presented at *Symposium on User's Experience with Elastic Plastic Fracture Toughness Test Methods,* Louisville, Ky., 20-22 April 1983.

[*10*] Joyce, J. A. and Gudas, J. P. in *Elastic-Plastic Fracture, ASTM STP 668,* J. D. Landes, J. A. Begley, and G. A. Clarke, Eds., American Society for Testing and Materials, Philadelphia, 1979, pp. 451-468.

[*11*] Saxena, A. and Hudak, S. J., Jr., "Review and Extension of Compliance Information for Common Crack Growth Specimens," Westinghouse Scientific Paper 77-9E7—AFCGR-P1, Pittsburgh, 1977.

[*12*] Merkle, J. G. and Corten, H. T., "A *J*-Integral Analysis for the Compact Specimen Considering Axial Force As Well As Bending Effects," *Journal of Pressure Vessel Technology, Transactions of the American Society of Mechanical Engineers,* Vol. 96, No. 4, Nov. 1974, pp. 286-292.

[*13*] Clarke, G. A. and Landes, J. D., "Evaluation of *J* for the Compact Specimen," *Journal of Testing and Evaluation,* Vol. 7, No. 5, Sept. 1979, pp. 264-269.

[*14*] Johnson, H. H., *Materials Research and Standards,* Vol. 5, 1965, pp. 442-445.

[*15*] Lowes, J. M. and Fearnehough, G. D., *Engineering Fracture Mechanics,* Vol. 3, 1971, pp. 103-108.

[*16*] Vassilaros, M. G. and Hackett, E. M. in *Fracture Mechanics: Fifteenth Symposium, ASTM STP 833,* R. J. Sanford, Ed., American Society for Testing and Materials, Philadelphia, 1984, pp. 535-552.

P. H. Davies[1] *and C. P. Stearns*[1]

Fracture Toughness Testing of Zircaloy-2 Pressure Tube Material with Radial Hydrides Using Direct-Current Potential Drop

REFERENCE: Davies, P. H. and Stearns, C. P., "**Fracture Toughness Testing of Zircaloy-2 Pressure Tube Material with Radial Hydrides Using Direct-Current Potential Drop,**" *Fracture Mechanics: Seventeenth Volume, ASTM STP 905*, J. H. Underwood, R. Chait, C. W. Smith, D. P. Wilhem, W. A. Andrews, and J. C. Newman, Eds., American Society for Testing and Materials, Philadelphia, 1986, pp. 379–400.

ABSTRACT: This paper addresses problems involved in measuring fracture toughness of thin pressure tube material which, in the presence of radial-axial hydrides, undergoes a significant brittle to ductile fracture transition. Compact tension specimens (~5 mm thickness) are machined from flattened tensile strips of Zircaloy-2 in which radial hydrides (30 to 100 ppm hydrogen) are produced by precipitation under stress. Axial fracture toughness is determined for the unirradiated material between room temperature and 300°C using the dc potential drop method. At low and intermediate temperatures crack growth is governed predominantly by the presence of the radial hydrides, and the potential drop is shown to underestimate crack extension due to short-circuiting across tight crack faces. In the upper shelf regime where crack extension is governed mainly by the flow properties of the matrix, the potential drop overestimates crack extension due to through-thickness yielding. It is shown that good, reproducible results can be obtained by careful data analysis using individual specimen calibrations.

KEY WORDS: elastic-plastic, fracture, Zircaloy-2, zirconium alloys, pressure tube, compact tension specimen, *J*-integral, potential drop, embrittlement, hydride

The defect tolerance of zirconium alloy pressure tubes is known to be reduced by high hydrogen concentrations due to precipitation of zirconium hydrides [*1*]. The fabrication route of CANDU (CANada Deuterium Uranium) reactor pressure tubes aims to minimize this effect by ensuring that hydride

[1]Research Engineer and Technologist, respectively, Ontario Hydro Research Division, Toronto, Ontario, Canada. Dr. Davies is currently at Atomic Energy of Canada Ltd., Chalk River Nuclear Laboratories, Chalk River, Ontario, Canada.

precipitation occurs in the circumferential-axial plane with platelet normals in the thickness direction. Under reactor operating conditions, however, hydrides in Zircaloy-2 pressure tubes could be reoriented to the radial-axial plane with platelet normals in the hoop direction. Therefore there is a need to study the effects of radial-axial hydrides that can lead to severe embrittlement [2,3].

In the present work axial fracture toughness (linear elastic fracture mechanics [LEFM] K_c) or axial crack growth resistance (elastic-plastic J_R-Δa) is determined for flattened, unirradiated Zircaloy-2 pressure tube material with radial hydrides from room temperature to reactor operating temperatures of 300°C. Thin compact tension specimens (~5 mm thickness) are tested using the d-c potential drop method for monitoring crack extension [4]. The problems of measuring the fracture toughness of thin pressure tube material exhibiting a significant brittle-to-ductile fracture transition are discussed.

Material and Specimen Preparation

All test specimens used material from the mid-length of Zircaloy-2 pressure tube 841, outer diameter approximately 114 mm, wall thickness from 5.0 to 5.4 mm. Rings approximately 60 mm wide were sectioned from the tube and flattened by a continuous reverse bending technique. After flattening, the strips were stress-relieved at 400°C for 24 h to reduce the residual stress from straightening. Hydride layers were electrolytically deposited on the surfaces of the strips using a 0.1*M* H_2SO_4 electrolyte at 90°C with a current density of 150 mA cm^{-2}. Strips containing 30, 40, 60, and 100 ppm hydrogen with radial hydrides were then produced by diffusion annealing followed by slow cooling under stress as indicated in Table 1 [5].

The typical coarse hydride morphology for the 60 ppm hydrogen strips is shown in Fig. 1. The hydride continuity coefficient (HCC) is defined as the fraction of the total length of a band along the tube radius that contains a projection of radial-axial hydrides [6]. This parameter was measured for each strip following the approach of Bell and Duncan [6]. Composite photographs of the hydride structure across the wall were made up at ×100 magnification

TABLE 1—*Hydriding parameters for Zircaloy-2 strips.*

Nominal Hydrogen Concentration, ppm	Annealing		Applied Stress, MPa	Cooling Rate to 100°C, °C min^{-1}
	Temperature, °C	Time, h		
30	250	120	250	2
40	270	90	250	2
60	300	50	250	2
100	350	21	220	2

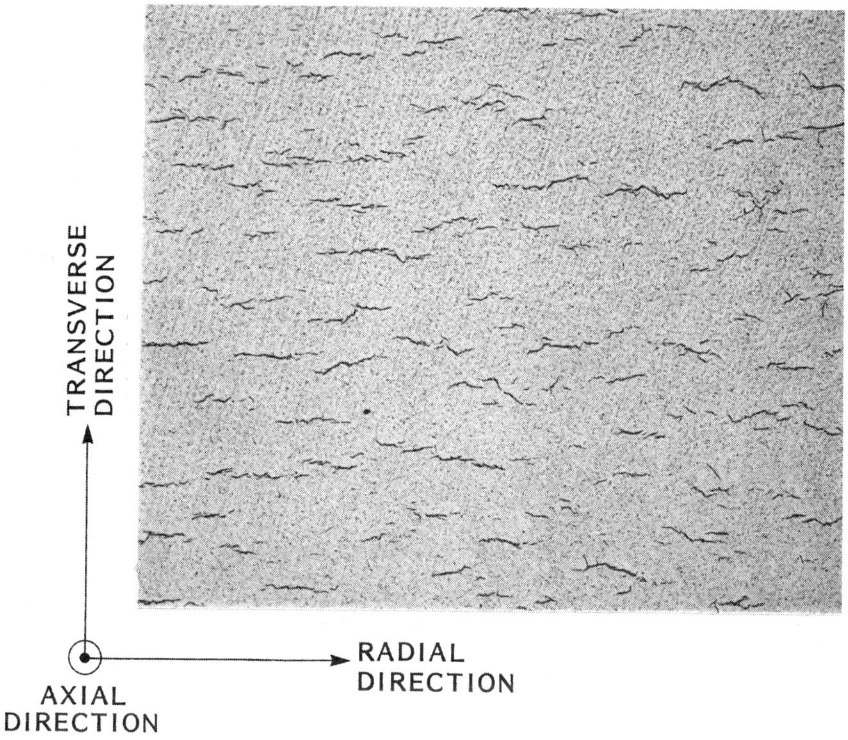

FIG. 1—*Photomicrograph of Zircaloy-2 with 60 ppm hydrogen, HCC = 0.5 to 0.6, × 100.*

in the radial-transverse plane. For the HCC measurement a band width of 0.1 mm × wall thickness (after removal of the surface hydride layer) was used. Material with room temperature HCC values <0.3 (30 ppm hydrogen), 0.3 to 0.45 (40 ppm hydrogen), 0.5 to 0.6 (60 and 100 ppm hydrogen), and 0.8 to 0.9 (100 ppm hydrogen) was produced [5].

Five compact tension and four transverse tensile specimens were machined from each strip. Enlarged end uniaxial test pieces (25.4 mm gage length and 3.2 mm diameter) were tested at a crosshead speed of 1 mm min^{-1}. Typical uniaxial tensile properties of the Zircaloy-2 pressure tube material are shown in Fig. 2.

The compact tension specimens were rough machined into plate specimens oriented for crack growth in the axial direction in the radial-axial plane. The final in-plane dimensions of each specimen were identical with a width, W, of 38.1 mm and initial slot length, a/W, of 0.43. The remaining in-plane dimensions were in the proportions recommended in ASTM Test for Plane-Strain Fracture Toughness of Metallic Materials (E 399) for a standard compact tension specimen. In order to prevent further increase in the total hydrogen

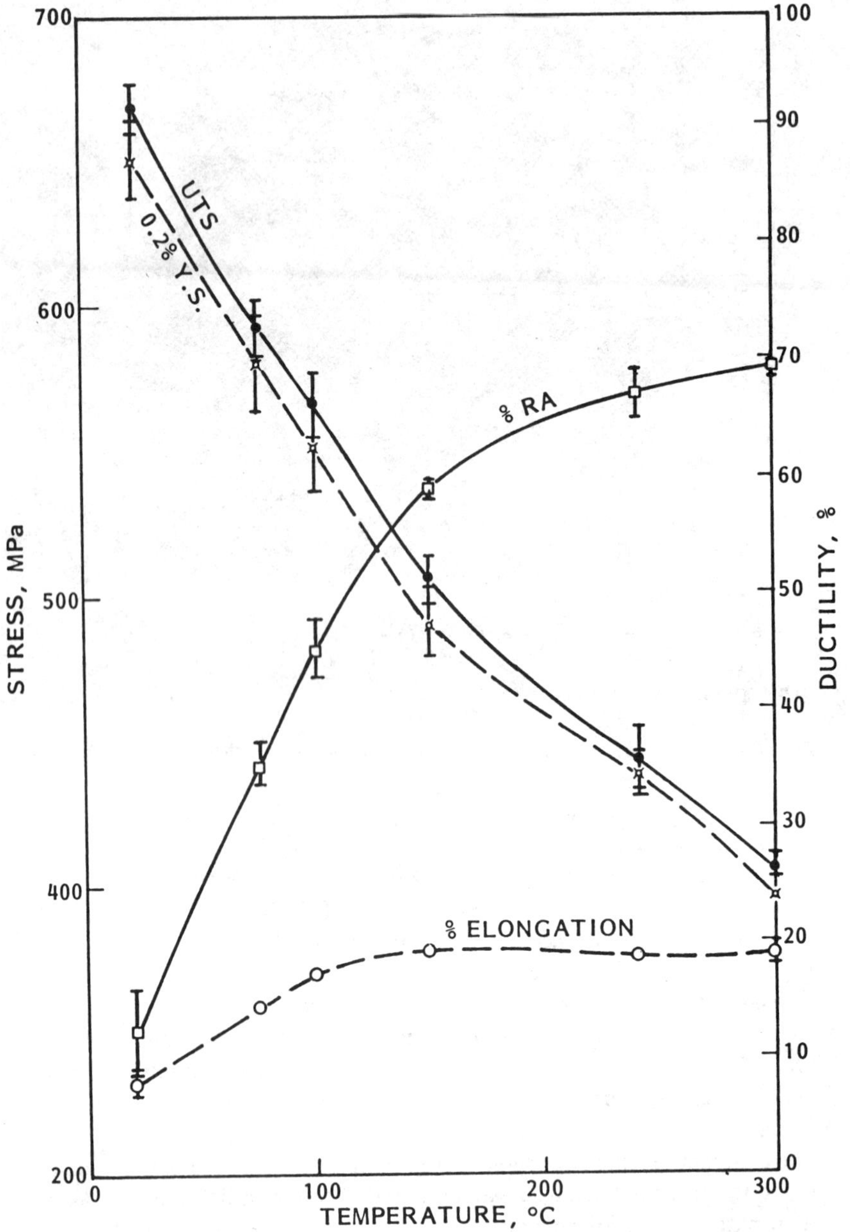

FIG. 2—*Transverse tensile properties versus temperature of Zircaloy-2 containing 60 ppm hydrogen with radial hydrides.*

concentration of the specimens at high temperatures, surface hydride layers were removed by grinding while keeping the reduction in specimen thickness to a minimum. A final plate thickness, B, of 4.7 to 5.1 mm was obtained.

Experimental Procedure

Testing was carried out in a servohydraulic-controlled testing machine with a 50 kN load cell on the 20% transducer range. Before testing, specimens were fatigue cracked at a maximum stress intensity factor of 12 $\text{MPa}\sqrt{\text{m}}$ to produce an initial a/W of 0.5.

During the fracture toughness tests, crack extension was monitored using the d-c potential drop technique with a constant current supply of 5 A. The current leads were stainless steel strips attached by screws to the front edge of the specimen above and below the crack mouth. Voltage leads of zirconium alloy wire, 0.6 mm diameter, were spot welded diagonally opposite each other across the crack mouth. Oxidized Zr-2.5 Nb loading pins were used to prevent short-circuiting through the load train [7]. The potential drop amplifier was a thermally stabilized differential amplifier with a filtered buffer and offset operating at ×2000 gain; the noise was 1.6 μV referred to the input [8]. A potential drop calibration factor, $(\Delta a/W)/(\Delta V/V_0)$, of 0.86 was obtained using fatigue cracked specimens for an initial crack length, a_0/W, of 0.50 and for crack lengths in the range $a/W = 0.5$ to 0.7.

Specimens were pin-loaded in stroke control at 1 mm min^{-1} ($\dot{K} \sim 1$ $\text{MPa}\sqrt{\text{m}}\ \text{s}^{-1}$). Analogue and digital recordings of load, stroke, and potential drop were made continuously during a test. The automated acquisition of potential drop data is described elsewhere [8].

Tests were carried out at room temperature and at temperatures up to 300°C using an electrically heated air furnace. Specimens were allowed to soak for at least 1 h before testing to ensure dissolution of the equilibrium concentration of hydrogen at the test temperature.

After testing, low temperature heat tinting at 200°C was used to mark any stable crack extension during loading. For consistency, the initial and final crack lengths were measured at nine locations across the thickness and averaged, as recommended in ASTM Tentative Test Procedure for Determining the Plane-Strain J_I-R Curve [9]. The average crack extension was then calculated and used to calibrate the potential drop method for each test.

Fracture toughness values were calculated from the load, load-line displacement (stroke corrected for machine and load-train compliance), and instantaneous crack size. The LEFM parameter K_{Ic} or K_c was used to characterize unstable crack growth following ASTM E 399. For consistency, in all other cases the elastic-plastic crack growth resistance curve J_R-Δa was obtained following the aforementioned ASTM tentative procedure [9].

Results and Analysis of Results

Crack Growth Behavior

The fracture toughness specimens exhibited three main types of macroscopic crack growth behavior:

1. Unstable crack growth with negligible surface yield.
2. Stable crack growth (tunnelling) with localized surface yield.
3. Stable crack growth with large-scale surface yield.

The crack growth behavior of each specimen is indicated in Table 2. Examples of surface plastic zones and fracture surfaces are shown in Figs. 3*a* to 3*f*.

The fast fracture exhibited by Specimen 21-1 at 23°C (Figs. 3*a* and 3*b*) was similar to unstable crack growth observed at temperatures up to 200°C for the 60 ppm hydrogen specimens and at temperatures up to 270°C for the 100 ppm hydrogen specimens. Scanning electron microscopy confirmed that these fractures occurred predominantly by the cleavage fracture of radial-axial hydrides with some ductile shear between hydrides on adjacent planes [5].

Crack tunnelling with localized surface yield (Figs. 3*c* and 3*d*) was observed at 100°C (30 ppm, 40 ppm), 120°C (40 ppm), 150°C (60 ppm), and 180°C (60 ppm). However, the angle between the surface shear bands and the crack plane increased with increasing temperature and decreasing hydrogen con-

TABLE 2—*Summary of crack growth behavior of small Zircaloy-2 compact tension specimens.*

Nominal Test Temperature, °C	Specimen No.	Thickness, mm	Crack Growth Behavior*
	30 ppm Hydrogen, HCC < 0.3		
23	19-2	5.04	*a*
100	19-3	4.91	*b*
100	19-5	4.78	*b*
150	19-1	5.06	*c*
150	19-4	4.76	*c*
	40 ppm Hydrogen, HCC = 0.3 to 0.45		
23	21-1	4.95	*a*
100	20-2	4.95	*b*
100	21-4	4.73	*b*
120	20-3	4.80	*b*
120	21-2	4.88	*b*
150	20-4	4.76	*c*
150	21-3	4.75	*c*
200	20-1	5.06	*c*
200	21-5	4.78	*c*
220	20-5	4.94	*c*

TABLE 2—*Continued.*

Nominal Test Temperature, °C	Specimen No.	Thickness, mm	Crack Growth Behavior*
	60 ppm Hydrogen, HCC = 0.5 to 0.6		
23	16-4	4.88	*a*
23	17-5	4.78	*a*
150	16-1	4.95	*b*
150	18-4	4.85	*b*
180	16-3	4.99	*b*
180	17-2	5.10	*a*
180	18-3	4.91	*b*
190	17-1	5.01	*a*
200	16-2	5.10	*a*
200	18-5	4.87	*c*
220	16-5	4.82	*c*
220	17-4	4.91	*c*
220	18-1	5.14	*c*
240	17-3	5.13	*c*
240	18-2	5.09	*c*
	100 ppm Hydrogen, HCC = 0.8 to 0.9		
23	9-5	4.98	*a*
150	9-3	4.91	*a*
150	10-2	4.80	*a*
200	14-2	4.86	*a*
240	9-2	4.88	*a*
240	10-3	4.96	*a*
240	15-3	4.69	*a*
250	14-4	5.04	*a*
260	14-1	4.87	*a*
260	15-5†	5.02	*c*
270	14-3	4.91	*a*
270	15-4†	4.91	*c*
280	14-5	5.07	*c*
280	15-2	4.76	*c*
300	9-1	5.02	*c*
300	10-4	5.04	*c*

**a*. Unstable crack growth with negligible surface yield.
b. Stable crack growth (tunnelling) with localized surface yield.
c. Stable crack growth with large-scale surface yield.
†HCC = 0.5 to 0.6.

centration and HCC. Full through-thickness yielding (Figs. 3*e* and 3*f*) was only observed at temperatures $\geq 200°C$ (40 ppm), 240°C (60 ppm), and $\geq 260°C$ (100 ppm). Some evidence of the cleavage fracture of radial hydrides was found in the tunnelled region of crack growth at intermediate temperatures, the proportion of cleavage fracture decreasing with increasing temperature [5]. For specimens exhibiting full through-thickness yield, however,

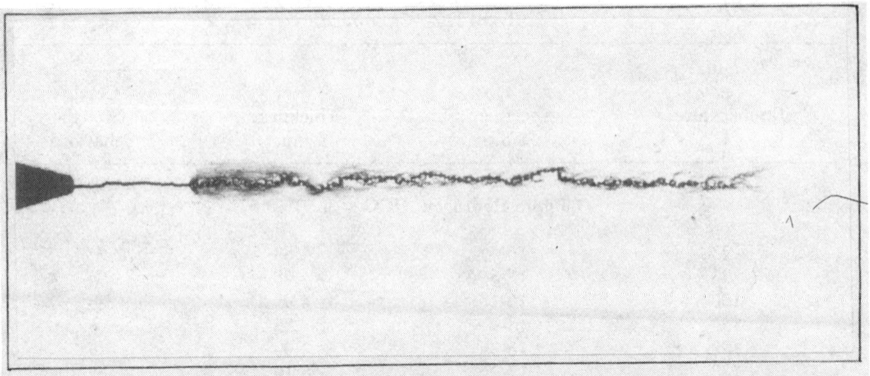

(a)

SURFACE PLASTIC ZONE (x6)

(b)

FRACTURE SURFACE (x 7.5)

FIG. 3—*Compact tension specimens of Zircaloy-2 with radial hydrides.* (a and b) *Specimen 21-1 (40 ppm hydrogen) tested at 23°C.*

failure was entirely by the formation and coalescence of equiaxed voids ahead of the crack tip (ductile tearing) [5].

Potential Drop Calibration

In general, the potential drop calibration for fatigue cracked specimens was not applicable to Zircaloy-2 with radial hydrides. This is shown in Fig. 4, which summarizes potential drop calibration factors based upon the mea-

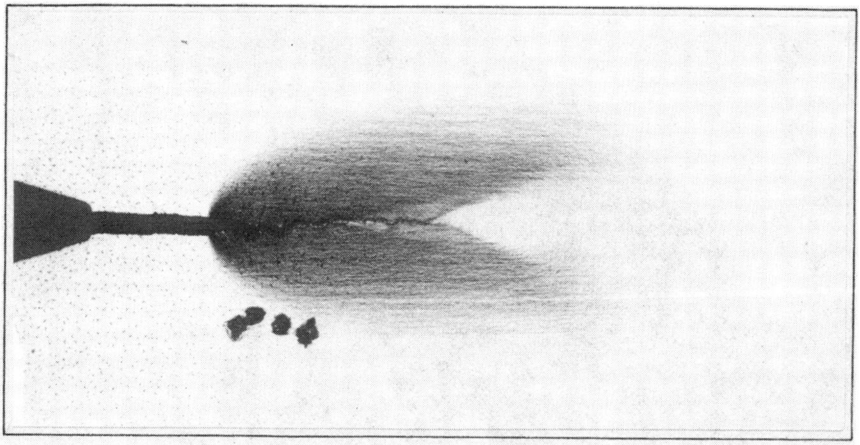

(c)
SURFACE PLASTIC ZONE (x6)

(d)
FRACTURE SURFACE (x 7.5)

(FIGS. 3c and 3d) *Specimen 19-3 (30 ppm hydrogen) tested at 100°C.*

sured nine-point average crack extension for each specimen exhibiting stable crack growth. At low temperatures (<150°C) the measured calibration factor was up to 60% higher than the fatigue crack calibration of 0.86. On the other hand, at high temperatures (>150°C) the measured calibration factor was up to 50% below 0.86. The actual calibration for any specimen also depended upon hydrogen concentration (at temperatures < 200°C) as well as instantaneous load and stroke.

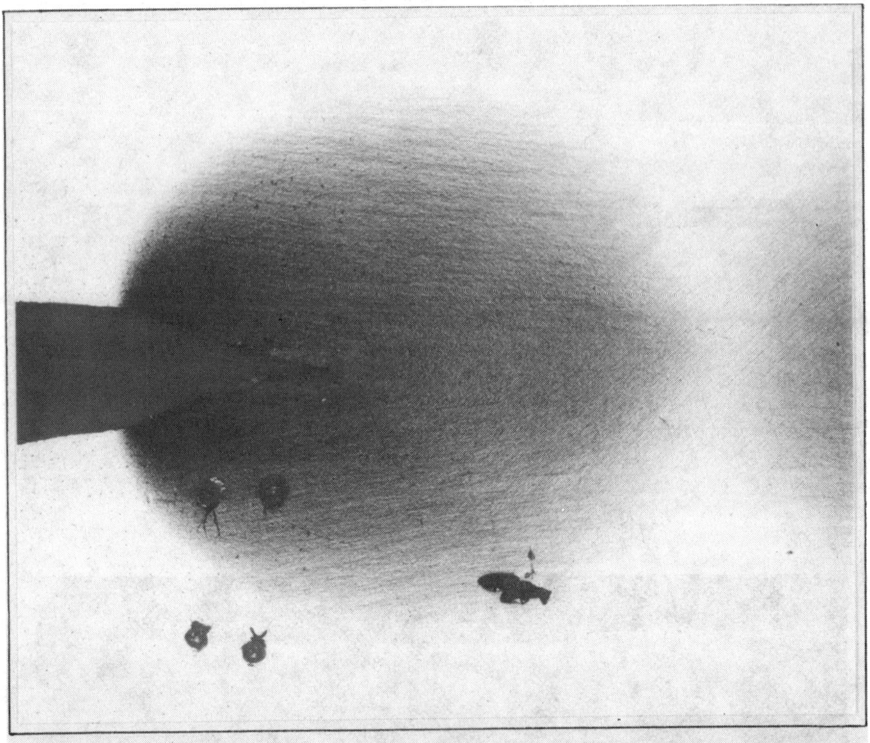

(e)

SURFACE PLASTIC ZONE (x6)

(f)

FRACTURE SURFACE (x7.5)

(FIGS. 3e and 3f) *Specimen 20-1 (40 ppm hydrogen) tested at 200°C.*

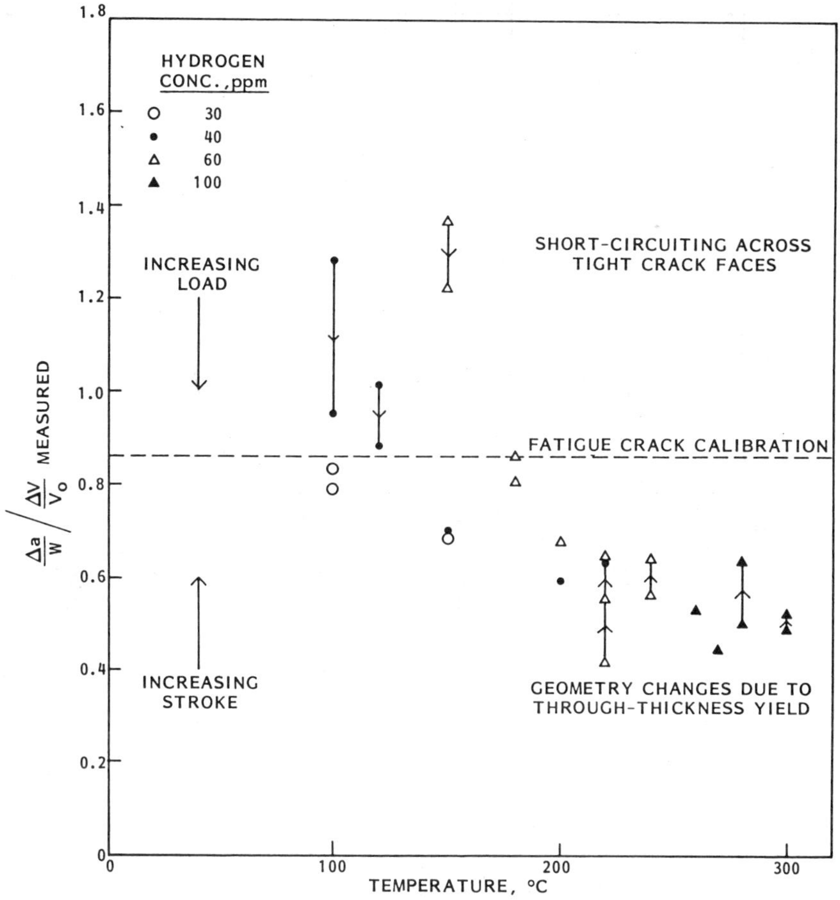

FIG. 4—*Comparison of d-c potential drop calibration based upon measured nine-point average crack extension with fatigue crack calibration for Zircaloy-2 compact tension specimens with radial hydrides.*

Figure 5 demonstrates the data analysis carried out for all specimens exhibiting stable crack growth at temperatures $\geq 200°C$. In Fig. 5*a* the linear portion of the J_R–Δa curve was extrapolated back to the abscissa in order to re-zero the potential drop data (i.e., determine a new reference voltage, V_0). This effectively eliminated the contribution of the initial surface contraction/crack tip blunting to the voltage signal. The resultant curve is shown in Fig. 5*b*. The potential drop data was then matched to the measured nine-point average crack extension (i.e., the potential drop calibration factor was determined). The resultant corrected data are shown in Fig. 5*c*.

For specimens exhibiting unstable crack growth any prior crack extension was generally less than 1 mm. For this situation the potential drop calibration factor of 0.86 was used.

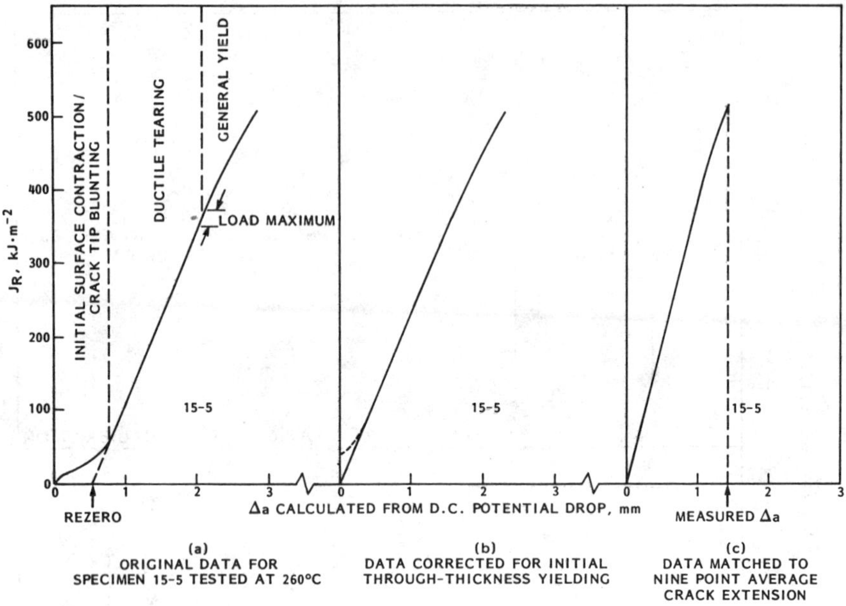

FIG. 5—*Typical data analysis used for compact tension specimens of Zircaloy-2 with radial hydrides exhibiting stable crack growth at temperatures ≥200°C.*

Fracture Toughness

Figure 6 summarizes the maximum load fracture toughness results for all specimens tested in terms of the LEFM parameter, K. For specimens exhibiting unstable fracture, K_{Ic} or K_c was used. For comparison, in all other cases values of J were converted to equivalent K values assuming plane stress crack tip conditions, (i.e., $K_{max} = \sqrt{EJ_{max}}$). The reduction in fracture toughness with increasing hydrogen concentration and HCC is clearly indicated.

For the 30 ppm hydrogen specimens (HCC < 0.3) the fracture toughness at maximum load, K_{max} increased gradually from 38 MPa$\sqrt{m}$ at room temperature to ~150 MPa$\sqrt{m}$ at 150°C. For the 40 ppm specimens (HCC = 0.3 to 0.45), K_{max} increased gradually from 33 MPa$\sqrt{m}$ at room temperature to ~170 MPa$\sqrt{m}$ at 200°C. The corresponding crack growth resistance curves for specimens exhibiting stable crack growth are given in Figs. 7*a* and 7*b*, the increase in fracture toughness with crack extension and temperature mainly reflecting increasing yield at the specimen surface.

For the 60 ppm hydrogen specimens (HCC = 0.5 to 0.6), K_{max} increased from 29 MPa$\sqrt{m}$ at room temperature to ~160 MPa$\sqrt{m}$ at 240°C. However, there was a tendency for some low toughness fast fracture behavior (K_c = 35 to 41 MPa$\sqrt{m}$) in the transition region between 180 and 200°C. Crack growth resistance curves for the 60 ppm specimens (Fig. 7*c*) indicated that stable

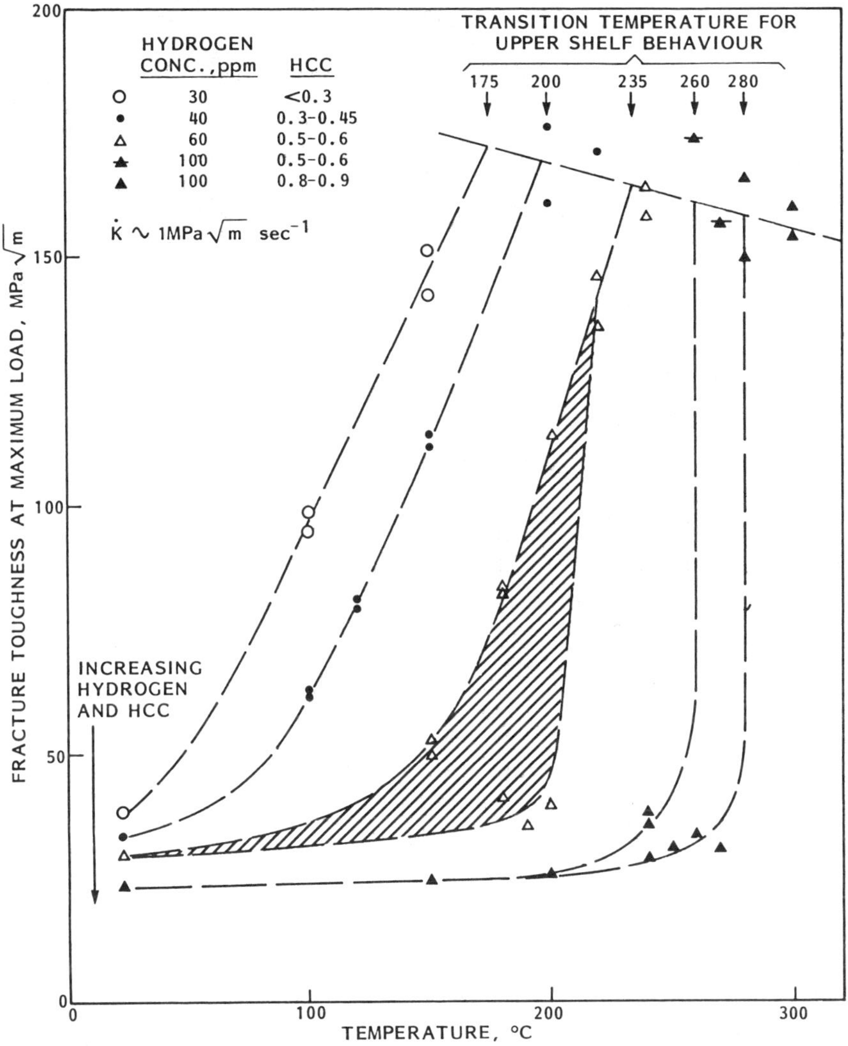

FIG. 6 *Fracture toughness versus temperature for Zircaloy-2 compact tension specimens with radial hydrides; 30 to 100 ppm hydrogen.*

crack growth at 150 and 180°C was governed predominantly by the presence of the radial hydrides, the tunnel-shape crack front limiting the extent of surface yield. However, at temperatures ≥200°C more widespread yielding could be achieved in the absence of fast fracture. Specimens 16-5, 18-1, and 17-4 tested at 220°C were loaded to displacements of 2 mm (load maximum at 2.5 mm), 4 mm, and 8 mm, respectively. The three J_R–Δa curves obtained at 220°C reflect the increasing potential drop calibration factors of 0.41, 0.65,

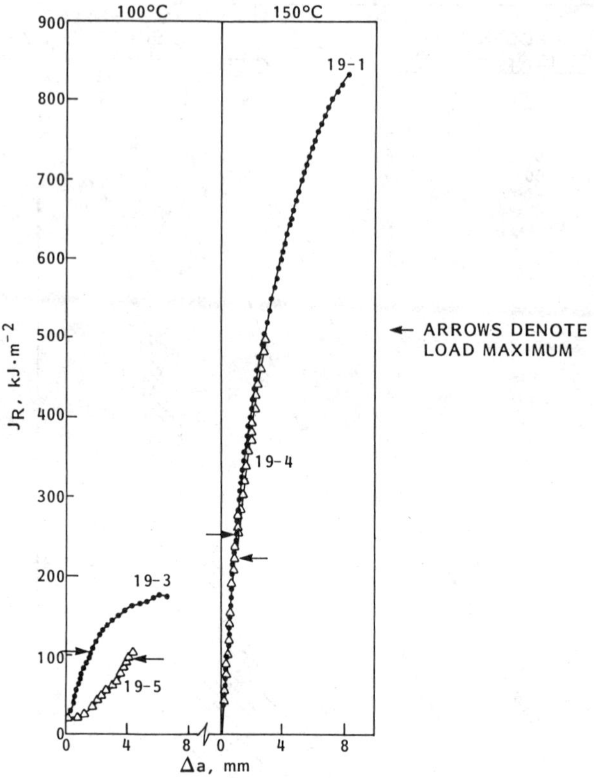

FIG. 7—*Crack growth resistance curves for Zircaloy-2 with radial hydrides.* (a) *30 ppm hydrogen.*

and 0.68 used for Specimens 16-5, 18-1, and 17-4, respectively, in order to match the total crack extension measured.

For the 100 ppm hydrogen specimens, a sharp well-defined fracture transition occurred at 260°C for Specimens 15-4 and 15-5 (HCC = 0.5 to 0.6) and at 280°C for Strip 14 (HCC = 0.8 to 0.9). At lower temperatures, the fracture toughness for fast fracture varied from 23 MPa$\sqrt{m}$ at room temperature to 38 MPa$\sqrt{m}$ at 240°C, some scatter occurring at 240°C just below the fracture transition. The J_R–Δa curves at temperatures above the transition temperature (Fig. 7*d*) were extremely steep being characteristic of full through-thickness yielding.

The fracture toughness in the upper shelf regime was relatively independent of hydrogen concentration and HCC, although there was considerable scatter in the K_{max} values achieved (Fig. 6). This was partly due to problems in defining a maximum load toughness in a region of rapidly increasing dis-

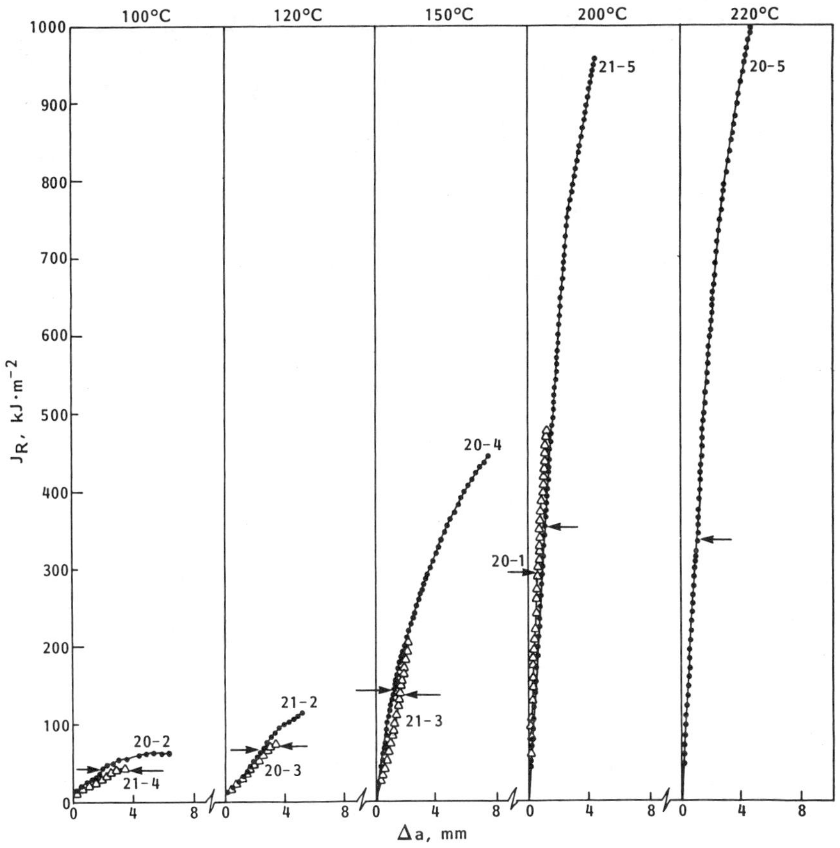

FIG. 7(b)—*40 ppm hydrogen.*

placement. Values shown are considered to be within ±4 MPa$\sqrt{m}$ of the maximum load toughness. In addition, specimens tested were of varying thickness (4.76 to 5.13 mm), resulting in variations in the displacement at the limit load (increasing displacement with decreasing thickness) and hence the maximum load toughness achieved during testing. A linear regression analysis of all the test results was used to determine the trend in the upper shelf regime; this is indicated by the broken line in Fig. 6. The decrease in K_{max} with increasing temperature is consistent with the toughness being governed by the flow properties of the matrix in the upper shelf.

Trend lines are also shown in Fig. 6 for the lower shelf and transition regions. A transition temperature for upper shelf behavior or full through-thickness yield, T_T, was defined as the intersection of the trend lines for the transition and upper shelf regions to the nearest 5°C. The increase in T_T with increasing hydrogen concentration and HCC can be seen.

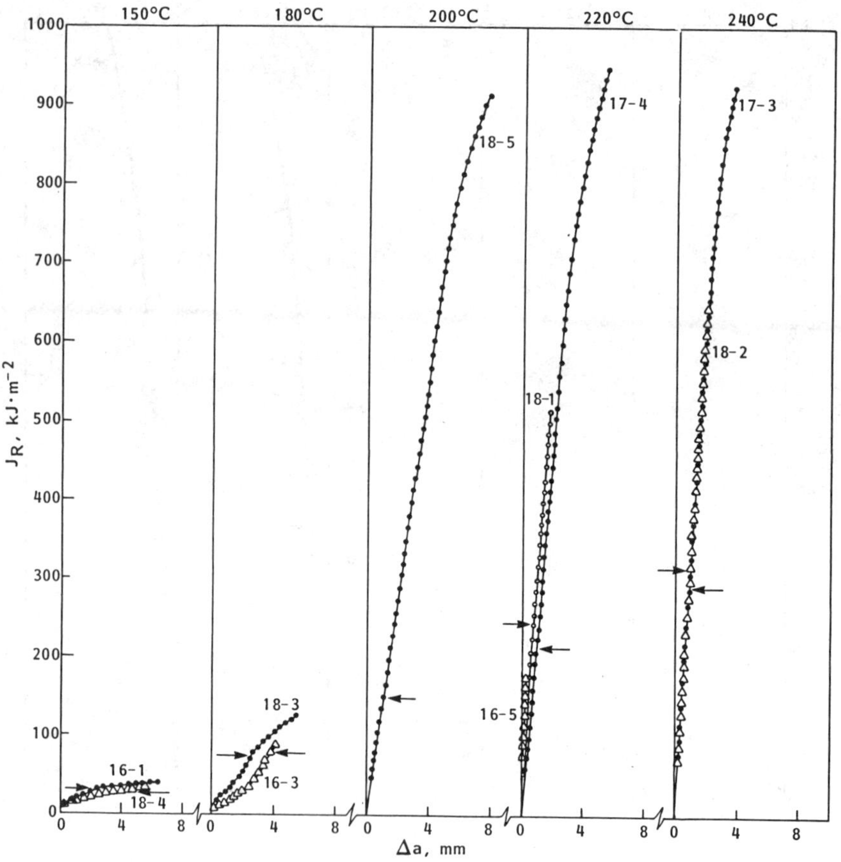

FIG. 7(c)—*60 ppm hydrogen.*

Discussion

The axial fracture toughness of unirradiated Zircaloy-2 pressure tube material has been shown to be significantly reduced by the presence of coarse radial-axial precipitates of zirconium hydride. Zirconium hydrides induce cleavage cracks ahead of the crack tip, resulting in a sharp brittle-ductile fracture transition for hydrogen concentrations ≥60 ppm (HCC ≥ 0.5 to 0.6) (Fig. 6). The solubility of hydrogen in zirconium increases with temperature being negligible below 150°C and increasing rapidly to 65 ppm at 300°C [*10*]. Therefore above 150°C the volume fraction of hydrogen present as radial hydride decreases rapidly with increasing temperature. This sharp increase in hydrogen solubility accounts for the steepness of the brittle-ductile fracture transition above 200°C, a region where there is only a small decrease in the uniaxial yield stress and tensile strength of Zircaloy-2 (Fig. 2).

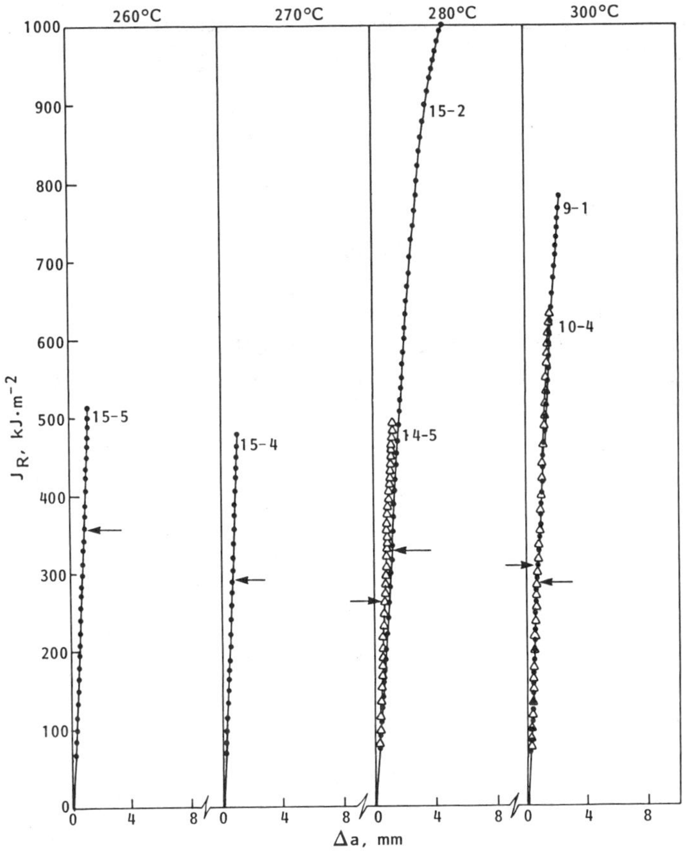

FIG. 7(d)—*100 ppm hydrogen.*

The sensitivity of the fracture transition to local crack tip conditions is also indicated in Fig. 6 by the duplex behavior of the 60 ppm specimens. Nominally identical specimens tested between 180 and 200°C exhibited either unstable fast fracture or ductile stable crack growth. A higher local hydrogen concentration and/or hydride continuity head of the crack tip would increase the probability of fast fracture. The statistical characterization of fracture in the transition region has been discussed by others [*11*].

Specimen Size Limitations

The lowest fracture toughness of 23 MPa$\sqrt{m}$ for Specimen 9-5 (100 ppm hydrogen, HCC = 0.8 to 0.9) exhibiting unstable fast fracture at room temperature was the only valid plane strain fracture toughness (K_{Ic}) measurement meeting the stringent requirements of ASTM E 399. These requirements include:

$$\left.\begin{aligned} a, W - a, B &\geq 2.5\left(\frac{K_{Ic}}{\sigma_y}\right)^2 \\ \text{and} \qquad & \\ \Delta a &\leq 0.02\,(W - a) \end{aligned}\right\} \qquad (1)$$

where σ_y is the 0.2% offset yield stress. However, the remaining fracture toughness values for fast fracture were valid LEFM plane stress fracture toughness (K_c) measurements for the specimen thickness tested, since the critical fracture load was less than 80% of the general yield load [*12*].

Only Specimen 19-3 (30 ppm hydrogen, HCC < 0.3) tested at 100°C met the proposed ASTM size requirements for plane strain *J*-controlled crack growth up to a load maximum [*9*]:

$$\left.\begin{aligned} a, W - a, B &\geq 20\,\frac{J_{max}}{\bar{\sigma}} \geq 20\,\frac{K_{max}^2}{E\bar{\sigma}} \\ \text{and} \qquad & \\ \Delta a_{max} &\leq 0.1\,(W - a) \end{aligned}\right\} \qquad (2)$$

where $\bar{\sigma}$ is a flow stress taken as the mean of the uniaxial yield stress and ultimate tensile strength. For example, for a typical specimen thickness of 5 mm the maximum *J* capacity is 160 kJ m^{-2} (K_{max} = 125 MPa$\sqrt{m}$) at room temperature decreasing to 100 kJ m^{-2} (K_{max} = 90 MPa$\sqrt{m}$) at 300°C. It is apparent that these values represent only a small proportion of the total load bearing capacity of thin pressure tube material deforming under plane stress crack tip conditions in the transition and upper shelf regions.

After a load maximum, general yield across the remaining ligament limits the maximum toughness that can be measured using small specimens. In the present case a maximum specimen width of 38.1 mm could be machined from the strips. The initial crack size was governed by the lower crack size limit recommended by ASTM for using the *J*-integral equation [*9*] ($a/W = 0.5$). Although the resultant crack growth resistance curves for Zircaloy-2 at temperatures above 200°C never achieved a limiting plateau level, it is likely that testing wider specimens with a larger remaining ligament would result in the initial steep portion of the resistance curves being maintained to larger crack extensions (Figs. 5 and 7). The curving over of the J_R-Δa curves after net section yielding produces more conservative crack growth resistance curves for flaw evaluation purposes.

Direct-Current Potential Drop

The d-c potential drop calibration based upon fatigue cracked specimens has been shown to be inadequate for testing Zircaloy-2 with radial hydrides. At low and intermediate temperatures where crack growth was controlled mainly by the presence of coarse radial hydrides, the potential drop signal underestimated the true crack extension. This is believed to be caused by bridging or short-circuiting across the rough crack surfaces which decreases with increasing load as the tight tunnel-shape crack faces open up. This effect has been observed previously during delayed hydride cracking tests in Zr-2.5Nb pressure tube material [*7*] as well as during the inspection of CANDU pressure tube rolled joint assemblies [*13*]. In comparison, at high temperatures where crack growth was governed by the flow properties of the matrix, the potential drop signal overestimated the true crack extension. This is a result of the significant specimen geometry changes (crack tip blunting, lateral contraction, yield at the back surface) which tend to decrease with increasing displacement. For example, the largest decrease in remaining cross-sectional area and hence increase in resistance occurred in the initial stages of loading prior to a load maximum. This effect was shown in Fig. 5*a*, the original uncorrected J_R–Δa curve for Specimen 15-5 at 260°C, as well as in Fig. 7*c* for Specimens 16-5, 18-1, and 17-4 at 220°C.

A major assumption of the potential drop technique is the linearity of the crack length/voltage signal relationship. Therefore the application of the method to a material such as Zircaloy-2 with radial hydrides for which this assumption is violated, is questionable. Unfortunately, the number of methods available for monitoring crack extension in thin specimens is limited. Unloading compliance tests on as-received Zr-2.5Nb [*14*] produced good estimates of the nine-point average crack extension for plane specimens at room temperature ($<7\%$ variation for $\Delta a > 1.5$ mm). For plane specimens at 150°C and short crack extensions ($\Delta a < 1.5$ mm), however, the compliance calculated crack length underestimated the true crack extension by approximately 20%, resulting in nonconservative crack growth resistance curves. Multiple specimen testing has some merit but provides no information concerning toughness variability from specimen to specimen, since only a single resistance curve is obtained. The toughness distribution of an inhomogeneous material, such as Zircaloy-2 with radial hydrides, can only be determined by testing a number of specimens to comparable crack extensions.

In view of the foregoing observations, a single specimen technique may be preferable, provided the results obtained can be shown to be conservative. Crack growth resistance curves for Zircaloy-2 specimens exhibiting large-scale surface yield have been shown to be increasingly conservative (shifted to larger crack extensions) with increasing stroke due to the increase in potential drop calibration factor (Fig. 7*c* at 220°C). The sensitivity of the slope of the crack growth resistance curve to the potential drop calibration factor in the upper shelf regime is shown in Fig. 8. Here the initial slope, dJ/da, and tearing modulus, $T_{mat} = (E/\bar{\sigma}^2)\ dJ/da$ [*15*], were measured from an expanded scale for crack extensions up to a load maximum. The reduction in dJ/da and T_{max} with increasing calibration function indicates the degree of conservatism of the potential drop method in estimating the crack growth resistance of thin specimens in the upper shelf. (It should be noted that previous tests on pressure tube material have confirmed that crack growth resistance curves are relatively insensitive to specimen thickness in the upper shelf regime [*14*].) Values of dJ/da between 290 and 430 MPa for Zircaloy-2 are similar to previous results of 200 to 400 MPa for as-received Zr-2.5Nb in the upper shelf [*14,16*]. Tearing moduli of 130 to 230 may be compared with a T_{mat} of over 200 for high toughness materials such as stainless steel [*17*].

In sharp contrast, inspection of crack growth resistance curves in the lower transition region for Zircaloy-2 with radial hydrides (Figs. 7*a*, 7*b*, and 7*c* at temperatures $\leq 180°C$) reveals more material inhomogeneity as well as nonconservative results (i.e., an apparent increase in slope with increased loading). This is the region where significant crack front tunnelling occurs, and under such circumstances both unloading compliance and potential drop are known to underestimate crack extension [*18,19*]. In this regime the number of test specimens should be increased in order to study the material variability. Additional specimens are also required for short crack extensions to determine the initial slope of the crack growth resistance curve.

Summary

Zircaloy-2 pressure tube material has been shown to undergo a sharp brittle-ductile fracture transition in the presence of radial-axial hydrides (≥ 60 ppm hydrogen, HCC $\geq$ 0.5 to 0.6). The d-c potential drop technique based upon a fatigue crack calibration is inadequate for measuring the axial fracture toughness of this material. At low and intermediate temperatures the presence of coarse radial-axial zirconium hydride precipitates produces bridging or short-circuiting across tight tunnel-shape crack faces. In this region the potential drop signal underestimates the measured nine-point average crack extension by up to 60%. At higher temperatures significant through-thickness yield and specimen configuration changes occur, and the voltage signal overestimates the average crack extension by up to 100%. Under such circumstances individual specimen calibration is necessary.

The resultant axial crack growth resistance curves have been shown to be reproducible and conservative under conditions of large-scale yield (upper shelf and upper transition region). Crack growth resistance curves in the lower transition region exhibit more variation and are nonconservative. In this region additional specimens are required for short crack extensions to determine the initial slope of the crack growth resistance curve. Further specimens are needed for studying material inhomogeneity in the transition regime.

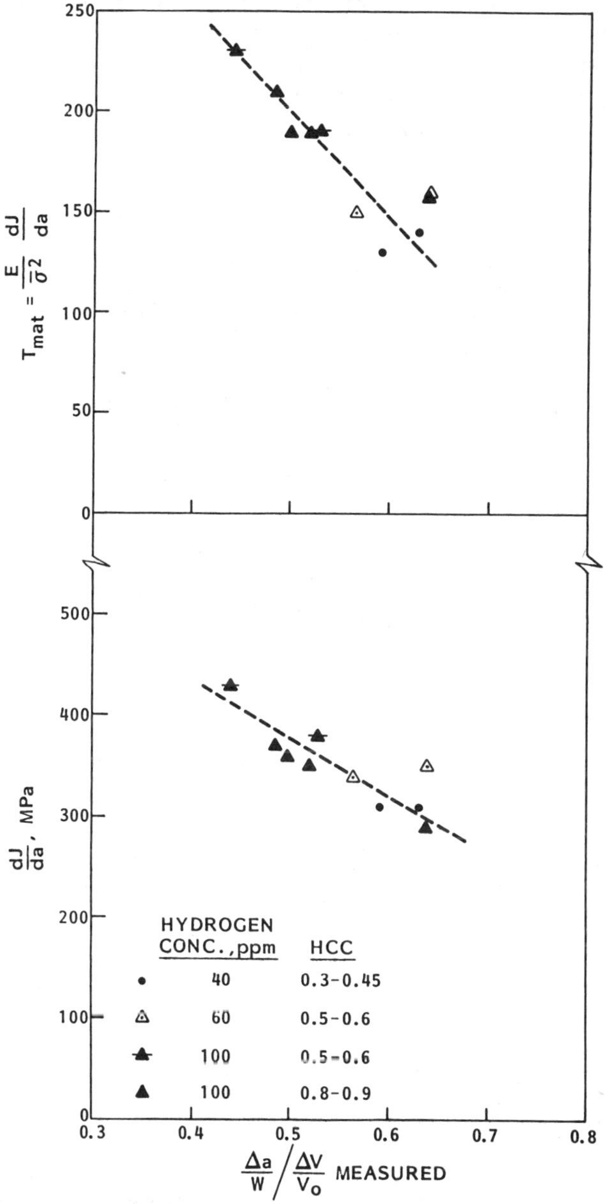

FIG. 8—*Tearing modulus and crack growth resistance slope versus measured d-c potential drop calibration in the upper shelf regime.*

Acknowledgments

The authors wish to thank Dr. M. Leger and Mr. G. Shek for assistance in initiating this study as well as in providing invaluable discussion on the fracture of zirconium alloys. Thanks are also due Mr. H. Seahra and Mr. P. Vesely for specimen preparation, metallography, and measurement of the uniaxial tensile properties of Zircaloy-2.

References

[1] Dawson, J. K., Asher, R. C., Watkins, B., Boulton, J., and Wanklyn, J. N. in *Proceedings*, 2nd Geneva Conference on Peaceful Uses of Atomic Energy, Paper No. 150, 1964, p. 461.

[2] Watkins, B., Cowan, A., Parry, G. W., and Pickles, B. W. in *Applications-Related Phenomena for Zirconium and Its Alloys, ASTM STP 458*, American Society for Testing and Materials, Philadelphia, 1969, p. 141.

[3] Pickles, B. W, Cowan, A., and Johnson, E. R. in *Zirconium in Nuclear Applications, ASTM STP 551*, American Society for Testing and Materials, Philadelphia, 1974, p. 385.

[4] Simpson, L. A. and Clarke, C. F. in *Elastic-Plastic Fracture, ASTM STP 668*, American Society for Testing and Materials, Philadelphia, 1979, p. 643.

[5] Davies, P. H., Shek, G. K., and Leger, M., Ontario Hydro Research Division Report, Toronto, in preparation.

[6] Bell, L. G. and Duncan, R. G., Atomic Energy of Canada Ltd., Report No. 5110, Whiteshell Nuclear Research Establishment, Pinawa, Manitoba, June 1975.

[7] Simpson, L. A. and Clarke, C. F., Atomic Energy of Canada Ltd., Report No. 5815, Whiteshell Nuclear Research Establishment, Pinawa, Manitoba, Oct. 1977.

[8] Stearns, C. P., Ontario Hydro Research Division, Report No. M83-183-K, Toronto, Dec. 1983.

[9] Albrecht, P., Andrews, W. R., Gudas, J. P., Joyce, J. A., Loss, F. J., McCabe, D. E., Schmidt, D. W., and Van Der Sluys, W. A., "Tentative Test Procedure for Determining the Plane Strain J_I-R Curve," *Journal of Testing and Evaluation*, Vol. 10, No. 6, Nov. 1982, p. 245.

[10] Kearns, J. J., *Journal of Nuclear Materials*, Vol. 27, 1968, p. 64.

[11] Landes, J. D. and Shaffer, D. H., in *Fracture Mechanics: Twelfth Conference, ASTM STP 700*, American Society for Testing and Materials, Philadelphia, 1980, p. 368.

[12] Srawley, J. E. and Brown, W. F., Jr., in *Fracture Toughness Testing, ASTM STP 381*, American Society for Testing and Materials, Philadelphia, 1965, p. 153.

[13] Hallett, J. B. and Licht, H., Atomic Energy of Canada Ltd., Report No. 7509, Chalk River Nuclear Laboratories, Chalk River, Ontario, Nov. 1981.

[14] Davies, P. H. and Stearns, C. P., Ontario Hydro Research Division, Report No. 82-411-K, Toronto, Oct. 1982.

[15] Paris, P. C., Tada, H., Zahoor, A. and Ernst, H. in *Elastic-Plastic Fracture, ASTM STP 668*, American Society for Testing and Materials, Philadelphia, 1979, p. 5.

[16] Simpson, L. A. and Wilkins, B. J. S. in *Mechanical Behaviour of Materials*, Vol. III, K. J. Miller and R. F. Smith, Eds., Pergamon Press, Oxford, 1980, p. 563.

[17] Bamford, W. H. and Bush, A. J. in *Elastic-Plastic Fracture, ASTM STP 668*, American Society for Testing and Materials, Philadelphia, 1979, p. 553.

[18] Clarke, G. A. in *Fracture Mechanics: Thirteenth Conference, ASTM STP 743*, American Society for Testing and Materials, Philadelphia, 1981, p. 553.

[19] Voss, B. and Blauel, J. G. in *Fracture and the Role of Microstructure*, Proceedings of Fourth European Conference on Fracture, Vol. 1, K. L. Maurer and F. E. Matzer, Eds., 1982, p. 47.

J. A. Kapp[1] *and M. I. Jolles*[2]

Assessment of *J-R* Curves Obtained from Precracked Charpy Specimens

REFERENCE: Kapp, J. A. and Jolles, M. I., **"Assessment of *J-R* Curves Obtained from Precracked Charpy Specimens,"** *Fracture Mechanics: Seventeenth Volume, ASTM STP 905*, J. H. Underwood, R. Chait, C. W. Smith, D. P. Wilhem, W. A. Andrews, and J. C. Newman, Eds., American Society for Testing and Materials, Philadelphia, 1986, pp. 401-411.

ABSTRACT: *J-R* curves were determined for five materials (7075-T651, 2024-T351, HY130, HY80, and A723, Class 1, Grade 4) using precracked Charpy specimens and standard size C(T) and SE(B) specimens. Crack growth in the Charpy specimens was estimated using the "load drop" method of analysis of the load-displacement trace, and crack extension in the C(T) and SE(B) specimens was determined using the electric potential method. The results show that physical crack extension in the larger specimen was not well estimated by the Charpy specimen results. If the crack extension is presented as relative crack growth (as a percentage of the uncracked ligament), however, the agreement between the two widely different specimen sizes is much better, although not exact. Except in the relatively brittle 7075-T651, the J corresponding to 0, 1%, and 2% crack growth was higher in the Charpy specimens than in the larger specimens. This was attributed to the inability of the "load drop" method to determine the exact location of the crack initiation. Although nonconservative, we believe the "load drop" method analysis of precracked Charpy data is adequate for quality control toughness testing, provided it is realized that J_{Ic} and *J-R* curves may be overestimated slightly.

KEY WORDS: J_{Ic} testing, precracked Charpy specimens, *J-R* testing, fracture mechanics

Recent work has demonstrated that adequate measurements of crack extension in precracked Charpy specimens can be made using their load-displacement characteristics alone [*1-2*]. The method has been called the

[1]U.S. Army Armament, Munitions, and Chemical Command, Armament Research and Development Center, Large Caliber Weapon Systems Laboratory, Benet Weapons Laboratory, Watervliet, NY 12189-5000. Currently Academic Visitor, Department of Mechanical Engineering, Imperial College of Science and Technology, London SW7 2BX, England.
[2]Naval Research Laboratory, Washington, DC 20375.

"load drop" method and is a simplified "key curve" analysis [*3*]. The purpose of this paper is to compare *J-R* curves generated using the precracked Charpy specimens with *J-R* curves generated with standard size compact tension C(T) or bend specimens SE(B) using accurate crack extension measurement methods.

To estimate crack extension using the "load drop" method, the basic assumption is that substantial plastic deformation occurs on the uncracked ligament, such that the maximum load (P_{max}) generated during the test approaches the limit load. In this case, crack initiation should occur very near P_{max}, and the "load drop" beyond P_{max} should be related to the amount of crack extension by [*1,2*]

$$\frac{P_{\Delta a}}{P_{max}} = \frac{b_{\Delta a}^2}{b_o^2} \tag{1}$$

where b_o is the original uncracked ligament dimension; $b_{\Delta a}$ is the uncracked ligament after an increment of crack extension, Δa, ($b_{\Delta a} = b_o - \Delta a$); and $P_{\Delta a}$ is the load beyond P_{max} where the estimate of crack extension is made. This method has proven an adequate approximation of crack extension in several materials [*1,2*].

To compare *J-R* curves generated using small specimens with larger specimen measurements, C(T) and SE(B) specimens were obtained from the same stock from which the Charpy specimens were made. The larger specimens were of standard planar dimensions: $W = 5.08$ cm (2.0 in.) and $B = 2.29$ cm (0.9 in.). Crack extension was determined using the direct-current electric potential method as outlined in Ref *4*.

J was calculated for the precracked Charpy specimen in accordance with the familiar form [*5*]

$$J = \frac{2A}{Bb_o} \tag{2}$$

where *A* is the total area under the load-displacement curve. Since small amounts of crack extension were encountered in these specimens, the correction for crack extension was also small and thus no correction for crack growth was made. The C(T) and SE(B) specimens were tested in an automated facility with computer-aided data processing capabilities. For these specimens, *J* was calculated incrementally accounting for crack extension. For the $(i + 1)$ increment, *J* was given as

$$J_{i+1} = \left[J_i + \left(\frac{f(a/W)}{b}\right)_i \frac{A_{i,i+1}}{B}\right]\left[1 - \left(\frac{\gamma}{b}\right)_i [a_{i+1} - a_i]\right] \tag{3}$$

where J_i is the total J calculated up to the previous increment; $(a_{i+1} - a_i)$ is the increment of crack growth that occurred between the $(i + 1)$ and (i) increments; $A_{i,i+1}$ is the area under the load-displacement trace between the $(i + 1)$ and (i) increments; and

$$\gamma_i = \begin{cases} 1 + 0.76(b_i/W) & \text{for C(T) specimens} \\ 1 & \text{for SE(B) specimens} \end{cases} \tag{4}$$

and

$$f(a/W) = 2 \qquad \text{for SE(B) specimens}$$

$$f(a/W) = 2[(1 + \alpha)/(1 + \alpha^2)] \qquad \text{for C(T) specimens} \tag{5}$$

where

$$\alpha = [(2a/b)^2 + 2(2a/b) + 2]^{1/2} - [2(a/b) + 1] \tag{6}$$

Equations 3 and 4 are based on the analysis of Ernst et al [*3*], and Eqs 5 and 6 are the tension component correction of the C(T) specimen from Clarke and Landes [*6*].

All specimens were precracked in accordance with the procedure outlined in ASTM Test for J_{Ic}, A Measure of Fracture Toughness (E 813). The theoretical nominal limit load was determined for each material, and the maximum load during fatigue precracking did not exceed 40% of the calculated limit load.

Materials

Five different materials were tested, three steels and two aluminum alloys. The aluminums were 2024-T351 and 7075-T651; the steels were HY80, HY130, and A723, Grade 1, Class 4 pressure vessel steel. All except the A723 steel were obtained in plate form, and specimens were obtained such that the T-L orientation was tested. Specimens of the A723 steel were obtained from thick hollow cylindrical forgings testing the C-R orientation. The mechanical properties of these materials are given in Table 1. These materials were chosen because of the wide range of properties they exhibit and their wide use in fracture critical applications.

Results and Discussion

The *J-R* curves developed are shown in Figs. 1 to 5. In all the figures, the symbols represent the curves developed using the larger specimens, and the continuous curves are average values from several (usually four) precracked Charpy specimens. The crack extension is represented in two ways: as physical crack extension and as a percentage of the original uncracked ligament.

TABLE 1—*Mechanical properties of the materials tested.*

Material	0.2% Offset Yield Strength, MPa	Ultimate Strength, MPa	Reduction of Area, %	Elongation, %
A723, Grade 1, Class 4	1310	1317	4	11
HY80	614	714	66	21
HY130	958	986	68	20
2024-T351	338	483	19	14
7075-T651	514	583	14	11

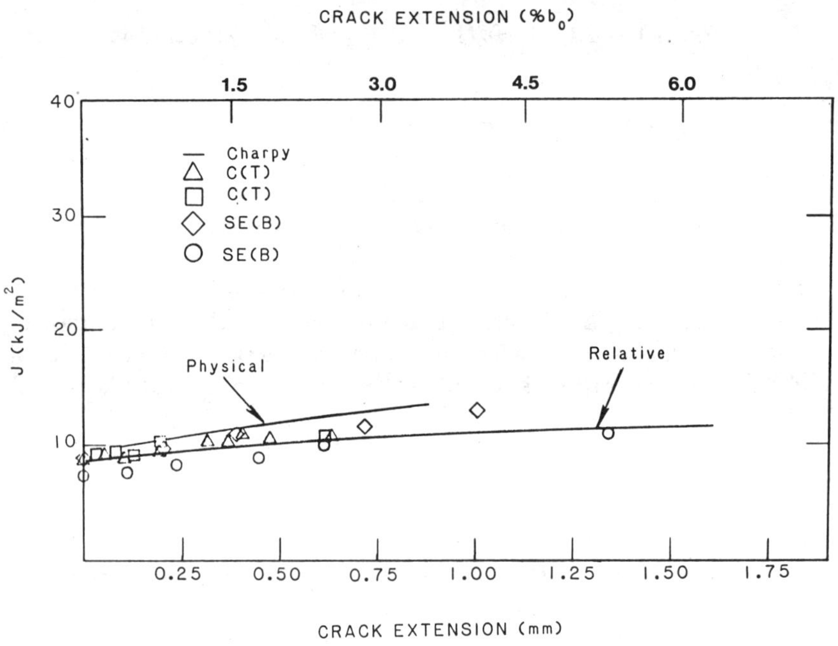

FIG. 1—J-R *curves for 7075-T651 aluminum.*

The scales were made so that the data for the larger C(T) and SE(B) specimens were at the same location on the plot.

The dual representation of the data was made because of the findings in a previous study [*1*]. In that study the value of J that resulted in about 1% crack extension in the precracked Charpy specimens compared favorably with K_{Ic} values in larger specimens. Since K_{Ic} corresponds to between 0 and 2% crack extension, the empirical observation was made that relative crack extension may be the common denominator when comparing toughness measurements in specimens of vastly different size. Although such an observation may have significant implications in the development of fracture test methods and anal-

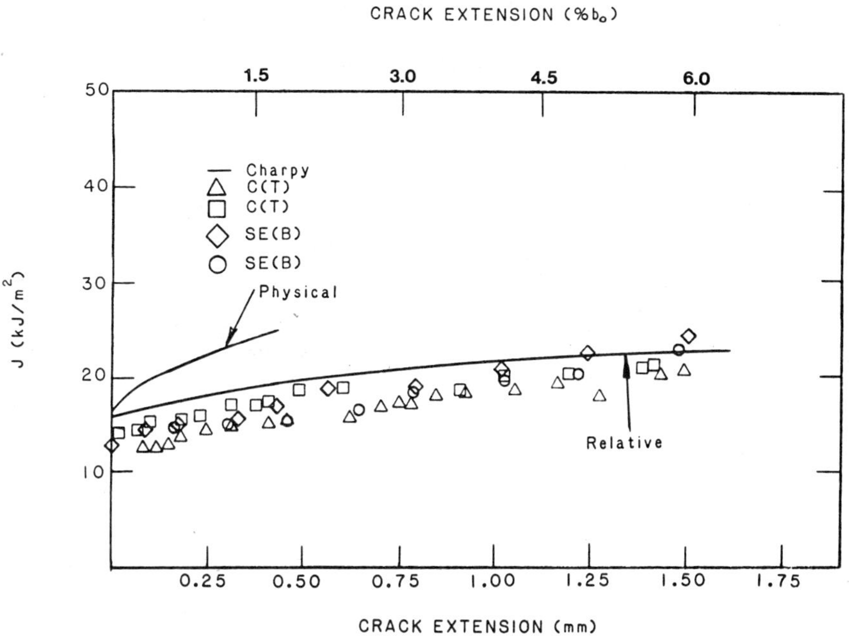

FIG. 2—J-R *curves for 2024-T351 aluminum.*

ysis, we make no claims as to its universal application. It merely seems to work in the testing of Charpy specimens using "load drop" analysis. The authors know of no continuum mechanics reason for such a specimen size dependence and caution against the use of "load drop" analysis or relative crack growth analysis to any structure other than precracked Charpy specimens without substantial experimental verification of its applicability.

The aluminum results are given in Figs. 1 and 2. The 7075-T651 curves (Fig. 1) show a very shallow slope, which suggests relatively brittle behavior even with very small precracked Charpy specimens. The initiation values of crack extension are well approximated using the "load drop" analysis of the Charpy specimens. Comparing the physical crack extension curves, we observe that the precracked Charpy results give a higher value of dJ/da than the larger specimens, but when considering crack extension as a percentage of the original uncracked ligament, either specimen size gives essentially the same curve. For 2024-T351 we find a substantially tougher material behavior than with 7075-T651. Both the initiation J values and the slopes of the R-curves are greater using both small and larger specimens. The agreement between large specimen results and precracked Charpy results is not so good as with the previous alloy. This is especially true with the physical crack extension results. For the small specimens, the initiation J value is somewhat

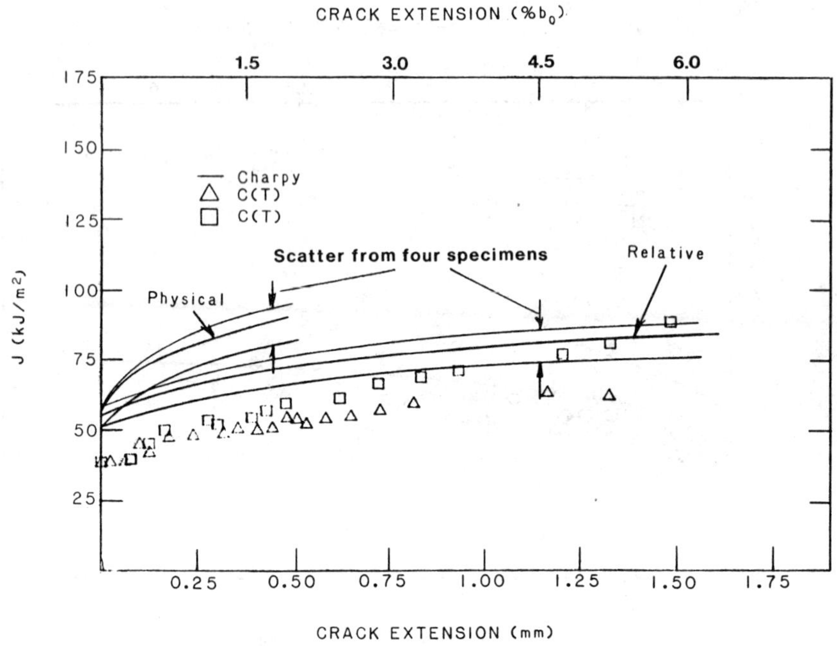

FIG. 3—J-R *curves for A723, Class 1, Grade 4 pressure vessel steel.*

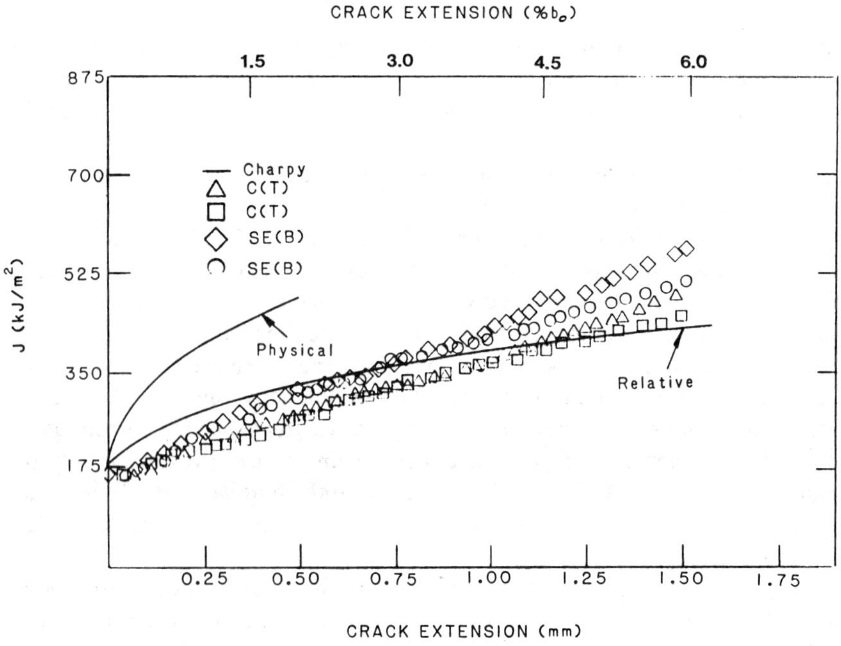

FIG. 4—J-R *curves for HY80 steel.*

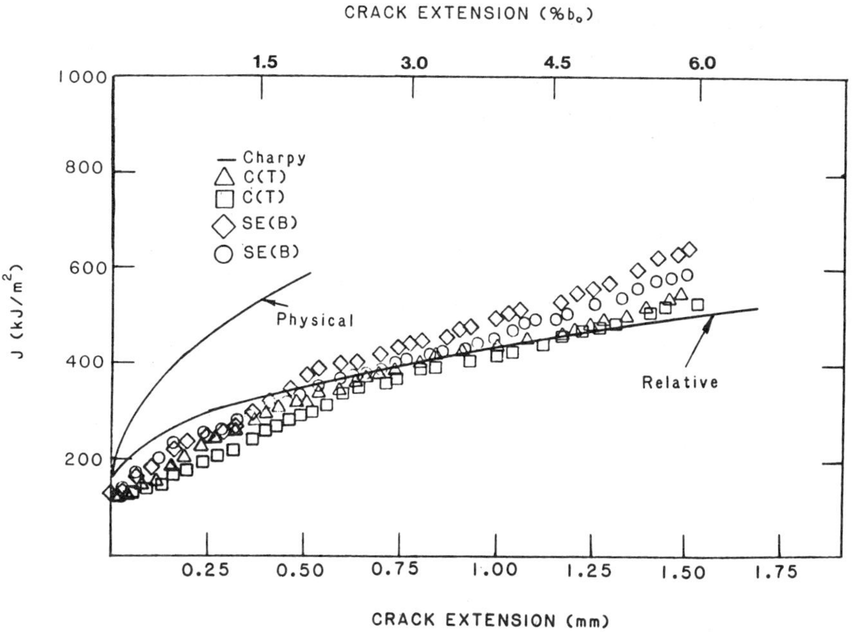

FIG. 5—J-R *curves for HY130 steel.*

higher and dJ/da is also much greater. When these same data are plotted as a percentage of the uncracked ligament, the R-curves are in much better agreement. The initial portion of the large specimen curve is overestimated, but once about 1.5 or 2.0% crack extension occurs, the agreement is quite good. Furthermore, the SE(B) data seem to give a somewhat greater dJ/da than the C(T) results. This was also seen in the steel materials. The fact that the precracked Charpy results agree better with the SE(B) than the C(T) data was expected, since the precracked Charpy specimen is also an SE(B) specimen of significantly smaller dimensions.

The steel results are given in Figs. 3 to 5. The A723 steel was originally a hollow cylindrical forging, and it was not possible to obtain large SE(B) specimens in the proper configuration. Thus only larger C(T) specimens were tested. As with the 2024-T351 aluminum, "load drop" analysis of precracked Charpy specimens gives a greater initiation J and a much steeper dJ/da when considering physical crack extension. Again, the agreement is improved when crack extension is given as a fraction of the uncracked ligament. Unlike the earlier results, the agreement does not become very good until about 3.5% crack extension. This may be somewhat deceiving because SE(B) specimens were not tested in the A723 alloy, while they were in the 2024-T351. A direct comparison of the precracked Charpy and C(T) results in 2024-T351 (Fig. 2) aluminum also shows that reasonable agreement was not achieved until about

5% relative crack extension. This suggests that if the same trend were observed in A723 steel between SE(B) and C(T) specimens, then even better agreement between the large specimen results and the precracked Charpy data may have been achieved if larger SE(B) specimens of this alloy had been tested.

The fact that the initiation J values are overestimated in both 2024-T351 and A723 is probably due to the assumption that crack extension begins at peak load. In relatively brittle materials, such as 7075-T651, it is likely that P_{max} is closely associated with the onset of crack extension because the load-displacement trace up to maximum load is nearly linear. This suggests that global elastic behavior and nonlinearity beyond P_{max} can indeed by only attributed to crack growth. On the other end of the scale, dealing with a very ductile material, where the entire uncracked ligament is subjected to plastic deformation, the drop in the load the specimen can support is either due to crack extension or necking. It is the case that falls between the two extremes where inaccuracy would be expected to be maximized. In that instance, crack extension commences when the uncracked ligament is partially plastic upon rising load. This may be the case for the 2024-T351 and A723 materials and will be discussed further below.

Returning to our *J-R* curves, we come to the HY80 results (Fig. 4). Again, comparing physical crack extensions, the precracked Charpy specimens give a much higher dJ/da property, but in this case the initiation values are well predicted using either specimen. The curves generated representing crack extension as a percentage of original uncracked ligament show that nearly the entire large specimen *J-R* curve can be very well approximated with the "load drop" precracked Charpy data. Similar results were obtained in HY130 (Fig. 5). The large specimen physical crack extension in HY130 was not well measured from physical crack extension of precracked Charpy specimens. Plotting relative crack extension again gives very good agreement between small and large specimens. For HY130, the initiation J value of the larger specimens were overestimated with the precracked Charpy specimens. In either HY80 or HY130, the best agreement on the relative crack extension *J-R* curves occurs between about 1.5 and 5.0% relative crack extension.

The original purpose of the "load drop" method was to generate a simple estimate of K_{Ic} using small specimens that has application as a quality control measure [*1*]. Since K_{Ic} is a measure of the stress intensity factor that results in between 1 and 2% crack extension, we can compare the large specimen and small specimen R-curves at these amounts of relative crack extension. The data reported here were generated using specimens that were precracked to a/W of approximately 0.5. Thus the relative amounts of crack extension $\Delta a/a_o$ and $\Delta a/b_o$ are approximately the same and can be determined directly from the R-curves (Figs. 1 to 5). These comparisons are given in Table 2 for all the materials tested. In the table a single value is given which is the average of four precracked Charpy specimens and the average of all the larger specimen results.

TABLE 2—*Average J values (kJ/m²) from small and large specimens at various amounts of relative crack extension.*

Material	$\Delta a/a_o = 0.0\%$		$\Delta a/a_o = 1.0\%$		$\Delta a/a_o = 2.0\%$	
	Charpy	SE(B) + C(T)	Charpy	SE(B) + C(T)	Charpy	SE(B) + C(T)
7075-T651	8.5	8.1	9.1	9.4	9.7	10.3
2024-T351	16	13	18	15	19	17
A732, Class 1, Grade 4	56	39	64	49	71	58
HY80	177	163	275	231	316	275
HY130	174	128	285	233	349	315

The first general comment that can be made is that the initiation J value is universally overestimated except in the case of the brittle 7075-T651 alloy. This can be explained by the assumption that the crack begins to propagate at maximum load. Probably small amounts of crack extension occur in the 2024-T351 and the A723 alloys before peak load. For the higher toughness HY80 and HY130, some crack extension could have occurred at the maximum load. Since both of these alloys strain-harden significantly, crack growth with a fully plastic remaining ligament may occur without the "load dropping" and thus we would not see it without "load drop" analysis.

At greater amounts of relative crack extension, "load drop" still overestimates the R-curve, although the absolute and relative differences became much less. For example, at 1% crack extension, J from the Charpy specimen is about 20 to 30% higher than the larger specimens for all the materials except the 7075-T651 where the difference is almost negligible. Similarly, at 2% crack extension, the differences are reduced to between 10 and 20% for the more ductile materials. If the J values are represented as their K equivalents, the relative differences are reduced by roughly one half (i.e., 10 to 15% at 1% crack extension and 5 to 10% at 2% crack extension).

Table 3 allows the examination of the discrepancies between the small

TABLE 3—*Toughness comparisons and validity considerations.*[a]

Material	$\frac{\Delta a}{a_o} = 0$		$\frac{\Delta a}{a_o} = 1.0\%$		$\frac{\Delta a}{a_o} = 2.0\%$		
	$\frac{J_{LD}}{J_{LS}}$	$\frac{25\,J_{LD}}{\sigma_f b_o}$	$\frac{J_{LD}}{J_{LS}}$	$\frac{25\,J_{LD}}{\sigma_f b_o}$	$\frac{J_{LD}}{J_{LS}}$	$\frac{25\,J_{LD}}{\sigma_f b_o}$	$\frac{P_{max}}{P_{LL}}$
7075-T651	1.05	0.08	0.97	0.08	0.94	0.09	0.51
2024-T351	1.23	0.20	1.20	0.23	1.12	0.24	0.85
A723, Class 1, Grade 4	1.44	0.21	1.31	0.24	1.22	0.27	0.84
HY80	1.09	1.28	1.19	1.99	1.15	2.29	1.15
HY130	1.36	0.90	1.22	1.47	1.10	1.81	1.19

[a]Subscript LD = "Load Drop" Charpy Specimens, LS = Larger Specimens.

specimen and large specimen data from a specimen size criterion viewpoint. For cracks to grow under *J*-controlled conditions, the guideline of *a*, *b*, and *B* dimensions of the specimen must be greater than 25 J/σ_f, with σ_f the arithmetic average of the yield strength and ultimate strength. For precracked Charpy specimens, the remaining ligament, *b*, is the important dimension, thus the column 25 $J/b_o\sigma_f$. When this quantity is less than one, *J*-controlled crack growth is assumed to be occurring; when the ratio is greater than unity, the specimen is too small for the *J* test. The larger specimens had b_o values about five times the b_o values of the Charpy specimens; thus if the quantity in the table exceeds five, *J* was not controlling in the larger samples. The final column in the table is the ratio of the average measured peak load (P_{max}) to the theoretical limit load (P_{LL}) of the precracked Charpy specimen [5]. This gives an indication of crack extension before peak load or any strain-hardening effects that would mark crack extension near peak load.

For 7075-T651, 2024-T351, and A723, the size validity criterion is met; the specimen was sufficiently large for *J*-controlled crack growth. Therefore, using precracked Charpy specimens, valid *R*-curves should result. The fact that the *R*-curves do not coincide for the 2024-T351 and the A723 materials is probably due to crack initiation occurring not at peak load but before it. This would have the effect of moving the entire precracked Charpy *R*-curve to the right or point-by-point addition of that amount of crack extension that occurred before maximum load. If an estimate of that amount of crack growth before peak load could be made, then better agreement would result. At this time a simple method of determining that small increment of crack growth is not available.

In the higher toughness HY80 and HY130 steels, crack growth did not occur under *J*-controlled conditions. According to the guidelines in ASTM E 813, the precracked Charpy *R*-curves cannot be considered as valid. What is interesting is that the agreement between small and large specimens of these materials was as good as the agreement between small and large specimens of the less tough 2024-T351 and A723 steel. This suggests that either the validity requirement is too restrictive or a coincidence has occurred. Further work on refining the validity criterion would answer this question. Also, in HY80 and HY130, it is clear that significant strain-hardening occurred. The effect of strain-hardening could be crack growth at peak load with no "load drop." This has the same result as the case of crack growth near but before peak load (i.e., "load drop" analysis underestimates crack extension). Real crack extension would move the entire *R*-curve to the right, thus giving better agreement with the large specimen data.

Summary and Conclusions

J-R curves were developed for five materials using standard specimens with well characterized methods of analysis and precracked Charpy specimens with "load drop" analysis. The results show that physical crack extension in the larger specimens is not well approximated with the precracked Charpy specimens and the *J* that results in the onset of crack extension is overestimated significantly with precracked Charpy specimens. If the crack extension data are presented as a fraction of the uncracked ligament, much better agreement is obtained. In this case the "load drop" analysis still overestimates the overall *R*-curve but to a much smaller degree. Comparisons between *J* that result in 1 and 2% crack extension show that "load drop" is not conservative by between 10 and 30% and was consistent even in specimens that were not valid according to presently recommended size requirements. The overestimate is attributed to the inability of the "load drop" method to sensitively determine the onset of crack growth and the inherent geometry dependence of *J-R* curves.

References

[1] Kapp, J. A. and Underwood, J. H., "Single Specimen *J*-Based Fracture Toughness Test for High Strength Steels," in *Fracture Mechanics: Fourteenth Symposium—Vol. II: Testing and Applications, ASTM STP 791*, American Society for Testing and Materials, Philadelphia, 1983, pp. II-402–II-414.

[2] Kapp, J. A., "*J-R* Curve Determination Using Precracked Charpy Specimens and the Load-Drop Method for Crack Growth Measurements," in *Fracture Mechanics: Sixteenth Symposium, ASTM STP 868*, American Society for Testing and Materials, Philadelphia, 1985, pp. 281–292.

[3] Ernst, H., Paris, P. C., Rossow, M., and Hutchinson, J. W., "Analysis of Load-Displacement Relationships to Determine *J-R* Curve and Tearing Instability Properties," in *Fracture Mechanics (Eleventh Conference), ASTM STP 677*, American Society for Testing and Materials, Philadelphia, 1975, pp. 581–599.

[4] Jolles, M. I., "Automated Technique for *R*-Curve Testing and Analysis," in *Automated Test Methods for Fracture and Fatigue Crack Growth, ASTM STP 877*, American Society for Testing and Materials, Philadelphia, 1985, pp. 248–259.

[5] Rice, J. R., Paris, P. C., and Merkle, J. C., in *Progress in Flaw Growth and Fracture Toughness Testing, ASTM STP 536*, American Society for Testing and Materials, Philadelphia, 1973, pp. 231–245.

[6] Clarke, G. A. and Landes, J. D., "Evaluation of the *J*-Integral for the Compact Specimen," *Journal of Testing and Evaluation*, Vol. 7, No. 5, Sept. 1979, pp. 264–269.

Ductile Fracture

Kazumi Hirano,[1] *Hideo Kobayashi,*[2] *and Hajime Nakazawa*[3]

A Single Specimen Determination of Elastic-Plastic Fracture Resistance by Ultrasonic Method

REFERENCE: Hirano, K., Kobayashi, H., and Nakazawa, H., **"A Single Specimen Determination of Elastic-Plastic Fracture Resistance by Ultrasonic Method,"** *Fracture Mechanics: Seventeenth Volume, ASTM STP 905*, J. H. Underwood, R. Chait, C. W. Smith, D. P. Wilhelm, W. A. Andrews, and J. C. Newman, Eds., American Society for Testing and Materials, Philadelphia, 1986, pp. 415–434.

ABSTRACT: A single specimen ultrasonic determination procedure of the empirical measure of elastic-plastic fracture resistance under monotonic and multiple unloading conditions was proposed. It was shown that the elastic-plastic fracture toughness J_{Ic} and the tearing moduli T_J (Δa_{max}) calculated from the slope of the determined J-Δa_{max} curves are independent of specimen size and side groove depth. It was also found that the advancing crack tip opening displacement δ_a is nearly constant during the stable crack growth. The multiple unloading R-curve was successfully determined and corresponded to the monotonic R-curve. In addition, the transition behavior from low cycle fatigue to monotonic stable crack growth characteristics was experimentally clarified.

KEY WORDS: elastic-plastic fracture, single specimen technique, ultrasonic method, JR-curve, tearing modulus, δR-curve, multiple unloading R-curve, low cycle fatigue crack growth

Nomenclature

a	Crack length
a_0	Initial crack length
a_i	Instantaneous crack length
Δa	Crack extension
Δa_{max}	Maximum crack extension at the center of specimen thickness

[1]Mechanical Engineering Laboratory, Agency of Industrial Science and Technology, MITI, Namiki 1-2, Sakura-mura, Niihari-gun, Ibaraki-ken, 305, Japan.
[2]Faculty of Engineering, Tokyo Institute of Technology, Ōokayama 2-12-1, Meguro-ku, Tokyo, 152, Japan.
[3]Faculty of Engineering, University of Chiba, Yayoi-cho 1-33, Chiba-shi, 260, Japan.

Δa_p	Physical crack extension (ASTM E 813)
Δa_{SZ}	Crack extension due to stretching
da/dN	Fatigue crack growth rate
$A_{i,i+1}$	Area under load versus load-line displacement curve between lines of constant displacement at Points i and $i + 1$
ΔA	Area under load versus load-line displacement curve during a rising load
A_{total}	Total area under load versus load-line displacement curve
b	Uncracked ligament, $W - a$
b_0	Initial uncracked ligament
b_i	Instantaneous remaining uncracked ligament
B_N	Net specimen thickness
C	Material constant relating da/dN and ΔJ
E	Young's modulus
ΔE_1	Top-on ultrasonic echo difference output voltage
ΔE_2	End-on ultrasonic echo difference output voltage
ΔH_2	End-on ultrasonic echo height difference
J	J-integral
J_{Ic}	Critical J-integral (elastic-plastic fracture toughness)
dJ/da	Slope of JR-curve
J_{total}	J-integral calculated using A_{total}
ΔJ	Cyclic J-integral
ΔK	Stress intensity factor range
ΔK_{eff}	Effective stress intensity factor range
k	Proportional constant relating δ and ΔH_2
P	Load
R	Load ratio
SZW	Stretch zone width
T_J	Tearing modulus
$T_J(\Delta a_p)$	Tearing modulus calculated using the slope of J-Δa_p curve
$T_J(\Delta a_{max})$	Tearing modulus calculated using the slope of J-Δa_{max} curve
V_L	Load-line displacement
ΔV_L	Change in load-line displacement during unloading
W	Specimen width
α	$= \sqrt{(2a_0/b_0)^2 + 2(2a_0/b_0) + 2} - (2a_0/b_0 + 1)$
γ	$= 1 + 0.766\ b/W$
η	$= 2 + 0.522\ b/W$
σ_{ys}	Yield stress
σ_u	Ultimate tensile strength
σ_f	Flow stress
δ	Crack tip opening displacement
δ_{Ic}	Critical crack tip opening displacement (elastic-plastic fracture toughness)
δ_o	Original crack tip opening displacement
δ_a	Advancing crack tip opening displacement

Recently, several attempts have been made to evaluate an elastic-plastic fracture resistance under cyclic loading condition and to combine elastic-plastic fatigue with monotonic crack growth resistance data [1-5]. The phenomenon of elastic-plastic fracture takes on a complex, three-dimensional aspect. The flat fracture near the midthickness develops under predominantly plane-strain condition, whereas the shear lips near the surface develop under plane-stress condition. It is therefore questionable how to define the crack extension in the experimental determination of the crack growth resistance curve that is meaningful and independent of specimen size and geometry. If the J-integral resistance curve is characterized in terms of maximum crack extension Δa_{max} at the center of specimen thickness where plane-strain flat fracture occurs, the experimentally determined J-Δa_{max} curve appears to be independent of specimen size and side groove depth, where the specimen sizes are geometrically similar. Since Δa_{max} is nondestructively and continuously measured by the ultrasonic monitoring technique, an empirical measure of plane-strain elastic-plastic fracture resistance curve can easily be determined from a single small specimen [6].

In this paper, a single specimen determination procedure of the elastic-plastic fracture resistance curves in terms of J-integral and crack tip opening displacement δ under monotonic and cyclic loading conditions is proposed. Multiple unloading elastic-plastic fracture resistance curves are successfully determined under multiple unloading condition according to the proposed procedure and are compared with those obtained under monotonic loading condition. The transition behavior from low cycle fatigue to monotonic stable crack growth characteristics is also discussed.

Ultrasonic Monitoring of Inner Crack Growth and Determination of Empirical Measure of Elastic-Plastic Fracture Resistance

An ultrasonic monitoring technique of inner crack growth in experimental fracture mechanics (Fig. 1) is fundamentally based on conventional ultrasonic nondestructive testing using the pulse echo technique. Transducers which both send and receive an ultrasonic pulse are placed on the top ("top-on" ultrasonic method [6]) and the end ("end-on" ultrasonic method [7]) of the compact (CT) specimen. The transducer placement variation does not require calibration, and the time-of-flight variation is easily calibrated and very versatile. An advantage of the ultrasonic monitoring technique of crack growth is that the directivity of the ultrasonic wave can provide inner crack length measurements where the crack tunneling occurs. In addition, the ultrasonic monitoring technique is applicable to any test specimen geometry that behaves in an elastic-plastic manner. Accordingly, not only fatigue crack growth but also fatigue precrack tip plastic blunting, onset of stable crack growth, or crack extension during an elastic-plastic fracture toughness testing can nondestructively and continuously be measured at the specimen midthickness under the predominantly plane-strain condition [8-14].

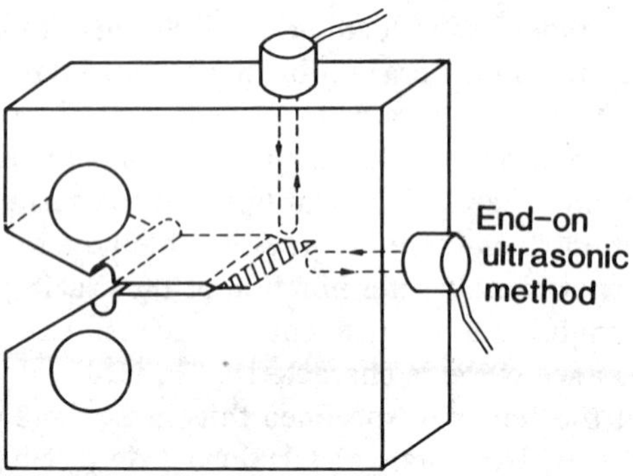

FIG. 1—*Schematic illustration of "top-on" and "end-on" ultrasonic method.*

The proposed single specimen determination procedure of the empirical measure of the elastic-plastic fracture resistance curve based on the ultrasonic method is summarized as follows:

1. As schematically illustrated in Fig. 1, transducers which both send and receive an ultrasonic pulse are placed on the top and end surface of the CT specimen.
2. The specimen is loaded in a displacement control mode. Continuous photographs of the end-on ultrasonic echo signal on a cathode ray tube (CRT) are taken during fracture toughness loading, in addition to the top-on and end-on ultrasonic echo difference output voltages (ΔE_1, ΔE_2) or the load versus load-line displacement ($P - V_L$) curve measurements.
3. The crack extension Δa is measured from the echo-shifted distance on the photograph of the end-on ultrasonic CRT presentation (Fig. 2*b*).
4. The J-integral versus Δa curve is determined from a single specimen by plotting the J values experimentally evaluated using the area under the $P - V_L$ curve against Δa (Fig. 2*a*). Elastic-plastic fracture toughness J_{Ic} is also determined at the rapid-increase or the maximum points in the ΔE_1, ΔE_2 versus V_L curves, respectively [*13*].
5. The original and advancing crack tip opening displacement (δ_o, δ_a) are nondestructively and continuously measured from the end-on ultrasonic echo height difference ΔH_2 at the original fatigue precrack tip or the advancing crack tip using the expression

$$\delta = k \cdot \Delta H_2 \tag{1}$$

where k is the proportional constant. The δ_o, δ_a versus Δa curves are also determined from a single specimen by plotting these δ_o and δ_a values against Δa (Fig. 2*a*).

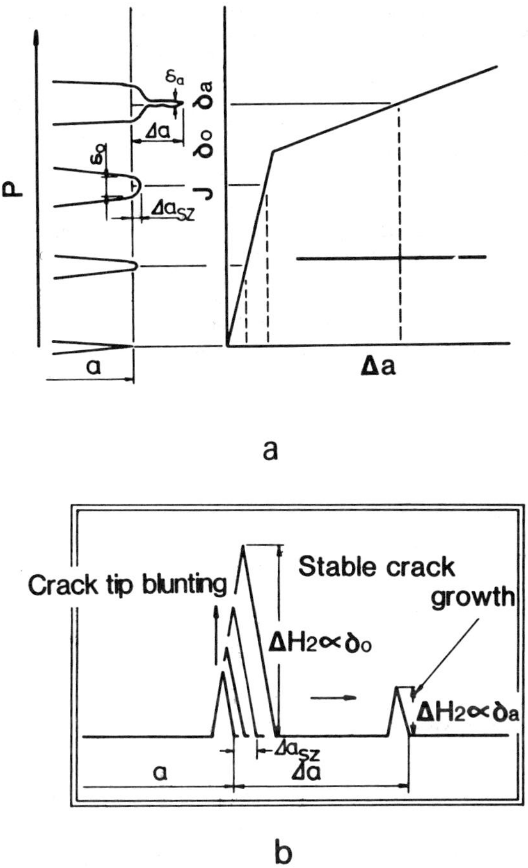

FIG. 2—*Single specimen determination procedure of an empirical measure of elastic-plastic fracture resistance.*

The multiple unloading elastic-plastic fracture resistance curve is also determined from a single specimen, since the ultrasonic monitoring technique proposed here is applicable to the cyclic loading tests.

Materials and Experimental Procedure

Materials and Test Specimens

The materials tested were intermediate-strength, high-toughness A533 (Grade B, Class 1) steel and 2½ Cr-Mo steel. Details of their chemical composition, heat treatments, and mechanical properties are given in Tables 1 and 2.

TABLE 1—*Chemical composition (weight percent).*

Material	C	Si	Mn	P	S	Ni	Cr	Mo	Cu	Al
A533B-1	0.19	0.27	1.27	0.006	0.008	0.63	...	0.52	...	...
2½Cr-Mo	0.14	0.17	0.61	0.009	0.015	0.12	2.45	1.01	0.10	0.02

TABLE 2—*Heat treatments and mechanical properties.*

Material	Heat Treatments	Yield Strength (σ_{ys}), MPa	Ultimate Tensile Strength (σ_U), MPa	Elongation (δ), %	Reduction of Area (ψ), %	Strain-Hardening Exponent (n)
A533B-1	1173 K for 50 min; air cooled 1203 K for 4 h; water quenched 933 K for 4 h; tempered	539	680	25.0	...	0.14
2½Cr-Mo	1193 K for 15.5 h; water quenched 933 K for 10 h; air cooled, tempered	510	659	28.8	75.9	0.10

The specimens used were 25-mm-thick compact type (1CT) (Fig. 3). Cracks were placed in the L-T orientation. All specimens were fatigue precracked in accordance with ASTM Test for Plane-Strain Fracture Toughness of Metallic Materials (E 399) to crack lengths corresponding to a/W of 0.5, where a is the crack length and W the specimen width. The 40-mm-thick compact type (2CT) specimen was sometimes used for the A533B-1 steel. After precracking, side grooves were machined into some specimens along the nominal crack plane to total section reductions of 12.5, 25, and 50%. Most specimens were side grooved with a 45-deg cutter with a root radius of approximately 150 μm.

Test Procedure

J-integral and δR curve tests were conducted at room temperature on a closed-loop electrohydraulic testing machine under the displacement-controlled condition at a speed of 0.5 mm/min. To study the effect of multiple unloading on crack growth behavior, tests also were carried out under displacement-controlled unloads with a constant ΔV_L (approximately 0.2 ~ 0.5 mm) as indicated in Fig. 4. Number of cycles were six and thirteen. Low cycle fatigue crack growth tests were carried out under load-controlled cyclic loading condition at a load ratio R (minimum load to maximum load) of zero. The ultrasonic measurements here were performed using a ceramic, normal trans-

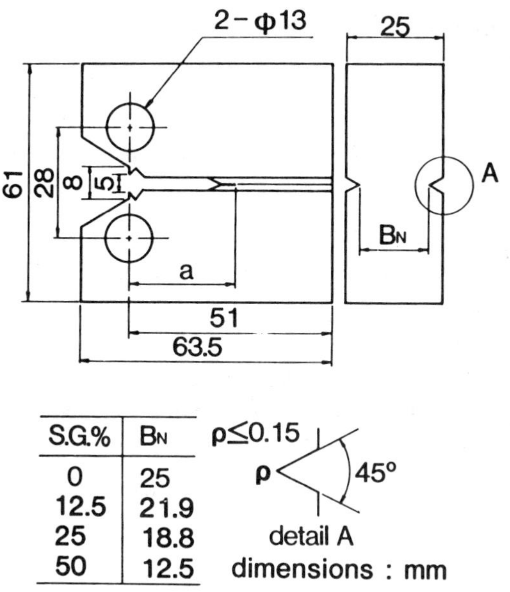

S.G.%	BN
0	25
12.5	21.9
25	18.8
50	12.5

FIG. 3—*Configuration and dimensions of test specimen (1CT).*

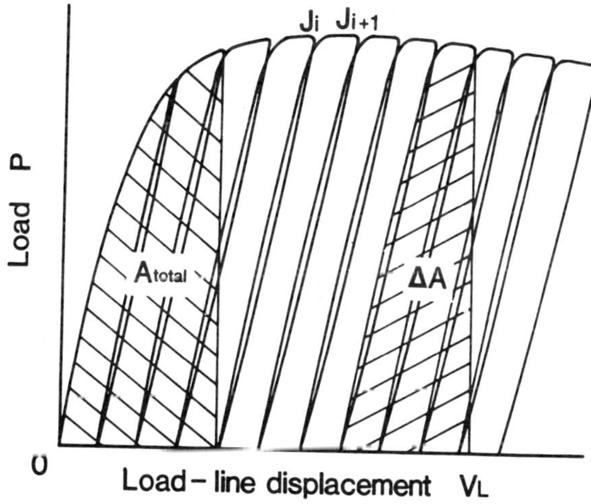

FIG. 4—*Estimation of cyclic J-integration.*

ducer operating at 5 MHz and a commercial ultrasonic flaw detector. The ultrasonic beam is focused on the center of the fatigue precrack front. The behavior detected by the ultrasonic method corresponds to stable crack growth which almost occurs at the specimen midthickness under the predominantly plane-strain condition.

The J-integral was calculated under the monotonic loading condition using the expression [15]

$$J_{i+1} = \left[J_i + \left(\frac{\eta}{b}\right)_i \frac{A_{i,i+1}}{B_N}\right]\left[1 - \left(\frac{\gamma}{b}\right)_i (a_{i+1} - a_i)\right] \tag{2}$$

where

$\eta = 2 + 0.522\ b/W$,
$\gamma = 1 + 0.766\ b/W$,
b_i = instantaneous remaining uncracked ligament,
B_N = net specimen thickness,
a_i = instantaneous crack length, and
$A_{i,i+1}$ = area under load versus load-line displacement curve between lines of constant displacement at Points i and $i + 1$.

The J-integral was calculated under the cyclic loading condition using the expression [16]

$$J_{i+1} = \frac{1}{b_i}[J_i \cdot b_{i+1} + (J_{\text{total},i+1} - J_{\text{total},i})b_0] \tag{3}$$

where

$J_{\text{total}} = \dfrac{1 + \alpha}{1 + \alpha^2} \cdot \dfrac{2A_{\text{total}}}{B_N \cdot b_0}$ [17],
$\alpha = \sqrt{(2a_0/b_0)^2 + 2(2a_0/b_0) + 2} - (2a_0/b_0 + 1)$,
a_0 = initial crack length,
b_0 = initial uncracked ligament, and
A_{total} = area indicated in Fig. 4.

Cyclic ΔJ were determined from area ΔA under load versus load-line displacement curve during a rising load as indicated in Fig. 4. The following expression was used to estimate ΔJ [18]:

$$\Delta J = \frac{1 + \alpha}{1 + \alpha^2} \cdot \frac{2 \cdot \Delta A}{B_N \cdot b} \tag{4}$$

Elastic-Plastic Fracture Resistance *R*-Curve

Δa *and* δ *Measurements*

End-on ultrasonic crack extension measurements taken at the termination of tests are compared with optical measurements on the fatigue post-fractured surface (Fig. 5). The test results show that the agreement between the ultrasonic measurements Δa (ultrasonic) and the actual maximum crack ex-

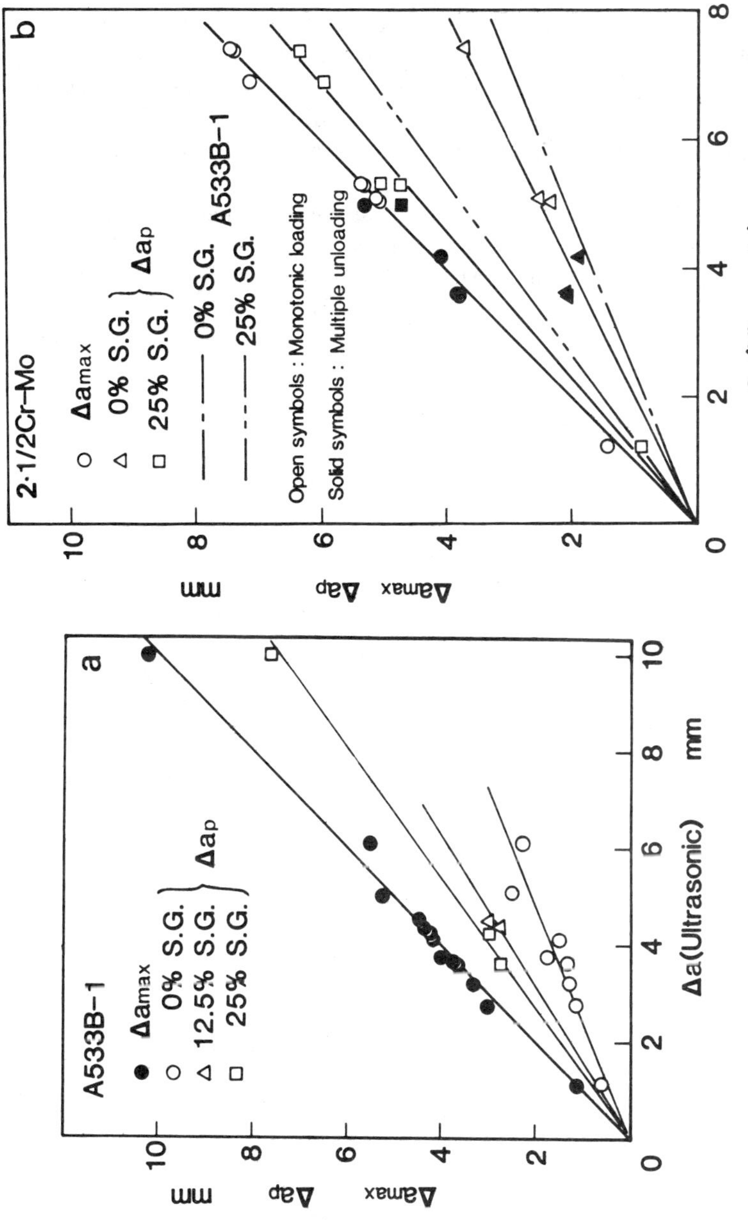

(a)

(b)

FIG. 5—*Ultrasonic measurements of crack extension.* (a) *A533B-1 steel.* (b) *2½Cr-Mo steel.*

tension Δa_{max} at the center of specimen thickness is excellent regardless of side groove depth. Within the limits of this experiment, the same linear relationships between Δa_{max} and the physical crack extension Δa_p in ASTM Test for J_{Ic}, A Measure of Fracture Toughness (E 813) exist in accordance with respective side groove depth under both monotonic and cyclic loading conditions. However, there is a difference in the curvature of a characteristic thumbnail-shaped crack front between A533B-1 and 2½Cr-Mo steels with nearly the same strength level, since the slope of Δa_{max} versus Δa_p curves varies.

The relationship between end-on ultrasonic echo height difference ΔH_2 and original crack tip opening displacement δ_o is shown in Fig. 6 for A533B-1 steel. In this figure, stretch zone width *SZW* fractographically obtained at the center of specimen thickness from the multispecimen test are also plotted against ΔH_2 for comparison. It is shown that there is a unique relationship

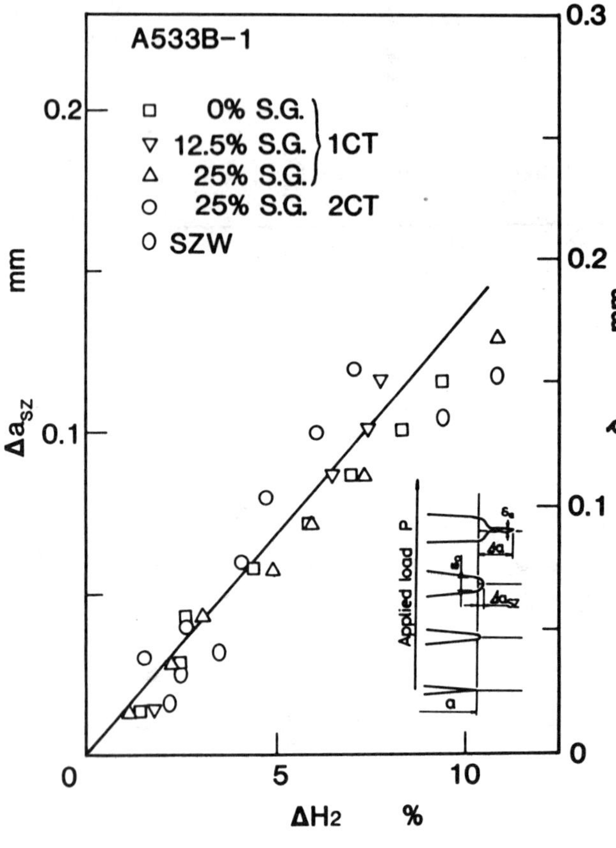

FIG. 6—*Relationship between end-on ultrasonic echo height difference* ΔH_2 *and original crack tip opening displacement* δ_o.

between ΔH_2 and δ_o as expressed by Eq 1, regardless of specimen size and side groove depth, and that this unique relationship corresponds to the multispecimen test results shown by the relationship between SZW and ΔH_2. From the test results described above, the ultrasonic monitoring technique proposed here is shown to give excellent estimates of actual maximum crack extension and crack tip opening displacement at the center of specimen thickness under plane-strain condition regardless of specimen size and side groove depth. It is therefore a useful tool in experimental determination of an empirical measure of plane-strain elastic-plastic fracture resistance curve.

Single Specimen Determination of JR- *and* δR-*Curves*

The J-integral versus Δa_{max} curves obtained according to the proposed single specimen determination procedure are shown in Fig. 7 for A533B-1 steel. The test results show almost no influence of specimen size and side groove depth on the J-Δa_{max} curve. Accordingly, if the J-integral resistance JR-curve is characterized in terms of Δa_{max} at the center of specimen thickness where

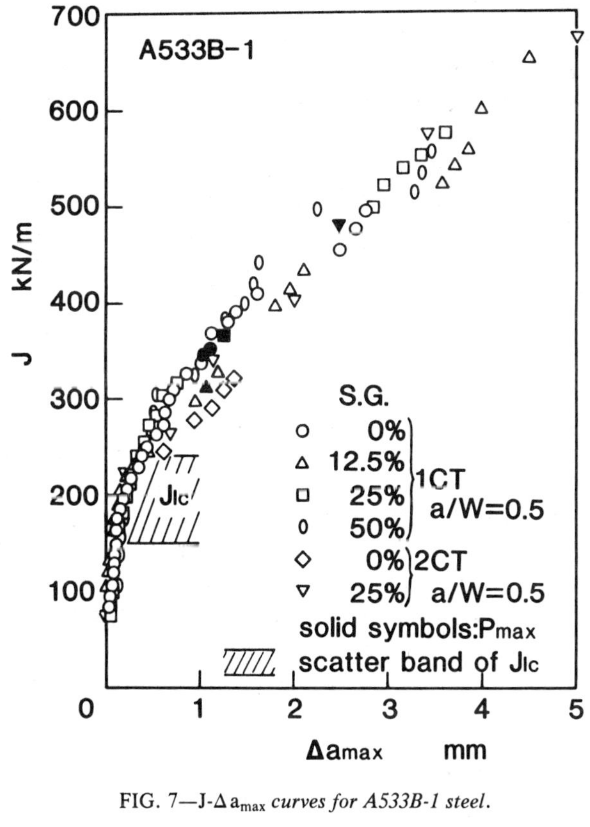

FIG. 7—J-Δa_{max} *curves for A533B-1 steel.*

predominantly plane-strain flat fracture occurs, the experimentally determined *JR*-curve appears to be independent of specimen size and side groove depth, where the specimen sizes are geometrically similar. Since Δa_{max} is nondestructively and continuously measured by the ultrasonic monitoring technique as described above, an empirical measure of the plane-strain elastic-plastic fracture resistance curve can easily be determined from a single small specimen. On the other hand, it can also be presumed from the relationship between Δa_{max} and Δa_p at each percentage of side groove depth shown in Fig. 5 that *J* versus Δa_p curves depend on the side groove depth, and that the slope of J-Δa_p curve decreases with increasing the percentage of side grooves [*19*], although these detailed results are omitted in this paper.

The tearing moduli are calculated using the expression [*20*]

$$T_J = \frac{E}{\sigma_f^2} \cdot \frac{dJ}{da} \tag{5}$$

where

T_J = tearing modulus,
dJ/da = slope of *JR*-curve,
E = Young's modulus, and
σ_f = flow stress $(= (\sigma_{ys} + \sigma_u)/2)$.

The dJ/da values in this investigation were calculated using the least-squares linear fit of *J* crack extension data which fell at least 0.15 mm beyond the measured blunting line to approximately 5 mm of crack growth. Plots of T_J (Δa_{max}) determined from the J-Δa_{max} curve against the percent of side grooves are presented in Fig. 8 for A533B-1 steel. The $T_J(\Delta a_{max})$ values are also nearly constant regardless of specimen size and side groove depth, although there is considerable scatter as shown by the hatched band in this figure. On the other hand, $T_J(\Delta a_p)$ decreases as the percent of side grooves increases, and $T_J(\Delta a_p)$ for 25 and 50% side-grooved specimens are nearly consistent with $T_J(\Delta a_{max})$.

Crack tip opening displacement resistance δ_o, δ_a versus Δa_{max} curves are shown in Fig. 9 for A533B-1 steel. It can be seen from this figure that the δ_o − Δa_{max} curve is represented by the plastic blunting line and fracture resistance curve after the onset of stable crack growth as well as the *JR*-curve [*21*]. These quantitative measurements also show that the advancing crack tip opening displacement δ_a increases immediately after the onset of stable crack growth and has a tendency to stay nearly constant during the stable crack growth (solid line), and that it is several times smaller than elastic-plastic fracture toughness δ_{Ic}, although there is considerable scatter. The test results correspond qualitatively to that obtained by Garwood and Turner [*22*] using the multispecimen test method.

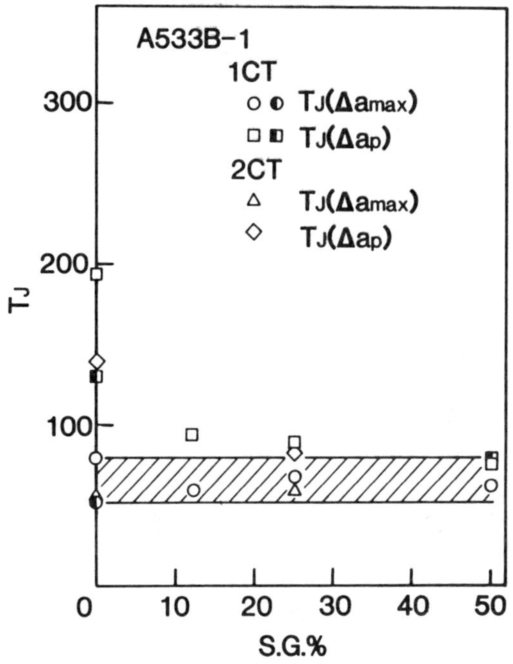

FIG. 8—*Effect of side groove depth on* T_J.

Multiple Unloading *R*-Curve Determination

Comparison of P versus V_L curves under monotonic and multiple unloading conditions is shown in Fig. 10 for 2½Cr-Mo steel. There is little difference between an envelope P-V_L curve under multiple unloading and a monotonic P-V_L curve, and the P-V_L curve is not influenced by the multiple unloading within the limits of this experiment. In other words, the distinct cyclic hardening and softening behavior do not appear on the P-V_L curve for this material, because a considerable crack extension occurs at each cycles.

J versus Δa_{max} curves developed under the cyclic loading condition are shown in Fig. 11 for 2½Cr-Mo steel. In this figure, the J versus Δa_{max} curve developed under the monotonic loading condition is also presented by the straight line for comparison. Although there is considerable scatter in comparison with the monotonic *JR*-curve, the test results show almost no influence of side groove depth and number of cycles on the multiple unloading *JR*-curve. The *JR*-curve represents an empirical measure of plane-strain elastic-plastic fracture resistance under multiple unloading condition which is side groove depth, multiple unloading cycles independent. The multiple unloading *JR*-curve corresponds to the monotonic *JR*-curve. Accordingly, it is

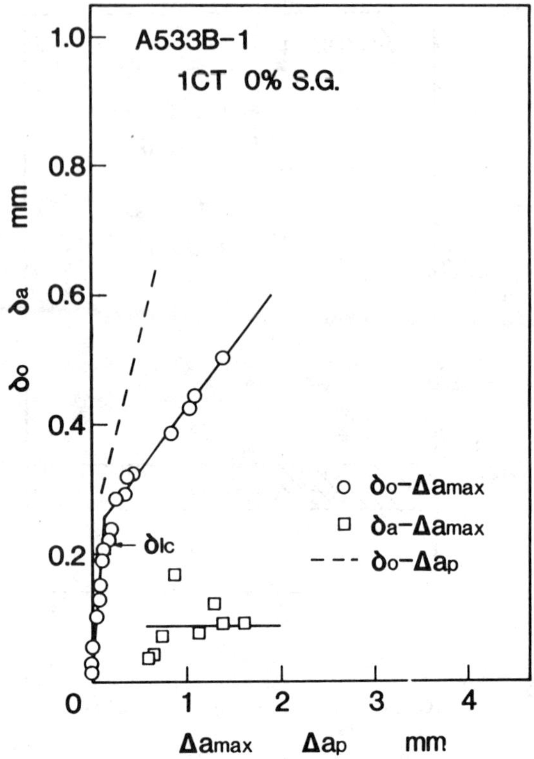

FIG. 9—δ_o, δ_a *versus* Δa_{max} *curves for A533B-1 steel.*

concluded that there is little influence of multiple unloading on empirical stable crack growth resistance characteristics.

Figure 12 shows δ_o, δ_a versus Δa_{max} curves developed under the multiple unloading conditions for 2½Cr-Mo steel. The test result under monotonic loading condition for δ_a is presented as the scatterband in this figure. It is found that the multiple unloading $\delta_o R$-curve is also represented by both the blunting line and fracture resistance curve, and that corresponds fairly closely to the monotonic $\delta_o R$-curve shown by the scatterband. Multiple unloading δ_a values fluctuate within the range of monotonic δ_a values shown by the hatched band, and have a tendency to be identical to monotonic δ_a values in the wide range of Δa_{max}, although there is considerable scatter. It is considered that there is no significant difference in an empirical measure of stable crack growth resistance in terms of δ under monotonic and multiple unloading conditions as well as *JR*-curve.

Stable crack growth resistance $T_J(\Delta a_{max})$ and fracture toughness J_{Ic} are summarized in Table 3. The $T_J(\Delta a_{max})$ values are nearly constant regardless of side groove (SG) depth. There is no significant difference in $T_J(\Delta a_{max})$ val-

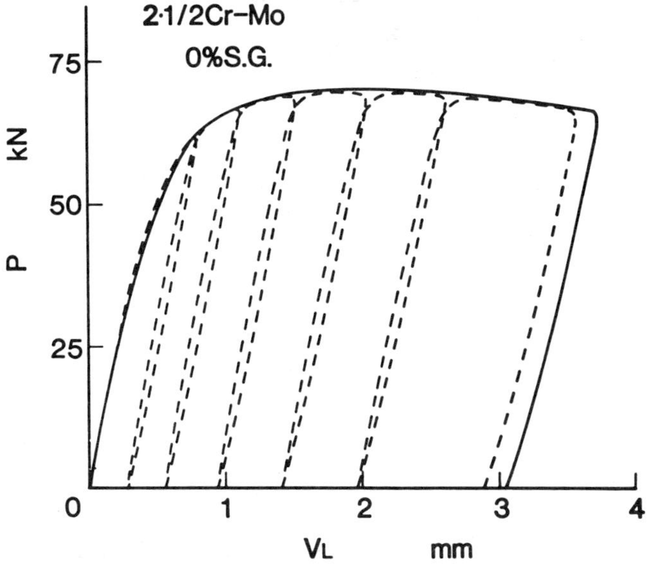

FIG. 10—*Comparison of* P *versus* V_L *curves under monotonic and multiple unloading conditions for* $2^1/_2$*Cr-Mo steel.*

ues under monotonic and multiple unloading conditions, since the multiple unloading *JR*-curve corresponds to the monotonic *JR*-curve. In addition, the J_{Ic} values determined according to the ultrasonic method are also independent of specimen size, crack length, and side groove depth within the valid specimen size requirements [*6,13,19*]. The J_{Ic} values show almost no influence of the multiple unloading. Comparison of elastic-plastic fracture resistances for $2^1/_2$Cr-Mo and A533B-1 steels shows that the material whose J_{Ic} value is large does not always have a large $T_J(\Delta a_{max})$. It is therefore necessary to evaluate the stable crack growth resistance as well as the elastic-plastic fracture toughness in order to assess the structural integrity.

Transition from Low Cycle Fatigue to Monotonic Stable Crack Growth Characteristics

Low cycle fatigue crack growth rates da/dN or Δa_{max} under the cyclic loading condition as a function of ΔJ are shown in Fig. 13 for $2^1/_2$ Cr-Mo steel. In this figure, the results from da/dN versus ΔK high cycle fatigue crack growth rate data were converted to a da/dN versus ΔJ format and also plotted for comparison. The dotted chain line represents the da/dN-ΔJ diagram converted from the da/dN-ΔK_{eff} relationship covering near-threshold fatigue crack growth rates [*23*]. The da/dN-ΔJ relationship gradually deviates from a linear relationship of $da/dN = C \cdot \Delta J$, where C is material constant, for

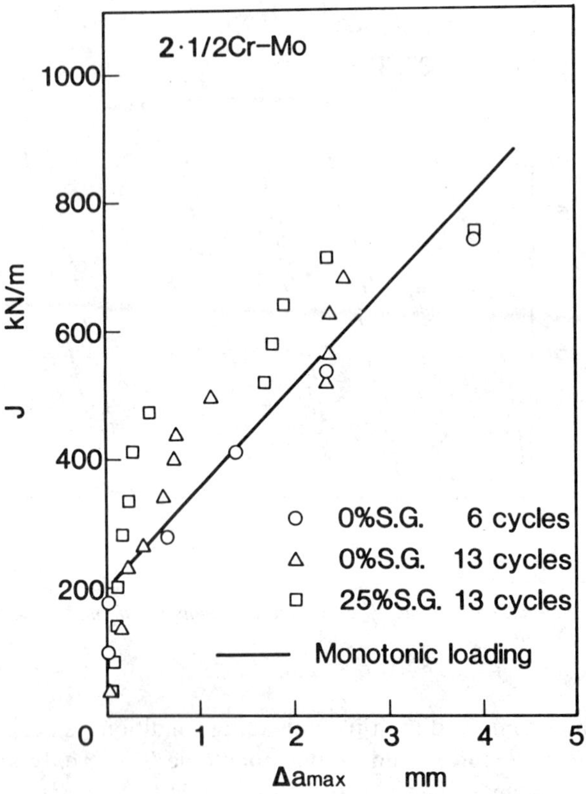

FIG. 11—*J versus* Δa_{max} *curves for 2½Cr-Mo steel under multiple unloading condition.*

high cycle fatigue crack growth characteristics under the predominantly linear elastic condition and connects smoothly to the Δa_{max}-ΔJ relationship under the cyclic loading condition. Beyond the J_{Ic} value, the Δa_{max}-ΔJ relationship gradually approaches the stable crack growth characteristics under the monotonic loading condition. The transition behavior also depends on the disappearance of crack closure phenomenon and the change of fracture micromechanism. In fact, recent test results [*24*] show that the low cycle fatigue crack growth rates at $R = -1.0$ using the center cracked tension (CCT) specimen still lie on the linear extrapolation of the da/dN-ΔJ relationship beyond the J_{Ic} value. To fully understand the transition region from low cycle fatigue to monotonic stable crack growth characteristics, many further tests with varying load ratio, loading or strain rate, and test specimen geometry will be required.

Conclusions

1. The maximum crack extension Δa_{max} at the center of specimen thickness under the plane-strain condition can nondestructively and continuously be

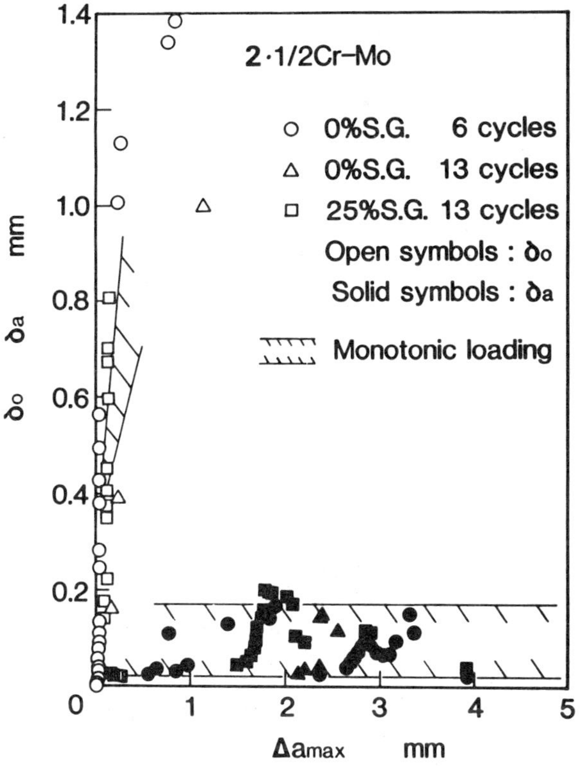

FIG. 12—δ_o, δ_a *versus* Δa_{max} *curves for* $2^1/_2$*Cr-Mo steel under multiple unloading condition.*

TABLE 3—*Stable crack growth resistance and fracture toughness.*

Material	Test Condition	Specimen Type[a]	$T_J(\Delta a_{max})$	J_{Ic}, kN/m
$2^1/_2$Cr-Mo	Monotonic loading	0% SG	103	174
		25% SG	88	169
	Cyclic loading	0% SG	87, 88	177
		25% SG	72	...
A533B-1	Monotonic loading	0% SG	57	180
		12.5% SG	49	152
		25% SG	52	197
		50% SG	50	194

[a]SG = side groove.

measured regardless of specimen size and side groove depth by the ultrasonic monitoring technique.

2. The crack tip opening displacement at the original fatigue precrack tip or the advancing crack tip (δ_o and δ_a) can nondestructively and continuously be measured at the center of specimen thickness under plane-strain condition

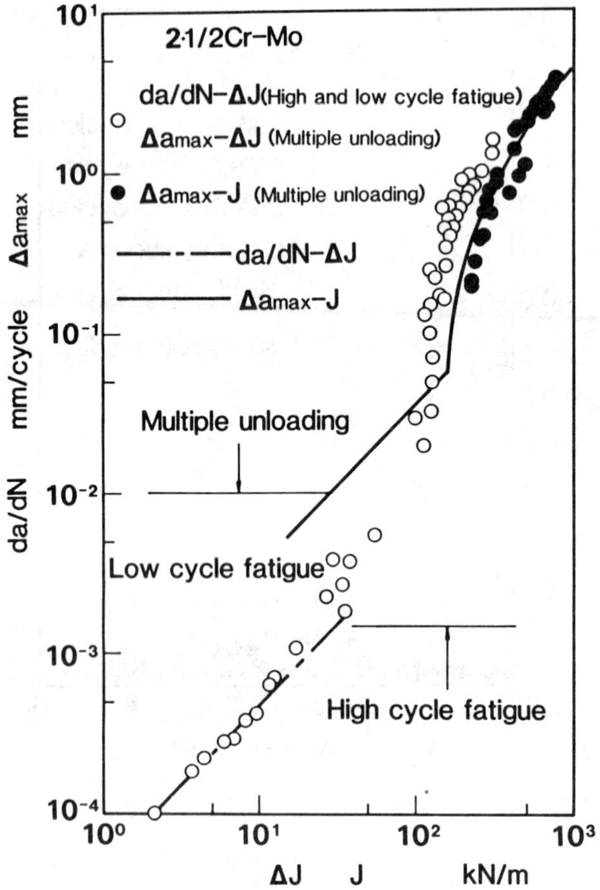

FIG. 13—*Transition from low cycle fatigue to monotonic stable crack growth characteristics.*

regardless of specimen size and side groove depth by the ultrasonic monitoring technique.

3. The measure of plane-strain elastic-plastic fracture resistance J-Δa_{max} curve can also be determined from a single small specimen according to the proposed procedure when crack tunneling occurs. The tearing moduli $T_J(\Delta a_{max})$ values calculated using the slope of the J-Δa_{max} curve are nearly constant regardless of specimen size and side groove depth. On the other hand, the tearing moduli $T_J(\Delta a_p)$ decrease as the percent of side grooves increases, and those for 25 and 50% side-grooved specimens are nearly consistent with $T_J(\Delta a_{max})$.

4. The material whose J_{Ic} value is large does not always have a large $T_J(\Delta a_{max})$ value. It is therefore necessary to evaluate the stable crack growth

resistance as well as the elastic-plastic fracture toughness in order to assess the structural integrity.

5. The Δa_{max}, δ_o and δ_a at the center of specimen thickness can nondestructively and continuously be measured under the cyclic loading condition by the ultrasonic monitoring technique. Multiple unloading R-curves can also be determined from a single specimen.

6. There is little difference between the monotonic and multiple unloading R-curves.

7. The transition behavior from low cycle fatigue to monotonic stable crack growth characteristics has been experimentally clarified.

Acknowledgments

We are pleased to acknowledge the considerable assistance of Mr. Ryuichi Kubota and Mr. Toshihiko Kusumoto, who contributed their efforts to the accomplishment of the experimental program.

References

[1] Wilhelm, D. P. and Ratwani, M. M., *Journal of Engineering Materials Technology, Transactions of ASME*, Vol. 100, 1978, pp. 416-420.

[2] Schwalbe, K. H., *Engineering Fracture Mechanics*, Vol. 11, 1979, pp. 331-342.

[3] Rhodes, D., Radon, J. C., and Culver, L. E., *Fatigue of Engineering Materials and Structures*, Vol. 4, No. 1, 1981, pp. 49-63.

[4] Kobayashi, H., Nakamura, H., and Nakazawa, H., *Mechanics of Fatigue*, American Society of Mechanical Engineers, Vol. 47, 1981, pp. 133-143.

[5] Landes, J. D. and McCabe, D. E. in *Elastic-Plastic Fracture: Second Symposium, Vol. II—Fracture Resistance Curves and Engineering Applications, ASTM STP 803*, American Society for Testing and Materials, Philadelphia, 1983, pp. II-723-II-738.

[6] Hirano, K., Kubota, R., Kobayashi, H., and Nakazawa, H., *Transactions of the Japan Society of Mechanical Engineers*, Vol. 49, No. 448, 1983, pp. 1556-1564 (in Japanese).

[7] Underwood, J. H., Winters, D. C., and Kendall, D. P., *The Detection and Measurements of Cracks*, The Welding Institute, Cambridge, 1976, pp. 31-38.

[8] Klima, S. J., Fisher, D. M., and Buzzard, R. J., *Journal of Testing and Evaluation*, Vol. 4, No. 6, 1976, pp. 397-404.

[9] Underwood, J. H., *Experimental Mechanics*, Vol. 18, No. 9, 1978, pp. 350-355.

[10] Hirano, K., Kobayashi, H., and Nakazawa, H. in *Proceedings*, 23rd Japan Congress on Materials Research, 1980, pp. 357-363.

[11] Hirano, K., Kobayashi, H., and Nakazawa, H. in *Proceedings*, 3rd Pan-Pacific Conference for Nondestructive Testing, 1980, pp. 145-153.

[12] Hirano, K., Kobayashi, H., and Nakazawa, H., in *Proceedings*, 24th Japan Congress on Materials Research, 1981, pp. 75-80, 94-98.

[13] Hirano, K., Kobayashi, H., and Nakazawa, H., *Journal of the Society of Materials Science (Japan)*, Vol. 32, No. 360, 1983, pp. 1018-1024 (in Japanese).

[14] Nakazawa, H. and Hirano, K., *Japanese Journal of Applied Physics*, Vol. 22, Supplement 22-3, 1983, pp. 8-13.

[15] Ernst, H. A., Paris, P. C., and Landes, J. D. in *Fracture Mechanics (13th Conference), ASTM STP 743*, American Society for Testing and Materials, Philadelphia, 1981, pp. 476-502.

[16] Garwood, S. J., Robinson, J. N., and Turner, C. E., *International Journal of Fracture*, Vol. 11, No. 3, 1975, pp. 528-530.

[*17*] Clarke, G. A. and Landes, J. D., *Journal of Testing and Evaluation*, Vol. 7, No. 5, 1979, pp. 264–269.
[*18*] Dowling, N. E. and Begley, J. A. in *Mechanics of Crack Growth, ASTM STP 590*, American Society for Testing and Materials, Philadelphia, 1976, pp. 82–103.
[*19*] Hirano, K., Kobayashi, H., and Nakazawa, H., *Journal of Testing and Evaluation*, Vol. 13, No. 5, 1985, pp. 356–362.
[*20*] Paris, P. C., Tada, H., Zahoor, A., and Ernst, H. *in Elastic-Plastic Fracture, ASTM STP 668*, American Society for Testing and Materials, Philadelphia, 1979, pp. 5–36.
[*21*] Shih, C. F., deLorenzi, H. G., and Andrews, W. R. in *Elastic-Plastic Fracture, ASTM STP 668*, American Society for Testing and Materials, Philadelphia, 1979, pp. 65–120.
[*22*] Garwood, S. J. and Turner, C. E., *International Journal of Fracture*, Vol. 14, 1978, pp. R195–R198.
[*23*] Hirano, K., Ishii, A., Kobayashi, H., and Nakazawa, H., Preprints for the Japan Society of Mechanical Engineers, No. 810-11, 1981, pp. 202–204 (in Japanese).
[*24*] Hirano, K., Iwasaki, K., Kobayashi, H., and Nakazawa, H., unpublished work.

Michael G. Vassilaros,[1] *Richard A. Hays,*[1] *and John P. Gudas*[1]

J-Resistance Curve Analysis for ASTM A106 Steel 8-Inch-Diameter Pipe and Compact Specimens

REFERENCE: Vassilaros, M. G., Hays, R. A., and Gudas, J. P., "***J*-Resistance Curve Analysis for ASTM A106 Steel 8-Inch-Diameter Pipe and Compact Specimens,**" *Fracture Mechanics: Seventeenth Volume, ASTM STP 905*, J. H. Underwood, R. Chait, C. W. Smith, D. P. Wilhem, W. A. Andrews, and J. C. Newman, Eds., American Society for Testing and Materials, Philadelphia, 1986, pp. 435–453.

ABSTRACT: An experimental investigation was performed to evaluate the applicability of using *J*-integral resistance (*J-R*) curves from laboratory specimens to describe the fracture behavior of circumferentially cracked 8 in. diameter ASTM A106 steel pipe. The approach of the study was to perform *J*-integral tests on 48 in. long, 8 in. diameter, Schedule 80 ASTM A106 steel pipes and compare them with *J*-integral tests performed on ½T, 1T plan, and 2T plan compact specimens machined from the pipe wall. The pipes were tested in four-point bending and were monitored for load, crack opening displacement, bend angle, and electrical potential drop. Elastic compliance and direct-current potential drop (DCPD) techniques were used simultaneously to predict crack length during the tests. The compact tension specimens were tested in a conventional screw-driven testing machine while being monitored for load, load-line deflection, and electrical potential drop. Elastic compliance and DCPD techniques were also used for crack length predictions during these tests.

The results of this series of *J*-integral tests indicate that similar *J-R* curves result from the two types of specimens. However, small specimen tests cannot easily be used to predict full *J-R* curve behavior for larger, full-sized specimens owing to the modest amount of crack extension available in laboratory-sized specimens.

KEY WORDS: fracture mechanics, ASTM A106 steel, pipe bend specimen, compact tension specimen, *J*-integral, *J*-resistance curve, elastic compliance, direct-current potential drop

The use of *J*-integral fracture mechanics concepts to characterize the ductile fracture properties of engineering materials has been shown to be valid and widely applicable [*1*]. Development of single-specimen *J*-integral test

[1]Research Metallurgist, Materials Engineer, and Branch Head, Fatigue and Fracture Branch, respectively, David Taylor Naval Ship R&D Center, Bethesda, MD 20084.

methodology [*2,3*] has broadened the application of this concept and has led to increased interest in the *J*-integral resistance curve. The theoretical foundation for the use of the *J*-integral beyond crack initiation was discussed by Hutchinson and Paris [*4*] and conditions for *J*-controlled crack growth were defined. Subsequently, analyses of the stability of crack growth in the *J*-controlled regime were conducted by Paris and co-workers [*5,6*]. Included in these investigations were the development of tearing instability relationships under fully plastic conditions for several specimen geometries and the introduction of the tearing modulus parameter which incorporates the slope of the *J*-resistance curve.

The objective of this research was to investigate the applicability of using *J*-resistance (*J-R*) curves from laboratory (compact) specimens to describe the *J*-integral fracture behavior of circumferentially precracked 8 in. diameter ASTM A106 steel pipe. The investigation was divided into two parts. The first phase involved *J*-integral characterization of 8 in. diameter Schedule 80 ASTM A106 steel pipe specimens subjected to four-point bend loading. The second phase of the investigation was to perform *J*-integral tests on ½T, 1T plan, and 2T plan compact tension specimens machined from the pipe wall. Results of these tests were then compared with the results from the tests on the full-sized pipe specimens.

Materials

The alloy used in this investigation was ASTM A106 Grade B steel. It was supplied in the form of four 6.1 m (20 ft) lengths of 219 mm (8.625 in.) outside diameter Schedule 80 pipe with a typical wall thickness of 14 mm (0.54 in.). The pipe was manufactured by the United States Steel Corporation and conformed to the requirements of ASME Section III Sub-Article NCA 3800. The nominal chemical composition of the steel is given in Table 1.

Twelve tensile tests were performed with three specimens from each of the lengths of pipe. The specimens were machined with the longitudinal axis of the specimen parallel to the longitudinal axis of the pipe which corresponds to the orientation of tensile properties which govern crack extension in the circumferentially cracked pipes. The 9.1 mm (0.357 in.) diameter tensile specimens with 35.6 mm (1.40 in.) gage lengths were tested in accordance with ASTM Methods of Tension Testing of Metallic Materials (E 8). The results of these tests are given in Table 2.

TABLE 1—*Chemical composition (wt%) of ASTM A106 GrB steel pipe.*

C	Mn	P	S	Si	Fe
0.23	0.81	0.0062	0.013	0.164	bal.

TABLE 2—*Room temperature mechanical properties of ASTM A106 GrB steel pipe.*

	Yield Strength, MPa (ksi)	Ultimate Tensile Strength, MPa (ksi)	Elongation % in 2 in.	Reduction in Area, % (L = 4*D*)
Average of 12 tests	28 (41.1)	495 (71.7)	38	64.5
Standard deviation =	0.921	1.62	3.8	0.522

Charpy V-notch specimens were machined from several sections of pipe and oriented with the plane of the notch perpendicular to the longitudinal axis of the pipe and with crack growth occurring in the circumferential direction. This orientation (LC) was the same as that used for the compact specimens and the full-scale pipe tests. ASTM Methods for Notched Bar Impact Testing of Metallic Materials (E 23) was followed when testing the Charpy specimens. An upper shelf Charpy impact toughness value on the order of 149 J (110 ft-lb) was attained at approximately 38°C (100°F).

Compact Specimen *J-R* Curve Testing

Three geometries of compact tension specimens were produced from the pipe. These included ½T, 1T plan, and 2T plan geometries (i.e., the 1T and 2T plan specimens had the same dimensions in the side plane as full-thickness 1T and 2T specimens) as shown in Figs. 1*a* and 1*b*. Compact specimens ranged from 10 mm (0.4 in.) to 13 mm (0.5 in.) thick. The ½T and 1T plan specimens were machined directly from blanks cut from the pipe. The 2T plan specimen blanks cut from the pipe wall were pressed flat before machining. This flattening procedure produced a prestrain of +6% or −6% across the specimen thickness. All specimens were machined such that crack growth occurred in the LC orientation.

J-integral fracture toughness tests performed on the compact tension specimens utilized two separate techniques for measuring crack extension. The first, used on all three geometries of compact specimens, was the elastic compliance technique described by Joyce and Gudas [*3*]. The second method utilized the direct-current potential drop (DCPD) technique to measure the crack extension during the *J*-integral fracture toughness tests. This method, described by Vassilaros and Hackett [*7*], monitored the internal resistance (IR) drop across the notch face of the compact specimen subjected to a constant direct current. The technique partitions the changes in electrical resistance into components resulting from plasticity and crack blunting, and that resulting from a change in crack length. The DCPD was used only on the 2T plan specimens with the data being gathered concurrently with elastic compliance data.

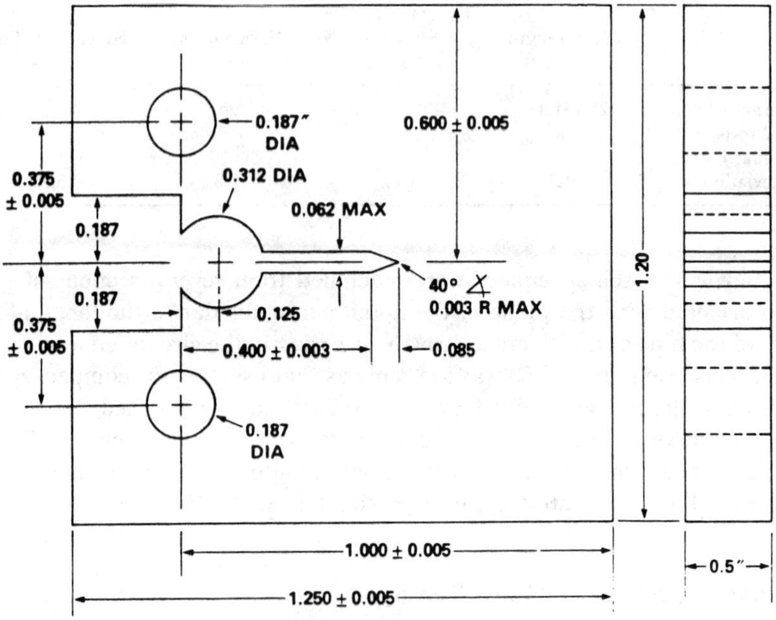

FIG. 1*a*—½*T compact specimen.*

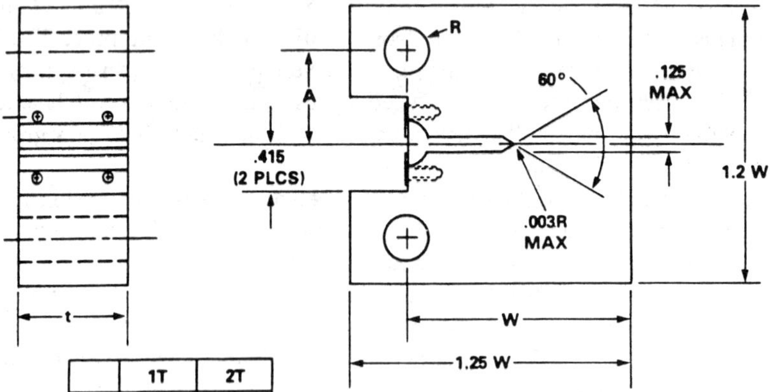

	1T	2T
W	2.00	4.00
A	0.80	1.30
R	0.375	1.00
t	0.475	0.470

FIG. 1*b*—*1T and 2T plan compact specimens.*

The J-integral values calculated for the compact specimen test results used the deformation theory formulation with the crack growth correction expression by Ernst et al [8]:

$$J_{(i+1)} = \left[J_i + \frac{\eta}{b_i} \times \frac{A_{i,(i+1)}}{B_N}\right] \times \left[1 - \frac{\gamma}{b_i}\,(a_{(i+1)} - a_i)\right] \tag{1}$$

where

η = 2 + (0.522) b/W for a compact specimen,
W = specimen width,
γ = 1 + (0.76)b/W,
b_i = instantaneous length of remaining ligament,
B_N = minimum specimen thickness,
a_i = instantaneous crack length, and
$A_{i,(i+1)}$ = area under the load versus load-line displacement record between lines of constant displacement at points i and $i+1$.

All specimen preparation and measurement procedures detailed in ASTM Test for J_{Ic}, A Measure of Fracture Toughness (E 813) were followed in this phase of testing. Fatigue precracks were grown to 0.65 a/W in all compact specimen geometries.

J-initiation was calculated using a modification of ASTM E 813. A least-squares linear regression analysis was performed on the first four valid points closest to the blunting line and then again after the addition of each qualified point representing the next increment of crack extension through the ASTM E 813 prescribed region of crack growth. The specific J-initiation value selected for each test was the point of intersection between the blunting line and the fit line from the set of data corresponding to the first peak in correlation of the least squares fit as a function of crack length (i.e., points were added as long as the addition increased the correlation of the least squares fit). The slope of data in the entire ASTM E 813 range was used to calculate the tearing modulus for each specimen test.

The compact specimens tested had thicknesses ranging from 10 to 13 mm (0.4 to 0.5 in.). This range of thicknesses imposes an upper bound valid J-integral value on the order of 155 kJ/mm^2 (900 in.-lb/in.^{2}) in order to maintain plane strain conditions in accordance with ASTM E 813. These conditions and the extensive shear apparent on the fracture surface indicate that all the fracture occurred under non-plane strain conditions.

J-R Curve Testing of Pipe Specimens

The J-integral fracture toughness tests were performed with ASTM A106 Grade B steel pipe configured in four-point bending. Each pipe specimen had an overall length of 1219 mm (48 in.). Figure 2 presents a schematic view of

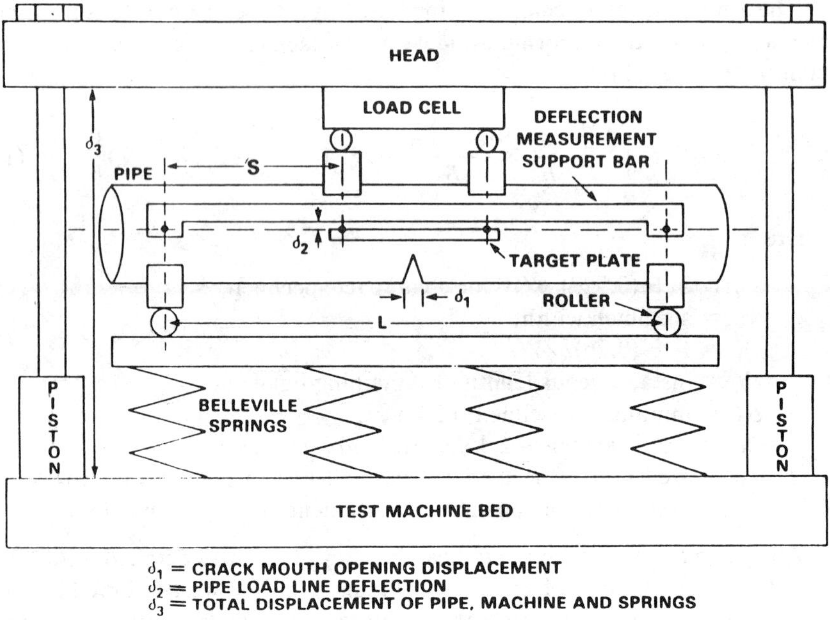

FIG. 2—*Schematic drawing of pipe test arrangement.*

the arrangement of the test apparatus as well as points of measurement included in the test. For all tests, the center span length was 305 mm (12 in.), while the moment arm length was either 381 or 457 mm (15 or 18 in.). Initial total fatigue precrack lengths ($2a$) measured on the mean circumference of the pipe ranged from 135 to 211 mm (5.3 to 8.3 in.), or 21 to 33% of the total circumference. Measurements taken during the tests included crack mouth opening displacement (δ_1), deflection of neutral axis of the pipe (δ_2), total system deflection (δ_3), and load (see Fig. 2). The DCPD crack length measurement current inputs were located at the extreme ends of the pipe; the potential output leads were located 51 mm (2 in.) apart and centered about the crack plane at the crack mouth. The pipes were insulated at the loading saddle blocks to avoid electrical grounding.

All precracking and testing were performed in a 1 300 000 N (300 000 lb) servo-hydraulic test machine in the deflection-controlled mode. Circumferentially oriented notches with 0.08 mm (0.003 in.) radius tips were machined into the specimens. Fatigue precracks were grown a minimum of 5 mm (0.2 in.) from the tip of each machined notched at a frequency of 0.17 Hz. The maximum ΔK applied during the final stages of precracking was 33 MPa$\sqrt{m}$ (30 ksi$\sqrt{in.}$). Strip heaters were placed around the test section to maintain a temperature in the range of 52 to 65°C (125 to 150°F), ensuring upper shelf behavior as measured in the Charpy impact toughness tests.

The J-integral tests performed on the full-sized pipe specimens utilized both the elastic compliance and DCPD techniques to measure crack extension. The elastic compliance technique used the slope of the crack mouth opening displacement measurements obtained during small elastic unloadings (15 to 20% of maximum load) performed during the tests. The compliance expression experimentally determined by Joyce [9] using 102 mm (4 in.) diameter aluminum 6061 pipe was modified for these tests using the elastic compliance measurements of initial fatigue cracks which were optically measured after the test. The DCPD technique for crack length measurement was similar to the compact specimen technique described by Vassilaros and Hackett [7]. However, the relationship between crack length and potential drop used for the pipe was obtained by fitting an exponential equation to the data published by Wilkowski and Maxey [10].

The J-integral values for these pipe tests were evaluated using two different expressions. The first expression was published by Tada et al [11]. This expression uses the pipe bend angle and the material flow stress to evaluate J:

$$J = \sigma_f R\phi \,[\sin(\theta/2) + \cos\theta] + K^2/E \qquad (2)$$

where

σ_f = flow stress,
R = mean radius,
θ = 1/2 total crack angle,
ϕ = total bend angle,
K = applied stress intensity (elastically calculated), and
E = elastic modulus.

The bend angle was calculated using the measured load-line deflection taken near the neutral axis divided by the span of the moment arm. This expression is an approximation to the actual J-integral assuming elastic-perfectly plastic material behavior without any crack growth correction.

The second J-integral expression was published by Zahoor and Kanninen [12]. This expression uses the actual bending moment and load-line deflection and employs a crack growth correction. The expression is

$$J = K^2/E + \beta \int_{\delta_0}^{\delta} (2P)\, d\delta + \int_{\phi_0}^{\phi} \gamma\, J\, d\phi \qquad (3)$$

where

K = stress intensity factor,
E = elastic modulus,
β = $-h'(\phi)/Rt\, h(\phi)$,
$2P$ = total bending load,
δ = plastic load-line deflection,

$\gamma = h''(\phi)/h'(\phi)$,
R = radius,
t = thickness,
ϕ = total crack angle, and
$h(\phi) = [\cos(\phi/4) - \frac{1}{2}\sin(\phi/2)]$.

The *J*-integral crack initiation values for the pipe bend tests could not be calculated using ASTM E 813 because of the insufficient number of data points in the required crack extension range. The crack initiation values reported for this study were obtained by calculating the *J*-integral value at the intersection of the blunting line ($J = 2\Delta a\sigma_f$) and a power law approximation to the *J-R* curve data beyond the blunting line as described by Vassilaros et al [*13*].

Compact Specimen Fracture Toughness Test Results

Figures 3 and 4 are *J-R* curves from 10 mm (0.4 in.) thick ½T and 12 mm (0.474 in.) thick 1T plan compact specimens, respectively, which were produced using the elastic compliance technique. These curves indicate that the A106 steel had a crack initiation toughness ranging from 363 to 694 kJ/m^2 (2073 to 3962 in.-lb/in.^{2}) and a high residual toughness as measured by the tearing modulus. Values of tearing modulus ranged from 222 to 396. Individual values for all tests are given in Table 3. The large scatter in the *J*-initiation

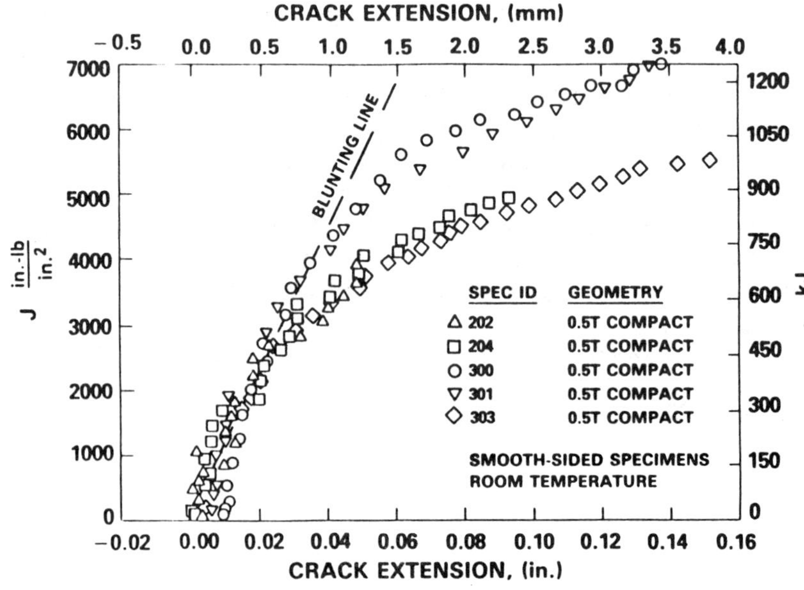

FIG. 3—J-R *curves for ½T compact specimens of ASTM A106 steel.*

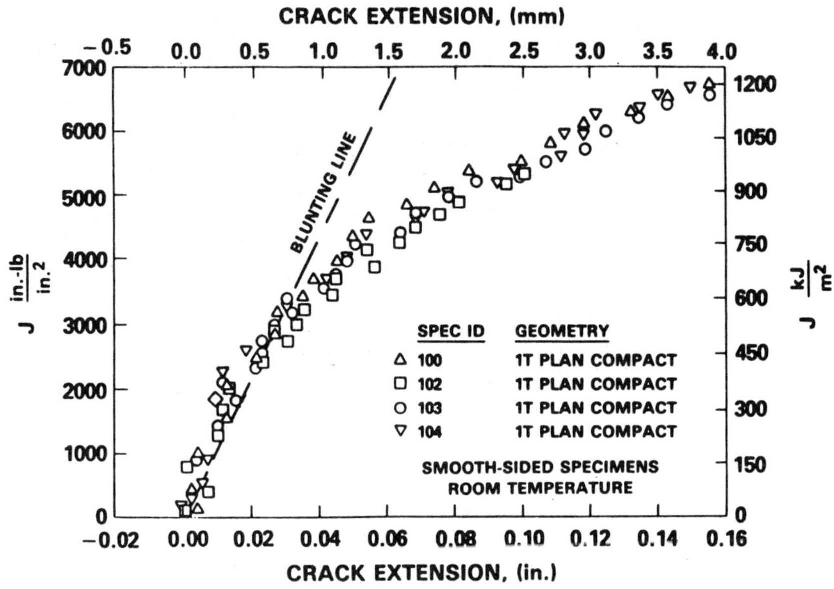

FIG. 4—J-R *curves for 1T plan compact specimens of ASTM A106 steel.*

values was due in part to normal material variability and additionally to the difficulties in evaluating *J-R* curves with high slopes. When *J-R* curves have high slopes, small variations in measured slopes can produce large variations in *J*-initiation values. The average *J*-initiation and tearing modulus values for the ½T plan specimens were 510 kJ/m² (2912 in.-lb/in.²) and respectively, while those for the 1T plan specimens were 558 kJ/m² (3185 in.-lb/in.²) and 361 respectively. The apparent differences between the two sizes of specimens are most probably a result of the difference in specimen thickness, with the thinner specimens exhibiting both a higher average initiation toughness and a higher average tearing modulus. The variation can be minimized by using side-grooved compact specimens, but the *J-R* curve would not be appropriate for modelling the *J-R* curve from pipe specimens. Another advantage of side-grooving is the tendency of the crack to resist tunneling and thus to minimize the error between the final crack length as predicted by the elastic compliance technique and the optically measured final crack length. For plane-sided compact specimens tested in this investigation, errors in predicted final crack length were as great as 50% (Table 3).

J-integral fracture toughness tests were also performed on 13 mm (0.5 in.) thick 2T plan specimens to increase the range of crack extension covered by the *J-R* curve. These specimens were tested using the elastic compliance technique and the DCPD technique simultaneously in the hope that the DCPD technique would eliminate the errors in crack extension measurement shown

TABLE 3—*J-initiation and tearing modulus from ½T and 1T compact specimen tests using the elastic compliance technique.*

Specimen No.	0.5-in.-Thick Compact Specimen Geometry	Test Tempera-ture	*J*-Initiation	Crack Extension Range to Calculated *J*-Initiation	Tearing Modulus	Crack Extension		
						Predicted	Measured	
			kJ/m² (in.lb/in.²)	mm (in.)		mm (in.)	mm (in.)	%
202	½T	RT	363 (2073)	1.22 (0.048)	263	1.24 (0.049)	1.68 (0.066)	50
204	½T	RT	470 (2684)	1.30 (0.051)	285	2.36 (0.093)	3.02 (0.119)	28
300	½T	RT	694 (3962)	1.55 (0.061)	222	4.90 (0.196)	not measured	...
301	½T	RT	660 (3220)	1.45 (0.057)	230	5.31 (0.209)	6.91 (0.272)	30
303	½T	RT	459 (2623)	1.47 (0.058)	260	3.86 (0.152)	4.45 (0.175)	15
100	1T plan	RT	632 (3606)	1.88 (0.074)	355	3.86 (0.152)	5.69 (0.224)	47
102	1T plan	RT	445 (2543)	1.91 (0.075)	396	2.44 (0.096)	3.35 (0.132)	37
103	1T plan	RT	551 (3149)	1.98 (0.078)	352	4.14 (0.163)	4.72 (0.186)	12
104	1T plan	RT	578 (3299)	1.80 (0.071)	340	3.78 (0.149)	3.25 (0.128)	14

in Table 3. *J-R* curves produced from testing two 2T plan compact specimens are shown in Fig. 5. The DCPD techniques produced resistance curves with lower initiation values and lower slopes than the elastic compliance technique. The lower initiation values may have been due to the higher sensitivity in measuring tunnelled cracks with DCPD, while the lower slope was due to the curve extending to the final measured crack length and not underestimating the final crack length as was the case with the elastic compliance method. The calculated *J* levels at crack initiation and tearing modulus results for these 2T plan specimens fall within the range of those calculated for the ½T and 1T plan specimens (Table 4).

Comparison of the full *J-R* curves of these three different specimen geometries does indicate some differences in fracture behavior. The ½T and 1T plan compact specimen *J-R* curves appear similar over the entire range of crack extension (Fig. 6). As is seen in Fig. 7, however, the divergence of the 1T plan and 2T plan compact specimen *J-R* curves is apparent. This difference in the curves begins at about 1.3 mm (0.5 in.) of crack extension and is manifested as a lower resistance curve from the 2T plan specimens. The lower *J-R* curve is most probably due to the prestrain introduced in the 2T plan specimen during flattening. Similar effects have been reported for mild steel [*14*] and HY-80 [*15*]. Therefore the extended crack growth portions of the

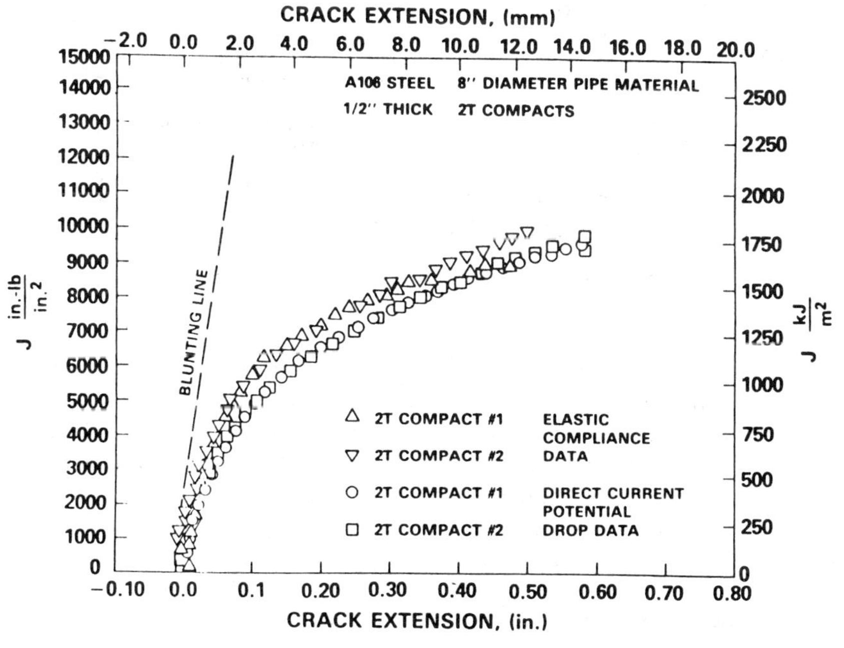

FIG. 5—J-R *curves for 2T plan compact specimens of ASTM A106 steel.*

TABLE 4—*J-initiation and tearing modulus results from compact specimens test using d-c potential drop technique.*

Specimen No.	0.5-in.-Thick Compact Specimen Geometry	Test Temperature	*J*-Initiation, kJ/m² (in.lb/in.²)	Crack Extension Range to Calculated *J*-Initiation, mm (in.)	Tearing Modulus
1a-PD	2T plan	RT	278 (1590)	1.83 (0.072)	291
2a-PD	2T plan	RT	385 (2197)	2.18 (0.086)	234
7a-PD	2T plan	125°F	391 (2234)	1.85 (0.073)	299
8a-PD	2T plan	125°F	500 (2860)	1.98 (0.078)	257

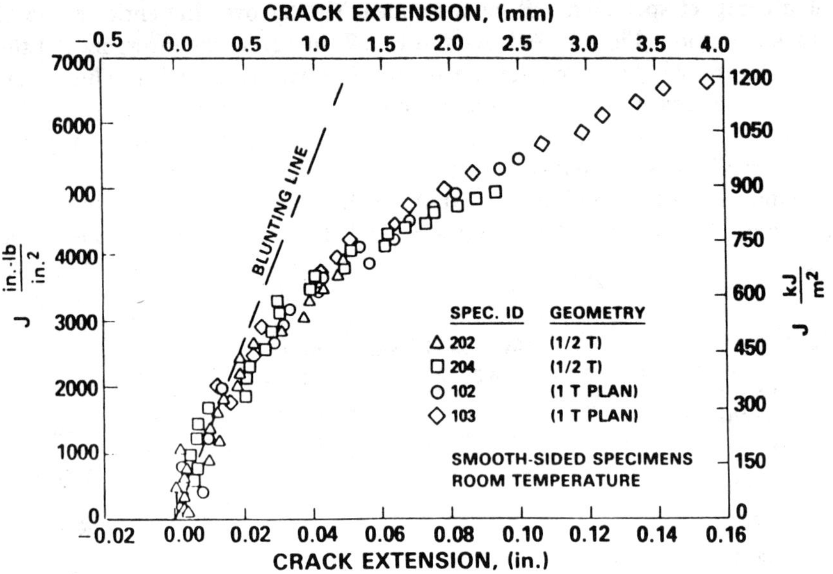

FIG. 6—J-R *curves from elastic compliance for 1/2T and 1T plan compact specimens.*

J-R curves from these specimens are not representative of the pipe material but instead underestimate the ductile fracture properties.

All the above results are from *J*-integral toughness tests that were conducted at room temperature. Two tests were conducted on compact specimens at the upper shelf temperature of 52°C (125°F). The results of these tests indicated that no change in ductile fracture properties resulted from the modest temperature increase of 28°C (50°F).

Pipe Specimen Fracture Toughness Test Results

J-integral tests were performed on nine ASTM A106 Grade B steel pipes loaded in four-point bending at 125°F using elastic compliance and DCPD

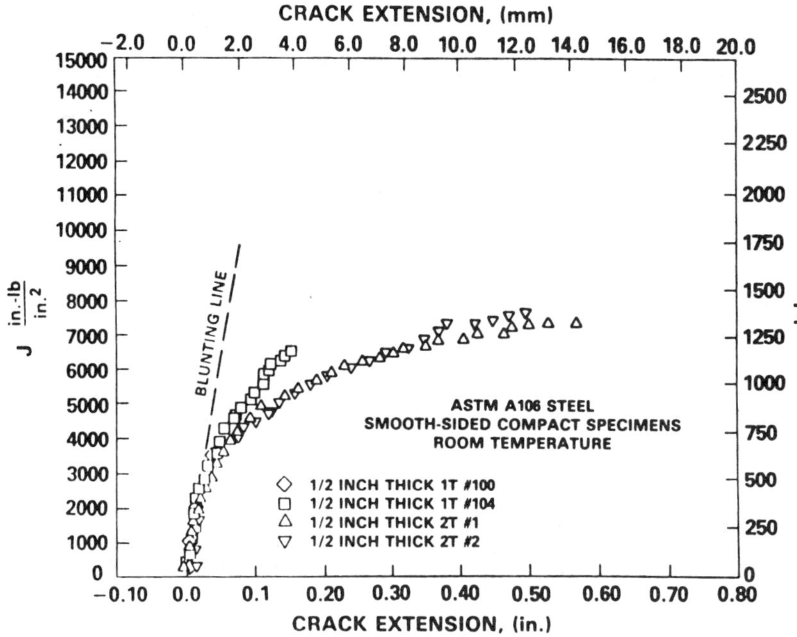

FIG. 7—J-R *curves from elastic compliance for 13 mm (½ in.) thick 1T and 2T plan compact specimens.*

techniques simultaneously. The results of these tests are shown in Figs. 8 and 9 which are plots of *J*-Zahoor (Eq 3) versus crack extension using elastic compliance and DCPD techniques, respectively. Had the figures been produced using *J*-Tada (Eq 2), they would have appeared very similar (within 5% of *J* value). An example of this similarity is given in Fig. 10 for Pipe 7. There are several reasons for the good agreement between the two formulations. By examining Eqs 2 and 3, it is apparent that both equations share the same elastic term. Also, because of the choice of short initial crack lengths (total crack angles ranging from 75 to 118°) in the investigation, the crack growth term (γ) in Eq 3 was small for these tests. Returning to Figs. 8 and 9, it is seen that the two sets of *J-R* curves follow the same trends, with the curves generated using the DCPD technique displaying less scatter than those generated using the elastic compliance technique. This is due to the use of measured initial and final crack lengths in the generation of DCPD crack extension data, whereas the unloading compliance technique cannot correct for errors in crack length.

The *J*-initiation values calculated using the above method range from 358 to 770 kJ/m² (2042 to 4397 in.-lb/in.²) and are listed in Table 5. Although the values have a wide range, they do agree with the *J*-initiation values obtained from elastic compliance tests performed on plane-sided ½T compact specimens. In fact, the average *J*-initiation values of 510 kJ/m² (2912 in.-lb/in.²)

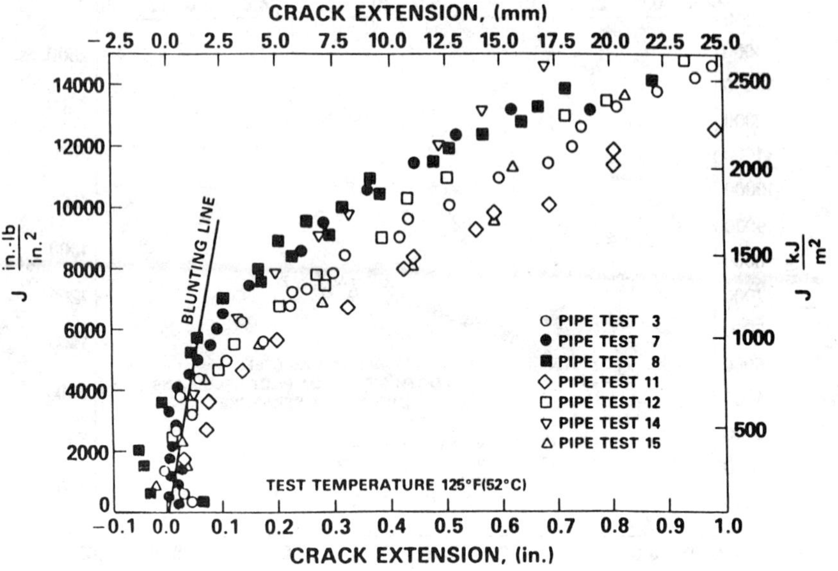

FIG. 8—J-R *curves from elastic compliance for four-point bend tests of ASTM A106 steel pipe specimens.*

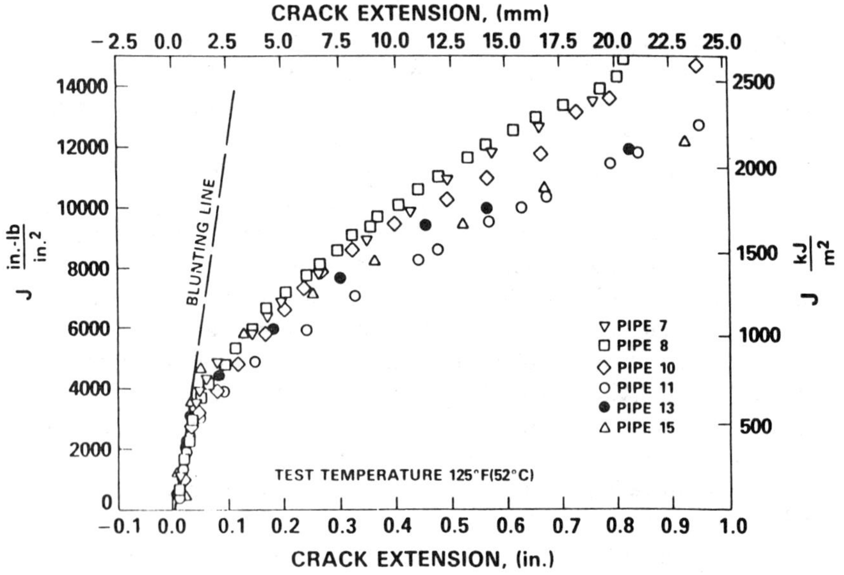

FIG. 9—J-R *curves from d-c potential drop for four-point bend tests of ASTM A106 steel pipe specimens.*

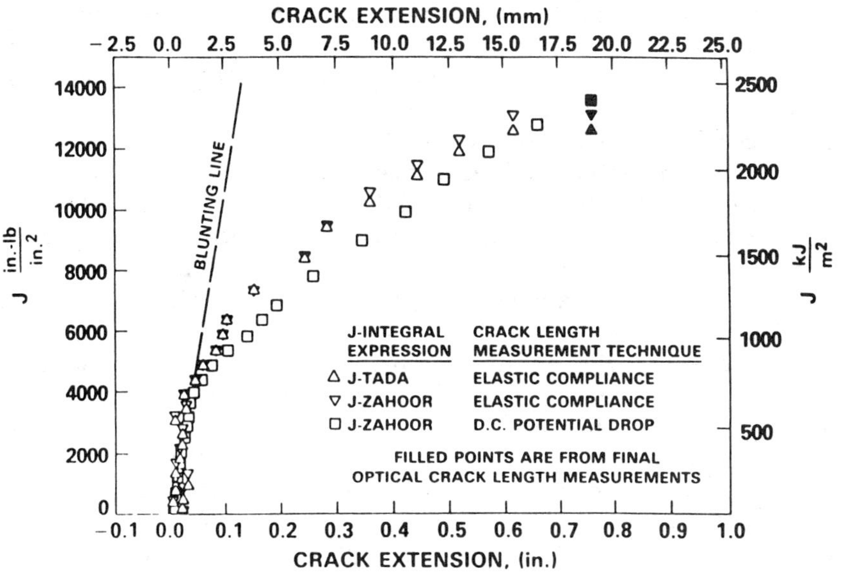

FIG. 10—J-R *curves for Pipe Test #7.*

TABLE 5—*J-initiation values from pipe tests.*

	J-Initiation, kJ/m² (in.lb/in.²)		Final Crack Extension Measurements		
Pipe Test No.	Elastic Compliance (Fig. 12)	DCPD (Fig. 13)	Elastic Compliance, mm (in.)	Optically Measured, mm (in.)	% Error
3	516 (2947)	...	28.0 (1.101)	35.8 (1.41)	28
7	769 (4397)	559 (3197)	14.6 (0.576)	19.2 (0.757)	31
8	697 (3985)	514 (2940)	18.7 (0.738)	20.4 (0.803)	9
10	...	443 (2530)	18.6 (0.733)	25.3 (0.995)	35
11	357 (2042)	411 (2349)	28.4 (1.120)	25.8 (1.017)	9
12	446 (2550)	...	29.2 (1.150)	36.2 (1.427)	24
13	...	503 (2873)	...	[a]	...
14	679 (3880)	...	...	[a]	...
15	437 (2496)	722 (4125)		[a]	...
Average	557 (3185)	525 (3002)	...	...	...

[a]No optical measurement.

for ½T compact specimens and 551 kJ/m² (3149 in.-lb/in.²) for 1T compact specimens agree very well with the average *J*-initiation value of 558 kJ/m² (3185 in.-lb/in.²) from the pipe tests.

In addition to the *J*-initiation values, the *J-R* curve behavior of ½T and 1T plan compact specimens appears similar to the *J-R* curve behavior of the pipe bend tests. When examining *J-R* curves which display an underestimation of crack extension, the real *J-R* curve behavior must be kept in mind by using the optically measured final crack length as a guide. Figure 11 shows *J-R* curves which represent the range of curves from the pipe specimen tests and the range of curves from the plane-sided ½T compact specimens. Although it appears from inspection of this figure that these compact specimens overpredict the *J*-integral fracture toughness of the ASTM A106 steel pipe, the final measured crack lengths indicate that the actual specimen behavior was similar to the behavior of the material in the pipe test. It must also be noted that the crack extension measured during the pipe test was also in error from +9% to −35% (Table 5).

Representative *J-R* curves from the 10 mm (0.4 in.) thick, plane-sided 1T plan compact specimens and the pipe specimens are shown in Fig. 12. Here again, the fracture resistance behavior of the laboratory compact specimens and that of the pipe specimens appear similar. However, the limited amount of crack extension attained in testing of the compact specimens precludes good comparison of the full *J-R* curves. The *J-R* curves from the 2T plan compact specimens display over 12.7 mm (0.50 in.) of crack extension and provide comparison of *J-R* curves from the pipe specimens over a larger range (Fig. 13). Figure 13 shows that the curves from the 2T plan compact specimens have a lower average initiation value and slope. The average initiation value of the 2T plan compacts was 489 kJ/m² (2795 in.-lb/in.²), which was

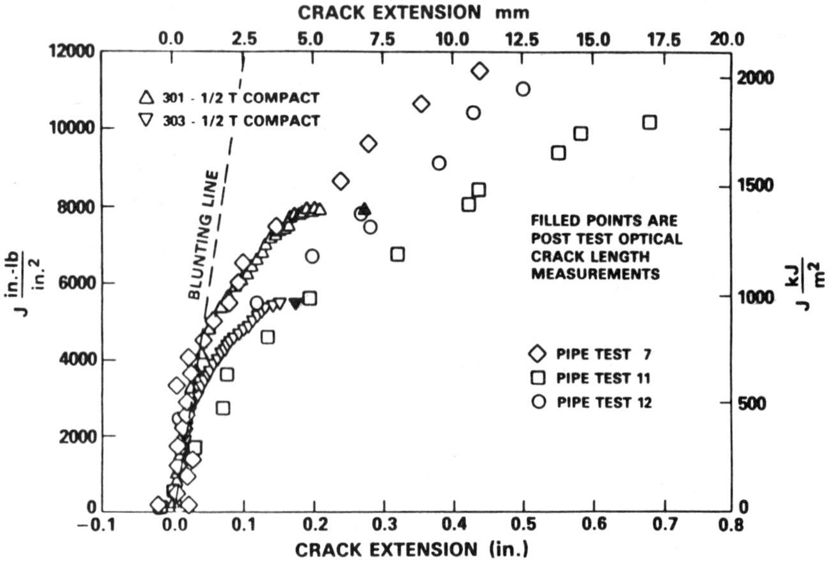

FIG. 11—J-R*curves for ½T compact specimens and pipe bend specimens of ASTM A106 steel.*

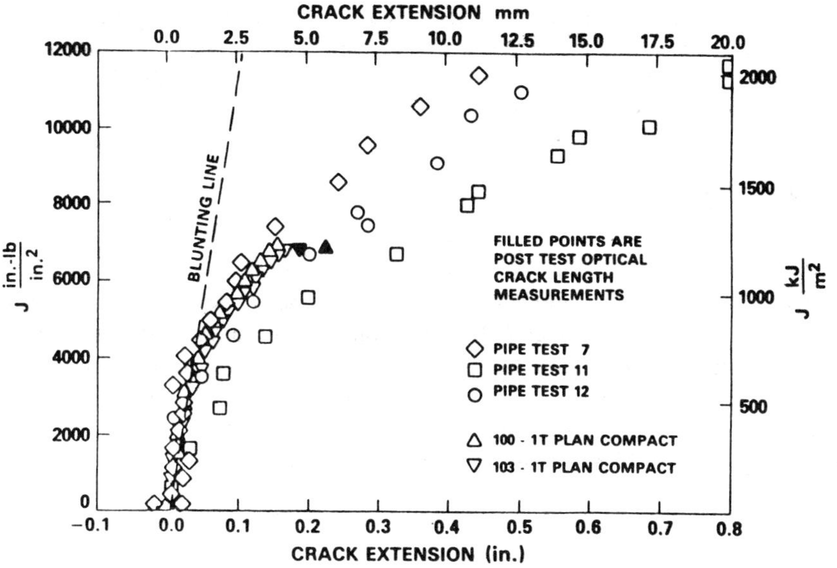

FIG. 12—J-R *curves for 1T plan compact specimens and pipe bend specimens of ASTM A106 steel.*

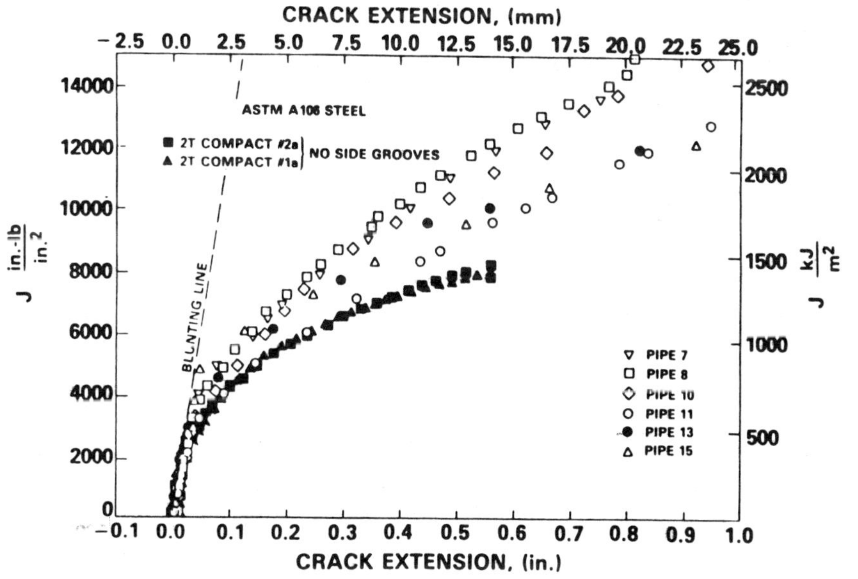

FIG. 13—J-R *curves for 2T plan compact specimens and pipe bend specimens of ASTM A106 steel.*

14% lower than that of the pipe specimens. The entire resistance curves of the 2T plan compact specimens fall below most of those from the pipe tests when comparing results from elastic compliance or DCPD. Inspection of tested 2T plan compact specimens and pipe specimens revealed similar fracture surfaces with extensive shear lips and crack tunneling.

Conclusions

The purpose of this research was to investigate the applicability of using small laboratory compact specimens to model the *J*-integral resistance curve behavior of 8 in. diameter ASTM A106 steel pipe. The results of this investigation appear to show that *J-R* curves from ½T, 1T plan, and 2T plan specimens can be used to predict *J-R* curves from 8 in. diameter pipe over the range of crack extension available in the compact specimens, provided the thicknesses of the compact specimen and the pipe wall are equal and that the specimen strain histories are the same.

Acknowledgments

The authors wish to acknowledge the support of Messrs. Milton Vagins and Jack Strosnider of the U.S. Nuclear Regulatory Commission. Additionally, the authors wish to thank those at the David Taylor Naval Ship R&D Center who have contributed to experimental phases of the program and to acknowledge the advice and guidance of Dr. Hugo Ernst, Mr. Gery Wilkowski, Professor Paul Paris, and Dr. Akram Zahoor.

References

[*1*] Begley, J. A. and Landes, J. D. in *Fracture Mechanics, ASTM STP 514*, American Society for Testing and Materials, Philadelphia, 1972, pp. 1-20.
[*2*] Clarke, G. A., Andrews, W. K., Paris, P. C., and Schmidt, D. W. in *Mechanics of Crack Growth, ASTM STP 590*, American Society for Testing and Materials, Philadelphia, 1976, pp. 27-42.
[*3*] Joyce, J. A. and Gudas, J. P. in *Elastic-Plastic Fracture, ASTM STP 668*, American Society for Testing and Materials, Philadelphia, pp. 451-468.
[*4*] Hutchinson, J. W. and Paris, P. C. in *Elastic-Plastic Fracture, ASTM STP 668*, American Society for Testing and Materials, Philadelphia, 1979, pp. 37-64.
[*5*] Paris, P. C., Tada, H., Zahoor, A., and Ernst, H. in *Elastic-Plastic Fracture, ASTM STP 668*, American Society for Testing and Materials, Philadelphia, 1979, pp. 5-36.
[*6*] Paris, P. C., Tada, H. Zahoor, A., and Ernst, H. in *Elastic-Plastic Fracture, ASTM STP 668*, American Society for Testing and Materials, Philadelphia, 1979, pp. 251-265.
[*7*] Vassilaros, M. G. and Hackett, E. M., "*J*-Integral *R*-Curve Testing of High Strength Steel Utilizing the Direct-Current Potential Drop Method," in *Fifteenth Symposium, Fracture Mechanics: ASTM STP 833*, American Society for Testing and Materials, Philadelphia, 1984, pp. 535-552.
[*8*] Ernst, H. A., Paris, P. C., and Landes, J. D., "Estimations of *J*-Integral and Tearing Modulus *T* from a Single Specimen Test Record," in *Fracture Mechanics; Thirteenth Conference, ASTM STP 743*, American Society for Testing and Materials, Philadelphia, 1981, pp. 476-502.

[9] Joyce, J. A., "Instability Tests of Compact and Pipe Specimens Utilizing a Test System Made Compliant by Computer Control," U.S. Nuclear Regulatory Commission Report, NUREG CR-2257, March 1982.

[10] Wilkowski, G. M. and Maxey, W. A., "Review and Applications of the Electric Potential Method for Measuring Crack Growth in Specimens, Flawed Pipes, and Pressure Vessels," in *Fracture Mechanics: Fourteenth Symposium—Volume II: Testing and Applications, ASTM STP 791*, American Society for Testing and Materials, Philadelphia, 1983, pp. II-266—II-294.

[11] Tada, H., Paris, P., and Gamble, R., "Stability Analysis of Circumferential Cracks in Reactor Piping Systems," U.S. Nuclear Regulatory Commission Report, NUREG CR-0838, June 1979.

[12] Zahoor, A. and Kanninen, M. F., "A Plastic Fracture Mechanics Prediction of Fracture Instability in a Circumferentially Cracked Pipe in Bending—Part I: *J*-Integral Analysis," *Journal of Pressure Vessel Technology, Transactions of ASME*, Vol. 103, No. 4, Nov. 1981.

[13] Vassilaros, M. G., Gudas, J. P., and Joyce, J. A., "Experimental Investigation of Tearing Instability Phenomena for Structural Materials," U.S. Nuclear Regulatory Commission, NUREG/CR-2570 Rev. 1 (Aug. 1982).

[14] Green, G., Smith, R. F., and Knott, J. F., "Mechanics and Mechanisms of Crack Growth," BSC Conference, M. J. May, Ed., Cambridge, England, April 1973, p. 58.

[15] Clayton, J. O. and Knott, J. F., *Materials Science*, Vol. 10, No. 63, 1976.

O. L. Towers[1] *and S. J. Garwood*[2]

Influence of Crack Depth on Resistance Curves for Three-Point Bend Specimens in HY130

REFERENCE: Towers, O. L. and Garwood, S. J., **"Influence of Crack Depth on Resistance Curves for Three-Point Bend Specimens in HY130,"** *Fracture Mechanics: Seventeenth Volume, ASTM STP 905*, J. H. Underwood, R. Chait, C. W. Smith, D. P. Wilhem, W. A. Andrews, and J. C. Newman, Eds., American Society for Testing and Materials, Philadelphia, 1986, pp. 454–484.

ABSTRACT: To obtain a better understanding of the behavior of shallow cracks in structures the influence of crack depth on fracture toughness has been studied. Fracture toughness tests were carried out on 50-mm thick three-point bend specimens sampling HY130 high strength steel. The mode of crack growth in all cases was tearing, and fracture toughness values at initiation of crack growth were evaluated as well as resistance (R) curves. Fracture toughness values were calculated in terms of both the crack tip opening displacement (CTOD) and the J integral. Crack extension was deduced from unloading or reloading compliance and by use of an alternating-current potential drop technique.

It was found that the fracture toughness values at initiation of crack growth, and the R-curve slopes, were consistently higher for crack depth to width (a/W) ratios of 0.3 or less than for the a/W ratio of 0.5 commonly used for routine fracture toughness tests. For very deep cracks, $a/W \simeq 0.8$, the fracture toughness at initiation of tearing was little different from that for $a/W \simeq 0.5$. The R-curve slope, however, was higher for $a/W \simeq 0.8$ than it was for $a/W \simeq 0.5$. The latter observation (and the observed increase in fracture toughness with decrease in crack length for a/W less than 0.3) is considered to be consistent with loss of constraint for very short and very deep cracks.

KEY WORDS: high strength steel, fracture toughness, crack length, crack initiation, crack growth, resistance curves, crack tip opening displacement, J-integral

Nomenclature

A	Area under load versus load point displacement
A_e, A_p	Elastic and plastic components of A, respectively
a	Crack length

[1]Manager, Process Modelling, Rolls-Royce Limited, Derby, England; formerly at The Welding Institute, Abington Hall, Abington, Cambridge, England.

[2]Head of Fracture, The Welding Institute, Abington Hall, Abington, Cambridge, England.

a_0	Original fatigue crack length
ac/pd	Alternating current/potential drop
b	Original uncracked ligament $= W - a_0$
B	Specimen thickness
B_{eff}	Effective specimen thickness
B_N	Net section thickness for side-grooved specimens
C	Slope of relationship between voltage change and crack extension
CTOD	Crack tip opening displacement
E	Elastic modulus
E'	Effective, plane strain, elastic modulus ($= E/1 - \nu^2$)
J	J-integral
J_{Ic}	J value at initiation of crack growth measured to ASTM E 813 procedures
J_Q	J value at initiation of crack growth
K	Mode I stress intensity factor
K_{Fmax}	Maximum value of K imposed in fatigue cycle
K_{Ic}	Plane strain fracture toughness for Mode I loading
$K_{nominal}$	Nominal K calculated neglecting side grooves
LVDT	Linear variable displacement transformer
m	"Constant" relating J-integral value to CTOD value
P	Applied load
P_L	Limit load
P_0	Load to yield uncracked specimen
q	Load point displacement
r	Rotation factor
S	Loading span
T	Tearing modulus
V	Clip gage displacement
V_1, V_2	Values of V for lower and upper clip gages, respectively
V_p	Plastic component of V
W	Specimen width
x	Distance of rotation point from crack tip
Y	Geometric correction factor for calculating K values
Z	Nondimensional elastic compliance
z	Knife edge height above specimen surface
z_1, z_2	Values of z for lower and upper clip gages, respectively
δ	Crack tip opening displacement (CTOD)
δ_i	CTOD at initiation of tearing
Δa	Crack extension
ΔK	Range of K imposed in fatigue cycle
ΔV	Change in voltage for ac/pd traces
ν	Poisson's ratio
η	Geometric factor used to calculate J values from A
η_e	Elastic component of η

η_{max} Overall value of η
η_p Plastic component of η
σ Stress
σ_f Flow stress
σ_u Ultimate or tensile strength
σ_Y Yield, or 0.2% proof, stress

Introduction

This investigation was prompted by the fact that most fracture toughness tests are performed on deeply notched specimens, largely for convenience of analysis. Cracks, or defects in structures, however, are generally shallow relative to the thickness of the section in which they lie. It is therefore necessary to establish the behavior of shallow cracks in order to ensure that suitable analysis procedures are adopted when assessing the significance of real, shallow, defects in structures.

The present test program sets out to establish the influence of crack depth on the resistance of the material to initiation and propagation of ductile fractures in HY130. The specimen adopted was loaded using the three-point bend configuration. This was used in preference to the compact tension geometry because of yielding around the loading holes in the latter geometry when the crack depth is shallow. Crack growth was monitored by indirect means, and both compliance and alternating-current potential drop (ac/pd) techniques were used for this.

Material

The results of a chemical analysis on a sample of the 52-mm-thick HY130 plate used for these studies are given in Table 1. The results of room temperature tensile tests (on specimens oriented both parallel and transverse to the rolling direction) are given in Table 2.

Test Program

The fracture toughness test program, which is summarized in Table 3, consisted of two phases. Phase I consisted of tests on 18 specimens with blunt notches of different notched depths to establish: (1) calibrations for the determination of crack length using compliance and ac/pd techniques, and (2) the rotation factors appropriate to different notch depths to indicate the best method for calculating CTOD values. Phase II consisted of tests on specimens with fatigue cracks prepared to different depths in order to develop crack growth resistance curves ("*R*-curves") for specimens of varying crack depth.

The Phase II tests involved the development of *R*-curves for specimens with

TABLE 1—*Chemical analysis of HY130 plate (Welding Institute Analysis No. S/82/168).*

Sample Ref. No.	Element (Weight Percent)															
	C	S	P	Si	Mn	Ni	Cr	Mo	V	Cu	Nb	Ti	Al	B	Sn	Co
2A74	0.12	0.007	0.007	0.25	0.77	4.81	0.53	0.41	0.07	0.16	0.005	0.007	0.052	0.0006	0.01	0.06

TABLE 2—*Results of tensile tests on HY130 performed at +20°C.*

Orientation Relative to Rolling Direction	0.2% Proof Stress, MPa (ksi)	Tensile Strength, MPa (ksi)	Elongation, %	Reduction of Area, %
Longitudinal	980 (142)	1026 (149)	23	61
Longitudinal	967 (140)	1029 (149)	23	59
Transverse	976 (142)	1032 (150)	20	58
Transverse	905 (131)	1023 (148)	22	58

TABLE 3—*Specimens tested.*

Phase No.	Specimen No.	Blunt Notch or Fatigue Crack?	a_0/W*	Purpose of Test
Phase I	M1-31	blunt notches	0	blunt notch calibrations
	M1-1 and 32		0.05	
	M1-2		0.10	
	M1-33		0.12	
	M1-3		0.15	
	M1-34		0.17	
	M1-4		0.20	
	M1-5 and 35		0.25	
	M1-6		0.30	
	M1-36		0.36	
	M1-7		0.40	
	M1-8		0.50	
	M1-37		0.59	
	M1-9		0.61	
	M1-10		0.71	
	M1-38		0.79	
Phase II	M1-14	fatigue cracks	0.60	single-specimen *R*-curves
	M1-15		0.58	
	M1-17		0.24	
	M1-18		0.21	
	M1-19		0.44	
	M1-20		0.45	
	M1-21		0.75	
	M1-22		0.80	
	M1-23		0.35	
	M1-24		0.38	
	M1-27		0.11	
	M1-28		0.09	
	M1-29		0.11	
	M1-30		0.18	
	M1-39		0.27	
	M1-43		0.23	

*a_0 = Initial notch or crack depth (being average of two surface measurements for blunt notch specimens or per ASTM E 813 for specimens with fatigue cracks), and
W = specimen width.

different fatigue crack depths. Sixteen specimens over a range of different nominal crack depths were tested to generate single-specimen R-curves using unloading compliance and ac/pd techniques.

Specimen Preparation

The fracture toughness test specimens listed in Table 3 were all prepared for three-point bend tests, with the specimen width, W, and specimen thickness, B, both being nominally 50 mm, and the specimen length being approximately 230 mm. The specimens were all extracted in T-L orientation to the notation of ASTM Test for Plane-Strain Fracture Toughness of Metallic Materials (E 399).

The overall dimensions of the specimens conformed to the requirements for the subsidiary specimen to the British Standard Methods for Crack (Tip) Opening Displacement (CTOD) Testing (BS 5762: 1979). All the test pieces except one, however, were side-grooved nominally by 20% of the specimen thickness (i.e., 5-mm-deep side-grooves on each side of the specimen). This is not specified or allowed for by BS 5762. The specimen geometry used is allowed by the ASTM Test for J_{Ic}, A Measure of Fracture Toughness (E 813) as an "alternative specimen", provided the net thickness (for side-grooved specimens) is greater than $25J_{Ic}/\sigma_f$, where σ_f is the "effective yield strength" or flow stress. ASTM E 813 allows side-grooves in specifying that "specimens may be side-grooved up to 25% of their original thickness". Side-grooves were adopted in the current test program because they help to ensure straight crack fronts during crack growth, and indeed the recently published tentative test procedure for determining the plane strain J_I-R curve points this out [1]. The square section three-point bend specimen adopted in this program also conforms to the requirements of the J_I-R curve test procedure [1], provided the various validity criteria of Section 9 of that document are conformed to.

Fatigue cracking was performed at room temperature in three-point bending over a 200-mm loading span with the maximum stress intensity in the fatigue cycle, K_{Fmax}, being nominally limited to 32 MPa$\sqrt{m}$. The R-ratio (the minimum load in the fatigue cycle divided by the maximum load) was kept to 0.1 or less throughout the fatigue operation.

With one exception, the side-grooves were machined into the specimens with fatigue cracks after the fatigue cracking operation. Specimen M1-43, however, was side-grooved to a 2.5 mm depth each side of the specimen (i.e., to 10% of the specimen thickness, B) with a 3.18-mm (1/8-in.) radius cutter before fatigue cracking. After fatigue cracking, the specimen was side-grooved to the full 20% of B, using the V-tipped milling cutter with a 60° included angle which was used to machine all the side-grooves. This procedure was used in an attempt to obtain a straighter fatigue crack front than was obtained without the use of side-grooves during pre-cracking. Six specimens with shallow cracks were produced by fatigue cracking with a 60-mm specimen width. The top 10 mm of the specimen was then machined off.

In Table 4 the initial fatigue crack lengths and the crack extensions are collated for the fatigue precracked specimens based on the requirements of the CTOD testing standard (BS 5762), the J_{Ic} testing standard (ASTM E 813), and the tentative procedure for determining the J_I-R curve [1]. The methods for averaging the initial fatigue crack length, a_0, measurements are different in these codes, which accounts for the differences apparent in Table 4. As the crack fronts are bowed, with the fatigue crack growth being greatest near the center of the specimen, the largest initial crack length values are obtained by BS 5762 procedures and the smallest by the tentative J_I-R curve test procedures, with the a_0 value of ASTM E 813 lying between the two. Again, if the crack extension is bowed, the BS 5762 procedures will give the largest values of crack extension and the tentative J_I-R curve procedure the smallest. This difference is accentuated by the requirement in BS 5762 to exclude the stretch zone from the measurement of crack extension. If the crack growth is greater at the root of the side-groove than it is in the center of the specimen, as clearly occurs for the side-grooved specimens with the shortest cracks, then the order of the crack growth value is inverted (i.e., the tentative J_I-R curve procedure gives the largest crack extensions and BS 5762 the smallest).

The validities of the fatigue crack shapes to BS 5762, ASTM E 813, and the tentative J_I-R curve test procedure are summarized in Table 4. All specimens were valid to the requirements for fatigue crack shape of BS 5762, but very few were valid to ASTM E 813 or the tentative J_I-R curve test procedure. This is because the maximum variations in the individual fatigue crack length measurements are limited to a given percentage of the specimen width in BS 5762, but to a given percentage of the average crack length (be it weighted or otherwise) in the other two cases. Clearly, for a given crack front shape, the requirements based on limiting the variations in fatigue crack length measurements to a proportion of the crack length become more arduous the shorter the crack length. For the shortest crack lengths the requirements of ASTM E 813 and the tentative J_I-R curve test procedure are too restrictive to be practical. It is also apparent from Table 4 that the uniformity of crack extension in the specimens was rarely within the requirements of the tentative J_I-R curve test procedure.

Instrumentation

Tests were performed in three-point bending using servohydraulic test machines. The loading span was normally 200 mm, but for the four shallow notched specimens (M1-27 to M1-30) a 210 mm loading span was used to reduce their maximum load level. For certain tests an alternating-current potential drop (ac/pd) technique was utilized. The ac/pd equipment used is a developed version of the equipment used by Okumura et al [2]. A 1-kHz fre-

TABLE 4—*Fatigue crack length and crack extension details for Phase II fatigue precracked specimens.*

	Initial Fatigue Crack Length, a_0, in mm to:			Crack Extension, Δa, in mm to:			Valid Fatigue Crack to:			Uniformity of Crack Extension. Valid to:
Specimen No.	BS 5762	ASTM E 813	Tentative Procedure for J_I-R Curves	BS 5762	ASTM E 813	Tentative Procedure for J_I-R Curves	BS 5762?	ASTM E 813?	Tentative Procedure for J_I-R Curves?	Tentative Procedure for J_I-R Curves?
14	30.27	29.93	29.79	6.35	6.16	5.97				X[c]
15	29.38	28.96	28.78	7.18	7.18	7.13				X[c]
17	12.20	11.85	11.71	31.39	31.69	31.89		X[a]	X[b]	X[c]
18	10.34	10.23	10.19	34.40	34.45	34.55		X[a]	X[b]	X[c]
19	22.44	21.97	21.78	2.62	2.66	2.65		X[a]	X[b]	X[c]
20	22.85	22.38	22.19	14.11	13.83	13.59		X[a]	X[b]	X[c]
21	37.61	37.37	37.26	3.95	3.94	3.90				
22	40.13	39.91	39.81	2.21	2.22	2.19				
23	17.79	17.36	17.17	22.14	22.40	22.56		X[a]	X[b]	X[c]
24	19.40	18.98	18.81	16.62	16.63	16.59		X[a]	X[b]	X[c]
27	5.93	5.60	5.47	1.82	2.49	2.97		X[a]	X[b]	X[c]
28	5.00	4.70	4.53	2.66	3.18	3.52		X[a]	X[b]	X[c]
29	5.69	5.46	4.97	2.09	2.76	3.24		X[a]	X[b]	X[c]
30	9.70	9.14	8.91	34.19	34.69	35.02		X[a]	X[b]	X[c]
39	14.25	13.87	13.70	1.25	1.49	1.65		X[a]	X[b]	X[c]
43	11.70	11.57	11.51	1.37	1.53	1.62		X[a]	X[b]	X[c]

[a]Individual crack length measurement different from average to ASTM E 813 by >7%.
[b]Individual crack length measurement different from average to tentative procedure for J_I-R curve by >7%.
[c]Crack extension at center of specimen differs from that at side groove root by >2% of W.

quency was used for the tests with a 1.6 A current. The specimens were electrically insulated from the test machine.

Two clip gages were mounted across the notch mouth of every specimen tested (Fig. 1). The arrangement used resulted in the lower and upper knife edges being 7.1 mm and 54.0 mm above the top surface of the specimen, respectively. The spacing between the knife edges was therefore 46.9 mm. For all tests a "comparator bar" arrangement was used to measure load point displacement [3]. Where ac/pd was used the two clip gages and the comparator bar arrangement were electrically insulated from the specimen.

Experimental Procedures

Data Logging

Continuous records of applied load versus lower clip gage displacement and of potential drop or load point displacement versus lower clip gage displacement were taken on an *x-y-y* chart recorder for all tests. In addition, the variables of applied load, upper and lower clip gage displacement, crosshead displacement, load point displacement (from the comparator bar), and elapsed time were recorded on an Intercole System Limited Compulog III Computer. Initially the data logging rate was approximately one reading of

FIG. 1—*Typical test specimen showing detail of double clip gage arrangement and ac/pd voltage leads in foreground.*

all six variables every second during loading and three every second during unloading. Later, however, the logging rate was reduced to approximately one complete scan every second throughout.

Testing Sequence

Force was applied to the specimens up to a level generally corresponding to a stress intensity factor, K, of 32 MPa$\sqrt{m}$, which was calculated using the nominal crack length to width ratio, the nominal specimen width of 50 mm, and the nominal gross-section thickness of 50 mm. This force level was chosen because it corresponds to the nominal maximum used for fatigue cracking. For the specimen tested with a notch depth of zero, M1-31, the calculation of a stress intensity factor is not relevant, and the initial load for this specimen was based on a similar proportion of the limit load, P_L, to that for the specimen with the shallowest notch depth. The initial load levels were generally in the range of 18 to 25% of the nominal limit load calculated using the expression given in ASTM E 813 and Ref *1*. On reaching the initial load level referred to above, the specimens were unloaded by 25% of that load (i.e., to 75% of the initial level). The specimen was then reloaded to the initial load and unloaded again. This sequence was repeated so that at least three unloadings were performed. For later tests these three initial unloadings were by 75% of the initial loading (i.e., to 25% of the initial load). This was done on the basis of the recommendation of Clause 8.3.2. in Ref *1* that the initial unloadings should be from 0.4 to 0.1 P_L. Following the initial unloadings the blunt notch specimens were either loaded until the maximum load point was exceeded or until the force capacity of the machine was reached. The specimens were then unloaded, heat tinted at approximately 300 to 400°C, and broken open after cooling to around −196°C in liquid nitrogen. The specimen with a zero notch depth, however, was left unbroken.

Subsequent to the initial unloadings the tests on fatigue precracked specimens were continued with unloadings being performed at regular intervals during the test. In all unloadings subsequent to the initial three, the unloadings were to a load level of 75% of the level at the start of the unloading (after that level had been allowed to settle on the test machine's digital load readout). At the end of the test the specimens were finally unloaded, heat tinted at approximately 400°C, and broken open after cooling in liquid nitrogen. Early tests were performed in crosshead displacement control, with a displacement rate during loading of 0.5 mm/min and during unloading of 0.2 mm/min. Before each unloading the digital reading of load on the test machine was monitored, and the unloading was not begun until this reading had settled at a constant level. For some of the specimens, however, partial instabilities occurred after passing the maximum load point. In these cases, the clip gage displacement increased very rapidly prior to the crack restabilizing. For this reason the later tests were performed in (lower) clip gage control to improve the stability of the test setup over that for crosshead control, with the lower clip gage displacement rate being 0.25 mm/min during loading and 0.1 mm/min during unloading.

Analysis of Test Results

Rotation Factors of Blunt Notch Specimens

Values of the crack tip opening displacement (CTOD) can be derived on the assumption that a bend specimen hinges about a point located in the ligament beneath the crack. (This is the assumption implicit in the BS 5762 procedure for calculating the plastic component of CTOD to that test procedure.) In BS 5762 the hinge is assumed to lie at a point distant from the crack tip by 40% of the remaining ligament ($W - a$), or the rotation factor, r, is assumed to be 0.4, where

$$r = \frac{x}{W - a} \tag{1}$$

where x is the distance of the rotation point from the crack tip (Fig. 2).

Assuming the crack flanks remain straight, the rotation factor can be derived from the values of upper and lower clip gage displacement [4]. Plots obtained of r versus lower clip gage displacement on the blunt notch specimen calibration are shown in Fig. 3.

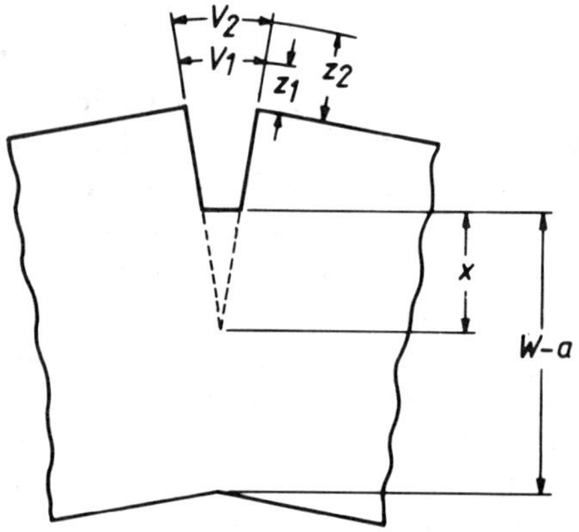

FIG. 2—*Schematic diagram of rotation point in bend specimen.*

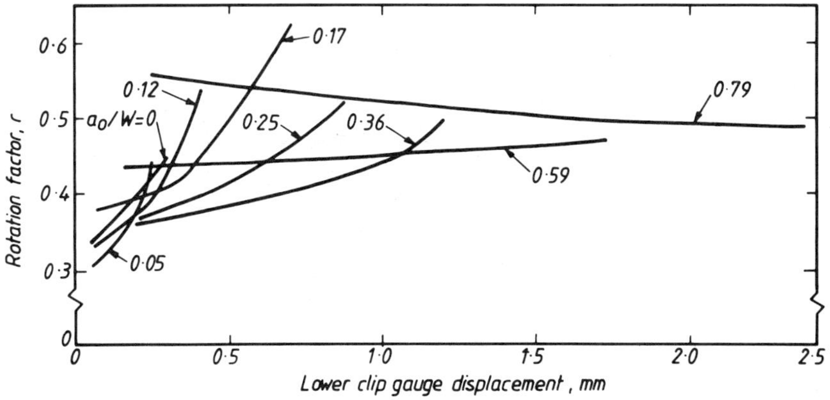

FIG. 3—*Rotation factor versus lower clip gage displacement for tests on blunt notch specimens.*

Derivation of Crack Extension and Resulting R*-curves for Fatigue Precracked Specimens*

Crack Extension from ac/pd—A schematic diagram from a typical ac/pd versus lower clip gage displacement trace is given in Fig. 4. This trace shape is typical for tests on ferritic steel specimens, and the factors that give rise to this shape are described in Refs *2* and *5*.

The investigations of Okumura et al [2] and Marandet and Sanz [5] indicate that the difference between the potential drop at a given stage of crack growth and the potential drop at the minimum in the trace (i.e., ΔV in Fig. 4) is close to being linearly related to the crack growth increment, Δa. In Fig. 5 this difference in potential drop is plotted against the ductile crack extension measured per ASTM E 813 procedures for the present test results. This plot gives the results for the blunt notch specimens, and the Phase II specimens for single specimen R-curves, but excluded are specimens that became unstable during the test. There is a clear trend for the crack extension to be closely related to the voltage difference, ΔV. To derive crack extensions from the plots of ac/pd versus lower clip gage displacement (V_1) for single specimen R-curves the crack extension is assumed to vary linearly with the difference of the potential drop from the minimum (ΔV in Fig. 4); that is, $\Delta a = C \cdot \Delta V$.

The slope of this relationship, C, was defined for each specimen as the crack extension measured for the final unloading point to ASTM E 813 (from the heat tinted broken specimen halves) divided by the voltage difference, ΔV, at the final unloading point. Thus the slope of the Δa versus ΔV relationship is fixed for each specimen at the minimum in the ac/pd versus V trace (assumed to correspond to $\Delta a = 0$) and at the final unloading point.

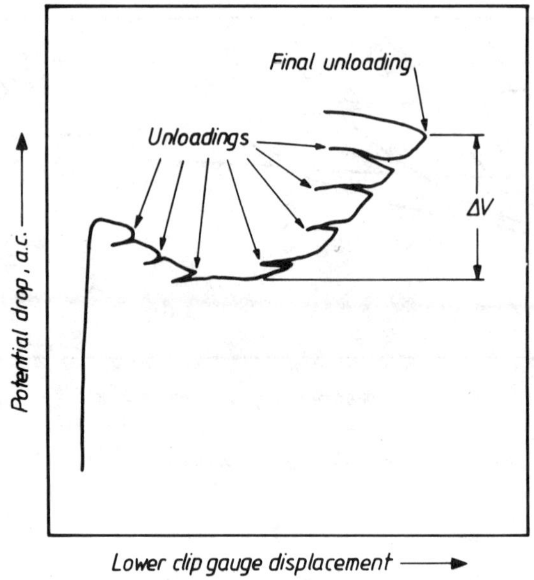

FIG. 4—*Schematic appearance of typical ac/pd versus lower clip gage displacement trace for single specimen* R-*curves.*

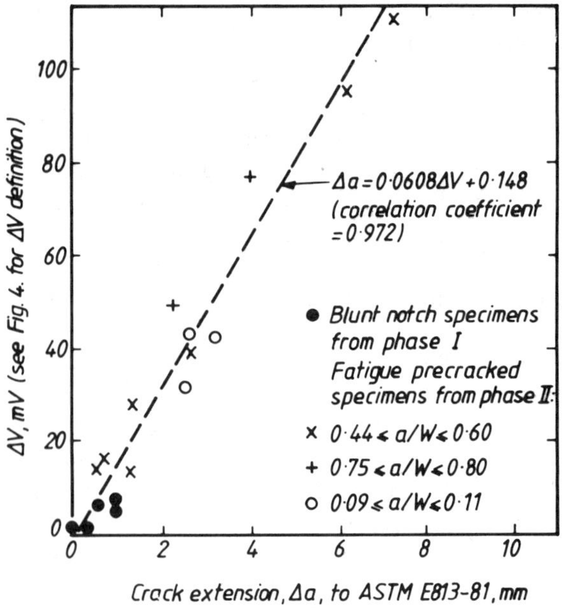

FIG. 5—*Plot of potential drop relative to minimum versus crack extension for final unloading points.*

Crack Extension from Compliance—The unloading (and reloading) compliance data from the load versus lower clip gage displacement traces were used to derive crack growth increments. Although there are published solutions that enable the crack length to be deduced from the compliance based on notch mouth opening for a three-point bend specimen, these solutions are usually for notch mouth openings measured on the top surface of the specimen, as for example is the case for the solutions given by Tada et al [*6*] and in the tentative J_I-R curve test procedure [*1*]. In practice, the mouth notch opening is often measured at some distance above the top surface, as is the case for the present tests (Fig. 3). Garwood and Willoughby [*7*] have published a modification to the approximation given in the tentative J_I-R curve procedure [*1*] to account for the measurement being made at a certain height above the top surface of the specimen. A comparison between this modified relationship and the results of finite element modelling of a three-point bend specimen on which the relationship is based indicated that the error in predicted compliance is within 3% for a knife edge height, z, to width, W, ratio of 0.12 and for $0.3 < a/W < 0.6$. The relationship proposed by Garwood and Willoughby has up to now only been validated relative to practical test results for fracture toughness specimens with a/W ratios of around 0.3 to 0.6 and z/W values of approximately 0.2 or less.

In Fig. 6 the relationship of Garwood and Willoughby [*7*] has been plotted in terms of normalized lower clip gage compliance, $WB_{\text{eff}}EV_1/PS$, versus a/W (Fig. 6*a* being the plot for a/W values in the range 0 to 1.0 and Fig. 6*b* being for $0 < a/W < 0.3$). For the fatigue precracked specimens, the original crack length to width, a_0/W, ratios are based on values determined per ASTM E 813 procedures. Clearly the relationship obtained by Garwood and Willoughby is close to the experimental data for relatively high a/W ratios, say >0.3, but for low a/W is not so accurate. Indeed, this loss of accuracy at low a/W ratios is not unexpected and was apparent when the relationship was compared with the finite element results on which the approximation is based. To provide a means of deriving crack length from lower clip gage compliance for low a/W values a polynominal was fitted to the data obtained for the initial and final unloadings performed on both the blunt notch and fatigue precracked test pieces for a/W values in the range 0 to 0.5, giving the relationship.

$$a/W = -0.25453 + 0.48441\alpha - 0.23854\alpha^2 + 0.068964\alpha^3 - 0.01059\alpha^4 + 0.00080063\alpha^5 - 0.000023319\alpha^6$$

where

$$\alpha = \frac{WB_{\text{eff}}E}{S}\frac{V_1}{P} \tag{2}$$

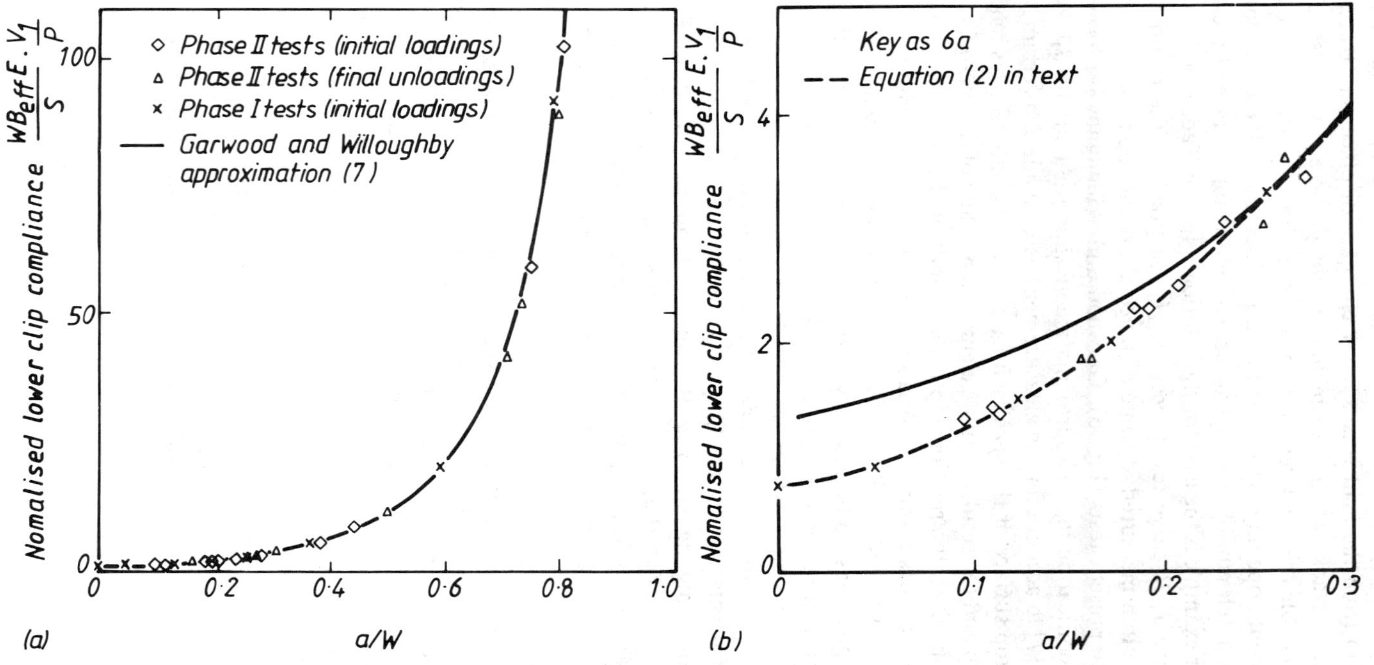

FIG. 6—*Normalized lower clip gage compliance versus a/W for initial unloadings and final unloading (a measured per ASTM E 813 procedure for fatigue precracked specimens:* E *taken as* 206×10^3 *MPa). (a)* $0 < a/W < 1.0$. *(b)* $0 < a/W < 0.3$.

and the variables a, B_{eff}, E, P, S, V_1 and W are as defined in the Nomenclature section. The curve fit "oscillates" over the range $0.3 < a/W < 0.5$. For this reason this relationship has only been used to derive crack extensions from unloading compliance over the range $0 < a/W < 0.3$. The crack extension for a given unloading was obtained by subtracting the original crack length (deduced from the unloading compliance for the initial unloadings) from the current crack length (deduced from the unloading compliance for the particular unloading of interest). The same procedure was followed to obtain the crack extensions based on reloading compliance. (The main reason for using the reloading compliance data was that unloading compliance data were not always available for the Phase II tests.) For a/W values of 0.3 or less the crack length was deduced from (unloading or reloading) compliance based on Eq 2. For $a/W > 0.3$ the relationship obtained by Garwood and Willoughby was used.

Calculation of Fracture Toughness Values

Derivation of CTOD Values—The CTOD values were calculated from the load versus lower clip gage displacement traces for each unloading using the general procedure of BS 5762. Some assumptions made when calculating CTOD values differed from the procedures of BS 5762; these differences are described below:

(*a*) *Crack length* (a)—This was taken as the fatigue crack length throughout and was the weighted average calculated per ASTM E 813 procedures (to be consistent with the J estimation procedure).

(*b*) *Specimen thickness* (B)—The specimen thickness B was taken as a B_{eff} for side-grooved specimens, where

$$B_{eff} = B - \frac{(B - B_N)^2}{B} \tag{3}$$

per ASTM E 813 procedures. BS 5762 does not provide for side-grooved specimens. Although this should not influence the plastic component of CTOD (because the specimen thickness does not enter the equation), the method for calculating the stress intensity factor, K, for the elastic component of CTOD needs to allow for the specimen being side-grooved. Freed and Krafft [*8*] proposed that the plane strain fracture toughness, K_{Ic}, for a side-grooved specimen could be deduced from the nominal K, $K_{nominal}$, calculated neglecting the side-grooves by the use of the relationship

$$K_{Ic} = K_{nominal}\left(\frac{B}{B_N}\right)^m \tag{4}$$

where m lies somewhere between 0.5 and 1.0.

The B_{eff} values used in place of B as given by Eq 3 are higher than would be proposed by Freed and Krafft [*8*], and more pessimistic (i.e., lower) values of K and therefore CTOD will be derived.

(*c*) *Stress intensity factor* (K)—The procedure for K calculation in BS 5762 is based on an approximation that is inaccurate for a/W ratios of 0.6 or greater. For this reason the procedure given in the K_{Ic} testing standard (ASTM E 399), which is expected to be accurate to within $\pm 0.5\%$ for $0 < a/W < 1.0$ and for a span-to-width ratio of 4, was used to calculate Y for Specimens M1-21 and M1-22 which had initial a/W ratios of 0.75 and 0.80, respectively (Table 3).

Derivation of J—J values were derived using the relationship

$$J = \frac{\eta A}{B_N b} \tag{5}$$

where

A = area under the load versus load point displacement trace,
B_N = net section thickness,
b = original uncracked ligament = $W - a_0$,
W = specimen width,
a_0 = original fatigue crack length, and
η = geometric correction factor = $f(a_0/W)$ per ASTM E 813 terminology.

To comply with ASTM E 813, η is taken as 2 for the three-point geometry. This value, however, is only appropriate for relatively deep cracks (e.g., $a/W > 0.45$). For shallower cracks a lower value for η is appropriate. Various estimates of the variation of η with a/W ratio are presented in Fig. 7. This includes results for the elastic η factor η_e which can be derived using linear elastic fracture mechanics [*9*], an overall factor, η_{max} or η [*10*], and for the plastic η factor, η_p, derived for the limit load situation [*10*]. The continuous plot of η_p versus a/W was derived from the limit load solutions for four-point bending given by Haigh and Richards [*11*] which are based on slip-line field analysis. The details of this derivation are given in Appendix I. Based on the η values plotted in Fig. 7 it was decided to calculate J values with η being taken as equal to the η_e values calculated using the procedures described in Ref *9* for the three-point bend geometry, the derivation being described in Appendix II. Since the η_e values are less than those η values deduced for other regimes for a/W approximately less than 0.4, this should yield pessimistic estimates of the J values. For $a/W > 0.4$ the J values calculated based on η_e will be slightly higher, by at most 11%, than J values calculated per ASTM E 813 procedures (for which η is taken as equal to 2).

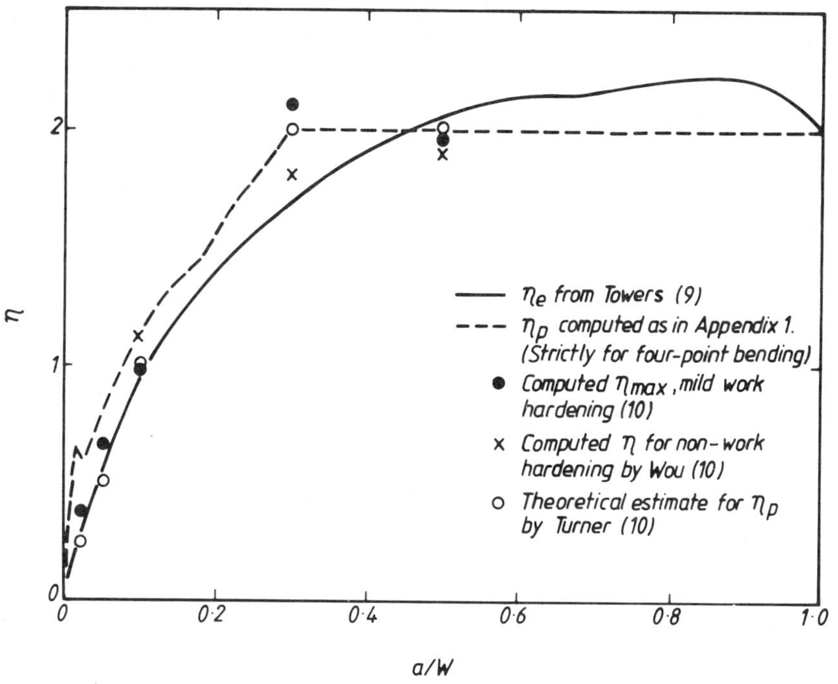

FIG. 7—*η values computed for the three-point bend geometry with a span-to-width ratio of 4.*

Test Results

J values nominally corresponding to initiation of tearing (i.e., J_Q values) are plotted versus original crack length to width ratio, a_0/W, in Fig. 8*a*. Values of the "tearing modulus" [*12*], T, where

$$T = \frac{\partial J}{\partial a} \frac{E}{\sigma_f 2} \tag{6}$$

are plotted versus a_0/W in Fig. 8*b*. The CTOD values nominally corresponding to initiation of tearing (i.e., δ_i values) are plotted versus a_0/W in Fig. 9*a*; Fig. 9*b* shows the slopes of the CTOD R-curves as a function of a_0/W. In Fig. 10 the J versus crack extension, Δa, plots have been grouped into discrete ranges in a_0/W. For the R-curves derived from compliance, plots were obtained by first plotting the fracture toughness values (i.e., J or CTOD values) against Δa values derived by subtracting the current estimate of crack length from the crack length derived from the average compliance for the initial unloadings. In many cases, however, the resulting R-curves displayed apparent

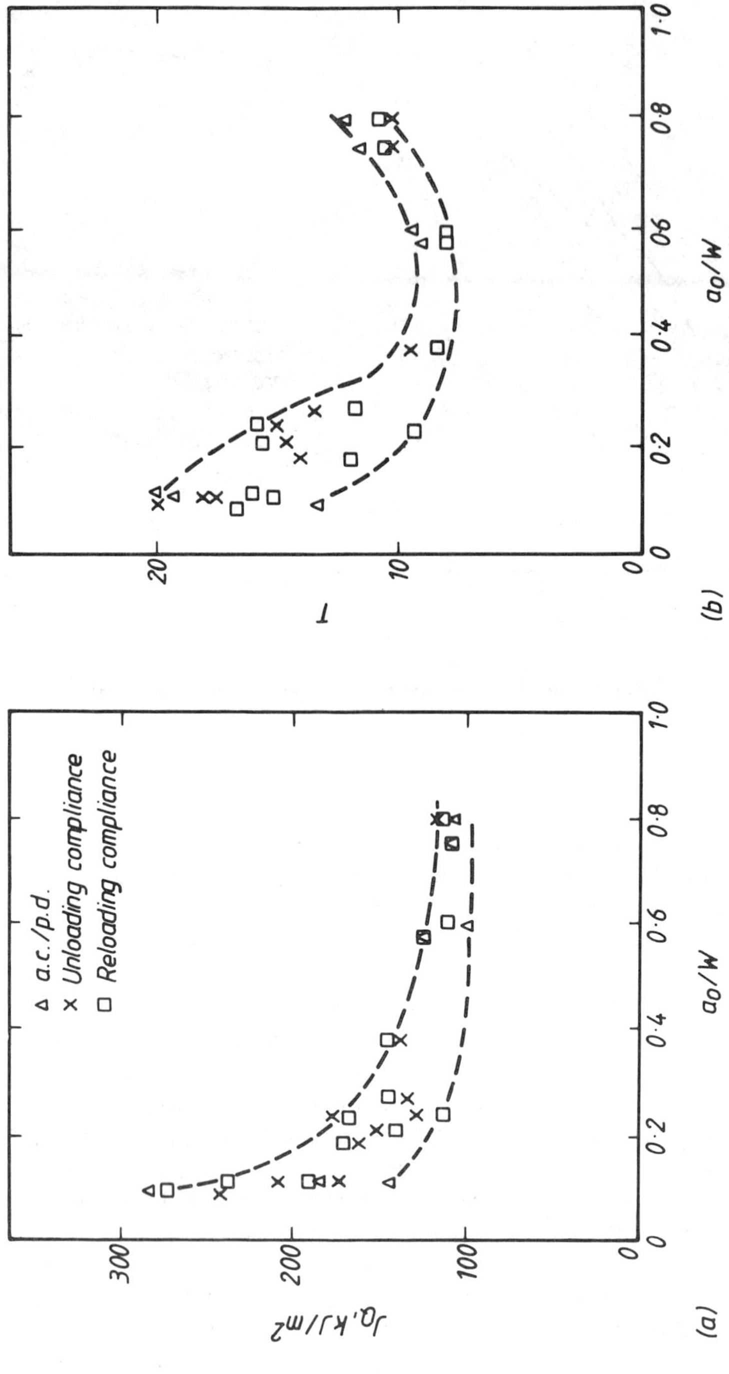

FIG. 8—*Data derived from linear regression fits to the single-specimen* J-R-*curves plotted against original crack length to specimen width ratio* a_0/W. (a) J *values at intercept between* R-*curves and blunting line* (*i.e*, J_Q). (b) T-*parameter.*

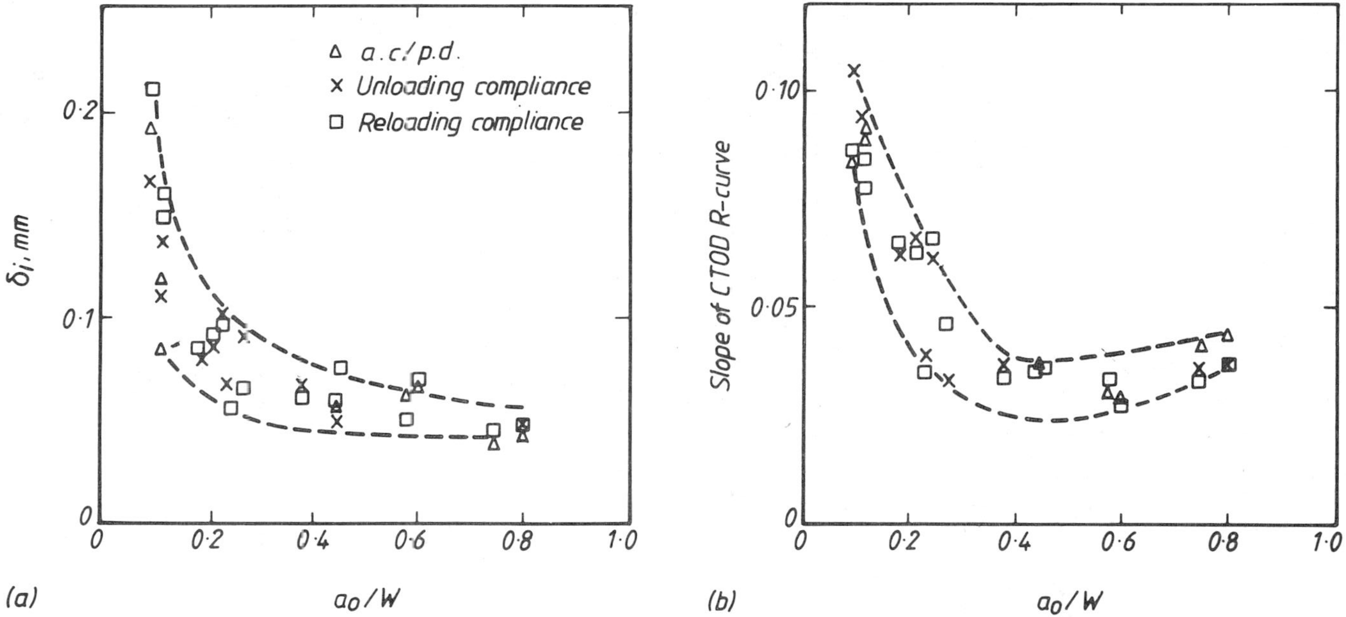

FIG. 9—*Data derived from linear regression fits to the single-specimen CTOD* R-*curves plotted against original crack length to specimen width ratio* a_0/W. (a) *CTOD values at intercept between* R-*curve and blunting line (i.e.,* δ_i*).* (b) *Slopes of CTOD R-curves.*

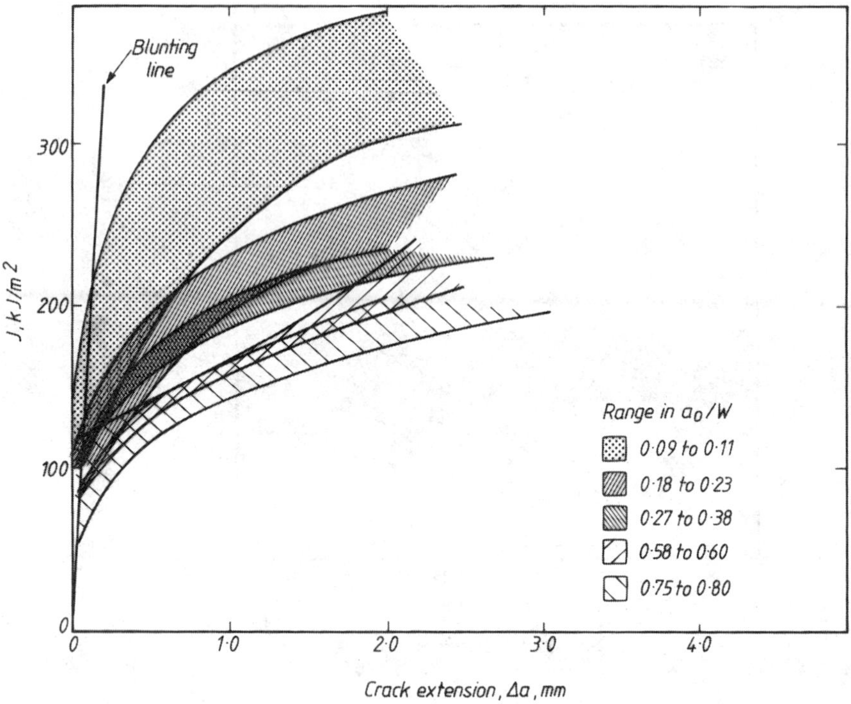

FIG. 10—*Summary of* J R-*curves for specimens of varying crack lengths.*

"negative crack growth" in the initial portion of the R-curve. To deal with this phenomenon, which is not unusual [*13,14*], curve fits were made manually to the overall R-curve and the point of zero crack extension was taken as the minimum value of Δa on the curve fit (i.e., the origin of the plot was moved so that negative crack growth was no longer apparent). It is possible that this practice could lead to overestimates of the toughness at initiation of tearing [*13*]. "Negative crack growth", however, was observed for specimens over a wide range of a_0/W ratios, and presumably the comparisons between results for specimens with varying a_0/W will not be affected (because overestimates of the initiation toughness will occur independently of the a_0/W ratio).

The fracture toughness at initiation of tearing, plotted in terms of J in Fig. 8 and in terms of CTOD in Fig. 9, was obtained by performing a linear regression analysis on three fracture toughness values taken from the curve fits to the R-curves at 0.5, 1.0, and 1.5 mm crack extension. The J_Q and δ_i values were then calculated at the intersection between this regression line and a blunting line (i.e., in a similar manner to ASTM E 813). For the J versus Δa plots the blunting line was taken as $J = 2\sigma_f \Delta a$, where σ_f, the "flow stress", was taken as 985 MPa based on the tensile test data in Table 2.

Discussion

The tendency for the tearing resistance to be greater at shallower crack depths, as indicated in Figs. 8 to 10, has an important implication with respect to fracture mechanics based assessments; that is, the testing of deeply cracked bend specimens with a/W of nominally 0.5 would appear to produce conservative values of initiation toughness and tearing modulus when applied to shallow surface cracks. At the same time it may be possible to rely on higher values of fracture toughness if shallow surface cracks are being assessed in a structure. It is necessary to be certain that the trends apparent in the reported data are real and not a consequence of the use of inappropriate formulae for the calculation of fracture toughness values. This is discussed further below.

Determination of J *for Shallow Cracks*

The J values reported here were calculated from the area under the load versus load point displacement trace, A, using the relationship of Eq 5. This relationship has been commonly adopted for deeply cracked bend specimens ($a/W = 0.5$) with η being taken as 2. For shallow cracked test pieces, however, there has been little work carried out to establish an appropriate method for estimating J values. In this report η has been taken as equal to the elastic value, η_e, and has thus been assumed to be independent of the extent of plasticity. The data shown in Fig. 7 indicate that this is a conservative procedure for shallow cracks in that η_e is lower than the computed values of η, which are more appropriate to situations with developed plasticity. Indeed, the use of η_e values for calculation of J results in slightly higher J values for deeply cracked test pieces than would be the case if η were taken as 2. Despite this, and the tendency for η_e to decrease rapidly with decreasing a_0/W ratio, there is a clear trend for the initiation toughness, J_Q, and the slope of the J-R curve, of which the tearing modulus is a measure, to be higher the shorter the original crack length (Fig. 8). (The increasing modulus evident in Fig. 8 for very deep cracks, $a/W > 0.6$, is discussed later.)

Determination of CTOD for Shallow Cracks

The CTOD values plotted in Fig. 9 were calculated based on the procedures of BS 5762, although some changes were necessary to allow for the present specimens being side-grooved, the fatigue crack lengths being measured to ASTM E 813 and the stress intensity factor solution in BS 5762 being limited in its applicability. The main limitation to the use of the formula in BS 5762 for shallow notches is that it is based on a hinge model for plastic displacements. This may not be adequate for shallow notched specimens. You and Knott [*15*] in their investigation into crack depth effects in high-strength steels opted for araldite resin impressions of the crack shape as a means of obtaining CTOD values for shallow cracks. Comparison of the rotation factors, *r*, plotted in Fig. 3 with the value of 0.4 assumed in BS 5762 for plastic displacements indicates that 0.4 is reasonable for deeply cracked test pieces, although in some cases a value nearer 0.5 is approached at large displacements. The tendency for the *r* values to rise rapidly with increased displacement for shallow notched test pieces can be explained in terms of plasticity development. For low displacements where plasticity is very limited, the apparent rotation point will be reasonably close to the middle of the ligament (i.e., $r = 0.5$) and (in the extreme) for no crack at all, the rotation point would presumably be very close to the middle of the ligament. (Having said that, it seems that the *r* value for Specimen M1-32, which has zero crack depth, starts from around 0.3. Possibly the three-point bending configuration affects the rotation point position relative to pure bending; also, the presence of side-grooves may alter the rotation point position.) As plasticity develops, the yielding will not just be confined to the ligament beneath the crack, it can also spread back from the crack tip and flanks to the top surface, and the situation presumably becomes such that the crack is increasingly contained within a yielded skin in a fairly uniform stress field (depending on the work hardening characteristics of the material). This leads to higher values of *r* than occur with a hinge mechanism. The rapidly changing *r* value with displacement for the shallow cracked specimens indicates that the assumption of a single value of 0.4 is inappropriate in this situation. Deduction of CTOD values from the rotation factors measured using the two clip gages, however, may itself be inappropriate because a rotation model assumes that the crack flanks remain straight. For shallow cracks the crack flanks will tend to be curved due to yielding occurring along them. Thus it is not appropriate to use a hinge model to calculate CTOD values for such cases if the measured rotation factor is used, because large overestimates of the crack tip opening displacement may occur.

By using the rotation factor of 0.4 (implicit in the BS 5762 formula) for shallow cracks the rotation point is taken closer to the crack tip than it is measured to be. This may compensate for the fact that the crack flanks are curved, but to an unknown extent. There is experimental evidence that the BS 5762 formula is adequate for a/W ratios as low as 0.15 [*16*]. Indeed Archer [*17*] found that the formulae being used to estimate CTOD values at that time were adequate for calculating CTOD values for a/W ratios down to 0.1. For very shallow cracks, $a/W = 0.05$, Archer's data indicate that the actual CTOD will be overestimated using the BS 5762 formula, but in the present investigation the lowest a_0/W ratio for fatigue precracked specimens was 0.09.

Influence of a/W *Ratio on Initiation Toughness and Tearing Resistance*

The above discussions of the methods used for fracture toughness calculation indicate that the trends shown in Figs. 8 to 10 are not solely caused by the method for calculating fracture toughness values. You and Knott [*15*] also found that for two high strength steels of approximate yield strengths of 700 and 1300 N mm^{-2}, (which they called HY80 and HY130, respectively), the CTOD value at initiation of ductile crack growth, δ_i, was higher the lower the a/W ratios. For "HY80" they found that δ_i was 164% higher for an a/W ratio of 0.1 relative to an $a/W = 0.5$. For "HY130" the equivalent increases in δ_i were 175% ($a/W = 0.1$) and 25% ($a/W = 0.2$). For the present data given in Fig. 9*a*, the average increases in δ_i relative to the δ_i for an a/W ratio of 0.5 are 151% ($a/W = 0.1$) and 56% ($a/W = 0.2$). The equivalent increases in J_Q obtained from Fig. 8*a* are 71% ($a/W = 0.1$) and 31% ($a/W = 0.2$). Thus the trends obtained in the present investigation are similar to those observed by You and Knott despite the fact that You and Knott measured δ_i for the shallow cracks by infiltration of the crack rather than by use of the BS 5762 formulation.

The difference in magnitude of the effect that the a/W ratio has on δ_i and J_Q values for the present investigation may be a result of the J estimation procedure underestimating the J values for short cracks or the CTOD estimation procedure overestimating CTOD, or a combination of the two. Alternatively the magnitude of the a/W effect may be different for the two parameters because the interrelation between the two parameters may also be dependent on a/W; that is, if

$$J = m\sigma_Y\delta \tag{7}$$

then m may be a/W dependent. Work reported by Dawes [*3*], however, indicates that m is little different for $a/W = 0.2$ relative to $a/W = 0.5$.

The increased resistance to crack initiation and growth that is evident for low a/W values (i.e., <0.3) relative to a/W values of ~ 0.5 has also been observed for ferritic steels in the ductile-to-brittle transition region. Thus an increased resistance to initiation of cleavage fracture with reduced a/W was found by Dawes [*3*]. Sumpter [*18*], and Anderson [*19*]. This can be explained, as can the present results, in terms of reduced constraint for shallow cracks, and this is discussed with respect to tearing in Ref *15* and with respect to cleavage fracture in Ref *19*.

A qualitative indication of a reduction in constraint in specimens with low a/W ratios can be obtained from the fracture faces. For relatively deep cracks, say $a/W = 0.5$, it can be seen in Fig. 11*a* that the crack extension is relatively uniform across the specimen thickness. This indicates that the constraint present at the midthickness is similar to that at the roots of the side-grooves. For the very shallow cracks, with $a/W = 0.1$, it can be seen in Fig. 11*b* that the crack extension occurring at the roots of the side-grooves is much greater than that occurring at the midthickness. This indicates that the constraint at the midthickness of the shallow-notched specimens is less than that present for the deeply cracked specimens. If the purpose of side-grooving is to

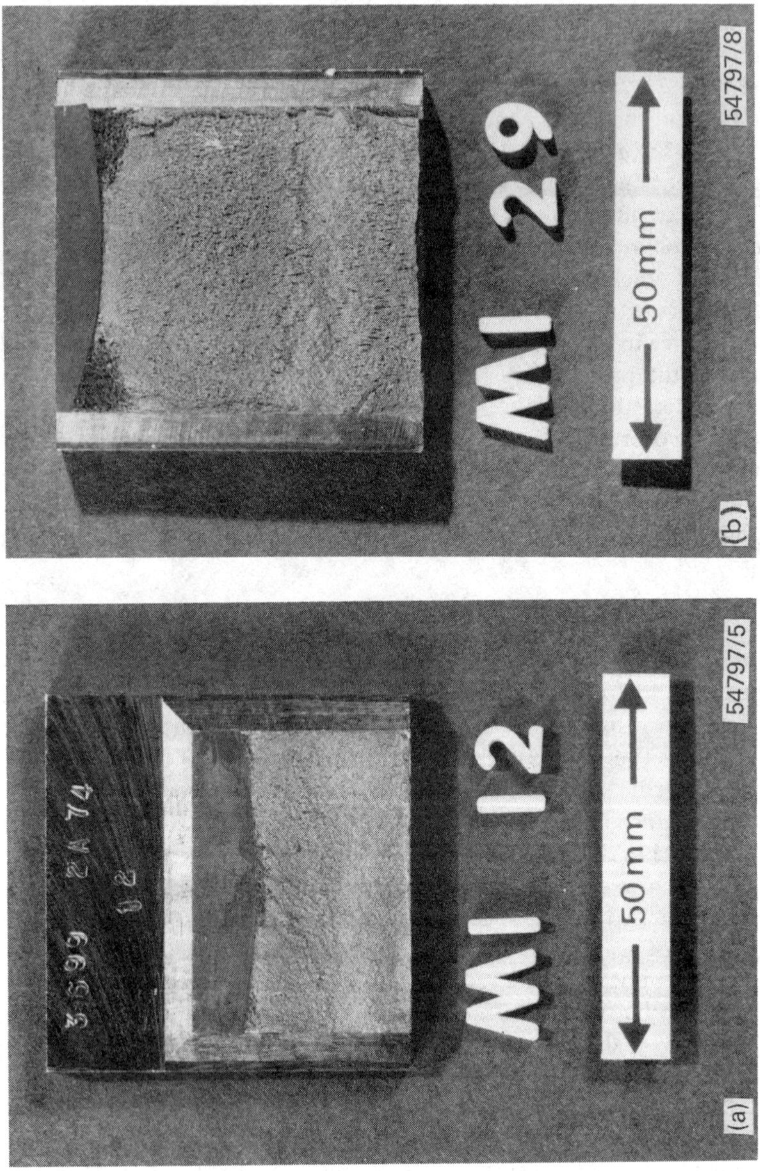

FIG. 11—*Typical fracture faces of fatigue precracked specimens.* (a) *Specimen M1-12 ($a_0/W = 0.56$).* (b) *Specimen M1-29 ($a_0/W = 0.11$).*

produce straight-fronted crack growth (both to improve crack growth predictions and to produce fairly uniform constraint across the crack front) it seems sensible to reduce the severity of the side-grooves for shallow cracks.

The slope of the resistance curves is apparently a/W dependent. Thus the tearing modulus, T, is 97% higher for $a/W = 0.1$ and 62% higher for $a/W = 0.2$ than it is for $a/W = 0.5$ (Fig. 8*b*). At the same time T is 36% higher for $a/W = 0.8$ than it is for $a/W = 0.5$ (Fig. 8*b*). The equivalent results for the slopes of the CTOD R-curves are that the slope is 180% higher for $a/W = 0.1$, 85% for $a/W = 0.2$, and 28% for $a/W = 0.8$ than it is for $a/W = 0.5$ (Fig. 9*b*). Again the variable magnitude of these effects for the J and δ R-curves may be either due to J being underestimated, δ being overestimated, or m being a/W dependent. The increased slope of the R-curves as a/W increases from 0.5 to 0.8 has also been experienced for 25-mm-thick compact tension specimens sampling HY130 by Gudas et al [*20*] and for 50-mm-thick three-point bend specimens sampling "HY130" by Etemad [*21*]. Vassilaros et al [*22*] also found that the tearing modulus (i.e., the slope of the J-R curve) increased as a/W increased from ~ 0.72 to ~ 0.82 in 25-mm-thick compact tension specimens sampling A533B. The fracture toughness for initiation of crack growth was apparently little affected by the change in a/W ratio from 0.5 to 0.8 in the present investigation (Figs. 8 and 9). If anything, the initiation toughness was slightly lower for the deeper cracks. Etemad also found that the initiation toughness, or at least the toughness at low values of crack extension, was lower for $a/W = 0.8$ than for 0.5. Gudas et al [*20*] and Vassilaros et al [*22*] found little consistent effect of a/W ratio on the initiation toughness. These observations appear to tie up with results reported by Anderson, where the effects of geometry on the ductile-to-brittle transition for ferritic structural steel were studied using three-point bend specimens [*20*]. Anderson found that for low values of fracture toughness, specimens with a/W ratios of ~ 0.75 gave lower values of fracture toughness than for a/W of ~ 0.52. For high values of fracture toughness (not necessarily after initiation of tearing), however, the specimens with deeper cracks ($a/W \sim 0.75$) produced higher values of fracture toughness than for $a/W \sim 0.52$. These observations can be explained in terms of constraint. For low values of fracture toughness a crack depth of ~ 0.7 to 0.8 appears to induce slightly greater crack tip constraint than for shallower cracks. As plasticity develops, however, (and the fracture toughness values correspondingly increase) loss of constraint occurs for the very deep cracks ($a/W \sim 0.7$ to 0.8) because the ligament is relatively small. A further complicating factor occurs for R-curves. The shallower the ligament, the steeper the stress and strain gradient into which the crack propagates. How this affects the situation, however, is difficult to assess quantitatively.

Conclusions

A clear trend has been found for the fracture toughness at initiation of ductile crack growth to be higher the shorter the crack length, particularly for "short cracks" with a/W values less than 0.3. This trend has been observed for fracture toughness values calculated both in terms of the J-integral and the crack tip opening displacement.

The slope of the J and CTOD resistance curves was found to increase as a/W was reduced from 0.5 to lower values. The tearing modulus, T, which is a measure of the slope of the J resistance curve, was 97% higher for $a/W = 0.1$ and 62% higher for $a/W = 0.2$ than it was for $a/W = 0.5$. These increases in resistance curve slope can be explained in terms of loss of constraint as the crack length decreases. A qualitative indication of this was provided by the shape of crack growth development. For very shallow cracks, $a/W = 0.1$, the crack growth was more extensive at the roots of the side-grooves than at the specimen midthickness, whereas for deeper cracks the crack growth was more uniform.

For a/W ratios of 0.75 to 0.80 the slopes of the R-curves were found to be higher than for $a/W = 0.5$ by 36% and 28% for the J and CTOD R-curves, respectively, whereas the values for initiation of crack growth were similar, if not slightly lower for the deeper cracks. This behavior can be attributed to a progressive decrease in triaxial constraint with plasticity development for the very deep cracks ($a/W = 0.8$).

Acknowledgments

This work was carried out at The Welding Institute for The David Taylor Naval Ship Research and Development Center as part of a research program on Tearing Instability. The authors wish to acknowledge Dr. H. H. Vanderveldt of the Naval Sea Systems Command (SEA 035) for supporting this project, and J. P. Gudas and M. G. Vassilaros of DTNSRDC for their helpful comments.

The experimental work was carried out by the staff of the Brittle Fracture Laboratory under the supervision of Mr. M. J. Neaves. Software development and data analysis were performed by Messrs. E. K. Briggs and B. Sibley, who are gratefully acknowledged. Finally, thanks are due Dr. A. A. Willoughby, who helped to instigate this project.

APPENDIX I

Derivation of η_p for Three-Point Bend Specimens

The most common means of estimating the J-integral for a specimen is from the area, A, under the load versus load point displacement trace. Thus

$$J = \frac{\eta A}{Bb} \tag{8}$$

where

B = specimen thickness,
b = original uncracked ligament = $W - a_0$,
W = specimen width,
a_0 = original fatigue crack length, and
η = geometric correction factor = $f(a_0/W)$ per ASTM E 813 terminology.

Apart from being dependent on test geometry, the eta factor, η, may also be dependent on the degree of deformation and also, possibly, on the material work hardening characteristics. As a result of the dependence on degree of deformation, the estimate of J is sometimes split into elastic and plastic components [*23*]; that is,

$$J = J_e + J_p = \frac{\eta_e A_e}{Bb} + \frac{\eta_p A_p}{Bb} \tag{9}$$

η_e can be derived from published stress intensity factor, K, solutions and/or compliance solutions because

$$J_e = \frac{K^2}{E'} = \frac{\eta_e A_e}{Bb} \tag{10}$$

where

K = stress intensity factor, and
E' = plane strain Young's modulus.

For the limit load condition (i.e., rigid plasticity), η_p can be simply derived as [*23*]

$$\eta_p = \frac{-(W - a)\,\partial P_L/\partial a}{P_L} \tag{11}$$

where P_L is the limit load.

Haigh and Richards [*11*] have given limit load solutions for four-point bend specimens where

$$P_L = mP_0 \tag{12}$$

where

P_0 = load to yield uncracked specimen, and
m = $1.26\,(1 - a/W)^2$ for $a/W \geq 0.295$,
= $f(1 - a/W)^2$ for $a/W \leq 0.295$.

Individual values of f for a/W values less than 0.295 are also given by Haigh and Richards [*11*]. A curve fit to the f versus a/W data was carried out using spline functions. Substitution of the approximations to P_L given by Haigh and Richards (and the spline function curve fit to their data) into the above equation for η_p produced the η_p versus a/W plot shown in Fig. 7.

APPENDIX II

Derivation of η_e for Side-Grooved Three-Point Bend Specimens

The elastic component of the η factor, η_e, is given by

$$J_e = \frac{K^2}{E'} = \frac{\eta_e A_e}{Bb} \tag{13}$$

where the variables have been defined as in Appendix I.

The above equation can be used to derive the relationship

$$\eta_e = \frac{2(1 - a/W)Y^2}{Z} \tag{14}$$

where Y is given by

$$Y = \frac{KBW^{1/2}}{P} \tag{15}$$

and Z, the nondimensional compliance, is given by

$$Z = \frac{qBE'}{P} \tag{16}$$

The background to this derivation is described in Ref *9*.

Values for Y, which for a given span is a function of the crack length to specimen width ratio, a/W, only, are given in Ref *9* and in the fracture toughness testing standards. Values for Z are given in Ref *9* for non-side-grooved specimens. Z can be divided into two components: that component due to the crack's presence, Z_c, and that component which is independent of the crack's presence, Z_{nc}. For a side-grooved specimen the variable B in the above equations becomes an effective thickness, B_{eff}. To be consistent with the rest of this report B_{eff} has been taken as

$$B_{eff} = B - \frac{(B - B_N)^2}{B} \tag{17}$$

where B is the gross section thickness and B_N is the net section thickness for a side-grooved specimen.

Substitution of B_{eff} in place of B in Eq 16 implies that the "no crack" compliance, Z_{nc}, is different for a side-grooved specimen, being a factor B_{eff}/B times the Z_{nc} value for a side-grooved specimen. This is in fact not the case. The "no crack" displacements occur by bending and shear deflections in the specimen arms. Hence the "no crack" displacements are the same whether or not side-grooves are introduced in the crack plane. (This neglects the slight increase which would occur in the "no crack" compliance due to the presence of the side-grooves.) To compensate for the "no crack" compliance, Z_{nc}, being unaffected by side-grooving whereas the "crack" com-

TABLE 5—η_e *values for three-point bend specimens* (S/W = *4*) *in steel* (ν = *0.3*) *for plane-sided and side-grooved specimens.*

	η_e Values	
a/W	Plane-Sided	Side-Grooved by 20% ($B_N/B = 0.8$)
0.05	0.556	0.579
0.10	0.931	0.968
0.15	1.196	1.240
0.20	1.395	1.443
0.25	1.556	1.605
0.30	1.692	1.740
0.35	1.809	1.855
0.40	1.909	1.950
0.45	1.989	2.026
0.50	2.049	2.081
0.55	2.092	2.118
0.60	2.120	2.141
0.65	2.140	2.157
0.70	2.158	2.170
0.75	2.177	2.186
0.80	2.198	2.204

pliance is, a factor B_{eff}/B can be entered into the Z_{nc} component; thus for a side-grooved specimen:

$$Z = \frac{qB_{eff}E'}{P} = Z_{nc}\frac{B_{eff}}{B} + Z_c \tag{18}$$

The resulting values for η_e, when Eq 18 is substituted into Eq 14 and the values for Y, Z_{nc} and Z_c are taken from Ref *9*, are given in Table 5 for specimens with B_N/B ratios of 0.8 (i.e., 20% side-grooved) and 1 (plane-sided) and for a/W ratios between 0.05 and 0.18.

References

[*1*] Albrecht, P. et al, "Tentative Test Procedure for Determining the Plane Strain J_I-R Curve," *Journal of Testing and Evaluation*, Vol. 10, No. 6, Nov. 1982, pp. 245-251.

[*2*] Okumura, N., Venkatasubramanian, T. V., Unvala, B. A., and Baker, T. J., "Application of the A. C. Potential Drop Technique to the Determination of *R*-curves of Tough Ferritic Steels," *Engineering Fracture Mechanics*, Vol. 14, No. 3, 1981, pp. 617-625.

[*3*] Dawes, M. G., "Elastic-Plastic Fracture Toughness Based on the COD and *J*-contour Integral Concepts," in *Elastic-Plastic Fracture, ASTM STP 668*, J. D. Landes, J. A. Begley, and G. A. Clarke, Eds., American Society for Testing and Materials, Philadelphia, 1979, pp. 307-333.

[*4*] Willoughby, A. A. and Garwood, S. J., "On the Unloading Compliance Method of Deriving Single Specimen *R*-curves in Three Point Bending," in *Elastic-Plastic Fracture: Second Symposium, Vol. II—Fracture Resistance Curves and Engineering Applications, ASTM STP 803*, C. F. Shih and J. P. Gudas, Eds., American Society for Testing and Materials, Philadelphia, 1983, pp. II-372 to II-397.

[5] Marandet, B. and Sanz, G., "Experimental Verification of the J_{Ic} and Equivalent Energy Methods for the Evaluation of the Fracture Toughness of Steels," in *Flaw Growth and Fracture, ASTM STP 631*, American Society for Testing and Materials, Philadelphia, 1977, pp. 462–476.

[6] Tada, H., Paris, P. C., and Irwin, G. R., "The Stress Analysis of Cracks Handbook," Del Research Corp., St. Louis, Mo., 1973, p. 2.17.

[7] Garwood, S. J. and Willoughby, A. A., "Fracture Toughness Measurements on Materials Exhibiting Stable Ductile Crack Extension," paper presented to International Conference on Fracture Toughness Testing, London, 9–10 June 1982, Paper 32.

[8] Freed, C. N. and Krafft, J. M., "Effect of Side Grooving on Measurements of Plane-Strain Fracture Toughness," *Journal of Materials*, Vol. 1, No. 4, Dec. 1966, pp. 770–790.

[9] Towers, O. L., "Tests for Fracture Toughness and Fatigue Assessment: A Compilation of Stress Intensity Factors, Compliances and Elastic η Factors," The Welding Institute, Cambridge, England, 1985.

[10] Turner, C. E., "Methods for Post-Yield Fracture Safety Assessment," in *Post-Yield Fracture Mechanics*, D. G. H. Latzko, Ed., Applied Science Publishers, London, 1979, pp. 23–210.

[11] Haigh, J. R. and Richards, C. E., "Yield Point Loads and Compliance Functions of Fracture Mechanics Specimens," Central Electricity Research Laboratories Memorandum RD/LM461, 3 May 1974, Leatherhead, Surrey, England.

[12] Paris, P. C., Tada, H., Zahoor, A., and Ernst, H., "The Theory of Instability of the Tearing Mode of Elastic-Plastic Crack Growth," in *Elastic-Plastic Fracture, ASTM STP 668*, J. D. Landes, J. A. Begley, and G. A. Clarke, Eds., American Society for Testing and Materials, Philadelphia, 1979, pp. 5–36.

[13] Voss, B. and Mayville, R. A., "The Use of the Partial Unloading Compliance Method for the Determination of J_I-R curves and J_{Ic}," in *Elastic-Plastic Fracture Test Methods: The User's Experience, ASTM STP 856*, E. T. Wessel and F. J. Loss, Eds., American Society for Testing and Materials, Philadelphia, 1985, pp. 117–130.

[14] Neale, B. K. and Priest, R. H., "The Unloading Compliance Method for Crack Length Measurement Using Compact Tension and Precracked Charpy Specimens," in *Elastic-Plastic Fracture Test Methods: The User's Experience, ASTM STP 856*, E. T. Wessel and F. J. Loss, Eds., American Society for Testing and Materials, Philadelphia, 1985, pp. 375–393.

[15] You, C. P. and Knott, J. F., "Ductile Fracture in High-Strength Structural Steels," in *Proceedings*, 4th European Colloquium on Fracture, EGF4, Pergamon Press, Oxford, 1982, Vol. 1, pp. 23–29.

[16] Slatcher, S., "The Accuracy of COD Values Calculated to BS 5762," *International Journal of Fracture*, Vol. 22, No. 2, June 1983, pp. R53–R54.

[17] Archer, G. L., "The Relationship Between Notch Tip and Notch Mouth Displacements in SENB Fracture Toughness Specimens," Welding Institute Research Report E/63/75, Cambridge, England, March 1975.

[18] Sumpter, J. D. G., "The Effect of Notch Depth and Orientation on the Fracture Toughness of Multi-Pass Weldments," *International Journal of Pressure Vessels and Piping*, Vol. 10, 1982, pp. 169–180.

[19] Anderson, T. L., "The Effects of Crack-Tip Region Constraint on Fracture in the Ductile-to-Brittle Transition Region," Ph.D. dissertation, Colorado School of Mines, Golden, CO, Sept. 1983.

[20] Gudas, J. D., Joyce, J. A., and Davis, D. A., "Investigation of Specimen Geometry Modifications to Determine the Conservative J_I-R Curve Tearing Modulus Using the HY-130 Steel System," in *Fracture Mechanics, ASTM STP 677*, C. W. Smith, Ed., American Society for Testing and Materials, 1979, pp. 474–485.

[21] Etemad, M. R., "Use of Resistance Curves in Post-Yield Fracture Mechanics," Ph.D. thesis, Imperial College, London, 1983.

[22] Vassilaros, M. G., Joyce, J. A., and Gudas, J. P., "Effects of Specimen Geometry on the J_I-R Curve for ASTM A533B Steel," in *Fracture Mechanics: Twelfth Conference, ASTM STP 700*, American Society for Testing and Materials, Philadelphia, 1980, pp. 251–270.

[23] Turner, C. E. in *Fracture Mechanics: Twelfth Conference, ASTM STP 700*, American Society for Testing and Materials, Philadelphia, 1980, pp. 314–337.

M. R. Etemad[1] *and C. E. Turner*[1]

An Investigation of the *I* and *dJ/da* Concepts for Ductile Tearing Instability

REFERENCE: Etemad, M. R. and Turner, C. E., "**An Investigation of the *I* and *dJ/da* Concepts for Ductile Tearing Instability,**" *Fracture Mechanics: Seventeenth Volume, ASTM STP 905*, J. H. Underwood, R. Chait, C. W. Smith, D. P. Wilhem, W. A. Andrews, and J. C. Newman, Eds., American Society for Testing and Materials, Philadelphia, 1986, pp. 485–502.

ABSTRACT: Unstable ductile tearing was found in a series of tests on notched bend-bars of HY130 steel tested in a screw-driven machine. *R*-curves were measured and ductile tearing stability analyzed according to the energy rate, *I*, and *dJ/da* tangency concepts. The effect of compliance on the energy rate available and an appropriate estimation of work dissipation rate for the *I* theory are described. The *dJ/da* concept is used in both its "exact" form and as originally stated in terms of *T*, with certain compliance terms modified. Subject to choice of appropriate compliance terms, the experimental behavior is predicted satisfactorily by both *I* and *dJ/da* methods, although the restriction of this analysis to a configuration very similar to that on which the *R*-curve was measured must be recalled.

KEY WORDS: fracture mechanics, crack propagation, *J*-integral, mechanical properties, *R*-curves, elastic-plastic fracture

An energy rate balance concept was proposed [*1,2*] for predicting unstable ductile tearing, separate from the tearing modulus or *dJ/da* tangency concept derived by Paris et al [*3*] and Hutchinson and Paris [*4*]. The energy rate balance, *I*, was seen as a special case of the well-known Orowan second derivative of energy condition for ductile instability [*5*] and therefore fundamentally correct, but difficulties were apparent in attempting to evaluate the terms required. The object of the present paper is to summarize the energy rate, *I*, theory as presently formulated and to compare results with the *T* or *dJ/da* theories by applying them to some experimental cases of ductile tearing instability obtained from HY130 steel on certain bending tests.

[1] Post-Doctoral Research Assistant and Professor of Materials in Engineering, respectively, Mechanical Engineering Department, Imperial College, London, England.

R-Curves, Work Rate, and Instability

Prior to initiation, J can be related to the area under the load deformation curve, A, the ligament area, Bb, and η, which is a function of geometry and perhaps of the hardening exponent, n, [*2,6,7*]:

$$J = \eta\ A/Bb \tag{1}$$

interpreted with subscripts "el" for the elastic case, "pl" for strictly plastic behavior, and "o" for the overall elastic-plastic case. For deep notch bending $\eta = 2$ and for many test pieces η depends but little on degree of deformation (or hardening exponent), but in general η_{el} is a function of length or span, whereas, at least for deep notch low hardening cases, η_{pl} is not. Nevertheless, estimates of η can be made for linear elastic fracture mechanics (LEFM) and fully plastic behavior, so that in practice a satisfactory estimate can usually be made for other cases. Prior to initiation, A can be interpreted either as work done, U, or internal energy, w (not necessarily recoverable). After initiation there is some doubt whether dJ should be related to change of internal energy, dw, or to work increment, dU, early workers using dU, later proposals [*8,9*] advocating dw. More recent work [*10*] points out that dU seems appropriate in the rigid plastic limit; other proposals [*11,12*] suggest a compromise between dU, which suits the fully plastic case, and dw, which is necessary for LEFM.

It is beyond the scope of this paper to discuss the merits of these various terms beyond noting the relationship

$$dw = dU - BJda \tag{2}$$

Equations 1 and 2 were used [*8*] to define a particular R-curve, J_r, later known as the deformation theory model, in which dJ is related to dw, itself derived by "correcting" dU according to Eq 2 in an incremental procedure to give

$$\frac{dJ_r}{da} = \frac{d}{da}(\eta w/Bb) \tag{3a}$$

$$= \frac{\eta}{Bb}\frac{dw}{da} + f(\eta)\frac{J_r}{b} \tag{3b}$$

$$= \frac{\eta}{Bb}\frac{dU}{da} + g(\eta)\frac{J_r}{b} \tag{3c}$$

where $f(\eta) \equiv 1 + (b/\eta)(d\eta/da)$ and $g(\eta) = f(\eta) - \eta$. This is closely similar to, though not quite identical with, ASTM Test for J_{Ic}: A Measure of Fracture Toughness (E 813). If, however, a parameter to suit [*10*] is followed, here termed J_U, it seems

$$\frac{dJ_U}{da} = (\eta/Bb)\frac{dU}{da} \tag{4}$$

This implies that, for unstable behavior at fixed displacement (i.e., no increment of external work), in the rigid plastic limit:

$$dJ_U/da = 0 \tag{5}$$

The question of whether either definition of a *J-R* curve is geometry independent is not further pursued here, since the limited objective is to discuss unstable behavior of the bend bars of similar configuration to those in which the *R*-curves are measured (i.e., differing only by a/W, if at all). To that extent, as Kaiser and Carlsson [*13*] have pointed out, what is being examined is the ability to estimate correctly the terms that affect unstable behavior in the laws of mechanics, rather than the complete fracture mechanics problem of translation from one configuration to another.

Statement of the *I* Theory

The energy rate per unit thickness available from an elastic-plastic body with (linear) elastic unloading (EPE material) was termed *I* [*1,2*]. Expressed as a multiple of *G* (see Appendix):

$$I = G[2(\eta_o/\eta_{el}) - 1] \tag{6a}$$

If expressed as an additive term

$$I = G + G_p \tag{6b}$$

it is seen that *G* is the same "free" elastic energy as is available for LEFM, whereas the additional term, G_p, is a "bound" potential energy that arises because of an additional change of displacement, *sr*, on unloading (see Appendix, Fig. 5), which is both $-dq_{el}$ and $+dq_{pl}$, since dq_{total} is zero. This extra term is included in the evaluation of the energy rate available, *I*, because the dissipation term, dU/Bda, that has to be exceeded for unstable behavior itself includes both local "fracture" work and remote plasticity. The balance of energy rates is thus between overall available and dissipative energy rate terms, because that is how they are conveniently measured or estimated. It is recognized that, in principle, the plasticity term could be subtracted from both sides if a method were available for separating the surface (fracture) from the more general plasticity dissipation. This may yet prove feasible [*14*], but is not pursued here.

The I theory has been restated [*15*] following an examination of the proposal in light of a series of experiments that gave unstable ductile tearing in bend bars of HY130 steel. In summary, two factors greatly influenced the value of the various terms. The first was the assessment of η_o/η_{el} in the term for I as $(\eta_o/\eta_{el})_s$ for the whole system, where not only η_{el} (as previously recognized) but also η_o was strongly dependent on system compliances. The expressions obtained (see Appendix) are

$$\eta_{o,s} = (\eta_{el,t} - X\eta_{pl,t})/(\alpha + X) \tag{7a}$$

$$\eta_{el,s} = \eta_{el,t}/\alpha \tag{7b}$$

where subscript "t" refers to just the test piece for which η_{el} is known [*2,6*] and subscript "s" refers to the whole system of test piece plus machine or structure; $\alpha = 1 + (\Phi_m/\Phi_t)$ where Φ_m is compliance of machine or structure and Φ_t is compliance of the test piece; X is the ratio of overall plastic work to test piece elastic work, $U_{pl,s} \equiv U_{pl,t} \equiv XU_{el,t}$ since it is supposed there is no dissipation except in the test piece (i.e., the machine or structure is purely linear elastic). The second factor was the evaluation of the dissipative term,[2] itself taken as $dU_{(dissipation)} = dU_{(total)} - dU_{(elastic)}$. For fixed displacement dU_{el} is taken as $-BIda$; for fixed load $dU_{el} = +BIda$. The term $dU_{(total)}$ is taken from the definition of the R-curve, for which the choice of formula is not important whilst instability of only the same configuration as used in determining the curves is under discussion. Thus from Eq 3*c*:

$$\frac{dU}{Bda}(\text{total}) = \frac{b}{\eta}\frac{dJ_r}{Bda} - \frac{J_r}{\eta}g(\eta) \tag{8}$$

and for stability

$$I > \frac{dU}{Bda}(\text{diss}) \tag{9a}$$

$$> \frac{b}{\eta}\frac{dJ_r}{da} - \frac{J_r}{\eta}g(\eta) - (\pm I) \tag{9b}$$

For fixed displacement over the length on which the load-displacement diagram is measured, the minus sign applies and instability would occur when

[2] Recall that "dissipation" here includes both fracture surface energy and more remote plastic work; "total" also includes elastic work.

$$\frac{b}{\eta}\frac{dJ_r}{da} = \frac{J_r}{\eta} g(\eta) \tag{10a}$$

which from Eqs 4 and 3*c* corresponds to dU/da, or dJ_U/da, being zero (i.e., no external work contributed to the component, even by the machine).

For a sustained load on the component the plus sign applies, so instability occurs when

$$2I > \frac{b}{\eta}\frac{dJ_r}{da} - \frac{J_r}{\eta} g(\eta) \tag{10b}$$

that is,

$$\frac{dU}{Bda}\,(\text{diss}) = \left\{\frac{b}{\eta}\frac{dJ_r}{da} - \frac{J_r}{\eta} g(\eta)\right\}\Big/2 \tag{10c}$$

Equation 2 can be arranged as $dU - dw = BJda$. The left hand side (l.h.s.) is $-dP$ where P is potential energy. The right hand side (r.h.s.) refers to nonlinear elastic (NLE) behavior. For LEFM the difference between dU and dw is $BGda$. If it is argued that for EPE material the term should be $BIda$, and $g(\eta)$ would be $f(\eta) - (I/J)\eta$, then Eq 9*a* further reduces to

$$I > \frac{dU}{Bda}\,(\text{diss}) = \frac{b}{\eta}\frac{dJ_r}{da} - \frac{J_r}{\eta} f(\eta) \tag{10d}$$

that is, at instability $dU_{(\text{diss})} = dw$.

Statement of the dJ/da or T Theories

The mathematically rigorous derivation of $\left.\frac{dJ}{da}\right|_{\text{app}}$ is set out in Ref *4*. The expressions are in terms of load point displacement, q, and machine compliance, Φ_m:

$$T_{\text{app}}\frac{\sigma_0^2}{E} = \frac{dJ}{da} = \left.\frac{\partial J}{\partial a}\right|_Q - \left.\frac{\partial J}{\partial Q}\right|_a \left.\frac{\partial q}{\partial a}\right|_Q \left[\Phi_m + \left.\frac{\partial q}{\partial Q}\right|_a\right]^{-1} \tag{11}$$

or

$$= \left.\frac{\partial J}{\partial a}\right|_q - \left.\frac{\partial J}{\partial q}\right|_a \left.\frac{\partial Q}{\partial a}\right|_q \left[K + \left.\frac{\partial Q}{\partial q_c}\right|_a\right]^{-1} \tag{12}$$

where q_c is the displacement due to the crack alone and K is the stiffness of the machine plus uncracked test piece. The various differential terms in Eqs

11 and 12 were evaluated by assuming fully plastic behavior and using Eqs 1 and 3*c*. Using Eq 1 in the form

$$J = \frac{\eta \int Q \Delta q}{Bb}$$

Hence

$$\left.\frac{\partial J}{\partial q}\right|_a = \frac{\eta Q}{Bb} \tag{13}$$

Similarly, using Eq 3*c* and keeping q constant, gives

$$\left.\frac{\partial J}{\partial a}\right|_q = g(\eta)\, J/b \tag{14}$$

Furthermore, using $A = f(Q, q)$ and

$$J = -\frac{1}{B}\left.\frac{\partial A}{\partial a}\right|_q$$

or the complementary term

$$\left.\frac{\partial J}{\partial q}\right|_a = -\left.\frac{\partial Q}{B\partial a}\right|_q \quad \text{and} \quad \left.\frac{\partial J}{\partial Q}\right|_a = \left.\frac{\partial q}{B\partial a}\right|_Q \tag{15a,b}$$

For deep notch bending, $\eta = 2$, L (the constraint factor) $= 1.4$. Substituting Eqs 15, 14, and 13 into Eq 12 gave [4]

$$T_{\text{app}}\frac{\sigma_o^2}{E} = \frac{dJ}{da} = \frac{4Q^2}{b^2B}\left[K + \left.\frac{\partial Q}{\partial q_c}\right|_a\right]^{-1} - \frac{J}{b} \tag{16a}$$

If Eq 12 is used with Eq 15 but η is not taken as 2, and Eq 14 is not yet introduced, then provided the variables of geometry and deformation are separable,

$$T_{\text{app}}\frac{\sigma_o^2}{E} = \frac{dJ}{da} = \left.\frac{\partial J}{\partial a}\right|_q + \left[\left.\frac{\partial J}{\partial q}\right|_a\right]^2 \left[\frac{B}{\partial Q/\partial q_c|_a + K}\right] \tag{16b}$$

which is identical to Eq 20 [11].

A simple expression for T_{app} in deep notch bending at limit load was given [3] as

$$T_{\text{app}} = \frac{2b^2 S_e}{W^3} - \frac{\Theta E}{\sigma_y} = \frac{2b^2 S_e}{W^3} - \frac{JE}{b\sigma_y^2} \tag{17a,b}$$

where Θ is the angle of bend and S is an effective span, $S = S[1 + (\Phi_m/\Phi_u)]$, Φ_m is compliance of machine or structure and Φ_u is compliance of the unnotched test bar. In Ref *4* Eq 17*b* was given, obtained from degenerating Eq 11, using $g(\eta) = -1$, and by an estimate of the limit load, $Q_L = L\sigma Bb^2/S$, but was restricted to a rigid machine. The difference implied for a compliant system is discussed later in terms of a different effective span.

Experimental Program

A series of tests was conducted on three-point bend bars of HY130 steel to determine R-curves over a wide range of geometric proportions and to find unstable ductile behavior [*15*]. Only one relevant R-curve is reported here, together with a series of tests where $B = 50$ mm, $W = 48$ mm, $S = 200$ mm, and $a/W \simeq 0.3$. Of the five nominally similar tests, three clearly went unstable; one was stopped probably just prior to instability and for one the record went off scale at the point of interest so that it is uncertain whether the instability occurred. All tests were conducted in a screw-driven machine at a constant cross-head speed of 2 mm/min. Clip gage mouth opening and cross-head movement were recorded, the latter with a correction for local indentation of the loading rollers. All pieces were fatigue pre-cracked but not side-grooved. The relevant machine and material properties are given in Table 1. A typical cross-head test record showing unstable behavior is shown in Fig. 1. The R-curve measured on other (stable) pieces, $a/W = 0.5$, and on some of the present tests, is shown in Fig. 2. The solid line R-curve is based on a second-order polynominal best-fit used [*15*] for the $a/W \simeq 0.3$ data; the dotted R-curve is a by-eye fit to the combined data. From the five pieces, several possible records are made up in Table 2 allowing for uncertainty in the exact amount of crack growth at instability. The limits are the amounts observed on the five pieces at the end of the test, not more than 6.5 mm. In view of the recorded loads, that amount of growth is impossible for an occurrence prior to instability, since the highest maximum possible limit load for the remaining ligament would be exceeded. However, five plausible conditions are given in Table 2 for the purpose of analysis. It should be clear these do not represent five successive stages during a test, but five possible combinations of load, actual crack length, and other data representative of the conditions at instability in the five tests.

TABLE 1—*Test data.*

Material: HY130; test direction: TS; Modulus $E = 210 \times 10^3$ MN/m^2			
$\sigma_y = 896$;	$\sigma_{fl} = 933$;	$\sigma_u = 970$ (units MN/m^2)	
$J_i = 0.17$;	$J_{Ic} = 0.18(0.21)$*	based on $J = 2\sigma_{fl}\Delta a$ (units MN/m)	
Nominal dimensions (mm): $B = 50$;	$W = 48$;	$S = 200$;	$a_o/W = 0.25$
Machine compliance (mm/kN): $\Phi_m = 28 \times 10^{-4}$;	Φ_u (un-notched) $= 15 \times 10^{-4}$;		
	Φ_n (notched) $= 27 \times 10^{-4}$		

*Values in parentheses refer to the dotted R-curve, fitted "by-eye" in Fig. 2.

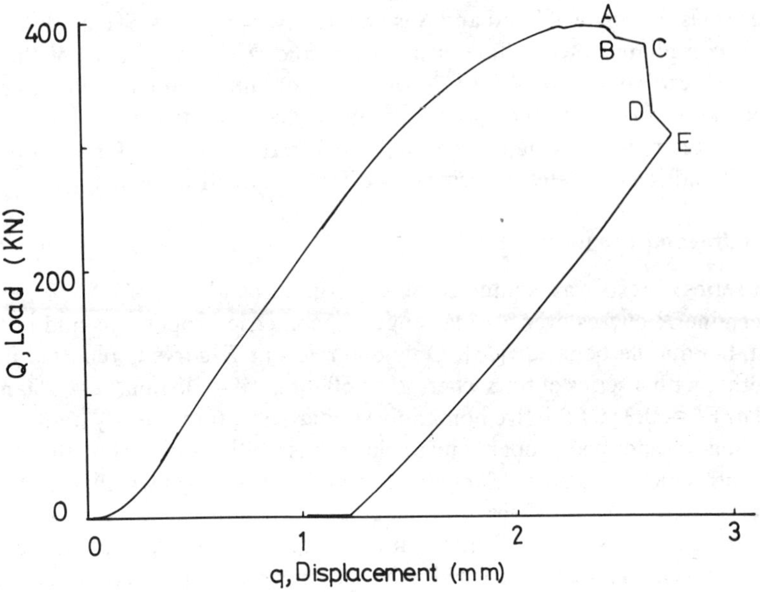

FIG. 1—*Typical load-displacement diagram showing unstable ductile behavior.*

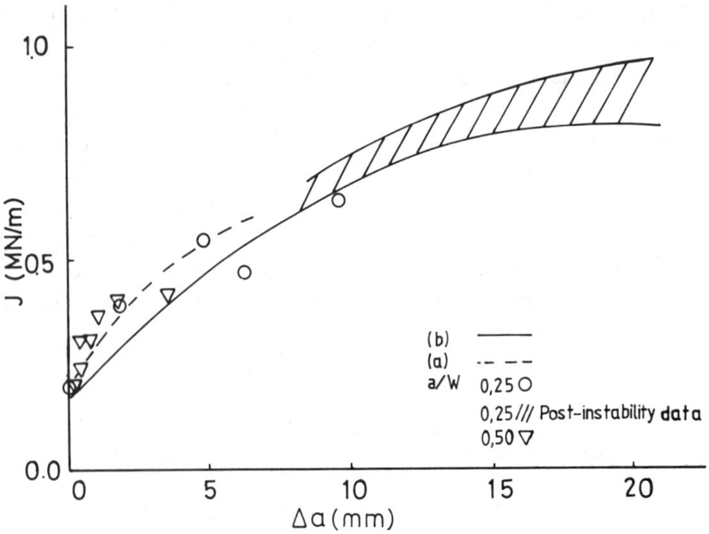

FIG. 2—R-*curve data for HY130 from three-point bend tests, with two particular estimates, curves* (a) *and* (b) *used in the analyses.*

TABLE 2—*Test data for onset of instability.*

(A) *Mean data representative of the five nominally similar tests.*

Q_{max}*	Q_{inst}*	Q_{lim}†	q_{inst}	Δa_{inst}
398	386	300 to 454	1.4	3.5 to 6.5

(kN and mm units)

(B) *Five plausible sets of data to represent the tests, as used in the analyses.*

	Δa	a/W	$\Delta a/b$	J_r‡	dJ/da‡	ω	Q_{inst}	σ§	Y	G
M0	0	0.25	0	0.17(0.17)	0.066(0.11)	13.5	386	1.01	1.73	0.167
M2	2.0	0.29	0.059	0.30(0.38)	0.060(0.070)	6.8	386	1.01	1.83	0.217
M3	3.5	0.32	0.108	0.39(0.48)	0.052(0.050)	4.6	369	0.966	1.88	0.232
M4	4.5	0.34	0.148	0.46(0.54)	0.048(0.043)	3.3	357	0.954	1.92	0.252
M6	6.5	0.39	0.220	0.54(0.58)	0.043(0.038)	2.5	308	0.645	2.08	0.152

(kN and mm units)

*There is some uncertainty whether the difference between Q_{max} and Q_{inst} is real or a recording error (see Fig. 1). Note also that inertia in the machine affects region *BC* of Fig. 1.

†Q_{lim} is based on σ_{fl} with constraint factor $L = 1$ for plane stress (lower limit) and $L = 1.155 \times 1.30 = 1.5$ for plane strain (upper limit). For the deeper notches, the geometric constraint is increased to 1.33 (for M4) and 1.35 (for M6) to give $L = 1.53$ and 1.56 respectively.

‡Values in parentheses refer to the dotted *R*-curve, fitted "by-eye" in Fig. 2.

§The gross section or LEFM stress at Q_{inst} and corresponding value of *G*.

The *I* Energy Rate Balance Method

Equation 10*d* is used for predicting instability, since the load here is well maintained. This instability will occur if

$$I > (b/\eta)(dJ_r/da) - (J_r/\eta)f(\eta) \tag{18}$$

In estimating the energy dissipation rate (r.h.s.) there is some uncertainty whether the geometric factors relate to the test piece or the whole system. The dissipation occurs in the former, but the instability relates to the latter; both sets of numbers are shown. Three sample cases are given in Table 3. To estimate the energy rate available (l.h.s.), the expression for *I* (Eq 6*a*) is used, $I = G[(2\eta_o/\eta_{el}) - 1]$, where both η_o and η_{el} refer to the whole system, including the effect of compliance. The term $\alpha = \{1 + (\Phi_m/\Phi_t)\} = 2.04$ (see Footnote 3). The ratio of plastic work to elastic test piece work, *X*, is required to evaluate η_o for the structure, and that is estimated from the loading diagram. Plane strain conditions are used for *G* because the value of maximum load

[3] Since Φ_t is the compliance of the notched component, a rather larger value of α might be appropriate to case M4 and M6 where the notch depth is significantly greater than for M0 and M2.

seems more compatible with that case. Table 3 data are in kiloNewton and millimetre units.

The results for J_r, I, and dw/Bda are plotted in Fig. 3. The choice of R-curve and η values in Table 3a affects the conclusions in this case by only about 1 mm on the crack length predicted for instability.

The *dJ/da* Analysis

Equation 12 is used in the dJ/da analysis with the same five cases postulated in Table 2. The compliance C is taken as the sum of machine plus unnotched values [*4*]. The term $(\partial Q/\partial q_c)|_a$ has been set zero, following the suggestion in Ref *4* that it may be possible to neglect it for near limit load behavior. The several other terms in Eq 12 are evaluated as in Eqs 13 to 15. The results are plotted in Fig. 4, showing instability, or nearly so, for the test data. This, in effect, is using Eq 16b with η appropriate to the notch depth in question, rather than using Eq 16b with $\eta = 2$ as in Ref *3*. In Eq 16b, $\partial J/\partial q|_a = \eta Q/Bb$, and in Ref *11* an experimental value of load, Q, is used rather than the limit load estimate of Ref *3*.

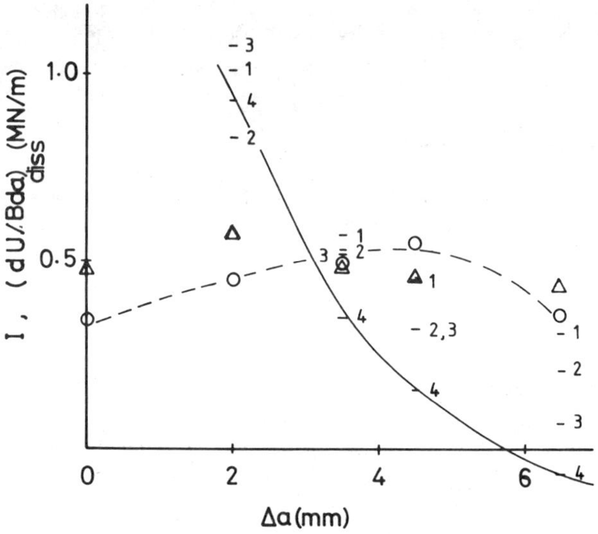

I dashed curve ○ (Eqs 6a, 7a,b)
dU_{diss}/Bda Eq 10c Δ (using η_{ot} and R-curve (b))
Eq 10d 1 (using η_{ot} and R-curve (a))
Eq 10d 2 (using η_{ot} and R-curve (b))
Eq 10d 3 (using η_{os} and R-curve (a))
Eq 10d 4 (using η_{os} and R-curve (b))
(Solid line indicates trend of 4).

FIG. 3—*Stability predictions by* I *analysis.*

TABLE 3—*Data used in* I *analysis.*

(A) *Dissipation (r.h.s.)*

	Δa	$a + \Delta a$	b_{inst}	$(a + \Delta a)/W$	η_{ot}	$(d\eta_o/da)_t$	J_r*	dJ_r/da*	dw/Bda*
M3	3.5	15.5	32.5	0.32	1.32 1.41†	0.038 0.060	0.39(0.48)	0.052(0.050)	0.57(0.53) 0.54(0.34)
M4	4.5	16.5	31.5	0.34	1.86 1.50†	0.028 0.059	0.46(0.51)	0.048(0.43)	0.45(0.32) 0.33(0.15)
M6	6.5	18.5	29.5	0.39	1.90 1.58†	0.017 0.059	0.54(0.58)	0.043(0.038)	0.43(0.39) 0.07(−0.07)

(B) *Available (l.h.s.)*

	σ	Y	a	G	X	η_{os}	η_{els}	$\left.\frac{2\eta_o}{\eta_{el}}\right\|_s$	I	Result*
M3	0.966	1.88	15.5	0.232	1.75	1.41	0.90	3.13	0.50	stable (stable) stable (unstable)†
M4	0.954	1.92	16.5	0.252	2.67	1.45	0.91	3.19	0.55	unstable (unstable) unstable (unstable)†
M6	0.645	2.08	18.5	0.152	3.00	1.58	0.95	3.32	0.35	stable (marginal) unstable (unstable)†

(kN and mm units)

Thus instability is predicted to occur for case M3 or M4, within the range of uncertainty of the appropriate values of Δa in the five tests conducted.

*Values in parentheses refer to the dotted *R*-curve, fitted "by-eye" in Fig. 2.

†For the first set of results for each case the values used in estimating the dissipation are η_o and $d\eta_o/da$ for the test piece, whereas for the second set the values are for the whole system.

Since ductile instability occurs at or after maximum load, it is supposed $\partial Q/\partial q_c|_a$ should be negative. As a limit to its value, the slope *AC* has been measured from Fig. 1, but if, as believed, the crack is already growing, the true value is between zero (limit state slope) and this measured value. Finally, Eq 21 in Ref *11* is used where $\partial Q/\partial q_c|_a$ is re-expressed in terms of T_{mat} to give

$$T_{app} = \frac{E}{\sigma_y^2}\left\{\left.\frac{\partial J}{\partial a}\right|_q + \left(\frac{\partial J}{\partial q}\right)_a^2\left(\frac{1}{K + \dfrac{dQ}{dq} + \dfrac{(\partial J/\partial a)^2}{(dJ/da)_{mat} - (\partial J/\partial a)_q}}\right)\right\} \tag{19}$$

All these results are shown in Fig. 4.

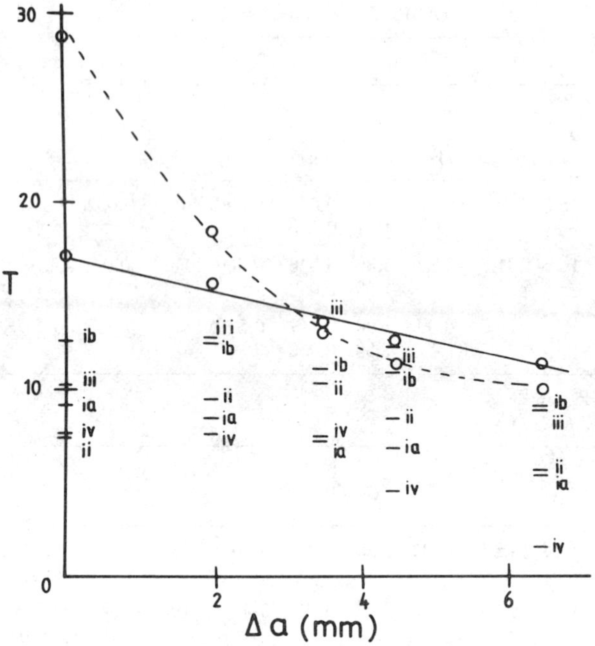

T_{mat} ○ dashed uses R-curve (a)
○ solid uses R-curve (b)
T_{app} (i) Eq 17b (a) refers to Eq 20
(b) refers to Eqs 21 and 22
(ii) Eq 16b $\partial Q/\partial q_c|_a = 0$
(iii) Eq 16b $\partial Q/\partial q_c|_a$ from Fig. 1
(iv) Eq 19 (based on the dashed R-curve)

FIG. 4—*Stability prediction by* T *analysis.*

Limit Case *T* Analysis

Equation 17b is used with S_e retained as an effective span to give a simple analysis for near limit state. Because the notch ratio is $a/W \simeq 0.3$, the constant 2 (which is based on $\eta = 2$ for deep notches) is modified to suit the values of η and constraint, already listed Tables 1 and 2. The effective span, S_e, was given in Ref *3* as $S[1 + (q_m/q_u)]$. This is interpreted as

$$S_e = S[1 + (\Phi_m/\Phi_u)] \tag{20}$$

where subscript "m" relates to machine (including spring bars in Ref *3* but not relevant here) and "u" to the un-notched test piece. The data for that case are shown in Fig. 4. Another curve is shown, defined using

$$\bar{S}_e = S\Phi_e/\Phi_u \tag{21}$$

where Φ_e is the effective compliance defined by

$$\Phi_e = C/[1 - (\Phi_m/\Phi_r)] \tag{22}$$

where $C = \Phi_m + \Phi_u$ and $1/\Phi_r = [\partial Q/\partial q_c|_a]$ (i.e., the numerical value of the negative slope of the term in the denominator of Eq 19). For limit state $1/\Phi_r \rightarrow 0$, and $\Phi_e \rightarrow C$ in Ref *3*. Equation 22 simply treats the denominator of the last term in Eq 16*b* as an effective compliance Φ_e, so the restriction in Ref *4* to a rigid machine in order to relate Eq 12 to Eq 17*b* does not seem necessary. It appears that, for the deep notch bend case, Eq 12 degenerates to Eq 17*b* with S_e as in Ref *3* if $1/\Phi_r = 0$, and to Eq 17*b* using $\bar{S}_e$ (Eq 21) rather than S_e (Eq 20) if $1/\Phi_r$ has some non-zero value, such as measured for Fig. 1. Thus four cases are shown in Fig. 4, all in the T notation, $(\sigma_y^2/E)(dJ/da)$:

1. T from Eq 17*b* but using appropriate value of η; (a) using S_e (Eq 20) (b) using $\bar{S}_e$ (Eqs 21 and 22).
2. dJ/da from Eq 16*b* with $\partial Q/\partial q_c|_a = 0$.
3. dJ/da from Eq 16*b* with $\partial Q/\partial q_c|_a = -1/16$ MN/mm (or $\Phi_r = 160 \times 10^{-4}$ mm/kN).
4. dJ/da from Eq 19 with $dQ/dq = -1/16$ MN/mm.

Discussion

Despite the general apparent validity of both I and dJ/da methods, a number of uncertainties remain. Two possible interpretations of dU were noted, Eqs 10*c* and 10*d*. The former is clearly more consistent in terms of J; the latter mixes use of J and I for physical arguments not strictly mathematically supported. Equation 10*d* is also used in four variations: choice of R-curve (a) or (b) in Fig. 2 and choice of η_{ot} or η_{os} in Table 2. The authors find it difficult to make entirely self-consistent choices between these possibilities, but, as seen in Fig. 4, the trend of Eqs 10*c* and 10*d* is rather different although actual prediction of instability differs but little. It is also noted that the sustained load case (Eq 10*b*) is used despite the inclusion of compliance effects in evaluating I. It is possible to formulate dw_{el}/Bda to include the effects of partial unloading, but it is not clear how to break down the system compliance, since the instability statement $dq_t = 0$ only implies $\Sigma d\Phi = 0$ for all the components (machine, un-notched length, the notch itself).

The justification for the use of I from these data could rest on the near limit state behavior equivalent to taking $\partial Q/\partial q_c|_a$ as zero in the dJ/da analyses. A more general agreement is that the R-curve on which the dissipation is based is itself derived at constant or slightly rising load, and contains the increase of elastic energy that has to be subtracted from Eq 9*b*. The concept is that the R-curve (including both plastic and elastic terms) is a material property, which is then scaled to the component by the relevant b and overall η terms to give the energy absorbed from which the elastic term is then subtracted to give the dissipation. If that logic is correct, it might be preferable to derive dissipative R-curves only which would include fracture and plastic work.

Although the theories differ in that the I method attempts to balance energy rates whereas the dJ/da method balances characteristic severities, there is a close similarity. This can be seen in outline in that Ref *1* starts from the Orowan second-derivative statement for instability and Ref *16* leads to it. It can be seen in closer detail because the dJ/da method now explicity accepts the variables separable formulation [*6,7*] that was advocated for the I method from the start [*17*], so that terms such as Eq 14 are now identical in both theories, and functions of the R-curves are identical (for a given definition of the J-R curve which is indeed common to the work under discussion), although they might not be if definitions such as J_u [*12*] or J_m [*11*] were used. There remains an apparent difference in that expressions for I and T_{app} are not self-evidently identical, though in practice must clearly be closely similar. Both terms admit of some uncertainty, the I theory in evaluating η_{os} through the compliance ratio α and plastic-to-elastic work ratio X, and the dJ/da theory in determining $\partial Q/\partial q_c|_a$. These compliance dependent terms dominate the final predictions. Note that in Table 3, estimates according to η_o (test piece) or η_o (structure) may change the prediction when it is well balanced. The latter are judged more correct here. In the dJ/da analysis (Fig. 4), curves (ia) and (ii) based on $\partial Q/\partial q_c|_a = 0$ predict stability, whereas those using $\partial Q/dq_c|_a =$ slope AC (Fig. 1) $\simeq -(1/16)$ MN/mm predict instability. The authors admit to some uncertainty over the compliance values used in which the inclusion or exclusion of extraneous terms (such as the mainly elastic indentation of the loading rollers and of shear deflection in theoretical estimates) could alter the values quoted by perhaps 1 mm/MN or even more in certain cases (i.e., a value comparable to Φ_u itself).

The amount of crack growth at instability is in the range of 2.5 to 4.5 mm, just at or beyond that normally associated with J-controlled growth [*4*]. It may be remarked that maintenance of a near pre-initiation crack tip stress state was not originally seen as a limit to the argument for the energy balance analysis which depends mainly on the relationship between J and work (Eq 2). Although it is accepted that the best definition of J for a geometry-independent R-curve is still uncertain, that uncertainty is only relevant to the present tests over the range of notch ratios $a/W = 0.25$ to 0.5, within the three-point bend configuration. Such experimental difference as there may be are encompassed by the two sets of R-curve data from Fig. 2 and used in Figs. 3 and 4.

If these niceties are overlooked and the generally satisfactory nature of both sets of predictions are accepted, it may be noted that in Ref *18* a value of $T_{mat} = 10$ was chosen for HY130 from deep-notched ($a/W = 0.8$) compact tension tests with side grooves, for crack growth up to 5 mm. Such a value, used here, would clearly give a prediction of instability. In Ref *19* a value of $T_{mat} = 20$ was suggested for tests at $a/W = 0.65$, no side grooves. Such a value used here would clearly give a prediction of stability. Hence it is the decrease of T_{mat} with a few millimetres of growth that has the dominant effect in any of the prediction methods, $T_{mat} = 20$ or 10 being roughly upper and lower bounds to the values of T at (say) 1 or 5 mm growth, according to the present data in Fig. 2. Thus prediction of stability, or not, lies within the uncertainty of published data.

Conclusions

1. Experiments on HY130 steel bend bars, nominally $a/W \simeq 0.30$, $W = B = 50$ mm, $S/W = 4$, showed unstable ductile behavior when tested in a conventional screw-driven testing machine. The instabilities occurred close to the limit state after some 2.5 to 4.5 mm of crack growth, implying $\Delta a/b_o$ between 6 and 12%, just at or beyond the conventional limits for J-controlled growth.

2. Both I [*1,2*] and dJ/da [*3,4*] theories of instability have been applied satisfactorily to these tests, within the uncertainties of the data. For these notch bend pieces the more recent dJ/da theories degenerate to the original T theory [*3*] if the values of η and effective span are selected appropriately.

3. In applying any of the theories as currently developed, the dominant effect is that of compliance, which introduces uncertainties into the term $\partial Q/\partial q_c|_a$ for dJ/da theories (or the effective span in Ref *3*) and into the estimates of η_{os} for the I theory.

4. The success of this application must be measured against the restricted range of test circumstances, namely only a small variation in notch depth ratio between the R-curve data and the unstable test geometry. The wider issue of use of the R-curve data from one configuration to prediction of behavior of another has not been discussed and doubt remains both in terms of the geometry independence of the R-curve and in the evaluation of some of the terms in the theoretical formulations.

APPENDIX

The Energy Release Rate I

The energy release rate in an elastic-plastic-elastic (EPE) system (i.e., with linear elastic unloading) was derived [*1,2*] but is re-stated here since it is central to the present analyses. The term derived [*1,2*] is then modified to allow for the effect of compliance and plasticity. These derivations are given more fully elsewhere [*12*]. The η terms are conveniently stated as

$$\eta = -(b/Q)(\partial Q/\partial a)|_q \tag{23}$$

with subscripts "el," "pl," or "o" denoting elastic (linear), plastic, and elastic-plastic behavior respectively. If the variables of geometry and deformation are truly separable, η does not vary from elastic to plastic state [7]. For most configurations, however, that ideal is not reached, so that $(\partial Q/\partial a|_{q,\,pl} > (\partial Q/\partial a|_{q,\,el}$. Thus in Fig. 5 where, on unloading, $sr = -dq_{el} = +dq_{pl}$, the energy release rate for thickness B, crack increment da, is

$$BIda = EDs + DCrs = BGda + BGpda \tag{24}$$

where G is the conventional LEFM term of the "free" internal energy; G_p is the "bound" potential energy that is only available when $-dq_{el}$ is in fact balanced by $+dq_{pl}$ (i.e., G_p is only "available" as part of an elastic-plastic interaction. I can be expressed as

$$\begin{aligned} BIda &= BGda + q_{el}(\partial Q/\partial a - \partial Q_{el}/\partial a)|_q \\ &= BGda + \left[\frac{\eta_o - \eta_{el}}{\eta_{el}}\right]\frac{q_{el}\,Q\eta_{el}}{b} \end{aligned} \tag{25}$$

Referring to Fig. 5, however, $Q\eta_{el}/b = -\partial Q_{el}/\partial a|_q = ED$ and $q_{el} = sn$ and $ED(sn) = 2(EDs) = 2BGda$. Therefore

$$I = G + G_p = G[2(\eta_o/\eta_{el}) - 1] \tag{26}$$

Clearly, if $\eta_o \equiv \eta_{el}$, then $I \equiv G$. If load is held constant, the increase of work, area $EHmn$ in Fig. 5, is $2IBda$.

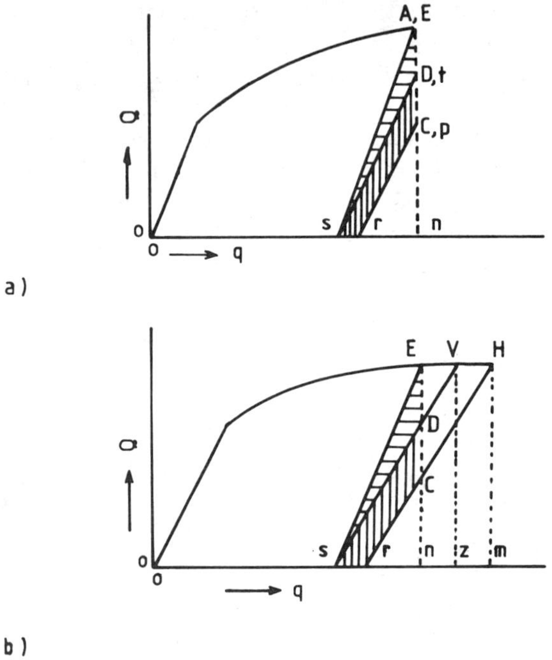

FIG. 5—*Elastic energy release rate,* I, *for elastic-plastic material with linear elastic unloading.* (a) *At fixed displacement.* (b) *At fixed load.*

Effect of Compliance on *I*

Let the compliance of an un-notched component be Φ_u and for a notched body, Φ_n. (Note Φ_n relates Q and q for the notched test piece and is not the same as the separate "crack component" used [4,11].) If there is a machine or structure of compliance Φ_m attached to the notched member, the compliance for the uncracked system is $\Phi_s = \Phi_u + \Phi_m$. These terms can be used to provide the constant of integration for the denominator if η_{el} is evaluated from the LEFM shape factor, Y:

$$\eta_{el} = bY^2a/\int Y^2 a da \tag{27}$$

For the whole system (elastic),

$$\eta_{el,s} = (\eta_{el,n})/[1 + (\Phi_m/\Phi_n)]$$

$$= (\eta_{el,n})/\alpha \tag{28}$$

where α is defined as $1 + (\Phi_m/\Phi_n)$.

If there is plastic work U_{pl} in the system overall (but only in the notched component), let

$$U_{pl} \equiv U_{pl,n} = X(U_{el,n}) \tag{29}$$

where X expresses the ratio of plastic work to the elastic work in the notched test piece. Writing

$$BbJ = \eta_{o,s} U_{o,s} = \eta_{o,s}(\alpha + X) U_{el,n}$$

$$= \eta_{el,s} U_{el,s} + \eta_{pl,n} U_{pl,n}$$

Then

$$\eta_{os} = [(\eta_{el,s})\alpha + (\eta_{pl,n})X]/(\alpha + X) \tag{30}$$

In the present work $(d\eta/da)_{o,s}$ is evaluated from $\eta_{o,s}$ for several values of Δa, although it can be found by differentiating Eq 30 to give

$$\left(\frac{b}{\eta}\frac{d\eta}{da}\right)_{o,s} = \left(\frac{b}{\eta}\frac{d\eta}{da}\right)_{el,n} + \frac{\Phi_m}{\Phi_n}\frac{\eta_{el}}{\alpha + X} - \eta_{el}(2 + X)\left(\frac{1}{1 + X} - \frac{1}{\alpha + X}\right) \tag{31}$$

References

[1] Turner, C. E. in *Fracture Mechanics (11th Conference), ASTM STP 677*, American Society for Testing and Materials, Philadelphia, 1979, pp. 614-628.

[2] Turner, C. E. in *Post-Yield Fracture Mechanics*, D. G. H. Latzko, Ed., Applied Science Publishers, London, 1979, pp. 23-211.

[3] Paris, P. C., Tada, H., Ernst, H., and Zahoor, A. in *Elastic-Plastic Fracture, ASTM STP 668*, American Society for Testing and Materials, Philadelphia, 1979, pp. 5-36, 251-265.

[4] Hutchinson, J. W. and Paris, P. C. in *Elastic-Plastic Fracture, ASTM STP 668*, American Society for Testing and Materials, Philadelphia, 1979, pp. 37-64.

[5] Orowan, E., *Journal of Applied Physics*, Vol. 26, No. 7, pp. 900-902.

[6] Turner, C. E. in *Fracture Mechanics (12th Conference), ASTM STP 700*, American Society for Testing and Materials, Philadelphia, 1980, pp. 314–337.

[7] Paris, P. C., Ernst, H., and Turner, C. E. in *Fracture Mechanics (12th Conference), ASTM STP 700*, American Society for Testing and Materials, Philadelphia, 1980, pp. 338–351.

[8] Garwood, S. J., Robinson, J. N., and Turner, C. E., *International Journal of Fracture*, Vol. 11, 1975, pp. 528–530.

[9] Ernst, H. A., Paris, P. C., and Landes, J. D. in *Fracture Mechanics (13th Conference), ASTM STP 743*, American Society for Testing and Materials, Philadelphia, 1981, pp. 476–502.

[10] Rice, J. R., Drugan, W. J., and Sham, T. L. in *Fracture Mechanics (12th Conference), ASTM STP 700*, American Society for Testing and Materials, Philadelphia, 1980, pp. 189–221.

[11] McCabe, D. E. and Ernst, H. A. in *Fracture Mechanics: Fourteenth Symposium*, Vol. I—*Theory and Analysis, ASTM STP 791*, American Society for Testing and Materials, Philadelphia, 1983, pp. I-561–I-584.

[12] Latzko, D. G. H., Turner, C. E., Landes, J. D., McCabe, D. E., and Hellen, T. K. in *Post-Yield Fracture Mechanics*, 2nd ed., Elsevier, The Netherlands, 1984.

[13] Kaiser, S. and Carlsson, A. J. in *Elastic-Plastic Fracture: Second Symposium, Volume II—Fracture Resistance Curves and Engineering Applications, ASTM STP 803*, American Society for Testing and Materials, Philadelphia, 1983, pp. II-58–II-79.

[14] Cotterell, B. and Mai, Y. W., *International Journal of Fracture*, Vol. 20, 1982, pp. 243–250.

[15] Etemad, M. R., "Use of Resistance Curves in Post-Yield Fracture Mechanics," Ph.D. thesis, University of London, 1983.

[16] Ernst, H. A. in *Elastic-Plastic Fracture: Second Symposium, Volume II—Fracture Resistance Curves and Engineering Applications, ASTM STP 803*, American Society for Testing and Materials, Philadelphia, 1983, pp. II-133–II-155.

[17] Turner, C. E., NUREG/CP-0010, CSNI Rept. 39, CSNI Specialist Meeting on Plastic Tearing Instability, St. Louis, Sept. 1979.

[18] Vassilaros, M. G., Joyce, J. A., and Gudas, J. P. in *Fracture Mechanics: Fourteenth Symposium, Vol. I—Theory and Analysis, ASTM STP 791*, American Society for Testing and Materials, Philadelphia, 1983, pp. I-65–I-83.

[19] Joyce, J. A. and Vassilaros, M. G. in *Fracture Mechanics (13th Conference), ASTM STP 743*, American Society for Testing and Materials, Philadelphia, 1981, pp. 525–542.

J. E. Carifo,[1] *J. L. Swedlow,*[1] *and Chul-Won Cho*[2]

Computation of Stable Crack Growth Using the *J*-Integral

REFERENCE: Carifo, J. E., Swedlow, J. L., and Cho, C.-W., **"Computation of Stable Crack Growth Using the *J*-Integral,"** *Fracture Mechanics: Seventeenth Volume, ASTM STP 905*, J. H. Underwood, R. Chait, C. W. Smith, D. P. Wilhem, W. A. Andrews, and J. C. Newman, Eds., American Society for Testing and Materials, Philadelphia, 1986, pp. 503–511.

ABSTRACT: Crack growth of almost 8 mm in a standard C(T) specimen is computed in ten steps by tracking a *J*-resistance curve for 7475-T7351 aluminum. Some discussion is presented as to why the far-field *J* is preferable to near-field quantities for this procedure. The analysis did not track the companion force-displacement curve.

KEY WORDS: *J*-integral, crack growth, node release, finite element

Computation of stable crack growth via finite elements is operationally straightforward; one need only release nodes at the crack front and beyond, allowing the material to relax or unload as each node is released. However, there remains the critical issue of determining the condition for nodal release. In fact, a number of prior investigators [*1–4*] have treated this problem but, for reasons to be discussed in the next section, we take issue with their respective conditions for nodal release.[3]

After some discussion of the condition for nodal release, we return to the problem statement and exhibit the physical data used in the present analysis. Results are then presented, followed by discussion and conclusions.

Conditions for Nodal Release

It would be ideal if one could decide when to release the nodes at the crack front itself. Candidates for this procedure include strain at the crack tip, as

[1]Department of Mechanical Engineering, Carnegie-Mellon University, Pittsburgh, PA 15213. Mr. Carifo is currently with Martin Marietta Aerospace, Baltimore Division, MD 21220.

[2]ALCOA Laboratories, Alcoa Center, PA 15069.

[3]Owing to the fact that the literature in this area is so widespread, the list of references herein is not intended to be comprehensive.

used by Newman [2], crack-tip opening displacement (CTOD) [3], and crack-tip opening angle (CTOA) [1,4]. However, there is some question regarding the accuracy with which any of these quantities can be computed; one has only to review the various round-robins on finite element analysis of crack problems to see the considerable disparity from one analyst to the next [5–8]. Hence the ideal of using a *near-field* quantity to decide upon nodal release is regarded as inoperational at best, and we have sought an alternative.

Any option for an alternative condition must involve a *far-field* quantity. For example, Liebowitz et al [9] use experimental records of load versus crack length, and Du and Lee [10] use the same information or a plastic energy versus crack size condition. A similar technique was followed by one of the investigators reported in [8] but, as the participants are anonymous, we cannot say who.

In seeking an appropriate condition for our work, we observe that some of the early investigators using a near-field condition produced results highly reminiscent of an R-curve.[4] It was inescapable that we consider just this type of information to derive our analysis. In the end, we chose to use a J-resistance curve as a suitable *far-field* condition. Certain disadvantages are obvious: J derives from deformation theory of plasticity while we perforce use flow theory, and J theory does not account for unloading while we must. The problems are addressed by Hutchinson and Paris [12], who go on to perform analytical versions of similar problems. On the other hand, J can be measured satisfactorily far from the crack tip in a finite element analysis, and there exists an established procedure for obtaining a J-resistance curve in the laboratory. We might also note that, when J was computed in the far field, the various round-robins [5–8] showed stable results, even though the calculations were mostly for nongrowing cracks. Hence, subject to certain constraints noted below, we chose the J-integral as the condition for crack growth, in the form of the J-resistance curve.

Procedure

The basic components of this work comprise a plane strain computer program, experimental data, and a specific model, in addition to the selection described above.

The computer program has been documented elsewhere [13] and neither the theory of incremental plasticity nor finite element methods need to be repeated here. Suffice it to say that the program utilizes the von Mises criterion together with the Prandtl-Reuss equations for an isotropically hardening material, and that it is coded to facilitate unloading as well as loading. The program uses constant strain triangles only, and certain modifications were made in order to handle nodal release at the crack front.

[4]For the reader not familiar with this form of data, perusal of Ref *11* is suggested.

As to experimental data, we were able to obtain for 7475-T7351 aluminum both a flow curve (Fig. 1) and a *J*-resistance curve (Fig. 2) obtained from two separate experiments. One of these was for limited crack growth, just under 4 mm, and the other went fully through the ligament. A piece-wise linear curve was fitted to the data (Fig. 2) so that we could interpolate between individual data points during computation. Owing to the fact that we were making a small deformation analysis, the blunting line was ignored.

The model involved a compact specimen (Fig. 3) having the proportions of a standard C(T) specimen. The element model is shown in Figs. 4 and 5. The *J*-integral paths are shown in Fig. 4, and mapping local to the crack tip in Fig. 5. The specimen is 127 mm wide and 152 mm high, and a_0 is nominally 63.5 mm. The crack was extended in ten steps of $a_0/80$ for a total extension of $a_0/8$.

Load is introduced via displacement control at the positions indicated in Figs. 3 and 4, alternately with crack growth. That is to say, when excitation at these positions led to a *J* value indicating crack growth, displacement at these positions was fixed and the appropriate node (in Fig. 5) was released. Then the far-field displacement was increased until *J* had risen sufficiently to indi-

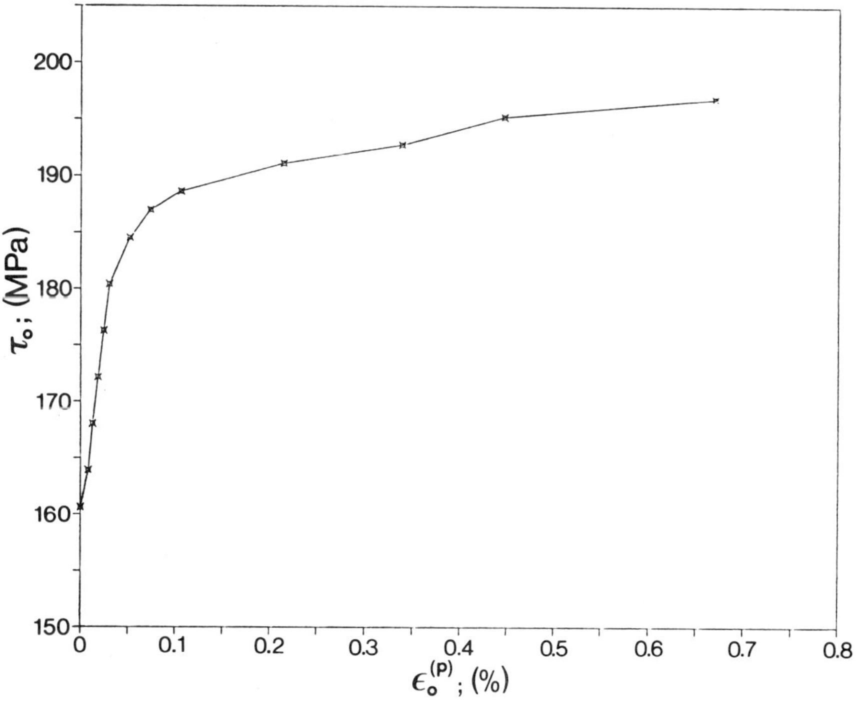

FIG. 1—*Experimental flow curve in terms of octahedral stress versus octahedral plastic strain.*

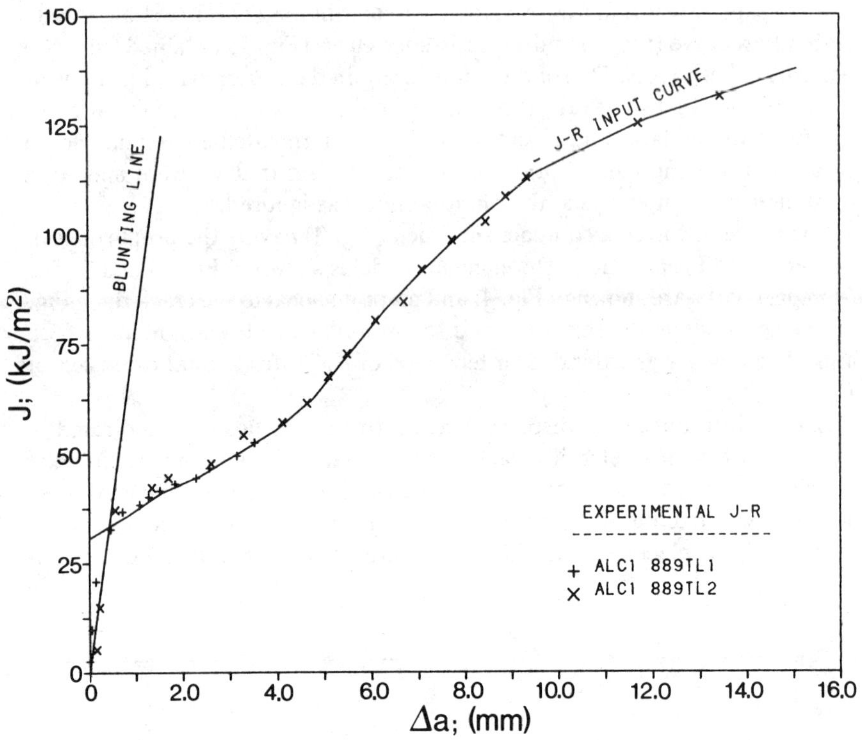

FIG. 2—*Experimental* J*-resistance curve.*

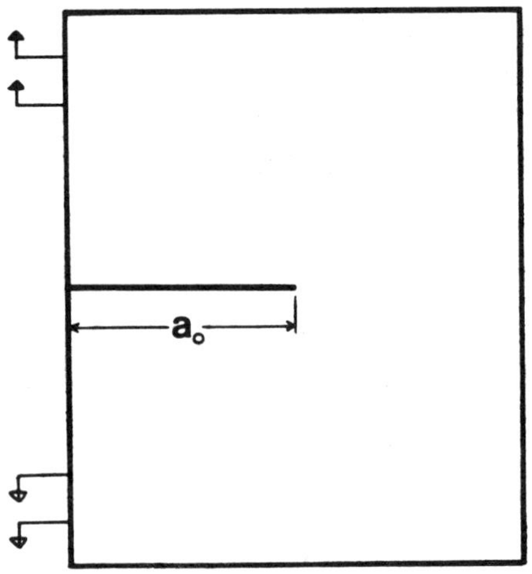

FIG. 3—*Compact test specimen geometry and the points of far-field displacement control.*

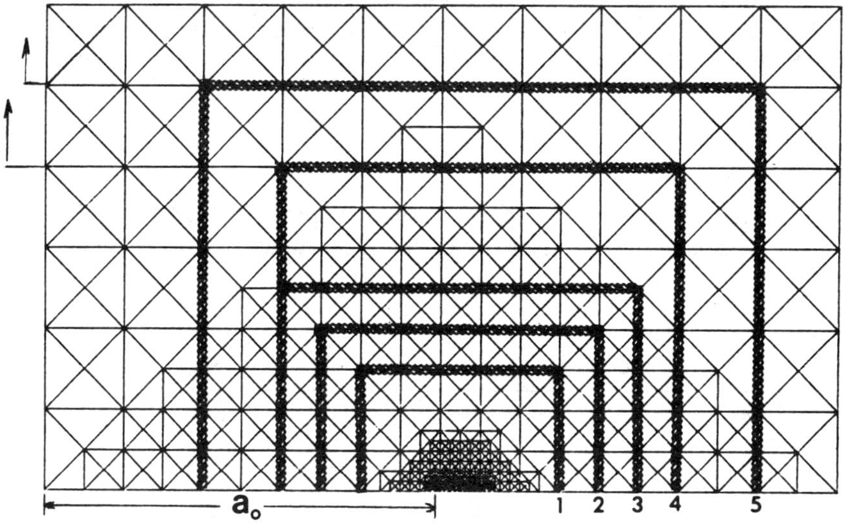

FIG. 4—*Map of the upper half of the specimen, the J-integral paths, and the points of far-field displacement control.*

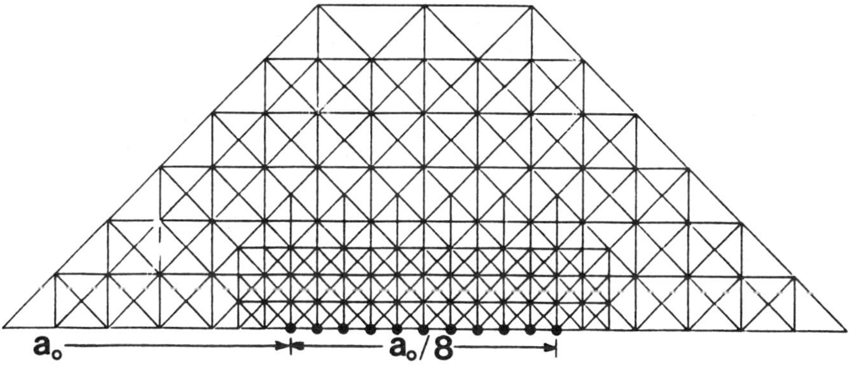

FIG. 5—*Gridding of map near and around crack tip where (●) denotes successive crack tip positions.*

cate further nodal release, and the whole process repeated until all nodes indicated in Fig. 5 had been released. We note that, since a_0/W is 0.5, greater than the threshold value of 0.4 noted by Newman [*14*], results here are not likely to be sensitive to the means for introducing load into the specimen.

A further word on nodal release is in order. When the computation begins, the ten nodes ahead of the original crack tip (see Fig. 5) are on a line of symmetry, and are pinned against any vertical motion. Naturally, vertical forces accrue to hold them in position. By *releasing* a node, we mean that the reaction force holding that node in place is relaxed in eight fractional steps.[5] Naturally, the load distribution ahead of the crack shifts to the right and, as we shall see, the force at the point of displacement control drops slightly, as does *J*.

Results

The computation described above is summarized in Table 1.[6] We observe that the *J* calculations are quiescent in that the standard deviation among the values is quite small.[7] Hence we regard our *J* values as virtually path-indepen-

TABLE 1—*J values calculated during node releases.*

Far Field Displacement, mm	*J* (Path Average), kJ/m²	Standard Deviation of *J*	Crack Length, mm	Far Field Net Force, kN/cm	Node Release?
0.404368	29.30	0.25	62.706	19.003	no
0.424586	32.28	0.28	62.706	19.931	yes, 1
0.424586	31.60	0.26	63.500	19.392	no
0.445821	34.81	0.29	63.500	20.337	no
0.468097	38.34	0.32	63.500	21.329	yes, 2
0.468097	37.50	0.31	64.294	20.723	no
0.491515	41.30	0.34	64.294	21.731	yes, 3
0.491515	39.94	0.32	65.088	21.099	no
0.516077	44.48	0.36	65.088	22.122	no
0.541884	48.97	0.39	65.088	23.188	yes, 4
0.541884	47.91	0.37	65.881	22.521	no
0.568985	52.75	0.41	65.881	23.607	yes, 5
0.568985	51.59	0.39	66.675	22.906	no
0.597433	56.81	0.43	66.675	24.013	yes, 6
0.597433	55.58	0.42	67.469	23.311	no
0.627304	61.19	0.46	67.469	24.425	no
0.658673	67.34	0.50	67.469	25.586	yes, 7
0.658673	65.91	0.48	68.262	24.838	no
0.691642	72.54	0.53	68.262	26.020	yes, 8
0.691642	71.04	0.50	69.056	25.269	no
0.726186	78.18	0.55	69.056	26.469	no
0.762508	86.02	0.61	69.056	27.714	yes, 9
0.762508	84.18	0.58	69.850	26.882	no
0.800608	92.64	0.63	69.850	28.150	yes, 10
0.800608	90.72	0.60	70.644	27.323	no

[5]This was the minimum number of steps that we could use with the software and hardware in hand without distorting the unloading-node opening curve. Of course, other codes might perform differently.

[6]The computation was performed on Carnegie-Mellon's DEC System Tops-20 and required 3.5 h of CPU time. This time is longer than one usually anticipates; it is a reflection of the iterative nature of unloading [*13*].

[7]Note that, although the standard deviation *grows* in absolute magnitude as loading progresses, it *reduces* as a percentage of *J*.

dent, and the potential disadvantages of having chosen to work with *J* as not having manifested themselves. In addition to this tabular information, it is important to see that the computation tracked the input *J*-resistance curve; this information appears in Fig. 6. (Here, each set of three nodal release points (*) represents the *J* calculated at the third, fifth, and seventh of the eight steps noted above.) We observe that the data are not precisely on the *J*-resistance curve but, in looking at the numbers, we find also that the computed data are virtually within one standard deviation of the curve.

It is also of interest to compare the computed and experimental force-displacement curves (Fig. 7). We are not at all surprised to observe the computed results as stiffer than the experimental values since (1) the test specimen was 43 mm thick while the analysis was done in plane strain (see Ref *15*, Fig. 12, for a related comparison), and (2) the finite element array is intrinsically stiffer than a continuum. Still, the difference is not so much as to be shocking, although it is greater than reported in Ref *8*. We see also that, beyond about 6 mm when unstable tearing begins, the discrepancy grows. It was not our expectation to be able to replicate unstable tearing.

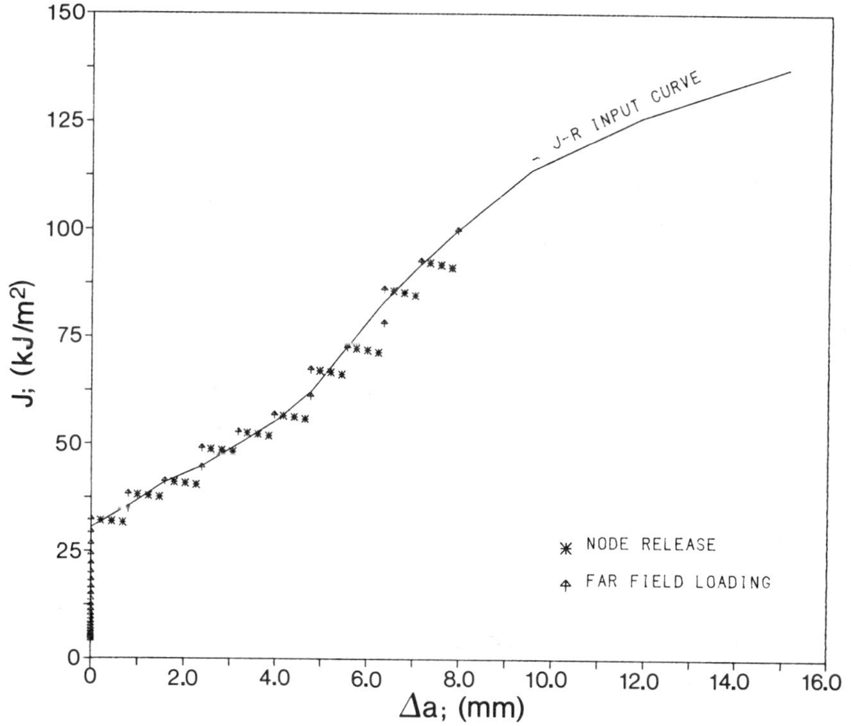

FIG. 6—*Tracking of the* J*-resistance curve. Note unstable tearing at 6 mm.*

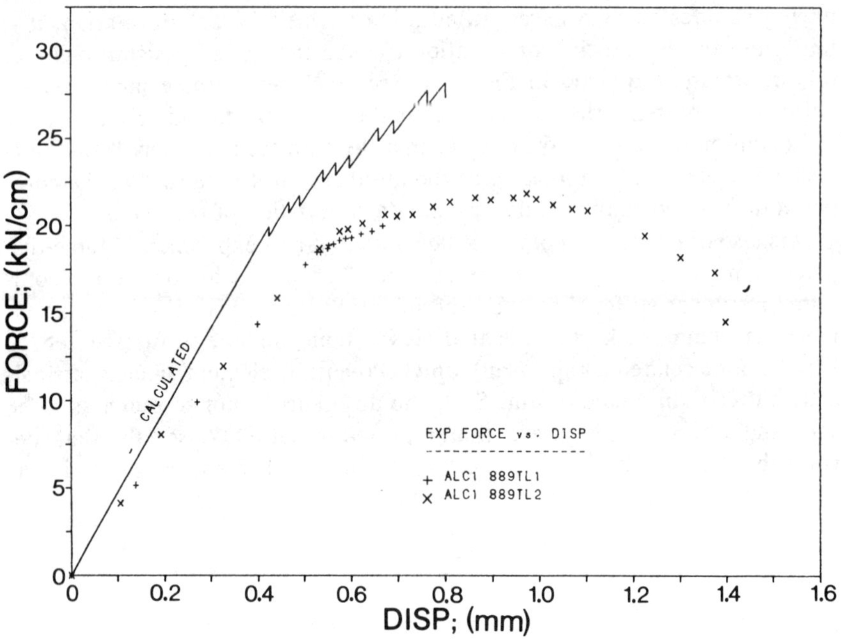

FIG. 7—*Computed and experimental force-displacement curves.*

Concluding Remarks

We observe that a *J*-based stable crack growth computation is certainly feasible. It is no problem to track the *J*-resistance curve but, with the coarseness of the finite element map used here, we were not able at the same time to track the force-displacement curve. By implication, if we *had* tracked the force-displacement curve, we would *not* have tracked the *J*-resistance curve.

In a more philosophical vein, proceeding as we have done has allowed the foregoing experimental evaluation. That is to say, had we used CTOD, say, as a criterion for nodal release, we would not have as clear a picture as to where the computation succeeded and where it had not.

Acknowledgments

This work was supported by a Carnegie Institute of Technology Fellowship awarded to JEC, and funded in major portion by Du Pont. Computing on the University's facilities is appreciated. In the early stages of this work, we benefited greatly by discussions with Professor F. B. Prinz.

References

[1] Anderson, H., *Journal of the Mechanics and Physics of Solids*, Vol. 21, 1973, pp. 337-356.
[2] Newman, J. C., Jr., in *Cyclic Stress-Strain and Plastic Deformation Aspects of Fatigue Crack Growth, ASTM STP 637*, American Society for Testing and Materials, Philadelphia, 1977, pp. 56-80.
[3] Newman, J. C., Jr., "Finite-Element Analysis of Initiation, Stable Crack Growth, and Instability using a Crack-Tip-Opening Displacement Criterion," NASA Report TM-84564, National Aeronautic and Space Administration, 1982.
[4] de Koning, A. U., Report NLR MP75035U, National Aerospace Laboratory, The Netherlands, 1975.
[5] Wilson, W. K. and Osias, J. R., *International Journal of Fracture*, Vol. 14, 1978, pp. R95-R108.
[6] Shiratori, M. and Miyoshi, T. in *Numerical Methods in Fracture Mechanics*, Pineridge Press, Swansea, England, 1980, pp. 417-431.
[7] Larsson, L. H., *International Journal for Pressure Vessels and Piping*, Vol. 11, 1983, pp. 207-228.
[8] Bleackley, M. H. and Luxmoore, A. R., *International Journal of Fracture*, Vol. 22, 1983, pp. 15-39.
[9] Liebowitz, H., Lee, J. D., and Subramonian, N. in *Nonlinear and Dynamic Fracture Mechanics*, AMD-35, American Society of Mechanical Engineers, New York, 1979, pp. 157-169.
[10] Du, S. and Lee, J. D., *Engineering Fracture Mechanics*, Vol. 16, 1982, pp. 229-245.
[11] *Fracture Toughness Evaluation by R-Curve Methods, ASTM STP 527*, American Society for Testing and Materials, Philadelphia, 1974.
[12] Hutchinson, J. W. and Paris, P. C. in *Elastic-Plastic Fracture, ASTM STP 668*, American Society for Testing and Materials, Philadelphia, 1979, pp. 37-64.
[13] Swedlow, J. L., *Computers and Structures*, Vol. 2, 1973, pp. 879-898.
[14] Newman, J. C., Jr., in *Fracture Analysis, ASTM STP 560*, American Society for Testing and Materials, Philadelphia, 1974, pp. 105-121.
[15] Riccardella, P. C. and Swedlow, J. L. in *Fracture Analysis, ASTM STP 560*, American Society for Testing and Materials, Philadelphia, 1974, pp. 134-154.

E. M Hackett,[1,2] *P. J. Moran,*[2] *and J. P. Gudas*[1]

Evaluation of Environmentally Assisted Cracking of a High Strength Steel Using Elastic-Plastic Fracture Mechanics Techniques

REFERENCE: Hackett, E. M., Moran, P. J., and Gudas, J. P., **"Evaluation of Environmentally Assisted Cracking of a High Strength Steel Using Elastic-Plastic Fracture Mechanics Techniques,"** *Fracture Mechanics: Seventeenth Volume, ASTM STP 905,* J. H. Underwood, R. Chait, C. W. Smith, D. P. Wilhem, W. A. Andrews, and J. C. Newman, Eds., American Society for Testing and Materials, Philadelphia, 1986, pp. 512–541.

ABSTRACT: The objective of this investigation was to assess the feasibility of determining the environmentally assisted crack growth characteristics of a high strength 4340 steel during cathodic polarization in seawater using *J*-integral techniques. A secondary objective was to demonstrate the capability of using these techniques for investigating stress corrosion cracking–hydrogen embrittlement mechanisms in high strength steels.

It was concluded that stress corrosion–hydrogen embrittlement crack growth behavior in a high strength steel can be characterized using techniques developed for the study of stable elastic-plastic crack growth. In the case of environmentally assisted cracking, an added dimension results from the rate dependence of the process, giving results which are displacement-rate dependent. Cathodic polarization of 4340 steel specimens in seawater resulted in up to a four-fold decrease in the energy required for fracture initiation (J_{Ic}). The extent of this decrease was a strong function of both pre-exposure time and displacement rate. A good correlation was obtained between fracture morphology changes predicted from the material *J-R* curves and actual fracture surface measurements. The technique should therefore be a useful tool in mechanistic studies of environmentally assisted crack growth in high strength steels.

KEY WORDS: stress corrosion cracking, hydrogen embrittlement, fracture mechanics, K_{Ic}, J_{Ic}, strain rate, seawater, cathodic polarization, 4340 steel

In recent years it has become common to study crack growth in engineering materials using the concepts of fracture mechanics. More specifically, linear elastic fracture mechanics (LEFM) has been extensively applied in studies of

[1]David Taylor Naval Ship R&D Center, Code 2814, Bethesda, MD 20084.
[2]Corrosion and Electrochemistry Research Laboratory, The Johns Hopkins University, Baltimore, MD 21218.

crack growth under monotonic loading, fatigue crack growth, and stress corrosion cracking (SCC). Studies of stress corrosion cracking using LEFM have been directed at identifying the stress intensity associated with the onset of environmentally assisted cracking (K_{Iscc}). Two common tests for K_{Iscc} are the precracked cantilever beam test [*1*] and the modified wedge-opening-load (WOL) test [*2*]. The precracked cantilever beam test involves the application of a deadweight load and therefore an increasing *K* field for crack growth. The WOL test is conducted at a constant deflection and therefore involves crack extension to arrest in a decreasing *K* field. Recent studies [*3,4*] have departed from these techniques by evaluating compact specimens under increasing load and displacement to determine the stress corrosion cracking susceptibility of high strength steels. Still more recent research [*5,6*] has focused on employing elastic-plastic fracture mechanics (EPFM) and the *J*-integral fracture parameter to study stress corrosion crack growth.

During the past decade the *J*-integral elastic-plastic fracture parameter has been extensively utilized to characterize the ductile fracture properties of structural alloys. However, very little attention has been focused on using EPFM to study environmentally assisted crack growth. The use of *J*-integral techniques has been validated over a broad range of elastic-plastic behavior [*7*], and reliable single-specimen tests for the determination of the critical value of *J* at the onset of crack growth (J_{Ic}) and the *J* versus crack growth resistance (*J-R*) curve have been developed [*8-10*]. In contrast to the LEFM approach, these techniques require an active crack length measurement system which is capable of resolving very small (approximately 0.003 mm [0.001 in.]) increments of crack growth. This capability has focused increased attention on crack growth behavior beyond J_{Ic} [*11*].

A recent investigation by Abramson and co-workers [*6*] concluded that, with certain restrictions, the *J*-integral is applicable to the analysis of stress corrosion cracking. This conclusion was reinforced by the recent work of Anderson and Gudas [*5*], who employed *J*-integral techniques to determine the SCC susceptibility of a high strength titanium alloy. Results from their investigation were shown to correlate well with complementary LEFM SCC tests employing precracked cantilever beam specimens.

The objective of this investigation was to assess the feasibility of determining the environmentally assisted crack growth characteristics of a high strength steel during cathodic polarization in seawater using *J*-integral techniques. A secondary objective was to demonstrate the capability of using these techniques for investigating SCC/hydrogen embrittlement (HE) mechanisms in high strength steels.

Test Methodology for Environmentally Assisted Fracture

Whether classified as hydrogen embrittlement or stress corrosion cracking, mechanisms for environmentally assisted cracking have been extensively studied using the concepts of linear elastic fracture mechanics [1-4]. A brief review of some of this work follows.

The LEFM studies of environmentally assisted crack growth can be divided into two categories: those performed under an increasing stress intensity (*K* increasing) and those performed under a constant deflection using a decreasing stress intensity (*K* decreasing). The most straightforward *K* increasing test was that described initially by Brown and Beachem [1], who utilized precracked cantilever beam specimens which were deadweight loaded. The lowest stress intensity at which no environmentally assisted crack growth occurred was taken to be the threshold stress intensity for subcritical crack growth (K_{Iscc}). This technique requires a number of specimens for the determination of (K_{Iscc}) and generally involves lengthy test periods. Largely for the advantages of portability, ease of loading, and specimen economy, Novak and Rolfe [2] proposed a modified bolt-loaded WOL specimen for the determination of K_{Iscc}. This specimen maintains a constant deflection and therefore a decreasing stress intensity for crack growth. The specimen is loaded to a value estimated to be greater than K_{Iscc}, then placed in an aqueous environment where subcritical crack growth initiates and proceeds until arrest. The arrest point is taken to be K_{Iscc}. This test has the advantage of providing a K_{Iscc} determination for each specimen tested, but may also require lengthy test periods if either the initiation time is long or the crack growth rate is slow.

A recent study by Clark and Landes [4] utilized compact specimens of AISI 4340 steel under increasing load and displacement to determine K_{Iscc} for the case of seawater. They concluded that rising load K_{Iscc} testing for this case can be used for rapidly screening high strength structural alloys with regard to their susceptibility to environmentally induced cracking in hydrogen-bearing environments. They also concluded that loading rate can have a significant effect on the apparent K_{Iscc} measured under rising load conditions and that there appeared to be no significant prior load history effects on the K_{Iscc} values determined for AISI 4340 steel using this technique.

Anderson and Gudas [5] extended this technique into the realm of elastic-plastic fracture mechanics (EPFM) by evaluating the environmental cracking susceptibility of a high strength titanium alloy in seawater using *J*-integral techniques. Their results showed that the stress corrosion susceptibility of this alloy was clearly identifiable in *J*-integral tests using the *J-R* curve extrapolation method as described in ASTM E 813. Importantly, these tests would not have provided valid linear elastic results due to specimen size constraints. Abramson et al [6] have recently evaluated the application of the *J*-integral to cases of environmentally assisted cracking of a magnesium alloy in an NaCl solution and concluded that with suitable modifications to allow for steady-state crack growth, *J*-integral techniques should apply to SCC. One of their more interesting results is the occurrence of drastically reduced *J-R* curves obtained from air tests conducted on specimens which were previously tested for K_{Iscc} at a constant deflection. They suggested that stable mechanical cracking proceeding from an SCC crack is easier than that occurring from a fatigue precrack. Another possibility is that the embrittling species was still present.

An important aspect of the *J-R* curve determination technique used in the present investigation is the capability of measuring crack extension during testing. This enables changes in fracture behavior predicted from the *J-R* curve to be correlated with fracture surface features, providing a useful tool for investigating environmentally assisted cracking.

Experimental Details

Material

Material for this investigation was provided in the form of 2.5-cm (1-in.)-thick plate of AISI 4340 steel. The plate was heat treated to obtain the mechanical properties shown in Table 1. Chemical composition of the plate is provided in Table 2.

Experimental Procedure

Compact specimens 2.5 cm (1 in.) thick were machined from the plate in accordance with Fig. 1. The notch orientation for all specimens was T-L. In this orientation the crack runs parallel to the rolling direction of the plate. Figure 1 shows a modified version of the standard compact specimen which incorporated a longer machined notch. The increased length notch was intended to allow the minimum possible length of fatigue precrack beyond the notch tip consistent with the criteria for fatigue precracking described in ASTM E 813. The short precrack was, in turn, desired to facilitate monitoring and control of electrochemical conditions close to the fatigue crack tip.

A cathodic polarization cell was constructed around the specimen as shown in Fig. 2. This cell maintained all specimens tested at −1.0 V versus a satu-

TABLE 1 — *Mechanical properties of 4340 steel.*

Ultimate Strength, ksi	0.2% Yield Strength, ksi	Elongation % in 50.8 mm (2 in.)
183.3	174.2	12.0

TABLE 2 — *Chemical composition (weight percent) of 4340 steel.*

C	Mn	P	Si	Cu	Ni	Cr	Mo	V
0.44	0.78	0.014	0.21	0.18	1.95	0.74	0.17	0.05

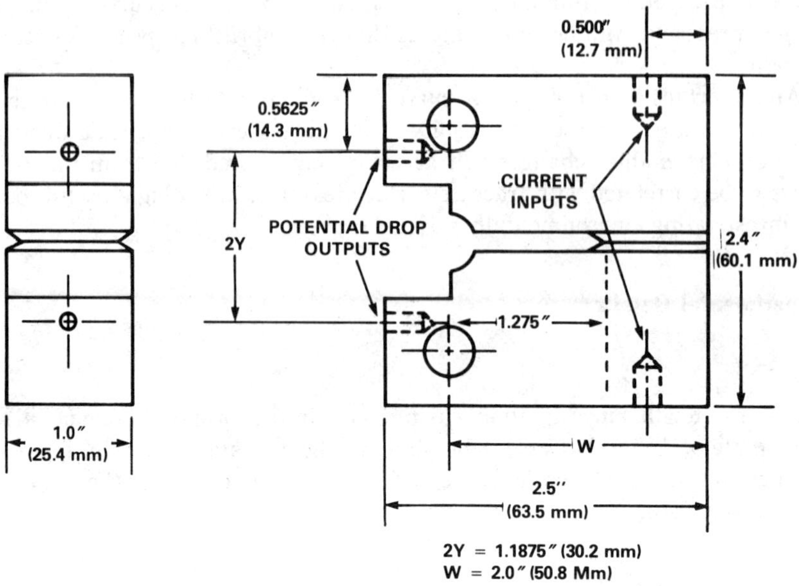

FIG. 1—*IT compact specimen.*

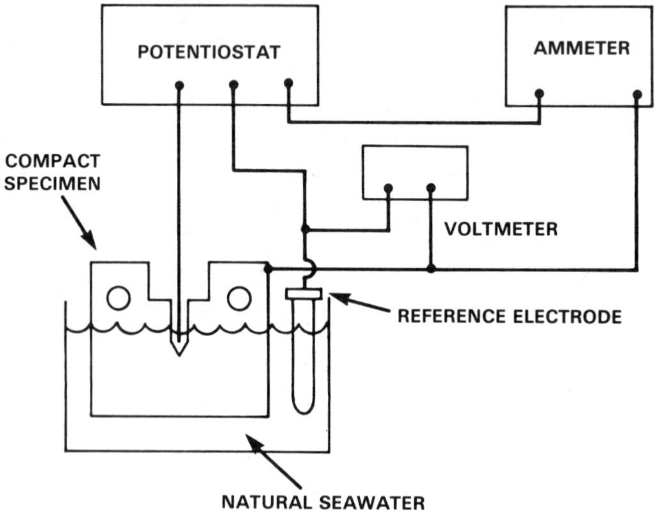

FIG. 2—*Schematic of cathodic polarization cell.*

rated calomel electrode (SCE) using potentiostatic control. Natural seawater shipped from the LaQue Center for Corrosion Control Technology was used as the electrolyte. The cell bath was supported from below the test machine, eliminating sealing difficulties around the cell. The palladium foil counter electrode was isolated from the machined notch with Teflon strips. The entire specimen, except for the crack starter notch and the side grooves, was protected from the seawater with a brush-applied lacquer. The tests were run at room temperature with no aeration. Electrolyte was added periodically to replace that lost due to evaporation.

Following machining, all specimens were precracked to an initial overall crack length of 35.6 mm (1.4 in.). This provided a fatigue precrack which was approximately 2.5 mm (0.1 in.) from the machined notch tip. In all cases the delta stress intensity used for fatigue precracking was below 33 MPa$\sqrt{m}$ (30 ksi$\sqrt{in.}$).

Fracture testing consisted of basically two variations. The longer time tests (greater than two days' exposure in the environment) were performed by loading the compact specimens in the test machine to a pre-load of 454 kg (1000 lb) corresponding to a stress intensity level of 21 MPa$\sqrt{m}$ (19 ksi$\sqrt{in.}$) for this geometry. This was below the expected K_{Iscc} for this alloy. A zinc wedge was then inserted into the notch area and the specimen was immersed in seawater for the pre-exposure time. It should be noted that this pre-load was probably not entirely maintained during the pre-exposure period due to relaxation of the zinc wedges. All specimens pre-exposed in this manner were connected to a coil of zinc in order to provide a favorable area ratio for cathodic polarization. The measured cell potential for these specimens was −1.0 V versus a saturated calomel reference electrode.

Following pre-exposure, the longer time specimens were removed from the seawater and inserted into the test machine and cell arrangement shown in Fig. 2 in a manner consistent with minimizing the time that the specimen was not exposed to the environment. This time was typically less than 5 min.

The shorter time tests (less than two days' exposure in the environment) were pre-exposed in the test setup and tested after the desired pre-exposure period. These specimens did not have to be transferred to the test machine following pre-exposure.

All fracture testing was performed in a screw-type tensile test machine which was translated in the horizontal plane to facilitate environmental testing. Tests were run in displacement control at two different displacement rates: 0.25 and 0.05 mm/min (0.01 and 0.002 in./min).

The *J-R* curve tests were performed both in air and the environment at the two displacement rates. For the environmental testing, cathodic polarization was continuously maintained during testing. The *J-R* curve test procedure employed for the majority of the testing was the computer-interactive, unloading compliance procedure of Joyce and Gudas [*9*]. In selected cases, a direct-current potential drop (DCPD) system, described by Vassilaros and Hackett [*10*], was employed for the crack extension measurements. The value of the J-integral was calculated using the formulation of Ernst [*12*]:

$$J_{(i+1)} = J_i + \frac{\eta}{b_i}\frac{A_{i,i+1}}{B_N}\left[1 - \frac{\gamma}{b_i}(a_{i+1} - a_i)\right]$$

where

η = 2 + (0.522) b/W for compact specimens,
W = specimen width,
γ = 1 + (0.76) b/W,
b_i = instantaneous length of remaining ligament,
B_N = minimum specimen thickness,
a_i = instantaneous crack length, and
$A_{i,i+1}$ = area under the load versus load-line displacement record between lines of constant displacement at points i and $i + 1$.

Following testing, specimens were fatigued at a low stress intensity to mark the final crack extension and then fractured completely in half. The fracture surfaces of selected specimens were examined in a scanning electron microscope. Fracture surface crack morphology and changes with propagation were correlated with crack behavior during propagation as described by the J-R curve tests.

Results and Discussion

Form of Presentation of Results

The results of this investigation will be presented primarily in the form of specimen load versus crack opening displacement (COD) records, J-integral R-curves, and scanning electron microscope (SEM) fractographs. The first two of these will be briefly discussed.

The specimen load versus COD records serve as input to the J-integral analysis. Calculation of the J-integral for a specimen involves the measurement of load-line displacement which, for the case of the compact specimen, is coincident with COD. The area (A) under the load-COD record is a measure of the energy absorbed by the specimen during fracture. Load-COD records for specimens with identical initial crack lengths can provide a qualitative measure of material toughness. All the specimens used in this investigation were precracked to the same initial crack lengths. When coupled with an on-line crack extension measurement system such as unloading compliance, the crack initiation point on a load-COD record can be defined. This was done for all of the specimens tested.

Salient features of a "typical" *J-R* curve are presented in Fig. 3. The blunting line is a mathematical representation of apparent crack extension due to crack tip blunting and is based on the material flow stress. J_{Ic} is a measure of the energy absorbed by the specimen at or near crack initiation. It is defined as the intersection of the blunting line, with a least squares line fit through data over a specified range beyond the blunting line. Details of this analysis are contained in ASTM E 813. If the validity requirements in this standard are met, the J_{Ic} value is a material property.

In addition, the slope of the *R* curve beyond initiation, normalized by the material elastic modulus divided by the material flow stress squared, is defined as the tearing modulus, *T*, ($T = dJ/da * E/\sigma \text{ flow}^2$), where *T* is a measure of the material's resistance to stable ductile tearing.

For the cases of environmentally assisted crack growth to be described below, the *J*-integral analysis cannot be rigorously applied. Details pertaining to this argument will be presented later. Despite the limitations on using the *J*-integral to rigorously analyze environmental crack growth, the extension of the testing and analysis techniques employed for *J-R* curve testing provides a useful tool for evaluating environmental damage mechanisms.

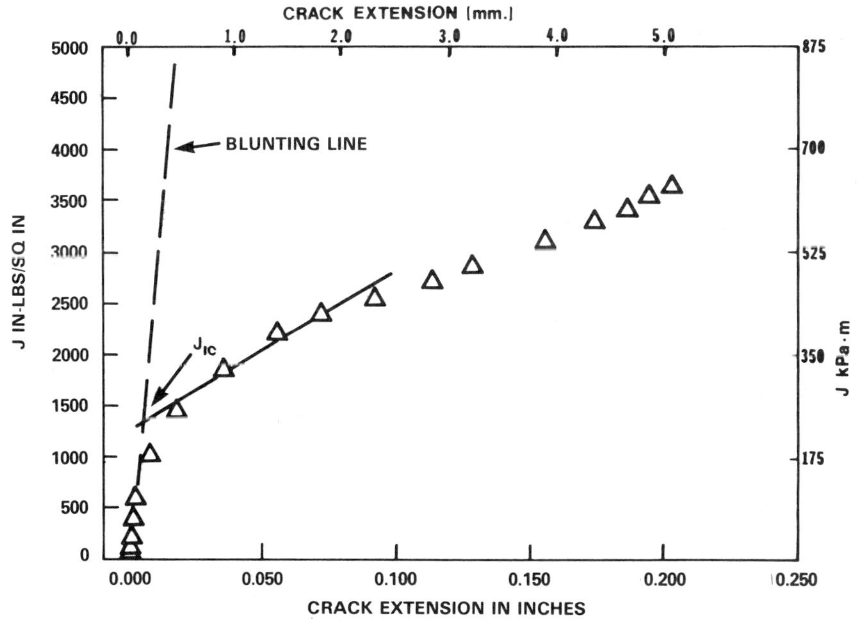

FIG. 3—*"Typical" J-integral R-curve.*

J-R *Curve Testing*

To serve as a baseline against which to measure the effects of environmentally assisted crack growth, *J-R* curve tests were performed in air at 0.25 mm/min (0.010 in./min) and 0.05 mm/min (0.002 in./min) displacement rates. The former is the standard test rate for air *J-R* curve testing; the latter represented a lower practical limit for use of the unloading compliance technique for *J-R* curve determination. Figure 4 shows a comparison of the load versus COD records for two air *J-R* curve tests. This plot shows the similarity of the two records and the lack of displacement rate effects in the air tests. The arrows indicate the crack initiation points in each case as determined from the unloading compliance technique. This figure also illustrates the predominantly linear-elastic behavior of the material in the air tests. Also evident is the difference between the magnitude of the unloadings for the two displacement rates. This is intentional since more data are acquired at the slower displacement rate, therefore requiring smaller unloadings to take the same amount of data. The time during the unloads is important since it has a direct bearing on the environmental interactions. Perhaps the most important feature shown in Fig. 4 is the occurrence of crack initiation at or near the maximum load achieved during the test followed shortly by catastrophic failure of the test specimens. Experience in performing *J-R* curve tests on different materials of similar geometry has shown that crack initiation most usually occurs

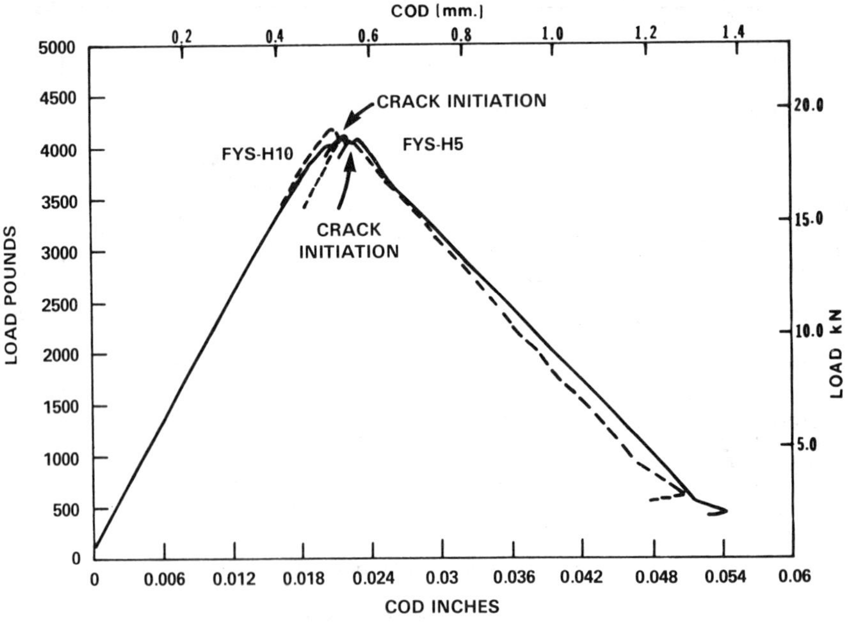

FIG. 4—*Load versus COD for 4340 steel air tests (FYS-H5 and FYS-H10).*

at or near the maximum load. This fact will be of importance in the subsequent discussion of the environmental tests.

The "*J-R* curves" for the air test specimens are presented in Fig. 5. "*J-R* curves" appears in quotes for two reasons. Firstly, the 4340 steel results shown in Fig. 5 are predominantly linear-elastic and therefore display no *R*-curve behavior in the conventional sense. Secondly, for the environmental tests, nonproportional loading occurs during environmentally assisted crack growth. This violates assumptions made in the derivation of the deformation theory *J*-integral, and it is therefore incorrect to call the results *J-R* curves in the conventional sense as applied to stable ductile tearing under proportional loading. The "*J-R* curves" for both of these air specimens indicate a crack initiation point (as determined by a deviation from the blunting line) of approximately 31 kPa·m (175 in.-lb/in.^{2}). Only a small amount of crack growth (less than 0.5 mm [0.020 in.]) was evident in these tests before catastrophic fracture, typical of a material which behaves in a linear elastic manner.

Figure 6 presents the load versus COD records for two seawater tests conducted in accordance with the shorter pre-exposure time procedure described previously (see Experimental Details section). One of the specimens (FYS-H8) was tested at the 0.25 mm/min (0.010 in./min) displacement rate, while the other specimen was tested at the 0.05 mm/min. (0.002 in./min) displacement rate. The specimens were geometrically identical and had the same ini-

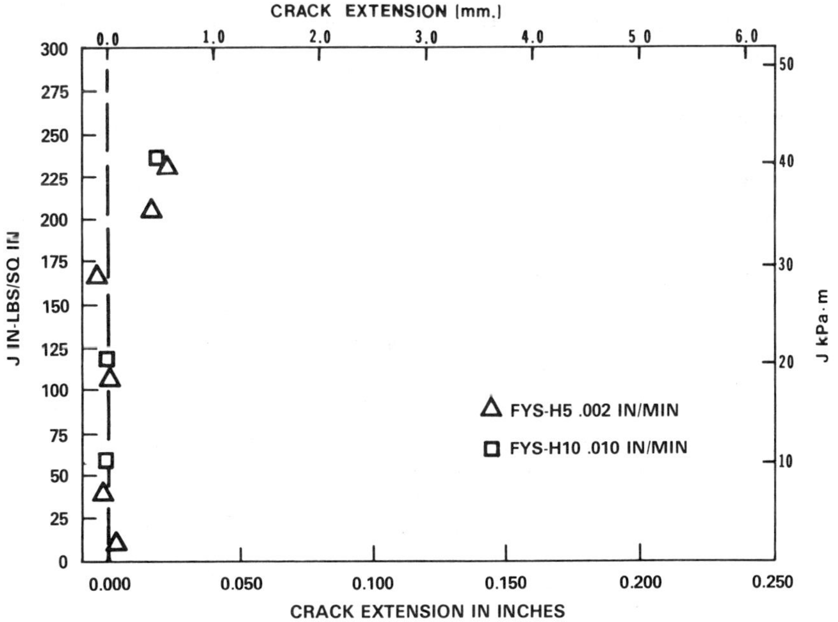

FIG. 5—"J-R *curves" for 4340 steel air tests (FYS-H5 and FYS-H10).*

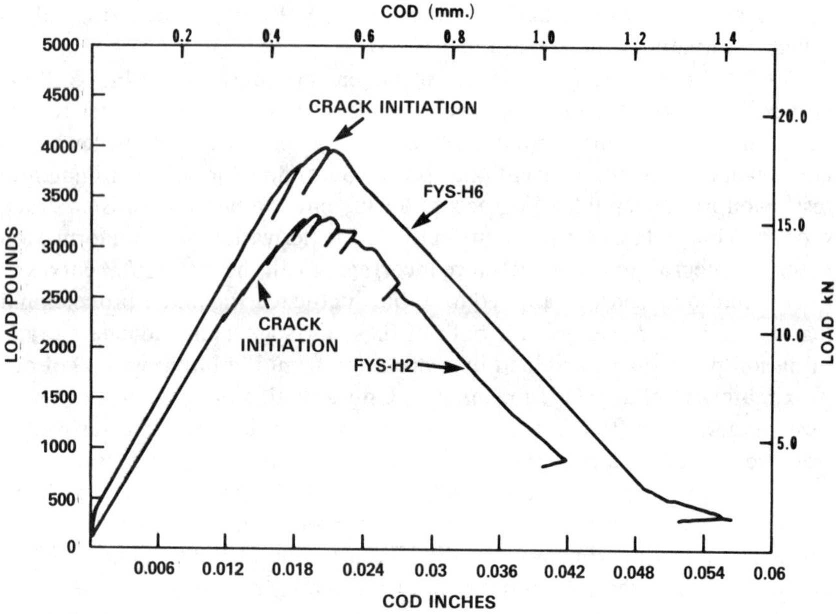

FIG. 6—*Load versus COD for 4340 steel seawater tests.*

tial precrack length. Figure 6 shows the effect of displacement rate on the fracture behavior of the 4340 steel under cathodic polarization in seawater. The arrows in the figure indicate the approximate crack initiation points, as determined from unloading compliance, for each of the two cases. The slower displacement rate used on FYS-H2 resulted in both a reduced maximum load and an earlier crack initiation point than the faster test. The movement of the crack initiation point back from near the maximum load is noteworthy. A crack initiation point at the location shown for FYS-H2 in Fig. 6 is not commonly encountered in routing *J-R* curve testing and is most likely the result of subcritical cracking under the influence of the environment. Noteworthy, also, is the fact that FYS-H8, although pre-exposed for a longer period than FYS-H2 (48 versus 14 h) exhibits a load-versus COD behavior that is virtually identical to that observed for the air tests in Fig. 4. This indicates the overriding importance of displacement rate in the seawater environment. This point will be discussed further in a later section.

The "*J-R* curves" for Specimens FYS-H2 and FYS-H8 are presented in Fig. 7. The similarity of the behavior of Specimen FYS-H8 and the two air test results is again evident, with FYS-H8 exhibiting a *J* of approximately 31 kPa·m (175 in.-lb/in.^{2}) near crack initiation. The *J* value near crack initiation for Specimen FYS-H2 is, however, much reduced in comparison to FYS-H8 or the two air tests (75 versus 175 in.-lb/in.^{2}). Also important is the abrupt

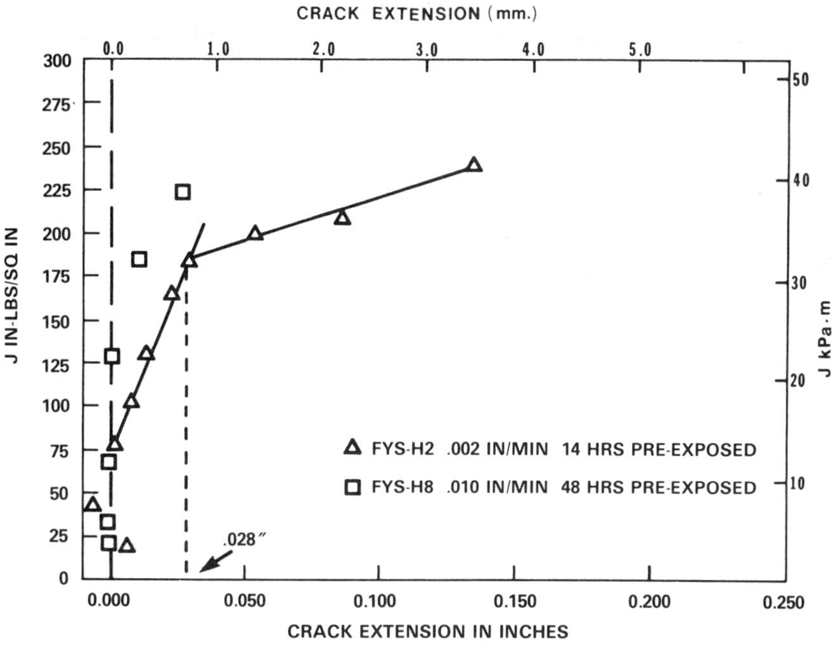

FIG. 7—"J-R *curves" for 4340 steel seawater tests.*

change in the "*J-R* curve" slope for FYS-H2 at approximately 0.7/mm (0.028 in.) of crack extension. While the *R*-curve in this case is not rigorously defined (as discussed previously), the change in slope should be indicative of a change in energy required to cause progressive fracturing. For this result, the slope change implies a higher energy fracture process occurring first followed by a lower energy mode. This result was exhibited by all the specimens that were tested in seawater with cathodic polarization at the lower displacement rate. A change in fracture surface morphology could be expected at the slope change point. Correlations between the "*J-R* curve" predictions and the specimen fracture surfaces were developed and will be described in a later section.

Figures 8 to 11 illustrate the effect of longer pre-exposure times on the fracture behavior of the 4340 steel in seawater with cathodic polarization. The arrows on the load versus COD records again indicate the approximate crack initiation points. The longer pre-exposure times caused a reduction in the *J* near initiation to below 9 kPa·m (50 in.-lb/in.^{2}). The effect of pre-exposure time, however, appears to have reached a plateau or limit in the range of 20 to 26 days, since there was no substantial difference in the fracture behavior of the two specimens (FYS-H6 and FYS-H12) pre-exposed at 19 days and 26 days respectively. Both of these specimens also exhibited the change in slope of the "*J-R* curve" noted previously for FYS-H2. An interesting feature of the

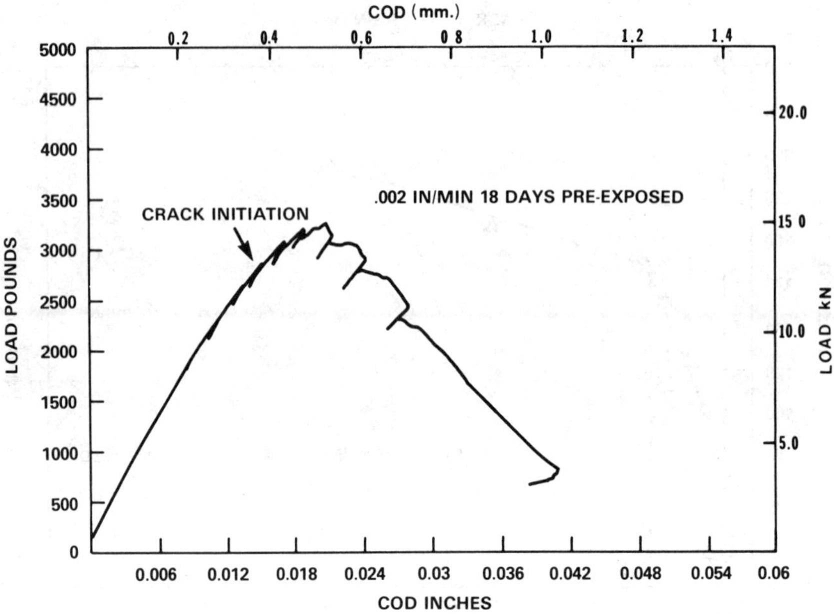

FIG. 8—*Load versus COD for FYS-H6 4340 steel seawater test.*

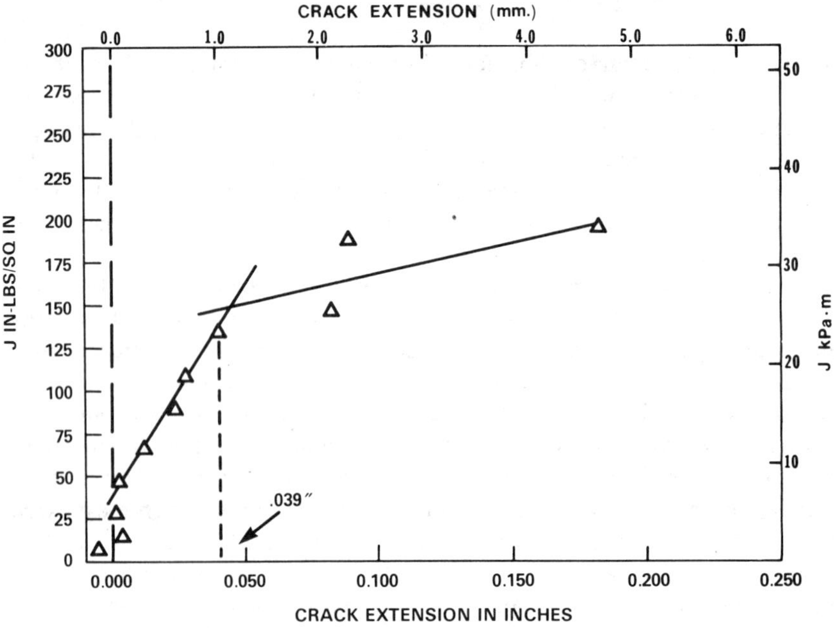

FIG. 9—"J-R *curve" for FYS-H6 at 0.002 in./min (19 days pre-exposed).*

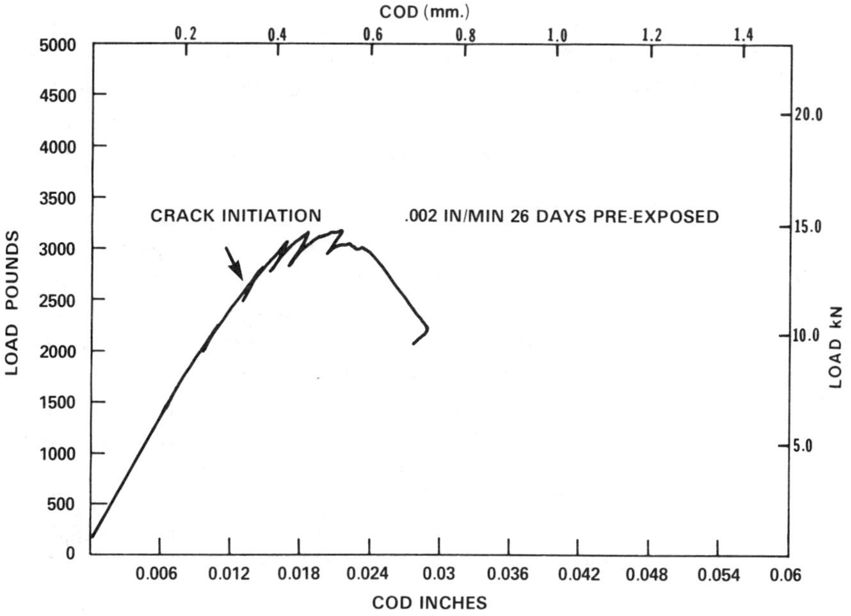

FIG. 10—*Load versus COD for FYS-H12 4340 steel seawater test.*

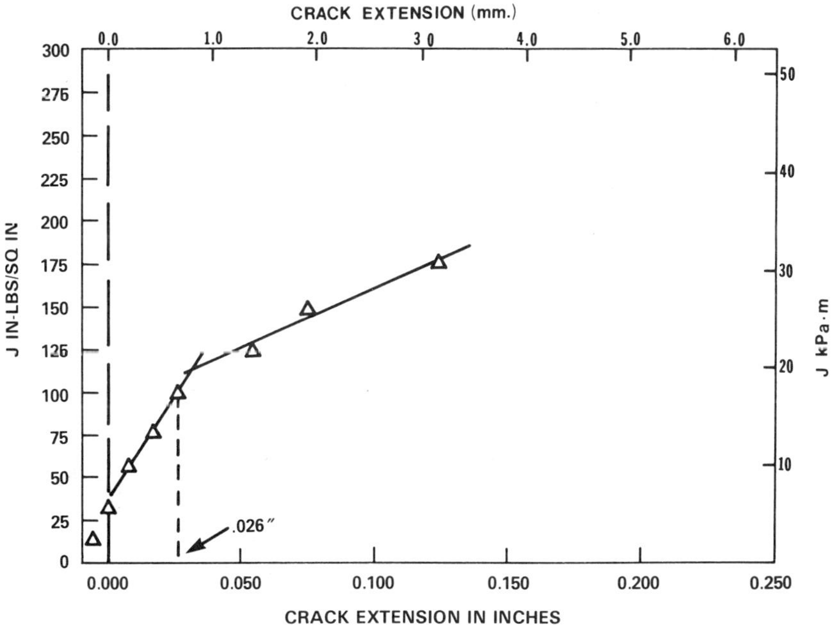

FIG. 11—"J-R *curve*" *for FYS-H12 at 0.002 in./min (26 days pre-exposed).*

load versus COD records of these two specimens (Figs. 8 and 10) is the "sawtooth" behavior beyond maximum load associated with the unloadings. This is indicative of environmental cracking occurring during some portion of the unload-reload cycle and can cause serious errors in crack length estimation from compliance. This result has also been seen previously by Anderson and Gudas [*5*].

A compilation of the "*J-R* curves" for the 4340 steel specimens tested at 0.05 mm/min (0.002 in./min) is presented in Fig. 12. The specimens pre-exposed for the longest periods (FYS-H6 and FYS-H12) exhibited approximately a four-fold decrease in the energy required for fracture initiation compared with the air test cases. Increasing the pre-exposure time did not seem to increase the value of crack extension at which the slope change in the "*J-R* curves" occurred. The limited amount of data, however, precludes a conclusive determination of this point. The initial slope of the "*J-R* curve" for Specimen FYS-H2 (14 h pre-exposed) was steeper than that obtained for either FYS-H6 (19 days pre-exposed) or FYS-H12 (26 days pre-exposed). This would seem to imply that the resistance to progressive fracturing was higher, initially, for FYS-H2 compared with FYS-H6 or FYS-H12. The initial "*J-R* curve" slopes for Specimens FYS-H6 and FYS-H12 were, however, very similar, implying little difference in the energy required to cause progressive fracturing between these specimens.

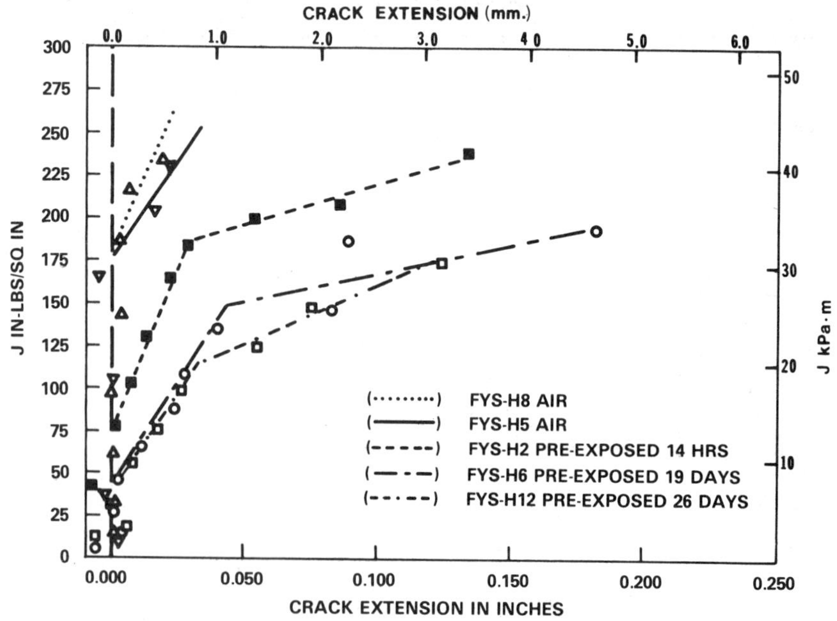

FIG. 12—"J-R *curves" for 4340 steel at 0.002 in./min.*

The direct-current potential drop (DCPD) technique for crack extension measurement was applied to specimen FYS-H13 with the results shown in Figs. 13 and 14. Notable in Fig. 13 is the absence of the unloadings required for the compliance-based crack extension estimation technique. The initiation point determined from DCPD is shown by the arrow in Fig. 13. Specimen FYS-H13 was tested in seawater with cathodic polarization at the 0.05 mm/min (0.002 in./min) displacement rate. Pre-exposure time for this specimen was only 2 h, the intent being to determine at what time environmental effects begin to control the fracture process. As shown in Fig. 13, the crack initiation point has moved downward from the maximum load, but the load-COD behavior is still similar to that observed for the air tests (see Fig. 4). The "*J-R* curve" for Specimen FYS-H13 is shown in Fig. 14. This figure indicates a J near initiation of approximately 28 kPa·m (160 in.-lb/in.^{2}). While somewhat less than the comparable J values obtained for the air test specimens, the "*R* curve" did not exhibit the features associated with the other seawater tests. This would seem to indicate that the 2 h pre-exposure time provides an effective lower bound for the observance of environmentally assisted cracking in the 4340 steel.

The K_{Iscc} values for the seawater compact specimens were calculated from the apparent J near crack initiation (J_{Iscc}) and compared with K_{Iscc} values obtained with cantilever beam specimens tested in similar environments. The

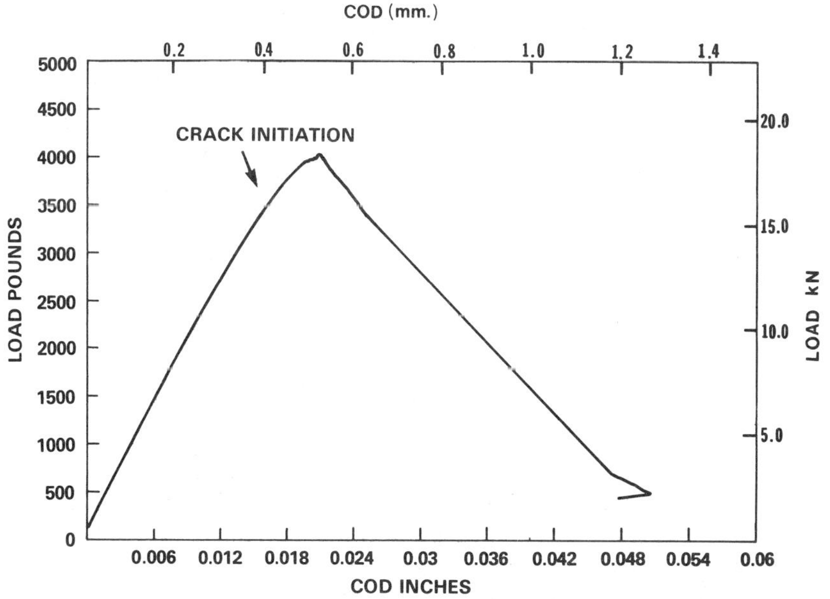

FIG. 13—*Load versus COD for FYS-H13 at 0.002 in./min (2 h pre-exposed).*

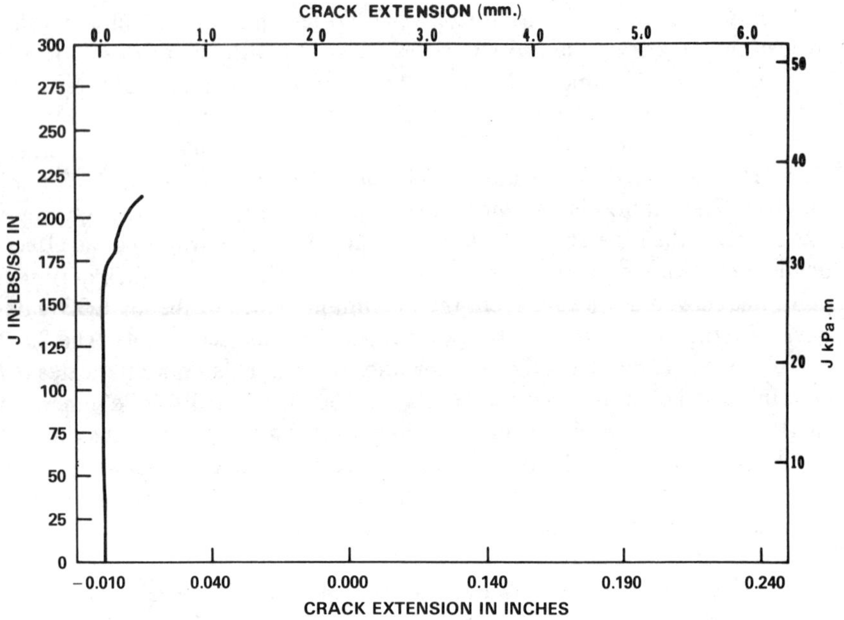

FIG. 14—"J-R *curve" for FYS-H13 4340 at 0.002 in./min (2 h pre-exposed), DCPD.*

cantilever beam results were obtained in a separate investigation on the same heat of 4340 steel by Hauser and Caton [*14*]. The results of these comparisons are presented in Table 3. This table shows K_{Iscc} values from the cantilever beam specimens ranging from 22 to 55 MPa$\sqrt{m}$ (20 to 50 ksi$\sqrt{in.}$). The higher values were from specimens that had been precracked in excess of 22 MPa$\sqrt{m}$ (20 ksi$\sqrt{in.}$), while the lower values were obtained from specimens that had been precracked at less than 22 MPa$\sqrt{m}$ (20 ksi$\sqrt{in.}$). The lower values are therefore more indicative of the actual K_{Iscc} determined with the cantilever beam tests. The lowest value for K_{Iscc} calculated from "J_{Iscc}" was 34 MPa$\sqrt{m}$ (31 ksi$\sqrt{in.}$) for Specimen FYS-H12, the specimen pre-exposed for the longest duration (26 days). The calculated K_{Iscc} values from this investigation also increased with decreasing pre-exposure time.

Fractographic Examinations

The fracture surfaces of selected specimens were examined in a scanning electron microscope to characterize the microfracture process both in air and in the seawater environments. The primary aim of the fractographic investigation, however, was to perform measurements on the fracture surfaces to correlate with the predicted fracture mode changes from the specimen "*J-R* curves".

TABLE 3—*Correlation of results of fracture tests with cantilever beam results.*

		Estimated J_{Ic}		$K(J_{Ic})$*		K Cantilever Beams	
Specimen	Environment	kPa·m	(in.-lb/in.^{2})	MPa$\sqrt{m}$	(ksi$\sqrt{in.}$)	MPa$\sqrt{m}$	(ksi$\sqrt{in.}$)
FYS-H5	air	31	(175)	83	(75)		
FYS-H2	SW+CP† 14 h	13	(75)	54	(49)	22 to 55	(20 to 50)
FYS-H6	SW+CP† 19 days	9	(50)	44	(40)		
FYS-H12	SW+CP† 26 days	5	(30)	34	(31)		

*$K = \sqrt{J \cdot E/(1-\nu^2)}$.
†SW = seawater; CP = cathodic polarization.

As expected, the fractographic examinations showed that a fracture mode change was occurring in the seawater tests, with the region immediately ahead of the fatigue precrack being almost completely intergranular, while further fracturing occurred by microvoid coalescence. This is shown for Specimen FYS-H12 in Figs. 15 and 16. Figures 17 and 18 show the interface between the fatigue precrack and fracture regions for Specimens FYS-H12 and FYS-H2 respectively. Both specimens show intergranular fracture beginning at the interface. The interfranular fracture in Specimen FYS-H12 appears to be more distinct, while there is more evidence of ductility for Specimen FYS-H2 (Fig. 18). This is perhaps attributable to the longer pre-exposure time for FYS-H12 compared with FYS-H2 (26 days versus 14 h).

The fatigue precrack/fracture interface for an air test specimen (FYS-H5) is shown in Fig. 19. This figure shows the immediate start of fracture to be by microvoid coalescence with evidence of significant ductility.

The zones of intergranular fracture at the precrack tip on Specimens FYS-H12 and FYS-H2 are shown in Figs. 20 and 21 respectively. The intergranular fracture zone was estimated as shown in the figures, and measurements were taken at eleven evenly spaced points across the surfaces at the points indicated. The two edge points were averaged to give a total of a ten point average of the intergranular zone across the specimen half.

The two sets of fractographs in Figs. 20 and 21 constitute SEM stereo pairs which were taken at an angle of separation of 6°. When aligned correctly and viewed with a stereo viewer, a three-dimensional image of the fracture surface is obtained. The difficulty in separating the intergranular cracking (IGC) zone from the ductile fracture region necessitated the use of this technique. The lines marking the extent of the IGC zone in Figs. 20 and 21 were drawn

FIG. 15—*SEM fractograph of Specimen FYS-H12 showing intergranular fracture region immediately ahead of the precrack.*

with the aid of a stereo viewer which helped enhance the delineation between the two regions due to the widely different three-dimensional textures involved. Even with the aid of a stereo viewer, the delineations are subject to different interpretations and are, at best, reasonable engineering approximations of the extent of the IGC zones.

The ten point measured average of the IGC zone for FYS-H12 was 0.59 mm (0.0231 in.) The fracture mode change predicted from the "*J-R* curve" (see Fig. 11) was 0.66 mm (0.026 in.). Therefore the *J-R* curve prediction overestimated the IGC zone measurement by 11% in this case. The fracture mode change predicted from the "*J-R* curve" for FYS-H2 was 0.71 mm (0.028 in.) (see Fig. 7). The ten point measured average of the IGC zone for FYS-H2 was 0.67 mm (0.0263 in.). Again, the *J-R* curve prediction overestimated the fracture surface measurement; this time, however, by only 6%.

Considering the uncertainties associated with the fracture surface measurements, these correlations are excellent and are on the same order of accuracy of the compliance estimation of the final measured crack extension in air *J-R* curve tests. The significance of these correlations will be discussed in the following section.

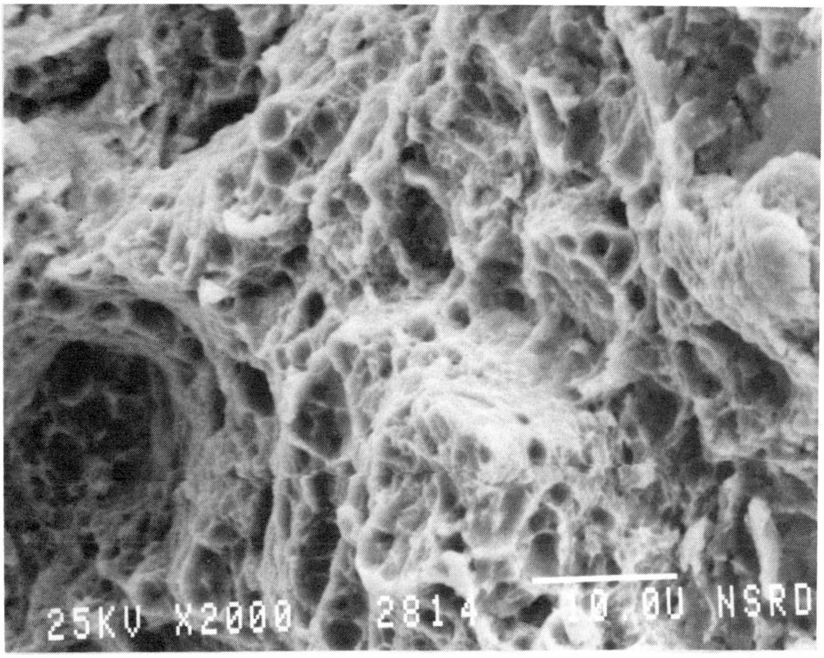

FIG. 16—*SEM fractograph of Specimen FYS-H12 showing microvoid coalescence in the center of the specimen.*

Figures 22 and 23 serve to further illustrate the differences between the fracture surfaces of a seawater specimen (FYS-H12) and an air test specimen (FYS-H10) at a low magnification. While similar overall, Specimen FYS-H12 has the IGC zone separating the fatigue precrack and the ductile tearing ridges (Fig. 22), whereas Specimen FYS-H10 exhibits ductile tearing ridges extending to the tip of the fatigue precrack (Fig. 23).

Discussion

Environmentally induced intergranular cracking in high strength steels is most often rationalized with a hydrogen embrittlement mechanism. The accumulation of atomic hydrogen ahead of a crack tip can be related to the total charge passed through the vicinity of the crack tip using the constant charge criterion proposed by Scully [*13*]. This criterion maintains that the charge in the crack tip region must exceed a minimum value for damage to occur. This charge is obviously dependent both on the strain and strain rate at the crack tip, since the rate of creation of fresh metal surfaces at the crack tip is dependent on the strain rate.

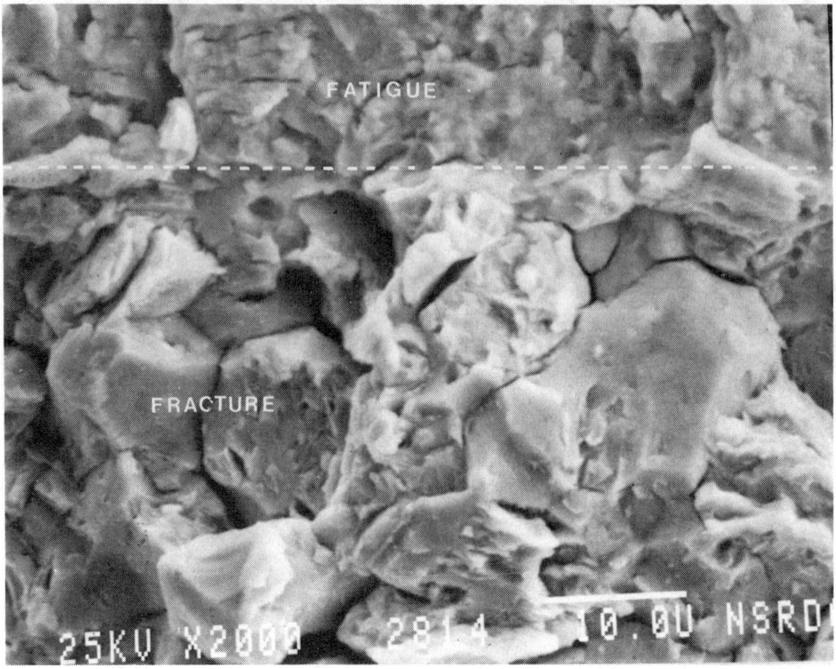

FIG. 17—*SEM fractograph of Specimen FYS-H12 showing the fatigue pre-crack/fracture interface.*

Results such as those shown in Fig. 14 for FYS-H2 (SW, 0.002 in./min, 14 h pre-exposed) and FYS-H8 (SW, 0.010 in./min, 48 h pre-exposed) can be at least partially rationalized with the above criterion. In the case of the slower displacement rate specimen (FYS-H2), 14 h of pre-exposure time combined with the slower displacement rate was sufficient to cause significant environmental degradation of the fracture properties. The faster displacement rate specimen (FYS-H8), however, exhibited no deleterious environmental effects despite having been pre-exposed for a longer time (48 h). These data clearly support the necessity of achieving both a critical strain and strain rate at the crack tip to cause environmental damage.

The lower limit on pre-exposure time for the occurrence of environmentally assisted fracture at the slower displacement rate was defined from Specimen FYS-H13 as 2 h. The results therefore indicate a required pre-exposure time for the occurrence of environmental damage of between 2 and 14 h at the slower displacement rate. This is again consistent with the above criterion in that a minimum time is required for the accumulation of a critical amount of hydrogen ahead of the crack tip. The rate determining step for this accumula-

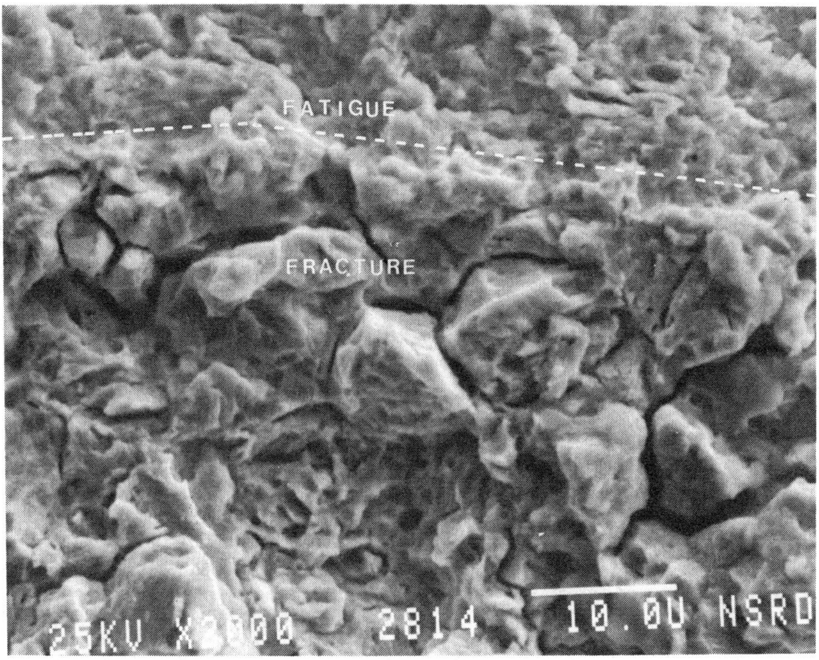

FIG. 18—*SEM fractograph of Specimen FYS-H2 showing fatigue pre-crack/fracture interface.*

tion could be surface absorption of hydrogen, absorption into the metal, bulk or grain boundary transport, or a combination of these factors.

All the specimens exhibiting environmentally assisted cracking were observed to switch fracture modes from intergranular to microvoid coalescence within 1.0 mm (0.040 in.) from the fatigue precrack tip. This implies that the dual criteria of critical strain and strain rate are no longer maintained beyond this distance. Under the constant displacement conditions imposed for this investigation, the strain rate at the crack tips of the compact specimens actually decrease with increasing crack length. The reason for cessation of environmental damage therefore appears to be due to the lack of a critical accumulation of hydrogen ahead of the crack tip. This most definitely involves hydrogen transport processes and is a certain function of the pre-exposure period. The exact relationship between the extent of the IGC zone and the pre-exposure period is not clear from this study. Beyond 14 h of pre-exposure time the data for Specimens FYS-H2, H6, and H12 indicate no dependence of the extent of this zone on the pre-exposure period.

The excellent correlations obtained between the physically measured extent

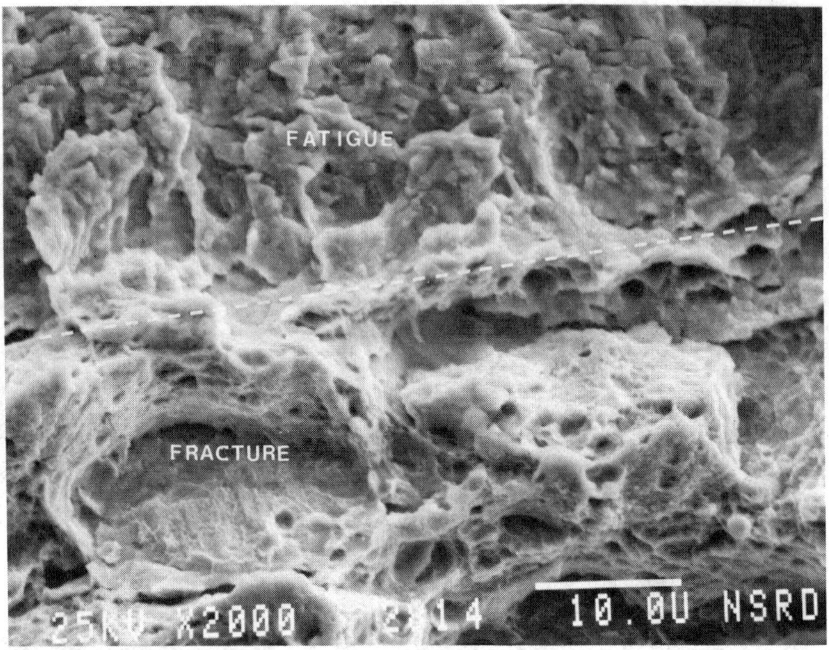

FIG. 19—*SEM fractograph of Specimen FYS-H5 showing fatigue pre-crack/fracture interface.*

of the IGC zone and the on-line predictions during the *J-R* curve testing provide a useful tool for evaluating environmental damage mechanisms in high strength steels. As an example, if a hydrogen damage mechanism is assumed to be operative, diffusible hydrogen can be measured at different positions on the fracture surface after a test is completed and correlations could be made with the value of the *J*-integral. This would provide a correlation between stable hydrogen content and the energy required to produce fracture.

In all the tests exhibiting environmental damage the "*R*-curves" exhibited relatively steep slopes just beyond crack initiation followed by a distinct change to lesser slopes with progressive fracturing. The initial steep slopes indicate that a higher energy fracture mode was operative in that region. The fracture morphology was determined to be intergranular (IG) in the initial region, followed by microvoid coalescence (MVC). This seemed contradictory since IG usually implies a lower energy for fracture than MVC. This result can be explained, however, through consideration of the crack tip mechanics. For the IGC zone the crack tip is highly branched. This results in a loss of constraint at the crack tip and, consequently, a higher "*R*-curve" slope in that region.

A few words regarding the DCPD and unloading compliance (UC) crack length estimation techniques are in order. As mentioned previously, the UC technique places a lower practical limit on the displacement rate that can be used for testing. Also, the occurrence of crack growth during the unloading portions of the tests (Figs. 8 and 10) casts doubt upon the validity of the measurements. Finally, the unloadings increase the amount of time possible for environmental interactions and can create a pumping action of the electrolyte at the crack tip with unknown consequences for environmental damage.

The DCPD technique was employed in this investigation primarily to validate its usefulness in environmental testing and to address this final point made above. The DCPD method requires no unloadings and thus can be used to evaluate the effect of the unloadings in a UC test on environmental damage. The single DCPD test performed for this investigation did not have a sufficient pre-exposure period to initiate environmental damage and therefore did not address the effect of the unloadings. The method was, however, proved to be workable in aqueous environments.

This research effort was focused on AISI 4340 steel heat treated to the 1241 MPa (180 ksi) yield strength level. At this strength level the alloy is highly susceptible to environmentally assisted fracture in aqueous environments. This alloy was chosen for this research effort for exactly that reason, since the objective of this research was to evaluate a new approach to assessing environmental cracking susceptibility. The J-integral approach employed in this work, however, enables lower strength alloys to be tested in small specimen sizes and can address the key issue of environmental interactions in the case of large scale plasticity.

Conclusions

Stress corrosion/hydrogen embrittlement crack growth behavior in a high strength steel can be characterized using techniques developed for the study of stable elastic-plastic crack growth. The results in the case of environmentally assisted fracture are test rate dependent and also dependent on the initial pre-exposure period. Assuming a hydrogen embrittlement mechanism, the results of this research support the constant charge criterion of Scully in that both a critical strain and strain rate are necessary to sustain environmentally assisted cracking.

In the case of the 4340 steel tested in this investigation, environmental degradation of the fracture properties in seawater with cathodic polarization resulted in up to a four-fold decrease in the energy required for fracture initiation (J_{Iscc}).

Good correlations were obtained between predicted fracture morphology changes from the material "J-R curves" and actual fracture surface measurements made after test completion. This provides a useful tool for mechanistic studies of hydrogen embrittlement/SCC in high strength steels.

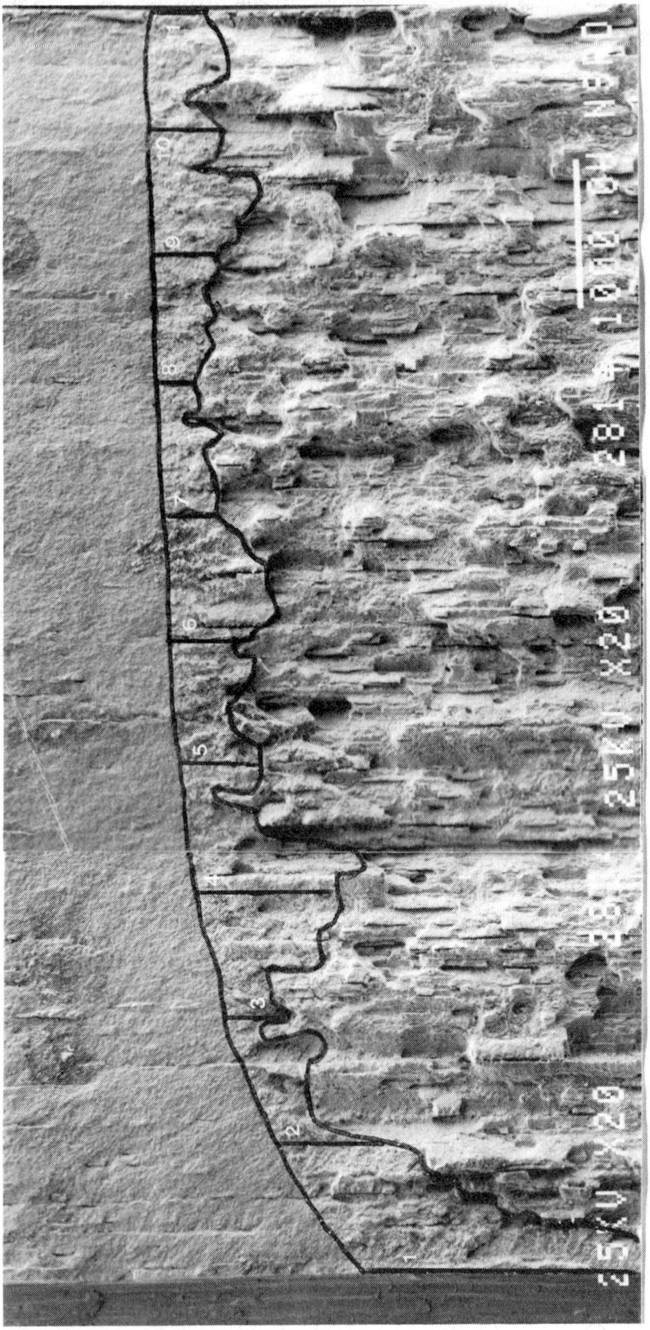

FIG. 20—*SEM stereo fractographs of specimen FYS-H12 showing IGC zone measurements.*

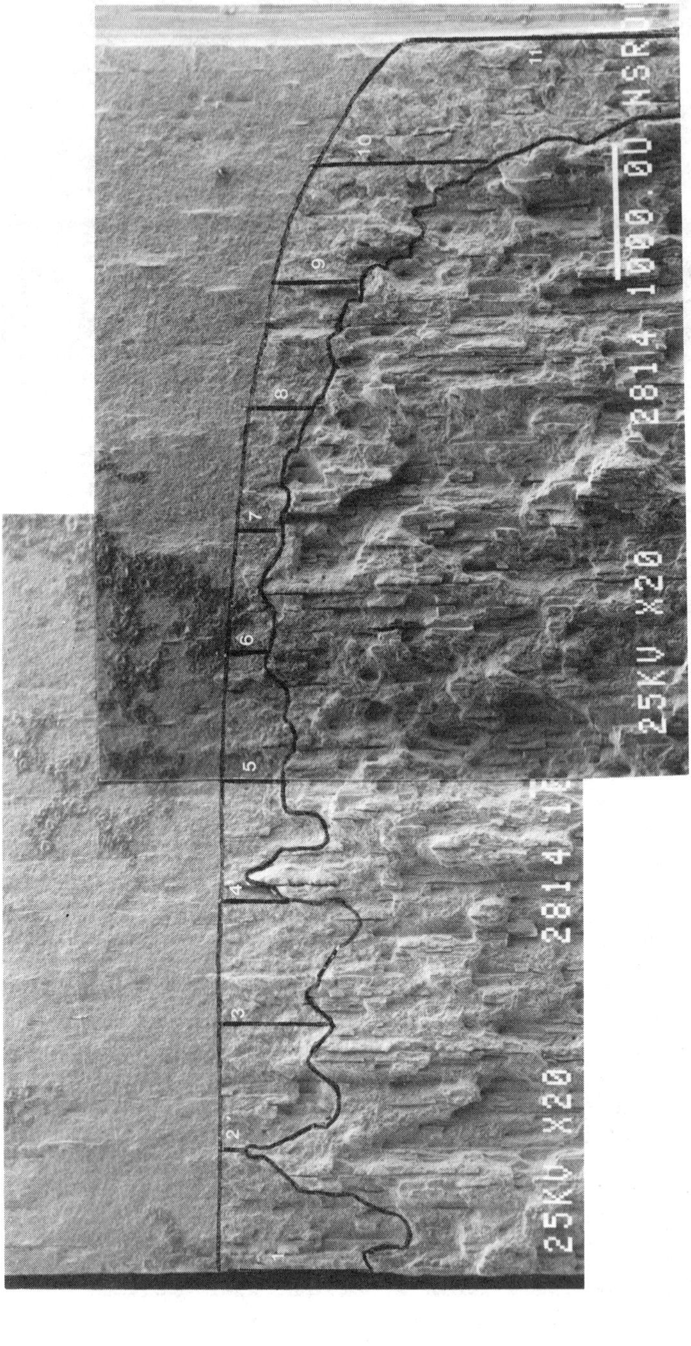

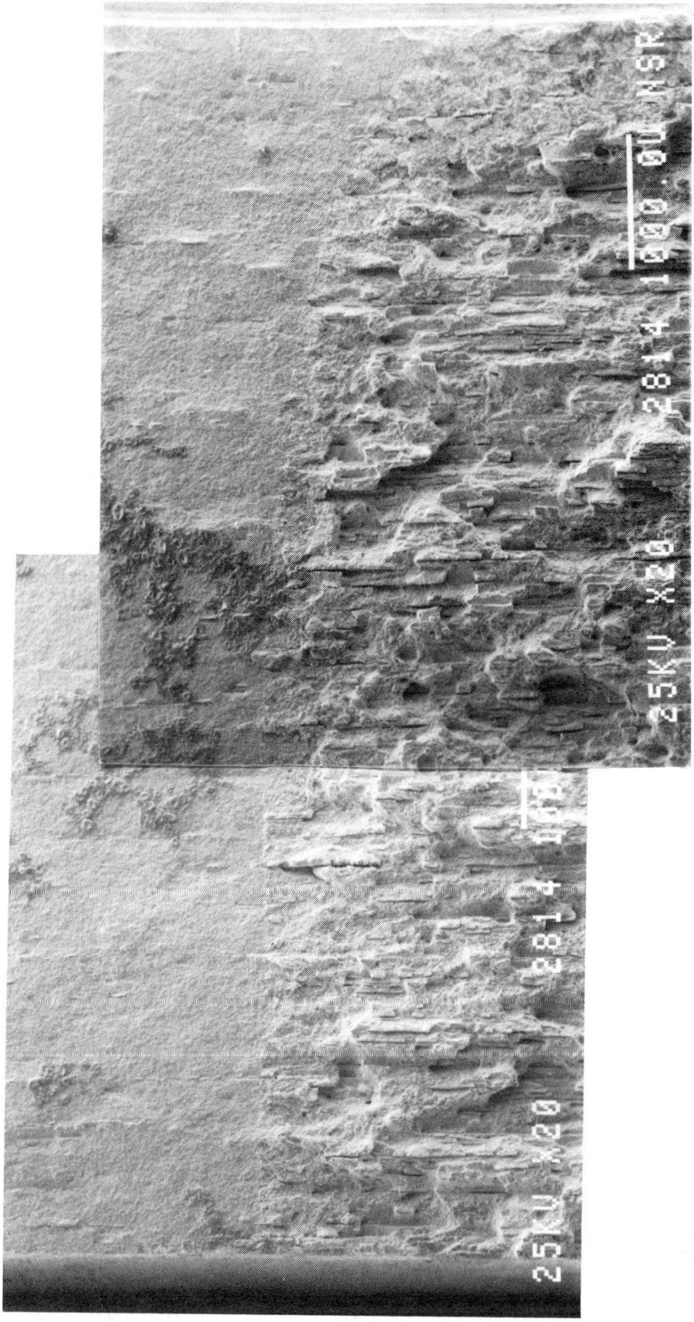

FIG. 21—*SEM stereo fractographs of Specimen FYS-H2 showing IGC measurements.*

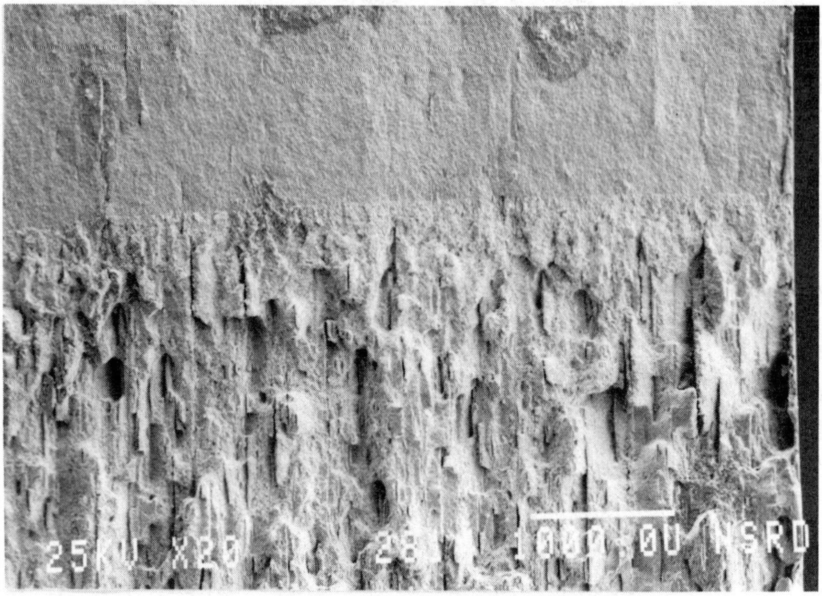

FIG. 22—*SEM fractograph of Specimen FYS-H12 showing tearing ridges separated from the fatigue pre-crack by the IGC zone.*

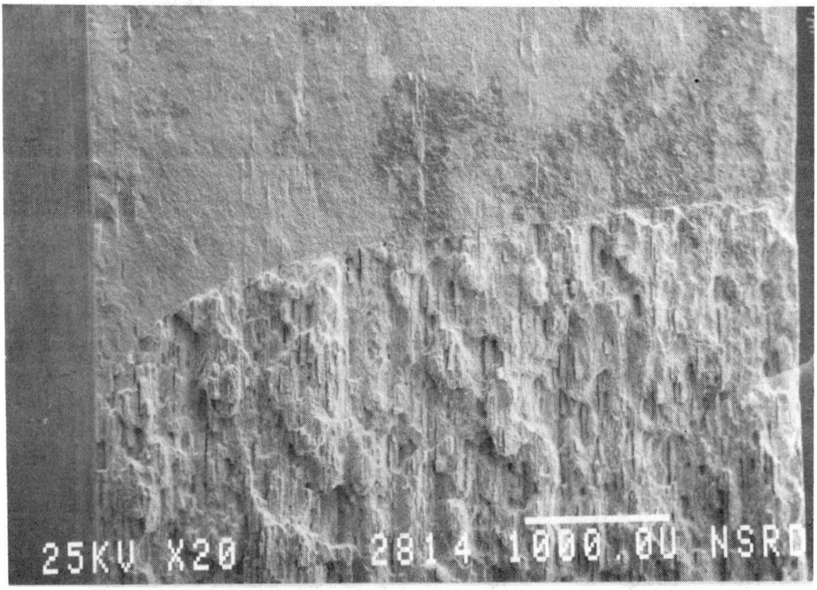

FIG. 23—*SEM fractograph of Specimen FYS-H10 (air) showing tearing ridges extending to the fatigue pre-crack.*

The K_{Iscc} derived from J_{Iscc} in rising load tests on 4340 steel was found to correlate closely with K_{Iscc} results from a separate investigation on the same plate using cantilever beam specimens. It should be pointed out, however, that the incubation time for the rising load tests is a critical factor in producing a good correlation between these two types of tests.

Acknowledgments

The authors wish to express thanks to Messrs. M. G. Vassilaros and J. R. Scully of the David Taylor Naval Ship R&D Center for many helpful discussions during the course of this work.

References

[1] Brown, B. F. and Beachem, C. D., "A Study of the Stress Factor in Corrosion Cracking by Use of the Precracked Cantilever-Beam Specimen," *Corrosion Science*, Vol. 5, 1965.

[2] Novak, S. R. and Rolfe, S. T., "Modified WOL Specimen for K_{Iscc} Environmental Testing," *Journal of Materials*, Vol. 4, 1969, pp. 701-728.

[3] Kobayashi, H. et al, "Influence of Hydrogen on the Fracture Toughness of High Strength Steels," in *Advances in Fracture Research (Fracture '81)*, Vol. 4, Cannes, France, 1981, pp. 1909-1916.

[4] Clark, W. G. and Landes, J. D. "An Evaluation of Rising Load K_{Iscc} Testing," in *Stress Corrosion—New Approaches, ASTM STP 610*, American Society for Testing and Materials, Philadelphia, 1976, pp. 108-127.

[5] Anderson, D. R. and Gudas, J. P., "Stress Corrosion Evaluation of Titanium Alloys Using Ductile Fracture Mechanics Technology," *Environment-Sensitive Fracture: Evaluation and Comparison of Test Methods, ASTM STP 821*, American Society for Testing and Materials, Philadelphia, 1984, pp. 98-113.

[6] Abramson, W. G., Evans, J. T., and Parkins, R. N., "The Application of the *J*-Integral to Stress Corrosion Cracking," University of Newcastle upon Tyne, Department of Metallurgy and Engineering Materials Report produced under Grant 2043/074 RAR, April 1982.

[7] Begley, J. A. and Landes, J. D. in *Fracture Mechanics, ASTM STP 514*, American Society for Testing and Materials, Philadelphia, 1972, pp. 1-20.

[8] Clarke, G. A. et al in *Mechanics of Crack Growth, ASTM STP 590*, American Society for Testing and Materials, Philadelphia, 1976, pp. 27-42.

[9] Joyce, J. A. and Gudas, J. P. in *Elastic-Plastic Fracture, ASTM STP 668*, American Society for Testing and Materials, Philadelphia, 1979, pp. 451-468.

[10] Vassilaros, M. G. and Hackett, E. M. "*J*-Integral *R*-Curve Testing of High Strength Steels Utilizing the Direct-Current Potential Drop Method," in *Fracture Mechanics: Fifteenth Symposium, ASTM STP 833*, American Society for Testing and Materials, Philadelphia, 1984, pp. 535-552.

[11] Hutchinson, J. W. and Paris, P. C. in *Elastic-Plastic Fracture, ASTM STP 668*, American Society for Testing and Materials, Philadelphia, 1979, pp. 37-64.

[12] Ernst, H. A. et al, "Estimations on *J*-Integral and Tearing Modulus *T* from a Single Specimen Test Record," in *Fracture Mechanics: Thirteenth Conference, ASTM STP 743*, American Society for Testing and Materials, Philadelphia, 1981, pp. 476-502.

[13] Scully, J. C., *Corrosion Science*, Vol. 15, 1975, p. 207.

[14] Hauser, J. A., Caton, T. C., et al, "Significance of Experimental Factors in Fracture Mechanics Test Techniques for Marine Stress Corrosion Cracking," to be presented at the International Conference and Exhibition on Fatigue, Corrosion Cracking, Fracture Mechanics and Failure Analysis, Salt Lake City, Utah, 2-6 Dec. 1985.

Thomas J. Watson[1] *and Mitchell I. Jolles*[1]

Plastic Energy Dissipation as a Parameter To Characterize Crack Growth

REFERENCE: Watson, T. J. and Jolles, M. I., **"Plastic Energy Dissipation as a Parameter To Characterize Crack Growth,"** *Fracture Mechanics: Seventeenth Volume, ASTM STP 905*, J. H. Underwood, R. Chait, C. W. Smith, D. P. Wilhem, W. A. Andrews, and J. C. Newman, Eds., American Society for Testing and Materials, Philadelphia, 1986, pp. 542–555.

ABSTRACT: The global plastic energy dissipation for crack initiation and slow growth was experimentally determined for monotonic, quasi-static loading conditions utilizing compact tension and three-point bend specimens of HY-130 steel. For non-side-grooved specimens, the global plastic energy dissipation was linear with crack extension. For side-grooved specimens, however, the global plastic energy dissipation R-curves were not linear. Moreover, for non-side-grooved specimens, the values of the global plastic energy dissipation at crack growth initiation and the global plastic energy dissipation rate were found to be dependent on specimen size, geometry, and initial crack length. Although no tests were conducted to determine the sensitivity of side-grooved specimens to size or initial crack length, it was found that, for these specimens, the geometry dependence of the global plastic energy dissipation rate was minimized but not eliminated.

KEY WORDS: plastic energy dissipation, plastic work, plastic energy dissipation rate, elastic-plastic fracture, crack initiation, crack growth

Several candidate fracture parameters have been used as a means for quantifying crack growth in the elastic-plastic regime. These include crack tip field parameters such as the stress intensity factor, integral parameters such as J, energy release rates such as G^{Δ}, and geometric parameters such as crack tip opening displacement. A viable fracture parameter must:

(i) be representative of the magnitude of the crack tip field and able to quantify crack growth initiation, slow stable extension, and instability.

[1]Materials Research Engineer, and Head, Fracture Mechanics Section, respectively, Naval Research Laboratory, Washington, DC 20375-5000.

(ii) have an R-curve which is independent of specimen size, geometry, and type of loading.
(iii) have an R-curve which is independent of the initial crack length and the amount of crack extension.
(iv) be easily obtained from available experimental data and calculable for the structural problem of interest.
(v) be relatively insensitive to details of numerical modelling.

Each of the above fracture parameters fails to meet one or more of the requirements. That is, the stress intensity factor is no longer representative of near-field behavior because the analytic basis for this parameter is no longer valid when large-scale plasticity is obtained local to the crack tip. For crack growth in high toughness materials, the *J-R* curve has been found to be dependent on specimen size, geometry, and the amount of crack extension [*1,2*]. Furthermore, it has been shown that J is no longer path independent and that the energy release rate definition of J is not valid subsequent to crack extension [*3*]. When G^{Δ} is obtained from finite element analyses, it is found to be mesh size dependent so that this parameter must be obtained over a distance which is characteristic for a given material [*4,5*]. The crack tip opening displacement, as obtained in British Standard BS 5762, cannot be used to quantify crack growth because it has no physical meaning beyond crack initiation. Also, from a practical point of view, this crack tip opening displacement definition is very limited in that it does not easily allow for a prediction of the fracture resistance of a real structure. Clearly, there is a need for a new approach to the elastic-plastic fracture problem.

Until recently, the idea of using plastic energy dissipation as a parameter to characterize crack growth has attracted little interest. To date, the two main approaches to plastic energy dissipation are based on either a local or a global criterion. The phenomenon of recrystallization has been used to study the local plastic energy dissipation as a function of crack length in specimens subjected to monotonic, quasi-static loading conditions [*6–9*]. It was found that the local plastic energy dissipation was linear with crack extension, and it was claimed that the local plastic energy dissipation rate was independent of specimen size and geometry [*8,9*]. It has even been noted that the behavior of specimens subjected to cyclic loading can be quantified in terms of the local plastic energy dissipation [*10*]. Moreover, through the use of the finite element technique to model crack growth in center-cracked tension specimens subjected to monotonic, quasi-static loading, it was found that the global plastic energy dissipation was linearly related to crack length [*11*].

Both criteria, as epitomized by the above studies, have their advantages and limitations. The main advantage of the local plastic energy dissipation criterion is that it is based solely on the plastic deformation behavior of material local to the crack tip. Hence variables such as specimen size and geometry do not enter into the determination of this parameter. The main limitation of this criterion is that there is no analytic basis to establish it as a viable approach to the fracture problem. On the other hand, the main advantage of the global plastic energy dissipation is that it has a sound analytical foundation [12]. Specifically, the global plastic energy dissipation rate (dU_p/dA) is directly related to the nonlinear strain energy release rate ($\bar{G}$) through the equation

$$\bar{G} = -\left(\frac{dU_e}{dA} + \frac{dU_p}{dA}\right) \quad (1)$$

where dU_e/dA is the elastic strain energy release rate. Note that when the material local to the crack tip approaches the perfectly plastic state, dU_e/dA approaches zero so that Eq 1 becomes

$$\bar{G} = -\frac{1}{B}\frac{dU_p}{da} \quad (2)$$

where B is the thickness and a the crack length for a given geometry. The main limitation of this parameter is its direct dependence on the global plastic deformation behavior of the structure of interest. That is, if the global plastic deformation behavior differs for two unlike geometries, it would be expected that the globally measured plastic energy dissipation would be both size and geometry dependent. Nevertheless, if a majority of the global plastic deformation could be contained within a region somewhat local to the crack tip, it is logical to expect that a measurement of this quantity would be independent of size and geometry, provided that the crack tip dominated the formation of plastic deformation. Therefore the purpose of this study was to determine if the global plastic energy dissipation could meet the requirements as stated above and, in so doing, become a viable candidate for the quantification of crack initiation and slow growth.

Materials and Methods

The test specimens for this investigation had the T-L orientation and were machined from 25.4-mm-thick plate of HY-130 steel. The tensile properties for this material are given in Table 1 and were obtained in accordance with ASTM Methods of Tension Testing of Metallic Materials (E 8). The specimen geometries (Fig. 1) were compact tension and three-point bend. Specimens that were side grooved 20% had the same nominal dimensions as the non-side-grooved specimens shown in Fig. 1. Side grooves were machined in a dry environment after precracking.

All specimens were fatigue precracked in accordance with specifications outlined in ASTM Test for J_{Ic}, A Method of Fracture Toughness (E 813) and had initial crack length to width ratios of 0.5, 0.55, or 0.8. All tests were

TABLE 1—*Room temperature tensile properties for HY-130 steel.*

Strain Rate, min^{-1}	Tensile Test Temperature, °C	Young's Modulus, MPa	0.2% Offset Yield Strength, MPa	Tensile Strength, MPa	Tensile Elongation, %[a]	Reduction of Area, %[b]
0.0008	26	207 000	987	1036	20	68

[a]Gage length = 50.8 mm.
[b]Nominal diameter = 12.8 mm.

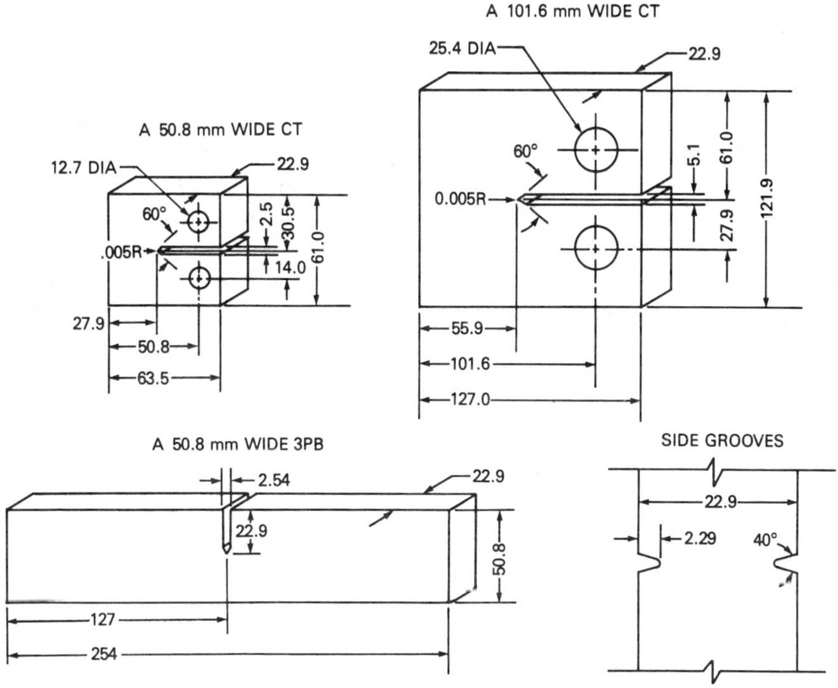

FIG. 1—*Specimen geometries used to determine the plastic energy dissipation (all dimensions in millimetres).*

performed through the use of a servohydraulic load frame and an automated technique for *R*-curve testing and analysis [*13*].

In order to analyze the experimentally determined data [*13*], the amount of physical crack extension was measured by heat tinting. All specimens were heat tinted at a temperature of 315°C for 30 min and then cooled to the temperature of liquid nitrogen. These specimens were then loaded at a high rate so that final fracture would occur by cleavage. For a given specimen, the initial and final crack length was obtained on a graphics tablet by digitization of a photograph of the fracture surface.

The global plastic energy dissipation was obtained by subtracting the elastic strain energy from the total energy under the load-displacement record (Fig. 2). For no crack growth, the global plastic energy dissipation (U_p) is given by

$$U_p = U_T - U_e \tag{3}$$

where U_T is the total energy absorbed by the specimen and U_e is the elastic strain energy. For crack growth, the global plastic energy dissipation is given by

$$U_p = U_T - U_e - U_{cg} \tag{4}$$

where U_T and U_e were previously defined and U_{cg} is the work done at the crack tip through a differential displacement du_i as the crack extends. Since the crack tip work is small relative to the total energy [3], this term may be neglected in the equality as an approximation and the global plastic energy dissipation may be determined by Eq 3. To make this calculation, the elastic compliance must be obtained for a given crack length. For the compact tension specimen, the elastic compliance (C) can be obtained by the equation [14]

$$C = \frac{1}{B_e E} f(a/W) \tag{5}$$

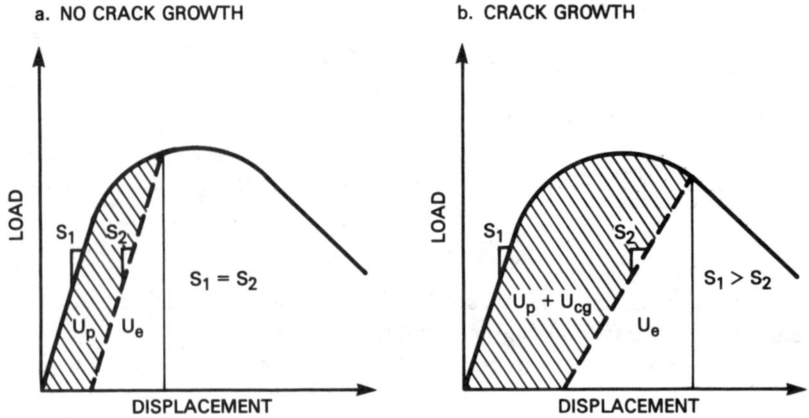

FIG. 2—*Two experimental load-displacement records that illustrate the determination of the plastic energy dissipation.*

where

$$f(a/W) = \left(\frac{1 + a/W}{1 - a/W}\right)^2 [2.163 + 12.219\,(a/W) - 20.065\,(a/W)^2 - 0.9925\,(a/W)^3 + 20.609\,(a/W)^4 - 9.9314\,(a/W)^5] \quad (6)$$

and B_e is the effective thickness [*15*], E is the effective modulus, and W is the specimen width. The elastic compliance for the three-point bend specimen is given by [*16,17*]

$$C = \frac{16}{B_e E}\,[1.195 + 1.5\,g(a/W)] \quad (7)$$

where

$$g(a/W) = \left(\frac{a/W}{1 - a/W}\right)^2 [5.58 - 19.57\,(a/W) + 36.82\,(a/W)^2 - 34.94\,(a/W)^3 + 12.77\,(a/W)^4] \quad (8)$$

and B_e, E, a, and W were as defined earlier. Having obtained the elastic compliance for the geometry of interest, the elastic strain energy can be calculated through the equation

$$U_e = \frac{1}{2}\,P^2 C \quad (9)$$

where P is the current load on the load-displacement record. The global plastic energy dissipation can then be calculated by using Eq 3.

The global plastic energy dissipation rate was obtained through the use of linear regression. If the graph of the global plastic energy dissipation versus crack extension were linear, linear regression was performed on all the data points. If the graph of these quantities were nonlinear, the global plastic energy dissipation rate was calculated by performing linear regression on successive groups of 15 data points.

Results and Analysis

The purpose of this study was to determine if the global plastic energy dissipation could meet the requirements of a viable fracture parameter. Of the five requirements listed above, only two were studied in this investigation. These were requirements (ii) and (iii), and the variables that were addressed included the effect of specimen size, geometry, and initial crack length.

The effect of specimen size on the global plastic energy dissipation was determined by testing non-side-grooved 50.8 (small) and 101.6 mm (large) wide compact tension specimens. Both sizes had the same initial crack length to width ratio and, as shown in Fig. 1, the same thickness. The average initiation values for the global plastic energy dissipation (Table 2) were found to be higher for the large compacts. The values of the global plastic energy dissipation rate for the large specimens were found to be greater than those obtained for small specimens (Fig. 3).

To investigate the effect of specimen geometry on the global plastic energy

TABLE 2—*Average initiation values of the plastic energy dissipation for HY-130 steel.*

Geometry	Width, mm	Initial Crack Length to Width Ratio	Side Grooves, %	U_p at Initiation, J/m
CT	50.8	0.55	0	287.4
CT	101.6	0.55	0	321.2
CT	50.8	0.55	0	287.4
3PB	50.8	0.55	0	318.0
CT	50.8	0.55	0	287.4
CT	50.8	0.80	0	187.7
CT	50.8	0.50	20	367.6
3PB	50.8	0.50	20	385.0

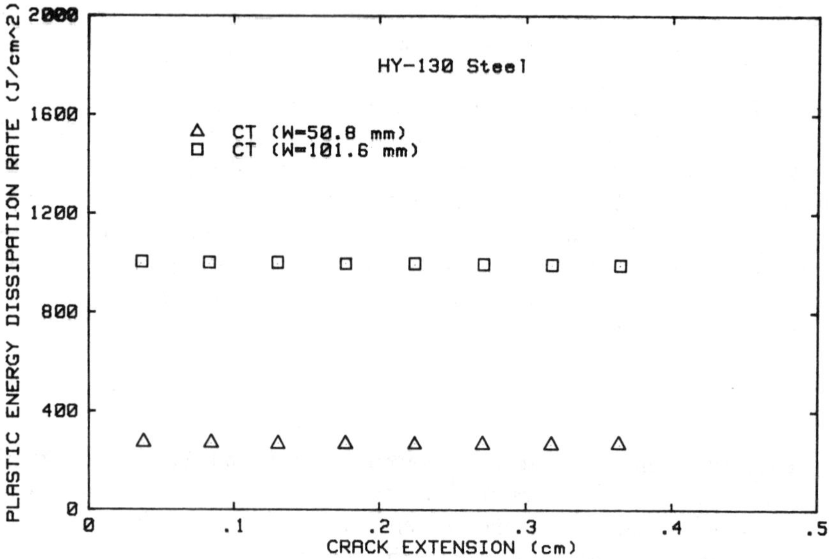

FIG. 3—*Graph of global plastic energy dissipation rate versus crack extension for small and large non-side-grooved compact tension specimens.*

dissipation, non-side-grooved 50.8-mm-wide compact tension and three-point bend specimens were tested. The values for the thickness, width, and initial crack length to width ratio for both geometries were the same. The average initiation values for the global plastic energy dissipation were found to be greater in the three-point bend specimens when compared with those obtained for the small compacts (Table 2). The values of the global plastic energy dissipation rate (Fig. 4) were found to be greater for the three-point bend specimens.

Finally, the effect of different initial crack lengths on the global plastic energy dissipation was determined by changing the initial crack length to width ratio in non-side-grooved 50.8-mm-wide compact tension specimens from 0.55 to 0.8. As shown in Table 2, it was found that the average values of the global plastic energy dissipation at initiation were lower for an initial crack length to width ratio of 0.8. The values of the global plastic energy dissipation rate (Fig. 5) were greater for the initial crack length to width ratio of 0.55.

Although one may be tempted to scale the results by introducing geometric normalizing factors such as W or B/b, there is no analytic justification for such an attempt, and the experience of the authors suggests general success will not be attained.

In an effort to explain the above behavior, measurements of the degree of shear lip formation through the thickness were obtained and plotted versus

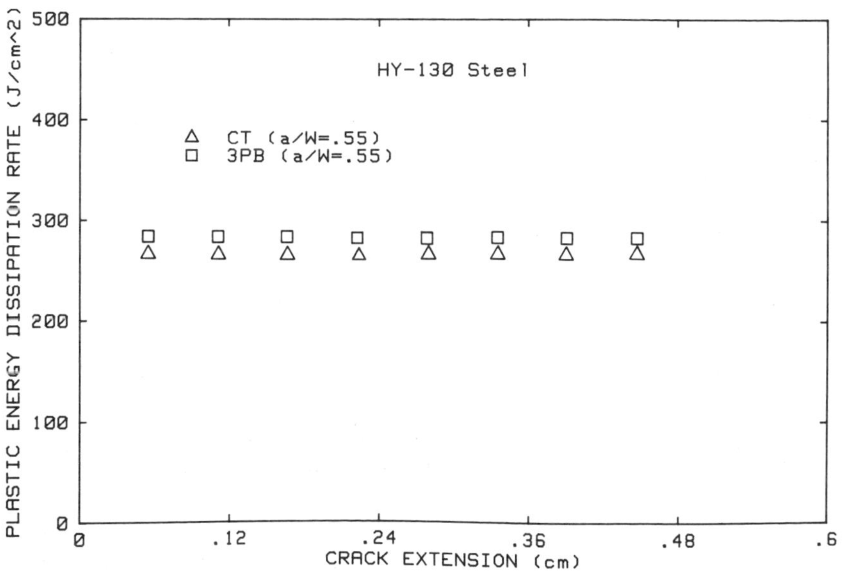

FIG. 4—*Graph of global plastic energy dissipation rate versus crack extension for non-side-grooved 50.8-mm-wide compact tension and three-point bend specimens.*

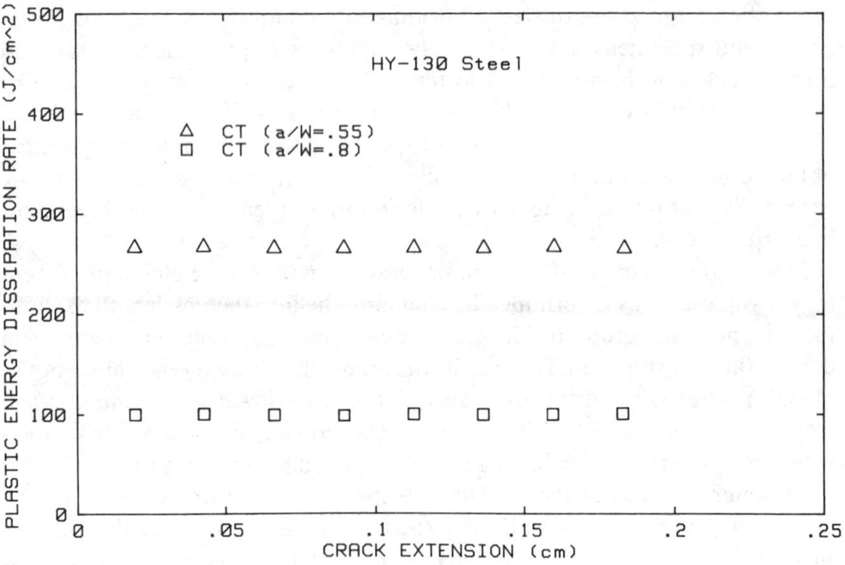

FIG. 5—*Graph of global plastic energy dissipation rate versus crack extension for non-side-grooved 50.8-mm-wide compact tension specimens precracked to an initial crack length to width ratio of 0.55 and 0.8.*

the distance from the fatigue precrack at the center of each specimen. Because the shear lip formation was the same for each side of a given specimen, only one side was measured. It was found that there was a direct relation between the global plastic energy dissipation and the degree of shear lip development when either the specimen size or initial crack length was changed. However, this correlation was not obtained for the different geometries.

The effect of specimen size and initial crack length on the global plastic energy dissipation rate is shown in Figs. 6 and 7. Note in Fig. 6 that for the same centerline distance, the value of the shear lip distance is greater in the large compact. This result seems to suggest that for the same amount of crack extension, the constraint is lower in the large compact. As shown in Fig. 7, this same line of reasoning can be used to explain the lower global plastic energy dissipation rate for the small compact specimens precracked to an initial crack length to width ratio of 0.8.

It is interesting to note that the effect of specimen size and initial crack length on constraint, and therefore, the global plastic energy dissipation rate, can be thought of in terms of the ratio of the thickness to the initial remaining ligament. This is significant because shear lip development during crack extension is not found in all materials, particularly aluminum. In general, constraint is often thought of in terms of the thickness and the in-plane dimension. That is, if the thickness is increased while holding the in-plane

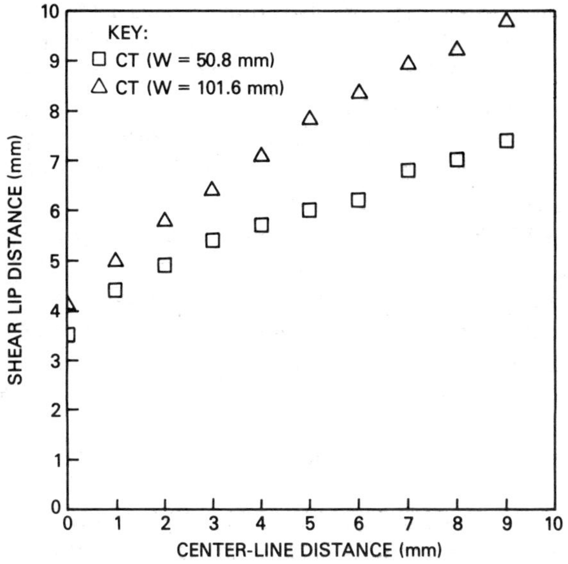

FIG. 6—*Shear lip distance versus center-line distance for small and large non-side-grooved compact tension specimens.*

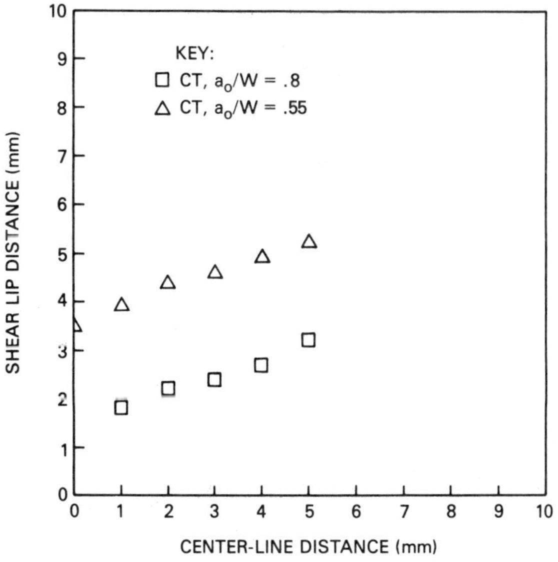

FIG. 7—*Shear lip distance versus center-line distance for non-side-grooved 50.8-mm-wide compact tension specimens precracked to an initial crack length to width ratio of 0.55 and 0.8.*

dimension constant, it is said that the material is more constrained. This is also true for the case where the in-plane dimension is decreased and the thickness is held constant. For both cases, the degree of constraint can be quantified by the ratio of the thickness to the in-plane dimension. For the compact specimen, the initial remaining ligament would be the characteristic in-plane dimension. Therefore note that the constraint is directly proportional to the thickness to initial remaining ligament ratio. Generally, when a material is more constrained, the plastic energy dissipated will be less. This is graphically illustrated in Fig. 8 where the global plastic energy dissipation rate *R*-curves in Figs. 3 and 5 are plotted in terms of the thickness to initial remaining ligament ratio. Of course, for this investigation, the effect of constraint on the global plastic energy dissipation rate is well characterized by the degree of shear lip formation.

For the geometries tested, as mentioned previously, there was no correlation between the global plastic energy dissipation rate and the development of shear lips. As shown in Fig. 4, the global plastic energy dissipation rate is greater for the three-point bend specimens. Yet, Fig. 9 shows that the degree of shear lip development is greater for the small compacts. It was felt that the explanation for the geometry dependence of the global plastic energy dissipation rate lies with the fact that the global plastic deformation behavior for non-side-grooved three-point bend and compact tension specimens is funda-

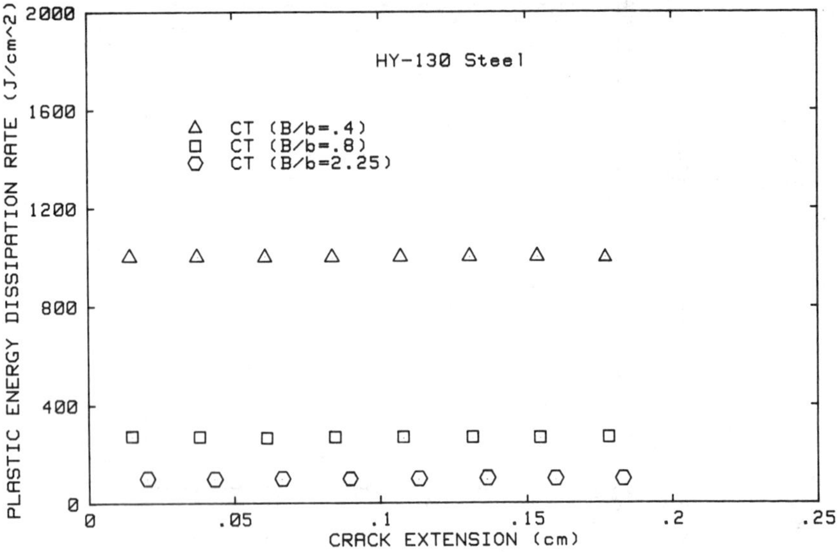

FIG. 8—*Graph of global plastic energy dissipation rate versus crack extension for non-side-grooved compact tension specimens with various thickness to initial remaining ligament ratios.*

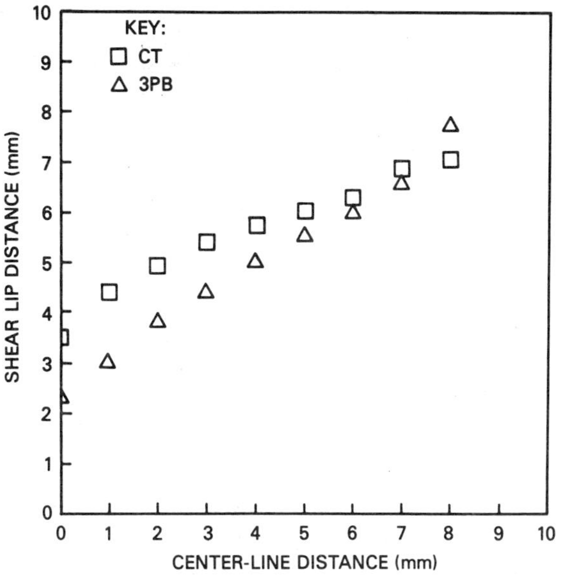

FIG. 9—*Shear lip distance versus center-line distance for non-side-grooved 50.8-mm-wide compact tension and three-point bend specimens.*

mentally different. That is, it has been found that the yield patterns in the three-point bend specimen are radically affected by the point load at the center of the specimen [*1*]. Indeed, the point load accounts for a significant percentage of the plastic deformation within the three-point bend specimen. For this geometry, then, it would be expected that the degree to which the material ahead of the crack tip has yielded would be different. Clearly, the growth of cracks into material that has achieved different values of plastic strain must be different. In an attempt to isolate the global plastic deformation behavior as a variable, 50.8-mm-wide compact tension and three-point bend specimens were side grooved 20% to increase the constraint and localize the plastic deformation to the crack tip region. By comparing Figs. 4 and 10, it was found that for the same amount of crack extension, the values of the global plastic energy dissipation rate for non-side-grooved specimens were greater than those obtained for side-grooved specimens. Moreover, Fig. 10 shows that the differences between the global plastic energy dissipation rate for the two geometries are minimized but not eliminated. This is not surprising since there is still a certain amount of yielding which is due to the point load in the side-grooved three-point bend specimen. Thus it is clear that for the side-grooved compact and three-point bend specimens tested in this study, the globally measured plastic energy dissipation rate is sensitive to geometry.

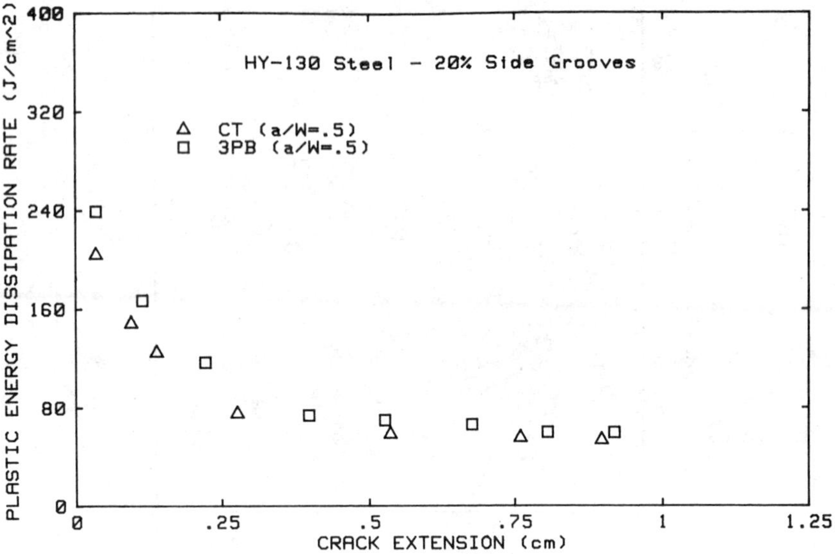

FIG. 10—*Graph of global plastic energy dissipation rate versus crack extension for side-grooved 50.8-mm-wide compact tension and three-point bend specimens.*

Conclusions

The global plastic energy dissipation at initiation and the global plastic energy dissipation rate were experimentally determined for a variety of specimen sizes, geometries, and initial crack lengths. The following is a summary of the conclusions drawn from this investigation:

1. The global plastic energy dissipation rate is affected by side grooves. For non-side-grooved specimens, the global plastic energy dissipation rate is constant. For side-grooved specimens, however, the global plastic energy dissipation rate decreases and approaches a constant value as the crack extends.

2. For non-side-grooved specimens, the values of the global plastic energy dissipation at initiation and the global plastic energy dissipation rate were found to be dependent on specimen size, geometry, and initial crack length. Through measurements of the degree of shear lip development, it was shown that the specimen size and initial crack length effects were due to differences in through-thickness constraint. The geometry dependence of these parameters was attributed to differences in the global plastic deformation behavior for the geometries tested.

3. For side-grooved specimens, the geometry dependence of the global plastic energy dissipation rate was minimized but not eliminated. Side grooves minimized the geometry dependence of the global plastic energy dissipation rate because the global plastic deformation becomes more localized with respect to the crack.

References

[1] Jolles, M. I., Schroedl, M. A., and Watson, T. J., "On the Geometry Dependence of *R*-Curves," presented at the 21st Annual Meeting of the Society of Engineering Science, Blacksburg, Va., 15-17 Oct. 1984.

[2] Hutchinson, J. W. and Paris, P. C., "Stability Analysis of *J*-Controlled Crack Growth," in *Elastic-Plastic Fracture, ASTM STP 668*, American Society for Testing and Materials, Philadelphia, 1979, pp. 37-64.

[3] Schroedl, M. A. and Jolles, M. I., "Crack Tip Integral Parameters for Elastic-Plastic Fracture," presented at ASTM 17th National Symposium on Fracture Mechanics, Albany, N.Y., 7-9 Aug. 1984.

[4] Kfouri, A. P. and Rice, J. R., "Elastic/Plastic Separation Energy Rate for Crack Advance in Finite Growth Steps," in *Fracture 1977*, Vol. 1, ICF4, University of Waterloo Press, Canada, 1977, pp. 49-59.

[5] Rousselier, G., "Numerical Treatment of Crack Growth Problems," in *Advances in Elasto-Plastic Fracture Mechanics*. Applied Science Publishers, Essex, England, 1979, pp. 165-189.

[6] Shoji, T., "Crack-Tip Blunting and Crack-Opening Displacement Under Large-Scale Yielding," *Metal Science*, Vol. 10, No. 5, May 1976, pp. 165-169.

[7] Shoji, T., Takahashi, H., and Suzuki, M., "Significance of Crack Opening Displacement and Crack Tip Plastic Strain Energy in Fracture Initiation," *Metal Science*, Vol. 12, No. 12, Dec. 1978, pp. 579-586.

[8] Shoji, T., "Determination of Crack Tip Energy Dissipation and Elastic-Plastic Fracture Toughness Parameter with Ductile Crack Extension," *Journal of Testing and Evaluation*, Vol. 9, No. 6, Nov. 1981, pp. 324-334.

[9] Saka, M., Shoji, T., Takahashi, H., and Abe, H., "A Criterion Based on Crack-Tip Energy Dissipation in Plane-Strain Crack Growth Under Large-Scale Yielding," in *Elastic-Plastic Fracture: Second Symposium, Vol. I—Inelastic Crack Analysis, ASTM STP 803*, American Society for Testing and Materials, Philadelphia, 1983, pp. I-130-I-158.

[10] Bodner, S. R., Davidson, D. L., and Lankford, J., "A Description of Fatigue Crack Growth in Terms of Plastic Work," *Engineering Fracture Mechanics*, Vol. 17, No. 2, 1983, pp. 189-191.

[11] Lee, J. D. and Liebowitz, H., "Considerations of Crack Growth and Plasticity in Finite Element Analysis," *Computers & Structures*, Vol. 8, 1978, pp. 403-410.

[12] Mura, T., *Micromechanics of Defects in Solids*, Martinus Nijhoff Publishers, The Netherlands, 1982, pp. 262-267.

[13] Jolles, M., "Automated Technique for *R*-Curve Testing and Analysis," in *Automated Test Methods for Fracture and Fatigue Crack Growth, ASTM STP 877*, American Society for Testing and Materials, Philadelphia, 1985, pp. 248-259.

[14] Saxena, A. and Hudak, S. J., Jr., "Review and Extension of Compliance Information for Common Crack Growth Specimens," *International Journal of Fracture*, Vol. 14, No. 5, Oct. 1978, pp. 453-468.

[15] Shih, C. F., Delorenzi, H. G., and Andrews, W. R., "Elastic Compliance and Stress Intensity Factors for Side Grooved Specimens," *International Journal of Fracture*, Vol. 13, 1977, pp. 544-548.

[16] Tada, H., Paris, P. C., and Irwin, G. R., *The Stress Analysis of Cracks Handbook*, Del Research Corp., Hellertown, Pa., 1973, p. 217.

[17] Roark, R. J. and Young, W. C., *Formulas for Stress and Strain*, McGraw-Hill, New York, 1975, p. 185.

Analysis and Mechanisms

S. L. Pu[1]

Stress Intensity Factors for a Circular Ring with Uniform Array of Radial Cracks of Unequal Depth

REFERENCE: Pu, S. L., "**Stress Intensity Factors for a Circular Ring with Uniform Array of Radial Cracks of Unequal Depth,**" *Fracture Mechanics: Seventeenth Volume, ASTM STP 905*, J. H. Underwood, R. Chait, C. W. Smith, D. P. Wilhem, W. A. Andrews, and J. C. Newman, Eds., American Society for Testing and Materials, Philadelphia, 1986, pp. 559-572.

ABSTRACT: The plane problem of a uniform array of unequal depth radial cracks originating at the internal boundary of a pressurized circular ring is considered. The 12-node quadrilateral isoparametric elements with collapsed singular elements around crack tips are used to compute stress intensity factors at crack tips.

In a previous study of equal depth radial cracks, the weakest configuration (having highest values of stress intensity factors) was a ring with two diametrically opposed cracks. The current study shows that if for any reason one of the two cracks should grow a little faster than the other, the stress intensity factor at the tip of the longer crack increases at a much faster rate to enhance the faster growth of the longer crack.

Numerical results are also obtained for cases of three and four radial cracks. They show the same trend: that once one or two cracks grow a little more than the rest, the stress intensity factors at these deeper cracks will be increased progressively higher to keep the faster pace of growth. This explains why failure caused by a single major crack has been observed most frequently.

The finite element results show that the relationship between the stress intensity ratio and the crack depth ratio is approximately linear. This approximation enables us to estimate stress intensity factors at unequal depth cracks by a simple numerical method. The estimations thus obtained are close to stress intensity factors computed from the finite element method.

KEY WORDS: stress intensity factors, multiple cracks, uneven cracks, fracture mechanics, thick-walled cylinders, multiply cracked cylinders

[1]Mathematician, U.S. Army Armament, Munitions, and Chemical Command, Armament Research and Development Center, Large Caliber Weapon Systems Laboratory, Benet Weapons Laboratory, Watervliet, NY 12189.

Nomenclature

a	Crack length normalized with respect to the inner radius of a ring
$c = a/t$	Dimensionless crack length, referring to equal crack depth configuration
$c_i = a_i/t$	Dimensionless crack length of the ith crack
$\rho_i = c_i/c$	Crack depth ratio
K_e	Mode I stress intensity for equal crack depth configuration
K_i	Mode I stress intensity for the ith crack
$N_i = K_i/K_e$	Stress intensity ratio for the ith crack
m	Slope of a straight line approximation in a graph of N_i versus ρ_i
m'	Coefficient in quadratic approximation
n	Number of radial cracks
p	Internal pressure loading
r	Radial distance from the center of a ring
r_c	Radial distance to the tip of a crack
R_1, R_2	Inner and outer radii of a ring, respectively
$t = R_2 - R_1$	Wall thickness
$W = R_2/R_1$	Wall ratio of the circular ring
Δ	Increment (e.g., $\Delta\rho_1$ = increment in ρ_1)
$\sigma_\theta(r)$	Tangential stress at r

Introduction

A series of efforts has been made to estimate stress intensity factors of radial cracks in a cylindrical pressure vessel. In order to facilitate the finite element computations of such a cracked cylinder, a collapsed 12-node triangular element with nodes shifted to $1/9$ and $4/9$ locations was developed in Ref *1*. The use of these elements around a crack tip has helped in computing stress intensity factors for an array of uniformly distributed radial cracks of equal depth emanating from the inner surface of a thick-walled tube. Our finite element results, reported in Ref *2*, are in good agreement with results obtained by other methods [*3–5*]. To reduce stress intensity factors at these cracks so the fatigue life of the pressure vessel may be prolonged, it becomes a common practice to use the autofrettage process. The computations of stress intensity factors at radial cracks become involved if a cylinder undergoes a partial autofrettage. One of the difficulties is the discrepancy among predictions on residual stress distribution by different investigators based on different assumptions. (References *6* to *13* represent a partial list of works on this subject.) Another difficulty is the lack of an existing general purpose finite element computer code that has the feature of treating an arbitrary initial stress distribution. To develop a method of computing stress intensity factors in this situation, the closed-form expressions for residual stresses for an incompressible, elastic-ideally plastic material [7] are chosen. In addition, an equivalent thermal load was found in Ref *14* that can be prescribed at the nodes of the existing computer codes NASTRAN and APES [*15*] to avoid the need of modification of computer codes in order to handle initial stresses. A summary of these methods and some results in stress intensity computations for cracks in a partially autofrettaged cylinder are given in Ref *16*. Other efforts on this subject are given in Refs *17* and *18*. Further extension of the method developed for radial cracks in cylindrical pressure vessels of elastic-perfectly plastic materials to similar cracks of strain-hardening materials is published in Ref *19*.

Thus far the stress intensity factors have been obtained for uniformly distributed radial cracks of equal depth. The weakest configuration was found to be two diametrically opposite cracks in non-autofrettaged cylinders. For autofrettaged cylinders, the weakest configuration is dependent on the degree of autofrettage and the ratio of bore pressure to yield strength. It remains true in general that two diametrically opposite cracks make a cylinder the least resistant to fracture. However, from the actual observations of fracture of thick-walled tubes, it is a single major crack that has caused the failure most frequently. The objective of this study is to compute the change in stress intensity factors at different crack tips when they grow from an equal depth configuration to an unequal depth configuration. To reach the goal we first use the finite element method to compute changes in stress intensity factors at various crack tips due to depth variation of only one of the cracks. Then an approximate numerical method is devised to estimate stress intensity factors for a general crack configuration of unequal depths. A separate section is devoted to the discussion of approximate methods for very shallow cracks, since the interaction among shallow cracks is weak and the accurate finite element results of stress intensity factors are not available. In the final section, conclusions are drawn.

Finite Element Results of Stress Intensity

The 12-node quadrilateral isoparametric elements were used in this study. The shape functions and the mathematical formulation of the element stiffness matrix are given in Ref *1*. The crack tip elements are formed by collapsing the quadrilateral elements into triangular elements around the crack tip. On each side of a triangle emitting from the crack tip, the locations of intermediate nodes are shifted from their usual $1/3$ and $2/3$ of the length of the side to $1/9$ and $4/9$ of the length measured from the tip. This simple technique gives the required singularity of the strain field at the crack tip [*1*]. The stress intensity factors can be computed quite accurately from certain nodal displacements [*1*]. A general purpose computer program having isoparametric elements such as NASTRAN may be used for our study; however, the computer code APES was chosen due to many convenient features of the program.

Using similar notations to those in Ref *2*, n denotes the number of radial cracks, W is the wall ratio, R_2/R_1, where R_1 and R_2 are inner and outer radii of the ring, respectively. The thickness of the wall, $t = R_2 - R_1$, is used to normalize the crack depth a. The dimensionless crack depth a/t is denoted by c. A subscript i, $i = 1, 2, \ldots, n$, is used to name a crack. Let the Cartesian coordinates be chosen such that the origin is at the center of the ring and a radial crack lies on the positive x-axis. This crack is assigned the number one. The crack number increases counterclockwise from the x-axis. In this study our finite element computations are limited to $W = 2$ and $n = 2, 3$, and 4 (Fig. 1). Starting from equal depth crack configuration, we proceed to unequal depth configurations by a systematic increment of crack depth of one, two, or three cracks. For a given n, let c and K_e be the crack depth and stress intensity factor, respectively, when all cracks are of the same depth. These quantities are used to normalize the crack depth (c_i) and the Mode I stress intensity factor (K_i) for the ith crack when cracks become unequal depths. New notations ρ and N are introduced for these dimensionless quantities:

$$\rho_i = c_i/c, \qquad N_i = K_i/K_e, \qquad i = 1, 2, \ldots, n \tag{1}$$

It is true that some cracks are under mixed Mode I and Mode II conditions when unequal depth crack configurations are considered. In this study, however, the effect of the shear mode is assumed to be negligible and only the dominant Mode I stress intensity factors are considered.

The idealizations used in this study for a ring with two, three, and four cracks are shown in Fig. 2. Only the upper half of a ring is shown due to symmetry consideration in finite element computations. Equilateral triangles are used around a crack tip and these collapsed singular elements are surrounded by a layer of quadrilateral elements (Details A and B of Fig. 2). The finite element results of K_e using the refined meshes (Fig. 3) are very close to results previously obtained in Ref *2* using slightly different meshes around a crack tip.

For two diametrically opposed radial cracks, stress intensity factors are computed for various lengths of c_1 with a fixed value of $c_2 = c = 0.2$. In a graph of N_i, $i = 1$ and 2, versus ρ_1, finite element results are shown as dots (Fig. 4). It can be seen that the dots may be joined approximately by straight

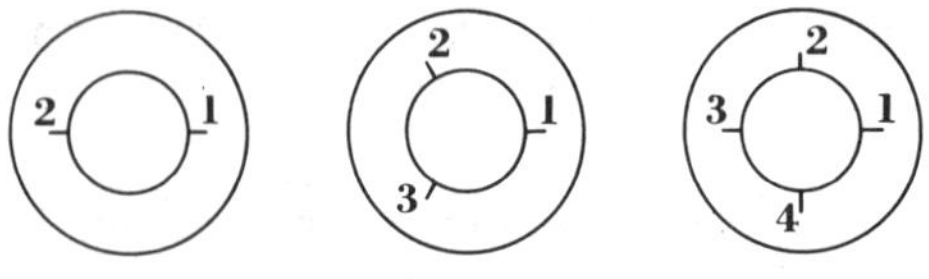

FIG. 1—*Schematic of a thick-walled cylinder containing two, three, or four radial cracks.*

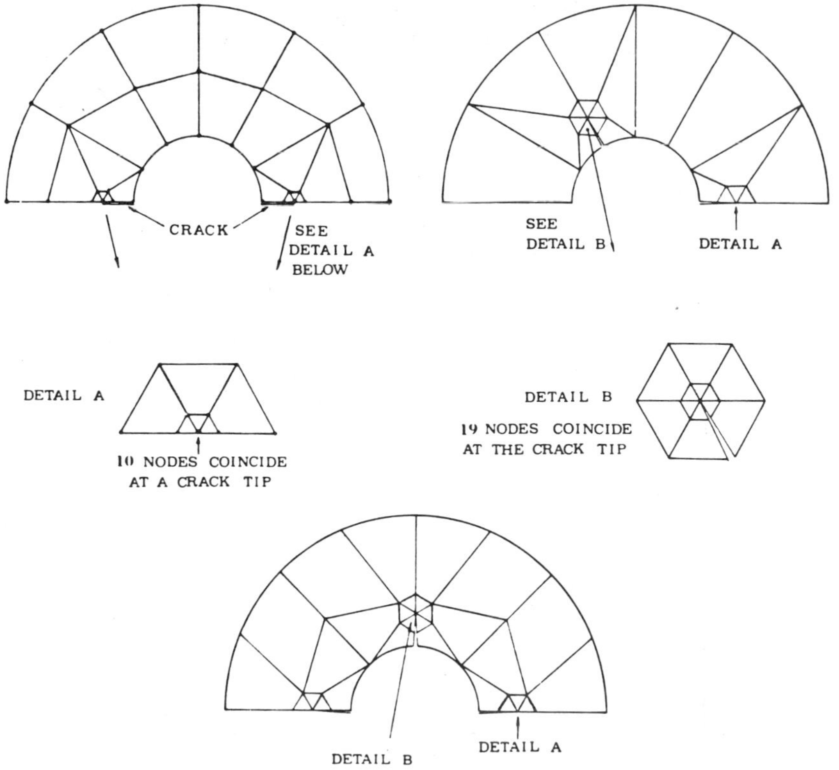

FIG. 2—*Finite element idealizations of a cylinder with two, three, or four cracks of unequal lengths.*

lines. Similar results are also obtained and shown in the same graph for $c_2 = c = 0.3$.

For three radial cracks, Cracks 2 and 3 remain the mirror image of each other so that the problem is symmetric with respect to the x-axis. Two cases are considered. In the first case, the crack depths of both Cracks 2 and 3 are fixed at $c_2 = c_3 = c = 0.2$ or 0.3. Stress intensity factors at crack tips of Cracks 1 and 2 are computed for various values of ρ_1. Finite element results are shown as dots for $c = 0.2$ and as crosses for $c = 0.3$ in Fig. 5. In the second case, the first crack has a fixed crack depth and the other two cracks grow simultaneously at the same rate. Finite element results for the latter case are used only for the purpose of checking our numerical approximations.

In the case of four radial cracks, Crack 4 is kept as the mirror image of Crack 2. This symmetry reduces the problem to the upper half of the ring with three cracks. Three cases have been studied. The first case is that only Crack 1 grows. In the second case, Cracks 1 and 3 are growing simultaneously at the same rate and Crack 2 has a fixed crack depth. The third case is opposite to

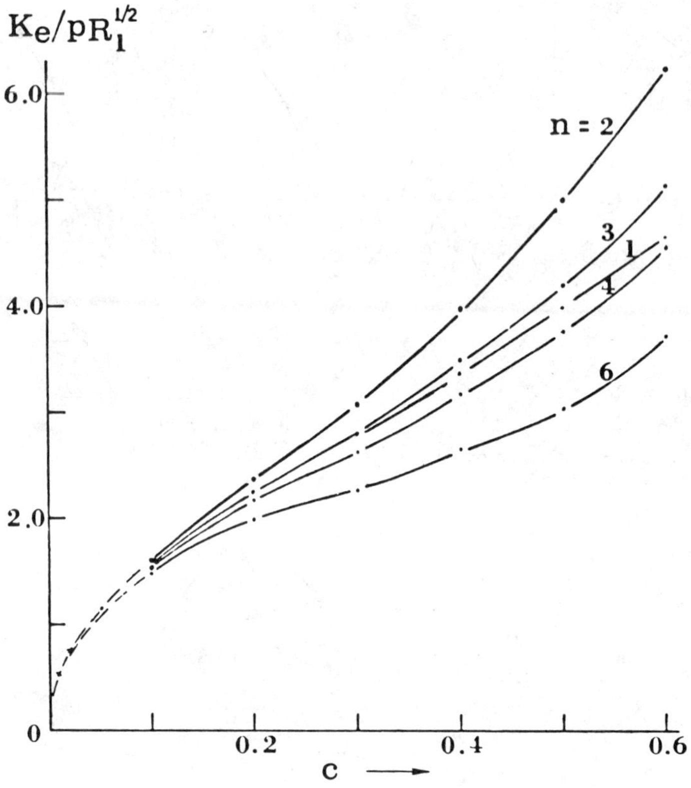

FIG. 3—*Stress intensity factors as function of crack depth for various numbers of radial cracks of equal depth.*

the first case. Crack 1 is fixed in length with Cracks 2 and 3 growing at the same rate. Using the idealization shown in Fig. 2, finite element results of stress intensity factors are obtained at each crack tip for various increments in crack depth of the growing crack or cracks. The results for the first case are shown as dots and crosses in Fig. 6. The numerical results for the other two cases are not given, since they are used only for checking the numerical approximations by total differentials.

Approximation by Total Differentials

We assume that the differential change in stress intensity factor at Crack i due to the differential change in length of Crack j may be given by

$$dN_i = \frac{\partial N_i}{\partial \rho_j} d\rho_j \tag{2}$$

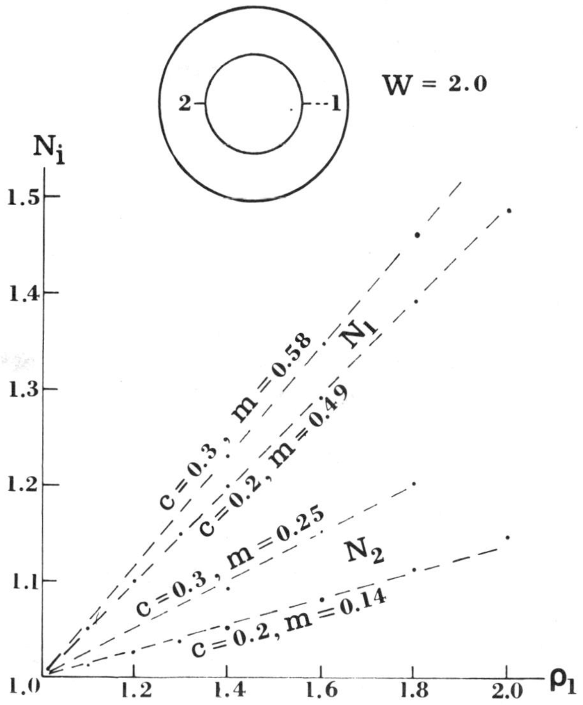

FIG. 4—*Stress intensity ratio,* $N_i = K_i/K_e$, *versus crack depth ratio,* $\rho_i = c_i/c$, *for two diametrically opposite cracks. Dots are finite element results that can be approximated by straight lines.*

The change in stress intensity factor due to a small change from ρ_j to $\rho_j + \Delta\rho_j$ can be expressed by

$$\Delta N_i = \int_{\rho_j}^{\rho_j+\Delta\rho_j} \frac{\partial N_i}{\partial \rho_j}\, d\rho_j \tag{3}$$

If the increment $\Delta\rho_j$ is divided into m_j intervals,

$$\Delta\rho_j = \Delta_1\rho_j + \Delta_2\rho_j + \dots + \Delta m_j\rho_j \tag{4}$$

and if in any interval ℓ there exists a quantity $(\partial N_i/\partial\rho_j)_\ell$ such that

$$\left(\frac{\partial N_i}{\partial \rho_j}\right)_\ell \Delta_\ell\rho_j = \int_{\rho_j+\Delta_{\ell-1}\rho_j}^{\rho_j+\Delta_\ell\rho_j} \frac{\partial N_i}{\partial \rho_j}\, d\rho_j \tag{5}$$

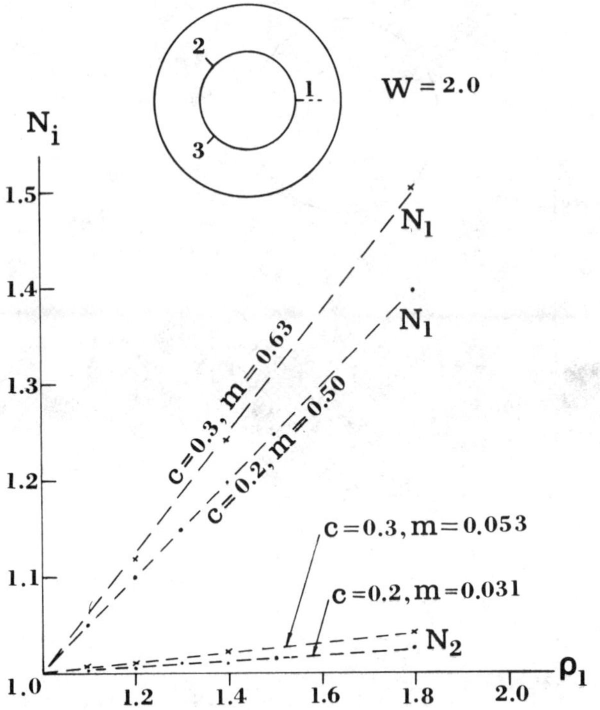

FIG. 5—*Stress intensity ratio*, $N_i = K_i/K_e$, *versus crack depth ratio*, $\rho_i = c_i/c$, *for symmetric three-crack cases. Dots and crosses are finite element results for* c = *0.2 and* c = *0.3, respectively.*

then the integral at the right-hand side of Eq 3 may be replaced by a summation

$$\Delta N_i = \sum_{\ell=1}^{m_j} \left(\frac{\partial N_i}{\partial \rho_j}\right)_\ell \Delta_\ell \rho_j \tag{6}$$

Furthermore, if the change in stress intensity factor at crack tip i due to simultaneous changes in length of two or more cracks may be approximated as the sum of changes due to each crack length change, then the total change of stress intensity factor at Crack i due to length changes of all cracks may be approximately expressed by

$$\Delta N_i = \sum_{j=1}^{n} \sum_{\ell=1}^{m_j} \left(\frac{\partial N_i}{\partial \rho_j}\right)_\ell \Delta_\ell \rho_j \tag{7}$$

The partial derivative $(\partial N_i/\partial \rho_j)_\ell$ may be calculated from the slope of the tangent to the curve of N_i versus ρ_j which is obtained by curve fitting of discreted

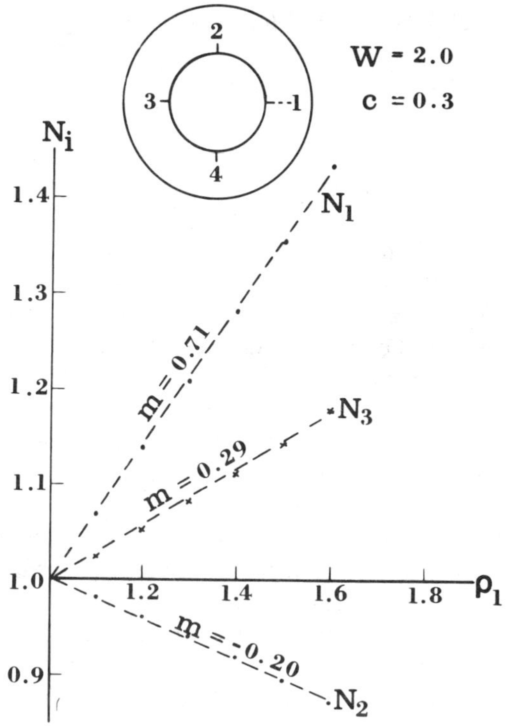

FIG. 6—*Stress intensity ratio,* $N_i = K_i/K_e$, *versus crack depth ratio,* $\rho_i = c_i/c$, *for symmetric four-crack cases. Dots and crosses are finite element results.*

points in the ρ-N plane from the finite element computations. The change in N_i must be restricted to that due to the change in ρ_j only. Figures 4 to 6 are used for this purpose with finite element values of N_i obtained for various values of ρ_1 with the crack depth of all other cracks fixed. A study of data points (ρ, N) in Figs. 4 to 6 shows that a straight line of the form

$$N - 1 = m(\rho - 1) \tag{8}$$

fits these points in a close approximation. The slope m of a least-squares regression is given by

$$m = \sum_i (N_i - 1)(\rho_i - 1) / \sum_i (\rho_i - 1)^2 \tag{9}$$

This slope may be used as the average value of the partial derivatives $\partial N_i/\partial \rho_i$ in a proper range of ρ_1 for a given set of values n and c. For $n = 2$, 3, and 4, the slopes m of linear regressions are computed and shown in Figs. 4 to 6 for $c = 0.2$ or $c = 0.3$ in a range $1 \leq \rho_1 \leq 1.8$.

For a given equal depth crack configuration we have

$$\frac{\partial N_i}{\partial \rho_i} = \frac{\partial N_j}{\partial \rho_j}, \qquad \frac{\partial N_i}{\partial \rho_j} = \frac{\partial N_j}{\partial \rho_i} \qquad (i, j = 1, 2, \ldots, n) \tag{10}$$

The stress intensity factor at any crack tip of unequal radial cracks may be estimated by using Eqs 7 and 10. The approximate method is applied to Case 2 of $n = 3$ and Cases 2 and 3 of $n = 4$. The results are compared with finite element results, and the difference is in general less than 4%. Two examples of the comparison are given in Table 1. In both examples the depth of Crack 1 was held constant and the depth of all other cracks increased an amount of 40% of the depth of Crack 1.

In the previous example of the four-crack case, the stress intensity factor of 3.6 at Crack 2 or 4 is well above 3.3, the would-be value if Crack 1 did not arrest but grew at the same rate. The stress intensity factor at Crack 2 due to the presence of Cracks 1 and 3 is considerably lower than 4.2, the stress intensity factor for the two-crack case with $c = 0.42$. The faster increase in stress intensity factor at Cracks 2 and 4, however, tends to change the four-crack configuration to a crack configuration dominated by two cracks.

The approximate method may be used to obtain results for crack configurations that are more difficult to compute by the finite element method. For instance, for a ring with three radial cracks of different depths, say $c_1 = 0.2$, $c_2 = 0.26$, and $c_3 = 0.3$, the entire ring with three crack tips has to be considered by the finite element method. The method of total differentials can easily yield the following estimates: $N_1 = 1.025$, $N_2 = 1.166$, and $N_3 = 1.26$. Taking $K_e/p\sqrt{R_1} = 2.23$ from Fig. 3 for $n = 3$, $c = 0.2$, we have $K_1/p\sqrt{R_1} = 2.28$, $K_2/p\sqrt{R_1} = 2.60$, and $K_3/p\sqrt{R_1} = 2.81$.

For a given value of n, if we obtain $\partial N/\partial \rho$ for many values of c, then a plot of $\partial N/\partial \rho$ versus c may be obtained. Such a plot for $n = 3$ is shown in Fig. 7. It shows that $\partial N_1/\partial \rho_1$ is nearly a linear function of c. Let $\partial N_2/\partial \rho_1$ be a quadratic function of c of the form

$$\partial N_2/\partial \rho_1 = m'c^2 \tag{11}$$

TABLE 1—*Comparison of* $K_i/p\sqrt{R_1}$ *finite element results with approximations.*

No. of Cracks (n)	Crack Depth		$i = 1$		$i = 2$		$i = 3$	
	c_1	Other c	Finite Element	Approximation	Finite Element	Approximation	Finite Element	Approximation
3	0.2	0.28	2.26	2.29	2.69	2.70	2.69	2.70
4	0.3	0.42	2.52	2.52	3.60	3.50	3.02	2.97

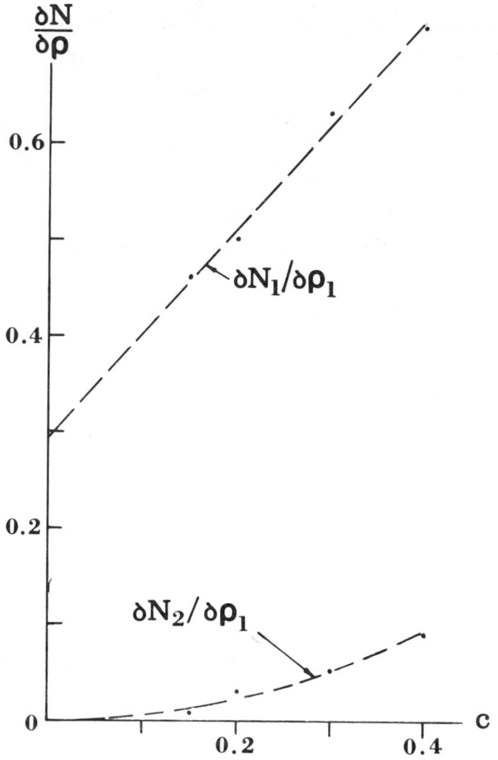

FIG. 7—*An approximate graph of* $\partial N_1/\partial\rho_1$ *and* $\partial N_2/\partial\rho_1$ *as a function of crack depth* c *for symmetric three radial cracks in a cylinder of wall ratio* w = 2.

The curve that best fits data in the least-squares sense is given by

$$m' = \sum_i c_i^2(\partial N_2/\partial\rho_1)_i/\sum_i c_i^4 \tag{12}$$

For the curve in Fig. 7, m' is approximately 0.577. The equation of the linear regression in Fig. 7 is

$$\partial N_1/\partial\rho_1 = 1.077c + 0.296 \tag{13}$$

A plot such as Fig. 7 or equations similar to Eqs 11 and 13 may be used to obtain approximate values of $\partial N_i/\partial\rho_1$ at a specific value of c for which no finite element results are available. Using $\partial N_1/\partial\rho_1 = 0.4037$ and $\partial N_2/\partial\rho_1 = 5.77 \times 10^{-3}$ for $n = 3$, $c = 0.1$ in Eqs 7 and 10 and taking $K_e/p\sqrt{R_1} = 1.56$ from Fig. 3, our estimations are $K_1/p\sqrt{R_1} = 1.563$, $K_2/p\sqrt{R_1} = 1.438$ and $K_3/p\sqrt{R_1} = 1.873$ for the three-crack case of $c_1 = 0.1$, $c_2 = 0.08$, and $c_3 =$

0.15. In our computation a negative value of -0.2 was used for $\Delta\rho_2$ in Eq 7 to indicate the decrease in length of Crack 2. The final estimation of stress intensity factors depends slightly on the judicious choice of c. In general, we take the smallest value of all crack lengths as c. In the previous example, if $c = 0.08$ is used, then c_3 would have a length increment of more than 80%. The estimations would have been $K_1/p\sqrt{R_1} = 1.538$, $K_2/p\sqrt{R_1} = 1.406$, and $K_3/p\sqrt{R_1} = 1.869$. The deviations in this example are relatively small. In some other cases the deviations become quite large when a length change is excessive. The choice of c for shallow cracks is more important and will be discussed in the following section.

Approximations for Shallow Cracks

The finite element method has some difficulty in calculating stress intensity factors for very shallow cracks due to the small size of crack tip elements required to solve such problems. For $c = 0.05$, finite element results are available in Ref *16* for various n. For shallow cracks with $c < 0.05$, no finite element results are available, since the accuracy of such computations is questionable. Some results for $c < 0.05$ were obtained by other methods. The most accurate results were reported in Ref *20* by the modified mapping collocation (MMC) method. However, the approximate methods for shallow cracks discussed in Ref *21* are easy to use and yield reasonably accurate results. The crosses in the range $0 \leq c < 0.05$ in Fig. 3 are obtained from the approximate formula [*22*] for a single crack

$$K = \{1.12\, \sigma_\theta(r = R_1) - 0.68[\sigma_\theta(r = R_1) - \sigma_\theta(r = r_c)]\}\sqrt{\pi c} \quad (14)$$

where $\sigma_\theta(r = R_1)$ and $\sigma_\theta(r = r_c)$ are hoop stresses of an uncracked cylinder at $r = R_1$ and $r = r_c$, respectively; r_c denotes the radius of the crack tip. Using them and finite element results for $c = 0.05$ as a guide, the curves in Fig. 3 are extended in broken lines to the region $c < 0.1$. The extended curves agree well with available MMC results.

Using the approximate method described in the preceding section and values from Figs. 3 and 7, stress intensity factors can be easily estimated for equally spaced shallow cracks. The judicious choice of c in shallow crack cases is very important. For small n and small c the crack interaction is weak. It is recommended to choose $c = c_i$ for the computation of K_i. For instance, $n = 3$, $c_1 = 0.01$, $c_2 = 0.02$, and $c_3 = 0.03$, we obtain $K_1/p\sqrt{R_1} = 0.52$ by assuming the final crack configuration is reached from $c_1 = c_2 = c_3 = c = 0.01$. In the computation of K_2, we assume the final crack configuration is reached from $c_1 = c_2 = c_3 = c = 0.02$. The result is $K_2/p\sqrt{R_1} = 0.73$. For K_3, initial crack length $c = 0.03$ is assumed, depths of Cracks 1 and 2 are decreased from 0.03 to $c_1 = 0.01$ and $c_2 = 0.02$, the computed result is $K_3/p\sqrt{R_1} = 0.90$. The corresponding results by using Eq 14 are $K_1/p\sqrt{R_1} =$

0.53, $K_2/p\sqrt{R_1} = 0.74$, and $K_3/p\sqrt{R_1} = 0.90$. Another easier alternative method for shallow cracks is the use of stress intensity factors for equal-depth cracks. For $n = 3$, when all crack depths are $c = 0.01$, the stress intensity factor at a crack tip is $K/p\sqrt{R_1} = 0.52$. Similarly $K/p\sqrt{R_1} = 0.73$ for $c = 0.02$ and $K/p\sqrt{R_1} = 0.90$ for $c = 0.03$. Using these values to the crack of the right crack depth in the unequal depth crack configuration, all three approximate methods for shallow cracks are useful up to $c = 0.1$. As a last example, for $n = 3$, $c_1 = 0.05$, $c_2 = 0.07$, and $c_3 = 0.09$, the first method gives $K_1/p\sqrt{R_1} = 1.122$, $K_2/p\sqrt{R_1} = 1.31$, and $K_3/p\sqrt{R_1} = 1.495$. Equation 14 gives $K_1/p\sqrt{R_1} = 1.15$, $K_2/p\sqrt{R_1} = 1.347$, and $K_3/p\sqrt{R_1} = 1.512$. The use of equal crack depth results gives $K_1/p\sqrt{R_1} = 1.12$, $K_2/p\sqrt{R_1} = 1.31$, and $K_3/p\sqrt{R_1} = 1.50$.

Conclusions

The 12-node quadrilateral isoparametric elements with collapsed singular elements around a crack tip can be used to efficiently compute stress intensity factors at tips of multiple radial cracks of unequal depths in a thick-walled cylinder. Based on finite element results of some selected crack configurations, the method of total differentials can estimate stress intensity factors accurately. The advantage of the approximate method is not only in time saving, but also in being able to give an estimate when the problem is not easily computed by the finite element method.

For shallow cracks, the stress intensity factor at a crack tip may be estimated from either equal-depth multiple-crack configuration or a single-crack configuration of that particular depth. A deeper crack will have a higher stress intensity factor which in turn will cause further growth of the deeper crack. This process will change a large number of shallow cracks into a crack configuration consisting of several dominant cracks, and eventually will end in a single crack domination.

Although no autofrettage residual stress has been considered in this study, the approximate method aided by the method summarized in Ref *23* may be used for partially autofrettaged, pressurized, multiply cracked cylinders when crack depths are unequal.

References

[*1*] Pu, S. L., Hussain, M. A., and Lorensen, W. E., *International Journal for Numerical Methods in Engineering*, Vol. 12, 1978, pp. 1727-1742.

[*2*] Pu, S. L. and Hussain, M. A. in *Fracture Mechanics (11th Symposium), ASTM STP 677*, American Society for Testing and Materials, Philadelphia, 1979, pp. 685-699.

[*3*] Bowie, O. L. and Freese, C. E., *Engineering Fracture Mechanics*, Vol. 4, 1972, pp. 315-321.

[*4*] Grandt, A. F., "Two Dimensional Stress Intensity Factor Solutions For Radially Cracked Rings," Technical Report AFML-TR-75-121, Air Force Materials Laboratory, 1975.

[*5*] Tracy, P. G., *Engineering Fracture Mechanics*, Vol. 11, 1979, pp. 291-300.

[6] Davidson, T. E., Barton, C. S., Reiner, A. N., and Kendall, D. P., "Overstrain of High-Strength Open-End Cylinders of Intermediate Diameter Ratio," in *Proceedings*, First International Congress on Experimental Mechanics, Pergamon Press, Oxford, 1963, pp. 335-352.

[7] Hill, R., *The Mathematical Theory of Plasticity*, Oxford at the Clarendon Press, 1950.

[8] Hodge, P. G. and White, G. N., *ASME Journal of Applied Mechanics*, Vol. 72, 1950, pp. 180-184.

[9] Weigle, R. E., "Elastic-Plastic Analysis of a Cylindrical Tube," Technical Report WVT-6007, Watervliet Arsenal, Watervliet, N.Y., 1960.

[10] Chen, P. C. T., "Numerical Prediction of Residual Stresses in an Autofrettaged Tube of Compressible Material," in *Proceedings*, 1981 Army Numerical Analysis and Computer Conference, pp. 351-362.

[11] Davidson, T. E. and Kendall, D. P. in *Mechanical Behavior of Materials Under Pressure*, H. L. P. Pugh, Ed., Elsevier, New York, 1970.

[12] Chu, S. C., *Journal of the Franklin Institute*, Vol. 294, 1972, pp. 57-65.

[13] Chen, P. C. T., "The Finite Element Analysis of Elastic-Plastic Thick-Walled Tubes," in *Proceedings*, Army Symposium on Solid Mechanics, The Role of Mechanics in Design-Ballistic Problems, 1972, pp. 243-253.

[14] Hussain, M. A., Pu, S. L., Vasilakis, J. D., and O'Hara, P., *ASME Journal of Pressure Vessel Technology*, Vol. 102, 1980, pp. 314-318.

[15] Gifford, L. N., Jr., "APES—Second Generation Two-Dimensional Fracture Mechanics and Stress Analysis by Finite Elements," DTNSRDC Report 4799, 1975.

[16] Pu, S. L. and Hussain, M. A. in *Fracture Mechanics: 14th Symposium—Volume 1: Theory and Analysis, ASTM STP 791*, American Society for Testing and Materials, Philadelphia, 1983, pp. I-194-I-215.

[17] Kapp, J. A. and Eisenstadt, R., in *Fracture Mechanics (11th Symposium), ASTM STP 677*, American Society for Testing and Materials, Philadelphia, 1979, pp. 746-756.

[18] Parker, A. P., "Stress Intensity and Fatigue Crack Growth in Multiply Cracked, Pressurized, Partially Autofrettaged Thick Cylinders," *Fatigue of Engineering Materials and Structures*, Vol. 4, No. 2, 1982.

[19] Pu, S. L. and Chen, P. C. T., *ASME Journal of Pressure Vessel Technology*, Vol. 105, 1983, pp. 117-123.

[20] Parker, A. P. and Andrasic, C. P., "Stress Intensity Prediction For a Multiply-Cracked, Pressurized Gun Tube With Residual and Thermal Stresses," U.S. Army Symposium on Solid Mechanics, AMMRC MS 80-5, 1980, pp. 35-39.

[21] Pu, S. L., "Stress Intensity Factors for Radial Cracks at Outer Surface of a Partially Autofrettaged Cylinder Subjected to Internal Pressure," ARRADCOM Technical Report No. ARLCB-TR-82003, Benet Weapons Laboratory, Watervliet, N.Y., May 1982.

[22] Underwood, J. H. and Throop, J. F., "Surface Crack K-Estimates and Fatigue Life Calculations in Cannon Tubes," in *Part-Through Crack Fatigue Life Prediction, ASTM STP 687*, American Society for Testing and Materials, Philadelphia, 1979, pp. 195-210.

[23] Pu, S. L., "A Functional Stress Intensity Approach to Multiply-Cracked, Partially Autofrettaged Cylinders," in *Transactions*, 28th Conference of Army Mathematicians, ARO Report 83-1, 1983, pp. 263-283.

George T. Sha[1] *and Chien-Tung Yang*[1]

Weight Functions of Radial Cracks Emanating from a Circular Hole in a Plate

REFERENCE: Sha, G. T. and Yang, C.-T., "**Weight Functions of Radial Cracks Emanating from a Circular Hole in a Plate,**" *Fracture Mechanics: Seventeenth Volume, ASTM STP 905*, J. H. Underwood, R. Chait, C. W. Smith, D. P. Wilhem, W. A. Andrews, and J. C. Newman, Eds., American Society for Testing and Materials, Philadelphia, 1986, pp. 573–600.

ABSTRACT: Explicit nodal weight functions for a pair of symmetric radial cracks emanating from a hole in a plate are presented with special emphasis on the load-independent characteristics of the explicit weight functions that can obviate repeated calculations of the Mode I stress intensity factors (K_I) under different loading conditions. An analytical expression, which relates the explicit crack-face weight function component for Mode I deformation to the radial distance from the crack tip along the crack face, is also provided to facilitate K_I evaluation by combining the uncracked stress field and the explicit crack-face weight functions through the use of the linear superposition principle. The utilization of the explicit weight functions, which are obtained from a simple loading, for calculating K_I under complex loading conditions such as biaxial loading and pin-joint loading with and without interference pressure, is uniquely demonstrated by combining the superposition principle and the weight function concept.

KEY WORDS: radial cracks, K_I solution, weight function, Green's function, biaxial loading, pin-joint loading, interference pressure

An efficient finite element evaluation of explicit weight functions has been established recently [*1*] at all locations within the structure of interest. This needed computational efficiency is achieved by coupling the virtual crack extension (VCE) technique [*2,3*] with singular elements [*4,5*]. The application of finite element techniques to a nodal weight function evaluation without using singular elements has also been demonstrated by Parks and Kamentzky [*6*], Vanderglas [*7*], and Vainshtok [*8*] with the VCE technique. Because of the load-independent characteristics, the availability of the explicit weight

[1]Staff Research Engineer and Senior Project Engineer, respectively, Allison Gas Turbine Division, General Motors Corporation, Indianapolis, IN 46206-0420.

functions for a given crack geometry can obviate the unnecessary repeated finite element calculations of Mode I stress intensity factors (K_I) under different loading conditions for the same geometry. The prime purposes of this paper are to present the characteristics of explicit weight functions for a pair of symmetric cracks emanating from a circular hole in a plate along the crack face, the inner hole perimeters, and the external traction application boundaries. An analytical expression, which relates the Mode I crack-face weight function component to the radial distance (r_s) from the crack tip along the crack face, is also provided for the concerned crack geometries to facilitate the K_I calculations by combining the uncracked stress field with the predetermined explicit weight functions. The use of the explicit weight functions along the traction application locations, which are obtained from a simple loading, for the K_I evaluations under complex loading conditions such as biaxial and pin-joint loading with and without interference pressures, is demonstrated extensively by coupling the linear superposition principle with the weight function concept.

There is no simple mechanics principle other than the weight function concept [*9,10*] that permits K_I calculation from one loading to another loading condition for a given geometry. Bueckner's weight function [*9*] is, in fact, the Green's function of the stress intensity factors of a cracked body. According to Rice's displacement derivative representation, the explicit weight function vector at (x_i, y_i) location with crack length (a) is defined as

$$h_{Ii}(x_i, y_i, a) = \frac{H}{2K_I^*}\frac{\partial u_i(x_i, y_i, a)^*}{\partial a} \tag{1}$$

where

H = effective modulus with $H = E$ (Young's modulus) for plane stress state, and $H = E/(1 - \nu^2)$ for plane strain stress-state with ν = Poisson's ratio,

K_I^* = Mode I stress intensity factor under * loading system, and

$\partial u(x_i, y_i, a)^*/\partial a$ = partial displacement derivative with respect to the virtual crack extension, da, under * loading system.

The Mode I explicit weight function vector, $h_I(x_i, y_i, a)$, consists of $h_{Ix}(x_i, y_i, a)$ and $h_{Iy}(x_i, y_i, a)$ components respectively along the global x and y axes direction, which are defined as

$$h_{Ix}(x_i, y_i, a) = \frac{H}{2K_I}\frac{\partial u_x(x_i, y_i, a)}{\partial a} \tag{2a}$$

$$h_{Iy}(x_i, y_i, a) = \frac{H}{2K_I}\frac{\partial u_y(x_i, y_i, a)}{\partial a} \tag{2b}$$

where u_x and u_y are the displacement components along the x and y axes respectively.

Following Rice [10], the Mode I stress intensity factor (K_I) is expressed as a sum of work-like product between the applied tractions and the explicit weight function (h_I) at their points of application as

$$K_I = \int_s t \cdot h_I ds + \int_v f^b \cdot h_I dv \tag{3}$$

where

s = boundary with traction t applied, and
f^b = body force within volume v.

Without body-force loading, only the surface integral is involved in the K_I calculation (Eq 3). However, by applying the linear superposition principle, the K_I evaluation for a given crack geometry can be obtained from the surface integral alone by combining the equivalent uncracked stress field $\sigma(r_s)$, which includes body force loading, thermal loading, and applied tractions at crack faces and other locations, with the explicit crack-face weight functions, $h_{Ir_s}(r_s, a)$, as

$$K_I = \int_0^a \sigma(r_s) \cdot h_{Ir_s}(r_s, a) dr_s = \Sigma f_i(r_s) \cdot h_{Ir_s}(r_s, a) \tag{4}$$

where

r_s = radial distance from the crack tip along the crack face,
h_{Ir_s} = crack-face weight function component for Mode I deformation, and
f_i = consistent nodal forces equivalent to $\sigma(r_s)$ applied stress.

It is found that the explicit crack-face weight function component under Mode I deformation, $h_{Ir_s}(r_s, a)$, of the symmetric radial cracks emanating from a circular hole in a plate can be represented accurately by

$$h_{Ir_s}(r_s, a) = \sum_{n=1}^{N} A_n(a) r_s^{(n/2-1)} \tag{5}$$

where

N = maximum number of terms needed for accurate representation of $h_{Ir_s}(r_s, a)$, and
$A_n(a)$ = least-square regressed coefficients that are functions of crack length (a) only.

The regressed coefficients A_n for any crack length of symmetric radial cracks emanating from a hole in a plate can be obtained with cubic spline interpolation technique from an adequate set of crack-face weight functions of discrete crack lengths. Consequently, the stress intensity factor at any crack length can be evaluated through Eq 4 with the interpolated explicit crack-face weight functions and coupling with the equivalent uncracked stress field. This approach of evaluating K_I for radial cracks emanating from a circular hole in a plate resulted in a total loss of accuracy of less than 1%.

The nodal weight functions, which depend on geometry and constraints, can be evaluated explicitly for the entire structure of interest with Eq 1 from finite element analysis with a relatively simple loading. From these predetermined explicit weight functions, the evaluation of the Mode I stress intensity factors (K_I) under complex loading conditions can be obtained by combining the linear superposition principle with the weight function concept, as illustrated in this paper. The explicit weight functions at the traction application locations other than the crack faces are extensively used for K_I evaluation under complex loading conditions. Examples of K_I calculations for the symmetric radial cracks emanating from a hole in a plate subjected to remote biaxial loading and pin-joint loading with and without interference pressure are illustrated to show the powerfulness of using the weight function concept for K_I determination.

Explicit Weight Functions for Symmetric Radial Cracks

The technical details for the explicit weight functions for the entire structure can be found in the paper by Sha [*1*] through the VCE technique coupled with singular elements. The characteristics of the explicit weight functions for the crack geometry of the symmetric radial cracks emanating from a hole in a plate (Fig. 1) are reported in this paper. Because of the symmetric crack geometry with symmetric loading conditions, only one fourth of the entire geometry is modeled with a typical finite element mesh (Fig. 2). If the symmetric crack geometry is subjected to asymmetric loading, which can potentially induce the mixed fracture mode, one half of the geometry needs to be idealized. The efficient finite element evaluation of the decoupled explicit weight functions (h_I and h_{II}) for the mixed crack problem has also been established recently by Sha and Yang [*11*] by combining the symmetric mesh in the crack-tip neighborhood with the VCE technique. The numerical results of Mode I stress intensity factors (K_I) and the Mode I explicit nodal weight functions (h_I) are confined geometrically for $D/W = 0.25$, $0.256 \leq 2a/W \leq 0.85$, and $H/W = 1$ or 2. By using the load-independent characteristics of the weight functions for a given geometry, the explicit nodal weight functions, which are determined with simple remote tension, are applied to complex loading conditions as demonstrated in this paper for K_I calculations. Once the explicit weight functions along the traction application locations are predetermined explicitly, the K_I calculation can be made either from Eq 3 of line integral only in the absence of body force loading or Eq 4 by combining the crack-face weight functions and the equivalent uncracked stress field, which

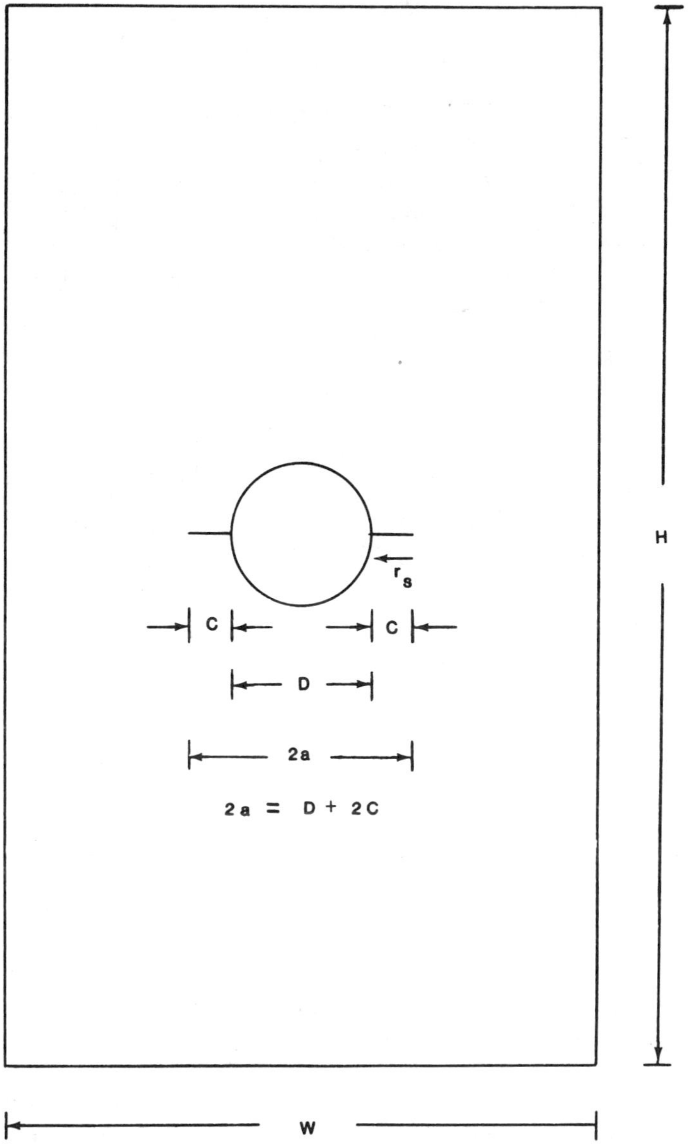

FIG. 1—*Nomenclature used for symmetric radial cracks emanating from a circular hole in a plate.*

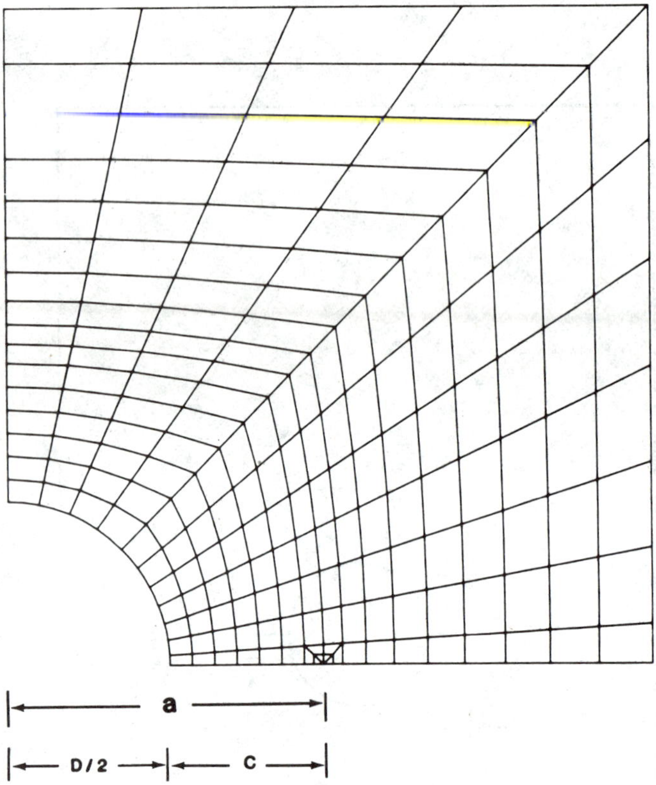

FIG. 2—*Typical finite element mesh for symmetric radial cracks with* D/W = *0.25,* 2a/W = *0.49, and* H/W = *1.0.*

can include body force loading and other loading conditions. The weight functions along the crack face and at other traction application boundaries are presented in this paper for the convenience of the numerical calculation of K_I and ΔK_I for predicting the crack propagation life purposes.

The normalized stress intensity factors ($K_I/\sigma\sqrt{\pi a}$), obtained in our finite element efforts for the symmetric cracks emanating from a hole in a plate with remote tension compared well with Bowie [*12*], Newman [*13*], and Cartwright and Parker [*14*] as shown in Fig. 3 for $D/W = 0.25$, $0.256 \leq 2a/W \leq 0.85$, and $H/W = 1$ and 2. The typical crack-face explicit Mode I weight functions component $h_{I r_s}$, as a function of r_s, as shown in Fig. 4 for the symmetric radial crack, can be accurately represented by Eq 5 with discrete crack lengths with $0.256 \leq 2a/W \leq 0.85$. $N = 4$ of Eq 5 is satisfactory with maximum error between the fitted and the predetermined crack-face nodal weight functions not more than 0.5% for the problem of a cracked hole in a plate. The coefficients, $A_n(a)$, which are a function of crack length (a) as

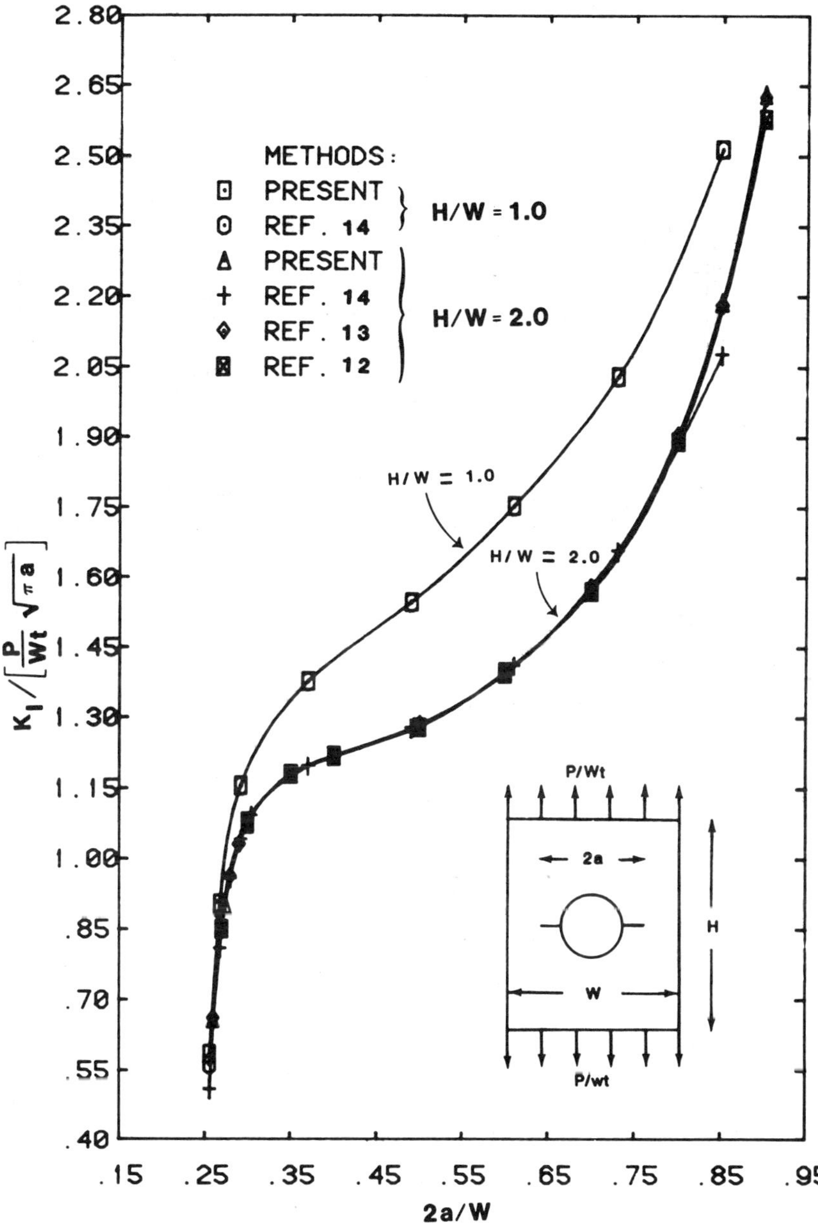

FIG. 3—*Normalized* K_I *values as a function of* 2a/W *ratios for symmetric radial cracks with* D/W = *0.25 and* H/W = *1.0 or 2.0.*

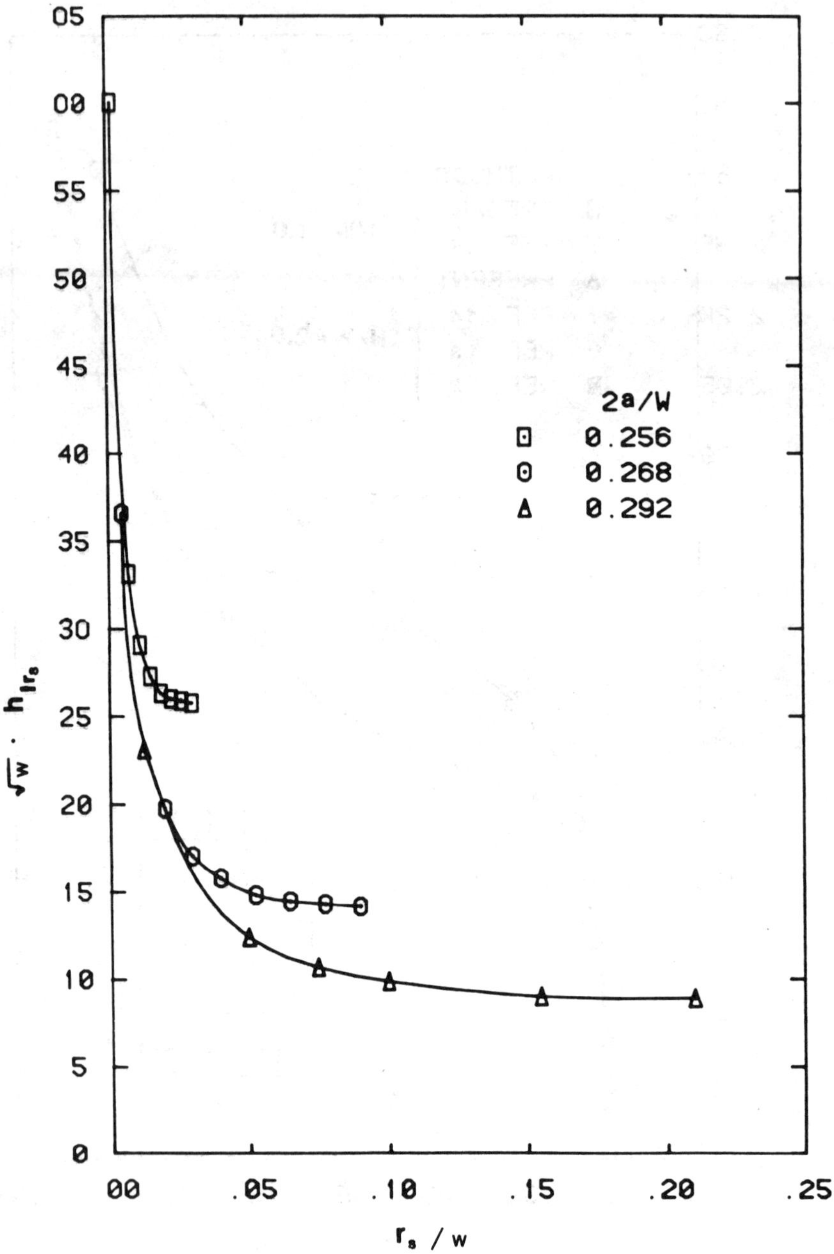

FIG. 4—*Dimensionless crack-face Mode I explicit weight functions,* $\sqrt{W} \cdot h_{Ir_s}(r_s, a)$ *as a function of* r_s/W *ratios for symmetric radial cracks emanating from a hole with different crack lengths in a plate.*

shown in Eq 5 can be obtained from the least-square regression analyses of the crack-face weight functions obtained from the finite element analyses.

These regressed coefficients for different crack lengths are given in Table 1 for the symmetric cracks emanating from a hole in a plate with $D/W = 0.25$ and $0.256 \leq 2a/W \leq 0.85$. It is found that these coefficients $A_n(a)$ for any crack length $(2a)$ can be obtained from the cubic spline interpolation technique from the adequate crack-face weight functions of different crack lengths. Consequently, the stress intensity factor K_I and ΔK_I at any crack length and loading can be obtained for the life prediction purposes through Eq 4 with the interpolated crack-face explicit Mode I weight functions and with the equivalent uncracked stress field based on the superposition principle. As a result of the interpolative feature of the coefficients $A_n(a)$ of Eq 5 for the crack-face weight functions, the cycle-by-cycle crack propagation life analyses become economically feasible for calculating K_I and ΔK_I of the progressive crack enlargement until reaching final fracture instability under cyclic fatigue loading.

The implicit weight function approach, which takes advantage of the load-independent characteristics of the weight function for K_I evaluation without involving explicit determination of the Mode I weight function component (h_I), has been adopted by Grandt [*15,16*], Hsu and Rudd [*17*], Andrasic and Parker [*18*], and Pu [*19*]. It can be divided into two groups depending on whether the partial displacement derivative with respect to crack length $(\partial u/\partial a)$ at the crack-face location is approximated. The former group is represented by Grandt [*15,16*] and Andrasic and Parker [*18*] with a $\partial u/\partial a$ approximation, which is made either from the conic section given by Orange [*20*] or from the numerical approximation of crack-face displacement data of different crack lengths. The latter group is typified by Pu [*19*] and Hsu and Rudd [*17*], with no assumptions on u and $\partial u/\partial a$, using the finite element method to compute K_I values associated with each different polynomial crack-face pressure distribution. The present work on explicit Mode I weight function vectors differs from the previously mentioned implicit weight function publications [*15–19*] in that both explicit nodal weight function vectors $h_{Ii}(x_i, y_i, a)$ and $\partial u/\partial a$ values for the entire structure of interest are deter-

TABLE 1—*Regressed least-square coefficients,* A_n, *of Eq 5 for crack-face weight functions.*

$2a/W$	A_1	A_2	A_3	A_4
0.256	0.816 68	−2.107 58	150.314 00	1663.544 98
0.268	0.819 77	−1.290 12	49.042 00	240.305 73
0.292	0.815 20	−0.543 59	13.633 20	85.891 84
0.370	0.804 88	−0.067 80	5.356 20	14.847 49
0.490	0.800 92	0.017 44	4.974 40	6.502 49
0.610	0.798 31	0.070 70	5.455 00	4.659 97
0.730	0.798 13	0.043 47	7.351 20	1.797 80
0.850	0.793 90	0.079 69	12.171 00	−4.606 30

mined efficiently with the finite element VCE technique. This is done through the following two-step process:

- Obtain the displacement derivative with respect to the crack extension du/da at the perturbed locations for the entire structure of interest from the perturbations of elemental stiffness and nodal forces of a few elements as a result of virtual crack extension.
- Convert the displacement derivatives from the perturbed locations to the unperturbed locations. Mathematically, this is equivalent to changing the total displacement derivatives du/da to partial displacement derivatives $\partial u/\partial a$.

Mathematically rigorous procedures [*1,6,7*] are adopted in carrying out this two-step process within the framework of finite element methodology.

The extensive use of these predetermined weight functions at locations other than crack faces for evaluating the needed Mode I stress intensity factors under complex load conditions is illustrated in the following sections to show the usefulness of the weight function concept along with the explicit weight function characteristics at these locations. These complex loading conditions, including biaxial loading and pin-joint loading with and without mandrel interferences, are exercised for symmetric radial cracks emanating from a circular hole in a plate for calculating K_I by combining the weight function concept with the linear superposition principle.

Biaxial Loading Condition

By applying the superposition principle to the biaxial remote loading condition, the Mode I stress intensity factor for Case 5a is the sum of Cases 5b and 5c (Fig. 5) with biaxiality ratio $\alpha = \sigma_{xx}/\sigma_{yy}$. The evaluation of K_I^{5b} and K_I^{5c} require the weight function components $h_{Iy}(x_i, y_i, a)$ along the top face (AB) and $h_{Ix}(x_i, y_i, a)$ along the side face (BC) respectively. The dimensionless explicit weight function components (Figs. 6 and 7, respectively) are $\sqrt{W} \cdot h_{Iy}(x_i, y_i, a)$ along the top face and $\sqrt{W} \cdot h_{Ix}(x_i, y_i, a)$ along the side face for the symmetric radial cracks of discrete crack lengths ($0.256 \leq 2a/W \leq 0.85$) emanating from a hole, as shown in Fig. 1 with $D/W = 0.25$ and $H/W = 1.0$. The normalized stress intensity factors as a function of a wide range of $\alpha(-2 \leq \alpha \leq 2)$ are shown in Fig. 8, which is obtained from Eq 3 in the absence of body force loading by combining the appropriate explicit weight function components with the corresponding load vector at the load application locations. The comparison of our data, which are obtained from the weight function concept, with those of Cartwright and Parker [*14*], which are obtained from the boundary collocation technique, show excellent agreement, as shown in Fig. 9 for $\alpha = 0$ and 1. With these predetermined Mode I weight function components (h_{Iy} along the AB face and h_{Ix} along the BC face, as shown in Fig. 5), the K_I value at any biaxiality ratio α value can be

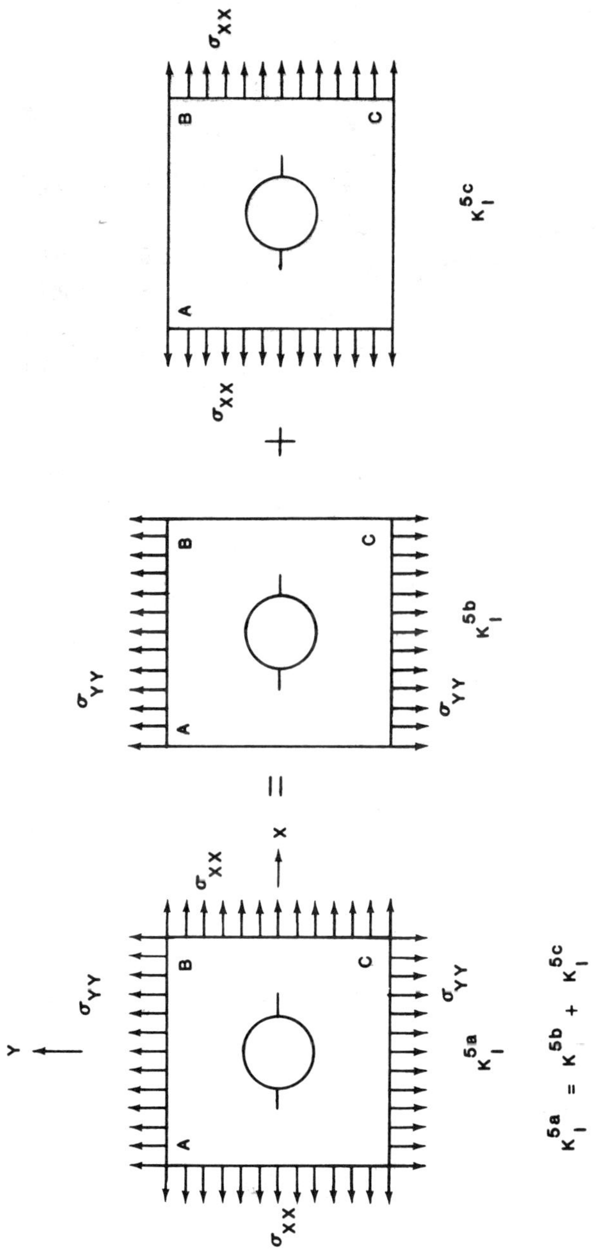

FIG. 5—*Stress intensity factor calculations under biaxial loading by coupling the linear superposition principle and the weight function concept.*

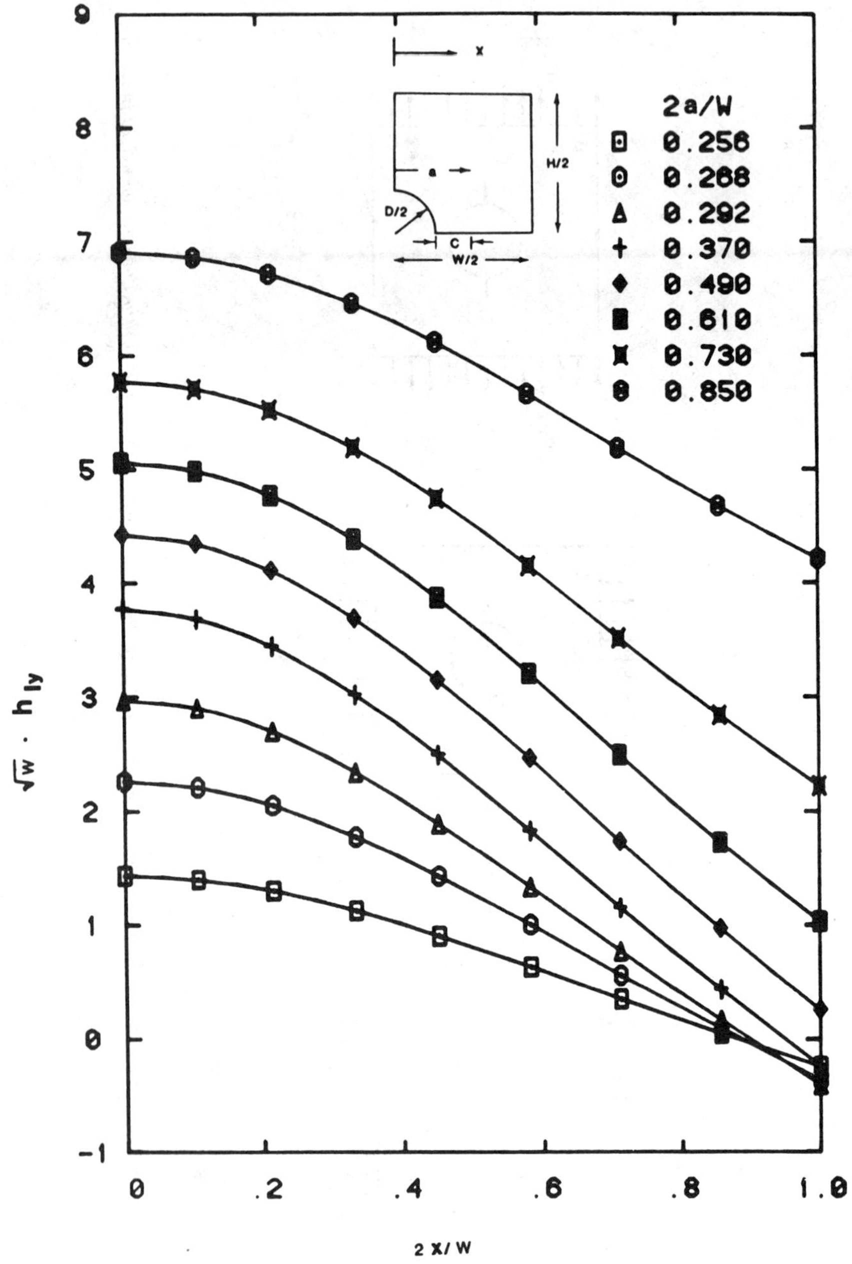

FIG. 6—*Dimensionless weight function components,* $\sqrt{W} \cdot h_{Ix}$, *along the top-face as a function of* 2x/W *ratios with* $0.256 \leq 2a/W \leq 0.850$, D/W = *0.25, and* H/W = *1.0.*

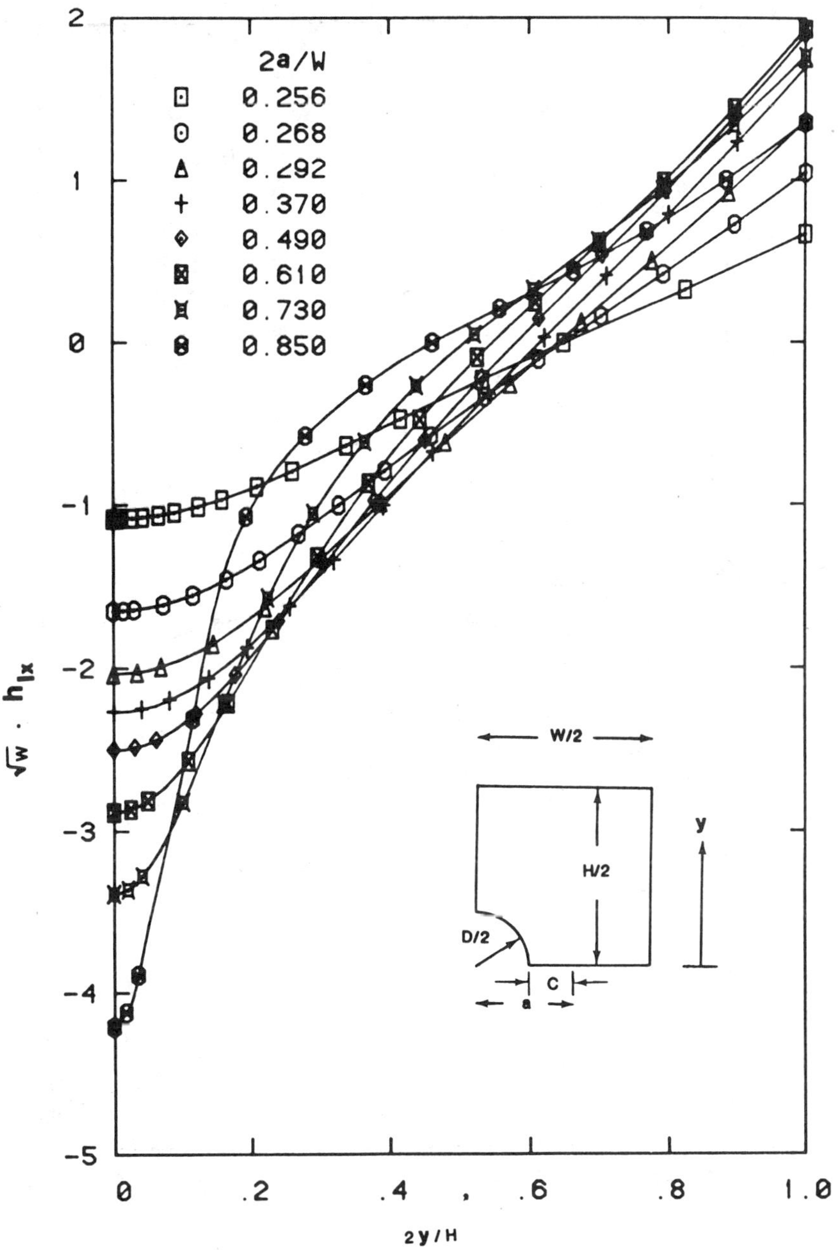

FIG. 7—*Dimensionless weight function components,* $\sqrt{W} \cdot h_{Ix}$, *along the side-face as a function of* 2y/H *ratio with* $0.256 \leq 2a/W \leq 0.850$, D/W = *0.25, and* H/W = *1.0.*

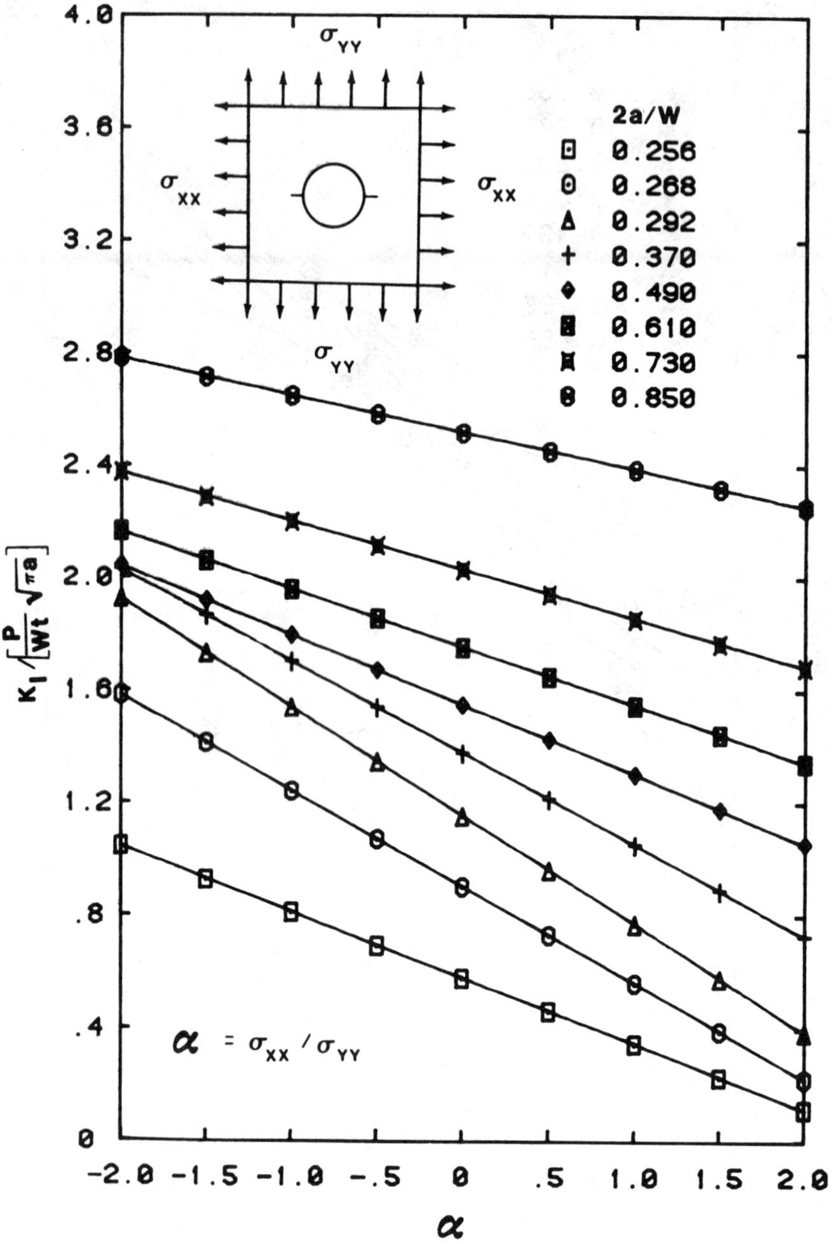

FIG. 8—*Normalized stress intensity factors* K_I *of different crack lengths as a function of biaxial stress ratios* (α).

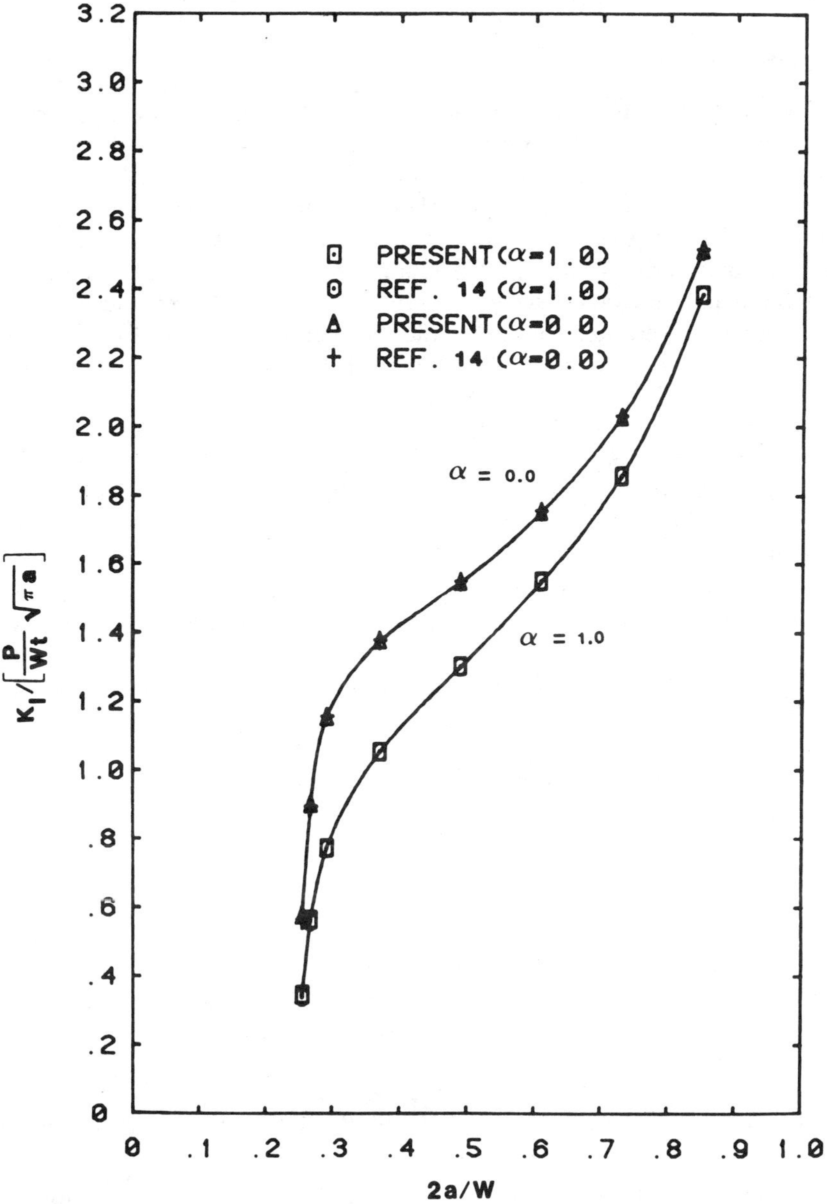

FIG. 9—*Normalized* K_I *values as a function of* a/W *ratios for* $\alpha = 0$ *and 1 with* D/W = *0.25 and* H/W = *1.0.*

obtained efficiently and accurately by coupling the weight function concept with the superposition principle.

Pin-Joint Loading

The use of the superposition principle for the pin-joint loading problem is shown in Fig. 10. The stress intensity factor of the pin-joint load (K_I^{10a}) can be obtained with the sum of 1/2 K_I^{10b} with remote tension and 1/2 K_I^{10c} with symmetric concentrated point-force P acting on each half of the hole boundary above and below the crack line in the direction normal to the crack line. Four potential distributions with equal resultant force P (Fig. 11) are investigated with the weight function concept to study the effect of different distributions of the identical resultant force P at the hole perimeter on the K_I values of symmetric radial cracks with different crack lengths ($0.256 \leq 2a/W \leq 0.85$ and $D/W = 0.25$). These four different load distribution studies are (1) concentrated point force P at $\theta = \pm\pi/2$ location ($K_I = P/2 \cdot h_{Ir}(\theta)|_{\theta=\pi/2}$) (2) sine-pressure distribution ($K_I = t \cdot \int_0^{\pi/2} P_{max} \sin\theta \cdot h_{Ir}(\theta) r d\theta$ with $P_{max} = 4P/\pi Dt$ and t = plate thickness), (3) square sine pressure distribution ($K_I = t \cdot \int_0^{\pi/2} P'_{max} \sin^2\theta \cdot h_{Ir}(\theta) r d\theta$ with $P'_{max} = 3P/2Dt$), and (4) uniform pressure distribution ($K_I = t \cdot \int_0^{\pi/2} P \cdot h_{Ir}(\theta) r d\theta$ with $p = P/Dt$). The effect of load distributions on the calculated K_I values with the weight function concept of the last three distributions requires the explicit weight function component in the radial direction $h_{Ir}(r, \theta, a)$ along the hole perimeter with θ as designated in Fig. 11. Since the Mode I explicit weight functions (h_I) are vectors, these Mode I weight functions in radial directions along the hole perimeter, $h_{Ir}(r, \theta, a)$, can be obtained with the coordinate transformation from $h_{Ix}(x, y, a)$ and $h_{Iy}(x, y, a)$ as

$$h_{Ir}(r, \theta, a) = h_{Ix}(x, y, a) \cos\theta + h_{Iy}(x, y, a) \sin\theta \tag{6}$$

The dimensionless Mode I radial explicit weight functions, $\sqrt{W} \cdot h_{Ir}(r, \theta, a)$, along the hole perimeter as a function of $2\theta/\pi$ are shown in Fig. 12 for $0.256 \leq 2a/W \leq 0.85$ with $D/W = 0.25$. It is found that the dimensionless radial explicit weight function component of $2a/W \geq 0.3$ with $D/W = 0.25$ can be accurately represented by the Fourier series as

$$\sqrt{W} h_{Ir}(r, \theta, a) = \sum_{n=0}^{3} (a_n \sin n\theta + b_n \cos n\theta) \tag{7}$$

These Fourier coefficients (a_n and b_n) are shown in Table 2. The inability of using Fourier representation for $h_{Ir}(r, \theta, a)$ data at the hole perimeter with very short cracks ($2a/W \leq 0.3$ and $D/W = 0.25$) is not completely understood. It is found that the refined mesh model with short cracks is needed to eliminate the oscillatory $h_{Ir}(r, \theta, a)$ results as a function of θ. It is noted that

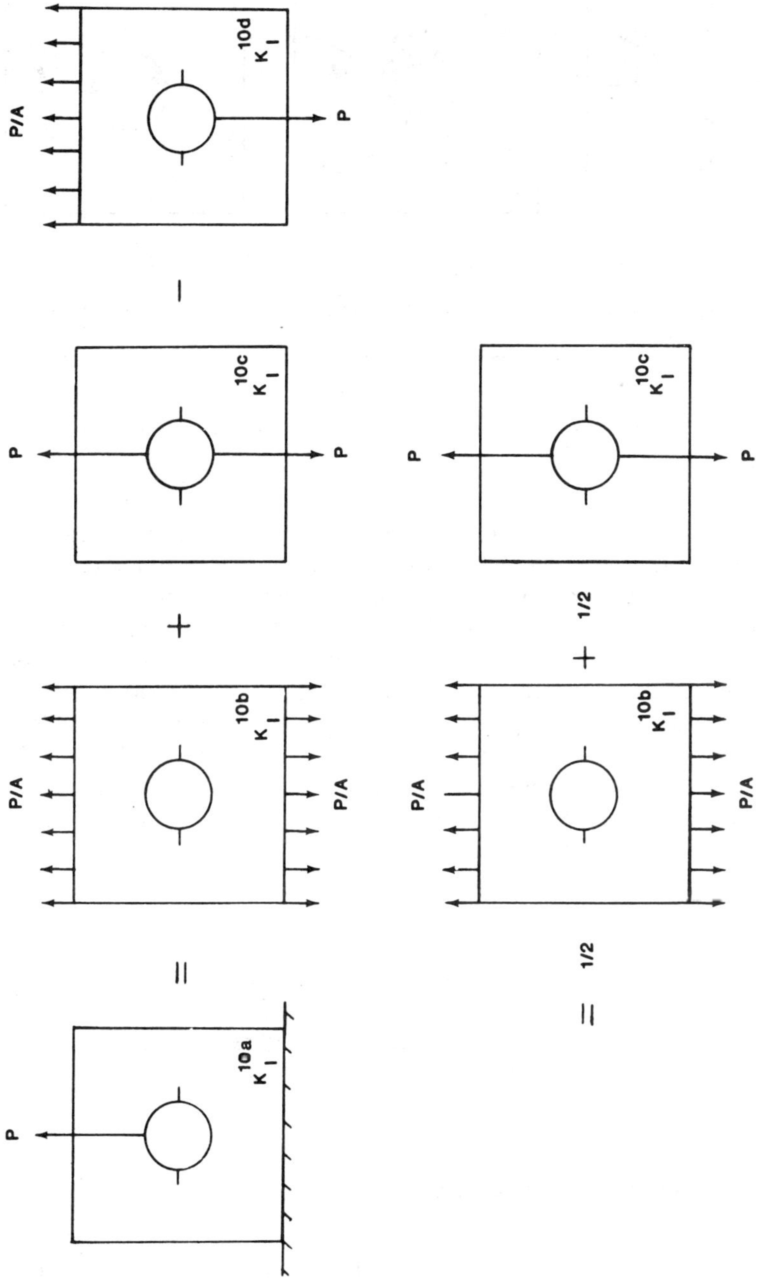

FIG. 10—*Applying superposition principle to pin-joint load problems.*

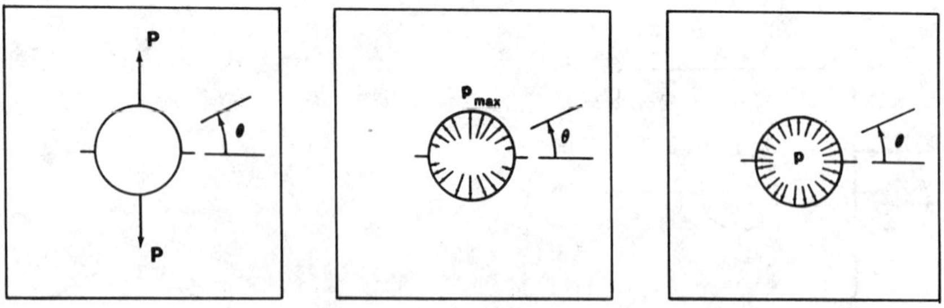

(a) concentrated point-load (b) sine pressure loading (d) uniform pressure

(c) square sine pressure loading

FIG. 11—*Symmetrical radial cracks at a center hole in a square panel with identical resultant loading for all four cases.* (a) *Concentrated point-load.* (b) *Sine pressure loading.* (c) *Square sine pressure loading.* (d) *Uniform pressure.*

the dimensionless explicit Mode I radial weight function component $\sqrt{W} \cdot h_{I_r}(r, \theta, a)$ at $2\theta/\pi = \pm 1$ locations is weighted more than any other location in the K_I evaluation (Fig. 12). This explicit Mode I weight function information at the traction boundaries is very useful to fracture mechanics practitioners and designers when they need to decide on a traction application for a given traction vector that produces the least stress intensity factor. As indicated in Eqs 3 and 4, the physical meaning of the nodal value of the Mode I explicit weight function can be interpreted as the contribution to the K_I value if a unit traction is applied at that nodal location in the Mode I deformation sense. For example, with unit tractions applied at the hole perimeter in the radial direction (Fig. 11),

$$K_I|_\theta = h_{I_r}(r, \theta, a)|_\theta \tag{8}$$

As indicated in Fig. 12, by applying the radial traction at the $\theta = \pi/2$ location, the hole perimeter will produce the maximum K_I value. With the radial traction at $\theta \leq 5$ deg, crack closure (negative K_I) will result, as shown in Fig. 12 for $0.256 \leq 2a/W \leq 0.85$ with $D/W = 0.25$ and $H/W = 1.0$.

Pin-Loaded Hole with Interference Pressure

The interference fitted fastener hole with pin-joint load can be simulated structurally by combining the uniform internal pressure at the hole perimeter with the pin-joint load described in the previous section. Therefore the stress intensity factor for the symmetric radial crack emanating from the interfer-

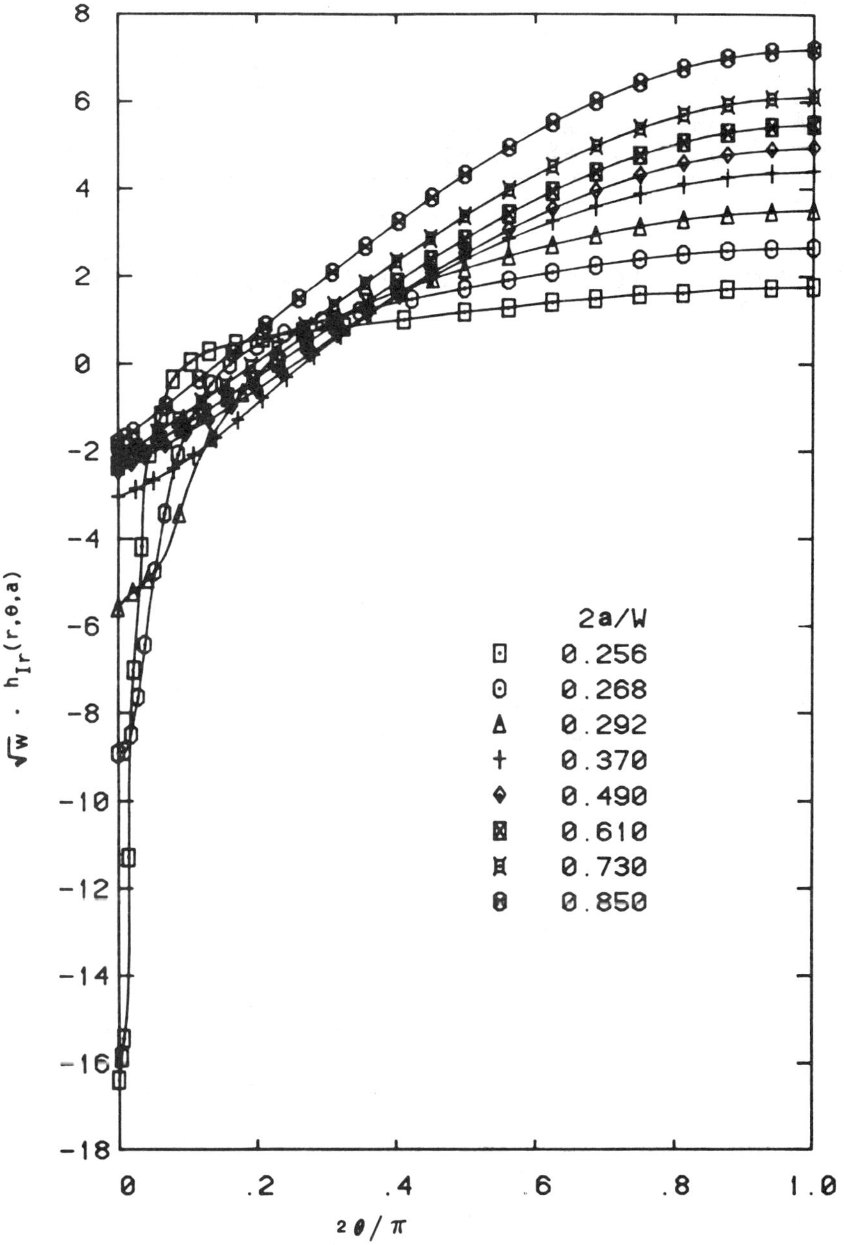

FIG. 12—*Dimensionless explicit radial weight function component,* $\sqrt{W} \cdot h_{Ir}$, *at the hole perimeter* $(0.0 \leq 2\theta/\pi \leq 1.0)$ *with* D/W = *0.25 and* H/W = *1.0.*

TABLE 2—*Regressed Fourier coefficients of Eq 7.*

$2a/W$	a_1	a_2	a_3	b_0	b_1	b_2	b_3
0.370	−75.6762	41.2543	−1.4819	62.1085	−57.1447	−16.4837	8.4889
0.490	19.2775	−9.9830	1.9096	−14.8487	16.1530	−2.4249	−1.3395
0.610	15.2589	−5.8561	0.8068	−9.6091	8.8730	−0.6355	−0.9709
0.730	12.1929	−3.3986	0.4410	−6.1492	5.0667	−0.5237	−0.5892
0.850	11.2789	−2.2680	0.3362	−4.3011	3.4156	−0.5523	−0.3825

ence fitted hole in a plate with a pin-joint load can be obtained by the superposition principle as

$$K_I = {}^1\!/_2(K_I^{10b} + K_I^{10c}) + K_I^{11d} \tag{9}$$

The uniform internal pressure, which represents the interference pressure, can be expressed as

$$p = \frac{\beta P}{Dt} \tag{10}$$

where $\beta \geq 0$ with $\beta = 0$ representing the pin-joint hole without interference fitting, and P is the resultant symmetric pin-joint force.

The effect of interference pressure on the K_I values with $0 \leq \beta \leq 2$ is shown in Fig. 13, which expresses the normalized $K_I(K_I/\sigma\sqrt{\pi c})$ as a function of $2c/D$ for various β values with sine pressure distribution assumed for K_I^{10c} calculation.

Conclusions

Because of the load-independent characteristics of the explicit weight functions for a given geometry, the evaluation of the explicit weight function can eliminate the unnecessary repeated calculation of the Mode I stress intensity factors under different loading conditions. In addition, there is no other known mechanics principle besides the weight function concept to permit the stress intensity factor calculation from one loading condition to another loading condition for a given crack geometry. An extensive utilization of the load-independent characteristics of the explicit weight functions is demonstrated. These explicit weight functions for the entire structure of interest of a given crack geometry are obtained first from a simple remote tension loading with an efficient finite element methodology and then are applied successfully to K_I evaluation under complex loading conditions illustrated for the symmetrical radial cracks emanating from a circular hole in a plate.

An analytical expression of Eq 5, which relates the Mode I explicit crack-face weight functions as a function of radial distance from crack tip along the

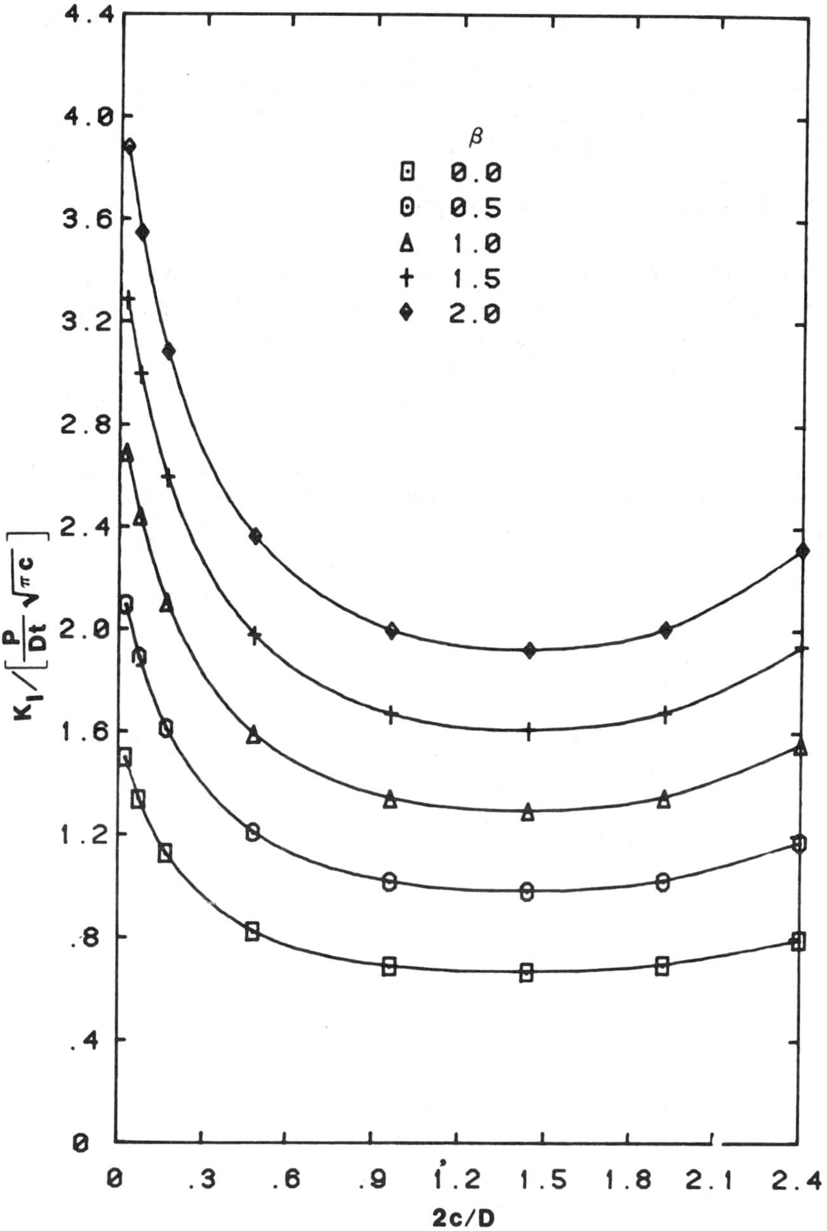

FIG. 13—*Effect of interference pressure loading on the normalized* K_I *values for pin-joint hole with sine pressure distribution assumption from* K_I^{10c} *contribution.*

crack face, can facilitate K_I evaluation with Eq 4. This is done by combining the crack-face weight functions with the equivalent uncracked stress field, which includes body force load, thermal loading, and incorporation of residual stresses.

The coefficient $A_n(a)$ of Eq 5 can be interpolated for any crack length by means of the cubic spline interpolation technique. Because of this interpolative feature, the cycle-by-cycle life analyses are economically feasible for permitting the needed K_I and ΔK_I values for predicting crack propagation life with the interpolated explicit crack-face weight function and with an uncracked stress field.

By coupling these predetermined $h_{Ir}(r, \theta, a)$ data with four assumed load distributions at the hole perimeter in the radial direction with the identical resultant force P, the normalized K_I values as a function of $2a/W$ ratios can be evaluated numerically according to Eq 3 as shown in Fig. 14 in the absence of body force loading. The different assumptions involved in the load distributions at the hole perimeters are the result of the large variations in K_I values despite having identical resultant loading P. The concentrated point force P gives the highest K_I value (K_I^{11a}); in contrast, the uniform pressure distribution leads to the lowest K_I value (K_I^{11d}) among four distributions studies in our investigation with the following ranking of $K_I^{11a} > K_I^{11c} > K_I^{11b} > K_I^{11d}$.

The effect of different pressure distributions at the hole perimeter on the stress intensity factor (K_I) is shown in Fig. 15, expressing the normalized K_I ($K_I/\sigma\sqrt{\pi C}$) as a function of $2C/D$ ratios. The effect is reduced gradually as $2C/D$ increases. The elevation in the normalized K_I values for the long cracks (Fig. 15) is caused by the proximity effect of back free face. This pronounced effect of the pressure distributions on the calculated K_I values for short cracks ($2c/D < 1$) can have important implications in crack propagation life predictions of pin-load joints because most of the fatigue life is spent while the crack is short. Finally, the normalized K_I values ($K_I/\sigma\sqrt{\pi a}$) are obtained with the weight function concept and the superposition principle as a function of $2a/W$ ratios with four pressure distributions at the hole perimeter with identical resultant force P (Fig. 16). As anticipated, the problem of the effect of pressure distribution at the hole perimeter with identical resultant force (P) on the calculated K_I value of pin-joint loading remains but with less pronounced intensity. The calculated K_I values with the weight function concept under the sine pressure and the uniform pressure distribution for the pin-joint problem compare well with those of Cartwright and Parker [*14*] obtained with the boundary collocation technique (Fig. 17).

By combining the explicit weight function concept and the linear superposition principle, the calculation of K_I under complex loading conditions is demonstrated extensively with the deliberate use of explicit weight functions at the load application boundaries other than crack-face locations. The complex loading condition of biaxial remote loading and pin-joint loading with and without interference pressures is demonstrated by combining the weight

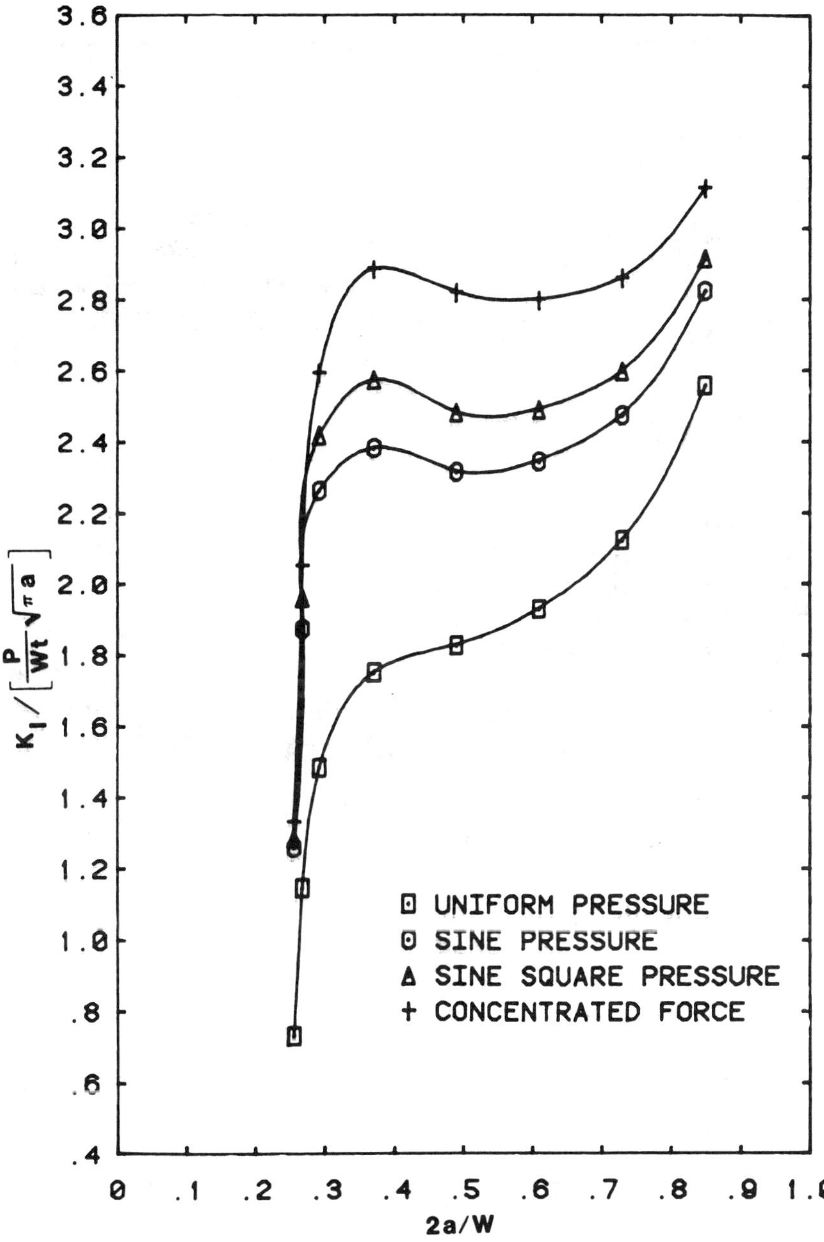

FIG. 14—*Effect of pressure distributions at the hole perimeter on the normalized* K_I *values for symmetric radial cracks emanating from hole with* D/W = *0.25,* H/W = *1.0, and 0.256* ≤ 2a/W ≤ *0.85.*

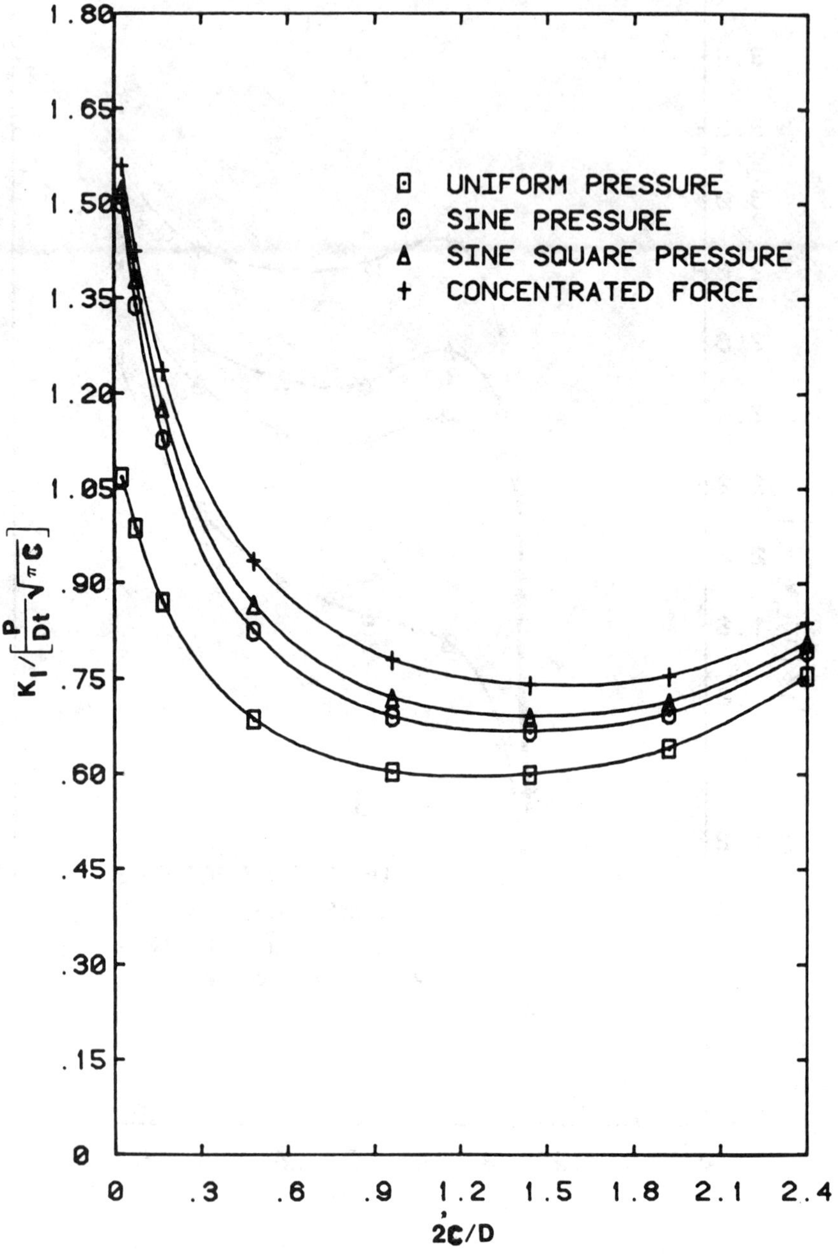

FIG. 15—*Effect of pressure distribution on the normalized* $K_I/\sigma\sqrt{\pi C}$ *at the hole perimeter values as a function of* 2C/D *for the pin-joint load problem with* D/W = *0.25 and* H/W = *1.0.*

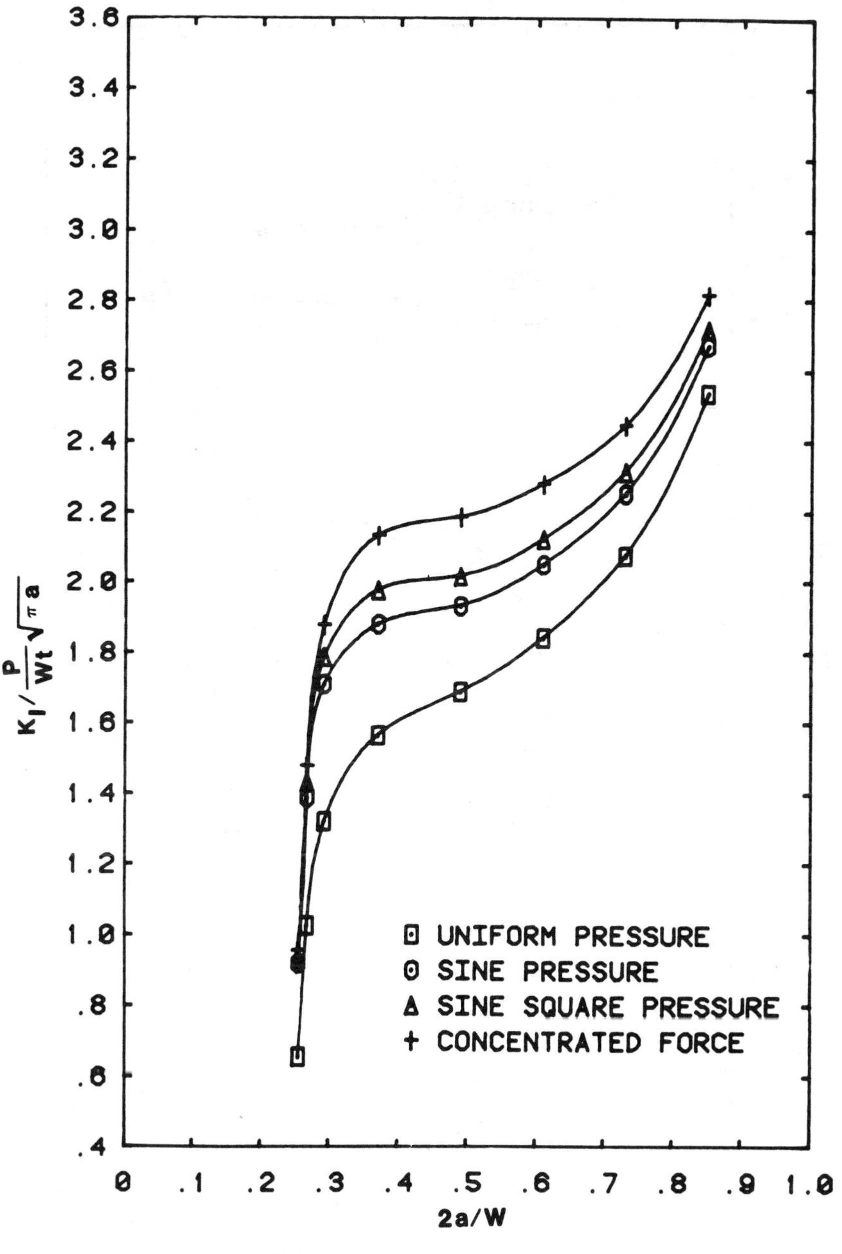

FIG. 16—*Normalized* $K_I/\sigma\sqrt{\pi a}$ *values as a function of* 2a/W *for pin-joint with four pressure distributions studied.*

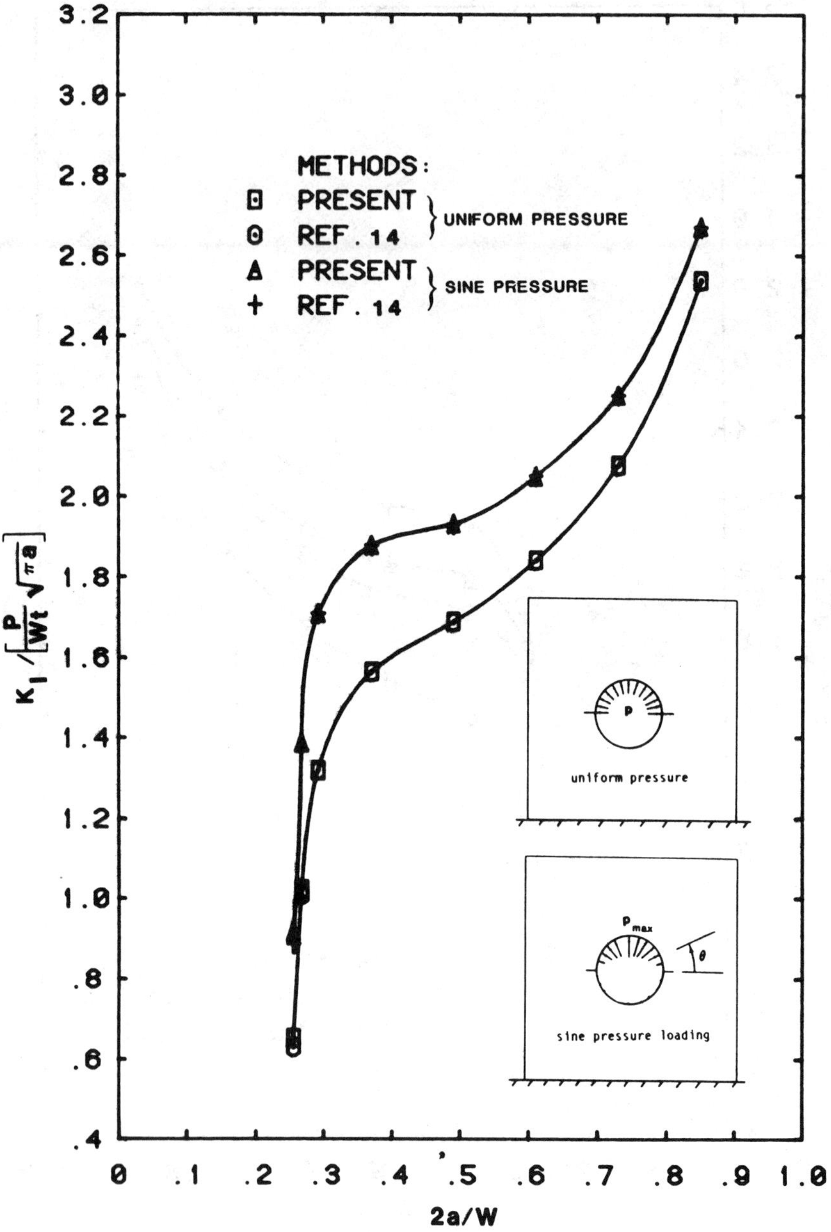

FIG. 17—*Normalized* K_I *values are compared between the weight function concept and the boundary collocation results for pin-joint loading with sine and uniform pressure distributions.*

function concept with the linear superposition principle for the symmetrical radial cracks emanating from a hole in a plate. The numerical K_I results obtained from the weight function concepts are found to be in excellent agreement with the well-established literature results for the symmetric radial cracks problem as studied in this paper.

The characteristics of the explicit Mode I weight function component for the symmetric radial cracks emanating from a hole in a plate (Fig. 1) are also provided at the key traction application boundaries. The Mode I explicit nodal weight function value can be considered as the stress intensity value under the unit traction load at the concerned nodal location in the direction normal to the traction boundary (Eq 8). Therefore the Mode I explicit weight function components along the traction application locations are extremely useful to fracture mechanics practitioners and designers regarding how to apply traction that will produce the reduced stress intensity values.

It is found that different pressure distributions at the hole perimeter can affect the Mode I stress intensity factors, with a more pronounced effect on short radial cracks than on long cracks emanating from a pin-joint loading hole. The study of the effect of different pressure distributions at the hole perimeter with the identical resultant force P in the y direction requires the weight function component in the radial direction. The radial weight functions at the hole perimeter can be obtained from the coordinate transformation (Eq 6). The effect of different pressure distributions at the hole perimeter on the Mode I stress intensity factor for short radial cracks ($2C/D < 1$) of the pin-joint loaded hole can provide an important implication in predicting the crack propagation life, since most fatigue life is spent while the crack is short. Except for very short radial cracks, the radial weight function component at the hole perimeter can be accurately represented as a Fourier series (Eq 7). The refined finite element mesh is needed to avoid the oscillatory ill-behaved radial weight function component at the hole perimeter for the problem of a pin-joint loaded hole.

In conclusion, the additional efforts needed to produce the explicit weight function components, as indicated in Ref *1*, are not much beyond the normal finite element calculation of stress intensity factors. However, the pay-off as indicated in this paper is far greater than the required effort to generate the load independent explicit weight functions. The deliberate use of the explicit weight functions along the traction application boundaries is applied extensively in this paper to calculate the K_I values under complex loading conditions. This is accomplished by combining the weight function concept with the linear superposition principle.

References

[*1*] Sha, G. T., "Stiffness Derivative Finite Element Technique to Determine Nodal Weight Functions with Singularity Elements," *Engineering Fracture Mechanics*, Vol. 19, No. 4, 1984, pp. 685–699.

[2] Parks, D. M., "A Stiffness Derivative Finite Element Technique for Determination of Crack Tip Stress Intensity Factors," *International Journal of Fracture*, Vol. 10, 1974, pp. 487-501.

[3] Hellen, T. K., "On the Method of Virtual Crack Extensions," *International Journal of Numerical Methods in Engineering*, Vol. 9, 1975, pp. 187-207.

[4] Henshell, R. D. and Shaw, K. G., "Crack Tip Elements Are Unnecessary," *International Journal of Numerical Methods in Engineering*, Vol. 9, 1975, pp. 495-507.

[5] Barsoum, R. S., "On the Use of Isoparametric Finite Elements in Linear Fracture Mechanics," *International Journal of Numerical Methods in Engineering*, Vol. 10, 1976, pp. 25-37.

[6] Parks, D. M. and Kamentzky, E. M., "Weight Functions from Virtual Crack Extension," *International Journal of Numerical Methods in Engineering*, Vol. 14, 1979, pp. 1693-1706.

[7] Vanderglas, M. L., "Weight Functions for Plane and Axisymmetric Crack Problems," in *Numerical Methods in Fracture Mechanics*, D. R. J. Owen and A. R. Luxmoore, Eds., Proceedings of 2nd International Conference held at University College, Swansea, 7-11 July 1980, pp. 51-65.

[8] Vainshtok, V. A., "A Modified Virtual Crack Extension Method of the Weight Functions Calculation for Mixed Mode Fracture Problems," *International Journal of Fracture*, Vol. 19, 1982, pp. R9-R14.

[9] Bueckner, H. F., "Field Singularities and Related Integral Representations," in *Mechanics of Fracture I: Method of Analysis and Solutions of Crack Problems*, G. C. Sih, Ed., Noordhoff, Leyden, The Netherlands, 1972.

[10] Rice, J. R., "Some Remarks on Elastic Crack-Tip Stress Fields," *International Journal of Solids and Structures*, Vol. 8, 1972, pp. 751-758.

[11] Sha, G. T. and Yang, C.-T., "Weight Function Calculations for Mixed Mode Fracture Problems with the Virtual Crack Extension Technique," *Engineering Fracture Mechanics*, Vol. 21, No. 6, 1985, pp. 1119-1149.

[12] Bowie, O. L., "Solutions of Plane Crack Problems by Mapping Technique," in *Mechanics of Fracture I: Method of Analysis and Solutions of Crack Problems*, G. C. Sih, Ed., Noordhoff, Leyden, The Netherlands, 1972.

[13] Newman, J. C., "An Improved Method of Collocation for the Stress Analysis of Cracked Plates with Various Shaped Boundaries," NASA TN D-6376, National Aeronautics and Space Administration, Washington, DC, 1971.

[14] Cartwright, D. J. and Parker, A. P., "Opening Mode Stress Intensity Factors for Cracks in Pin-Loads Joints," *International Journal of Fracture*, Vol. 18, 1982, pp. 65-78.

[15] Grandt, A. F., "Stress Intensity Factors for Some Through-Cracked Fastener Holes," *International Journal of Fracture*, Vol. 11, 1975, pp. 283-294.

[16] Grandt, A. F., "Stress Intensity Factors For Cracked Holes and Rings Loaded with Polynomial Crack Face Pressure Distributions," *International Journal of Fracture*, Vol. 14, 1978, pp. R221-229.

[17] Hsu, T. M. and Rudd, J. L., "Green's Function for Thru-Cracks Emanating from Fastener Hole," in *Proceedings Fourth International Conference on Fracture*, Vol. 3, Waterloo, Ontario, June 1977, pp. 139-148.

[18] Andrasic, C. P. and Parker, A. P., "Dimensionless Stress Intensity Factors for Cracked Thick Cylinders under Polynomial Crack Face Loadings," *Engineering Fracture Mechanics*, Vol. 19, No. 1, 1984, pp. 187-193.

[19] Pu, S. L., "A Functional Stress Intensity Approach to Multiply Cracked, Partially Autofrettaged Cylinders," in *Transactions of the 28th Conference of Army Mathematicians*, ARO Report 83-1, 1983, pp. 263-283.

[20] Orange, T. W., "Crack Shapes and Stress Intensity Factors for Edge-Cracked Specimens," in *Stress Analysis and Growth of Cracks, ASTM STP 513*, American Society for Testing and Materials, Philadelphia, 1972, pp. 71-78.

J. F. Yau[1]

An Empirical Surface Crack Solution for Fatigue Propagation Analysis of Notched Components

REFERENCE: Yau, J. F., **"An Empirical Surface Crack Solution for Fatigue Propagation Analysis of Notched Components,"** *Fracture Mechanics: Seventeenth Volume, ASTM STP 905*, J. H. Underwood, R. Chait, C. W. Smith, D. P. Wilhem, W. A. Andrews, and J. C. Newman, Eds., American Society for Testing and Materials, Philadelphia, 1986, pp. 601-624.

ABSTRACT: A stress intensity factor solution for a surface crack in a finite solid subjected to an arbitrary stress field is presented. The solution was developed based on the superposition principle, the weight function technique, and a referenced finite body correction factor. An additional surface factor is identified when the solution is applied to the reduction of data from tests on surface flaw specimens.

Verification of the solution is first made by comparison with published results for problems with non-uniform stresses. The examples include a surface crack under pure bending load and a corner crack at a circular hole.

Another assessment was achieved by verifying the solution against the test results of ASTM round-robin test data for cracked fastener holes. Good correlation was obtained on all aspects of experimental observations such as the backtracked stress intensity factors, the measured residual lives, and the observed variation in crack aspect ratio.

It has been further demonstrated, from Air Force supported test programs, that the solution is accurate in predicting crack propagation behavior and residual life for subcomponent specimens under complex loadings. The test specimens are made of four different superalloys in various shapes of geometric discontinuities. The test conditions include various combinations of temperature, stress level, mean stress, and cycle profiles. The results have shown that the great majority of the measured life/predicted life ratios are within a factor of 1.5. It has also been demonstrated that the capability of the methodology is independent of the stress levels, specimen types, materials, and cycle profiles used in the study.

KEY WORDS: surface crack, stress intensity factor, fatigue crack propagation, residual life prediction, stress analysis, superalloy, weight function, superposition principle, finite body effect, plastic zone

[1]Senior Project Engineer, Aircraft Engine Business Group, General Electric Company, Evendale, OH 45215.

Nomenclature

- a Crack depth of the semi-elliptical surface crack
- c Half surface crack length of the semi-elliptical surface crack
- ϕ Angular orientation along the crack front of the semi-elliptical surface crack
- W Width of a rectangular structural component
- H Thickness of a rectangular structural component
- σ_j Coefficients of polynomial representing stress distribution
- σ Adjusted stress in the surface crack solution expression accounting for stress gradient
- I_j Influence function for *i*th term of polynomial stress distribution
- m_j Back surface modification on the influence coefficients for the solution at the depth of the surface crack
- A Scaling factor in the Paris expression of crack growth rate curve
- n Exponential coefficient in the Paris expression of crack growth rate curve

The stress intensity factor is a fracture mechanics parameter representing the strength of the stress and displacement fields surrounding the crack tip, and it is considered to be the driving force for fatigue crack growth. The essence of this parameter is that it defines the unique crack tip stress field, regardless of structural geometry, crack dimensions, and loading conditions. Thus the baseline fatigue crack growth data obtained from laboratory tests using simple specimens can be utilized to predict the crack growth behavior and residual life of more complicated structural components.

In general structural components, the part-through surface flaw has been the most frequently observed crack configuration. For the purposes of design and life analysis, it is mandatory that an engineer be able to confidently deal with this crack geometry. Unfortunately, most surface crack problems are rather complex and exact solutions are usually not available. Various approximate solutions, obtained by pseudo-analytical, experimental, or numerical methods, have been shown to be in considerable disagreement for even the simplest case of a surface crack subjected to a uniform stress field [*1*]. This discrepancy is expected to be intensified for cases involving complicated structural geometries with steep non-uniform stress gradients. For these problems, good solutions often have to rely on a rigorous finite element analysis of the cracked component. For predictions of crack growth behavior and residual life, a series of analyses is required for various possible crack dimensions and configurations throughout the entire propagation process. The approach is certainly impractical, especially for the surface crack problem in which the crack propagates in multiple directions.

An empirical stress intensity factor solution for a surface crack in a finite body subjected to an arbitrary stress field is proposed in this paper. The solution was developed based on the weight function technique [*2*] and the superposition principle, and the finite element solution by Newman and Raju [3] for a surface crack in a finite solid subjected to uniform tensile stress field. The solution is expressed in a series of closed-form equations that appears attractive for application to crack propagation analysis and residual life prediction. This paper will first describe the analytical formulations of the solution. Comparisons are made of the proposed solution with some published results. Procedures for reducing surface flaw crack growth data and determining a surface factor are illustrated in the next section. This is followed by a brief discussion on numerical procedures for fatigue damage accumulation predictions. Finally, the capability of the solution in predicting fatigue propagation behavior and residual life is verified by correlation with notched specimen test data from various sources.

Analytical Formulation

Consider a semi-elliptical surface crack in a finite body (Fig. 1) subjected to arbitrary stress field in the depth direction. The stress intensity factor, K, at any point along the crack front can be expressed in a similar form to the Irwin part-through crack solution [4] as

$$K = F^* \hat{\sigma}^* \frac{\sqrt{\pi a}}{\phi} \tag{1}$$

where ϕ is the elliptical integral of the second kind, which can be approximated by the equations

$$\phi = \begin{bmatrix} 1 + 0.5707^* \left(\dfrac{a}{c}\right)^{1.42} & \text{for } \dfrac{a}{c} \leq 1 \\ 1 + 0.5707^* \left(\dfrac{a}{c}\right)^{-1.42} & \text{for } \dfrac{a}{c} > 1 \end{bmatrix} \tag{2}$$

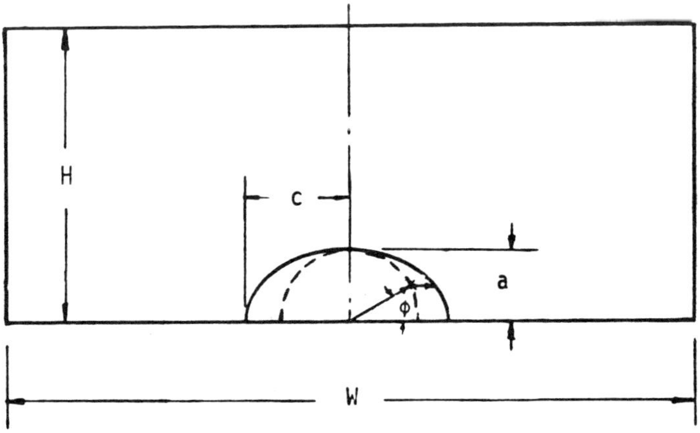

FIG. 1—*Configuration of a semi-elliptical surface crack.*

and F is the finite body correction factor for the crack subjected to uniform stress condition. The finite element solution by Newman and Raju [3] was adopted to define this factor, which is given by the equations

$$F = M_1 + M_2\left(\frac{a}{H}\right)^2 + M_3\left(\frac{a}{H}\right)^4 g f_\phi f_w \tag{3}$$

where f_w is a finite body correction factor defined as

$$f_w = \left[\sec\left(\frac{\pi c}{W}\sqrt{\frac{a}{H}}\right)\right]^{1/2} \tag{4}$$

For $a/c \leq 1$,

$$M_1 = 1.13 - 0.09\left(\frac{a}{c}\right) \tag{5}$$

$$M_2 = -0.54 + \frac{0.89}{0.2 + \frac{a}{c}} \tag{6}$$

$$M_3 = 0.5 - \frac{1}{0.65 + \frac{a}{c}} + 14\left(1 - \frac{a}{c}\right)^{24} \tag{7}$$

$$g = 1 + \left[0.1 + 0.35\left(\frac{a}{H}\right)^2\right](1 - \sin\phi)^2 \tag{8}$$

and

$$f_\phi = \left[\left(\frac{a}{c}\right)^2 \cos^2\phi + \sin^2\phi\right]^{1/4} \tag{9}$$

For $a/c > 1$,

$$M_1 = \sqrt{\frac{c}{a}}\left(1 + 0.04\,\frac{c}{a}\right) \tag{10}$$

$$M_2 = 0.2\left(\frac{c}{a}\right)^4 \tag{11}$$

$$M_3 = -0.11\left(\frac{c}{a}\right)^4 \tag{12}$$

$$g = 1 + \left[0.1 + 0.35\left(\frac{c}{a}\right)\left(\frac{a}{H}\right)^2\right](1 - \sin\phi)^2 \tag{13}$$

and

$$f_\phi = \left[\left(\frac{c}{a}\right)^2 \sin^2\phi + \cos^2\phi\right]^{1/4} \tag{14}$$

where ϕ is the angular orientation of the points along the crack front.

The stress term $\hat{\sigma}$ in Eq 1 is defined as the "adjusted" stress which is derived using the superposition principle and the weight (or influence) function technique. First, the distribution of the elastoplastic stresses normal to a presumed crack plane of an uncracked structural component is expressed by a polynomial of the form

$$\sigma(X) = \sum_{j=0}^{n} \sigma_j x^j \tag{15}$$

From the superposition principle, the crack solution is solved by applying this stress field to the crack surface of the cracked component [5]. Using the weight function technique, the adjusted stress $\hat{\sigma}$ is then defined as

$$\hat{\sigma} = \sum_{j=0}^{n} \sigma_j m_j I_j(\phi) a^j \tag{16}$$

where σ_j is the coefficient for the jth order term of the polynomial in Eq 15, and $I_j(\phi)$ is the corresponding polynomial influence function at the angular orientation along the crack front. The function I_j can be estimated by applying pressure $p(x) \sim |x|^j$ on the crack surface and computing the resulting stress intensity factors of a unit circular crack in an infinite body. It can be evaluated from the closed-form solution [6]

$$I_j = \frac{1}{2\pi a} \int_{-\pi}^{\pi} \int_0^a \left|\frac{x}{a}\right|^j \sqrt{\frac{2a \sin\theta}{\rho} - 1}\; d\rho d\theta \tag{17}$$

Numerical values of weight functions at the crack maximum depth ($\phi = 0$) and on the surface ($\phi = 90$) are listed in Table 1. The term m_j in Eq 16 is a back surface modification factor on the weight function coefficient, I_j, for

TABLE 1—*Polynomial influence functions,* I_j.

Order, j	Depth	Surface
0	1.000000	1.000000
1	0.718953	0.166704
2	0.600000	0.066669
3	0.528710	0.035716
4	0.479365	0.022224
5	0.442391	0.015153
6	0.413253	0.010990
7	0.389470	0.008334
8	0.369550	0.006537

computation of stress intensity factor at the maximum crack depth. The factors can be approximately determined by the equations

$$m_j = \begin{cases} 1 - f_j{*}\sin\left(\dfrac{\pi a}{2H}\right)^{1.7*}\left(1 - \dfrac{a}{c}\right) & \dfrac{a}{c} < 1 \\ 1.0 & \dfrac{a}{c} \geq 1 \end{cases} \tag{18}$$

where H is the thickness and the f_j values are scaling factors ($f_j = 0.0$, 0.418, 0.612, 0.701, 0.759, 0.793, 0.813, 0.832, 0.849 for $j = 0$ through 8 respectively). The limiting values of the equation (crack aspect ratio equals to zero) correspond to the relationship for the single edge crack. The factor is assumed to become unity when the aspect ratio of the crack is equal to or greater than one. The proposed stress intensity factor solution reduces to the solution form of Newman and Raju [*3*] in the case of uniform stress condition.

The solution for the corner crack is approximated by using the solutions at the surface locations of two mirror-imaged semi-elliptical surface cracks (Fig. 2) along with a modification factor,

$$f = \begin{cases} 1.085 - 0.075{*}\,c'/a' & c'/a' < 1.1 \\ 1.0 & c'/a' \geq 1.1 \end{cases} \tag{19}$$

where a' is the crack depth and c' is the half crack length of the imagined semi-elliptical surface crack.

Note that when the solution is applied in crack propagation analysis, the stress intensity factor at surface locations should be modified by a surface factor which accounts for the effects of free surface and surface conditions on the crack growth rate. The factor may vary with materials, machining processes, surface treatments, temperatures, etc. The procedures of experimentally determining this factor are described later.

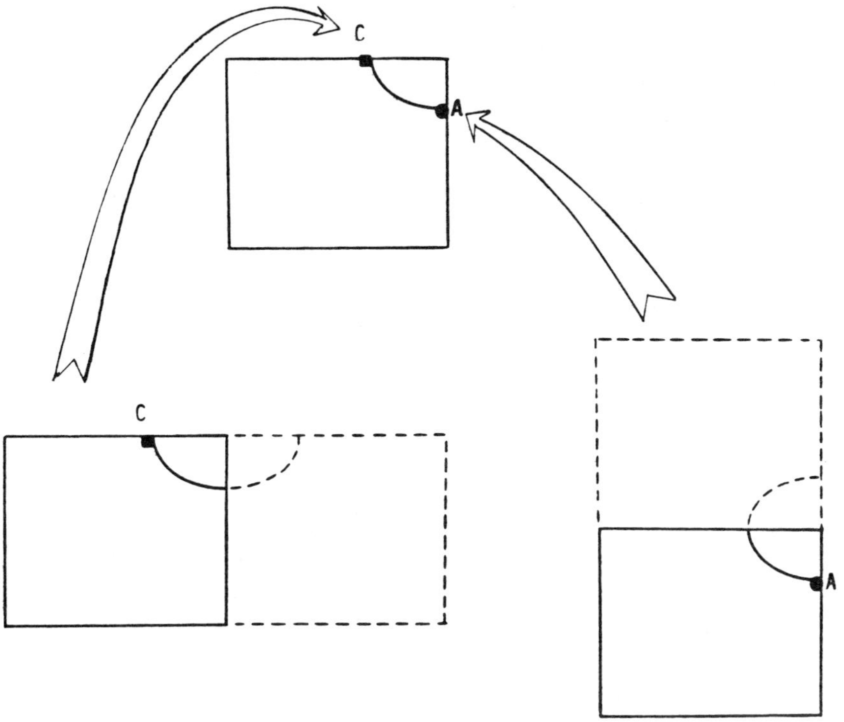

FIG. 2—*Mirror image approach for an approximate corner crack solution.*

It is also noted that the stress intensity factor was further modified by the Irwin plastic zone correction [*7*].

Numerical Results

Two example solutions from the open literature for non-uniform stresses are compared with the above approach.

Crack in a Finite Plate under Bending Load

The first example is a surface crack in a finite thickness plate under pure bending load. The solutions of the stress intensity factor at the maximum depth are compared in Fig. 3 for various crack aspect ratios. It is shown that the agreements with the referenced solutions [*3,8*] are excellent.

Corner Crack at Bolt Hole

A corner crack emanating from a circular hole in an infinite plate is another example of a more complex, non-uniform stress condition which was

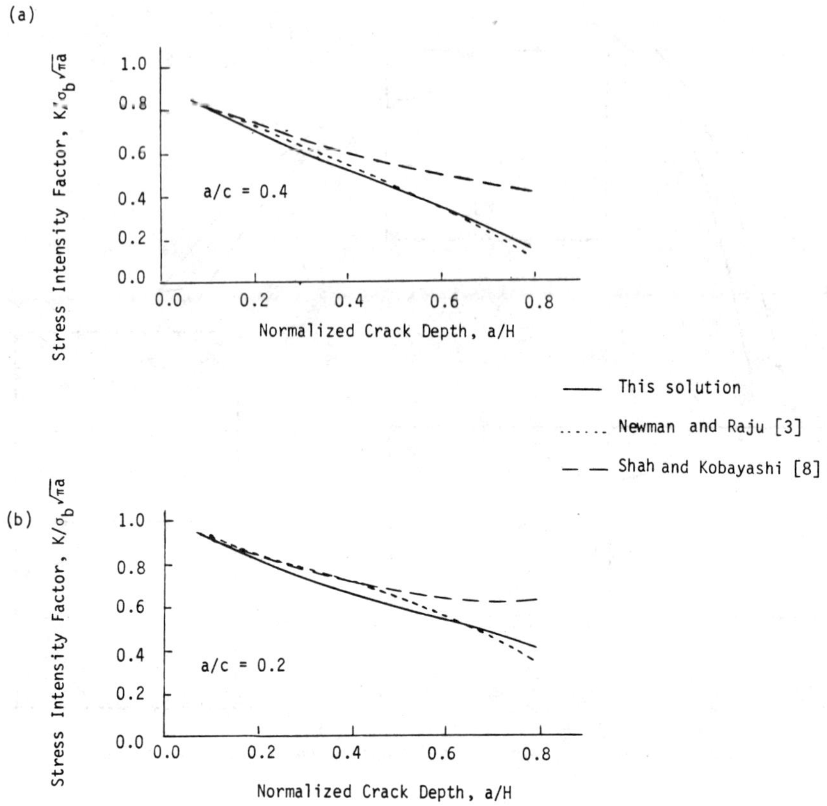

FIG. 3—*Comparison of stress intensity factor solutions for a surface crack under bending load.*

examined. In the approach, the stress distribution around a hole is first represented by the polynomial

$$\sigma(x) = 3 - 6.7x + 13.1x^2 - 15.56x^3 + 9.55x^4 - 0.84x^5 - 2.45x^6 + 1.33x^7 - 0.22x^8 \quad (20)$$

where x is the distance away from the edge of the hole. The crack configuration is assumed to be quarter circular. Following the procedures described in the previous section, the normalized stress intensity factor solutions of the crack at the two surface locations were obtained (assuming a surface factor of 0.88 for the reason described in the next section) and compared with a referenced solution [*9*] which was obtained from a finite element analysis. The stress intensity factor was normalized to the Bowie through-crack solution [*10*] of the same crack length. In Fig. 4, the agreement is shown to be good.

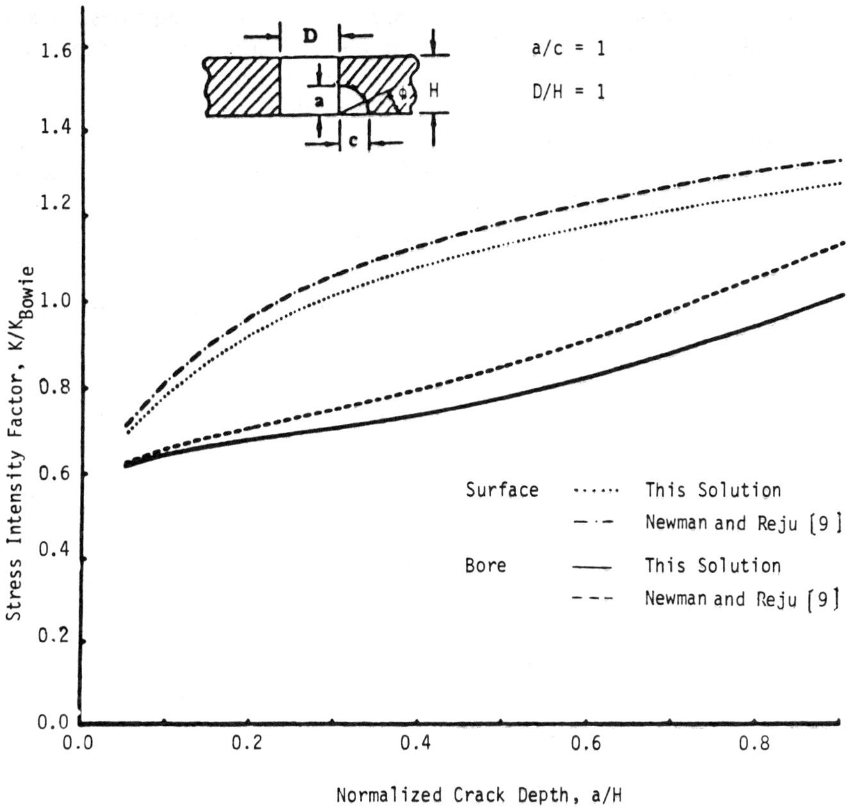

FIG. 4—*Comparison of stress intensity factor solutions for a corner crack at hole.*

Determination of Surface Factor

The surface condition (due to machining, treatment, etc.) and free surface response (plane stress as opposed to plane strain) have major effects on the fatigue crack growth behavior, especially at the surface locations of the crack. To account for this effect, a modification factor is applied to the stress intensity factor solution of the surface locations. The factor is to be determined in the data reduction process using the surface crack specimen of the same surface condition.

General Procedures for Data Reduction of Surface Crack Specimen

For the surface crack specimen, the surface crack length is the only crack dimension that can be measured during the test. Since the crack solution also depends on the crack depth dimension, it was necessary to obtain information on how the crack aspect ratio changes as the crack length progresses. This is accomplished by periodically heat tinting the specimen during the test and examining the produced color rings on the fracture surface after the test. Several test results were reviewed; it was found that the crack aspect ratio changes for the two materials studied (Inconel 718 and Ti-6Al-4V) were nearly independent of the test temperatures and *R*-ratios. Figure 5 presents an example for Inconel 718. A simple mathematical representation, as shown in the insert of Fig. 5, was used to obtain crack depth points as a function of cycles.

The growth rate data were reduced by a linear regression to the Paris equation

$$\frac{d\ell}{dN} = A(\bar{K})^n \tag{21}$$

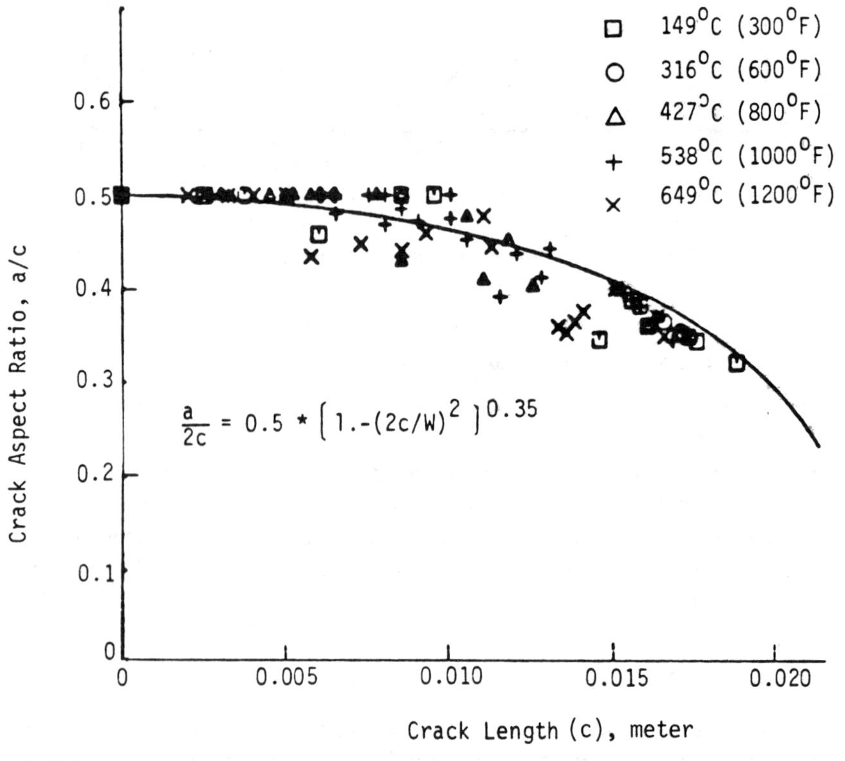

FIG. 5—*Variation of measured crack aspect ratios for Inconel 718 surface crack specimens* (R = *−1 to 0.5,* a/H = *0 to 0.8).*

where ℓ is either the crack length (c) or depth (a), and A and n are temperature-dependent functions. The term, $\bar{K}$, is the Walker equivalent stress intensity factor [*11*] defined as

$$\bar{K} = K_{max}(1 - R)^m \tag{22}$$

in which K_{max} is the stress intensity factor at the peak point of the cycle and m is a material parameter. Figure 6 gives an example of the derived crack growth rate curves for various test temperatures.

Both the surface crack length and depth data can be reduced independently to produce the crack growth rates at both locations of the crack front (i.e., dc/dN and da/dN). An example of these two curves from the Inconel 718 at 538°C (1000°F) is shown in Fig. 7. The surface modification factor on

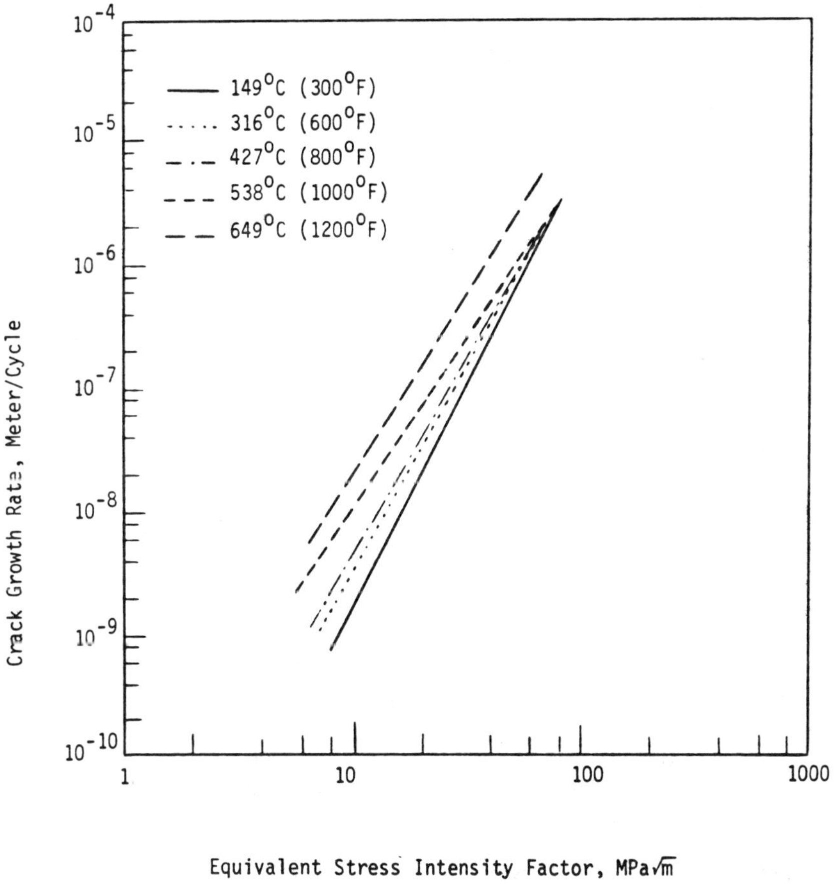

FIG. 6—*Crack growth rate curves of Inconel 718 at various temperatures.*

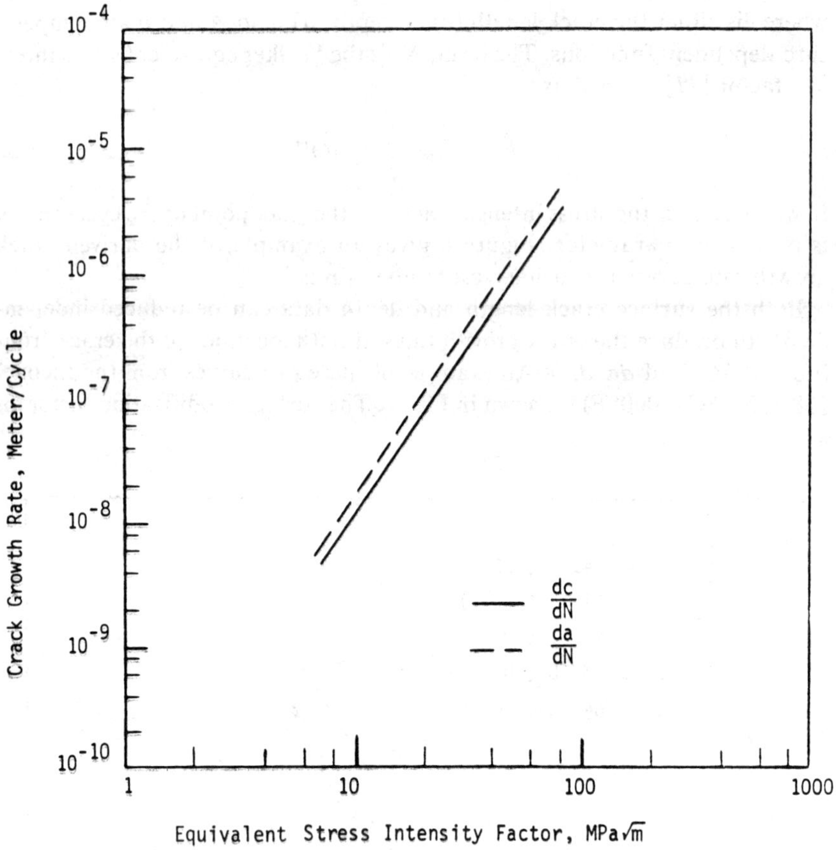

FIG. 7—*Crack growth rate curves reduced from both crack length and depth of Inconel 718 surface crack specimen at 538°C (1000°F).*

the surface stress intensity factor is the factor that shifts the *dc/dN* curve onto the *da/dN* curve. It was found that the factors (Table 2) for both the materials studied are nearly independent of material and temperature.

Once the crack growth rate curve has been established, it is necessary, for a fundamental assessment of the solution, to back predict and check the crack propagation life of each individual test using the generated curve. These results (Fig. 8) show that the back calculated lives were in excellent agreement with the measured lives. A great majority of the data points are within a factor of 1.5. A statistical plot of the data in terms of measured/predicted life ratio (Fig. 9) indicates that the results may be normally distributed. The standard deviation (slope) is found to be 0.109.

Fatigue Propagation Analysis and Life Prediction

The fatigue propagation life over a range of crack dimensions is evaluated by integrating the predicted crack growth rates determined by the computed

TABLE 2—*Surface modification factors of crack growth curves.*

Material	Temperature	Surface Factor
Inconel 718	300	0.88
	600	0.88
	800	0.89
	1000	0.89
	1200	0.89
Ti-6Al-4V (annealed)	300	0.90
	600	0.90
Ti-6Al-4V (double solution treated	75	0.90
and aged)	300	0.91
	600	0.91

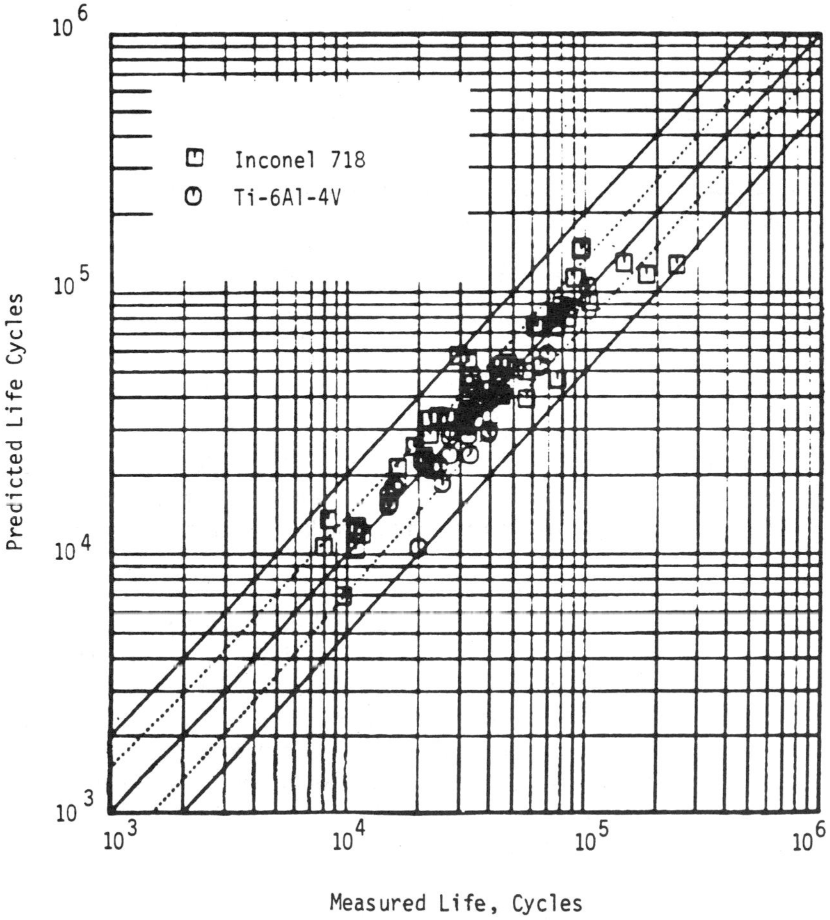

FIG. 8—*Comparison of back-predicted and tested cycles for Inconel 718 and Ti-6Al-4V surface crack specimens.*

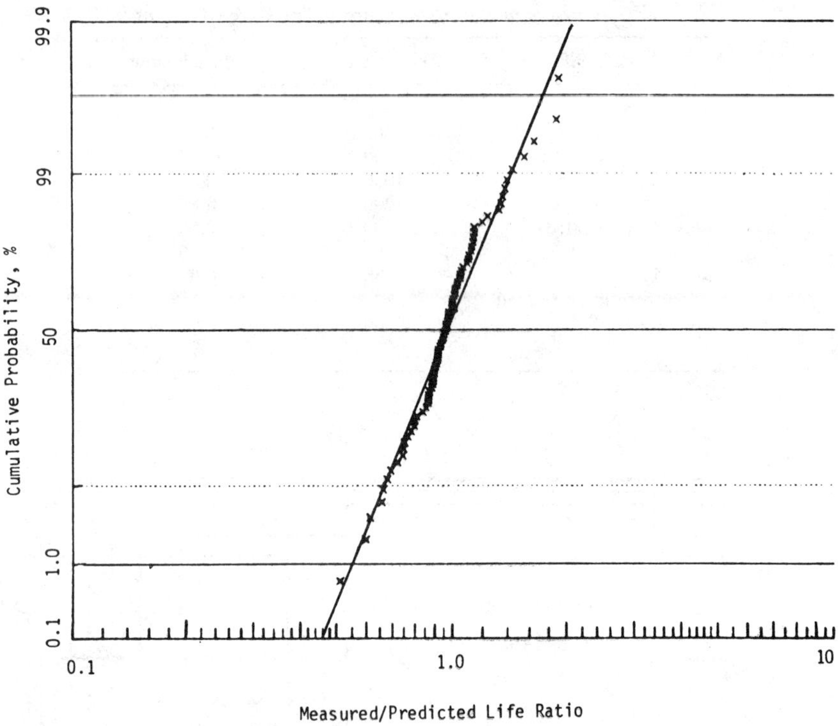

FIG. 9—*Normal distribution plot of measured/predicted life ratios for Inconel 718 and Ti-6Al-4V baseline surface crack specimens.*

stress intensity factors and the baseline crack growth rate curve. A Runge-Kutta routine was used for numerical integration. The cycle integration interval was initially based on the initial estimation of crack growth rate and is periodically adjusted relative to the current crack dimensions, stress intensity factors, and crack growth rates.

Both surface and corner cracks were treated as two-dimensional problems which consider only two locations of the crack front for analysis, namely, at the maximum crack depth and the free surface. The curve of the crack front line was assumed to remain in either semi-elliptical or quarter elliptical configuration for the surface or corner crack, respectively, before the crack broke through the width or thickness of the structural component. Once it broke through, abrupt transitioning to a one-dimensional crack was assumed, either to a single edge crack, a through-thickness internal crack, or, for hole specimens, to Bowie's solution [*10*] for a through-cracked hole problem. The finite body modification to the Bowie solution was from Isida's solution [*12*] for an off-center through-crack in a finite sheet.

For problems with complex cycles, the rain flow cycle counting algorithm

[13] was used for determination of the equivalent damaging cycles. The mean stress effect of the cycles was handled by Walker's relationship [11].

The numerical computational process was completed when a failure was predicted according to the following criteria:

1. Maximum stress intensity factor exceeds the fracture toughness.
2. Equivalent stress intensity factor that accounts for the mean stress effect exceeds 96% of the fracture toughness.
3. Net section stress of the remaining ligament is greater than the ultimate tensile stress.

Experimental Verification

Two major sources of experimental data were used to verify the capability of the solution for predicting fatigue crack growth behavior and residual life of notched specimens. The first data set was from a published source used in the round-robin fatigue crack life predictions by a task group of ASTM Subcommittee E24.06 on Fracture Mechanics Applications [14,15]. It involved sheet specimens with a cracked fastener hole subjected to constant amplitude cycle loading. The second data set was generated by the General Electric Company at Evendale, Ohio, for several Air Force supported verification test programs [16,17] conducted for verification of current applied fracture mechanics codes. The specimens include various geometric discontinuities and were subjected to both simple and complex loading cycles.

ASTM Round-Robin Predictions of Cracked Hold Specimens

The tests were conducted by the Air Force Materials Laboratory for specimens of polymethyl methacrylate (PMMA) plexiglass [18] and 7075-T651 aluminum alloy. Since the PMMA is a transparent material, it was possible to monitor the crack dimensions directly during the test and therefore determine the stress intensity factors that were achieved using the James-Anderson backtracking technique [19].

One difficulty in solving the problem by the present approach is the lack of information on the surface factors for these two materials. A value of 0.88 was assumed in the predictions based on the Inconel 718 experience. This factor should be verified experimentally using surface crack specimens of the same materials.

The crack propagation behavior and residual life for each individual test case were predicted using the provided baseline crack growth curve. Correlations of residual life predictions are presented in Fig. 10. The solid line in the figure represents a perfect correlation and the dotted lines represent a factor of two scatter bands. The correlations for the PMMA specimen are shown to be excellent. However, the predictions for the 7075-T651 specimens fall consistently on the conservative side. In fact, the majority of the round-robin par-

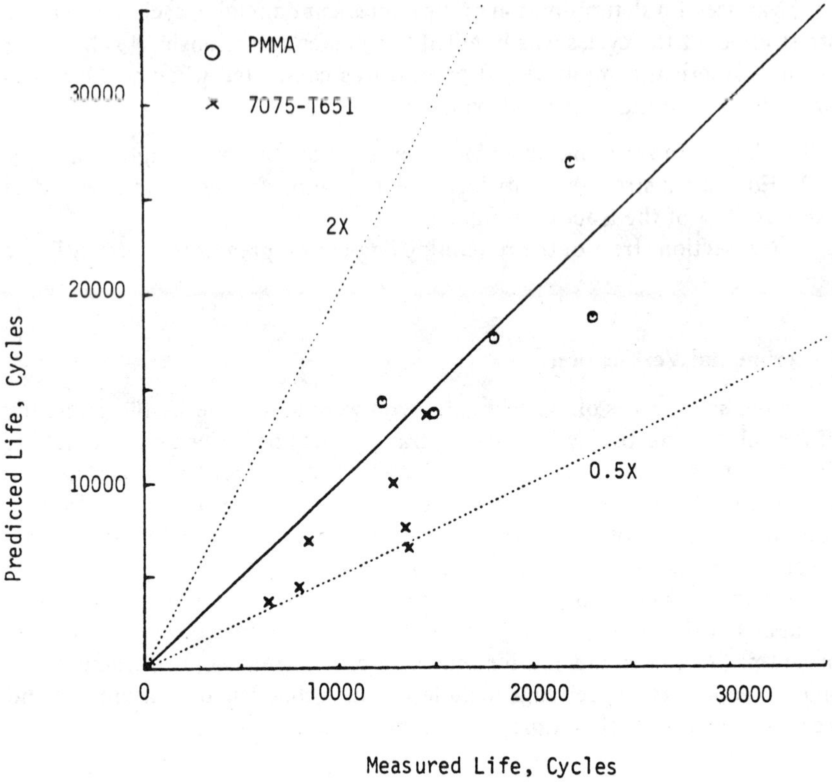

FIG. 10—*Predicted versus measured cycles for ASTM round-robin fatigue life prediction problems.*

ticipants experienced the same trend in their predictions of the 7075-T651 experiments.

For the PMMA tests, the stress intensity factors corresponding to the measured crack dimensions during each test have been computed and compared with the experimentally deduced stress intensity factors at both surface locations. The results are presented graphically in Figs. 11*a* and 11*b* for the bore of the hole and on the front surface locations, respectively. The dotted lines represent the 17% scatter bands which correspond to the scatter bands in the baseline data. The correlations are shown to be excellent at both locations.

The predicted and measured crack aspect ratio changes during each PMMA experiment have also been compared and reported elsewhere [*20*].

Air Force Supported Verification Test Programs

A series of subcomponent specimen tests has been performed in the past few years by General Electric in support of several Air Force funded engine

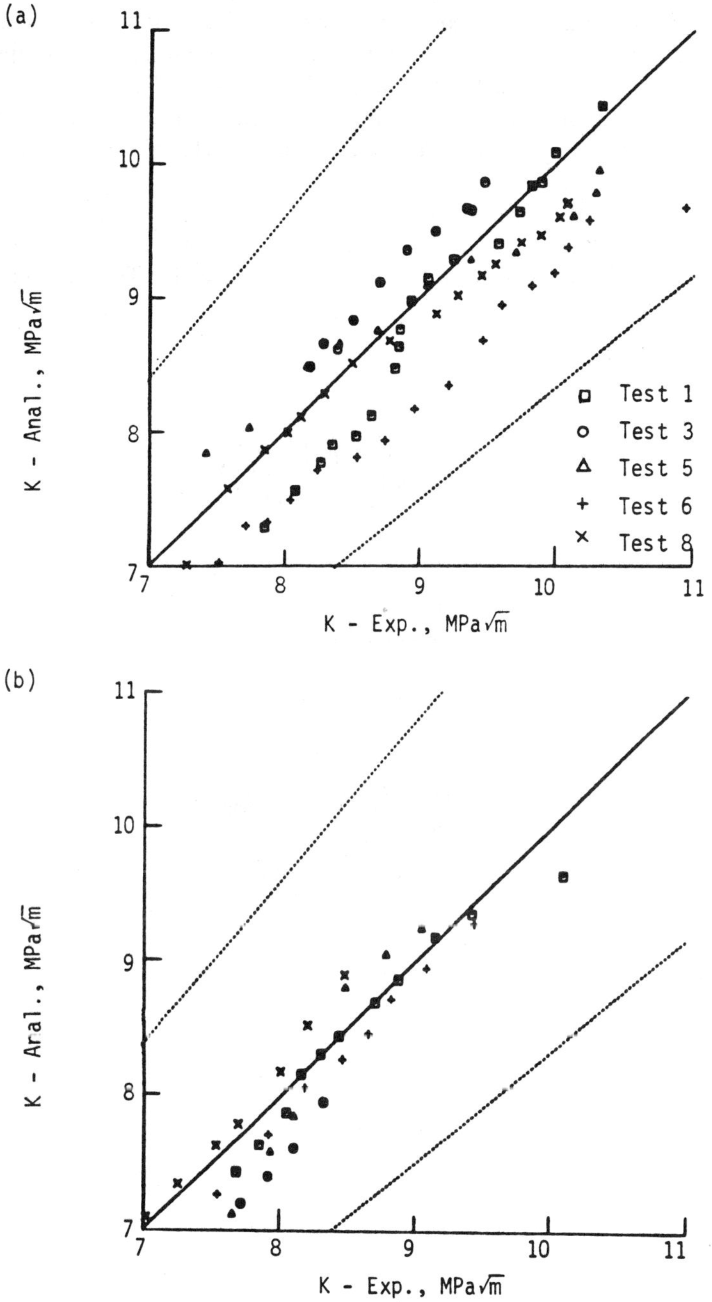

FIG. 11—*Predicted versus measured stress intensity factors for the PMMA specimens.* (a) *Bore.* (b) *Surface.*

programs. The effort was to assess validity of current applied fracture mechanics computer codes by comparing the test observations and analytical predictions of various notched specimens which simulated engine component geometries and were subjected to low cycle fatigue cycles. The materials of the specimens were typical alloys used in engine components. The test cycle profiles range from simple cycles to more complex cycle mixes and hold time effects.

In verification, the test data for the smooth bars subjected to complex cycle mix were first used to examine the adequacy of the approach in handling cycle mix problems. Typical cycle profiles are shown in Fig. 12. The effects of hold

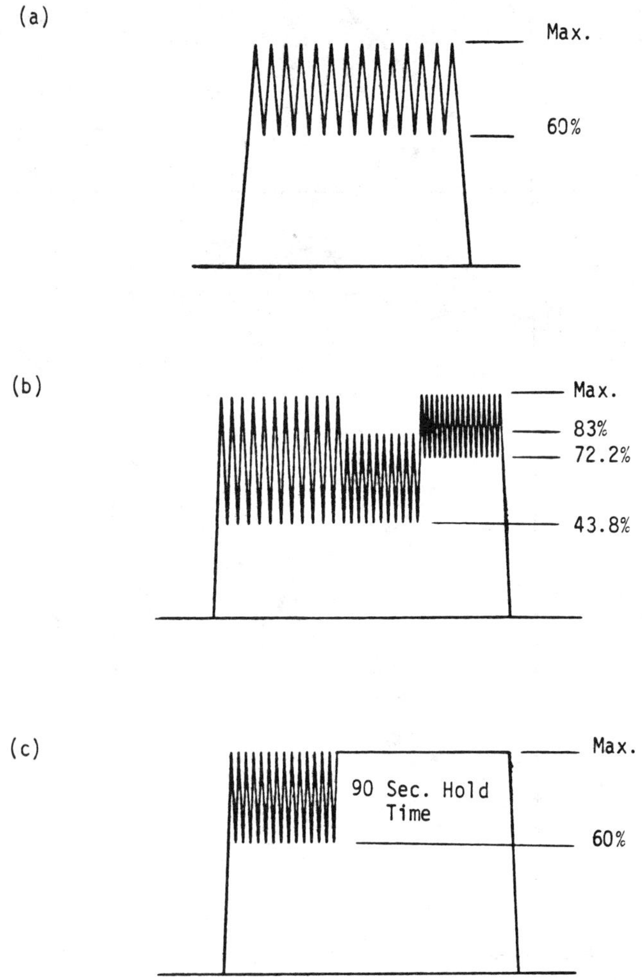

FIG. 12—*Sample test cycle profiles from one verification program.*

time and retardation were not modelled in the current analyses, since accurate and self-consistent methods for accounting for these effects have not yet been identified and the magnitude of these effects were found to be small for the studied range of experimental conditions. A good correlation of predicted versus observed lives for both Inconel 718 and Ti-6Al-4V specimens is shown in Fig. 13.

The notched specimens include double edge notch, fillet, and hole specimens (Fig. 14). A total of 53 specimens were tested. An electric discharge machining (EDM) starter notched was used to initiate a crack at a desired location relative to the geometric discontinuity. The details of specimen dimensions, crack location, cycle profile, stress level, etc., for each individual test case can be found in Ref *16*.

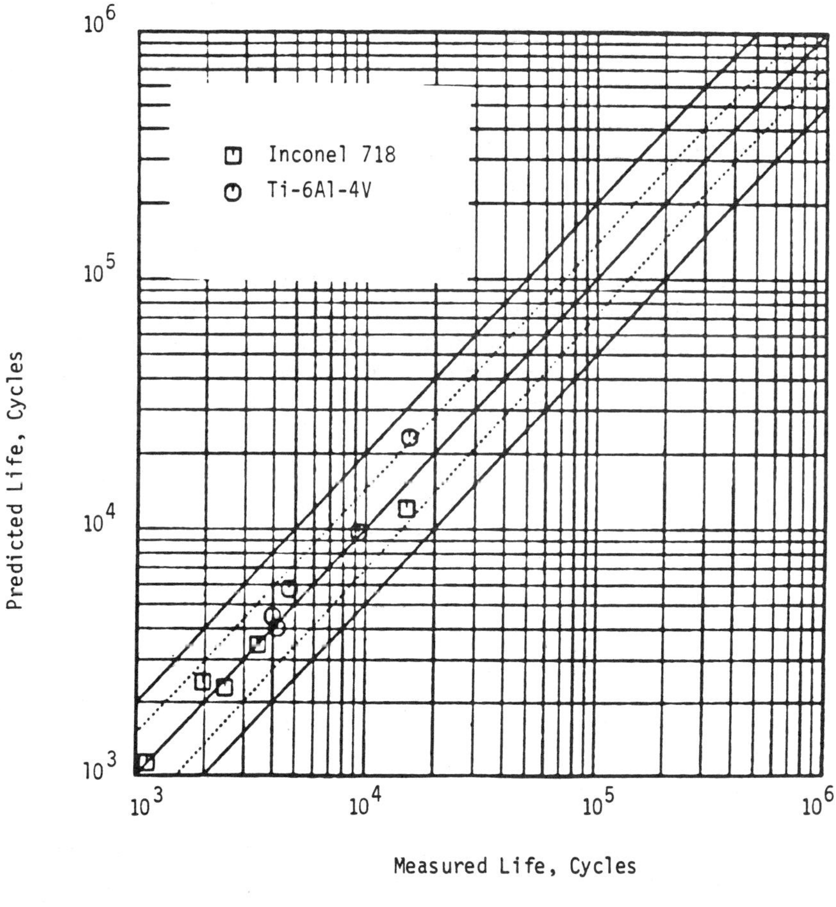

FIG. 13—*Predicted versus measured residual lives for Inconel 718 smooth specimens under complex cycle load.*

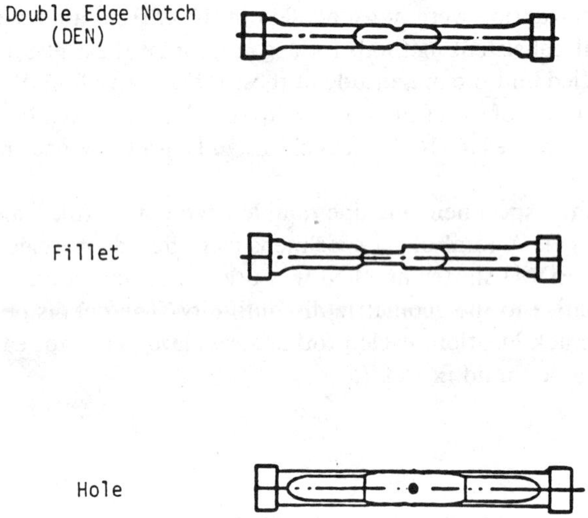

FIG. 14—*Sample configurations of complex specimens used in the verification programs.*

A comparison of the predicted and measured residual lives is shown in Fig. 15. As shown, the greatest majority of data points are within a factor of 1.5, which is very good for the residual life predictions.

Figure 16 shows the comparison of the measured/predicted residual life ratios versus the normalized maximum notch root stresses. The result of uniformly distributed data points demonstrates that the approach is independent of stress level, although the influence of localized plasticity and the potential limitations on linear elastic fracture mechanics need to be further evaluated. Figure 17 examines these life ratios with respect to different specimen types.

The statistical plot of the measured/predicted life ratios of the subcomponent tests in Fig. 18 shows good agreement with the baseline curve. The overall prediction for notched specimens appears to be slightly conservative (about 11%) compared with the baseline results, although the difference may not be statistically significance.

Conclusions

A stress intensity factor solution for a surface crack in a finite solid subjected to an arbitrary stress field has been presented. The solution was developed based on the superposition principle, the weight function technique, and a referenced finite body correction factor for uniform stresses. An additional factor is required to modify the solution at surface locations. The surface modification factor is to be determined through the reduction of data from tests on surface flaw specimens.

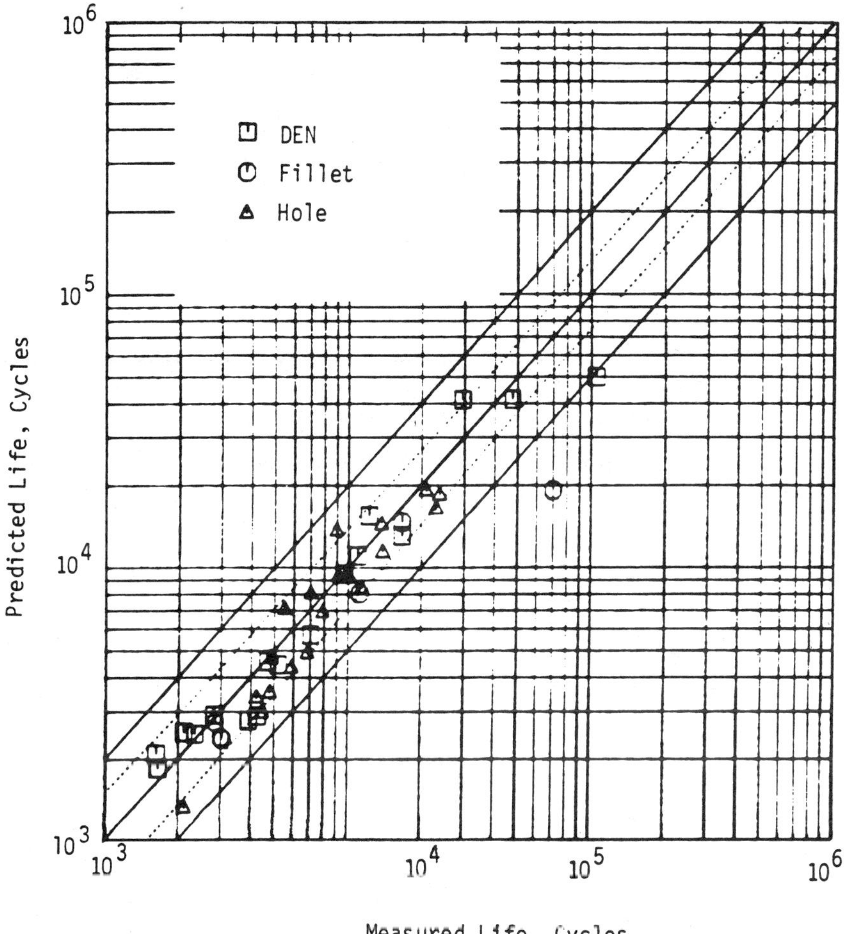

FIG. 15—*Predicted versus measured residual lives for Inconel 718 notched specimens under complex cycle load.*

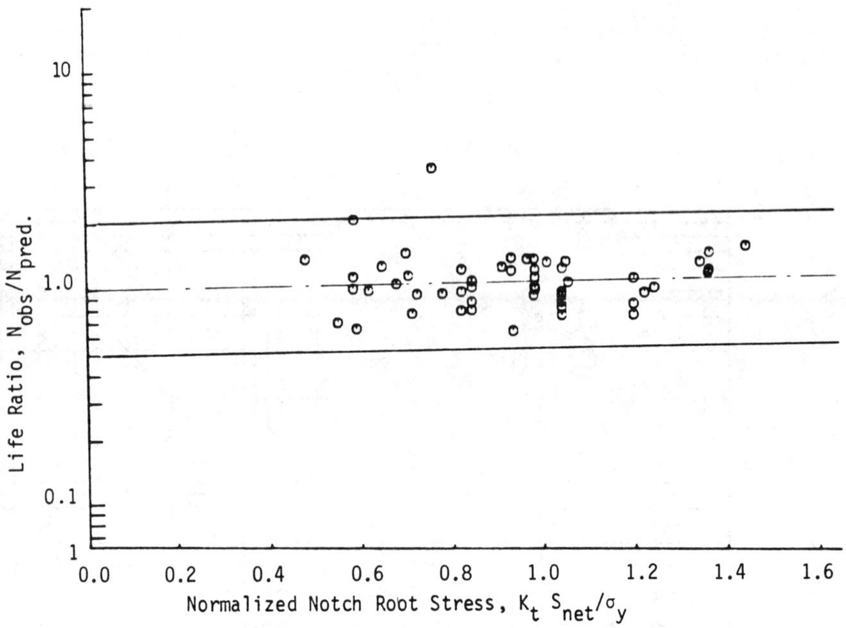

FIG. 16—*Measured/predicted residual life ratios versus normalized maximum notched root stresses of notched specimens.*

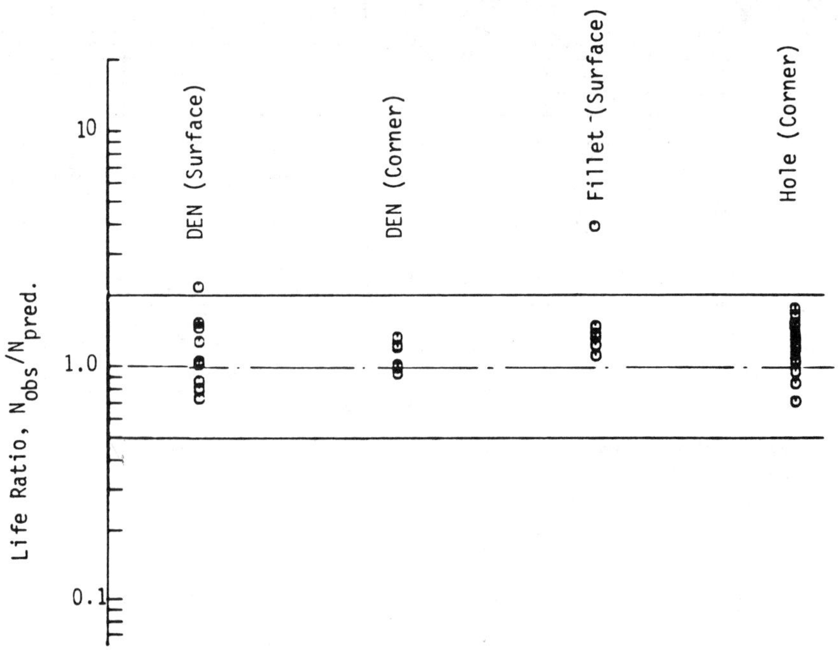

FIG. 17—*Comparison of measured/predicted residual life ratios for various notched specimen types.*

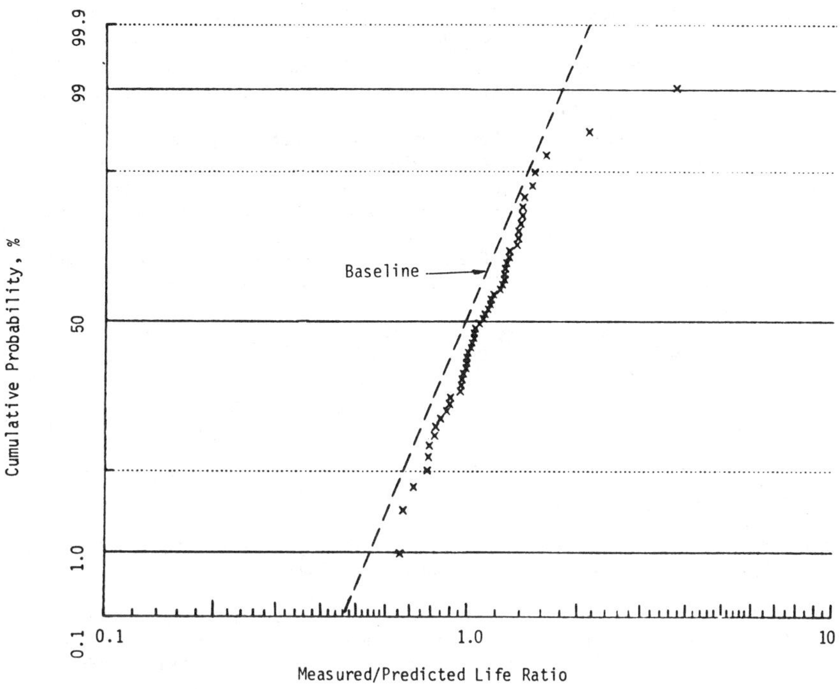

FIG. 18—*Normal distribution plot of measured/predicted residual life ratios for notched specimens.*

Numerical comparison for several non-uniform stress problems showed that the proposed solution agrees well with published solutions. Extensive experimental verification with test data from various sources has shown that the solution is accurate in predicting the crack propagation behavior and residual life of notched subcomponent specimens subjected to complex loadings.

Acknowledgments

The author acknowledges the Air Force at Wright-Patterson for their support of verification test programs (F33657-81-C-0007, F33657-81-2006) and their permission to publish part of the results. The author also expresses his appreciation to Drs. J. H. Laflen and P. A. Domas for their encouragement and counseling during the course of this work, and to Mr. O. C. Gooden and Ms. K. M. Helton for their assistance in preparing this manuscript.

References

[*1*] Newman, J. C., "A Review and Assessment of the Stress Intensity Factors for Surface Cracks," in *Part-Through Crack Fatigue Life Prediction, ASTM STP 687*, J. B. Chang, Ed., American Society for Testing and Materials, Philadelphia, 1979, pp. 16–42.

[*2*] Bueckner, H. F., "Weight Functions for the Notched Bar," *Zeitschrift für Angewandte Mathemetik and Mechanik*, Vol. 51, 1971, pp. 97–109.

[*3*] Newman, J. C., Jr., and Raju, I. S., "Analysis of Surface Cracks in Finite Plates under Tension or Bending Loads," NASA TP 1578, National Aeronautics and Space Administration, Washington, D.C., 1979.

[*4*] Irwin, G. R., "Crack Extension Force For a Part Through Crack in a Plate," *Journal of Applied Mechanics*, Vol. 29, No. 4, Dec. 1962, pp. 651–654.

[*5*] Grant, A. F., Jr., and Kullgren, T. E., "Stress Intensity Factors for Corner Cracked Holes under General Loading Conditions," *Journal of Engineering Materials and Technology*, Vol. 103, 1981, pp. 171–176.

[*6*] Heliot, J., Labbens, R. C., and Pellissier-Tanon, A., "Calculation of Stress Intensity Factors by the Boundary Integral Equations Method," presented at Eleventh National ASTM Symposium on Fracture Mechanics, Blacksburg, Va., June 1978.

[*7*] Irwin, G. R., "Plastic Zone Near a Crack and Fracture Toughness," in *Proceedings*, 7th Sagamore Conference, Syracuse University Research Institute, 1960, p. IV–63.

[*8*] Shah, R. C. and Kobayashi, A. S., "Stress Intensity Factor for an Elliptical Crack Approaching the Surface of a Plate in Bending," in *Stress Analysis and Growth of Cracks, ASTM STP 513*, American Society for Testing and Materials, Philadelphia, 1972, pp. 3–21.

[*9*] Newman, J. C., Jr., and Raju, I. S., "Stress Intensity Factor Equations for Cracks in Three-Dimensional Finite Body," NASA TM Report 83200, National Aeronautics and Space Administration, Washington, D.C., Aug. 1981.

[*10*] Bowie, O. L., "Analysis of an Infinite Plate Containing Radial Cracks Originating at the Boundary of an Internal Circular Hole," *Journal of Mathematics and Physics*, Vol. 35, No. 1, Apr., 1956, pp. 60–71.

[*11*] Walker, K., "The Effect of Stress Ratio During Crack Propagation and Fatigue for 2024-T3 and 7075-T6 Aluminum," in *Effects of Environment and Complex Load History on Fatigue Life, ASTM STP 462*, American Society for Testing and Materials, Philadelphia, 1970, pp. 1–14.

[*12*] Isida, M., "Stress-Intensity Factors for the Tension of an Eccentrically Cracked Strip," *Journal of Applied Mechanics, Transactions of ASME*, Vol. 33, 1965, p. 674.

[*13*] Matsuishi, M. and Endo, T., "Fatigue of Metals Subjected to Varying Stress," presented at Japan Society of Mechanical Engineers, Fukuoka, Japan, March 1968.

[*14*] Minutes of ASTM meeting, Subcommittee E24.06 on Fracture Mechanics Applications, 28 April 1982.

[*15*] Heckel, J. B. and Rudd, J. L., "Assessment of Stress Intensity Factors for Corner Cracks at Holes," AFWAL-TM-82-201-FIBE, Aug. 1982.

[*16*] Domas, P. A., Yau, J. F., and Ramaswamy, V. G., "TF34 Damage Tolerance Assessment Program," General Electric Company TIS Report R82AEB272, June 1982.

[*17*] Ramaswamy, V. G., Yau, J. F., and Laflen, J. H., "F101-GE-102 ENSIP Fracture Mechanics Verification Program Results," General Electric Company TIS Report R84AEB427, May 1984.

[*18*] Grandt, A. F. and Snow, J. R., "A Stress Intensity Factor Calibration for Corner Flaws at an Open Hole," AFML-TR-282, Air Force Materials Laboratory, Wright-Patterson AFB, May 1975.

[*19*] James, L. A. and Anderson, W. E., "A Simple Experimental Procedure for Stress Intensity Calibration," *Engineering Journal of Fracture Mechanics*, Vol. 1, No. 3, April 1969, p. 565.

[*20*] Yau, J. F., "Assessment of the MISKRA/RELIES Solutions for Corner Cracks at Holes," General Electric Company Class I TIS Report R83AEB496, Aug. 1983.

H. M. Müller,[1] *S. Müller,*[2] *D. Munz,*[1,2] *and J. Neumann*[1]

Extension of Surface Cracks During Cyclic Loading

REFERENCE: Müller, H. M., Müller, S., Munz, D., and Neumann, J., **"Extension of Surface Cracks During Cyclic Loading,"** *Fracture Mechanics: Seventeenth Volume, ASTM STP 905,* J. H. Underwood, R. Chait, C. W. Smith, D. B. Wilhem, W. A. Andrews, and J. C. Newman, Eds., American Society for Testing and Materials, Philadelphia, 1986, pp. 625–643.

ABSTRACT: The extension of semielliptical surface cracks in plates under cyclic tension and cyclic bending loading has been investigated. The analytical results compared different assumptions, applying local and weighted averaged ΔK values and the effect of crack closure. The experimental results showed that the developing crack front can be described fairly accurately by semiellipses. Applying local ΔK values for the prediction underestimates the developing crack aspect ratio a/c. Applying the weighted averaged ΔK or a crack closure factor improves the prediction.

KEY WORDS: surface crack, crack growth rate, fracture mechanics

The fundamental problem in practical application of fracture mechanics is the transformation of results from test specimens to the behavior of components. Fracture mechanics specimens usually are plates with through-the-thickness cracks. Real cracks are mostly two-dimensional cracks, which very often can be described as semielliptical surface cracks. These cracks extend in depth and length under external loading. For safety considerations, the crack growth rate and the change in the shape of these cracks have to be predicted. The crack shape developing during fatigue loading is important, especially for leak-before-break evaluation. The bases of the transformation of results of test specimens with one-dimensional cracks to the two-dimensional surface cracks are the experimentally obtained relation between the crack growth rates for one load cycle (da/dN) and the range of the stress intensity factor (ΔK), on the one hand, and the varying stress intensity factor along the crack front of the two-dimensional cracks, on the other hand.

[1]University of Karlsruhe, Karlsruhe, West Germany.
[2]Karlsruhe Nuclear Research Center, Karlsruhe, West Germany.

Several experimental results have been published on the change of geometry of semielliptical surface cracks for plates under tension and bending loads. Results obtained up to 1983 have been summarized in Ref *1*; some further results have been published since then [*2–6*].

A comparison between experimental results and predicted behavior showed good qualitative agreement. In some cases there was quantitative agreement, in other cases there was a more or less pronounced deviation from the prediction [*1*]. Different possible reasons for this, such as anisotropy, inhomogeneity of the material, residual stresses, and different stress state along the crack front, are mentioned in Ref *1*. According to Hodulak et al [*7*] the different stress state should affect the material resistance, according to Jolles and Tortoriello [*5,6*] it should affect crack closure.

In this paper predictions of the change of the shape of semielliptical surface cracks based on different approaches are compared with experimental results obtained for three materials.

Prediction of the Rate of Crack Growth in Depth and Length

Assumptions

The stress intensity factor of semielliptical surface cracks of depth a and length $2c$ in a plate of thickness t (Fig. 1) usually is written in the form

$$K = \frac{\sigma\sqrt{\pi a}}{\phi} F(a/c, a/t, \varphi) = K\sqrt{a}\, Y \tag{1}$$

where σ is a characteristic stress (e.g., surface stress in bending) and ϕ can be expressed as

$$\phi = [1 + 1.464(a/c)^{1.65}]^{1/2} \tag{2}$$

F is a function of the crack aspect ratio a/c, the relative crack depth a/t, and the location along the crack front, expressed by the angle φ and also depending on the type of loading. The varying stress intensity factor along the crack

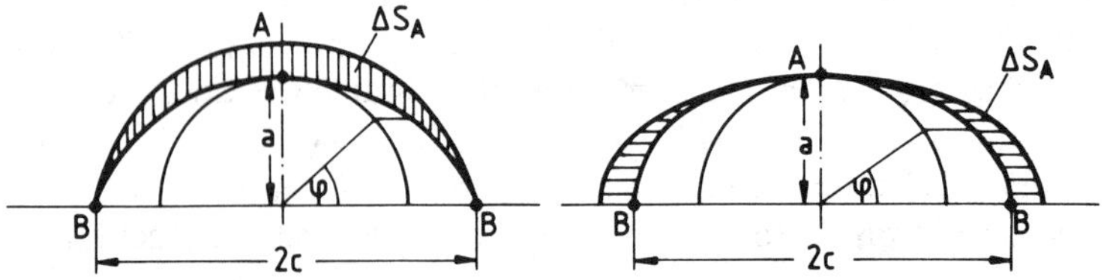

FIG. 1—*Incremental change in crack area for the calculation of averaged* K_A *and* K_B.

front has to be taken into account in predicting the change in shape of the crack front. The simplest assumptions made for this prediction are as follows:

1. The crack front can be described always as a semiellipse.
2. The material is homogeneous in a macroscopic sense.
3. No effect is exerted by the stress state (plane stress/plane strain).
4. The crack extension is calculated for the deepest point A and the surface points B only. Starting from an initial crack with the axes a_i and c_i, the crack extension Δa and Δc are calculated for the first load cycle, applying a material specific crack growth relation. Then a new $\Delta K(a/c, a/t, \varphi)$ is calculated with $a = a_i + \Delta a$ and $c = c_i + \Delta c$ and the extensions Δa and Δc are obtained for the next cycle. This procedure is repeated until wall penetration takes place.
5. The stress intensity factor responsible for crack extension are the local ΔK_A and ΔK_B at A and B, respectively.

Two modifications of this procedure are suggested in the literature. Cruse and Besuner [*8*] used, instead of local K values, averaged K values taking into account the changing K near Points A and B. These averaged values are given by (Fig. 1)

$$\bar{K}_A^2 = \frac{1}{\Delta S_A} \int_{\Delta S_A} K^2(\varphi) dS_A \tag{3a}$$

$$\bar{K}_B^2 = \frac{1}{\Delta S_B} \int_{\Delta B} K^2(\varphi) dS_B \tag{3b}$$

The incremental changes in crack area dS_A or dS_B are associated with crack growth in the a- or c-direction only.

Newman and Raju [*9*] introduced an empirical factor, reducing ΔK_B:

$$\Delta K_{B,\text{eff}} = 0.9\ \Delta K_B \tag{4}$$

This factor of 0.9 was introduced to verify experimental results by Corn [*10*], which showed that small semicircular cracks remain semicircular. Jolles and Tortoriello [*5*] interpreted this factor as differences in the crack closure behavior at the points A and B, introducing

$$\Delta K_{A,\text{eff}} = U_A \cdot \Delta K_A; \qquad \Delta K_{B,\text{eff}} = U_B \cdot \Delta K_B \tag{5}$$

Measurements of U_A and U_B have been performed by Fleck et al [*4*] with the results of $U_A = 0.85$ and $U_B = 0.75$. Thus the relative effects $U_B/U_A = 0.88$ is nearly identical to the empirical factor introduced by Newman and Raju. A general application of this factor, however, implies that U_B/U_A is independent of crack shape and crack size.

The calculations are performed for a power law crack growth relation:

$$da/dN,\ dc/dN = C(\Delta K)^n \tag{6}$$

Applying Eq 6, it can be seen that the relative change in crack length in the a- and c-directions is given by

$$\frac{da}{dc} = \left(\frac{Y_A}{Y_B}\right)^n \tag{7}$$

and thus is independent of the constant C and the amplitude of the applied stress.

Alternating Tension and Bending Loads

For tension and bending loads of plates Newman and Raju [9] published a closed-form solution for the function F in Eq 1 for $0 < a/c < 1$ and $0 < a/t < 0.8$, which can be extrapolated to $0.8 < a/t < 1.0$. The calculated change in crack geometry is presented as a/c-a/t curves (Figs. 2*a* and 2*b* and 3*a* and 3*b*). Figures 2*a* and 2*b* show the results for two different exponents n. For bending loading, a final crack aspect ratio a/c in the range of 0.1 to 0.2 is reached at wall penetration. Cracks with small initial values of a/c first grow in the direction of a semicircular crack, then the trend is reversed and a/c decreases.

For tension loading, a final a/c in the range of 0.7 to 0.8 is reached at wall penetration with the exception of long and deep cracks (a/c small, a/t large), which do not have a chance of reaching this limiting value before penetrating the plate.

The effect of n can also be seen from Figs. 2*a* and 2*b*. A greater n leads to a larger change in a/c. The difference between the use of local and averaged ΔK values and the effect of a crack closure factor of 0.9 is shown in Figs. 3*a* and 3*b*. Applying weighted averaged values results in less change of a/c than for local ΔK values. That means if a/c is increasing, it increases less, and if a/c is decreasing, it decreases less, than for the calculations with local values of ΔK. The crack closure factor results in less crack growth at the surface. The expected a/c ratio at wall penetration therefore is larger compared with that without applying this factor. This effect, however, is small for bending loading (Fig. 3*b*).

Experimental Results

Experimental Procedure

Three materials were tested: an aluminum alloy with the German designation AlZnMgCu 1.5, which is similar to 7475; a reactor-grade steel 20

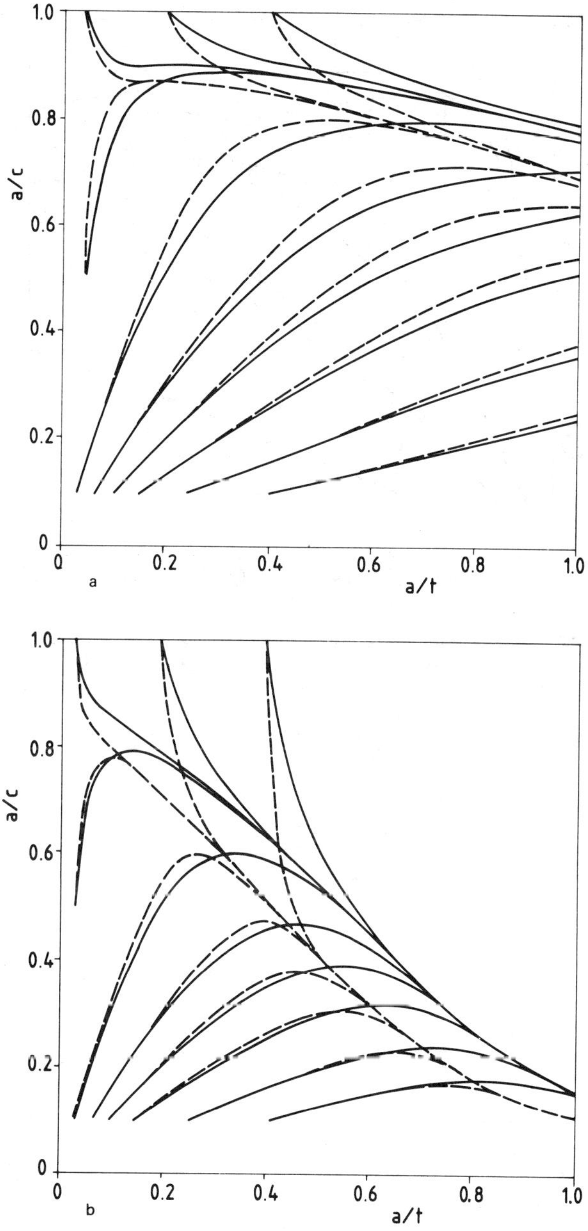

FIG. 2—*Change in crack geometry for* n = 2 (—) *and* n = 4 (---) *calculated with local* ΔK. (a) *Cyclic tension.* (b) *Cyclic bending.*

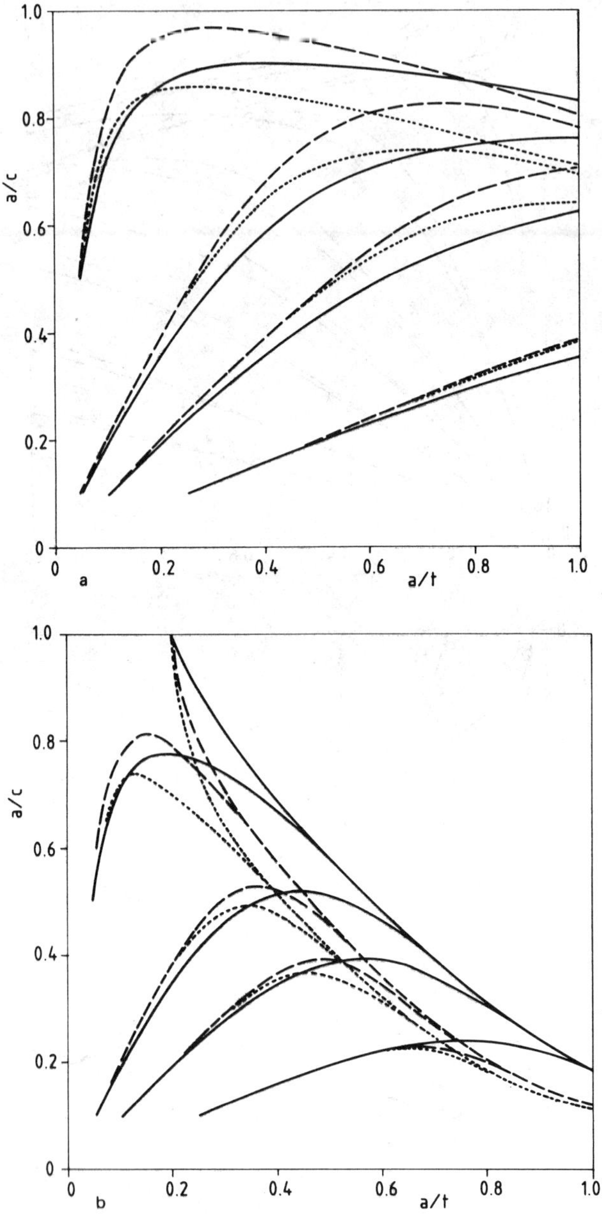

FIG. 3—*Change in crack geometry for* n = *3.726, calculated with averaged* (—) *and local* ΔK *with* (---) *and without* (...) *crack closure factor.* (a) *Cyclic tension.* (b) *Cyclic bending.*

MnMoNi 55, which is similar to ASTM A533B; and an austenitic stainless steel X6CrNi 1811, which is similar to AISI 304. The da/dN-ΔK curves were obtained with compact tension specimens per ASTM E 399 with a thickness $B = 25$ mm and a width $W = 50$ mm.

For the surface crack experiments, plates of a thickness $t = 20$ mm and a width $W = 80$ mm were used. The length of the plates was $L = 400$ mm for tension loading and $L = 180$ mm for bending loading. For tensile loading the specimens were fixed with hydraulic grips and the alignment was carefully controlled. The bending tests were performed under four-point loading. The outer span was 170 mm and the inner span 40 mm. The test frequency was 5 cycles/s. The R-ratio was 0.1 unless specified otherwise. Small semielliptical precracks were introduced by the electro-spark method. Two methods were applied to monitor the crack length. Beach marks were introduced by decreasing the stress amplitude with fixed upper stress level for about 1000 cycles. The amount of crack extension in depth and length could then be measured on the fracture surface. Some fracture surfaces are shown in Figs. 4*a* to 4*c*. In addition, the electrical potential method was applied. A direct current of 20 A was supplied and the potential drop measured between two points across the crack. The change in the potential drop is dependent on the shape of the developing crack. Therefore for each shape of an incipient crack a calibration curve has to be obtained using the beach marks. Figures 5*a* and 5*b* show examples. The potential drop method then can be used in subsequent tests with the same shape of the incipient crack without introducing beach marks.

Crack Growth Rate

The crack growth rate da/dN versus ΔK for compact specimens in the c-direction is shown in Figs. 6*a* to 6*c*. Within the measured range of ΔK the da/dN-ΔK relation can be described by a power law with the parameters C and n given in Table 1.

For the austenitic steel, crack growth rate measurements were also performed in the a-direction using four-point bend specimens. The crack growth rate was larger by a factor of about 3 (Fig. 6*c*).

Figures 7*a* and 7*b* show the rate of crack growth in depth da/dN and in length dc/dN for the aluminum alloy for surface cracks under cyclic tension and cyclic bending. If local ΔK values are used (Fig. 7*a*), there is a tendency for the rate of crack growth in the a-direction to be larger than in the c-direction. If the crack growth rate is plotted versus the averaged value ΔK (Fig. 7*b*), all crack growth rates measured for bending and tension and in the a- and c-directions are within a common scatter band and in agreement with the curve obtained for the compact specimens. Applying the idea of different crack closure (Eq 4) instead of averaged ΔK would also shift the data points of the a-direction closer to those of the c-direction.

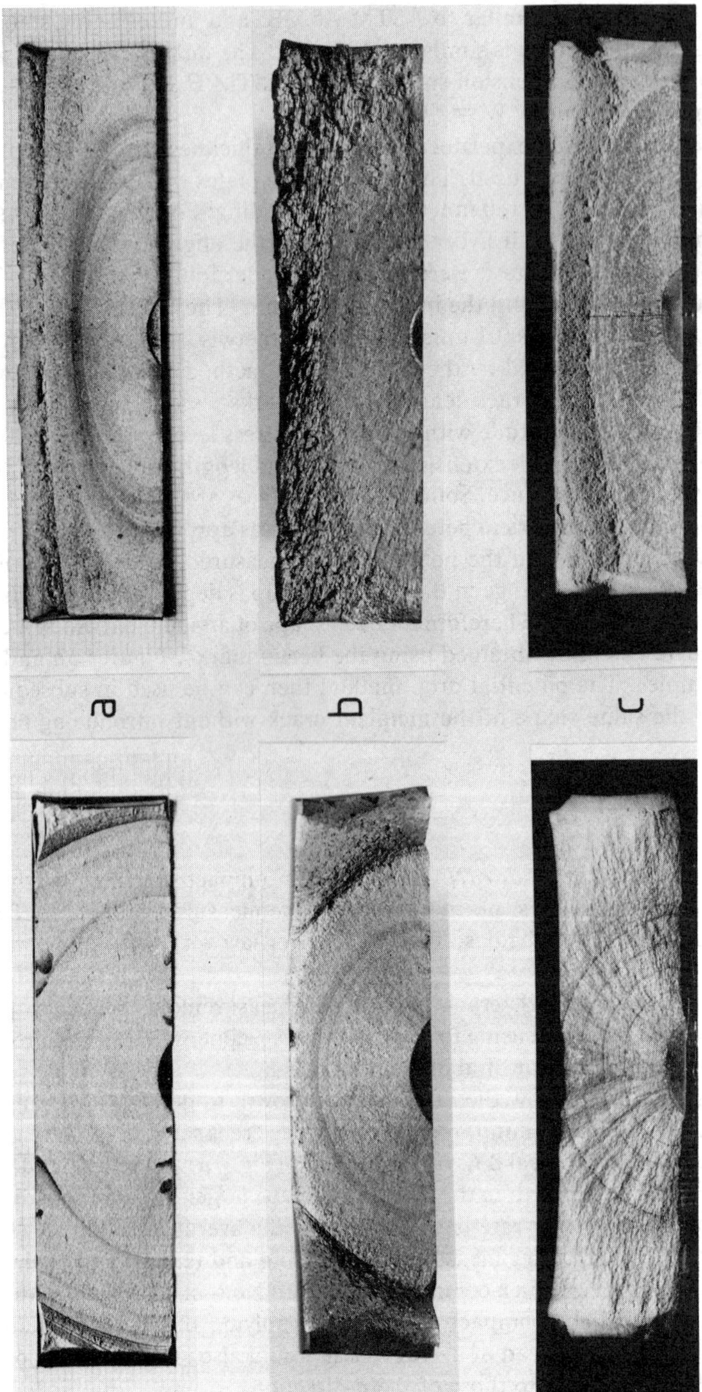

FIG. 4—*Fracture surfaces with beach marks for cyclic tension* (left) *and cyclic bending* (right). (a) *Austenitic steel.* (b) *Ferritic steel.* (c) *Al-alloy.*

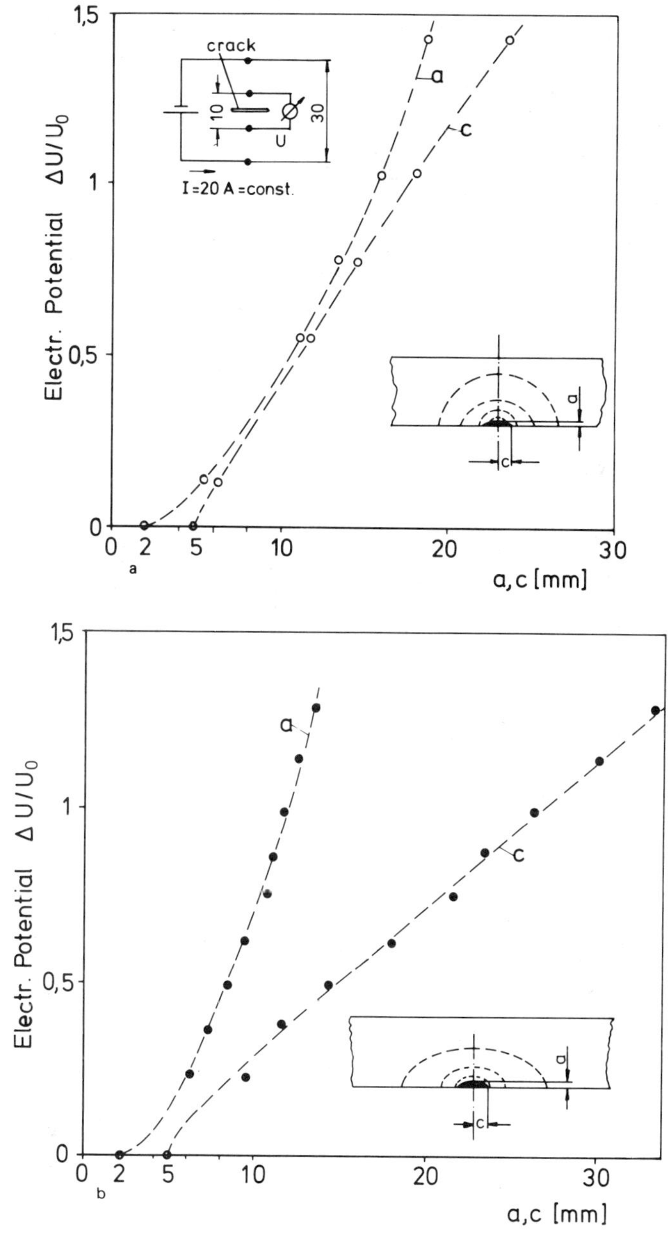

FIG. 5—*Electrical potential drop versus crack size.* (a) *Cyclic tension.* (b) *Cyclic bending.*

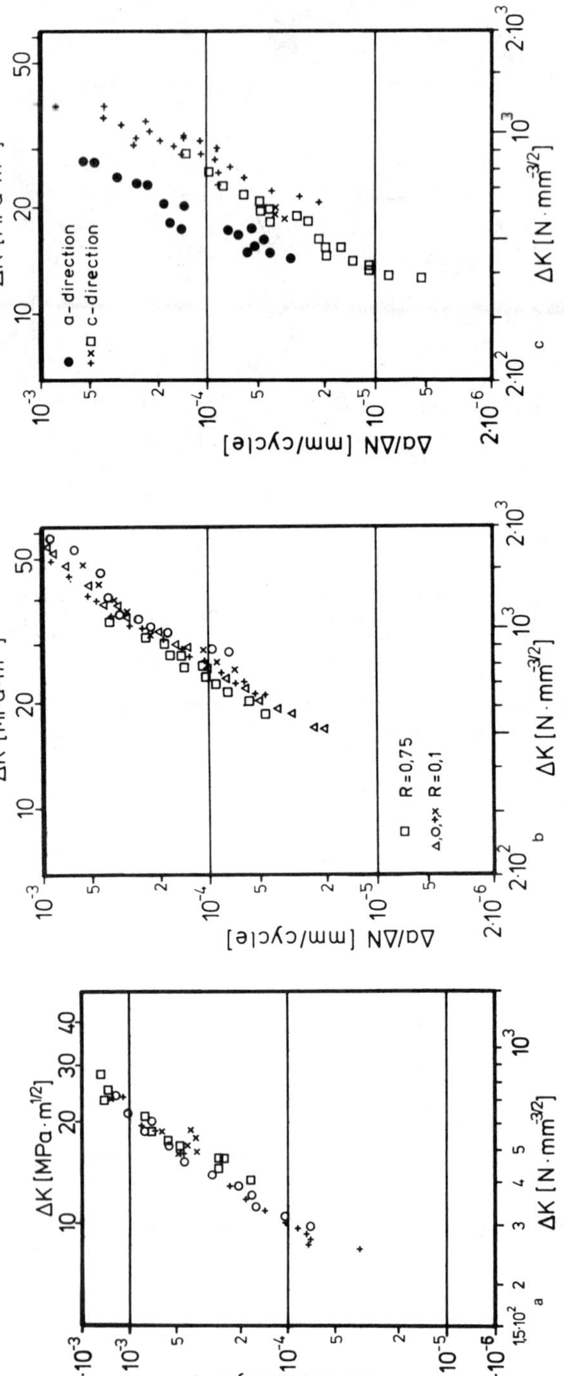

FIG. 6—da/dN-ΔK curves obtained with compact specimens (different symbols for different specimens). (a) *Al-alloy.* (b) *Ferritic steel.* (c) *Austenitic steel.*

TABLE 1—*Parameters of Paris relation.*

Material	n	C, N · mm	R	Crack Direction
AlZnMgCu 1.5	3	$3 \cdot 10^{-12}$	0.1	c
20 MnMoNi 55	3.3	$2 \cdot 5 \cdot 10^{-14}$	0.1	c
	3.3	$3 \cdot 5 \cdot 10^{-14}$	0.75	c
X6CrNi 1811	4	$3 \cdot 10^{-16}$	0.1	c
	4	$1 \cdot 10^{-15}$	0.1	a

The results for the austenitic stainless steel are shown in Fig. 8. There is a tendency for larger crack extension in the a- than in the c-direction; this may be the result of the anisotropic crack growth behavior. Applying the crack closure factor or the averaged ΔK values would also approach the data points for the a- and c-directions.

Change in Crack Shape

A prerequisite of the prediction of the change in the crack shape was that the crack contour could be described as a semiellipse. Figures 9*a* and 9*b* show as an example observed crack shapes and the adapted semiellipses. There are some deviations from the ideal semiellipse. The effect on the calculated ΔK, however, should be small. The changes in crack shape are plotted as a/c-a/t diagrams in Figs. 10 to 12. The diagrams also include the predictions for the specific n of the material using local and averaged ΔK and local ΔK with the "crack closure factor" of 0.9. A comparison of experimentally observed results with the predictions leads to the following conclusions:

1. The experimentally observed aspect ratios are generally in agreement with the predictions.
2. In agreement with the predictions, the crack aspect ratio at penetration of the plate was 0.8 to 0.9 for tension loading.
3. For bending loading the tests could not be continued until penetration of the plate. However, the aspect ratio followed the predicted curve, suggesting an a/c ratio of about 0.2 for $a = t$.
4. There is a tendency for the predictions applying local ΔK without the crack closure factor to underestimate the developing a/c ratio. A better, but not complete, agreement is obtained by applying the averaged ΔK or the crack closure factor.
5. For the austenitic stainless steel the increase of a/c in the range $0.2 < a/t < 0.5$ is larger than predicted.

The last effect can be explained by the anisotropic behavior of the steel plate (Fig. 6*c*). To reveal the effect of different crack growth rates, a/c-a/t

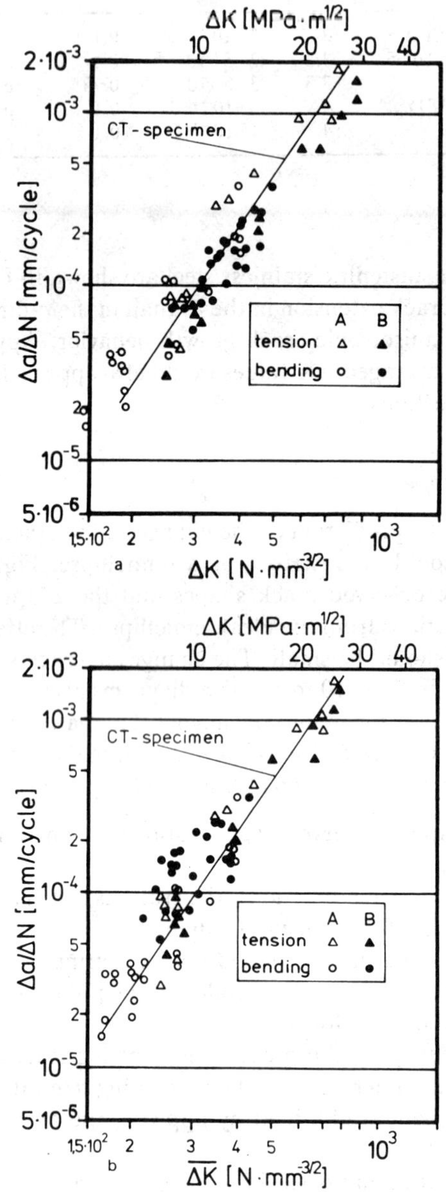

FIG. 7—*Al-alloy crack growth rates for surface cracks.* (a) *Local* ΔK. (b) *Averaged* ΔK.

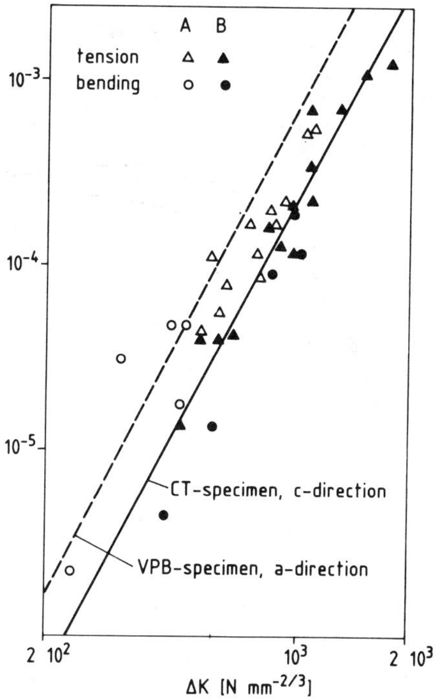

FIG. 8—*Austenitic steel crack growth rates for surface cracks.*

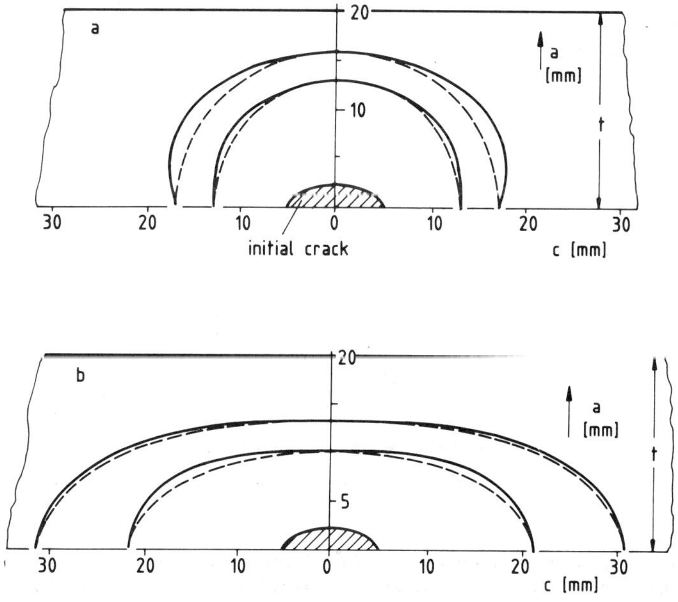

FIG. 9—*Observed crack-shapes* (solid line) *and adapted semiellipses* (dashed line). (a) *Cyclic tension.* (b) *Cyclic bending.*

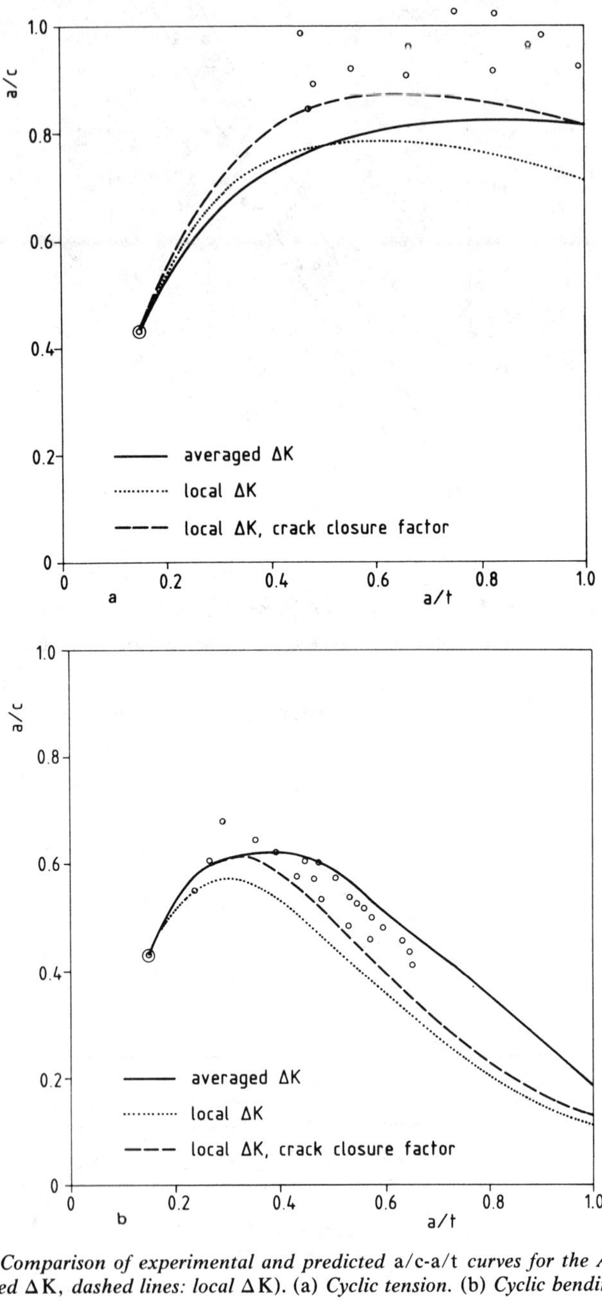

FIG. 10—*Comparison of experimental and predicted* a/c-a/t *curves for the Al-alloy* (*solid lines: averaged* ΔK, *dashed lines: local* ΔK). (a) *Cyclic tension.* (b) *Cyclic bending.*

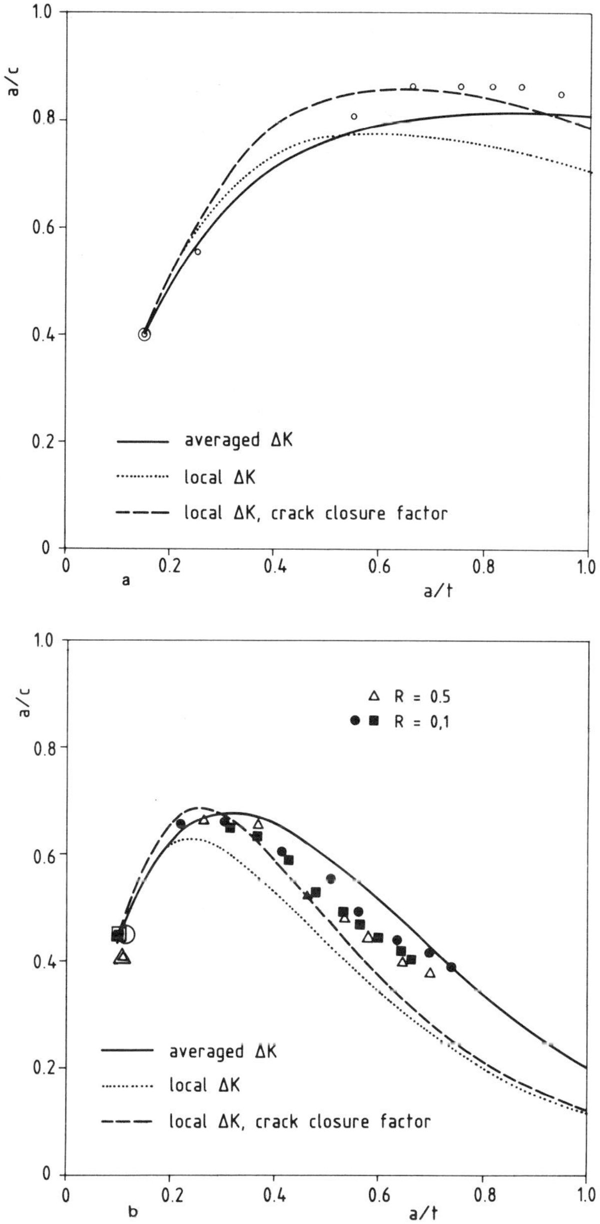

FIG. 11—*Comparison of experimental and predicted* a/c-a/t *curves for the ferritic steel (solid lines: averaged* ΔK, *dashed lines: local* ΔK). (a) *Cyclic tension.* (b) *Cyclic bending.*

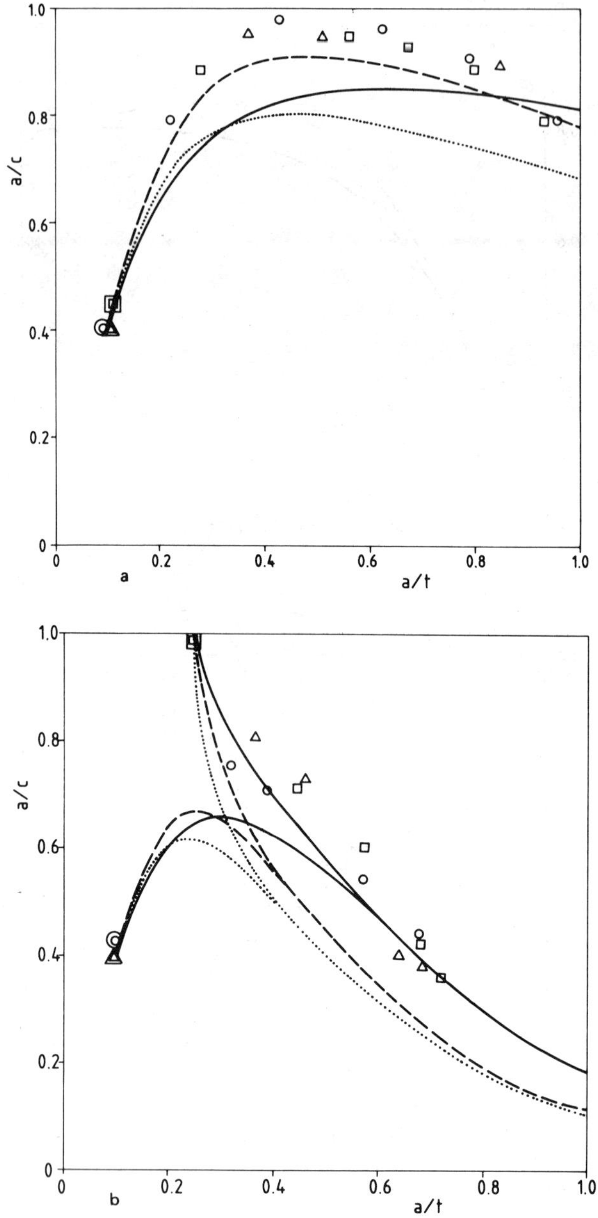

FIG. 12—*Comparison of experimental and predicted* a/c-a/t *curves for the austenitic steel (solid lines: averaged* ΔK, *dashed lines: local* ΔK). (a) *Cyclic tension.* (b) *Cyclic bending.*

Summary and Conclusions

The experimental and analytical results can be summarized as follows:

1. Experimental results for three materials showed that the crack developing under cyclic tension or cyclic bending loading can be described fairly accurately by semiellipses for $0.1 < R < 0.8$.
2. Crack growth rates in depth and in length directions are generally in agreement with results from compact specimens if plotted versus local ΔK. Plotting versus averaged ΔK or applying a crack closure factor may improve the agreement.
3. The developing crack shape generally is in agreement with the predictions. The application of local ΔK underestimates the developing a/c ratio somewhat. Using averaged ΔK or a crack closure factor improves the prediction without obtaining a complete agreement. Because of the combination of possible effects of anisotropy, crack closure, and scatter due to inhomogeneity of the materials, it is not possible to make an exact prediction of the developing crack shape. Without exact crack closure measurements it is also not possible to decide whether the predictions using local ΔK values can be improved applying weighted averaged ΔK values.
4. No systematic effect of mean stress (R-ratio) on the developing crack shape was observed.

References

[1] Görner, F., Mattheck, C., and Munz, D., *Zeitschrift für Werkstofftech.*, Vol. 14, 1983, pp. 11-18.

[2] Truchon, M. and Lieurade, H.-P. in *Advances in Fracture Research (Fracture '81)*, Vol. 1, Pergamon Press, Oxford, 1982, pp. 33-39.

[3] Smith, I. F. and Smith, R. A., *Fatigue of Engineering Materials and Structures*, Vol. 5, 1982, pp. 151-165.

[4] Fleck, N. A., Smith, I. F. C., and Smith, R. A., *Fatigue of Engineering Materials and Structures*, Vol. 6, 1983, pp. 225-239.

[5] Jolles, M. and Tortoriello, V. in *Fracture Mechanics: Fourteenth Symposium—Volume I: Theory and Analysis, ASTM STP 791*, American Society for Testing and Materials, Philadelphia, 1983, pp. I-297-I-307.

[6] Jolles, M. and Tortoriello, V. in *Fracture Mechanics: Fifteenth Symposium, ASTM STP 833*, American Society for Testing and Materials, Philadelphia, 1984, pp. 300-311.

[7] Hodulak, L., Kordish, H., Kunzelmann, S., and Sommer E. in *Fracture Mechanics, ASTM STP 677*, American Society for Testing and Materials, Philadelphia, 1979, pp. 399-410.

[8] Cruse, T. A. and Besuner, P. M., *Journal of Aircraft*, Vol. 12, 1975, pp. 369-375.

[9] Newman, J. C. and Raju, I. S., *Engineering Fracture Mechanics*, Vol. 15, 1981, pp. 185-192.

[10] Corn, D. L., *Engineering Fracture Mechanics*, Vol. 3, 1971, pp. 45-52.

Walter G. Reuter[1]

Comparison of Predicted Versus Experimental Stress for Initiation of Crack Growth in Specimens Containing Surface Cracks

REFERENCE: Reuter, W. G., "**Comparison of Predicted Versus Experimental Stress for Initiation of Crack Growth in Specimens Containing Surface Cracks,**" *Fracture Mechanics: Seventeenth Volume, ASTM STP 905*, J. H. Underwood, R. Chait, C. W. Smith, D. P. Wilhem, W. A. Andrews, and J. C. Newman, Eds., American Society for Testing and Materials, Philadelphia, 1986, pp. 644-660.

ABSTRACT: Fracture mechanics concepts were used to evaluate the ability to predict the stress for crack initiation for specimens containing surface cracks. The comparison is made between the predicted stress (σ_{pred}) and the experimentally determined stress (σ_{exp}). The predictions were made for conditions ranging from elastic to plastic. The predictions were based on suggestions by P. Paris for modifying an equation developed for center-cracked panels. Acoustic emission techniques were used to detect the load corresponding to crack initiation.

KEY WORDS: fracture mechanics, surface flaws, crack initiation, J_{Ic}

Nomenclature

a	Depth of surface flaw
a'	Half-crack length of a through crack
Δa	Change of flaw depth
b'	Half-width of plate
b	Remaining ligament
$2c$	Length of surface flaw
cF^2	Parameter quantifying the severity of a surface flaw
E	Young's modulus
ϵ'	True strain minus elastic strain
ϵ	True strain

[1]Principal Engineer, EG&G Idaho, Inc., Idaho Falls, ID 83415.

$\bar{e}$ Flow strain
e_{eng} Engineering strain
ϵ_{ys} Yield strain, σ_{ys}/E
$\{F(\sigma/\sigma_{ys})\}$ Stress correction term where σ and σ_{ys} are based on true stress
G_{ln} A coefficient in stress correction term of Eq 1
J J-integral
J_{Ic} Value of J corresponding to initiation of crack growth satisfying plane-strain conditions
K Applied stress intensity factor
K_{Ic} Plane-strain fracture toughness
M Correction factor for front and back faces
n Strain hardening term in Eq 4
P Load per unit thickness
P_0 Load per unit thickness corresponding to the perfectly plastic limit
Q Defect geometry correction factor
R^2 Regression correlation coefficient
t Thickness
T Tearing modulus
W Specimen width
α' Slope of line in Eq 4
$\bar{\sigma}$ Flow stress $= (\sigma_{ys} + \sigma_{uts})/2$
σ Applied true stress
σ_{exp} Experimentally determined engineering stress for crack initiation
σ_{pred} Predicted engineering stress for crack initiation
σ_{ys} Yield strength
σ_{uts} Ultimate tensile strength
$\propto$ Proportional to
Ψ^* A coefficient in stress correction term of Eq 1

Introduction

Fracture mechanics concepts may be used to assure structural integrity by (1) providing material procurement criteria, including welding procedures and techniques, and (2) evaluating effects of a defect on structural integrity. The latter is the subject of this paper.

In evaluating the effects of a defect on structural integrity, the most accurate approach is to conduct tests of actual components containing natural defects loaded in a manner to simulate the structure. The flaw type and typical dimensions used in tests should also simulate, as much as possible, the actual situation expected in service. This approach is expensive and its usefulness is probably limited to the conditions simulated unless adequate stress analyses are conducted. An appropriately designed and tested specimen can provide the same assurance of structural adequacy. This is the intended role of specimens containing surface cracks, which more closely represent structural behavior than any other type of specimen normally used to develop fracture toughness data.

Figure 1 shows a schematic of fracture mechanics concept approaches used to predict structural integrity. For linear-elastic conditions the general approach for predicting conservative crack initiation conditions for a structure containing a surface flaw is to use plane strain fracture toughness (K_{Ic}), obtained from standard fracture toughness tests (ASTM E 399), and analytical procedures such as those provided in Ref *1*. There has been sufficient testing of surface flawed specimens to verify the general approach although the accuracy has not been quantified.

For conditions of increased ductility, both elastic-plastic (E/P, constrained plasticity) and plastic, the test techniques provided in ASTM E 813 may be used to experimentally determine J_{Ic}, but there are no generally accepted methods for predicting the behavior (crack initiation, tearing, and instability) of a surface crack. Experimental research is being conducted at the Idaho National Engineering Laboratory (INEL) into the behavior of surface cracked specimens in E/P and plastic conditions. The research provides the following: (1) a basis for empirical correlations for predicting structural behavior based on results obtained from specimens containing surface cracks, and (2) a basis

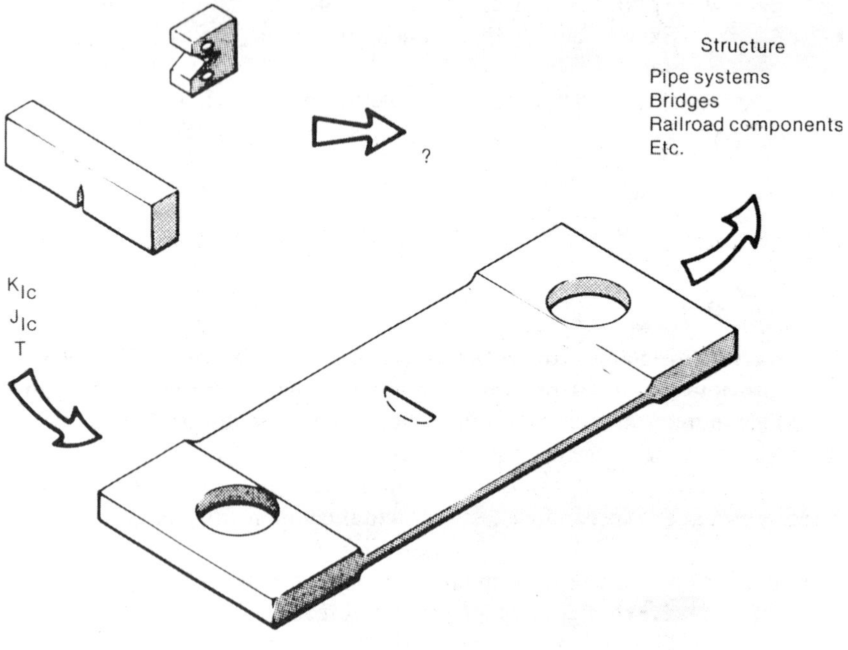

FIG. 1—*Schematic illustrating process for predicting structural integrity.*

for developing a correlation between results from standard fracture toughness specimens and surface cracked specimens.

This paper presents comparisons between stresses for crack initiation (σ_{pred}) predicted by standard techniques and experimentally determined stresses (σ_{exp}) obtained from specimens containing surface cracks.

Experimental Approach

The material tested was ASTM A-710 with the following chemistry (weight percent): 0.05 C, 0.74 Cr, 0.85 Ni, 0.47 Mn, 0.25 Si, 0.21 Mo, 0.010 P, 0.004 S, 1.20 Cu, and balance Fe. Tensile specimens were fabricated per ASTM E 8 with the long direction of the specimen parallel to the transverse direction of the plate and tested at temperatures ranging from 200 to 377 K. Compact specimens, fabricated and tested per ASTM E 399 and ASTM E 813, were used to measure K_{Ic} and J_{Ic} at 200, 255, and 297 K (see Table 1). Specimens containing surface cracks were tested at the same temperatures. The specimen dimensions, defect sizes, and stresses corresponding to crack initiation are given in Table 2. The loads corresponding to initiation of crack growth are based on acoustic emission techniques.

Analysis of Results

The stress corresponding to crack initiation was predicted using a modification of an equation developed for center-cracked plates [*2*]:

$$J = \bar{\sigma}\bar{\epsilon}a' \left\{ \Psi^* \left(\frac{\sigma}{\bar{\sigma}}\right)^2 + a' G_{\ln}^* \left(\frac{\sigma}{\bar{\sigma}}\right)^{n+1} \right\} \tag{1}$$

In Ref *3* it was noted that Paris had suggested that the addition of a geometry factor and values of ψ^* and $G_{\ln}^*$ corresponding to $a'/b' = 0$ could modify Eq 1 for use with pressure vessel cylinders containing a surface crack. These modifications are shown in Eq 2 along with changing $\bar{\sigma}$ to σ_{ys}:

$$J = \sigma_{ys}\epsilon_{ys}a' \left\{ F\left(\frac{\sigma}{\sigma_{ys}}\right) \right\} \left[\frac{M^2}{Q}\right] \tag{2}$$

TABLE 1—*Mechanical properties for A-710.*

Test Temperature, K	J_{Ic}, kJ/m^2	E, MPa	σ_{ys}, MPa	ϵ_{ys}
200	10.2	206.3×10^3	500	2.42×10^{-3}
255	18.6	201.5×10^3	467	2.32×10^{-3}
297	39.6	200.1×10^3	452	2.26×10^{-3}

TABLE 2—*Surface crack dimensions and experimental stress.*

Specimen Identification	Depth (a), mm	Length ($2c$), mm	Thickness (t), mm	Width (W), mm	Crack Initiation (σ_{exp}), MPa
		Tested at 200 K			
B-2	2.29	7.37	6.35	101.7	590.0
7	4.06	26.98	6.45	101.4	362.9
13	4.50	41.22	6.38	101.4	275.3
25	4.22	6.60	6.27	100.8	579.6
27	4.29	9.09	6.45	101.3	408.5
B-27	5.00	17.02	12.75	102.4	464.7
B-32	7.19	32.18	12.85	102.0	408.9
B-41	7.62	18.36	12.83	101.9	447.5
		Tested at 255 K			
14	4.62	39.50	6.35	101.35	316.7
22	2.97	7.11	6.50	101.4	505.1
30	4.42	8.89	6.32	101.0	529.9
B-1	1.90	6.73	6.36	101.7	623.1
B-14	1.27	2.54	6.43	101.8	621.7
B-28	4.19	13.31	12.70	102.2	610.0
B-31	5.77	26.80	12.80	102.1	331.6
B-36	1.65	3.66	12.57	101.7	549.9
B-39	5.21	11.81	12.78	101.7	469.2
B-44	7.11	17.78	12.72	101.9	400.2
		Tested at 297 K			
3	2.31	7.75	6.40	101.7	449.2
9	2.54	26.16	6.50	101.5	461.6
10	2.54	25.91	6.35	101.3	489.9
15	4.57	40.00	6.50	101.5	362.2
18	1.14	2.41	6.50	101.5	502.3
24	2.54	6.22	6.55	101.4	529.2
32	4.32	9.52	6.53	101.7	533.4
B-4	4.27	26.67	6.50	101.8	462.3
B-11	4.47	40.97	6.43	101.8	358.8
B-17	2.74	6.48	6.35	101.6	458.8
B-18	3.30	7.24	6.30	101.1	429.8
B-34	1.78	3.56	12.80	102.3	520.3
B-42	7.19	18.16	12.75	101.8	507.8
B-45	5.92	25.65	12.72	101.6	454.7

The stress correction term in braces in Eq 2 is separated into an elastic and a plastic component [*4*] consisting of

$$\left\{F\left(\frac{\sigma}{\sigma_{ys}}\right)\right\} = \frac{b'}{a'}\,\Psi\left[\frac{a'}{b'}, n = 1\right]\left[\frac{\sigma}{\sigma_{ys}}\right]^2 + \alpha'\,\frac{b'}{a'}\,G_{\ln}\left[\frac{a'}{b'}, n\right]\left[\frac{\sigma}{\sigma_{ys}}\right]^{n+1} \quad (3)$$

These components will be evaluated separately, and the constants solved for a very wide plate (i.e., $a'/b' \rightarrow 0$).

Solution for the Elastic Component

To obtain $b'/a' \ \Psi_1(a'/b' \rightarrow 0, n = 1)$, the approach in Ref *5* was used where

$$b'/a' \ \Psi_1 \left(\frac{a'}{b'}, \frac{P}{P_0}\right) = \frac{b'}{a'} \frac{a'}{b'} (1 - a'/b')\Psi g_1(a'/b', n = 1)$$

Because no plastic zone correction was considered, $\Psi = 1$, and for $a'/b' \rightarrow 0$:

$$(b'/a')\Psi_1 \left(\frac{a'}{b'}, \frac{P}{P_0}\right) = g_1(a'/b' \rightarrow 0, n = 1)$$

Data from Refs *6* and *7* were used to solve for the constant for the plane-strain condition by extrapolating the given values of g_1 to a value of g_1 at $a'/b' = 0$. The extrapolation techniques used were based on graphical or statistical means, as appropriate, and are provided in Table 1 of Ref *8*. The values were averaged to obtain $g_1 = 2.99$ for plane-strain conditions.

Solutions for the Plastic Component

It was necessary to solve the plastic component in two steps. The first solves for α' and n in the Ramberg-Osgood equation

$$\epsilon/\epsilon_{ys} = \frac{\sigma}{\sigma_{ys}} + \alpha' \left(\frac{\sigma}{\sigma_{ys}}\right)^n \tag{4}$$

using stress-strain data. To solve for the combination of n and α' that provides the best fit, Eq 4 was modified in Ref *8* by reducing it to the linear equation in Eq 5:

$$\epsilon'/\epsilon_{ys} = \alpha' \left(\frac{\sigma}{\sigma_{ys}}\right)^n \tag{5}$$

The selection of the best-fit values of n and α' was based on the linear fit of Eq 5 that had the least error where the natural logarithm of Eq 5 was taken to obtain:

$$\ln \epsilon'/\epsilon_{ys} = \ln \alpha' + n \ln \left(\frac{\sigma}{\sigma_{ys}}\right) \tag{6}$$

For each data point on the stress-strain curve, there is a linear equation:

$$Y_n = C_1 + C_2 X_n$$

where

$$C_1 = \ln \alpha',$$
$$C_2 = n,$$
$$X_n = \ln \sigma/\sigma_{ys}, \text{ and}$$
$$Y_n = \ln \epsilon'/\epsilon_{ys}.$$

For the linear equation of each data point, an error was calculated. The error for the nth data point was equal to

$$r_n = C_1 + C_2 X_n - Y_n \tag{7}$$

The n and α' constants that would provide the best fit to Eq 5 were selected by choosing the C_1 and C_2 combination with the minimum total data point error; for example,

$$\sum_{i=1}^{M} [r_n] = \text{minimize} \tag{8}$$

The second step is the approach suggested in Ref *5* to solve for $(b'/a')G_{\text{In}}$ $(a'/b', n)$, where

$$(b'/a')G_{\text{In}} = (1 - a'/b')g_1(a'/b' \rightarrow 0, n = n)$$

Because $a'/b' \rightarrow 0$:

$$(b'/a')G_{\text{In}} = g_1(a'/b' \rightarrow 0, n = n)$$

To solve the above, it was necessary to extrapolate g_1 at $a'/b' = 0$ for each value of n. Then, the value of g_1 at $a'/b' = 0$ for each n is plotted versus n. Data from Refs *6* and *7* were used to develop a correlation for g_1 at $a'/b' = 0$ for various values of n. For plane-strain conditions, the statistical techniques based on data from Ref *7* were selected (see Fig. 2 of Ref *8*).

Since Eq 4 was solved using the yield strength, yield strength was used in Eq 3. In Eq 4 the values of n ranged from 4.0 to 14.2 and for α' ranged from 5.0 to 7.9, making it difficult to select appropriate numbers. In a previous analysis [*8*] Paris had observed that $(b'/a')G_{\text{In}}$ was constant, which suggested that a plot of ϵ/ϵ_{ys} versus $\{F(\sigma/\sigma_{ys})\}$ would provide a more accurate means for estimating ϵ/ϵ_{ys} based on the stress correction term.

Figure 2*a* shows the power curve fit and the regression correlation (R^2) for tensile data for test temperatures ranging from 155 to 377 K. Since the stress correction terms of concern to this paper are less than 100, these data were used to more accurately fit the lower region of Fig. 2*a*, shown in Fig. 2*b*.

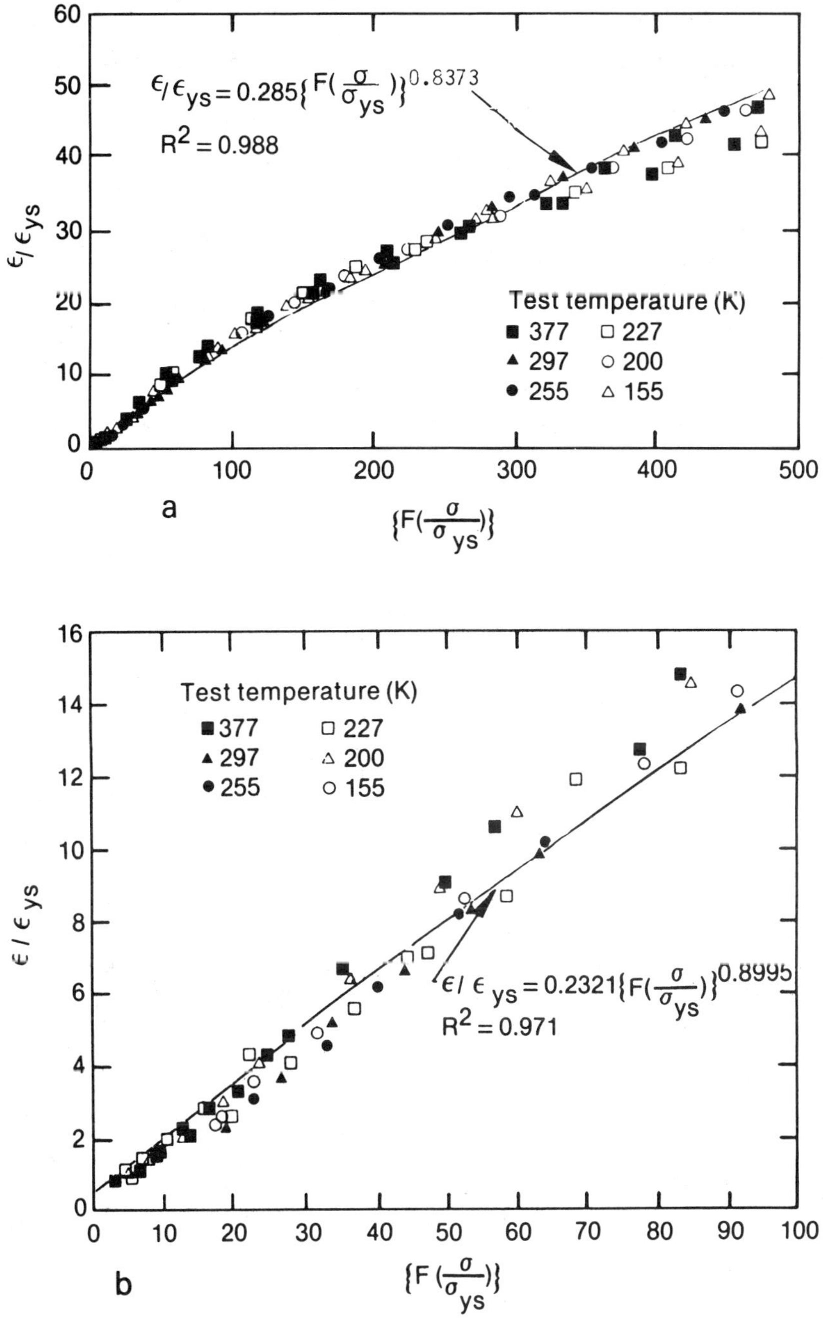

FIG. 2—*Plot of* ϵ/ϵ_{ys} *versus the stress function* $\{F(\sigma/\sigma_{ys})\}$ (a) *Full curve,* ϵ/ϵ_{ys} *to 52.* (b) *Initial portion,* ϵ/ϵ_{ys} *to 16.*

Calculated Results

Rearranging Eq 2 to solve for the stress correction term at $J = J_{Ic}$:

$$\left\{F\left(\frac{\sigma}{\sigma_{ys}}\right)\right\} = \frac{J_{Ic}}{\sigma_{ys}\epsilon_{ys}a\left(\frac{M^2}{Q}\right)} \tag{9}$$

The information provided in Tables 1 and 2 is used to calculate $\{F(\sigma/\sigma_{ys})\}$ (see Table 3). The stress correction term is related to ϵ/ϵ_{ys} by using Figs. 2*a* and 2*b*. Table 3 shows the engineering strain (e_{eng}) calculated from the true strain (ϵ) using ϵ_{ys} for the temperature of interest. The predicted engineering stress (σ_{pred}) corresponding to J_{Ic} is calculated using e_{eng} and the engineering stress-strain curve. Table 4 shows the comparison between σ_{pred} and σ_{exp}.

Discussion

The approach used to calculate σ_{pred} is aimed at conditions associated with the upper shelf, but is used in this analysis for the lower shelf and the lower region of the transition zone. On the lower shelf region (at 200 K) the specimens generally failed under elastic conditions, but for specimens containing sufficiently small defects, the net section stress exceeded the yield strength. This behavior provides an opportunity to evaluate the ability of the Paris approach to extend LEFM concepts from the elastic to the elastic-plastic region. Extending the approach to tests conducted at 255 and 297 K provides an opportunity for evaluating the success (or failure) of Eq 2 under even more ductile conditions. The ratios of $\sigma_{pred}/\sigma_{exp}$ are given in Table 4 along with cF^2 and are plotted in Figs. 3, 4, and 5.[2] Figures 3 to 5 each contain two figures which represent the lowest and highest ratio of predicted to experimental stress ($\sigma_{pred}/\sigma_{exp}$). These exist because of the geometry correction term, M, in Eq 9 which reflects the flaw severity at maximum depth and at the free surface. Predictions based on Eq 9 are depicted as solid symbols. For $\epsilon/\epsilon_{ys} < 0.7$, approaches based on LEFM are used and these results are depicted by open symbols. In every instance the predictions based on LEFM are higher than those based on Eq 9.

In all instances except one in Fig. 3*a* ($cF^2 = 1.70$ mm) and two in Fig. 3*b* ($cF^2 = 1.40$ and 2.11 mm), where $\epsilon/\epsilon_{ys} \geq 0.7$, LEFM-based calculations for the remaining specimens provide more satisfactory predictions than those based on Eq 9. Of these three, where $\epsilon/\epsilon_{ys} > 0.7$, the ratio of σ_{pred} to σ_{exp} ranged from 0.7 to 0.9, which means that the predicted values are conserva-

[2]Here cF^2 is a parameter used to quantify the severity of a surface flaw [*9*]. F is the combined net-section conversion, shape, and magnification factor for front and back face. F is related to the geometry correction term (M^2) [*1*] used in calculating K, for surface cracks ($K_I = \sigma\sqrt{(\pi a/Q)}M$) by $F = M\sqrt{a/cQ}(1 - (\pi ac/Wt))$.

TABLE 3—*Predicted stress (σ_{pred}) for crack initiation.*

Specimen Number	$F\left(\frac{\sigma}{\sigma_{ys}}\right)^{(1)}$ $\phi = 0°$	$F\left(\frac{\sigma}{\sigma_{ys}}\right)^{(1)}$ $\phi = 90°$	$\epsilon_t/\epsilon_{ys}^{(2)}$ $\phi = 0°$	$\epsilon_t/\epsilon_{ys}^{(2)}$ $\phi = 90°$	e_{eng} $\phi = 0°$	e_{eng} $\phi = 90°$	$\sigma_{eng} = \sigma_{pred}$ $\phi = 90°$ (MPa)	$\sigma_{eng} = \sigma_{pred}$ $\phi = 0°$ (MPa)
				Tested at 200 K				
B-2	5.794	4.725	1.127	0.938	0.00273	0.00227	441.6	431.2
7	2.381	1.099	0.506	0.105	0.00122	0.00025	251.8	51.8
13	1.818	0.667	0.397	0.161	0.00096	0.00039	198.0	78.7
27	2.433	3.616	0.516	0.738	0.00125	0.00179	258.1	369.2
B-27	2.574	2.014	0.543	0.436	0.00132	0.00106	272.6	218.7
B-32	1.447	0.946	0.324	0.221	0.00078	0.00054	160.8	111.1
B-41	1.404	1.744	0.315	0.383	0.00076	0.00093	156.6	191.8
				Tested at 255 K				
14	3.605	1.393	0.736	0.313	0.00171	0.00073	347.1	148.4
22	8.461	9.727	1.584	1.796	0.00368	0.00418	445.0	455.4
30	4.765	7.656	0.945	1.448	0.00220	0.00336	417.4	441.6
B-1	15.328	11.072	2.704	2.018	0.00629	0.00469	476.1	458.8
B-14	24.466	30.344	4.118	4.998	0.00960	0.01166	496.8	507.2
B-28	6.643	5.417	1.275	1.061	0.00296	0.00246	434.7	424.4
B-31	4.394	2.595	0.879	0.547	0.00204	0.00127	414.0	257.4
B-36	18.982	20.823	3.278	3.562	0.00763	0.00830	484.4	490.6
B-39	5.103	6.039	1.006	1.170	0.00234	0.00272	470.9	447.8
B-44	3.147	3.681	0.651	0.750	0.00151	0.00174	306.4	353.3
				Tested at 297 K				
3	26.494	20.727	4.424	3.547	0.01005	0.00805	458.5	478.2
9	35.031	9.051	5.687	1.684	0.01294	0.00381	500.2	441.6
10	34.128	8.942	5.555	1.665	0.01263	0.0037	498.9	438.8
15	8.735	3.235	1.631	0.667	0.00369	0.00151	438.8	302.2
18	61.491	71.775	9.434	10.842	0.02155	0.02481	523.7	530.6
24	23.688	25.703	4.00	4.305	0.00908	0.00978	483.7	486.4
32	10.956	15.615	1.999	2.750	0.00453	0.00623	4452.0	465.8
B-4	9.749	4.886	1.799	0.967	0.00407	0.00219	448.5	414.7
B-11	9.175	3.225	1.740	0.665	0.00394	0.00150	443.7	300.2
B-17	21.26	24.409	3.629	4.109	0.00824	0.00933	479.6	485.1
B-18	16.532	21.558	2.894	3.675	0.00656	0.00834	469.2	480.2
B-34	40.204	49.248	6.438	7.723	0.01466	0.01761	505.1	512.7
B-42	6.97	8.098	1.331	1.523	0.00301	0.00345	427.8	438.2
B-45	9.339	5.96	1.732	1.156	0.00392	0.00262	443.7	423.7

$\epsilon_t/\epsilon_{ys} = 0$

(1) $\epsilon_t/\epsilon_{ys} = 0.2321\, F(\sigma/\sigma_{ys})^{0.8995}$.
(2) $\epsilon_{ys} = 2.42 \times 10^{-3}$ @ 200 K.
$\epsilon_{ys} = 2.32 \times 10^{-3}$ @ 255 K.
$\epsilon_{ys} = 2.26 \times 10^{-3}$ @ 297 K.

TABLE 4—*Comparison between σ_{pred} and σ_{exp} for crack initiation.*

Specimen No.	σ_{exp} (MPa)	$\sigma_{pred}/\sigma_{exp}$		cF^2 (mm)		σ_{pred}⁽¹⁾ (MPa)	
		$\phi = 0°$	$\phi = 90°$	$\phi = 0°$	$\phi = 90°$	$\phi = 0°$	$\phi = 90°$
			Tested at 200 K				
B-2	590.0	0.75	0.73	1.40	1.70	...	...
7	362.9	0.69	0.14	2.67	5.79	433.3	239.9
13	275.3	0.72	0.29	2.62	7.37	391.9	233.2
27	408.5	0.63	0.90	3.15	2.11	440.2	...
B-27	465.1	0.59	0.47	2.95	3.76	494.7	439.5
B-32	409.2	0.39	0.27	4.32	6.60	372.6	301.5
B-42	447.8	0.35	0.43	5.03	6.07	366.4	407.8
			Tested at 255 K				
14	316.7	1.10	0.47	2.87	7.44	...	312.6
22	505.1	0.88	0.90	1.93	1.68	...	...
30	529.9	0.79	0.83	3.25	2.03	...	...
B-1	623.1	0.76	0.74	1.09	1.50	...	...
B-14	621.7	0.80	0.82	0.69	0.56	...	...
B-28	610.0	0.71	0.70	2.41	2.95	...	...
B-31	381.6	1.08	0.67	3.20	5.44	...	433.3
B-36	549.9	0.88	0.89	0.89	0.81	...	...
B-39	469.2	0.90	0.91	3.12	2.64	...	...
B-44	400.2	0.77	0.88	4.65	3.99	400.2	...
			Tested at 297 K				
3	449.2	1.09	1.06	1.40	1.78	...	...
9	481.6	1.04	0.92	0.94	3.63	...	...
10	489.9	1.02	0.90	0.96	3.66	...	...
15	362.2	1.21	0.83	2.72	7.34	...	431.9
18	502.3	1.04	1.06	0.64	0.53	...	...
24	529.2	0.91	0.92	1.58	1.45	...	...
32	533.4	0.85	0.87	3.20	2.26	...	...
B-4	462.3	0.97	0.90	2.97	5.94	...	...
B-11	358.8	1.24	0.84	2.56	7.32	...	427.8
B-17	458.8	1.05	1.06	1.75	1.52	...	...
B-18	429.9	1.09	1.12	2.21	1.70	...	...
B-34	520.3	0.97	0.99	0.96	0.79	...	...
B-42	507.8	0.84	0.86	4.72	4.06	...	...
B-45	454.7	0.98	0.93	3.43	5.36	...	...

⁽¹⁾Based on LEFM using Ref *1*.

tive. In Fig. 4*b*, all ten specimens had $\epsilon/\epsilon_{ys} > 0.7$ and the ratio $\sigma_{pred}/\sigma_{exp}$ ranged from 0.7 to 1.1. The ratio of 1.1 identifies some nonconservatism in the predicted stresses.

In all instances except two ($cF^2 = 7.32$ and 7.34 mm), Fig. 5*a* had $\epsilon/\epsilon_{ys} > 0.7$. Of the remaining twelve, the ratio of σ_{pred} to σ_{exp} ranged from 0.7 to 1.08. In Fig. 5*b*, all fourteen specimens had $\epsilon/\epsilon_{ys} > 0.7$, and the ratio $\sigma_{pred}/\sigma_{exp}$ ranged from 0.85 to 1.25. Those ratios in excess of 1.0 identify nonconservatively predicted stresses.

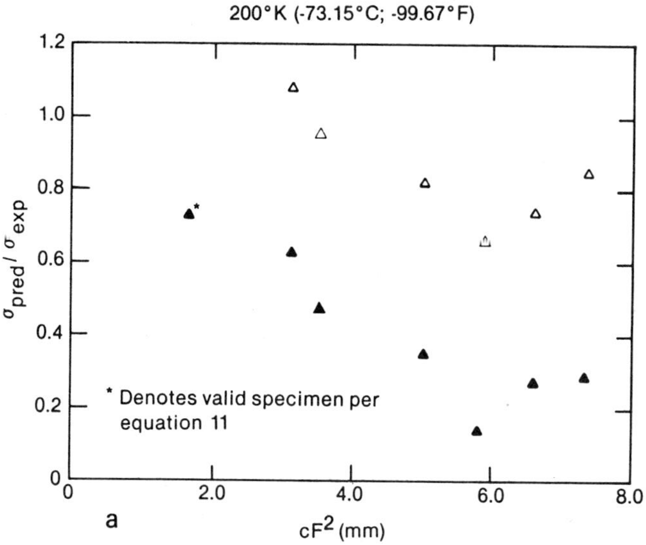

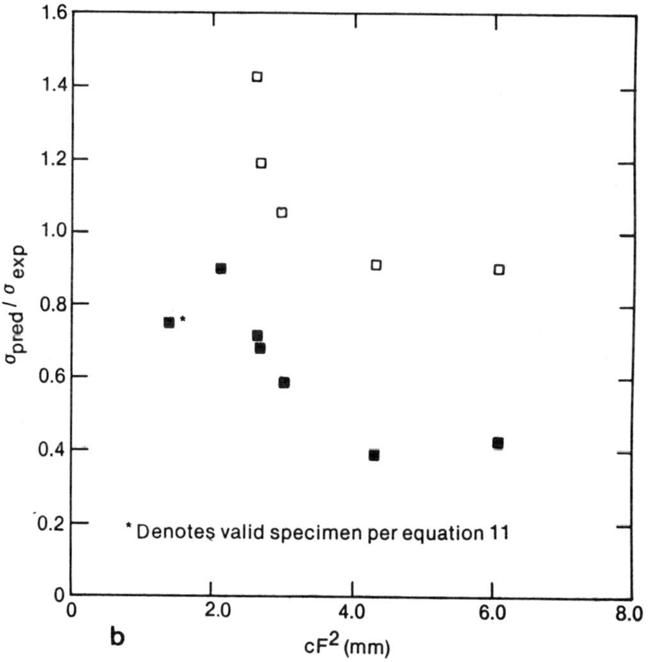

FIG. 3—*Plot of* $\sigma_{pred}/\sigma_{exp}$ *at 200 K versus* cF^2. (a) *Lowest ratio of* $\sigma_{pred}/\sigma_{exp}$. (b) *Highest ratio of* $\sigma_{pred}/\sigma_{exp}$.

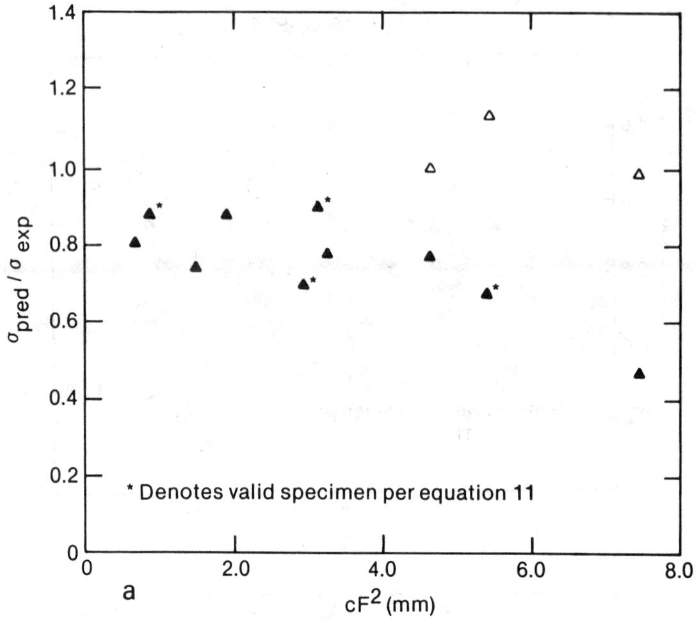

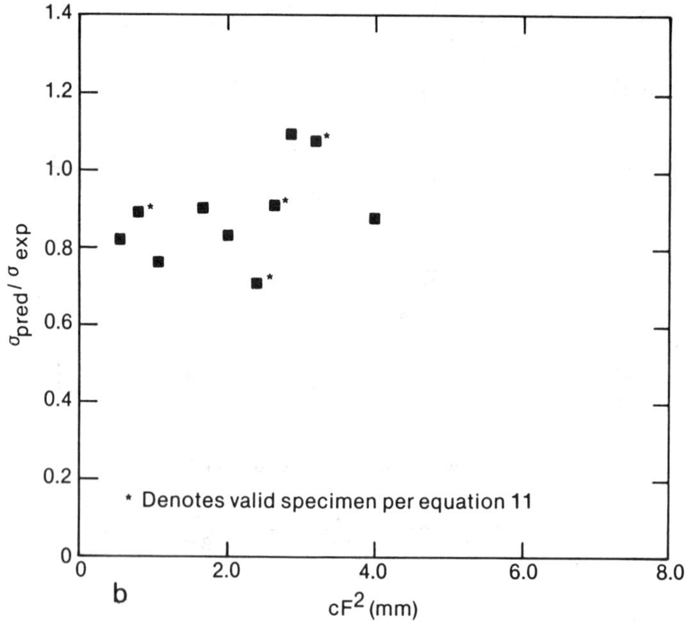

FIG. 4—*Plot of* $\sigma_{pred}/\sigma_{exp}$ *at 255 K versus* cF^2. (a) *Lowest ratio of* $\sigma_{pred}/\sigma_{exp}$. (b) *Highest ratio of* $\sigma_{pred}/\sigma_{exp}$.

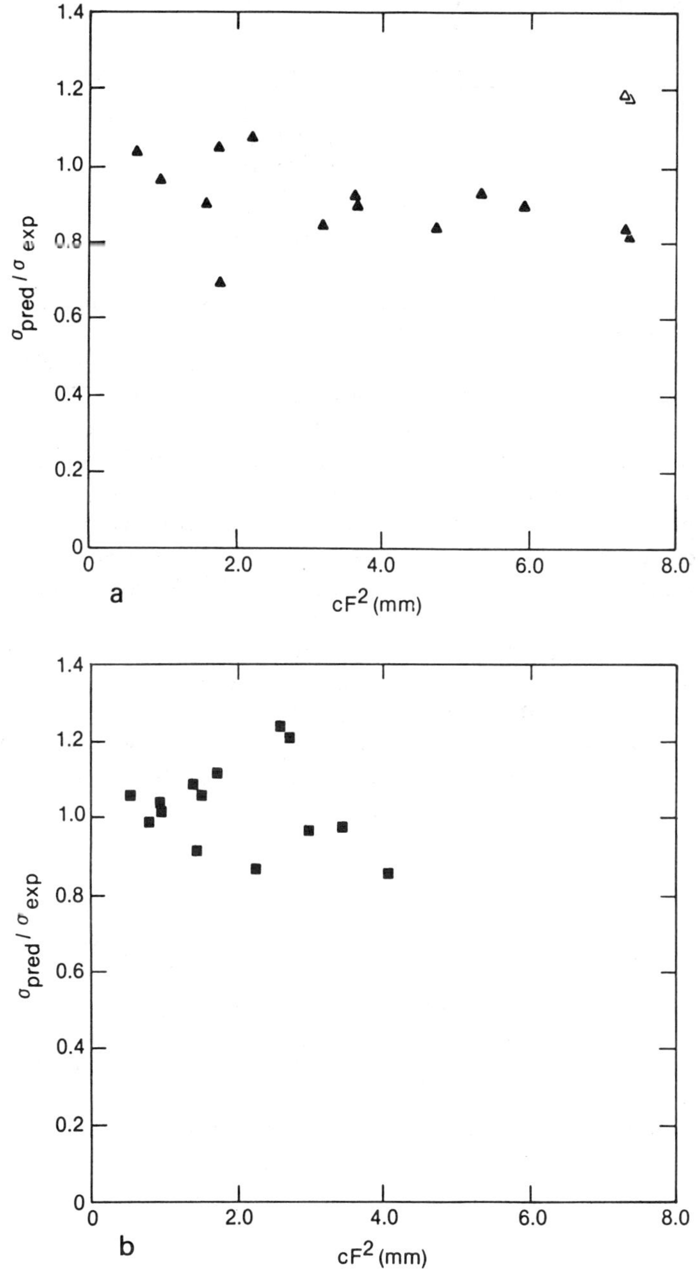

FIG. 5—*Plot of* $\sigma_{pred}/\sigma_{exp}$ *at 297 K versus* cF^2. (a) *Lowest ratio of* $\sigma_{pred}/\sigma_{exp}$. (b) *Highest ratio of* $\sigma_{pred}/\sigma_{exp}$.

An evaluation of Figs. 3 to 5 based on using LEFM (when $\epsilon/\epsilon_{ys} \leq 0.7$) or Eq 9 (when $\epsilon/\epsilon_{ys} > 0.7$) shows:

Figure No	$\sigma_{pred}/\sigma_{exp}$
3*a*	0.7 to 1.1
3*b*	0.75 to 1.15
4*a*	0.7 to 1.15
4*b*	0.7 to 1.1
5*a*	0.8 to 1.1
5*b*	0.85 to 1.25

These ratios suggest that the use of Eq 9 will provide reasonably accurate predictions of conditions for crack initiation.

The use of acoustic emission (AE) techniques to detect crack initiation is often suspect due to its being more of an art than a science. Therefore it is necessary to provide some verification that crack initiation has occurred. This verification has been done by stopping tests at various loads after crack initiation has been identified by AE, unloading and then using metallographic techniques. In all instances it was observed that crack initiation and growth had occurred. It is desirable for crack initiation as determined by AE to be actual crack initiation, but this is not necessary for the result to be useful. Crack initiation identified by AE is a function of the system sensitivity (e.g., set zero gain and catastrophic failure corresponds to crack initiation or set gain sufficient to detect plastic deformation and this becomes crack initiation). In the tests conducted to date the AE system gain, filtering, etc., have essentially been constant for a given test temperature and only minimally different than for the other test temperatures. Therefore AE crack initiation corresponds to a relatively consistent magnitude of crack growth. This provides the ability to detect trends between σ_{pred} and σ_{exp} but might tend to distort quantitative comparisons.

A potential source of error is the surface-flawed specimen's ability to satisfy conditions for plane strain and a *J*-controlled field. For elastic conditions, it is generally accepted that plane strain prevails when the ratio of plastic zone size (r_y) to thickness (t) is less than 0.02. For general plasticity it is much more difficult to specify conditions for plane strain. Therefore for this paper the usefulness of the approach is based on satisfying conditions for a *J*-controlled field.

It is expected that the requirement for a *J*-controlled field for crack initiation would fall somewhere between those given in Eqs 10 and 11:

$$t, b \geq \frac{25\,J}{\bar{\sigma}} \quad \text{for bending} \quad [10] \tag{10}$$

$$t, b \geq \frac{200\,J}{\bar{\sigma}} \quad \text{tensile loading of center-cracked panel} \quad [10] \tag{11}$$

If the specimen width (nominally 101.6 mm) is assumed to be the thickness, since the crack first penetrated the wall thickness before growing in the $2c$ direction, all the test specimens satisfy both Eqs 10 and 11. The requirements for b could only be satisfied for both the 6.4 and 12.7 mm thick specimens if Eq 10 is applicable. This is unlikely but substantial out-of-plane bending does occur due to the effect of the surface flaw, which suggests that Eq 10 could apply. Equation 11 is probably more applicable; one of the specimens tested at 200 K and four of the specimens at 255 K satisfy these requirements, whereas none of the specimens tested at 297 K satisfy the requirements. These requirements for b are summarized in Table 5.

Figures 3 and 4 have a notation adjacent to those specimens satisfying Eq 11. It is obvious that the data obtained from valid specimens are within the general data scatter for all specimens.

This leads to an observation that either the use of Eq 11 is substantially conservative in identifying conditions required to satisfy a J-controlled field or that the agreement between σ_{pred} and σ_{exp} is fortuitous. This latter possibility raises an interesting question, not answered in this paper, regarding the usefulness of J_{Ic} for structures such as thin-walled pipe where attainment of a J-controlled field is not possible. If a J-controlled field is not attained, then the use of J_{Ic} to predict crack initiation for structures containing surface flaws is of questionable value. A similar observation can be made for using the tearing modulus to predict crack growth.

Summary and Conclusions

Equation 1 for calculating J for center-cracked panels has been modified based on suggestions by Paris for estimating J for specimens containing surface cracks. Equation 9 has been used, when $\epsilon/\epsilon_{ys} > 0.7$, in conjunction with Fig. 2*b* and J_{Ic} to calculate the predicted stress for crack initiation for relatively thin surface-flawed specimens tested at 200, 255 and 297 K. When $\epsilon/\epsilon_{ys} \leq 0.7$, LEFM techniques were used to calculate σ_{pred}. An evaluation of Figs. 3 to 5 based on using Eq 9 (when $\epsilon/\epsilon_{ys} > 0.7$) shows the ratio $\sigma_{pred}/\sigma_{exp}$ to range from 0.7 to 1.25. The ratios in excess of 1.0 identify nonconservatively predicted stresses.

The specimen sizes were compared with requirements for satisfying J-con-

TABLE 5—*Requirements for* b.

Test Temperature, K	J_{Ic}, kJ/m^2	$\bar{\sigma}$, MPa	b Based on	
			Eq 10, mm	Eq 11, mm
200	10.2	591	0.43	3.45
255	18.6	548	0.84	6.81
297	39.6	524	1.88	15.11

trolled fields and only a relatively few specimens were in conformance with Eq 11. However, the data obtained from valid specimens are within the general data scatter for all specimens. Therefore it appears as if the Paris modification to the center cracked equation is useful for predicting crack initiation for ASTM A-710 tested at temperatures corresponding to the lower region of the transition zone.

Acknowledgments

This work was supported by the U.S. Department of Energy Assistant Secretary for Research, Office of Energy Research, Office of Basic Energy Sciences, under DOE Contract DE-AC07-76ID01570. The author would like to thank P. Paris for reviewing the manuscript and for his helpful comments.

References

[*1*] Newman, J. C., Jr. and I. S. Raju, *Analysis of Surface Cracks in Finite Plate under Tension or Bending Loads*, NASA Technical Paper 1578, Washington, D.C., 1979.

[*2*] Hutchinson, J. W., Needleman, A., and Shih, C. F., "Fully Plastic Crack Problems in Bending and Tension," in *Fracture Mechanics*, N. Perrine et al, Eds., University of Virginia Press, Charlottesville, 1978.

[*3*] Merkle, J. G., "Analysis, Explanation of Analytical Basis for Low Upper Shelf Vessel Toughness Evaluations," *Resolution of the Task A-11 Reactor Vessel Materials Toughness Safety Issue,* Appendix C, NUREG-0744, Vol. 2, Rev. 1, Oct. 1982, p. C-37.

[*4*] Tada, T. M. and Paris, P. C., *Tearing Instability Analysis Handbook*, USNRC Report NUREG/CR-1221, Jan. 1980.

[*5*] Zahoor, A., "Tearing Instability of Elastic-Plastic Crack Growth," Ph.D. dissertation, George Washington University, Washington, D.C., Aug. 1978.

[*6*] Shih, C. F. and Hutchinson, J. W., "Fully Plastic Solutions and Large Scale Yielding Estimates for Plane Stress Crack Problems," *Journal of Engineering Materials and Technology*, Oct. 1976, pp. 289–295.

[*7*] Kumar, V., German, M. D., and Shih, C. F., *Estimation Technique for the Prediction of Elastic-Plastic Fracture of Structural Components of Nuclear Systems*, SRD-88-094, Corporate Research and Development, General Electric Co., Schenectady, N.Y., 1 June 1980.

[*8*] Reuter, W. G. and Chung, D. T., "Resolution of Some Problems Related to NRC's Task Action Plan A-11," *Resolution of the Task A-11 Reactor Vessel Materials Toughness Safety Issue*, NUREG-0744, Vol. 2, Rev. 1, Oct. 1982.

[*9*] Reuter, W. G., Matthews, S. D., and Smith, F. W., "Critical Parameters for Ductile Fracture of Surface Flaws," *Engineering Fracture Mechanics*, Vol. 19, No. 1, 1984, pp. 159–179.

[*10*] Shih, C. F., "Methodology for Plastic Fracture," EPRI NP-1735, Electric Power Research Institute, Palo Alto, Calif., March 1981, pp. 5–11.

P. Balladon[1] *and J. Heritier*[1]

Comparison of Ductile Crack Growth Resistance of Austenitic, Niobium-Stabilized Austenitic, and Austeno-Ferritic Stainless Steels

REFERENCE: Balladon, P. and Heritier, J., **"Comparison of Ductile Crack Growth Resistance of Austenitic, Niobium-Stabilized Austenitic, and Austeno-Ferritic Stainless Steels,"** *Fracture Mechanics: Seventeenth Volume, ASTM STP 905*, J. H. Underwood, R. Chait, C. W. Smith, D. P. Wilhem, W. A. Andrews, and J. C. Newman, Eds., American Society for Testing and Materials, Philadelphia, 1986, pp. 661–682.

ABSTRACT: Fracture toughness and ductile crack growth properties are compared at room and elevated temperatures by using the *J* integral concept for three grades: (1) an austenitic stainless steel water-quenched from different temperatures, (2) a niobium-stabilized austenitic stainless steel containing niobium carbides and niobium carbonitrides in an austenitic matrix, and (3) an austeno-ferritic stainless steel that contains about 50% ferritic phase in an austenitic matrix and shows higher tensile properties compared with the two other grades.

The *J-R* curves of the two austenitic grades are mainly dependent upon the inclusion content and the direction of the crack growth plane. Some results show the influence of residual stress level on *J-R* curves.

The J_{Ic} and dJ/da values for the austeno-ferritic grade cannot be explained only by the roles of the inclusion content and the crack growth plane direction; a specific behavior of the ferritic phase in this duplex steel may decrease J_{Ic} compared with the two austenitic grades.

KEY WORDS: toughness, stainless steels, *J-R* curves, inclusions

Nomenclature

- Δa Total crack extension (current value)
- Δa_f Total crack extension (final value)
- Δa_e Crack extension without stretched zone (current value)

[1]UNIREC, Groupe USINOR, Centre de Recherches d'Unieux, B. P. 34, 42701 Firminy Cedex, France.

Δa_{ef} Crack extension without stretched zone (final value)
ΔV Potential drop (current value)
ΔV_i Potential drop at initiation
ΔV_f Potential drop (final value)
$\Delta V_{ef} = \Delta V_f - \Delta V_i$ Potential drop due to crack extension without stretched zone
δ Load-line displacement (current value)
δ_i Load-line displacement at initiation
δ_f Load-line displacement (final value)
$\sigma_f = 1/2$ (0.2% yield strength + ultimate tensile strength) Conventional flow stress
J J integral
J_i J value at initiation determined by potential drop
J_f J value at the end of the test (final value)
P Load

Austenitic and austeno-ferritic stainless steels are widely used in nuclear and chemical industries. Ductile tearing of these steels can occur from defects produced by fatigue crack growth, creep crack growth, stress corrosion cracking, or even defects pre-existing in the components.

Ductile tearing properties of Types 304, 316, and 316L austenitic stainless steel at room and elevated temperature have been widely investigated [*1–12*]. Previous work has shown the great dependence of ductile tearing resistance of austenitic stainless steels on microstructure and particularly on inclusion content and morphology with respect to crack orientation [*9,12*] as well as on monotonous and cyclic strain hardening effects [*11*].

The present work is a further investigation, using *J-R* curve methodology, of the influence of microstructure on the ductile crack growth resistance of stainless steels: namely grain size, presence of niobium carbides and carbonitrides, and presence of ferritic phase.

Grain size is known to affect flow stress in austenitic stainless steels at room and elevated temperatures. Consequently, tearing properties may be influenced by large variations in grain size.

Niobium carbides and carbo-nitrides may act as inclusions and therefore lower J_{Ic} and dJ/da.

Ferrite increases flow stress and decreases conventional ductility properties. Consequently, lower values of J_{Ic} and dJ/da than those obtained for austenitic stainless steels can be expected.

Three different steel grades were studied:

1. An austenitic stainless steel (Type 316L) water-quenched from different temperatures with the aim of evaluating the influence of austenitic grain size.
2. A niobium-stabilized austenitic stainless steel (Type 347) containing niobium carbides and niobium carbonitrides in an austenitic matrix.

3. An austeno-ferritic stainless steel containing about 50% ferritic phase in an austenitic matrix.

Experimental Program

Materials

The three steels studied were supplied as industrial products in the solution-annealed condition. Tables 1 to 3 give, respectively, for each grade, the type of product, the chemical analysis and inclusion content, and the mechanical properties at room temperature and 550°C (for Types 316L and 347 steels). The austeno-ferritic steel, containing about 50% ferrite, is generally used only at low temperature (lower than 250°C), so mechanical properties at high temperature are not given for this steel. To evaluate the influence of grain size on ductile tearing, samples of Type 316L steel have been heat treated at high temperatures between 1150 and 1300°C for 1.5 h and water quenched (WQ). Resulting grain sizes appear in Table 4, with estimated values of flow stress at room temperature and 550°C. This estimation is based on statistical analysis of mechanical properties of several heats of this kind of steel produced by Creusot-Loire [*13*].

Specimens

Specimens used for *J-R* curve determination were modified compact tension type of 25 mm thickness for Types 316L and 347 steels and 30 mm thickness for the austeno-ferritic steel. They were located at midthickness in the TL orientation for the 316L plate, in the LS orientation for the austeno-ferritic bar, and at midradius in the CR orientation for the 347 bar (TL, LS, and CR being crack plane orientation codes per ASTM Terminology Relating to Fracture Testing [E 616]). Specimens were 20% side-grooved after precracking. Side-grooving was done following the recommendations given in Ref *14*.

Experimental Procedure

Specimens were fatigue precracked in accordance with ASTM Test for J_{Ic}, A Measure of Fracture Toughness (E 813). The *J-R* curves were determined

TABLE 1—*Products.*

Steel	Product	Annealing Temperature, °C
Type 316L austenitic	30-mm-thick plate	1070
Type 347 austenitic	∅ 250-mm bar	1070
Austeno-ferritic	◻ 110-mm bar	1150

TABLE 2—*Chemical analysis and inclusion content.*

Steel	C	Mn	Si	S	P	Ni	Cr	Mo	Nb	Cu	N	Inclusion No. for 40 Zones per ASTM Practice E 45, Method D
Type 316L	0.025	1.81	0.49	0.002	0.027	12.22	17.22	2.38	...	...	0.074	57.5
Type 347	0.032	1.50	0.57	0.001	0.028	9.61	17.83	0.34	0.43	0.19	0.076	...
Austeno-ferritic	0.025	1.68	0.43	0.006	0.024	6.39	20.91	2.39	...	1.37	...	48.5

TABLE 3—*Mechanical properties.*

Steel	Tensile Properties at 20°C				Tensile Properties at 550°C				Impact Properties at 20°C
	0.2% Yield Strength, MPa	Ultimate Tensile Strength, MPa	Total Elongation, %	Reduction of Area, %	0.2% Yield Strength, MPa	Ultimate Tensile Strength, MPa	Total Elongation, %	Reduction of Area, %	Charpy U-Notch, J/cm²
Type 316L	255	535	67	82	130	445	48	64	270
Type 347	245	600	59	76	140	450	33	58	180
Austeno-ferritic	480	655	37	79	...	...	...	...	210

TABLE 4—*Influence of heat treatment on grain size and flow stress for Type 316L steel.*

Heat Treatment	Grain Size, μm	σ_f at 20°C, MPa	σ_f at 550°C, MPa
As-annealed	60	420	287.5
1150°C, 1.5 h, WQ	90	410	270
1200°C, 1.5 h, WQ	250	397.5	255
1250°C, 1.5 h, WQ	350	394	249
1300°C, 1.5 h, WQ	500	390	241

using both single specimen compliance and potential drop techniques. The single specimen compliance method has been described elsewhere [*11*]; this method was used for room temperature tests.

The potential drop method uses an apparatus giving impulses of current of constant amplitude. The specimen is insulated from the testing machine and the clip gage. Both load-displacement and potential-displacement curves are recorded. Figure 1 shows the principle of the method.

An example of the potential-displacement curves obtained for austenitic and austeno-ferritic steels is given in Fig. 2. One can see first a linear relationship between potential and displacement corresponding to blunting, then a deviation from the linearity. Crack growth initiation is assumed to take place at the onset of this deviation. This assumption has been verified by interrupted tests. A linear calibration curve between potential drop from this initiation point and the crack propagation (excluding the stretching zone) is assumed.

This calibration curve is drawn for each test knowing the potential drop from the initiation point to the last point before unloading and the crack ex-

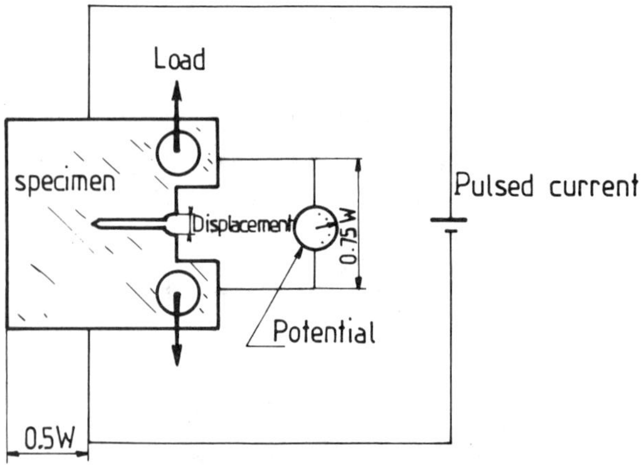

FIG. 1—*Principle of the potential drop method.*

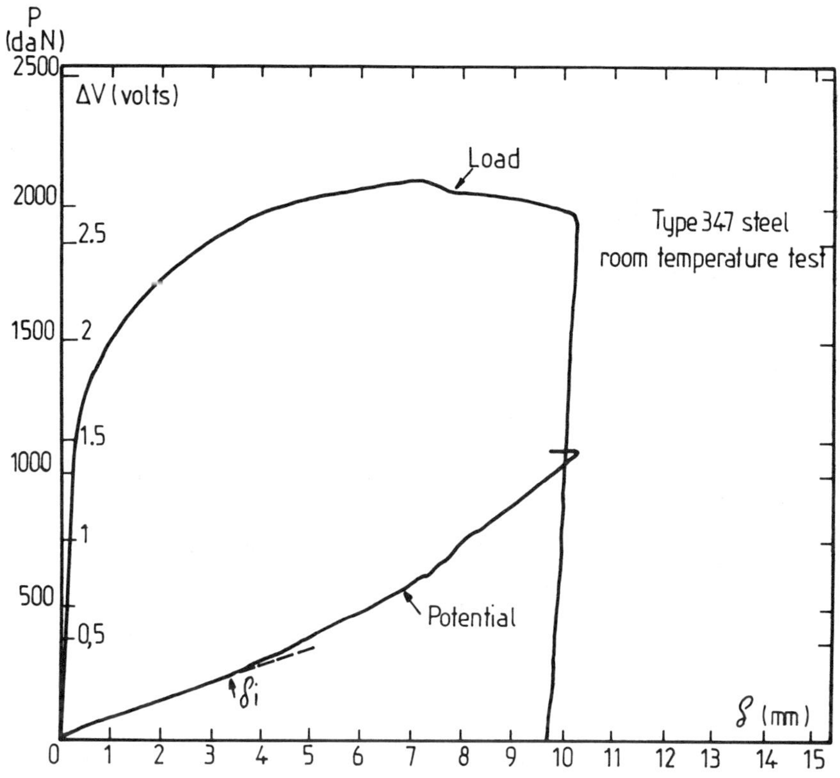

FIG. 2—*Example of potential/displacement curve with corresponding load/displacement curve.*

tension measured on the fracture surface after heat tinting and fracture of the specimen. Crack extension excluding the stretched zone can be determined either by direct measurement of the stretched zone, or by the formula $\Delta a_{sz} = J_i/4\sigma_f$, assuming the blunting line equation equal to $J = 4\sigma_f \cdot \Delta a$. This blunting line equation seems to be more realistic than $J = 2\sigma_f \cdot \Delta a$ for austenitic stainless steels, as discussed by several authors working on these materials [*8–10*]. Figure 3 describes the procedure used to determine the *J-R* curve by the potential drop method. Comparison of results obtained by both unloading compliance and potential drop methods at room temperature shows good agreement at room temperature (Fig. 4). From these results, potential drop was used at 550°C to determine *J-R* curves.

Results

J-Δa *Tests*

The *J-Δa* tests were performed at room temperature and 550°C for Types 316L and 347 steels and at room temperature only for the austeno-ferritic

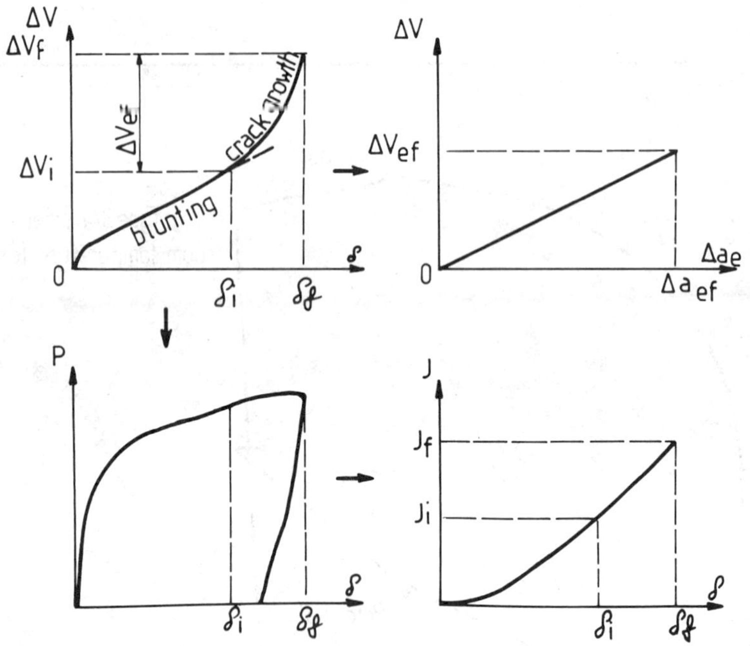

FIG. 3—*Procedure used to determine* J-R *curves by potential drop method.*

steel. The resulting *J-R* curves are shown in Figs. 5 and 6. J_{Ic} and dJ/da were determined following a modified procedure of ASTM E 813 using a blunting line equation equal to $J = 4\sigma_f \cdot \Delta a$, which has been found to be a better approximation of the experimental blunting line. J is estimated by the formula $J = (A/B_N b_0) f(a_0/W)$, where A is the area under the load displacement curve, B_N is the net thickness, b_0 the initial ligament size, a_0 the initial crack length, and $f(a_0/W)$ the Merkle-Corten correction.

The J_{Ic} and dJ/da values are given in Table 5. Results obtained for Type 316L steel at room and elevated temperatures are not valid following the conditions of ASTM E 813 for the ligament size and specimen thickness. Nevertheless, test results can be used to compare the toughness levels of these materials. In addition, dimensional measurements made after testing on specimens of heat-treated 316L steels, for room temperature tests, showed a light plastification of the arms of the specimens. This gives overestimated values of J_{Ic} at room temperature for specimens treated at 1150 and 1350°C. No plastification has been observed on as-annealed material, however, and it can be concluded that post-annealing treatments increase J_{Ic} values, even if these values are overestimated.

Micrographic Examination

Scanning electron microscopic examinations have been done on fracture surfaces in the stretched zone and in the crack extension zone.

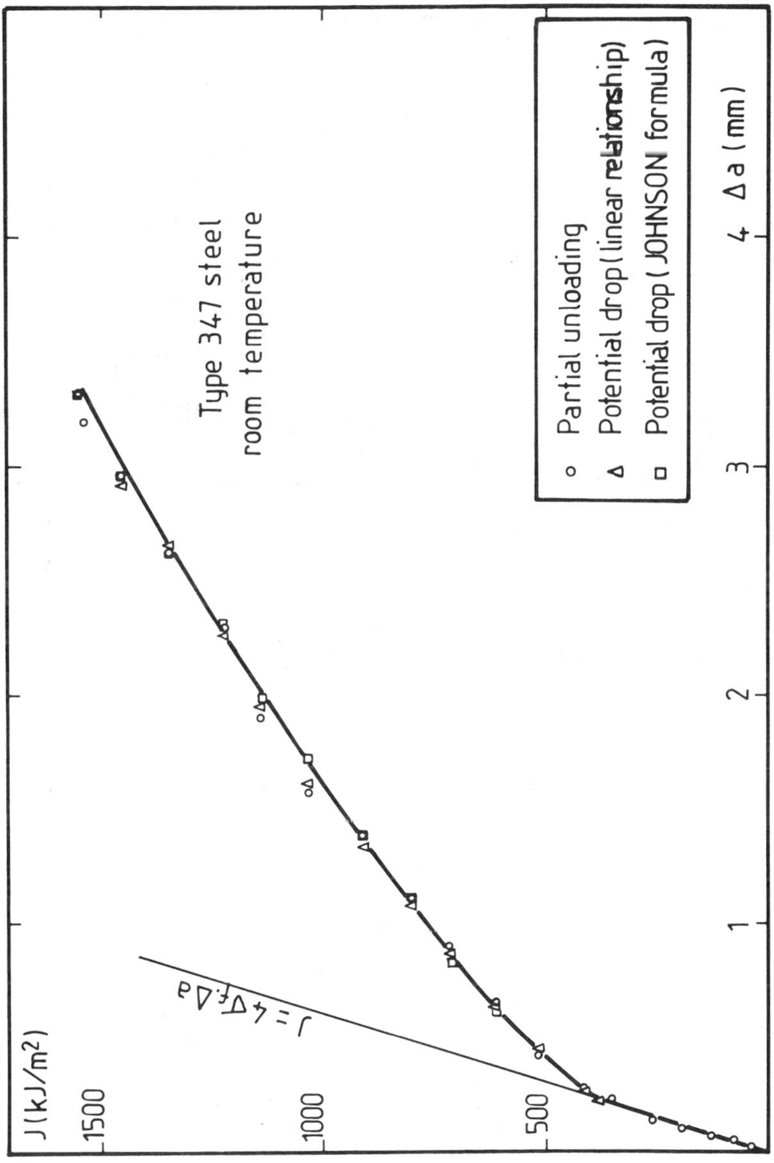

FIG. 4—*Comparison of partial unloading and potential drop methods.*

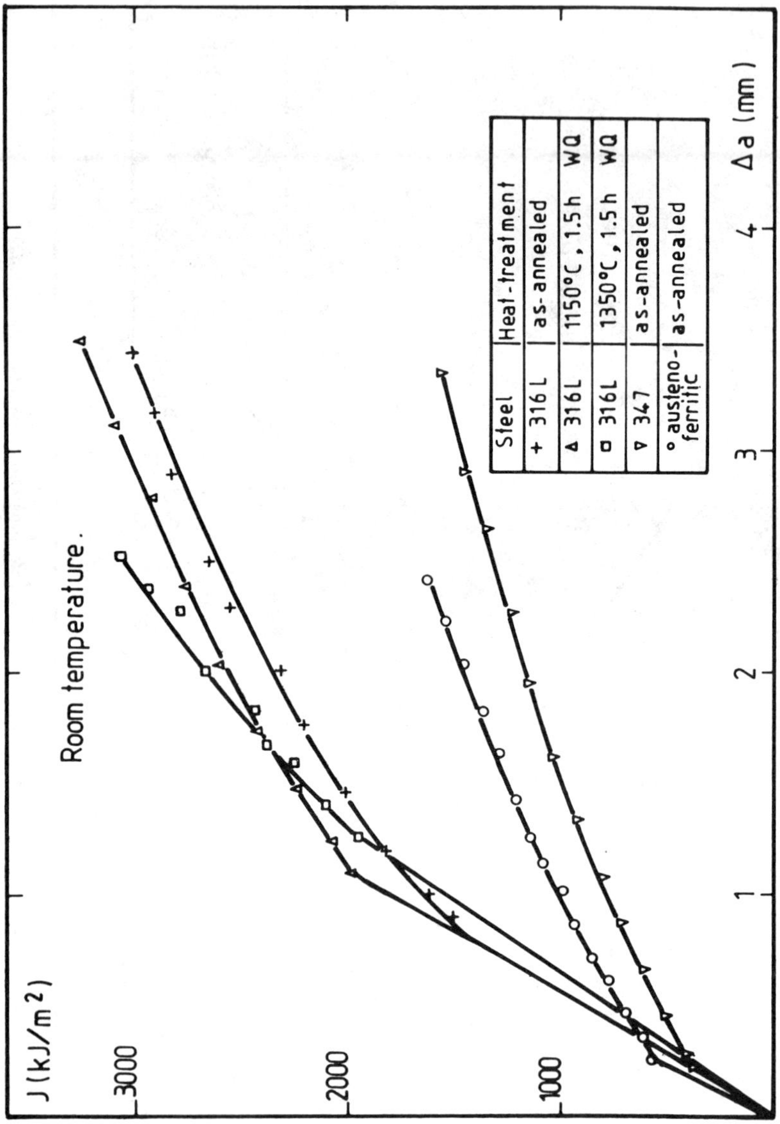

FIG. 5—J-R *curves at room temperature.*

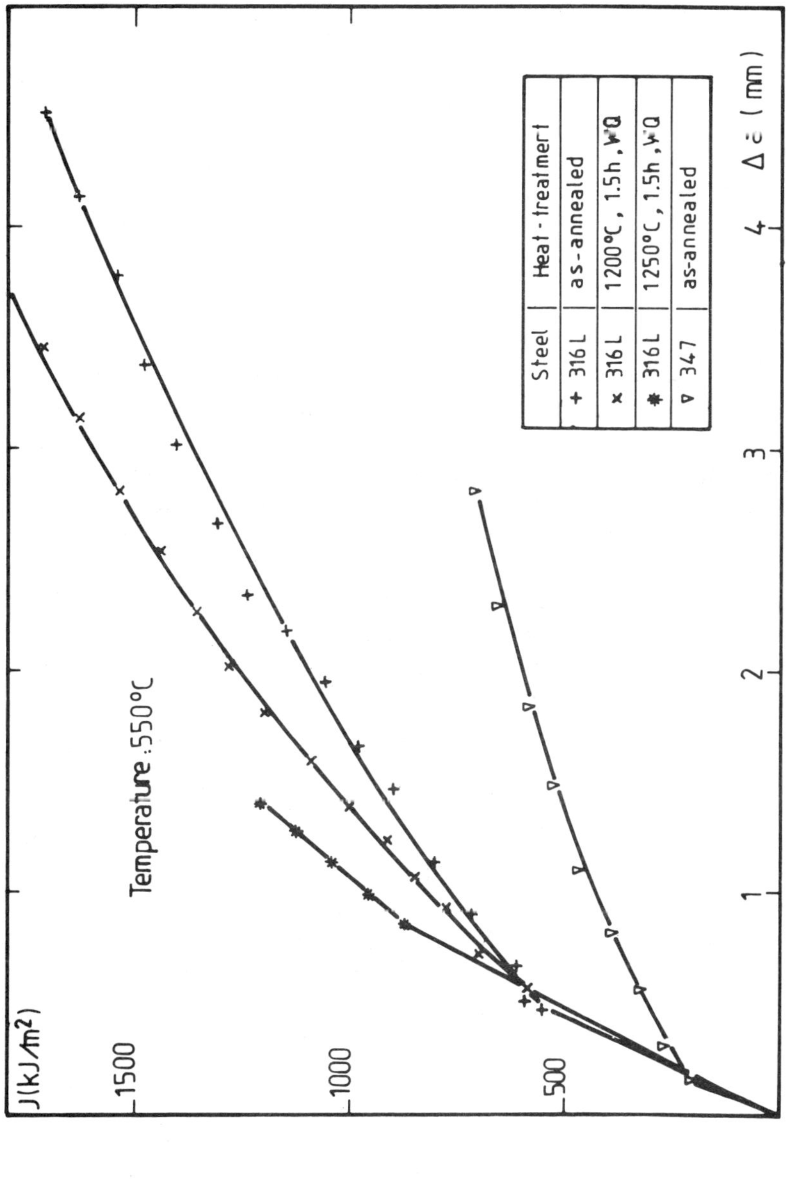

FIG. 6—*J-R curves at 550°C.*

TABLE 5—J_{Ic} *and* dJ/da *values.*

Steel	Heat Treatment	Test Temperature, °C	J_{Ic}, kJ/m^2	dJ/da, MPa	Observations
Type 316L	as-annealed	20	1660	580	not valid
	1150°C, 1.5 h, WQ	20	2040	520	not valid
	1300°C, 1.5 h, WQ	20	2000	1000	not valid
	as-annealed	550	600	340	not valid
	1200°C, 1.5 h, WQ	550	650	425	not valid
	1250°C, 1.5 h, WQ	550	870	600	not valid
Type 347	as-annealed	20	440	420	valid
	as-annealed	550	240	210	valid
Austeno-ferritic	as-annealed	20	640	480	valid

Type 316L Steel—Figure 7 compares stretched zones for as-annealed and 1150°C treated materials at room temperature. Figure 8 shows the crack extension zone for the same materials. Decohesions at inclusion-matrix interfaces are seen in the stretched zone. Increasing grain size seems to have no marked influence on the stretched zone morphology. On the contrary, the crack extension zone shows larger dimples for the post-annealed material, with few regions of very small dimples. The same observation has been made for specimens treated at 1300°C. Figure 9 gives an example of the fracture surface observed after testing at 550°C (specimen treated at 1200°C). The main difference between 550°C and room temperature specimens is the absence of very small dimples at 550°C.

Type 347 Steel—Figure 10 compares the stretched zone and the beginning of the crack extension zone for specimens tested at 20 and 550°C. Fracture aspect is the same at room and elevated temperatures. Decohesions are initiated at interface between matrix and niobium carbonitrides. The finer array of equiaxed dimples observed on the fractography of the specimen tested at 550°C is not statistically significant.

Fracture occurs first by coalescence between these decohesions with rather low plastic deformation due to the high density of carbonitrides, secondly by tearing of the ligaments remaining between the coalesced decohesions, giving rise to a very small dimpled structure.

Figure 11 shows at higher magnification cracks and dimples initiated on niobium carbonitrides.

Austeno-Ferritic Steel—Figure 12 gives an example of fracture surface observed near the stretched zone and at higher magnification in the crack extension zone. Besides the large dimpled structure already observed in Type 316L steel, which has been analyzed by X-ray (energy dispersive) as being the aus-

FIG. 7—*Stretched zones for Type 316L steel at room temperature.*

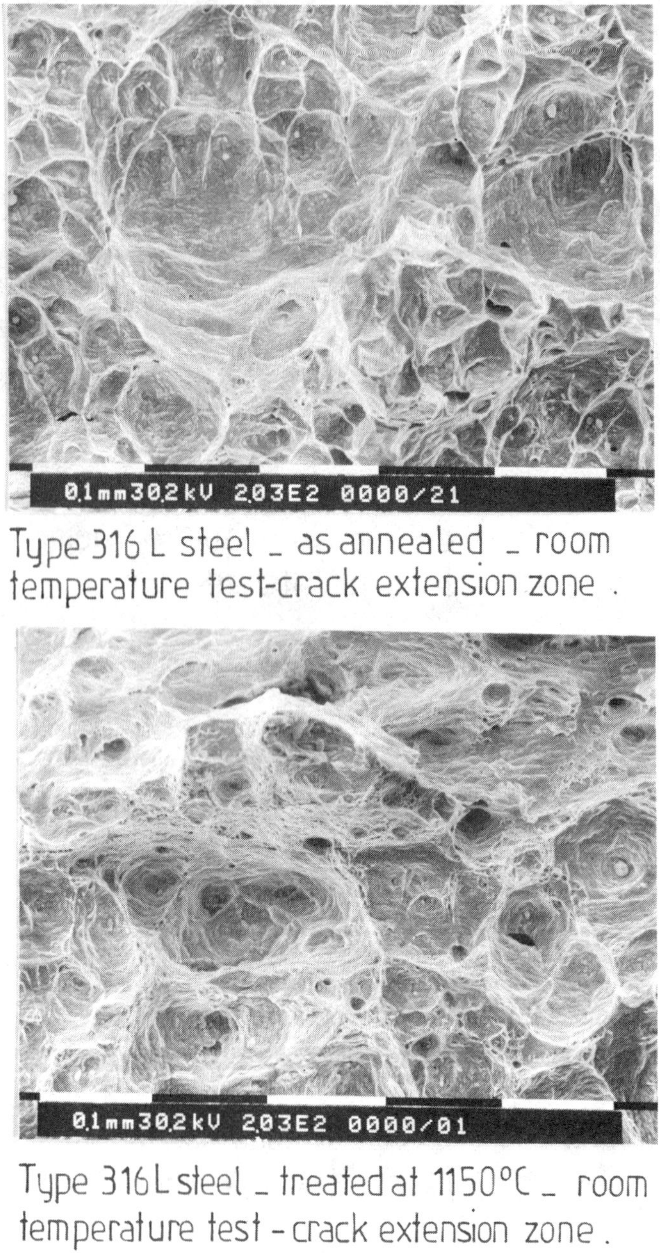

FIG. 8—*Crack extension zones for Type 316L steel at room temperature.*

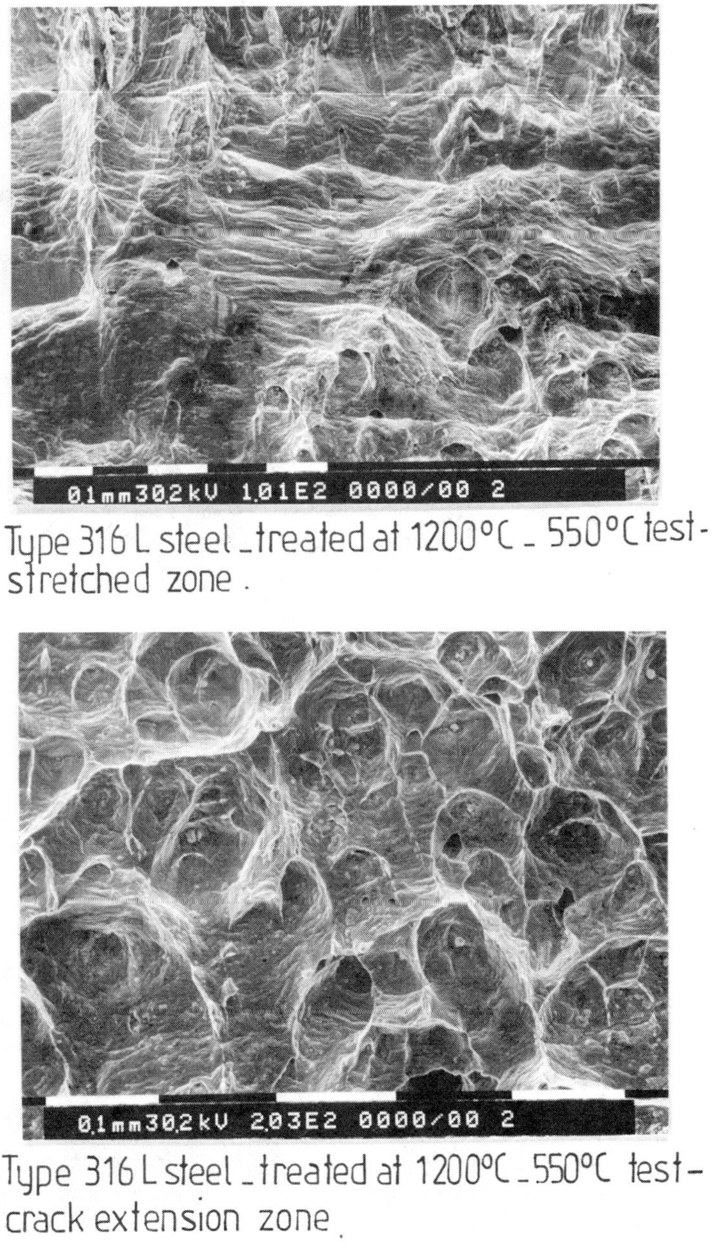

FIG. 9—*Stretched and crack extension zones for Type 316L steel at 550°C.*

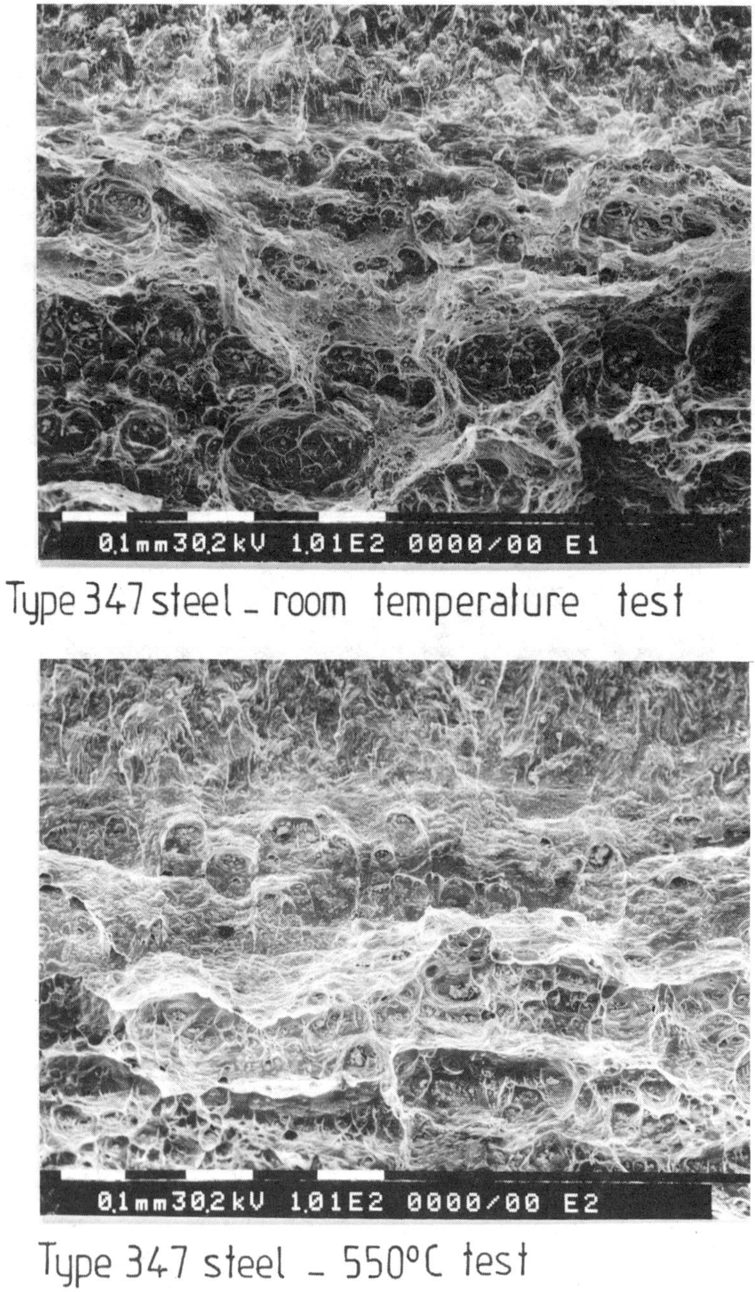

FIG. 10—*Stretched zones and crack extension zones for Type 347 steel at room temperature and 550°C.*

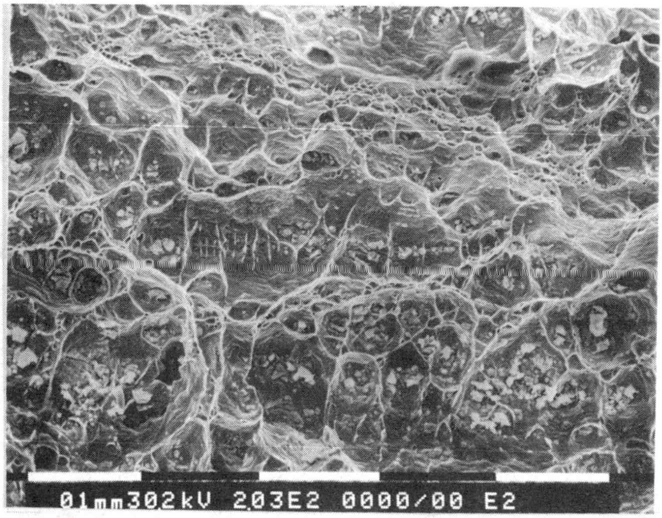

Type 347 steel - 550°C test - crack extension zone.

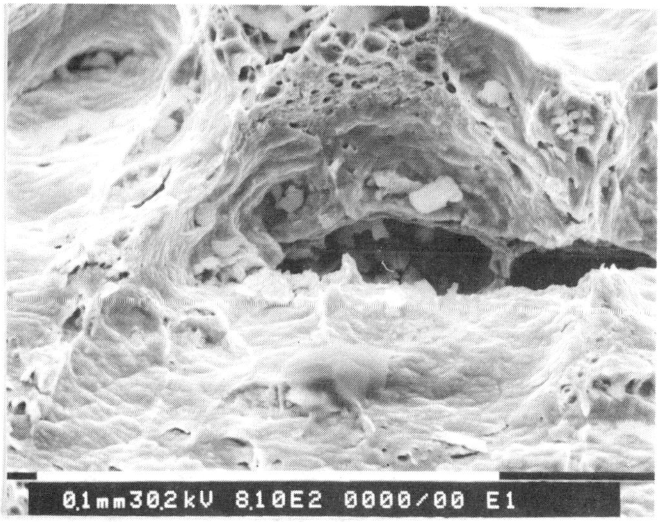

Type 347 steel - room temperature test- decohesion between niobium carbonitrides and austenitic matrix

FIG. 11—*Cracks and dimples initiated on niobium carbonitrides.*

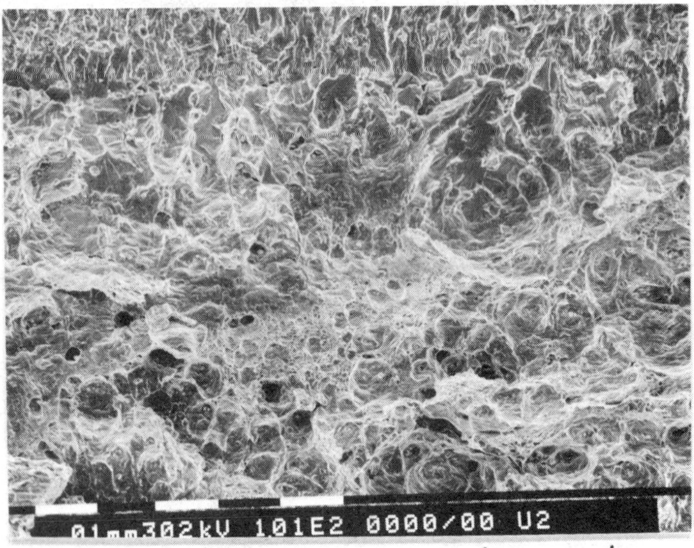

Austeno ferritic steel _ room temperature test-stretched and crack extension zones.

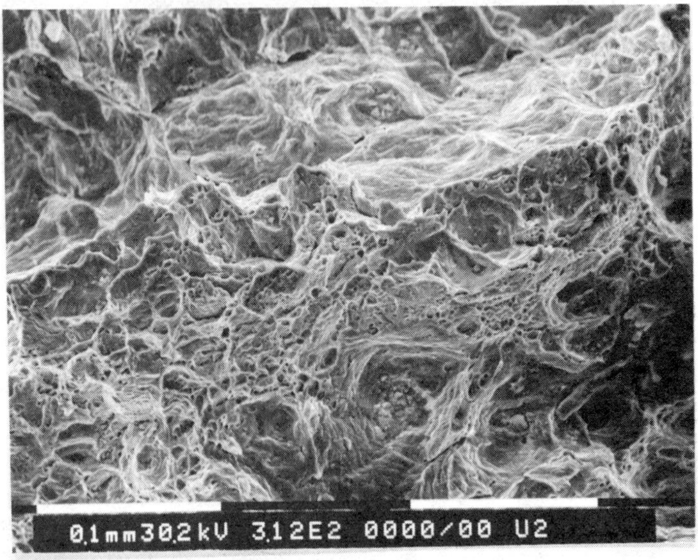

Austeno-ferritic steel _ room temperature test-crack extension zone.

FIG. 12—*Stretched and crack extension zones for austeno-ferritic steel.*

tenitic phase, small dimpled structures with rather flat surfaces can be seen. These regions have been analyzed as being the ferritic phase.

Discussion

Influence of Grain Size

At room temperature, the influence of grain size on ductile tearing properties of Type 316L steel is not clear. If we consider post-annealing treatments at 1150 and 1300°C, giving respectively grain sizes of 90 and 500 μm, no difference can be seen in J_{Ic} values, but dJ/da increases 100% when grain size increases from 90 to 500 μm. This influence can be related to the observation of a greater amount of small dimpled regions on fracture surfaces due to final tearing of ligaments between large dimpled regions. Comparison of results obtained on the as-annealed and the post-annealed at 1150°C materials, with grain sizes of respectively 60 and 90 μm, shows that J_{Ic} is increased about 20% between as-annealed and post-annealed material, but that dJ/da is not influenced. Microfractographic examinations do not show any difference between the two materials. This increase in J_{Ic} cannot be related to the small increase in grain size, and it can be assumed that other factors (for example, residual stresses in industrial annealed plates) may influence J_{Ic}. This assumption must be verified by further investigations. However, *J-R* curves obtained at room temperature for this steel are well in the experimental scatter band obtained with different low inclusion content Type 316L steels [*12*] (Fig. 13). At 550°C, both J_{Ic} and dJ/da increase with increasing grain size.

General improvement in ductile tearing properties observed at room temperature and 550°C by coarsening grain size is not in opposition with results obtained by tensile tests and Charpy U-notch tests on this kind of steel. Coarsening grain size affects tensile properties by lowering 0.2% yield strength and ultimate tensile strength and increasing total elongation to rupture and reduction of area. At the same time, Charpy U-notch properties are improved by lowering 0.2% yield strength and ultimate tensile strength [*13*].

Influence of Niobium Carbonitride

Results obtained on Type 347 steel show that niobium carbonitrides act as oxide and sulfur inclusions on Type 316L steel and are the main factor responsible for the lower ductile tearing properties compared with 316L steel with low inclusion content. The *J-R* curve obtained at room temperature for this steel is well in the experimental scatter band obtained with different high inclusion content Type 316L steels [*12*] (Fig. 13). The same trend is observed at 550°C.

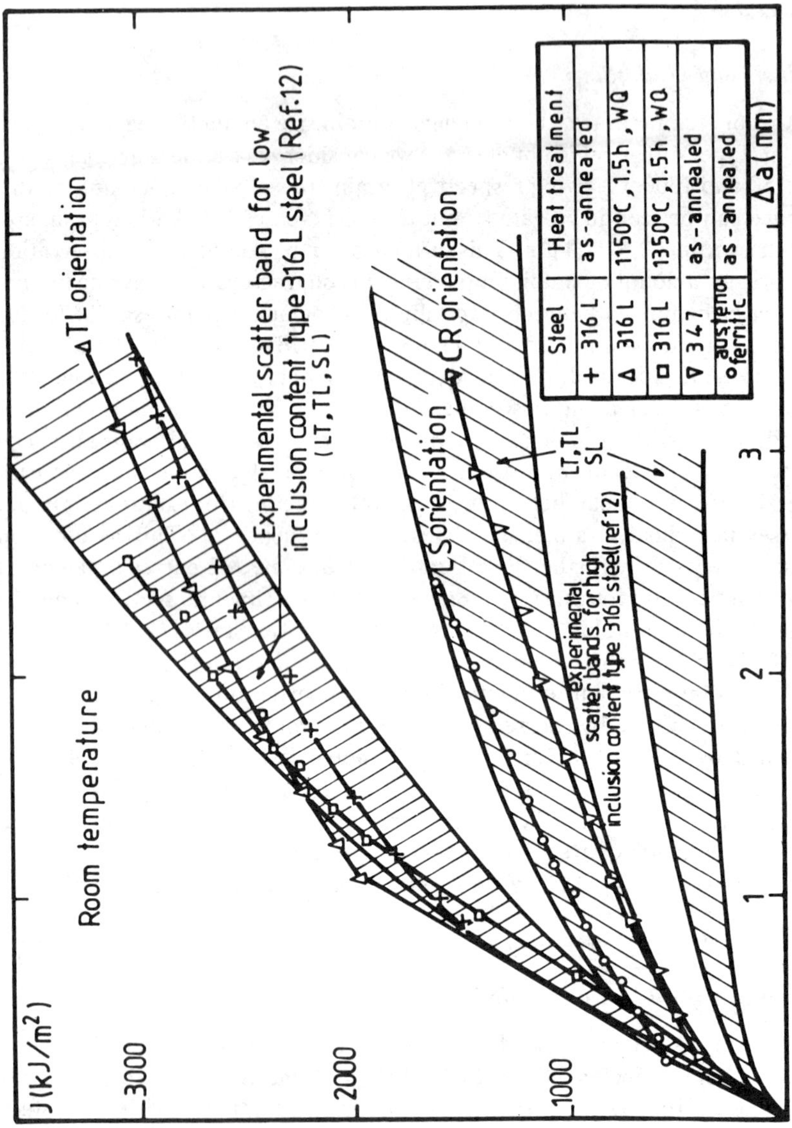

FIG. 13—*Comparison of* J-R *curves at room temperature with previous results obtained on Type 316L steel.*

Influence of Ferritic Phase

The *J-R* curve obtained on the austeno-ferritic steel can be compared with those obtained on different high inclusion content Type 316L steels (Fig. 13). Microfractographic examinations show that the fracture mode of the ferritic phase is quite different from that of the austenitic phase. The ferritic phase acts as a lower ductile tearing resistant phase. Ferrite thus plays a specific detrimental role in initiation of ductile crack growth; this leads to quite a lower level of J_{Ic} compared with a 316L grade (decrease of about 70%) for a similar inclusion number.

These results are in good agreement with other ductile properties such as total elongation to rupture and reduction of area in tensile tests, and Charpy U-notch properties, compared with those of Type 316L steel.

Influence of Test Temperature

As has been shown in Ref *9*, ductile crack growth resistance is much lower at 550°C than at room temperature. This lowering is relatively more important for very high toughness materials such as Type 316L steel than for lower toughness materials such as Type 347 steel. No clear evidence of different mechanisms can be seen from microfractographic examinations, except the lack of small dimpled regions related to tearing of ligaments between decohesions at the matrix/inclusion interface. However, these results are in agreement with the lower tensile properties, both flow stress and ductility, observed at 550°C.

Summary and Conclusions

1. Values of J_{Ic} and *dJ/da* for low sulfur content niobium-stabilized Type 347 steel are much lower than those obtained for low sulfur content Type 316L steel. The *J-R* curve for Type 347 is in the experimental scatter band of *J-R* curves for high inclusion content Type 316L. This result shows that carbides, nitrides, and carbonitrides of nobium in Type 347 act as other inclusions (oxides and sulfurs) in Type 316L.

2. In spite of its low inclusion content, the austeno-ferritic grade presents values of J_{Ic} and *dJ/da* close to those observed for high inclusion content Type 316L. This result indicates that the ferritic phase has a specific role in the duplex 50%-50% structure.

3. For the austenitic 316L grade, the increase of J_{Ic} (about 20%) at room temperature by post-annealing cannot only be explained by the small increase of grain size (60 to 90 μm).

References

[*1*] Hodkinson, P. H., "An Elastic-Plastic Fracture Mechanics Study of Crack Initiation in 316L Steel," Royal Military College of Science, Shrivenham, Swindon, U.K., May 1980.

[2] Chipperfield, G. C., *International Journal of Fracture*, Vol. 12, No. 6, Dec. 1976, pp. 873-886.
[3] Bamford, W. H. and Bush, A. J. in *Elastic-Plastic Fracture, ASTM STP 668*, American Society for Testing and Materials, Philadelphia, 1979, pp. 553-577.
[4] Zahoor, A. and Kanninen, M. F., *Journal of Pressure Vessel Technology, Transactions of ASME*, Vol. 103, Nov. 1981, pp. 359-365.
[5] Yagawa, G., Takahashi, Y., Ando, Y., and Tomoda, Y. in *Proceedings*, 6th SMIRT, Paris, 1981, F7/11.
[6] Tanaka, K. and Harrison, J. D., *International Journal of Pressure Vessels and Piping*, Vol. 6, 1978, pp. 177-200.
[7] Janssen, G. T. M., Eikhoff, J. V. D., and Van Rongen, H. J. M. in *Proceedings*, 6th SMIRT, 1981, Paris, L5/2.
[8] Mills, W. J., *Journal of Testing and Evaluation*, Vol. 9, No. 1, 1981, pp. 56-62.
[9] Balladon, P., Heritier, J., and Rabbe, P. in *Fracture Mechanics: Fourteenth Symposium—Volume II: Testing and Applications, ASTM STP 791*, American Society for Testing and Materials, Philadelphia, 1983, pp. II-496-II-513.
[10] Davies, P. H. in *Elastic-Plastic Fracture: Second Symposium, Volume II—Fracture Resistance Curves and Engineering Applications, ASTM STP 803*, American Society for Testing and Materials, Philadelphia, 1983, pp. II-611-II-631.
[11] Balladon, P., Heritier, J., and Jarboui, C. in *Fracture Mechanics: Sixteenth Symposium, ASTM STP 868*, American Society for Testing and Materials, Philadelphia, 1985, pp. 293-307.
[12] Balladon, P. and Heritier, J., "Ductile Crack Growth in Nuclear Austenitic Stainless Steel Plates: Influence of Crack Orientation and Steel Microstructure," in *Fracture Prevention in Energy and Transport Systems*, I. Le May and S. N. Monteiro, Eds., EMAS, Rio de Janeiro, 1983.
[13] Heritier, J. and Freycenon, J., Creusot-Loire Internal Report, 42701, Firminy Cedex, France, 1982.
[14] Albrecht, P. et al, *Journal of Testing and Evaluation*, Vol. 10, No. 6, Nov. 1982, pp. 245-251.

H. Homma,[1] D. A. Shockey,[2] and S. Hada[1]

Minimum Time Criterion for Crack Instability in Structural Materials

REFERENCE: Homma, H., Shockey, D. A., and Hada, S., "**Minimum Time Criterion for Crack Instability in Structural Materials,**" *Fracture Mechanics: Seventeenth Volume, ASTM STP 905*, J. H. Underwood, R. Chait, C. W. Smith, D. P. Wilhem, W. A. Andrews, and J. C. Newman, Eds., American Society for Testing and Materials, Philadelphia, 1986, pp. 683-696.

ABSTRACT: Cracks in 4340 steel, 1018 steel, and 6061 aluminum specimens were loaded by short pulses of various amplitudes and durations of about 20 to 80 μs to generate experimental results of critical stress amplitudes for crack instability. Stress intensity histories experienced by the cracks at instability were calculated by a finite element method. The results showed that it is necessary to introduce time effect into the instability criterion, such as in the minimum time theory proposed by one of the authors. Correlation between the minimum times obtained for the three materials and their mechanical properties was discussed to clarify the physical meaning of the minimum time. The experimental results showed that the minimum time becomes long for the ductile material.

KEY WORDS: dynamic crack instability, stress wave loading, fracture criteria, stress intensity histories, minimum time

Crack instability can be predicted by static linear elastic fracture mechanics concepts when a crack is loaded at a quasi-static loading rate and plastic deformation is limited in a small region at the crack tip so that the plane strain conditions are dominant throughout the specimen thickness. In such a situation, the crack begins to propagate in an unstable manner when the applied stress intensity equals or exceeds the fracture toughness value of the material. However, for rapid loading in which the stress field near the crack is significantly affected by the inertia, we lack reliable crack instability criteria and experimental results. One good example used for rapid loading is the impinging of stress waves upon a crack.

There are limited experimental data available on crack instability under

[1]Toyohashi University of Technology, Tempaku-cho, 440 Toyohashi, Japan.
[2]SRI International, Menlo Park, CA 94025.

stress wave loading [*1–3*]. Achenbach and Brock [*4*] postulated that a crack begins to propagate immediately when the dynamic stress intensity equals or exceeds the critical value. Lehnigk [*5*] assumed that a crack impinged by a square stress wave pulse becomes unstable when the crack tip opening displacement (CTOD) increases to a certain critical value. Since the dynamic stress and the displacement fields near the crack tip in a linear elastic media are expressed by equations using dynamic stress intensity similar to corresponding static equations [*6*], the CTOD criterion is equivalent to the stress intensity criterion as well as in the case of the quasi-static loading.

On the other hand, Kalthoff and Shockey [*7*] showed that experimental data of crack instability in epoxy plate by short tensile stress pulses could not be explained by the criteria described above. They proposed a minimum time criterion (i.e., that a crack becomes unstable when the dynamic stress intensity at the crack tip exceeds a critical value during a certain time defined as the *minimum time*) and explained the experimental results of the crack instability by that criterion. Homma et al [*8*] found that a crack in 4340 steel hardened to HRC 50 necessitated a 7 μs overshooting of the dynamic stress intensity from the critical value to be initiated.

The aim of this work is to clarify the applicability of the minimum time criterion for materials other than 4340 steel and to examine the material dependence of the obtained minimum time. In this paper the dependence of the minimum time on the material will be discussed in association with dynamic elasto-plastic stress histories at the crack tip in the materials.

Experimental Procedure

Materials and Specimens

Two kinds of materials, 1018 cold-rolled steel and 6061-T651 aluminum, were prepared for the experiments. Yield strength and static fracture toughness were 700 MPa and 56 MPa$\sqrt{m}$ for 1018 steel and 300 MPa and 26 MPa$\sqrt{m}$ for 6061 aluminum. Single-edge-notched specimens and a loading device identical to those used in previous experiments [*8*] were used in the experiments for the crack instability under a 80 μs duration pulse. A sinusoidal tensile stress pulse impinging on the crack was generated by the loading device. A new loading device was constructed to produce dynamic stress intensity histories with about a 20 μs duration in the specimens. The specimen geometry is shown in Fig. 1; the specimen is much smaller than that used previously. Edge notches were introduced into each specimen in the middle section by a saw. A fatigue crack was initiated from each notch root and grown to at least 3 mm by cyclic loading at a maximum stress intensity factor less than 18 MPa$\sqrt{m}$. The total depth of the notch and the fatigue crack was between 40 and 80% of the specimen width.

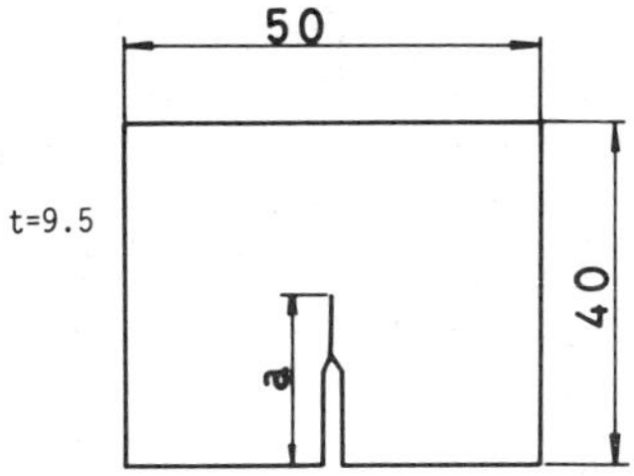

FIG. 1—*Specimen geometry*

Dynamic Loading Device

The new loading device is shown in Fig. 2. This consists of a launching device, load transfer rod, and bend fixture. A specimen was placed between the rod and the fixture. The launching device consists of a cylindrical barrel, 2000 mm long by 20 mm inner diameter, a reservoir for nitrogen gas, and a solenoid valve. A projectile, 20 mm long by 20 mm diameter, was positioned at the end of the barrel, and the reservoir was filled with nitrogen gas to the desired pressure. The projectile was launched by opening the solenoid valve between the barrel and the reservoir. The pressurized gas flew into the barrel at high speed and pushed the projectile to accelerate it against the load transfer rod. Impact velocity was controlled by the amount of gas pressure.

A strain gage was mounted on the midway position in the longitudinal direction of the load transfer rod to measure stress waves traveling towards the specimen after the collision of the projectile and returning from the interface between the rod and the specimen.

Results

Critical Amplitudes for Crack Instability

The method to determine the critical tensile pulse for the crack instability is the same as that used in a previous experiment [*8*]. The impact velocity of the projectile was gradually increased shot by shot. After each impact, the initial

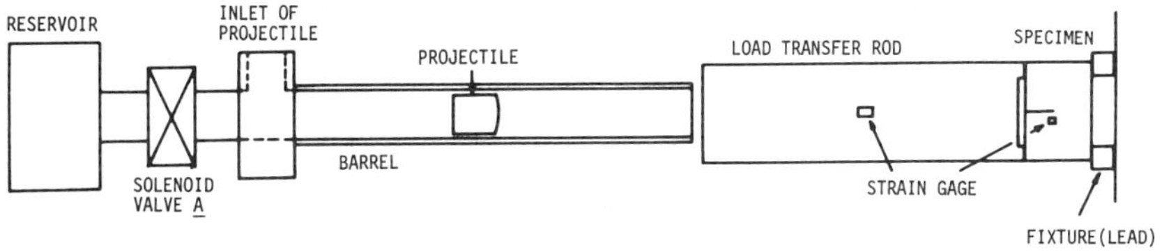

FIG. 2—*Loading device.*

fatigue crack tip was replicated with plastic tape, and the replica was examined at ×400 with an optical microscope to determine whether the crack had grown. A similar method was used to determine the critical magnitude of the impact bending load in the test described above.

Figures 3*a* and 3*b* show the experimental results on the two materials under 80 μs pulses and the results under 40 μs pulses obtained in the previous work [*8*]. Open circles indicate that no crack growth was observed; solid circles indicate crack instability. In all three materials, the critical pulse amplitudes with 40 and 80 μs durations decreased with increase of the crack length. On the other hand, the critical pulse amplitude with 20 μs duration for 4340 steel was independent of the crack length within the range investigated [*8*]. Results

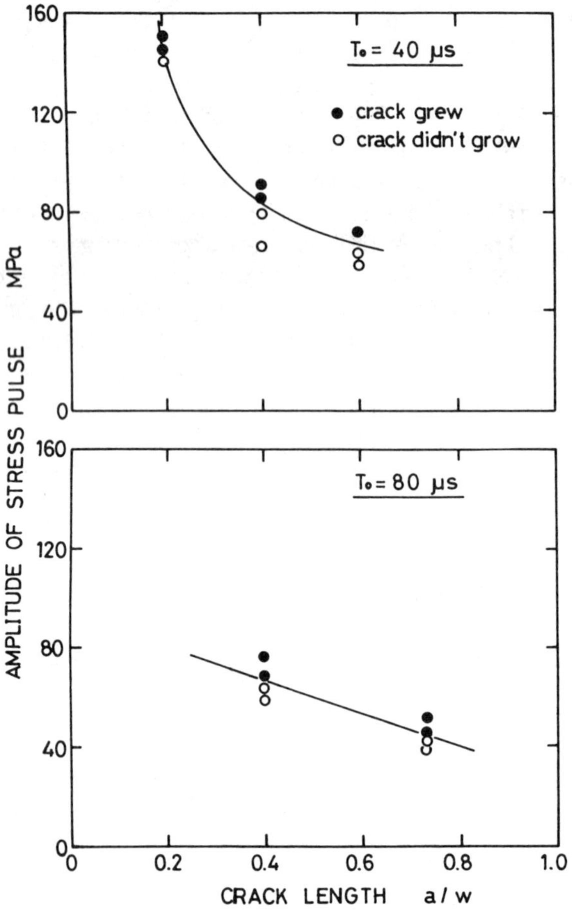

FIG. 3*a*—*Critical amplitude of stress pulse for crack instability as a function of initial crack length for 1018 steel.*

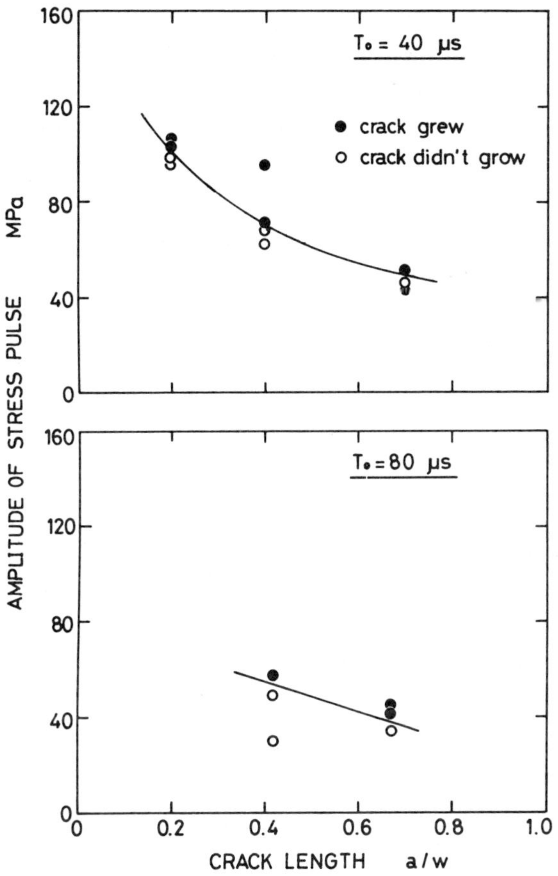

FIG. 3*b*—*Critical amplitude of stress pulse for crack instability as a function of initial crack length for 6061 aluminum.*

of the crack instability under 18 μs impact bending loads for 1018 steel were obtained in this experiment, but the results for 6061 aluminum could not be obtained due to insufficient capacity of the loading device. Since the crack tip stress intensity history was calculated not by the critical bending load but by the strain history measured by a strain gage mounted near the crack tip, the experimental results of the bending load are not shown here.

Critical Stress Intensity Histories

The crack tip stress intensity histories experienced by the cracks in the crack instability experiments shown in Figs. 3*a* and 3*b* were calculated to consider the results in light of dynamic fracture mechanics. Since the finite element code used in the calculation had been developed by one of the present authors and was explained in a previous paper [8], it is described briefly here. In the code, constant strain triangle elements were used and the Newmark technique in the finite difference method was utilized to solve the differential equation in respect to time. Dynamic stress intensity was calculated from the $\hat{J}$ integral proposed by Kishimoto et al [9].

Critical dynamic stress intensity histories for 1018 steel under the three durations are shown in Fig. 4. The history under the 18 microsecond bending load was one calculated from the measured strain history based on the stress intensity-strain near the crack tip-diagram calibrated under the static load. It is seen that the maximum values of stress intensity decreases with increase of the pulse duration.

Critical Stress Intensity and Minimum Time

Figure 5 plots the maximum values of the critical stress intensity as a function of the initial crack length for two materials and three durations. The values for a/w of 0.7 under 18 μs duration and for a/w of 0.2 under 40 μs duration are quite higher than the other results under each duration. Detailed examination of the strain histories measured by the strain gage mounted near the crack tip showed that the duration of the strain histories for a/w of 0.7 was about 13 μs, while the duration for the other a/w was about 18 μs. It may be that the higher value for a/w of 0.7 resulted from the decrease in

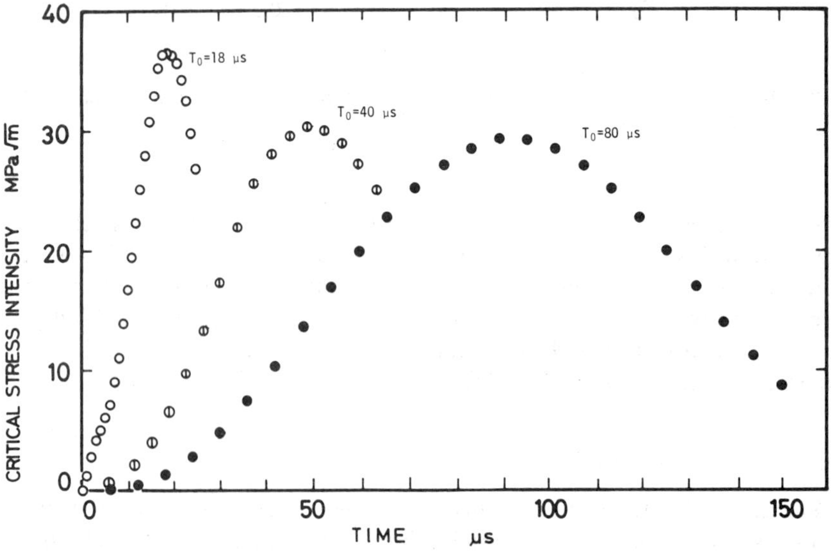

FIG. 4—*Critical dynamic stress intensity histories for 1018 steel under three pulse durations.*

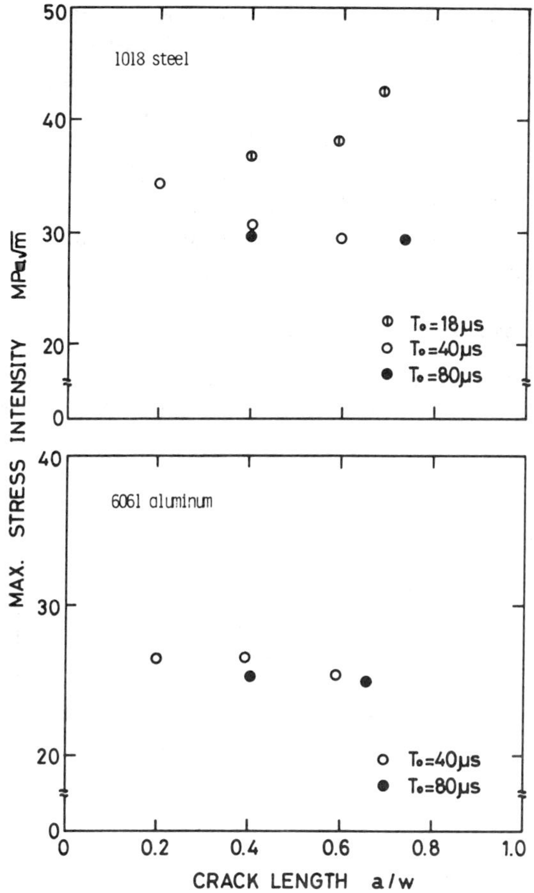

FIG. 5—*Maximum value of critical stress intensity history as a function of initial crack length.*

the duration of stress intensity history. On the other hand, there is no reason to be found for the results under the 40 μs duration. If the maximum values shown in Fig. 5 are taken as the dynamic fracture toughness values for the materials, that of 1018 steel decreases with increase of the pulse duration. The results for 6061 aluminum are too few for us to examine the dependence on the pulse duration.

Kalthoff and Shockey [7] proposed the minimum time criterion that a crack starts to propagate in an unstable manner when dynamic stress intensity exceeds a critical value for a certain time, called the *minimum time*. If the minimum time is postulated as a given time for a given material, the critical stress intensity value is calculated from the critical stress intensity history as shown in Fig. 4. The critical stress intensity values were obtained for several minimum times assumed as the trial. Those for the tested crack lengths and stress pulse durations deviated in the most narrow width for a certain time. The standard deviations of those values were 0.73 MPa$\sqrt{m}$ and 0.19 MPa$\sqrt{m}$ for 1018 steel and 6061 aluminum respectively. The dynamic fracture toughness value was determined as the average of those values and that time was taken as the minimum time. They were 29.7 MPa$\sqrt{m}$ and 11 μs for 1018 steel and 25.2 MPa$\sqrt{m}$ and 9 μs for 6061 aluminum. Homma et al [*8*] indicated that for hardened 4340 steel, the minimum time and the dynamic fracture toughness were 7 μs and 31.7 MPa$\sqrt{m}$. The dynamic fracture toughness values for the tested materials are shown in Fig. 6.

Elasto-Plastic Stress History near the Crack Tip

For one-dimensional elasto-plastic stress wave propagation in materials possessing the bilinear tensile stress-tensile strain relation, the plastic stress

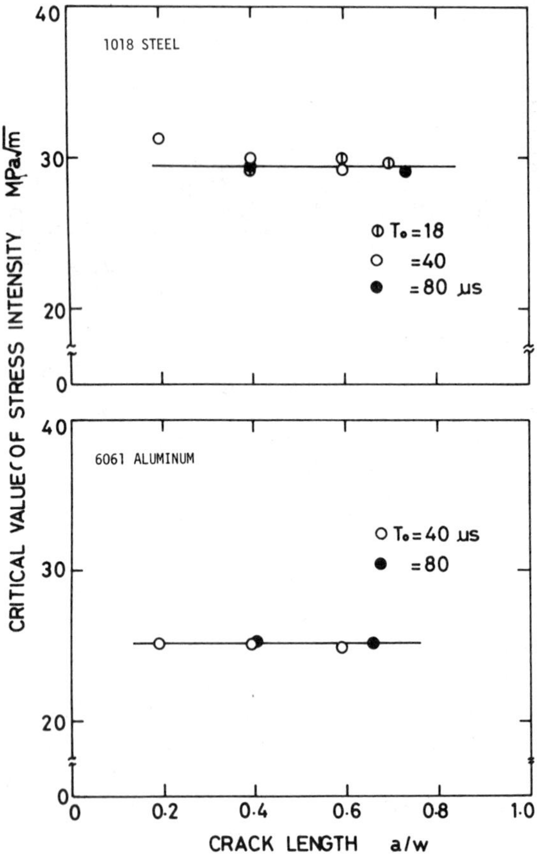

FIG. 6—*Critical stress intensity for crack instability based on the minimum time criterion.*

wave runs after the elastic stress wave at the velocity of $\sqrt{C/\rho}$, where C is the plastic modulus and ρ is the density, less than the elastic stress wave velocity [10]. Therefore the dynamic plastic deformation in the vicinity of the crack tip loaded by the stress pulse will develop later than the anticipated time based on quasi-static consideration from the elastic stress analysis results.

In order to examine the difference between the elastic and the elasto-plastic stress histories near the crack tip, dynamic stress analysis was carried out by finite element methods using constant strain triangular mesh. The constitution equation of the material was assumed by a linear elastic power law plastic strain hardening and the deformation theory was used for the plastic flow The initial stress method was used for solution of the nonlinear problem by the finite element method. The specimen, 88.9 mm wide, 914.4 mm long, and 9 mm thick, with a single-edge-crack in the midsection, was loaded by the pressure on the crack surface. This loading and the specimen geometry correspond to those for the results under 40 and 80 μs duration stress pulses for three materials and for the results of 4340 steel under 20 μs duration stress pulses. The crack length was 60% of the specimen width, and the pressure applied to the crack varied with time as

$$\sigma = \frac{1}{2}\sigma_0\left[1 + \sin\left\{\frac{\pi}{T_0}\left(t - \frac{T_0}{2}\right)\right\}\right] \tag{1}$$

where σ_0 is amplitude of the pressure, T_0 is time length at half of the amplitude (defined as the duration of the pulse in this paper), and t is time. The plastic zone size was estimated for the crack loaded by the critical stress pulse using the quasi-static formula

$$r_p = \frac{(1 - 2\nu)^2}{2\pi}\left(\frac{K}{\sigma_{ys}}\right)^2 \tag{2}$$

The size was 0.05 mm for 1018 steel and 0.2 mm for 6061 aluminum. Since very fine meshes are necessary to carry out the elasto-plastic stress analysis for the specimen in which the crack initiation took place, it is difficult to obtain reliable and stable numerical results by the finite element method. Therefore, in the calculation, the yield strength was reduced to about one half or one fourth of the actual yield strength, and the dependence of the elasto-plastic stress history on the yield strength was examined qualitatively. Mechanical properties of the materials used in the analysis are shown in Table 1.

Five kinds of materials are considered in the calculation. Steel 1 and Steel 2 have the same mechanical properties except the plastic modulus C and the strain hardening exponent n. The yield strength of Steel 3 is about one half of that of Steel 2. Although Steel 4 has the same yield strength as that of Steel 3, it behaves as a harder material than the latter. If Steel 3 were to be regarded as 1018 steel, then Steel 1 would correspond to 4340 steel.

TABLE 1—*Mechanical properties of test materials.*[a]

Material		Steel 1	Steel 2	Steel 3	Steel 4	Aluminum
Young's modulus, GPa		206	206	206	206	72.5
Poissons' ratio		0.3	0.3	0.3	0.3	0.3
Density, kg/mm^3		7.9×10^3	7.9×10^3	7.9×10^3	7.9×10^3	2.8×10^3
Yield strength, MPa		294	294	196	196	196
Plastic modulus, MPa		1176	196	196	1960	316
Strain hardening exponent		0.2	1.0	1.0	0.2	0.126
σ_{ep}/σ_e		0.86	0.85	0.63	0.97	0.84
Time lag	$T_0 = 18\ \mu s$	2.5	2.5	5.0	0.0	...
	$T_0 = 40\ \mu s$	2.5	2.5	...	...	5.0

[a] $\bar{\sigma} = C(A + \bar{\varepsilon}_p)^n$ = constitution equation for plastic deformation.

Figure 7 compares the stress histories at the crack tip mesh for Steels 1 and 2 under the stress pulse of $T_0 = 40$ μs with that obtained by elastic analysis for these steels. Before plastic deformation takes place at the crack tip, the stress-time curves for these materials coincide with the elastic analysis result. After yielding occurs at the crack tip mesh, the stress by elasto-plastic analysis increases at a lower rate and reaches the peak later than that by the elastic analysis. The time lag of the elasto-plastic stress histories must result from the slowness of the plastic stress wave propagation as mentioned above. The time lags for the five materials are listed in the bottom row of Table 1. It is seen that the time lag depends on the yield strength and the strain hardening prop-

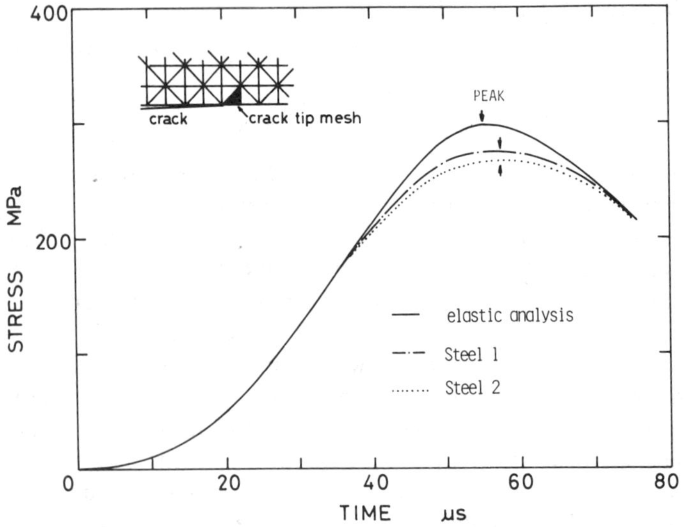

FIG. 7—*Stress histories at the crack tip mesh for Steels 1 and 2 under the crack surface pressure with 40 μs duration.*

erties. The ratio of peak stress obtained by elastic analysis to peak stress obtained by elasto-plastic analysis is considered to be a factor representing the strain hardening properties. High ratio means high strain hardening property. The ratios are also indicated in the seventh row from the top of Table 1. The time lag decreases with increase of the ratio.

Discussion

Crack instability is brought about by fracture of the material ahead of the crack tip. The fracture criterion depends on microscopic fracture mechanisms such as cleavage, dimple, and intergranular cracking. In any fracture mechanism, however, the fracture criterion may be related to the elasto-plastic stress at the crack tip. In order to examine the obtained minimum times for the tested materials, it is simply postulated that the crack instability occurs when the stress at the crack tip exceeds a certain value.

As mentioned in the previous section, the dynamic elasto-plastic stress histories at the crack tip are delayed in comparison with those obtained by the elastic analysis. Therefore, even if the dynamic stress intensity equals the quasi-static fracture toughness value for loading rate-insensitive materials, the crack does not start to propagate until the elasto-plastic stress at the crack tip exceeds the critical value. In other words, if the dynamic stress intensity immediately decreases just after it reaches the fracture toughness value, the crack remains restful because the stress at the crack tip does not reach the critical value.

Minimum time determined experimentally in this paper will be related to the time lag of the elasto-plastic stress history at the crack tip. The FEM dynamic elasto-plastic stress analysis carried out in this paper can allow us only to qualitatively consider that relationship. The calculation results indicate that the time lag of the elasto plastic stress history decreases with increase of the strain hardening. It coincides with the fact that the minimum time of 1340 steel is the shortest of the three tested materials. The minimum time of 6061 aluminum is shorter than that of 1018 steel. The ductility of 6061-T651 is about 10% for the elongation while that of 1018 is about 30%, and the difference between the tensile strength and yield strength is about 50 MPa for 6061-T651 and around 100 MPa for 1018 steel. Therefore the ratio of the strength difference to the ductility is larger in 6061 aluminum than in 1018 steel; thus 6061 aluminum has a higher strain hardening property than 1018 steel. This corresponds to the fact that the minimum time of the former is shorter than that of the latter.

We conclude that the minimum time is qualitatively related to the development of the dynamic elasto-plastic stress at the crack tip. When we discuss quantitatively the minimum time, we must take into account the relation between the microscopic fracture mechanism and the crack tip plasticity.

Conclusions

The following conclusions are drawn from tests on metallic materials such as hardened 4340 steel, cold-rolled 1018 steel, and 6061-T651 aluminum:

1. If a crack is loaded by a very short stress pulse so that the crack experiences the dynamic stress intensity history in less than 20 μs, a much greater dynamic stress intensity than that generated by the longer stress pulse (e.g., a 40-μs pulse) is necessary to cause crack initiation.
2. The minimum time criterion has been successfully applied to explain change of maximum dynamic stress intensity with stress pulse duration; minimum times were 7 μs for 4340 steel, 11 μs for 1018 steel, and 9 μs for 6061 aluminum.
3. Minimum time has been discussed in association with elasto-plastic stress histories at the crack tip and has been qualitatively related to the time lag for the development of the dynamic elasto-plastic stress at the crack tip in comparison with the stress obtained by the elastic analysis.

References

[*1*] Shockey, D. A. and Curran, D. R. in *Progress in Flaw Growth and Fracture Toughness Testing, ASTM STP 536*, American Society for Testing and Materials, Philadelphia, 1973, pp. 297–311.
[*2*] Costin, L. S., Duffy, J., and Freund, L. B. in *Fast Fracture and Crack Arrest, ASTM STP 627*, American Society for Testing and Materials, Philadelphia, 1977, pp. 301–318.
[*3*] Homma, H., Ushiro, T., and Nakazawa, H. in *Proceedings*, 2nd International Conference on Mechanical Behavior of Materials, Boston, 1976, p. 1453.
[*4*] Achenbach, J. D. and Brock, L. M. in *Dynamic Crack Propagation*, G. C. Sih, Ed., Noordhoff, The Netherlands, 1973, p. 529.
[*5*] Lehnigk, S. H. in *Dynamic Crack Propagation*, G. C. Sih, Ed., Noordhoff, The Netherlands, 1973, p. 529.
[*6*] Sih, G. C., Embley, G. T., and Ravera, L. M., *International Journal of Solids and Structures*, Vol. 8, 1972, p. 977.
[*7*] Kalthoff, J. F. and Shockey, D. A., *Journal of Applied Physics*, Vol. 43, No. 3, 1977, p. 986.
[*8*] Homma, H., Shockey, D. A., and Murayama, Y., *Journal of the Mechanics and Physics of Solids*, Vol. 31, No. 3, 1983, p. 261.
[*9*] Kishimoto, K., Aoki, S., and Sakata, M., Reprint 780-9, Japan Society of Mechanical Engineers, 1978, p. 75 (in Japanese).
[*10*] Johnson, W., *Impact Strength of Materials*, Edward Arnold, London, 1972, p. 212.

DISCUSSION

J. H. Giovanola[1] (*written discussion*)—The authors assume that the minimum time fracture theory is required to predict crack instability in structural

[1]SRI International, Menlo Park, CA 94025.

alloys loaded by stress pulses of durations less than 40 μs. Then, on the basis of experimental and analytical data reported in the paper and in Ref *1*, they postulate that the minimum time during which the crack must be loaded above the fracture toughness level can be interpreted as the time necessary to expand the plastic zone at the tip of the crack. The concept of a minimum time for dynamic crack initiation and the proposed interpretation of the minimum time are appealing. The reported experimental evidence, however, does not convincingly support the conclusions drawn by the authors.

Although a minimum time to fracture can be obtained from the data in the present paper and in Ref *1*, the minimum time fracture criterion may not be the only criterion for explaining these experimental results. One can argue, for instance, that the fracture data could also be satisfactorily explained in terms of the simple criterion

$$(K_I^{dyn})_{max} = K_{Id}$$

(where K_I^{dyn} is the applied stress intensity factor) once the scatter in the experimental data is considered.

Further, in fracture experiments, 1018 steel and 6061 aluminum exhibit rather large plastic zones. Therefore it is questionable whether the elastic stress intensity factor concept can be used to analyze the fracture data for these materials. The results of the finite element calculation presented in the paper may indicate the need for an elastic-plastic fracture analysis, rather than an interpretation of the minimum time.

A final remark pertains to the definition of the point of dynamic crack initiation. The authors measured the initiation toughness using the stress intensity required to extend the crack 20 to 50 μm. Their values for 4340 and 1018 steel are much smaller (about one half) than dynamic toughness values reported in the literature for these materials under comparable test conditions [*2–4*]. However, in Ref *2* to *4*, critical stress intensity for crack initiation was determined from strain gage measurements, which require a greater amount of crack extension (on the order of 500 μm) to indicate a discontinuity in the record. This may explain the difference in the measured fracture toughness values. Some operational definition of the point of crack initiation should be agreed upon for measuring dynamic fracture toughness values to allow valid comparisons of experimental data. A definition consistent with the procedure of ASTM E 399 would be desirable because it would allow comparison of quasi-static and dynamic toughness values.

H. Homma et al (*authors' closure*)—As pointed out, experimental K_{max} data for 1018 steel (Fig. 5) were widely scattered, but the reason why the result for a/w of 0.7 under 18 μs stress pulse was higher than those for other a/w was described in the paper. Since the amount of data is small, it is difficult to draw the conclusion that the K_{max} value for 1018 steel increases with decrease of stress pulse duration from only Fig. 5. Statistical analysis of the experimental data will give us a rather convincing argument in one sense. A statistical test was carried out to examine whether there is a significant difference among the mean values of K_{max} for the three stress pulse durations. It was deduced with 90% reliability from the test that the mean value of K_{max} for 18 μs duration significantly differs from the mean values for the other durations. Therefore, even if the scatter in the experimental data is considered, the pointed simple criterion can not be applied to explain the fracture data shown in Fig. 5.

The plastic zone sizes at the crack initiation roughly estimated from quasi-static formula were about 0.05 mm for 1018 steel and 0.2 mm for 6061 aluminum as described in the paper. These sizes satisfied by a wide margin the requirement for small-scale yielding. Therefore the elastic stress intensity concept is applicable. The computational results in the paper indicate the need for an elastic-plastic stress analysis to understand the dynamic fracture problems even if small-scale yielding conditions appear near the crack tip.

As pointed out, the discrepancy between the dynamic fracture toughness values was brought about by the method of detecting crack initiation. However, fracture toughness for 6061 aluminum determined by the present method almost agreed with the static plane strain fracture toughness because this material is loading rate insensitive. Also, dynamic fracture toughness of A533B steel obtained by the present method fell on an extrapolated line of the fracture-toughness/loading-rate curve in Ref *5*. To conclude, the authors believe that the present method for dynamic fracture toughness determination is quite reasonable.

Discussion References

[*1*] Homma, H., Shockey, D. A., and Murayama, Y., *Journal of the Mechanics and Physics of Solids*, Vol. 31, 1983, pp. 261-279.

[*2*] Giovanola, J. H., this publication, pp. 307-328.

[*3*] Costin, L. S., *Journal of Engineering Materials and Technology*, Vol. 101, 1979, pp. 258-264.

[*4*] Costin, L. S., Duffy, J., and Freund, L. B. in *Fast Fracture and Crack Arrest, ASTM STP 627*, G. T. Hahn and M. F. Kanninen, Eds., American Society for Testing and Materials, 1977, pp. 301-318.

[*5*] Homma, H., Watanabe, N., and Nakazawa, H. in *Proceedings*, 5th European Conference on Fracture, Lisbon, Vol. I, 1984, p. 241.

A. Shukla[1] *and S. Anand*[1]

Dynamic Crack Propagation and Branching under Biaxial Loading

REFERENCE: Shukla, A. and Anand, S., **"Dynamic Crack Propagation and Branching under Biaxial Loading,"** *Fracture Mechanics: Seventeenth Volume, ASTM STP 905*, J. H. Underwood, R. Chait, C. W. Smith, D. P. Wilhem, W. A. Andrews, and J. C. Newman, Eds., American Society for Testing and Materials, Philadelphia, 1986, pp. 697–714.

ABSTRACT: A photoelastic study of high-speed crack propagation and crack branching is conducted under biaxial loading conditions. Cross-type specimens fabricated from Homalite 100 are loaded in a specially designed loading fixture where loads parallel and perpendicular to the crack can be controlled independently. The experimental data obtained are analyzed using dynamic stress field equations and multiple points to obtain the instantaneous stress intensity factor K, the stress field parallel to the crack σ_{ox}, the crack velocity $\dot{a}$, the branching stress intensity factor K_{br}, and the branching angle θ. The results indicate a strong influence of the sign as well as the magnitude of σ_{ox} on both crack propagation and branching.

KEY WORDS: stress intensity factor, dynamic photoelasticity, crack branching, stress parallel to crack, Homalite 100

Dynamic fracture studies have received considerable attention in the past 20 years. Several investigators [*1–6*] have tried to characterize dynamic fracture in terms of the instantaneous value of the stress intensity factor $K(t)$ and crack velocity $\dot{a}$. Results obtained to date indicate that the crack arrests when $K(t) < K_{IA}$, where K_{IA} is a toughness dependent upon the material and temperature.

Several studies of crack branching have also been reported in the literature [*7–16*]. These studies have dealt primarily with the value of the strain energy release rate at the instant of branching, the branching angle, and the branching velocity. Although some progress has been made, the prediction of branching is not possible and the problem has not been formulated in sufficient detail to permit its solution.

[1]Associate Professor and Graduate Assistant, respectively, Department of Mechanical Engineering and Applied Mechanics, University of Rhode Island, Kingston, RI 02881.

There is disagreement in the literature about the possible existence of a branching strain energy release rate and the branching velocity as a material property independent of the size and shape of the fracture test specimen. Experiments performed by Irwin et al [2] and by Kobayashi et al [8] indicate that neither the branching stress intensity factor nor the branching strain energy release rate are unique material properties. Their results, however, are in contradiction to those obtained by Congleton and Petch [12] and Doll [16]. Recently Ramulu and Kobayashi [17,18] have proposed a dynamic branching criterion based on the value of K and a critical distance r_c. This is the first time a criterion has been proposed which includes higher order terms, since r_c depends on the stress field parallel to the crack, σ_{ox}. However, the criterion does not take into account the sign of σ_{ox} since both r_c and branching angle σ depend on $(\sigma_{ox}/K)^2$.

This study deals with high speed crack propagation and branching in a brittle polyester material, Homalite 100, under biaxial loading conditions. The magnitude and sign of σ_{ox} were systematically varied in a series of experiments and their effect on crack propagation and branching studied. Cross-type specimens were subjected to three kinds of loading, namely, uniaxial, biaxial tensile-compressive, and biaxial tensile-tensile. The nonsingular stress field coefficient σ_{ox} was controlled by varying the load parallel to the crack. Experimental results showed that K and σ_{ox} varied systematically as the crack propagated through the specimen. The stress intensity factor at branching did not show much variation, whereas the branching angle varied from 24 to 73° as σ_{ox} was changed from negative to highly positive values.

Dynamic Stress Field Representation

Irwin [19] has shown that, for a crack tip stress pattern translating in the positive x-direction at a fixed speed, the dilatation, Δ, and rotation, ω, can be expressed as

$$\begin{aligned}\Delta &= \frac{\partial u}{\partial x} + \frac{\partial v}{\partial y} = A(1 - \lambda_1^2)\,\mathrm{Re}\,\Gamma_1(z_1)\\ \omega &= \frac{\partial v}{\partial x} - \frac{\partial u}{\partial y} = B(1 - \lambda_2^2)\,\mathrm{Im}\,\Gamma_2(z_2)\end{aligned} \qquad (1)$$

where λ_1, λ_2, z_1, and z_2 are as defined in Fig. 1. Using Hooke's Law:

$$\begin{aligned}\sigma_{xx} &= \mu\,[\,A\,(1 + 2\lambda_1^2 - \lambda_2^2)\mathrm{Re}\Gamma_1 - 2B\lambda_2\mathrm{Re}\Gamma_2]\\ \sigma_{yy} &= \mu\,[-A\,(1 + \lambda_2^2)\mathrm{Re}\Gamma_1 + 2B\lambda_2\mathrm{Re}\Gamma_2]\\ \tau_{xy} &= \mu\,[-2A\lambda_1\mathrm{Im}\Gamma_1 + B(1 + \lambda_2^2)\mathrm{Im}\Gamma_2]\end{aligned} \qquad (2)$$

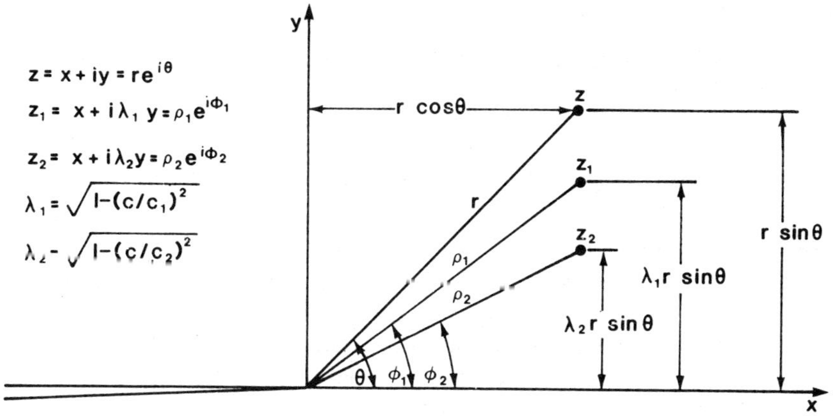

FIG. 1—*Coordinate system for crack tip stress analysis.*

where the constants A and B have to be evaluated, after selecting general stress functions, Γ_1 and Γ_2, so as to satisfy the boundary conditions for the crack problem of interest. For the opening mode case, one logical choice is

$$\Gamma_1 = \sum_{j=0}^{j=J} C_j\, z_1^{j-\frac{1}{2}} \quad \text{with} \quad \Gamma_2 = \sum_{j=0}^{j=J} C_j\, z_2^{j-\frac{1}{2}} \tag{3}$$

The leading coefficient, C_0, is related to the stress intensity factor, $K = C_0\sqrt{2\pi}$, and the leading term is the familiar inverse-square root stress singularity. The symmetry condition requires that $B = 2\lambda_1 A/(1 + \lambda_2^2)$.

An alternative choice is

$$\Gamma_1 = \sum_{m=0}^{m=M} D_m z_1^m \quad \text{with} \quad \Gamma_2 = \sum_{m=0}^{m=M} D_m z_2^m \tag{4}$$

which is required to completely describe the stress field in specimens with finite boundaries, in a manner analogous to that demonstrated to be the case in the static problem [*20*]. The boundary conditions on the crack faces require that $B = A(1 + \lambda_2^2)/2\lambda_2$. The leading term, D_0, gives rise to a superposed constant stress in the direction of crack propagation which is similar to the σ_{ox}-term in Irwin's static near-field equations [*21*]. This term can also be regarded as the far-field biaxiality correction factor studied extensively by Liebowitz and co-workers [*22*] for the static case.

Parameter Determination Using the Least-Squares Method

Equations 2 to 4 when combined with the stress-optic law can be used to relate the fringe order, N, at any point in an isochromatic field with the unknown real coefficients, C_j and D_m, through:

$$(Nf_\sigma/2t)^2 = \tau^2_{\max} = 1/4(\sigma_{yy} - \sigma_{xx})^2 + \tau^2_{xy} \tag{5}$$

where f_σ is the fringe sensitivity of the model material and t is the model thickness. The first step in the analysis of an isochromatic fringe pattern is to take a region around the crack tip from the experimental pattern being analyzed, extract a large number of individual data points, and determine the coordinates and fringe order at each point. These data points are then used as inputs to an overdetermined system of nonlinear equations of the form of Eq 5 and solved in a least-squares sense for the unknown coefficients. As a final check, the best-fit set of coefficients is used to reconstruct the fringe pattern over the region of data acquisition to ensure that the computed solution set does, in fact, predict the same stress distribution as that observed experimentally.

When analyzing dynamic stress patterns, the data acquisition region is usually restricted to that portion of the stress pattern that can be seen to translate with moderate changes in order to approximate the constant crack-speed assumption. In cases where the crack tip is approaching a specimen boundary, the data region should be restricted to no more than one quarter to one third of the distance from crack tip to the boundary. The number of coefficients necessary for an adequate representation of the stress field over the data acquisition region can be estimated by examining, as a function of the number of parameters, the average fringe order error, the values of the leading coefficients, and the reconstructed (computer-generated) fringe pattern corresponding to a given set of coefficients. Stability of the error term and leading coefficients, as well as good visual match between experimental and reconstructed patterns, indicates convergence to a satisfactory solution.

Experimental Procedure

The fracture specimen used in these studies was a cross-type specimen (Fig. 2). This type of specimen was chosen because it is easy to load in biaxial mode. The specimens were fabricated from a brittle polyester material, Homalite 100, whose properties have been characterized [*1*].

A starter crack was machined in the specimen and the crack tip blunted so that the models could be loaded to high values of K_Q (where K_Q is the stress intensity factor associated with the blunt starter crack tip). Loads were applied perpendicular and parallel to the crack with the help of hydraulic loading cylinders and recorded by using the quartz type of load cells. A special loading fixture was designed and fabricated which could be used for uniaxial

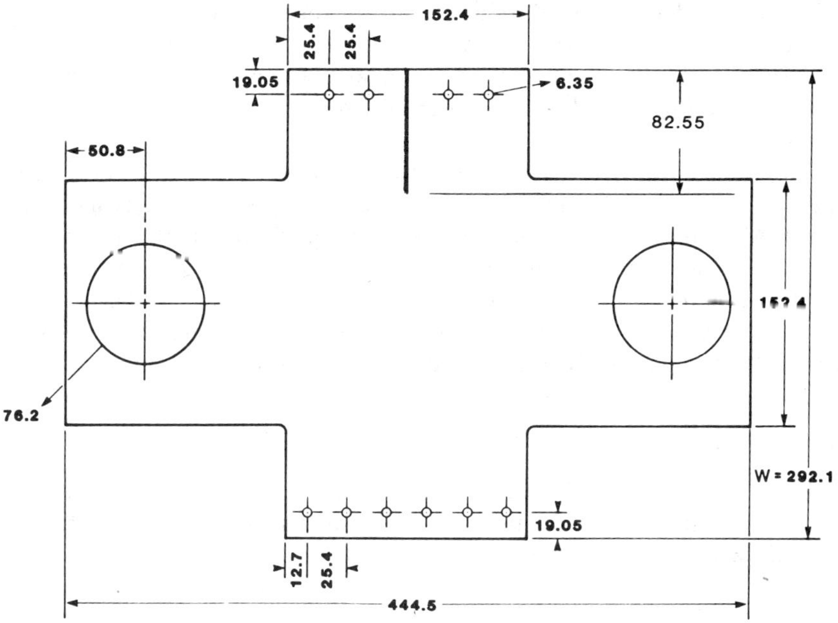

FIG. 2—*Specimen geometry (dimensions in millimetres).*

or biaxial loading. A schematic of the loading frame used for biaxial tension-tension experiment is shown in Fig. 3. The horizontal loading is accomplished by simply pulling the ram of the cylinder with a hydraulic pump. The vertical loading is accomplished by a system of pulleys and a steel rope. When the ram of the cylinder is moved down, it produces equal tension in the rope on either

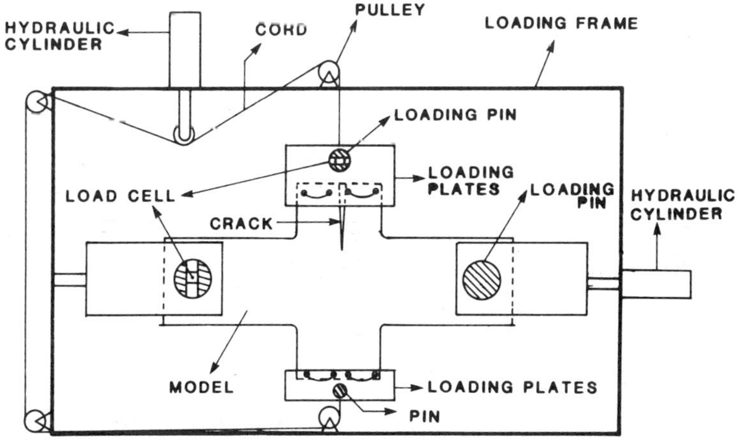

FIG. 3—*Loading frame along with the specimen for tension-tension loading.*

side of the model. In biaxial tension-compression experiments the hydraulic cylinder at the top is moved to the middle of the loading frame. The rope is removed and flat plates are kept at the top and bottom of the model and the ram of the cylinder pushed down vertically to achieve compressive parallel stress.

The model and the loading frame were kept in the optical bench of a high-speed multiple spark camera. The optical bench of this camera is shown in Fig. 4. After the models were loaded to specified loads, the crack was initiated with a solenoid actuated knife. As the crack began to propagate, it cut a line of silver-conducting paint and initiated the high-speed camera. The camera was set at a rate of about 100 000 frames per second and provided 20 isochromatic photographs at discrete times during the run-branching event.

Results and Discussion

Three different types of experiments were conducted, namely, uniaxial, biaxial tension-compression, and biaxial tension-tension. The results obtained from these experiments are discussed below.

Uniaxial Experiments

Three uniaxial experiments were conducted with normal stresses equal to 1.61, 3.85, and 4.52 MPa, respectively. The stresses in all the experiments are defined as the load measured by the quartz cell divided by the cross-sectional area of the arms of the cross specimens. Typical isochromatics obtained during the uniaxial experiments are shown in Fig. 5. In the first experiment the crack propagated after initiation at a constant velocity of 383 m/s. The instantaneous stress intensity factor showed an increasing trend with crack

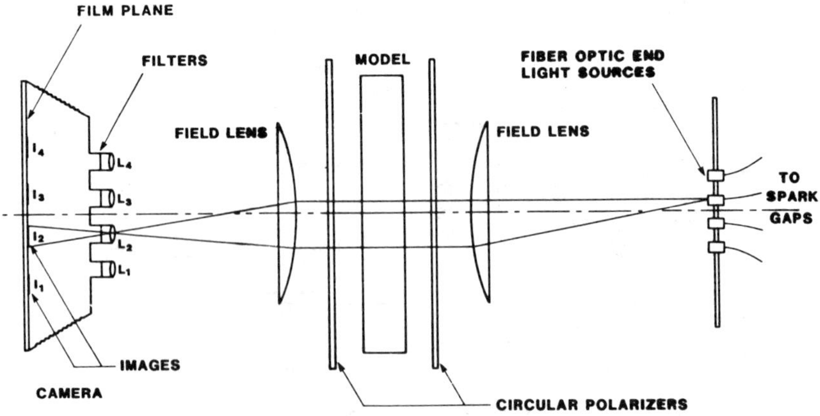

FIG. 4—*Optical bench of the high speed camera.*

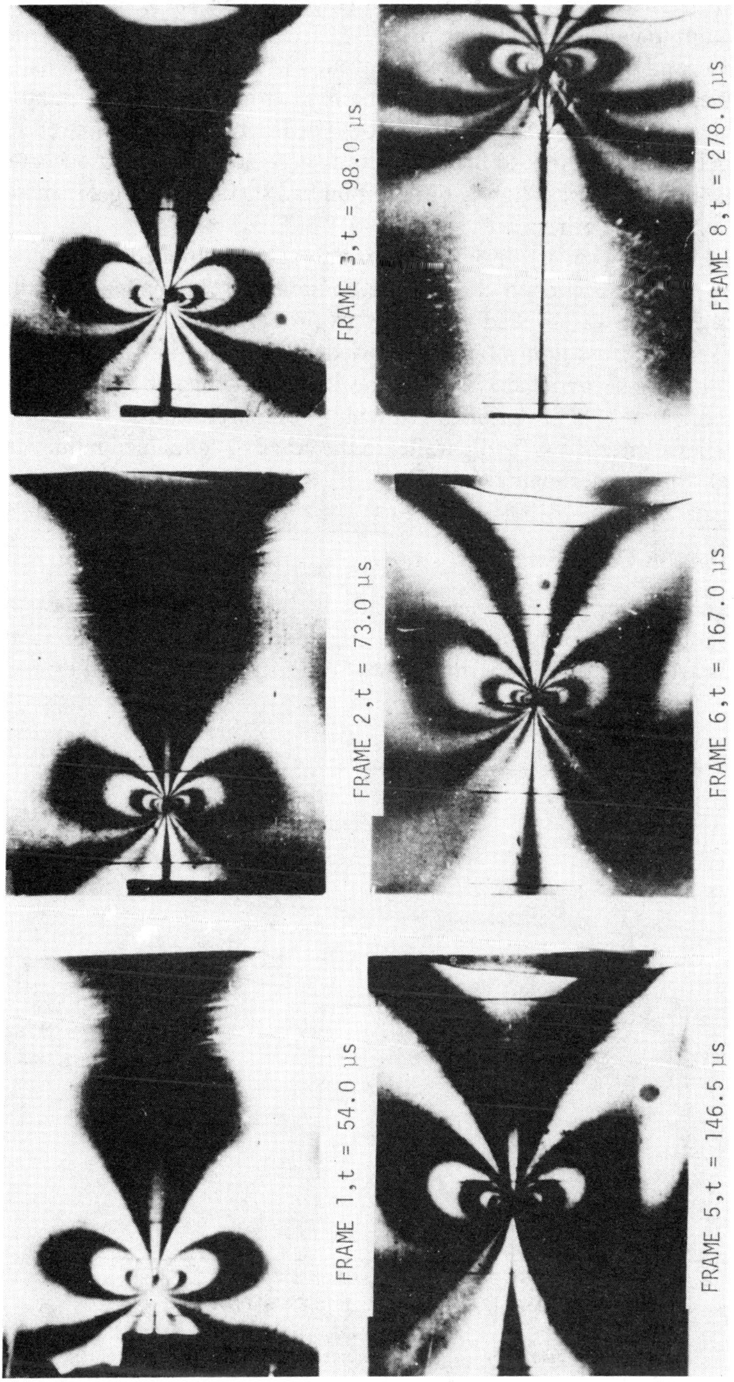

FIG. 5—*Typical isochromatic patterns obtained during uniaxial loading.*

length and varied from 1 to 1.6 MPa$\sqrt{m}$ (Fig. 6). This value of K was not large enough to cause crack branching.

In the second experiment the crack travelled at a slightly higher constant crack velocity of 400 m/s. Since the starting K_Q was higher, the instantaneous K varied from 1.1 to 1.84 MPa$\sqrt{m}$ (Fig. 6). The fracture surface showed several unsuccessful attempts to branch before successful branching occurred. The branching angle was slightly different on two surfaces, being 24° on one side and 22° on the other.

In the third experiment a little higher initial load was applied so that crack branching did not occur too close to the boundary of the model. An initial stress of 4.52 MPa was applied and the crack propagated at a constant velocity of 390 m/s. Stress intensity factor varied between 1.4 and 1.9 MPa$\sqrt{m}$. Crack branching occurred at 1.9 MPa$\sqrt{m}$ with a branching angle of 30°. The crack propagated 12.2 cm prior to branching compared with 17.8 cm in the second experiment. Stress field parallel to the crack σ_{ox} remained mildly negative in all three experiments (Fig. 7).

Biaxial Tension-Compression Experiments

In the next series of experiments an additional compressive load was applied parallel to the crack to observe its influence on crack propagation and branching. A set of three experiments was conducted. In these experiments

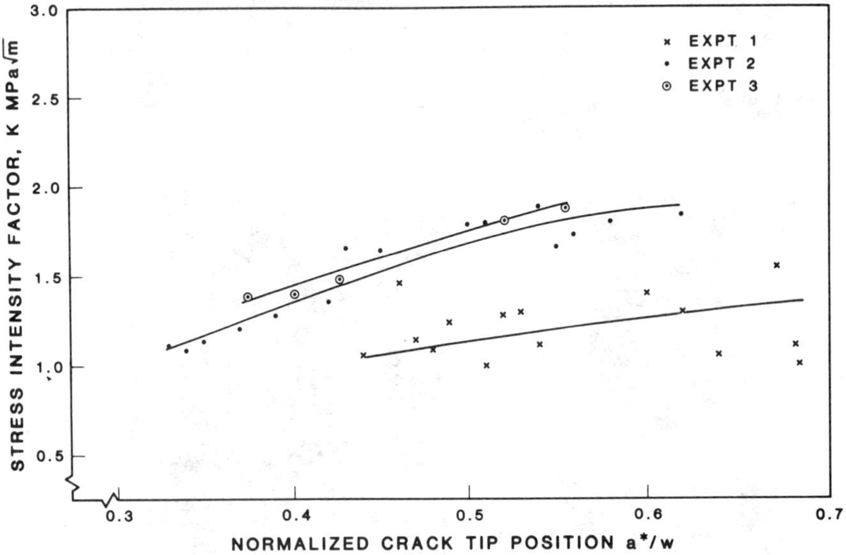

FIG. 6—*Stress intensity factor as a function of normalized crack tip position for uniaxial loading.*

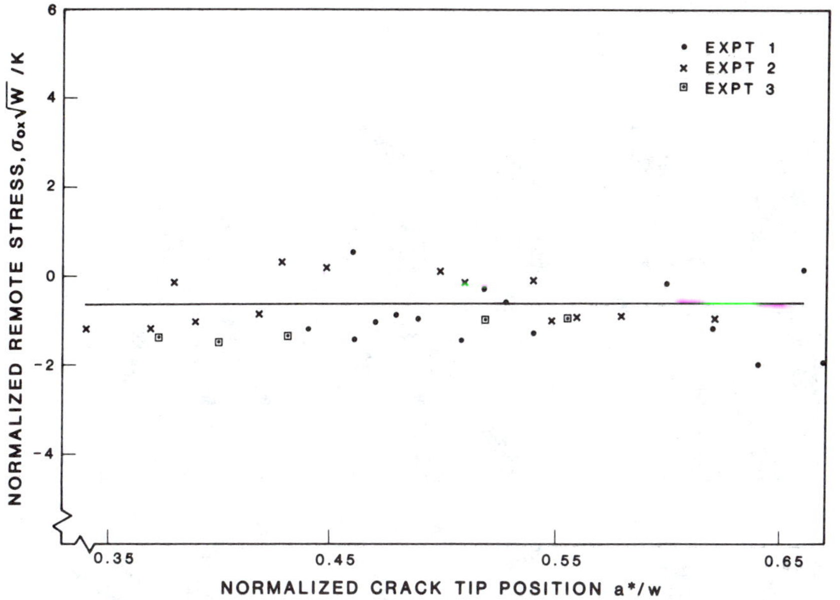

FIG. 7—*Stress parallel to the crack σ_{ox} as a function of normalized crack tip position for uniaxial loading.*

the stresses normal and parallel to the crack were systematically varied. A typical set of isochromatic fringes obtained during the experiment is shown in Fig. 8. The forward lean of the fringes indicates a negative value of σ_{ox}. These fringes were analyzed to obtain the value of K and σ_{ox} in all the experiments.

A tensile normal stress of 3.45 MPa and a compressive parallel stress of 1.24 MPa were applied in the first experiment. The crack did not branch in this experiment, but there was a conspicuous roughness of the fracture surface in the final stages of crack propagation, showing that the crack would have branched at a slightly higher initial K_Q. Like uniaxial experiments, K showed an increasing trend with values ranging from 0.98 to 1.77 MPa$\sqrt{m}$ (Fig. 9) and σ_{ox} oscillated about a mild negative mean of about −1.3 MPa (Fig. 10). The crack velocity was constant at 400 m/s.

In the second experiment the normal stress was increased to 4.6 MPa and the parallel stress was kept the same at 1.2 MPa. Stress intensity factor K and σ_{ox} obtained are shown in Figs. 9 and 10, respectively. Because of the higher K_Q, the K values obtained were fairly high, ranging from 1.7 to 2.01 MPa$\sqrt{m}$. Crack branching was achieved at a K of 2.01 MPa$\sqrt{m}$. Prior to branching, the crack travelled at a constant velocity of 385 m/sec. The branching angle was 29°.

In the third experiment the tensile normal stress was kept the same as in the second experiment (4.6 MPa) and a much higher compressive parallel

FIG. 8—*Typical isochromatic patterns obtained during tension-compression loading.*

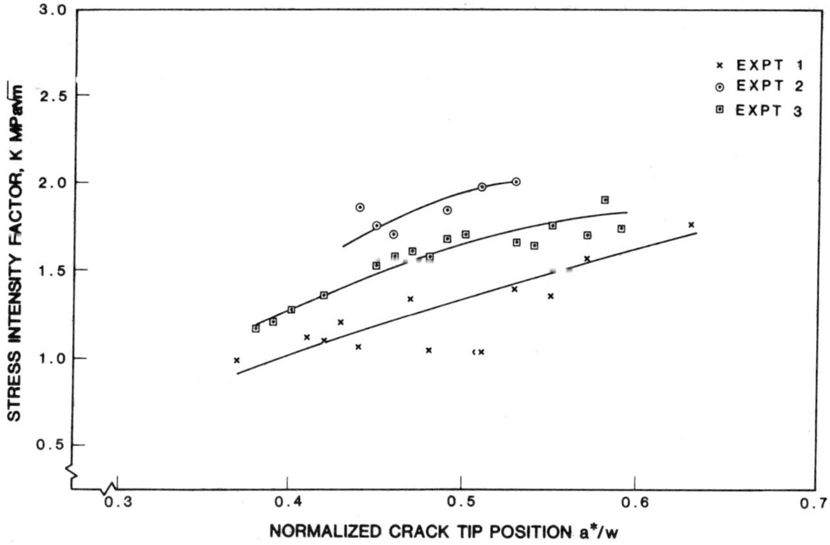

FIG. 9—*Stress intensity factor as a function of normalized crack tip position for tension-compression loading.*

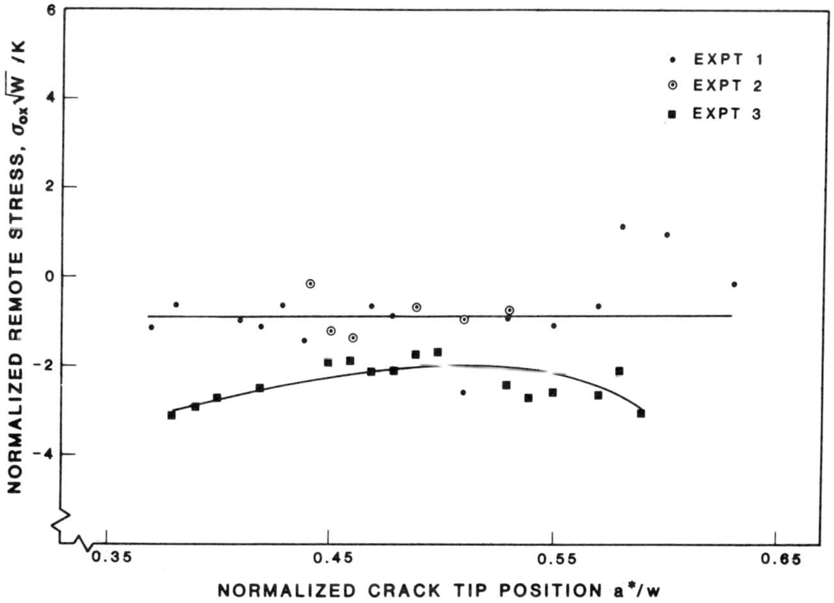

FIG. 10—*Stress parallel to the crack σ_{ox} as a function of normalized crack tip position for tension-compression loading.*

stress of 2.34 MPa was applied. It was observed that the K values for the same crack tip position were lower in this experiment than in the second experiment. σ_{ox} values were more negative due to high initial compression. K and σ_{ox} are plotted in Figs. 9 and 10, respectively. Crack branching occurred at a K value of 1.91 MPa$\sqrt{m}$ and the branching angle was 24.5°. Even though in this and the previous experiment the normal stress was more than the normal stress in the third uniaxial experiment, the pre-branching surface area was larger. This indicates that initial compression parallel to crack tends to suppress branching.

Biaxial Tension-Tension Experiments

A series of four experiments was conducted to observe the influence of tensile stress parallel to crack on propagation and branching. A typical set of isochromatic fringes obtained during the experiments is shown in Fig. 11. Owing to a large positive σ_{ox}, the fringes lean backwards. These fringes resemble the pattern obtained for DCB specimens in Ref *1*. Since σ_{ox} is positive, the instability of crack propagation direction is expected according to Cottrell's theory [*23*].

In the first experiment a low normal stress of 2.91 MPa and tensile parallel stress of 2.76 MPa were applied. Owing to low K_Q at initiation, branching was not obtained. The crack propagated at a constant velocity of 377 m/s. Stress intensity factor ranged from 1.27 to 1.68 MPa$\sqrt{m}$ and σ_{ox} oscillated around −0.1 MPa. K and σ_{ox} are shown in Figs. 12 and 13, respectively.

In the second experiment the normal stress was increased to 4.75 MPa and a low tensile parallel stress of 1.65 MPa was applied. Although the normal stress in this experiment was only 3% higher than Experiment 3 in the compression case, the stress intensity factors in the tension experiment for the same crack tip position were about 40 to 50% higher. In fact, the K values were so high that the crack branched very early. K varied between 1.81 and 2.21 MPa$\sqrt{m}$. Crack branching occurred at 2.21 MPa$\sqrt{m}$ with a branching angle of 30°. Owing to low tensile parallel stress, there was not much difference in the values of σ_{ox} (Fig. 13).

In the third experiment a normal stress of 4.1 MPa and a tensile parallel stress of 4.23 MPa were applied. The crack after initiation propagated at a constant velocity of 380 m/s and branched early in the experiment. Thus only two pictures could be taken prior to branching. The branching K was about 2 MPa$\sqrt{m}$. The post-mortem photograph of the model is shown in Fig. 14. The crack branching angle was 45°, which is quite high compared with the angles obtained during uniaxial or compressive parallel stress testing, thus indicating that a tensile normal stress increases crack branching angle. To further confirm this a fourth experiment was performed where the tensile parallel stress was much higher than the normal stress.

In the fourth experiment a normal stress of 3.71 MPa and a very high ten-

FIG. 11—*Typical isochromatic patterns obtained during tension-tension loading.*

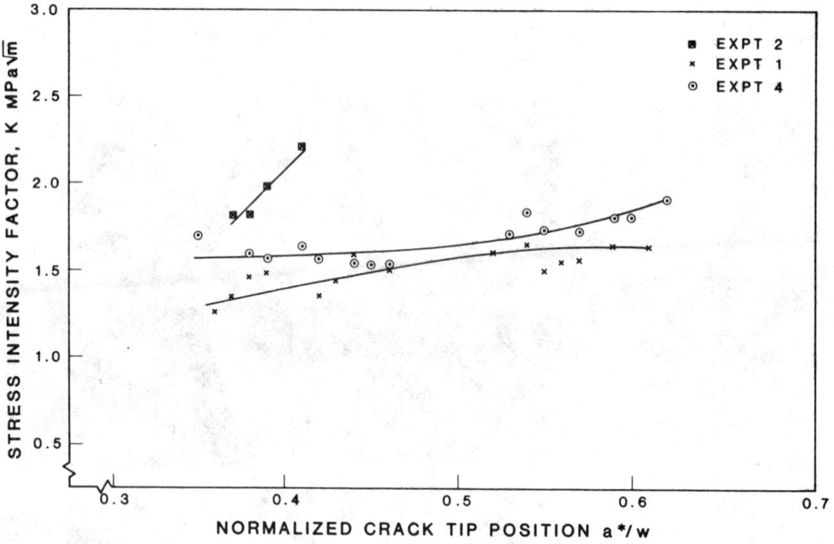

FIG. 12—*Stress intensity factor as a function of normalized crack tip position for tension-tension loading.*

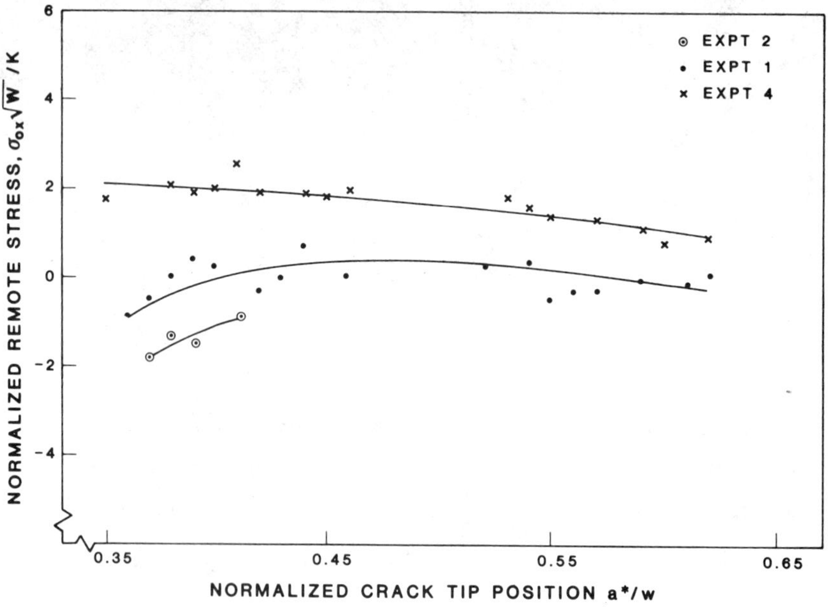

FIG. 13—*Stress parallel to the crack* σ_{ox} *as a function of normalized crack tip position for tension-tension loading.*

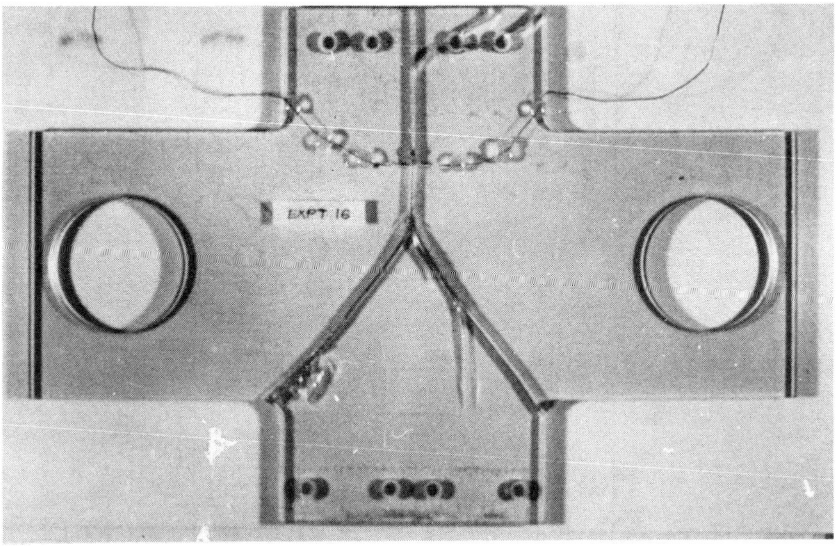

FIG. 14—*Post-mortem photograph from a tension-tension experiment.*

sile parallel stress of 6.4 MPa were applied. The crack propagated with a constant velocity of about 400 m/s. Owing to a large tensile parallel stress, σ_{ox} was highly positive and oscillated about 5 MPa (Fig. 13). Stress intensity factor varied between 1.52 to 1.91 MPa$\sqrt{m}$ (Fig. 12). Crack branching angle as obtained from the broken model was 73°, confirming that tensile parallel stress does increase branching angle.

The results from all the experiments are summarized in Table 1. For comparable values of normal stresses, tension-tension loading considerably enhances crack branching as is evident from the pre-branching fracture area. The fourth tension-tension experiment when compared with the second uniaxial experiment shows crack branching to occur with 40% less pre-branching area. Furthermore, comparison of the third uniaxial experiment with both the second and third tension-compression experiments shows that compressive parallel stress tends to suppress branching as the crack propagates longer distances before branching. These results are in qualitative agreement with the branching mechanism proposed by Ravichander and Knauss [*24*]. Since crack branching in a brittle material like Homalite 100 occurs by void growth around the running crack, a compressive load will tend to close these voids and thus suppress branching. On the other hand, a tension-tension loading will tend to open these voids more and thus enhance branching. The results from all the experiments did not show any marked difference in crack velocities or branching stress intensity factors. The crack propagated around a terminal velocity of 400 m/s and branched at a K value

TABLE 1—*Summary of results for all experiments.*

Experiment Type	Experiment No.	Normal Stress, MPa	Parallel Stress, MPa	Pre-Branching Fracture Area, cm^2	Branching Angle, °	Branching K, $MPa\sqrt{m}$	Normalized σ_{ox} at Branching, $\sigma_{ox}\sqrt{W}/K$
Uniaxial	1	1.61	0	...	...	...	−2.1
	2	3.85	0	17.8	23	...	−1.0
	3	4.52	0	12.2	30	...	−0.9
Tension-compression	1	3.45	−1.24	...	...	...	0.0
	2	4.60	−1.20	15.0	29	2.01	−0.7
	3	4.60	−2.34	12.4	25	1.91	−3.0
Tension-tension	1	2.91	2.76	...	...	...	0.0
	2	4.75	1.65	4.8	30	2.2	−0.8
	3	4.1	4.23	3.5	45	2.0	...
	4	3.71	6.4	10.8	73	1.91	1.0

of 2 MPa$\sqrt{m}$. The sign of the parallel stress σ_{ox} did have a marked influence on crack branching angles. The compressive load did not change the branching angles, whereas tension caused branching angles to increase from 23 to 73°.

Conclusions

The experimental results from the uniaxial and biaxial experiments indicated that:

1. The instantaneous stress intensity factor showed an increasing trend as the crack propagated through the model for all experiments.
2. The branching stress intensity factor and the crack velocity showed relatively no dependence on the nature of the remote parallel stress within the normal and vertical stress ratios of these experiments.
3. Tensile stress parallel to the crack, σ_{ox}, increases branching angles considerably. Branching angles as large as 73° were obtained for tension-tension loading in comparison to 25° for uniaxial and tension-compression loading.
4. Tensile stress σ_{ox} gives rise to higher stress intensity factors in comparison to the compressive case for the same crack tip positions and normal loads.
5. Tensile stress σ_{ox} enhances branching as seen by the fracture surface area data prior to branching in Table 1.
6. Experimental results indicate that any criterion predicting branching angle must take the sign of σ_{ox} into account.

Acknowledgments

The authors acknowledge the support of NSF under Grant MEA-8203277.

References

[1] Irwin, G. R. et al, "Photoelastic Studies of Crack Propagation and Arrest," NRC Report NUREG-0342, NRC 5, Nuclear Regulatory Commission, Washington, D.C., Oct. 1977.
[2] Irwin, G. R. et al, "Photoelastic Studies of Crack Propagation and Arrest in Polymers and 4340 Steel," NRC Report NUREG/CR-0542, Nuclear Regulatory Commission, Washington, D.C., Dec. 1978.
[3] Irwin, G. R. et al, "On the Determination of the *a-K* Relationship for Birefringent Polymers," *Experimental Mechanics*, Vol. 19, No. 4, 1979, pp. 121–128.
[4] Bradley, W. B. and Kobayashi, A. S., "An Investigation of Propagation Cracks by Dynamic Photoelasticity, *Experimental Mechanics*, Vol. 10, No. 3, 1970, pp. 106–113.
[5] Kobayashi, T. and Dally, J. W., "The Relation Between Crack Velocity and Stress Intensity Factors in Birefringent Polymers," in *Fast Fracture and Crack Arrest, ASTM STP 627*, American Society for Testing and Materials, Philadelphia, 1976, pp. 257–273.
[6] Kalthoff, J. F. et al, "Experimental Analysis of Dynamic Effects in Different Crack Arrest Test Specimens," in *Crack Arrest Methodology and Applications, ASTM STP 711*, American Society for Testing and Materials, Philadelphia, 1980, pp. 109–127.
[7] Clark, A. B. and Irwin G. R., "Crack Propagation Behaviors," *Experimental Mechanics*, Vol. 6, No. 6, 1966, pp. 321–330.

[8] Kobayashi, A. S. et al, "Crack Branching in Homalite 100 Sheets," *Engineering Fracture Mechanics*, Vol. 6, 1974, p. 81.
[9] Kalthoff, J. F., "On the Characteristic Angle for Crack Branching in Brittle Materials," *International Journal of Fracture Mechanics*, Vol. 7, 1971, pp. 478-480.
[10] Kitagawa, H. and Yuvki, R., "Analysis of Branched Cracks Under Biaxial Stress," in *Fracture 1977*, D. M. R. Taplin, III, Ed., Univ. of Waterloo Press, Waterloo, Ont., Canada, 1977, p. 201.
[11] Anthony, S. R. et al, "The Crack Branching Velocity," *Philosophical Magazine*, Vol. 22, 1970, pp. 1201-1216.
[12] Congleton, J. and Petch, N. J., "Crack Branching," *Philosophical Magazine*.
[13] Irwin, G. R., "Discussion of: The Dynamic Stress Distribution Surrounding a Running Crack—A Photoelastic Analysis," *Proceedings of SESA*, Vol. 16, 1958, pp. 93-96.
[14] Kerkhoff, F., "Wave Fractographic Investigations of Brittle Fracture Dynamics," in *Dynamic Crack Propagation*, G. C. Sih, Ed., Noordhoff, Leyden, The Netherlands, 1973, p. 3.
[15] Congleton, J., "Practical Applications of Crack Branching Measurements," in *Dynamic Crack Propagation*, G. C. Sih, Ed., Noordhoff, Leyden, The Netherlands, 1973, pp. 427-438.
[16] Doll, W., "Investigations of the Crack Branching Energy," *International Journal of Fracture*, Vol. 11, 1975, p. 184.
[17] Ramulu, M. and Kobayashi, A. S. "Dynamic Crack Branching—A Photoelastic Evaluation," ONR Technical Report UWA/DME/TR-82/43.
[18] Ramulu, M., Kobayashi, A. S., Kang, B. S. J., and Barker, D. B., "Further Studies on Dynamic Crack Branching," *Experimental Mechanics*, Vol. 223, No. 4, Dec. 1983, pp. 431-437.
[19] Irwin, G. R., "Constant Speed Semi-Infinite Tensile Crack Opened by a Line Force, *P*, at a Distance *b*, from the Leading Edge," Lehigh University Lecture Notes, Bethlehem, Pa.
[20] Sanford, R. J., "A Critical Re-Examination of the Westergaard Method for Solving Opening Mode Crack Problems," *Mechanics Research Communications*, Vol. 6, No. 5, 1979, pp. 289-294.
[21] Irwin, G. R., "Discussion of Photoelastic Stress Analysis for an Edge Crack in a Tensile Field," *Proceedings of SESA*, Vol. 16, No. 1, 1958, pp. 93-96.
[22] Liebowitz, H. et al, "Biaxial Load Effects in Fracture Mechanics," *Engineering Fracture Mechanics*, Vol. 10, 1978, pp. 315-335.
[23] Cottrell, B., "Notes on the Path and Stability of Cracks," *International Journal of Fracture*, Vol. 2, 1966, pp. 526-533.
[24] Ravichander, K. and Knauss, W. G., "Process Controlling the Dynamic Fracture of Brittle Solids," Workshop on Dynamic Fracture, Feb. 1983, sponsored by National Science Foundation and Army Research Office, Pasadena, Calif.

Ted L. Anderson[1] *and Steven Williams*[2]

Assessing the Dominant Mechanism for Size Effects on CTOD Values in the Ductile-to-Brittle Transition Region

REFERENCE: Anderson, T. L. and Williams, S., **"Assessing the Dominant Mechanism for Size Effects on CTOD Values in the Ductile-to-Brittle Transition Region,"** *Fracture Mechanics: Seventeenth Volume, ASTM STP 905*, J. H. Underwood, R. Chait, C. W. Smith, D. P. Wilhem, W. A. Andrews, and J. C. Newman, Eds., American Society for Testing and Materials, Philadelphia, 1986, pp. 715–740.

ABSTRACT: There has been some uncertainty in the field of fracture mechanics as to whether shifts in the ductile-to-brittle transition with specimen size are caused by statistical sampling effects or constraint effects. Fracture toughness data for three carbon-manganese steels were analyzed to examine the effectiveness of each model in predicting the results of different sized specimens. The data set used in this investigation contains nearly 500 crack tip opening displacement (CTOD) values for various geometries of single edge notched bend (SENB) specimens with thicknesses ranging from 10 to 100 mm. A limited amount of data for single edged notched (SENT) specimens and side grooved SENB specimens was also included.

It was found that the mechanism which accounts for size effects most effectively depends on the amount of plastic flow prior to fracture. Under conditions of small scale yielding, high constraint is maintained in both small and large specimens. In this region the Landes and Shaffer statistical sampling model appears to work well for explaining the higher average toughness of small specimens. However, when the ligament yields prior to fracture, the shifts in transition observed with specimen size cannot be explained by statistical sampling effects alone.

The temperature at which net section yielding first occurs tends to shift upward as (1) the specimen thickness increases, (2) the ligament length is decreased (in a bend specimen), or (3) the mode of loading is changed from tension to bending. When the ligament yields, the plastic deformation relaxes the crack tip constraint and the brittle-to-ductile transition becomes steep. The relaxation of crack tip constraint occurs more rapidly and at lower temperatures in smaller specimens. This gives rise to a steeper transition curve and a shift in transition temperature which cannot be accounted for by statistical effects. Thus as the critical CTOD increases the relative contribution of statistical sampling on size effects decreases and constraint effects tend to dominate.

[1]Assistant Professor, Mechanical Engineering Department, Texas A&M University, College Station, TX 77843.

[2]Shell U.K. Exploration and Production, Strand, London, England.

KEY WORDS: carbon-manganese steels, cleavage fracture, constraint, crack tip opening displacement, ductile-to-brittle transition, toughness, weakest-link statistics

Nomenclature

a	Crack length
B	Specimen thickness
b	Weibull scale parameter
c	Weibull shape parameter
E_p	Plastic modulus
F	Weibull distribution function
K_{Ic}	Plane strain fracture toughness (critical stress intensity factor)
L	Notch constraint factor
ℓ	Ligament length
M_{LS}	Landes and Shaffer predicted "mean"
n	Ratio of specimen thicknesses
P_0	Load at net section yield
r_p	Plastic rotational factor
W	Specimen width
X_c	Critical distance
x	Value of datum
δ_c	Critical CTOD
$\bar{\epsilon}$	Effective strain
$\bar{\sigma}$	Effective stress
σ_f	Fracture stress
σ_0	Yield strength
σ_x°	Initial stress in the x direction

Many steel structures operate in the ductile-to-brittle transition region of the material. Application of fracture mechanics is difficult in this region, because the cleavage fracture toughness is a function of test specimen size and geometry [*1–13*]. At a constant temperature and loading rate in the transition region, small specimens tend to have higher toughnesses than larger specimens. Thus nonconservative results can be obtained if small scale laboratory specimens are used to infer the fracture behavior of a large structure.

Although the existence of size effects in the transition region has been known for some time, there has been some disagreement as to the cause of these effects. Currently there are two possible explanations for size effects, one based on constraint and the other on statistical sampling effects. Crack tip constraint is a function of material flow properties, while statistical effects depend on microstructure. These two mechanisms are described briefly below.

The traditional explanation for size effects is that of crack tip constraint [*1–12*]. Large specimens have a high degree of stress triaxiality near the crack tip, which tends to promote cleavage fracture by making it easier for the material to reach the fracture stress. Many believe that constraint relaxes in smaller specimens, thereby shifting the transition to lower temperatures.

A second explanation for size effects was recently proposed by Landes and Shaffer [*14*]. Their model is based on the statistical nature of cleavage fracture. If fracture occurs by a weakest-link mechanism, the average toughness of a group of small specimens will be higher than the average toughness of larger specimens. Since more material is sampled along the crack front of a large specimen, there is a higher probability of obtaining a low toughness value. Landes and Shaffer used this concept to predict the toughness of large specimens from data for small specimens.

Both statistical and constraint models were examined in this investigation. The crack tip opening displacement (CTOD) data for several carbon-manganese steels were analyzed. The effectiveness of each model in predicting the results of different size specimens was investigated.

Materials and Test Specimens

Fracture toughness data for three carbon-manganese steel plates were used in this investigation. Some of these data have been published previously [*7,12*]. Table 1 summarizes the three materials. The chemical composition of each plate is given in Table 2. Each material has a yield strength of approximately 350 N/mm^2 (50 ksi) at 25°C. The composition and mechanical properties of the three steels are very similar; two of the plates were supplied in accordance with British Standard BS 4360 Grade 50D.

The data set for each material consists of CTOD values as a function of test temperature and specimen geometry. Most of the CTOD tests were performed in the ductile-to-brittle transition region. Most tests were carried out in accordance with BS 5762:1979, the British Standard for CTOD testing. However, a number of tests were performed on specimens with nonstandard geometries.

Many of the CTOD tests were performed in previous investigations; consequently the data were often not collected in a manner appropriate to the present study. This has frequently limited the way in which it was possible to

TABLE 1—*Summary of three C-Mn steel plates on which fracture toughness was measured.*

Material No.	Designation	Nominal Plate Thickness, mm	No. of Specimen Configurations	Approximate No. of CTOD Results
1	ABS EH36	25	8	85
2	BS 4360 50D	100	4	150
3	BS 4360 50D	70	21	250

TABLE 2—*Chemical composition (weight percent) of three steel plates.*

Material No.	C	S	P	Si	Mn	Ni	Cr	Mo	V	Nb	Ti	Al	Cu
1	0.12	0.006	0.015	0.38	1.39	0.03	0.05	0.007	[a]	[a]	[a]	[a]	0.05
2	0.16	<0.005	0.013	0.41	1.38	0.15	0.11	0.03	<0.01	0.032	<0.01	0.035	0.28
3	0.16	<0.005	0.10	0.41	1.33	0.06	0.05	0.02	<0.01	0.029	<0.01	0.026	0.07

[a] Data not available.

analyze each data set in this investigation. For example, the Landes and Shaffer statistical model [*14*] could not be applied to Materials 1 and 3 because there were insufficient data at each test temperature.

The background of each material and the corresponding CTOD data are summarized below.

Material 1

Material 1 is a 25 mm thick plate of ABS EH36 steel. The CTOD results, which were originally published by Anderson et al [*12*], are for five geometries of single edge notched bend (SENB) specimens and three geometries of single edge notched tension (SENT) specimens. These results were later used to assess a constraint-based model for the ductile-to-brittle transition region [*15*]. In the present investigation, the Material 1 toughness data are used to demonstrate the relationship between notch constraint and transition temperature.

Material 2

Material 2 is a 100 mm thick plate of BS 4360 Grade 50D steel. Some fracture toughness data for this material have been published by Pisarski [*7*]. These data include CTOD tests on 10, 50 and 100 mm thick SENB specimens in the transition region. In all cases, the specimen width, W, was equal to twice the thickness, B, and the crack length-to-width ratio, a/W, was nominally 0.5. The British CTOD standard, BS 5762:1979, refers to this configuration as the "preferred geometry".

In order to apply the statistical model [*14*] to this material, additional CTOD tests were performed in the present investigation. These extra tests were on 10 and 100 mm thick preferred geometry specimens as well as side grooved SENB specimens with a net thickness of 10 mm.

Material 3

Material 3 is a 70 mm thick plate of BS 4360 50D steel. Kamath [*16*] performed CTOD tests in the transition region on 21 geometries of SENB specimens. The results have not been published in the open literature. The scatter was relatively large in these fracture toughness data; this made comparisons between geometries difficult. The data for Material 3 were used in the present investigation to compare notch constraint in the various specimen geometries. Notch constraint data were obtained from ligament yield loads (defined later) at −50°C; these data were not as scattered as the fracture toughness data.

Statistical Model

Background

Landes and Shaffer [*14*] have proposed a statistical model to explain the differences in toughness results obtained from small and large specimens of the same material. The model is based on a "weakest link" theory (i.e., the initiation of unstable fracture is controlled by the region of lowest toughness along the crack front). The larger the specimen thickness, the more material is sampled by the crack front and the lower the toughness of the specimen is likely to be. According to the weakest link theory, if the probability that the toughness of a small specimen will exceed some particular value is α, then the probability that the toughness of a specimen n times larger will exceed the same value is α^n. This is because all n elements of the longer crack front must sample regions that have toughnesses greater than the required value.

Any distribution function that gives an adequate description of small specimen results can be used in the weakest link model to predict the results of larger specimens. In their original paper, Landes and Shaffer [*14*] used a two-parameter Weibull distribution [*17*] to describe the toughness distribution of small specimens. The Weibull cumulative distribution is given by

$$F_1(x) = 1 - e^{-(x/b)^c} \tag{1}$$

where b is a scale parameter, c is a shape parameter, and $F_1(x)$ is the probability of the value of a data point being less than x.

For specimens that are n times larger than a given size, the weakest link theory predicts that the cumulative distribution function is

$$F_n(x) = 1 - [e^{-(x/b)^c}]^n \tag{2a}$$

$$= 1 - e^{-\left(\frac{xn^{1/c}}{b}\right)^c} \tag{2b}$$

Landes and Shaffer approximated the mean value of the distribution as being equal to the scale parameter, b. Using this assumption, the predicted mean of large specimen results, M_{LS}, is given by

$$M_{LS} = \frac{b}{n^{1/c}} \tag{3}$$

The predicted toughness values resulting from the above approximation actually correspond to 63rd percentile values rather than arithmetic means. An alternative characterizing parameter is the median value, which is the value below which 50% of the results are expected to lie. The median is generally considered to be a better indication of the central position than the arithmetic mean when there is a large amount of scatter in the data, as is the case for fracture toughness results. The weakest link model predicts median toughness of large specimens (assuming a Weibull distribution) to be

$$x_{50} = \left(\frac{-\ell n\ 0.50}{n}\right)^{1/c} b \tag{4}$$

Similarly, the statistical model can predict 5% lower bound and 95% upper bound toughness values for large specimens:

$$x_{05} = \left(\frac{-\ell n\ 0.95}{n}\right)^{1/c} b \tag{5}$$

$$x_{95} = \left(\frac{-\ell n\ 0.05}{n}\right)^{1/c} b \tag{6}$$

Fracture Toughness Data

The weakest link statistical model was applied to CTOD data for Material 2, a 100 mm thick plate of BS 4360-50D steel (see Tables 1 and 2). Three types of test specimens were examined: full thickness 100 by 200 mm SENB specimens, 10 by 20 mm SENB specimens, and 15 by 20 mm specimens that were side grooved so that the net thickness at the fatigue crack was 10 mm. All specimens were through-thickness notched with the plane of the crack perpendicular to the rolling direction; a/W was nominally 0.5 in all cases.

Figure 1 illustrates how the subsize specimens were extracted from the 100 mm plate. Sets of eight plain (non-side-grooved) 10 by 20 mm specimens were taken from across the plate thickness. In addition, sets of three specimens were taken from the central 35 mm of the plate thickness. The side-grooved specimens were extracted from nine positions across the thickness in such a way as to sample as much of the plate thickness as possible.

The specimens were tested at various temperatures ranging from −125 to −50°C. Table 3 lists the number and type of specimen tested at each temperature. The fracture toughness results for all specimens are plotted in Fig. 2.

Analysis of the Data

The Landes and Shaffer statistical model assumes that the toughness distribution through the plate thickness is uniform and that low toughness regions are equally likely at all positions within the plate thickness. This assumption was examined by comparing the results from the central 35 mm of the plate thickness with those from the edges of the plate. The Mann-Whitney test [*18*] was used to determine whether or not there was a significant differ-

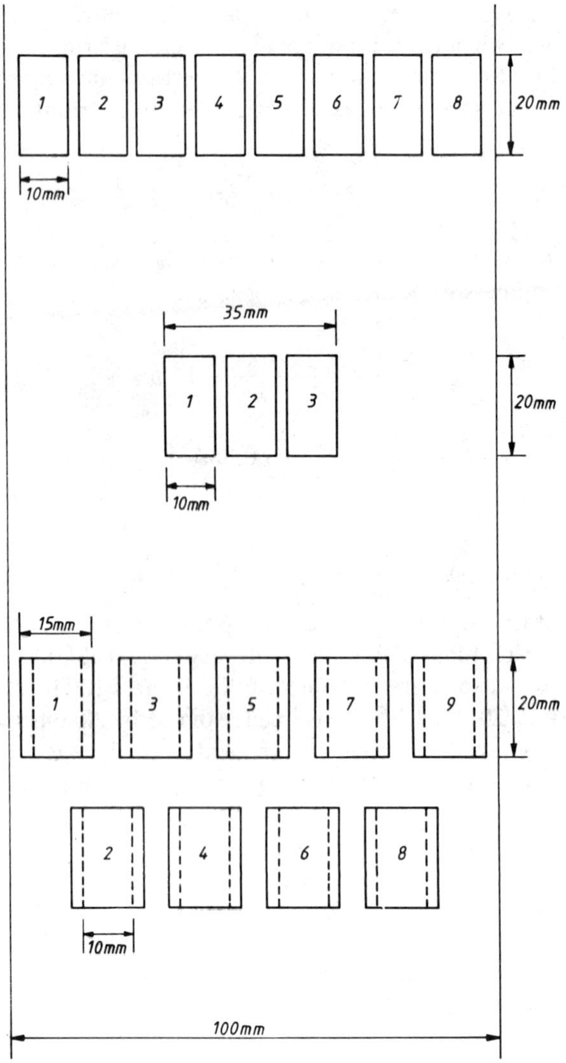

FIG. 1—*Relative position in thickness of small scale SENB specimens extracted from a 100 mm thick plate.*

ence in the median toughness values from the two locations. In every case but one (plain specimens tested at −125°C), there was a significant difference between median toughness values, with the lower results being obtained near the center of the plate. The CTOD results for side grooved specimens tested at −100°C are plotted against position in thickness (Fig. 1) in Fig. 3 to illustrate the variation in toughness with sampling location.

TABLE 3—*Fracture toughness data used for statistical characterization.*

Specimen Configuration	Number of Results at −125°C	−100°C	−86°C	−75°C	−50°C
100 by 200 mm	2	5	...	5	1
10 by 20 mm	18	21	4	...	...
15 by 20 mm (side grooved)	28	29	...	19	...

The Mann-Whitney test was also used to compare the toughness results of plain and side grooved specimens. At −125°C there was no significant difference between the toughness results from the two types of specimens. At −100°C, however, the toughness values from the side grooved specimens are lower than those from the plain specimens. Weibull and log normal distributions were fitted to the various sets of small specimen toughness data. In tests where a pop-in was observed or the test was stopped before fracture, the lowest estimate of the toughness was used to compute distributions. In most cases, both the Weibull and log normal distributions gave acceptable fits to the data, although the Weibull distribution yielded marginally better fits. A typical Weibull plot of fracture toughness data is shown in Fig. 4.

Although any distribution can be used in the weakest link statistical model, the Weibull distribution was used in the present study to predict toughness values for the large specimens. The Weibull distribution was chosen because it fitted the data slightly better than the log normal distribution and also because it was used by Landes and Shaffer in their original work.

The statistical model was used to estimate the median, the 5% lower bound, the 95% upper bound, and the approximate Landes and Shaffer mean values for the 100 mm thick specimens (Table 4). These estimates were made using Weibull distributions fitted to the small specimen data and Eqs 3 to 6. For each specimen type and test temperature, predictions were made using Weibull distributions which were fitted to all the available data as well as to data from the central 35 mm of the plate thickness. The use of the distributions fitted to all the available data corresponds to the Landes and Shaffer model which assumes that there is no systematic variation in toughness with sampling position. However, it has been shown that toughness tended to be lower near the center of the plate. When all the data across the thickness were used in the Weibull distribution, an n value of 10 was used in Eqs 3 to 6. This corresponds to the ratio of thicknesses of the small and large specimens. For the analyses based on the Weibull distribution fitted to the data from the center of the plate, it was assumed that the central region of the plate thickness controls the toughness and there is little risk of fracture initiation in the outer regions of the plate. In this case a value of n of 3.5 was used in Eqs 3 to 6.

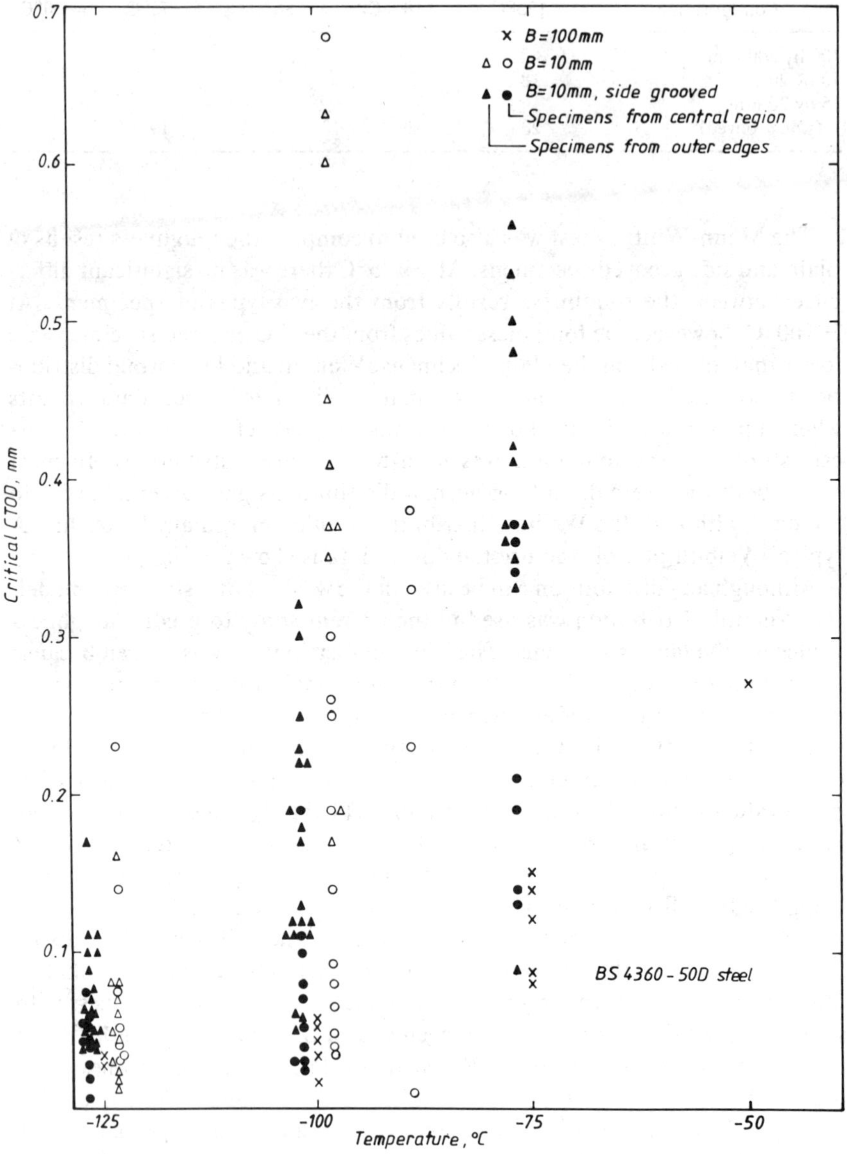

FIG. 2—*Fracture toughness data for Material 2.*

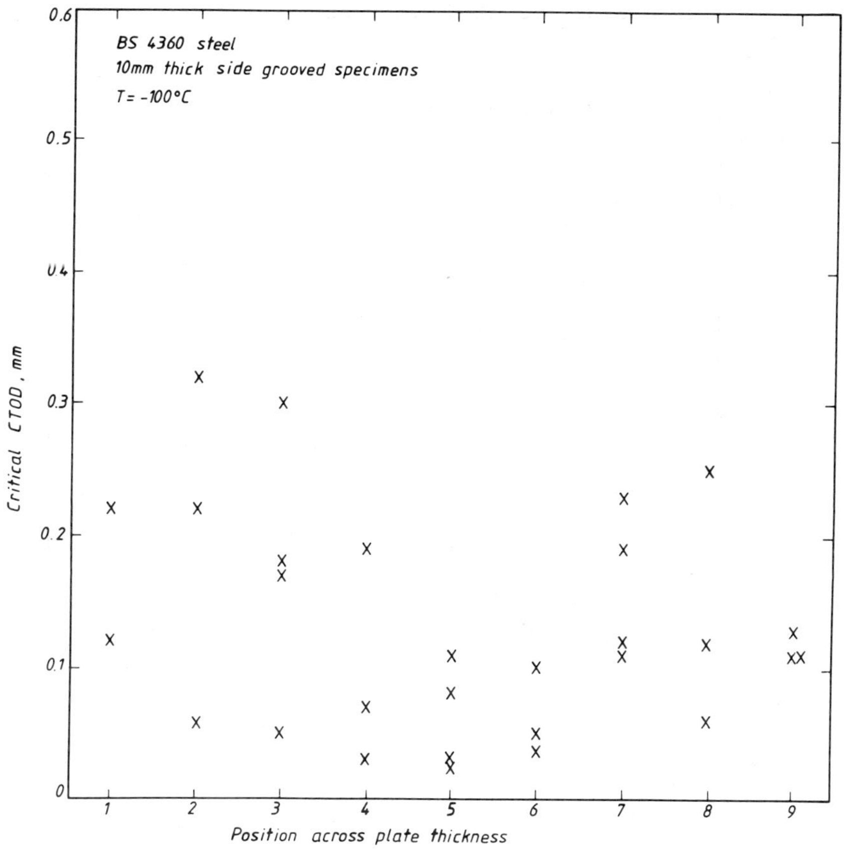

FIG. 3—*Fracture toughness of side grooved SENB specimens at −100°C as a function of position in thickness.*

The values estimated from Eqs 3 to 6 are listed in Table 4 and are plotted in Fig. 5 with the actual 100 mm data. At −125°C, the predictions from the statistical model, using either side grooved or plain specimens, are consistent with the 100 mm data. At −100°C, the predictions from the side grooved specimens are satisfactory, but the predictions from the plain specimens are slightly higher than the experimental test results. Only four toughness values for plain specimens were available at −86°C. These data include a very low pop-in value which considerably depressed the predicted large specimen toughness. Had more data been available at this temperature, the predicted toughness would probably have been higher. Even so, the predicted values lie above the interpolated values for the actual data and the predictions from side grooved specimens. At −75°C, the predictions from the side grooved specimens lie above the actual test results.

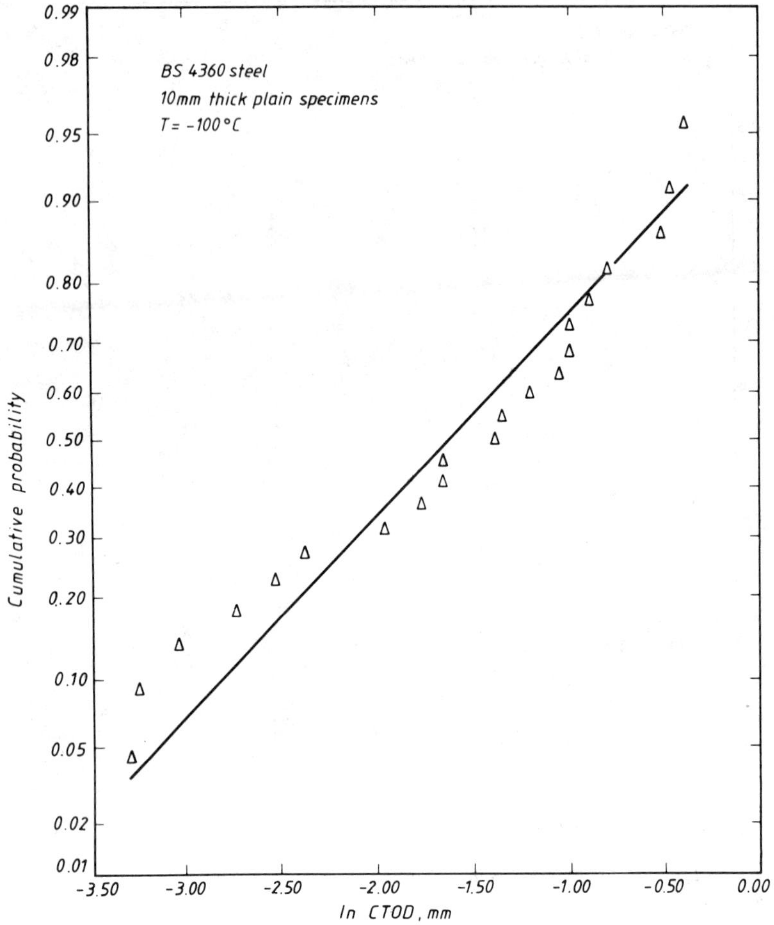

FIG. 4—*Cumulative probability plot for 10 by 20 mm SENB specimens at −100°C. Plots such as this were used to determine* c *and* b *in the Weibull distribution.*

It must be emphasized that the error bars in Fig. 5 do not represent scatter bands in the 10 mm data. Rather, they are predictions of the 95% upper bounds and 5% lower bounds of the 100 mm data (Eqs 5 and 6). It is worth noting that the predictions from specimens extracted from near the center of the plate (PC and SC) agree well with predictions from all subsize specimens (PA and SA), despite the fact that the toughness in the central region of the plate was consistently lower than in the outer regions (Figs. 2 and 3). For reasons stated earlier, n in Eqs 3 to 6 was taken as 10 for the SA and PA populations and 3.5 for SC and PC.

TABLE 4—*Summary of 100 mm thick specimen results predicted from small specimen test results (Material 2).*

Temperature, °C	Small Specimen Data Set [a]	5% Lower Bound CTOD, mm	Median CTOD, mm	95% Upper Bound CTOD, mm	Landes & Shaffer [*14*] "Mean" (M_{LS}) CTOD, mm
−75	SG,C	0.053	0.148	0.26	0.171
−75	SG,A	0.056	0.145	0.25	0.166
−86	P,C	0.015	0.088	0.235	0.112
−100	P,C	0.003	0.036	0.155	0.052
−100	P,A	0.007	0.044	0.125	0.057
−100	SG,C	0.004	0.026	0.072	0.034
−100	SG,A	0.006	0.029	0.070	0.036
−125	P,C	0.002	0.023	0.080	0.031
−125	P,A	0.001	0.008	0.028	0.011
−125	SG,C	0.006	0.021	0.043	0.026
−125	SG,A	0.005	0.018	0.038	0.022

[a] SG-Side grooved specimens.
P-Plain specimens.
C-Specimens from central 35 mm ($n = 3.5$).
A-Specimens across the thickness ($n = 10$).

Evaluation of Constraint Effects

According to Fig. 5, the statistical model works well at low toughness values but the 100 mm data and the statistically corrected data diverge as toughness increases. One possible explanation for this divergence is that constraint relaxes in the smaller specimens. Some support for this argument is given by the side grooved data which diverge from the plain specimen data as temperature increases. This suggests that the side grooved specimens maintain high constraint longer than the plain specimens. Additional evidence for the constraint mechanism as an explanation for size effects is given below.

Figures 6 to 8 show fracture toughness data for Material 1. The data in Figs. 6 and 7 were previously published by Anderson et al [*12*]. The ductile-to-brittle transition tends to shift to higher temperatures with increasing thickness, increasing crack length-to-width ratio (a/W), and changing from tensile loading to bending. The transition shifts in Figs. 7 and 8 are almost certainly due to constraint effects because the thickness is constant; the volume of material sampled along the crack front is the same for all specimen configurations.

The constraint in a given specimen geometry can be qualitatively measured by means of the notch constraint factor. The notch constraint factor, L, is defined by the limit load expression for SENB specimens:

$$P_0 = \frac{L\sigma_0(W - a)^2B}{4W} \tag{7}$$

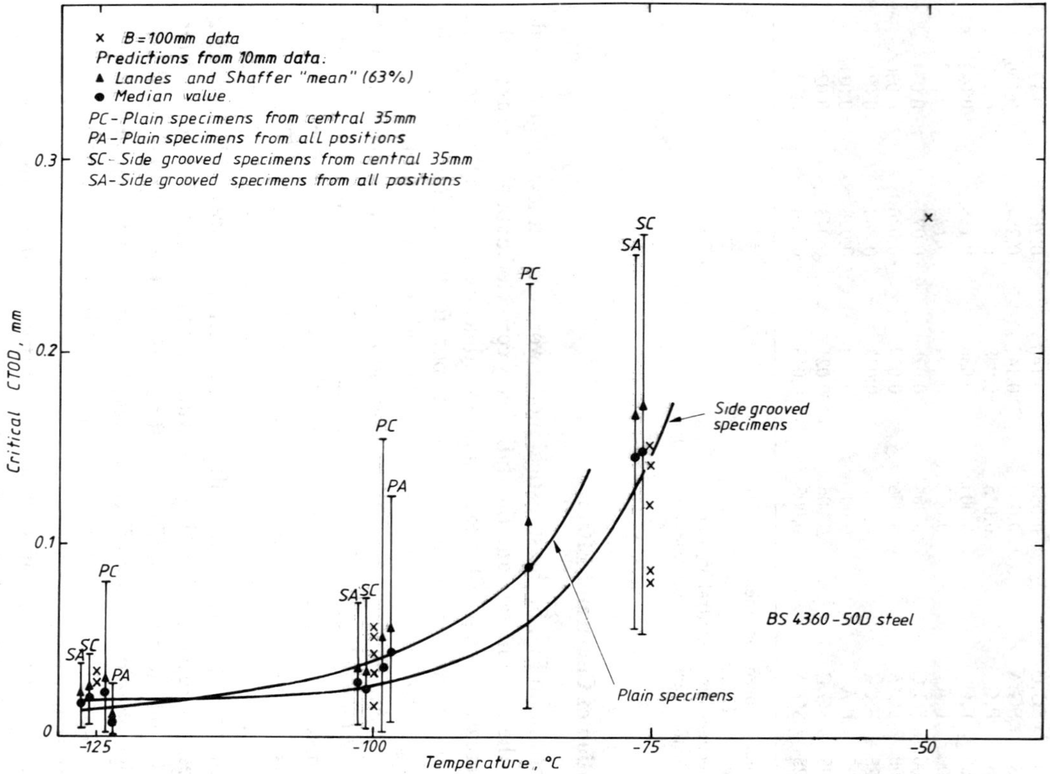

FIG. 5—*Toughness predictions from the Landes and Shaffer* [14] *model compared with actual data for 100 by 200 mm SENB specimens.*

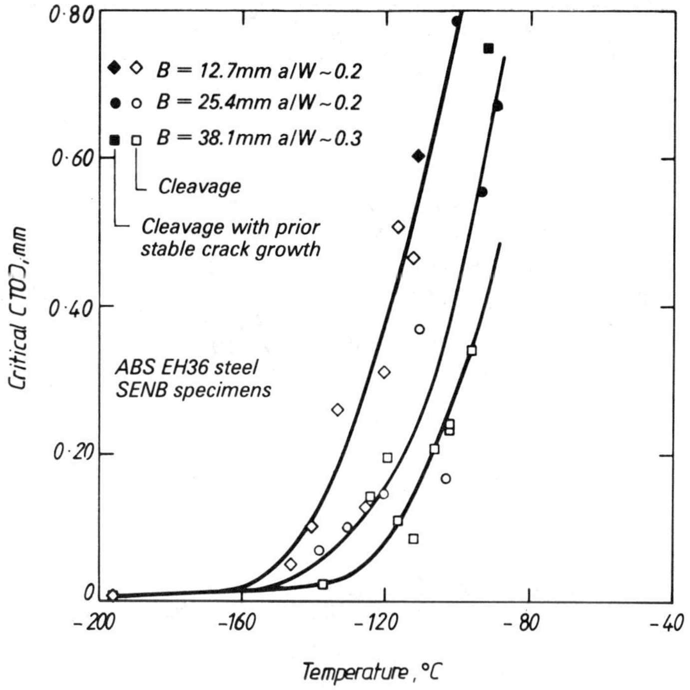

FIG. 6—*Fracture toughness as a function of specimen thickness in Material 1.*

where P_0 is the load at net section yield and σ_0 is the yield strength. The notch constraint factor is a dimensionless constant which is a measure of the elevation of the yield load due to the pressure of a crack. In the absence of a crack, $L = 1$. Notch constraint is a global constraint factor and should not be confused with crack tip constraint, which is a local quantity. (Crack tip con straint is discussed in detail in the next section.)

As notch constraint increases, the load at net section yield is elevated due to an increase in triaxiality. This is illustrated in Fig. 9 where dimensionless load (i.e., load normalized for yield strength and specimen dimensions) is plotted against displacement for three SENB specimens of Material 1. Dimensionless load, which is defined on the ordinate of Fig. 9, was obtained from a generalized version of Eq 7 where load is allowed to vary; when $P = P_0$, dimensionless load $= L$. Figure 9 shows that notch constraint increases with increasing thickness, a/W, or both.

For Material 1, L is related to the ductile-to-brittle transition temperature, defined at a critical CTOD of 0.1 mm. Figure 10 is a plot of L versus transition temperature. These data correspond to the transition curves for the five SENB geometries plotted in Figs. 6 and 7. The L values were obtained by constructing tangents to the elastic and fully plastic portions of dimensionless

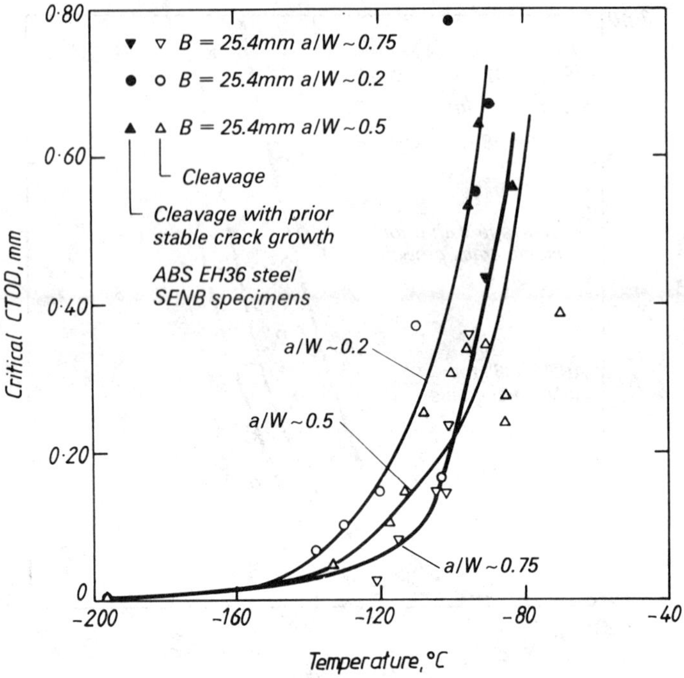

FIG. 7—*Fracture toughness as a function of crack length in Material 1.*

load-displacement curves (Fig. 9); L was defined at the intersection of the two tangent lines. As can be seen from Fig. 10, there is a direct relationship between notch constraint and transition temperature.

Data for Material 3 were used to determine the relationship between L and specimen dimensions. The data set for this material contained fracture toughness data for 21 configurations of SENB specimens which were machined from 70 mm thick plate. Although L is relatively temperature independent [*12*], values of L for Material 3 were all obtained from load-displacement records for specimens tested at $-50°C$. The intersection-of-tangents method, which is described above, was used to compute L.

Figure 11 is a plot of L as a function of crack length, specimen width, a/W, and ligament length (thickness = constant). The data lie on a common trend when L is plotted against ligament length. Apparently, $W-a$ is the governing dimension for notch constraint.

The trend in Fig. 11 would probably not apply to specimens with very short cracks or very short ligaments. Specimens with very short cracks would experience plastic relaxation to the crack mouth and a relaxation in constraint. As ligament length decreases, constraint would eventually relax as the plastic zone approached the back surface of the specimen. However, increases in

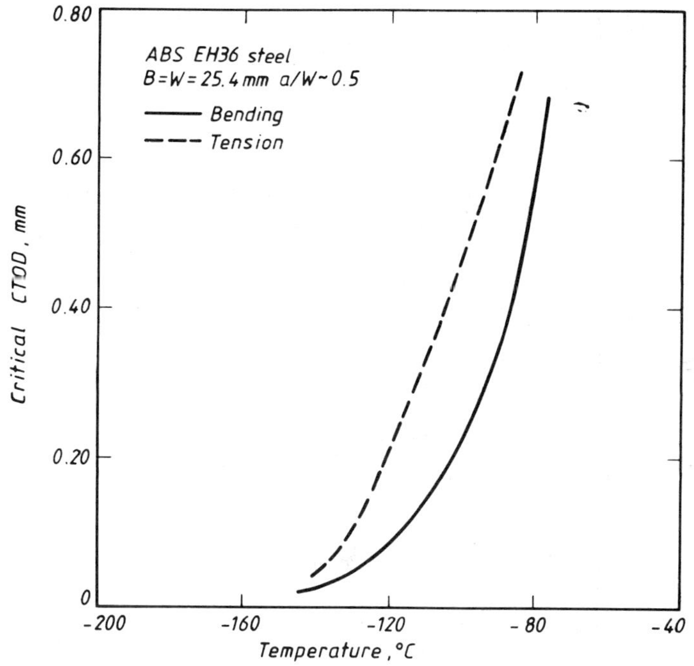

FIG. 8—*Comparison of fracture toughness for bending and tensile loading in Material 1.*

notch constraint with decreasing ligament length have been observed down to ligament lengths of 6 mm (see Figs. 7 and 10).

The effect of specimen thickness on notch constraint is shown in Fig. 12. For a constant width and crack length, L increases with thickness, as would be expected. However, for a constant geometry ($W = 2B$, $a/W = 0.5$), notch constraint decreases slightly with specimen size. In this set of preferred geometry specimens, thickness and ligament length were increased simultaneously. Apparently the tendency for notch constraint to increase with thickness was more than offset by decreases in constraint due to the increasing ligament length.

The data in Fig. 12 for preferred geometry specimens may appear to contradict previous statements about size effects. Figure 12 indicates that the 10 mm specimens may have had a slightly higher notch constraint factor than the 100 mm specimens, and yet the transition shift between 10 and 100 mm specimens (Fig. 5) has been attributed partly to constraint effects. This apparent contradiction may be resolved as follows. A small specimen with a high L value possesses a high degree of constraint under conditions of small scale yielding. However, when the ligament yields prior to fracture, the small specimen rapidly loses its constraint, resulting in a steep transition curve. Some evidence for this hypothesis is given in the following section.

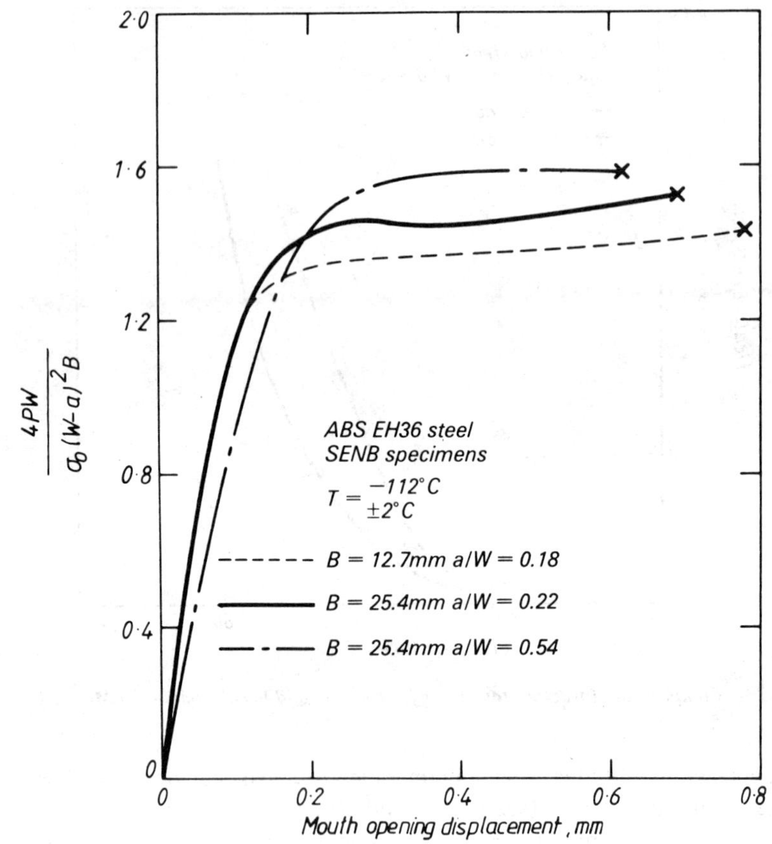

FIG. 9—*Nondimensional load-displacement curves for Material 1.*

Application of a Constraint Model

Plastic constraint at the crack tip and the resulting effect on fracture toughness in the transition region have been modelled by Anderson [*15*]. In the present investigation, this model was applied to Material 2 in order to predict the fracture toughness of the 10 and 100 mm specimens (Fig. 5).

Background

The constraint model stems from experimental measurements of crack tip region constraint (CTRC) in specimens of Material 1 [*15*]. These measurements were based on microhardness-crack tip strain correlations and the cleavage fracture model of Ritchie et al [*19*]; the latter states that fracture occurs when the fracture stress, σ_f, is exceeded over a critical distance, X_c, ahead of the crack tip. The fracture stress was measured by means of low

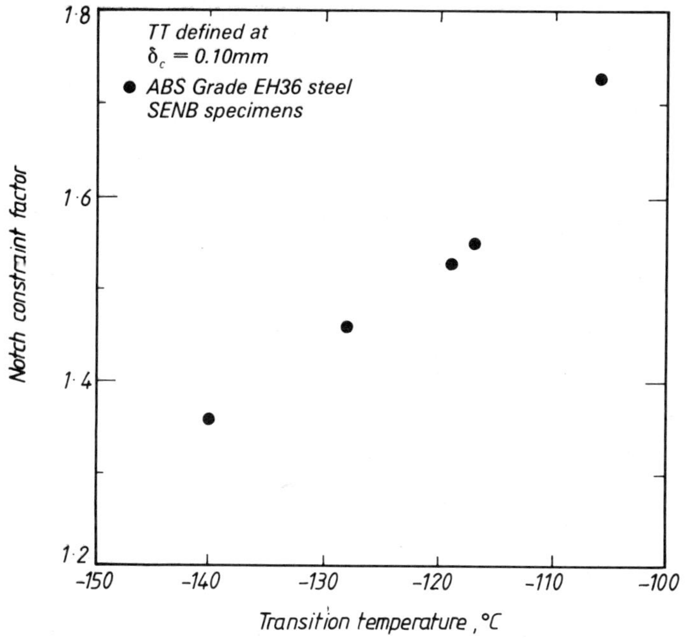

FIG. 10—*Notch constraint as a function of transition temperature (defined as* $\delta_c = 0.10$ *mm) for Material 1.*

temperature tensile tests, and X_c was inferred from Tracey's [*20*] finite element solutions for crack tip stress. Crack tip constraint was defined as

$$\text{CTRC} = \frac{\sigma_f}{\bar{\sigma}_c} \tag{8}$$

where $\bar{\sigma}_c$ is the critical effective stress at a distance X_c ahead of the crack tip.

The CTRC measurements indicated that constraint is initially high but relaxes with plastic deformation. The initial CTRC for small scale yielding was a function of specimen geometry. The rate at which constraint relaxed was also a function of specimen dimensions. The constraint relaxed most rapidly in specimens with the shortest ligaments. Also the ductile-to-brittle transition was steepest for the specimens with the shortest ligaments and the most rapid constraint relaxation. (For an example of this, see Fig. 7.)

The CTRC relaxation with plastic deformation has been modelled [*15*]. The critical stress perpendicular to the crack plane was assumed to be equal to σ_f at X_c ahead of the crack tip. As the crack tip blunts, the stress in the direction of crack propagation, σ_x, was assumed to relax from an initial value of σ_x°. The Tresca yield criterion was used in order to confine the model to two

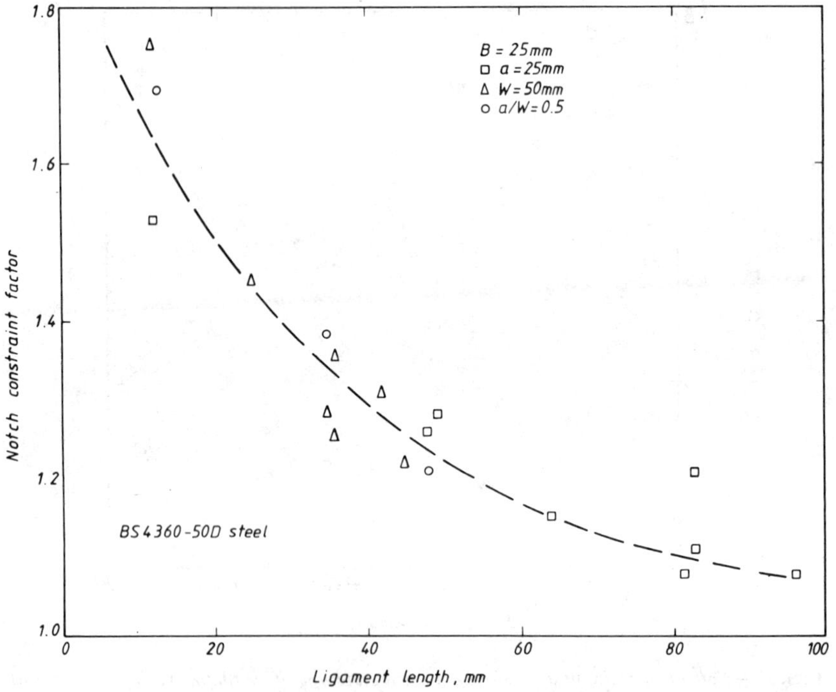

FIG. 11—*Notch constraint as a function of ligament length at a constant thickness (Material 3);* T = *−50°C.*

dimensions; it was further assumed that $\sigma_x < \sigma_z < \sigma_f$. The model produced the following equation for fracture toughness in the transition region:

$$\frac{\beta\delta_c^{0.875m}}{10X_c^{0.875}} - \frac{E_p\delta_c}{16(r_p\ell)^{0.125}X_c^{0.875}} = \sigma_f - \sigma_x^\circ - \sigma_0 \tag{9}$$

where δ_c is the critical CTOD, E_p is the plastic modulus, r_p is the plastic rotational factor, ℓ is the ligament length, and β and m are work hardening constants corresponding to

$$\bar{\sigma} = \sigma_0 + \beta\bar{\epsilon}^m \tag{10}$$

where $\bar{\sigma}$ is effective stress and $\bar{\epsilon}$ is effective strain.

Equation 9 expresses δ_c as a function of three stresses. The fracture stress is a material constant, independent of temperature and specimen geometry. The yield stress is independent of geometry but is temperature dependent. The geometry dependence is contained in σ_x°. This stress governs the temperature at which the upturn in the toughness-temperature curve occurs (i.e., the

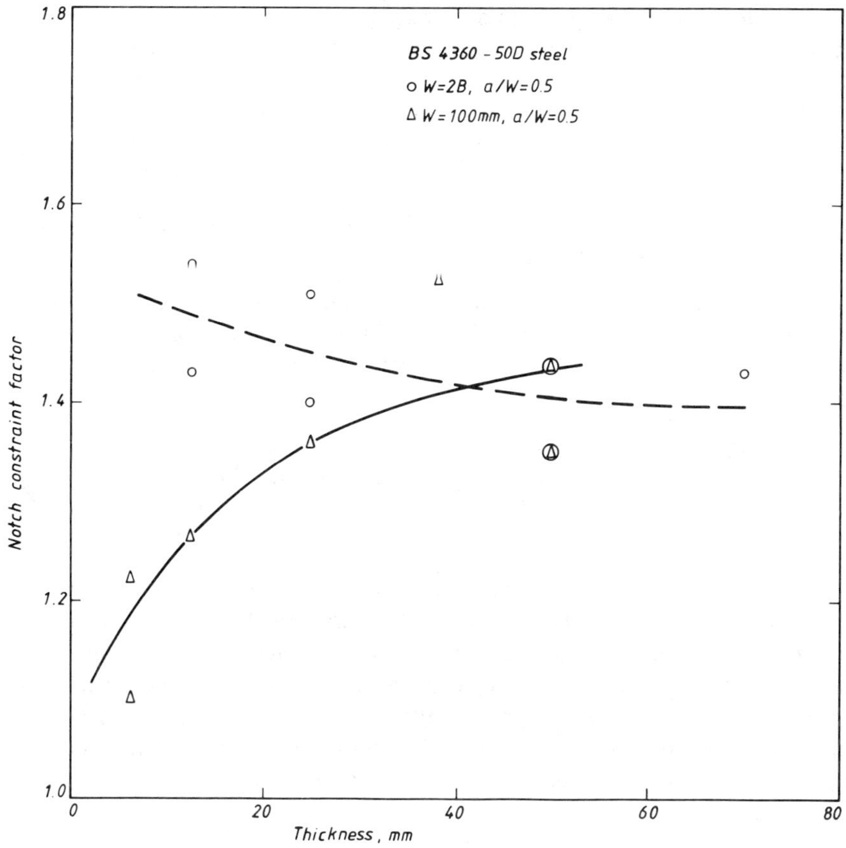

FIG. 12—*Notch constraint as a function of specimen thickness for Material 3;* T = *−50°C.*

border between the lower shelf and the transition region). An empirical relationship between σ_x° and L was obtained for Material 1 [*15*]:

$$\frac{\sigma_x^\circ}{\sigma_f} = 0.183L + 0.242 \qquad (11)$$

A notable feature of Eq 9 is that it predicts a steeper transition curve as ligament length decreases. This is consistent with experimental observations.

Predicting Transition Curves

The constraint model was used to predict fracture toughness values for 10 and 100 mm specimens of Material 2. The parameters assumed in the analysis are outlined below.

1. Fracture Stress—Since no data were available for this particular plate, a value of 1000 N/mm^2 was assumed for σ_f. This was based on typical values reported for similar materials [*21*]. It should be noted that the analysis is not sensitive to slight ($\leq$ 10%) variations in the assumed σ_f. Since σ_f and X_c are related through the Tracey curves [*20*], errors in σ_f are self compensating (e.g., a slight overestimate of σ_f will result in a slight underestimate of X_c).

2. Critical Distance—For $\sigma_f = 1000$ N/mm^2 and $K_{Ic} = 1110$ N mm$^{-3/2}$ at -196°C [*7*], $X_c = 0.245$ mm.

3. σ_x°—For both the 10 and 100 mm specimens, $L \simeq 1.45$. It was assumed that Eq 11 is valid for Material 2. This was considered to be a reasonable assumption, since this steel was very similar to Material 1, the steel for which Eq 11 was obtained. Thus $\sigma_x^\circ = 507$ N/mm^2.

4. Flow Properties—Yield strength and tensile strength of Material 2 have been measured as a function of temperature by Pisarski [*7*]. From these data, true stress-true strain curves were constructed. For this material $E_p \simeq 2500$ N/mm^2.

The predicted transition curves for the 10 and 100 mm thick specimens are plotted in Fig. 13. The 100 mm data and the statistically corrected mean values for 10 mm specimens are also plotted. The predicted curves lie above and to the left of the data. This discrepancy between prediction and experiment is probably due to simple assumptions adopted in the model. It is noteworthy, however, that the predicted temperature shift between the 10 and 100 mm data agrees reasonably well with the experimental data.

Discussion

The statistical model worked reasonably well at low temperatures where fracture was preceded by a small degree of plastic flow. At higher temperatures and toughnesses, the statistical model does not account for all the observed size effects on critical CTOD values. The following observations, based on this investigation and a previous study [*15*], led the authors to conclude that constraint effects contribute to size-related toughness shifts when cleavage fracture is preceded by large scale yielding.

1. Experimental measurements [*15*] of CTRC indicate that constraint relaxes when the remaining ligament yields. The rate of this relaxation increases with decreasing ligament length. Specimens with shorter ligaments also have steeper transition curves [*15*].
2. The initial value of CTRC (before ligament yield) is related to L [*15*] which in turn is related to transition temperature (Fig. 10). Specimen dimensions have a significant effect on L (Figs. 11 and 12).
3. There is a transition shift between the side grooved and plain subsize specimens. If there were no difference in constraint between the 10 and 100

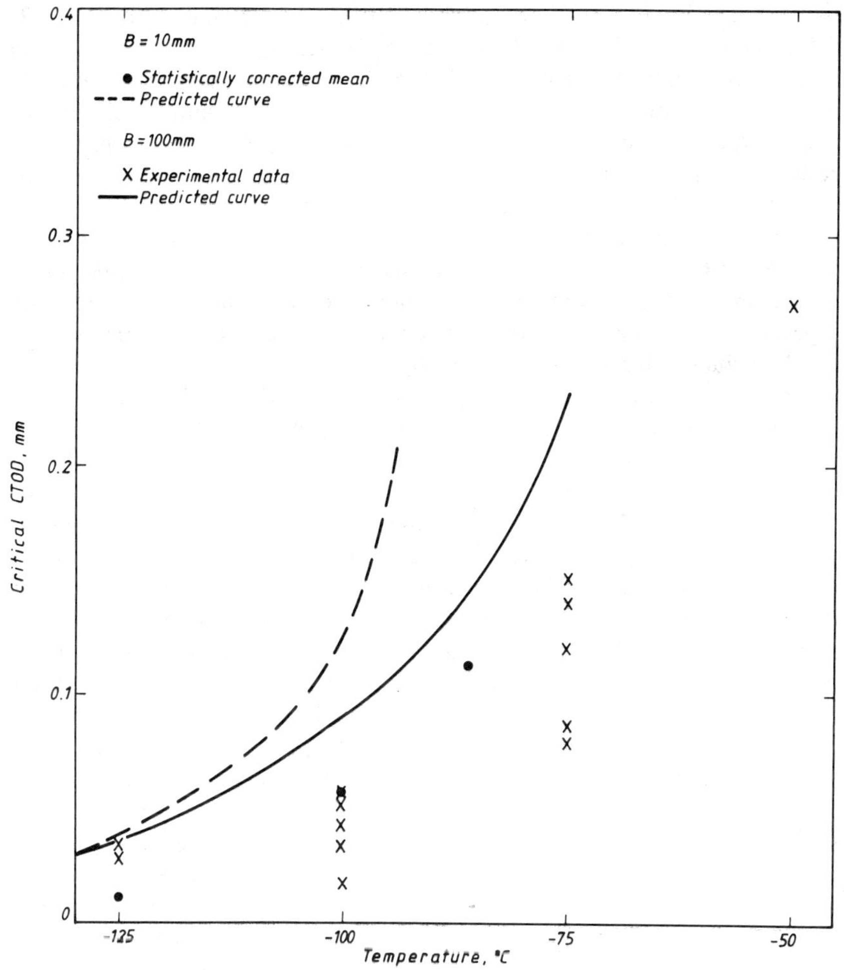

FIG. 13—*Predicted transition curves (Material 2) based on the constraint model* [15] *compared with experiment. The statistically corrected mean refers to the Landes and Shaffer "mean" (Eq 3) computed for the 10 mm plain specimens.*

mm specimens, the constraint in the 10 mm plain specimens would be at a maximum and side grooving would have no effect.

4. The constraint model works reasonably well for explaining the discrepancy between the 100 mm data and the statistically corrected 10 mm data. The predicted curves in Fig. 13 coincide at low temperatures but diverge at higher temperatures.

The interaction between constraint and statistical effects can be explained as follows. At low temperatures a small degree of plastic flow precedes fracture and a high level of constraint is maintained in both small and large specimens. Since L values are similar for the 10 and 100 mm specimens (Fig. 12), CTRC under contained yield should also be similar for the two specimen sizes. In this temperature range, all observed size effects can be explained by a statistical argument. As temperature increases, the amount of plastic flow prior to fracture increases and constraint relaxes rapidly in small specimens, resulting in a steep ductile-to-brittle transition. This relaxation in constraint was suppressed in the side grooved specimens, causing the transition to shift upwards in temperature.

Some of the small specimen data used to predict the large scale results (based on the statistical model) were upper shelf toughness values. This was especially true at higher test temperatures. Since upper shelf toughness is related to ductile tearing resistance, and transition region and lower shelf toughnesses correspond to cleavage resistance, statistical distributions that contain both types of toughness data are not strictly valid. One might argue that this fact, rather than constraint relaxation, causes the statistically corrected data to diverge from the large specimen data as temperature increases (Fig. 5). However, this conclusion is almost certainly incorrect. If tearing and plastic collapse did not occur in a material, the fracture toughness would increase without bound in the transition region. In real materials plastic collapse intervenes at higher temperatures, thereby imposing a ceiling on toughness. This has the effect of lowering the mean toughness from what it would be if the upper shelf were not present. The predicted large specimen toughnesses are also depressed from what they would be if there were no upper shelf. Thus, in the absence of an upper shelf, the predicted toughnesses would lie even further above the experimental data at higher temperatures.

It is apparent from this investigation that both constraint and statistical sampling effects can contribute to shifts in toughness with specimen size. Neither explanation for size effects applies to all situations. The relative contribution of each mechanism depends on the degree of plastic flow prior to fracture. Perhaps a model which considers both constraint and statistical effects would be appropriate for the ductile-to-brittle transition region.

Conclusions

1. The statistical sampling model works reasonably well for explaining size effects under conditions of small scale yielding.
2. Constraint relaxes with plastic flow. This relaxation occurs more rapidly and at lower temperatures in small specimens.
3. As critical CTOD increases, the relative contribution of statistical sampling effects decreases and constraint effects tend to dominate.

Acknowledgments

The authors would like to express their gratitude to colleagues in the Fracture Section of The Welding Institute who have shared their expertise and offered words of encouragement.

This paper was prepared under joint funding of the Research Members of The Welding Institute and the Materials, Chemicals and Vehicles Requirements Board in Great Britain. The fracture toughness data for Material 1 were collected while one of the authors (TLA) was employed by the National Bureau of Standards in Boulder, Colorado. The tests on this material were funded by the U.S. Department of Interior, Mineral Management Service. As stated in the text, most of the fracture toughness tests on Materials 2 and 3 were performed by Dr. H. G. Pisarski of The Welding Institute and Mr. M. S. Kamath (formorly with TWI).

References

[1] Dawes, M. G., "Elastic-Plastic Fracture Toughness Based on the COD and *J*-Contour Integral Concepts," in *Elastic-Plastic Fracture, ASTM STP 668*, American Society for Testing and Materials, Philadelphia, 1979, pp. 307–333.

[2] Stuber, A. B., "A Study of Crack Growth Behavior and Specimen Size Effects in the Crack (Tip) Opening Displacement Test," Ph.D. thesis, University of Kansas, Lawrence, 1980.

[3] Penelon, A., Bassim, M. N., and Dorlot, J. M., "Variation of Fracture Toughness with Specimen Geometry and Loading Conditions in Welded Low Alloy Steels," in *Fracture Mechanics, ASTM STP 677*, American Society for Testing and Materials, Philadelphia, 1979, pp. 449–462.

[4] Milne, I., "Fracture Toughness Testing of a Quenched and Tempered Medium Strength Steel in the Ductile-Brittle Transition Range," *Materials Science and Engineering*, Vol. 30, 1977, p. 243.

[5] Chell, G. G. and Spink, G. M., "A Post Yield Fracture Mechanics Analysis of Three-Point Bend Specimens and its Implications to Fracture Toughness Testing," *Engineering Fracture Mechanics*, Vol. 9, 1977, p. 101.

[6] Milne, N. and Worthington, P. J., "The Fracture Toughness of a Low Alloy Pressure Vessel Steel in the Post Yield Regime," *Materials Science and Engineering*, Vol. 26, 1976, p. 185.

[7] Pisarski, H. G., "Influence of Thickness on Critical Crack Opening Displacement (COD) and *J* Values," *International Journal of Fracture*, Vol. 17, 1981, p. 427.

[8] de Castro, P. M. S. T., Spurrier, J., and Hancock, P., "Comparison of *J* Testing Techniques and Correlation *J*-COD Using Structural Steel Specimens," *International Journal of Fracture*, Vol. 17, 1981, p. 83.

[9] Chell, G. G. and Davison, A., "A Post-Yield Fracture Mechanics Analysis of Single Edge Notched Tension Specimens," *Materials Science and Engineering*, Vol. 24, 1976, p. 45.

[10] Chell, G. G. and Gates, R. S., "A Study of Failure in the Post Yield Regime Using Single Edge-Notched Tension Specimens," *International Journal of Fracture*, Vol. 14, 1978, p. 223.

[11] de Castro, P. M. S. T., Spurrier, J., and Hancock, P., "An Experimental Study of the *a/W* Ratio Dependence of the COD Test Using Small-Scale Specimens," in *Fracture Mechanics, ASTM STP 677*, American Society for Testing and Materials, Philadelphia, 1979, pp. 486–497.

[12] Anderson, T. L., McHenry, H. L., and Dawes, M. G., "Elastic-Plastic Fracture Toughness Tests with Single-Edge Notched Bend Specimens," in *Elastic-Plastic Fracture Test Methods: The User's Experience, ASTM STP 856*, American Society for Testing and Materials, Philadelphia, 1985, pp. 210–229.

[13] Iwadate, T., Tanaka, Y., Ono, S., and Watanabe, J., "An Analysis of Elastic-Plastic Fracture Toughness Behavior for J_{Ic} Measurement in the Transition Region," in *Elastic-Plastic Fracture: Second Symposium, Volume II—Fracture Resistance Curves and Engineering Applications, ASTM STP 803*, American Society for Testing and Materials, Philadelphia, 1983, pp. II-531—II-561.

[14] Landes, J. D. and Shaffer, D. H., "Statistical Characterization of Fracture in the Transition Region," in *Fracture Mechanics: Twelfth Conference, ASTM STP 700*, American Society for Testing and Materials, Philadelphia, 1980, pp. 368–382.

[15] Anderson, T. L., "A Constraint-Based Model for Fracture in the Ductile-to-Brittle Transi-

tion Region," Welding Institute Report 248/1984, Abington Hall, Abington, Cambridge, England, Aug. 1984.
[*16*] Kamath, M. S., unpublished work, The Welding Institute, Abington Hall, Abington, Cambridge, England, 1979–80.
[*17*] Weibull, W., "A Statistical Distribution Function of Wide Applicability," *Journal of Applied Mechanics*, Vol. 18, 1953.
[*18*] Wine, R. L., *Statistics for Scientists and Engineers*, Prentice-Hall, Englewood Cliffs, N. J., 1964.
[*19*] Ritchie, R. O., Knott, J. F., and Rice, J. R., "On the Relationship Between Critical Tensile Stress and Fracture Toughness in Mild Steel," *Journal of the Mechanics and Physics of Solids*, Vol. 21, 1973, p. 395.
[*20*] Tracey, D. M., "Finite Element Solutions for Crack-Tip Behavior in Small Scale Yielding," *Journal of Engineering Materials and Technology*, Vol. 98, 1976, p. 146.
[*21*] Knott, J. F., "Micro-Mechanisms of Fracture and the Fracture Toughness of Engineering Alloys," in *Fracture 1977*, Vol. 1, ICF4, Waterloo, Canada, June 1977, p. 61.

J. A. Joyce[1] *and E. M. Hackett*[2]

Dynamic *J-R* Curve Testing of a High Strength Steel Using the Key Curve and Multispecimen Techniques

REFERENCE: Joyce, J. A. and Hackett, E. M., **"Dynamic *J-R* Curve Testing of a High Strength Steel Using the Key Curve and Multispecimen Techniques,"** *Fracture Mechanics: Seventeenth Volume, ASTM STP 905*, J. H. Underwood, R. Chait, C. W. Smith, D. P. Wilhem, W. A. Andrews, and J. C. Newman, Eds., American Society for Testing and Materials, Philadelphia, 1986, pp. 741-774.

ABSTRACT: *J*-integral *R*-curve tests were performed on three-point bend specimens of a 3-Ni steel at three loading rates: quasi-static, intermediate (25 mm/s), and drop tower (2.54 m/s). The key curve and multispecimen procedures were employed for the higher rate tests; this investigation is focused primarily on details of the test method development. The multispecimen and key curve techniques were found to yield upper shelf *J-R* curves which were in substantial agreement at the elevated loading rates. Numerical smoothing techniques required to apply a key curve method appear to separate the oscillatory high frequency component from the load-displacement record. For the 3-Ni steel tested for this investigation both J_{Ic} and T were found to be elevated with increasing loading rate.

KEY WORDS: *J-R* curve, key curve, dynamic loading, three-point bend specimen, J_{Ic}, tearing modulus (T), numerical smoothing, multispecimen technique

Over the past few years fracture characterization of elastic-plastic alloys has made extensive use of *J*-integral methods including the J_{Ic} parameter and *J-R* curve analyses. All uses of the *J*-integral for fracture resistance measurements derives from the solution by Hutchinson [*1*] and Rice and Rosengren [*2*] of the problem of a static tensile crack in a power law hardening nonlinear elastic material. The intensity of the stress and strain field is then defined by a single parameter which is a function of loading type, load intensity, and crack geometry. This situation was recognized by Landes and Begley [*3*] as similar

[1]Associate Professor of Mechanical Engineering, U.S. Naval Academy, Annapolis, MD 21401.
[2]Research Engineer, David Taylor Naval Ship Research and Development Center, Bethesda, MD 21402.

to the elastic stress intensity factor (K) and singularity which had been utilized to develop fracture criteria for elastic cracked bodies, and they experimentally demonstrated that J_{Ic} is a material parameter, although subject to certain size restrictions.

J_{Ic} is the value of the J-integral at or near crack initiation. The standard test procedure for the determination of J_{Ic} (ASTM E 813) describes a multispecimen approach which defines a J versus crack extension or J-R curve. J_{Ic} is then determined by an extrapolation technique, the applicability of which to certain materials is the subject of considerable debate.

In many applications the design of a structure to avoid all crack extension is impractical and the J_{Ic} parameter is considered overly conservative. This fact coupled with the recent development of tearing instability theory and the establishment of reliable single-specimen test procedures for the determination of J_{Ic} and the J-R curve [*4,5*] have centered attention on crack growth beyond initiation. Paris and co-workers [*6*] have presented a tearing instability theory defining the conditions for ductile instability in terms of a nondimensional tearing modulus (T). The applied tearing force is a function of the crack geometry and structure elastic compliance and the type of loading (i.e., load or displacement controlled), while the material tearing resistance is a function of the slope of the J-R curve beyond initiation; that is,

$$T_{mat} = dJ/da \times E/\sigma^2 \tag{1}$$

where

E = material elastic modulus, and
σ = material flow stress.

If $J_{applied}$ exceeds J_{Ic} and $T_{applied}$ equals or exceeds the material tearing resistance, a ductile instability is predicted by the Paris theory.

A considerable effort [*7–9*] has been directed toward estimations of the applied T for several important cases. It appears that estimation of T in complicated structural applications can be accomplished by finite element techniques.

Experimental verification of the Paris tearing instability theory has been demonstrated by several researchers [*10–13*], including a recent validation by Vassilaros and co-workers [*14*] for the case of a circumferentially cracked 8-in.-diameter pipe of an ASTM A106 steel used in the commercial nuclear industry.

The laboratory J-R curve determinations described above have all been conducted at slow (quasi-static) loading rates. Many of the anticipated loading events for the actual structures, however, are dynamic in nature. An example would be rapid pressurization or de-pressurization of a nuclear reactor vessel due to a loss of coolant accident (LOCA). A need exists, therefore, to extend the ductile fracture mechanics technology described above into the dynamic loading regime. This need was reinforced in a recent comprehensive review of dynamic fracture of metallic materials by the National Materials Advisory Board [*15*]. The committee recommended that "both experimental and mathematical analysis work need to be performed to identify and establish the appropriate material parameters needed for a dynamic elastic-plastic fracture mechanics methodology." Conventional single-specimen techniques using elastic-compliance for *J-R* curve determinations cannot be employed at higher loading rates. Techniques which can be utilized at elevated loading rates include multispecimen, direct-current potential drop (DCPD), alternating-current potential drop, and the key curve method.

The objective of this research was to develop methods of evaluating *J-R* curves for elastic plastic alloy steels at rates well above those that have been used to date. This report concentrates on details of test technique development for the key curve and multispecimen approaches, but also includes valuable data on an 3-Ni structural steel tested at three rates, including a drop tower rate. All testing was performed at room temperature, which is well up on the upper shelf for this material.

Experimental Program

Overview

The experimental program for this investigation involved tests of three-point bend fracture specimens at three rates: quasi-static, rapid (servo-hydraulic), and drop tower. The three-point bend fracture specimens were modified to allow for load-line displacement measurements. Tensile tests were performed at various loading rates in order to characterize the rate dependence of the material flow properties. Lastly, Charpy impact toughness tests were performed to identify the upper and lower shelves and transition temperature regions for the material.

Material

Material for this investigation was provided in the form of a 38.1 mm ($1^{1}/_{2}$ in.) thick 3-Ni steel plate. The chemical composition of this plate is given in Table 1.

Tensile Testing

Tensile tests were performed on standard 12.8 mm (0.505 in.) diameter and subsize 6.4 mm (0.252 in.) diameter tensile specimens. The subsize specimens were tested at loading rates of 0.25 mm/min (0.01 in./min), 50 mm/min (2 in./min), and 0.125 m/s (3.2 in./s) to characterize the rate dependence of the material flow properties.

TABLE 1—*Chemical composition of steel used for* J-R *curve testing.*

Identification	Chemical Composition, wt%										
	C	Mn	P	S	Cu	Si	Ni	Cr	Mo	V	Ti
3-Ni Steel (FYB)	0.153	0.33	0.012	0.013	0.033	0.18	2.55	1.66	0.37	0.003	less than 0.001

Impact Toughness Testing

Charpy V-notch impact specimens were machined from the 3-Ni steel plate with the notches in the T-L orientation. These specimens were tested in accordance with ASTM E 23 to characterize the transition temperature behavior of the material.

J-R *Curve Testing*

The basic geometry used for these tests was a 1T three-point bend fracture specimen (Fig. 1) common to ASTM fracture mechanics test procedures. For

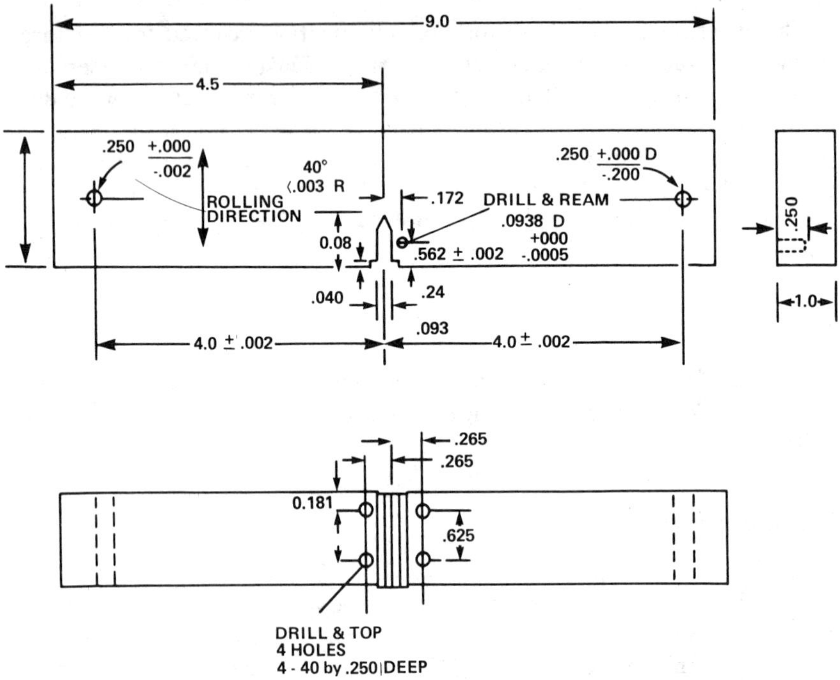

FIG. 1—*Three-point bend specimen 1T plan.*

the key curve method calibration specimens, a ½T bend specimen was also utilized. The notch orientation for all specimens was T-L. The specimens were precracked in accordance with the requirements in ASTM E 813. All specimens were side grooved to a total reduction of 20%.

Quasi-static *J-R* curve tests were used in this investigation to characterize the material used and to act as a baseline for comparison of rate effects found in the higher rate tests. All quasi-static tests were performed using the computer-interactive elastic compliance procedure of Joyce and Gudas [5] in accordance with the proposed *J-R* curve procedure [6] published recently. The apparatus used to measure load-line displacement (Fig. 2) consisted of a strain-gaged "flex bar" mounted directly on the specimen. This device was calibrated for load-line displacement on the bend specimen by simulating the deflection at the flex bar centerline with a micrometer.

Intermediate rate tests were conducted on a servohydraulic test machine using the stroke control mode with a ram velocity of 25 mm/s (1 in./s). The test apparatus described for the quasi-static tests was found to be adequate for these higher rate tests, except that a higher speed data acquisition system was required. The key curve and multispecimen procedures were employed for *J-R* curve determinations at intermediate rates, since the elastic compliance technique is not suitable for this test rate.

Drop weight tests were conducted in a drop tower at a striker (tup) velocity of approximately 2.54 m/s (100 in./s). The fixtures and electronic apparatus pertinent to these tests are shown in Fig. 3. The specimen is supported by flat, hardened anvils and is struck via a soft metal absorber. To avoid indenting the stop blocks, a flat striker geometry is used. Three transducer signals were acquired during these tests. Load is measured in two ways: (1) from the instrumented tup and (2) by two strain gages mounted on the specimen (Fig. 3). The strain gages mounted on the specimen constituted a half bridge, and their output was fed into a strain gage conditioning amplifier. This strain

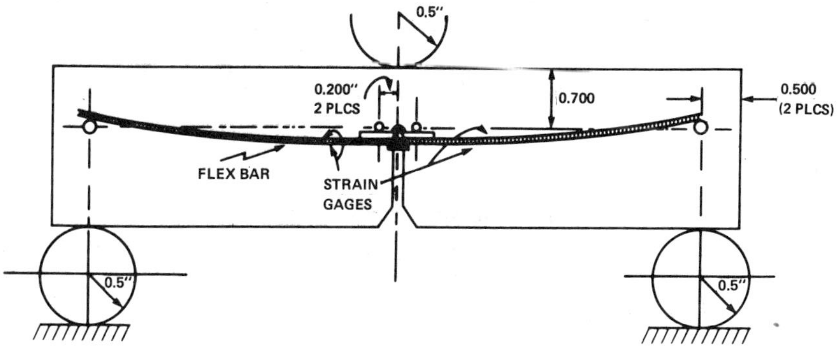

FIG. 2—*Diagram of three-point bend bar static test arrangement showing details of flex bar and specimen supports.*

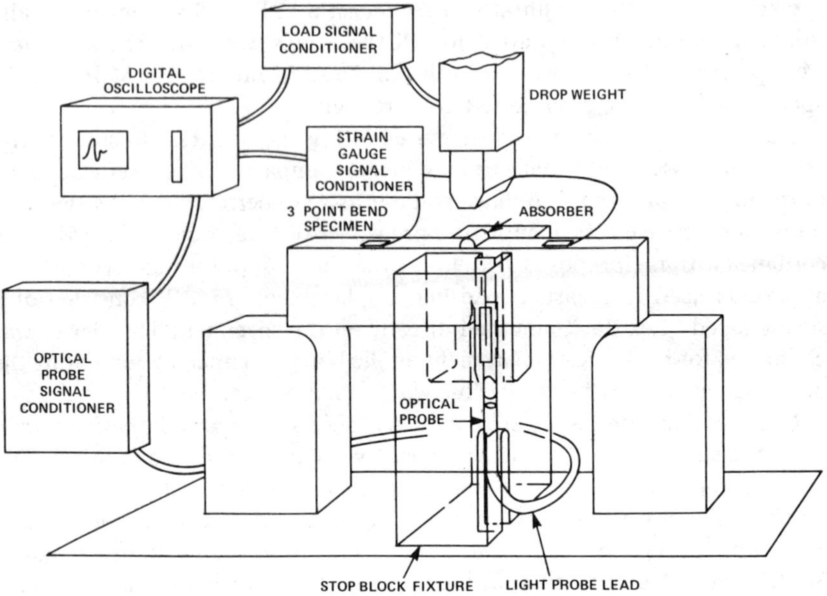

FIG. 3—*Experimental setup for drop weight tests.*

gage system was calibrated statically against the load cell in a screw-type tensile machine. Load-line displacement is measured using the fiber optic light probe transducer shown in Fig. 3. The linkage shown allows the specimen arm to rotate while requiring the piston facing the light beam to stay in line with the light beam, avoiding errors resulting from relative rotation to which this type of transducer is very sensitive. The offset of the attachment point for the linkage is exaggerated in Fig. 3; the actual distance is 3.75 mm (0.15 in.). This distance is approximately 2% of the beam span.

The expression used for calculation of the J-integral for the single-specimen quasi-static tests was that developed by Ernst et al [*17*]:

$$J_{(i+1)} = J_i + \frac{\eta}{b_i}\frac{A_{i,i+1}}{B_N}\left[1 - \frac{\gamma}{b_i}(a_{i+1} - a_i)\right] \quad (2)$$

where

η = 2 for three-point bend specimens,
W = specimen width,
γ = 1 for three-point bend specimens,
b_i = instantaneous length of remaining ligament,
B_N = minimum specimen thickness,
a_i = instantaneous crack length, and
$A_{i,i+1}$ = area under the load versus load-line displacement record between lines of constant displacement at points i and $i + 1$.

This work is directed towards high rate *J-R* curve testing; however, since considerable disagreement exists on dynamic expressions for *J*, static *J* expressions have been utilized here. To evaluate *J* using static equations requires only load-time and load point displacement-time records to be taken during the test for the multispecimen and key curve techniques.

Key Curve Method

The key curve method was first presented by Ernst et al [*18*]. They used dimensional analysis to show that for simple geometries in which plasticity is confined to the uncracked ligament region the load displacement relationship must have the form

$$\frac{PW}{Bb^2} = F1\left(\frac{\Delta}{W}, \frac{a}{W}, \frac{L}{W}, \frac{B}{W}, \text{material properties}\right) \tag{3}$$

where

P = applied load,
Δ = total load-line displacement,
a = crack length,
b = uncracked ligament,
B = specimen thickness, and
$W = a + b$ = specimen width.

Substitution of Eq 3 into the deformation plasticity theory formula for *J* gives an incremental formula for evaluation of *J* [*19*]:

$$dJ_n = \left[\frac{2b}{W}F1_n - \frac{b^2}{W^2}\left(\frac{\partial F1}{\partial(a/W)}\right)_n\right]d\Delta_n + \left[-\frac{2}{W}\int F1\,d\Delta + \frac{4b}{W^2}\int\left(\frac{\partial F1}{\partial(a/W)}\right)d\Delta + \frac{b}{W^3}\int\left(\frac{\partial^2 F1}{\partial(a/W)^2}\right)d\Delta\right]da_n \tag{4}$$

and an incremental formula for the evaluation of crack extension:

$$da_n = \frac{\dfrac{b^2}{W^2}\left(\dfrac{\partial F1}{\partial(\Delta/W)}\right)_n d\Delta_n - dP_n}{\dfrac{2b}{W}F1_n - \dfrac{b^2}{W^2}\left(\dfrac{\partial F1}{\partial(a/W)}\right)_n} \tag{5}$$

It was pointed out by Ernst et al [*18*] in the original work that deeply cracked bend bars should exhibit key curves which are independent of crack length (i.e., $P \alpha\, b^2$ for all applied displacements). Equations 4 and 5 can be reduced to the much simpler form that:

$$dJ = \frac{2b}{W} F1 d\Delta + \left[-\frac{2}{W} \int F1 d\Delta \right] da \tag{6}$$

and

$$da = \frac{\dfrac{b^2}{W^2} \dfrac{\partial F1 d\Delta}{\partial(\Delta/W)} - dP}{\dfrac{2b}{W} F1} \tag{7}$$

It will be shown in later sections that for the three-point bend bar identical key curves are obtained for various crack lengths, at least to within experimental accuracy, and that these simpler equations are applicable for the three-point bend geometry.

Key curves could be obtained by analytical methods and in fact are implicit in the J integral crack driving force diagrams obtained by Kumar et al [*20*]. Application of these diagrams to obtain key curves has not been successful, apparently due to an inability to accurately measure the material strain hardening exponent and to accommodate the fact that 25.4 mm (1 in.) thick specimens are intermediate between plane stress and plane strain in their behavior under load.

The technique used here is to obtain the desired key curve by experimental methods as done previously [*20–22*] for static and rapid hydraulic tests on compact specimens. Subsize and blunt notched specimens are generally used to retard the point of crack initiation, though as discussed below this did not prove very successful for three-point bend specimens.

The basic method used here then to develop J-R curves with the key curve method was to obtain the key curve $F1(\Delta/W)$ from tests on subsized geometrically similar specimens, then to obtain $dF1/(d(\Delta/W)$ using numerical differentiation techniques, and then to integrate Eqs 6 and 7 along the full scale specimens load displacement record to obtain J and Δa.

Multispecimen Method

The major positive feature of the multi-specimen method is its directness and simplicity. The multispecimen test method was developed for J-R curve testing by Begley and Landes [*3*] as a technique for J_{Ic} measurement. The multispecimen method used here follows the procedure of ASTM E 813. A series of specimens prepared so as to be geometrically identical is tested to different load point displacements and then heat tinted to mark the final crack extension. The specimens are broken open and the average crack extension is measured with an optical microscope. The value of J is estimated from each load displacement record to the individual termination point, and the resulting J-crack extension pairs are plotted to yield the J-R curve. Using the procedure defined in ASTM E 813 for quasi-static testing, a J_{Ic} point can then be identified for the material.

For the multispecimen tests conducted for this investigation J was determined from the following relationship derived by Rice [*23*]:

$$J = 2A/Bb \tag{8}$$

where:

A = area under the load/load-line displacement record to the point of termination,
B = net thickness of the specimen to the side grooves, and
b = initial ligament.

The multispecimen test procedure can be employed for higher loading rates if a rigid stop block fixture is utilized to stop the drop weight or pendulum after slightly differing load-line displacement values have been reached as described in Ref *24*.

Results and Discussion

Tensile Tests

Results from the tensile tests are presented in Table 2. As expected, the increased displacement rate tests exhibited slight elevation in the 0.2% yield and ultimate tensile stress compared with the quasi-static (0.25 mm/min) tests. Engineering stress-strain curves for the subsize tensile specimens at the three different displacement rates are provided in Fig. 4.

Impact Toughness Tests

Results of the Charpy V-notch impact toughness tests showed the 3-Ni steel to be clearly on the upper shelf at room temperature. The average upper shelf impact toughness was 95.6 J (70.3 ft-lb).

Quasi-Static J-R *Curve Testing*

A series of quasi-static tests was run in this study to act as a baseline for comparison with intermediate rate servohydraulic results and high rate drop weight results. A typical result is shown in Fig. 5 and demonstrates the excellent correspondence obtained between unloading compliance, key curve, and

TABLE 2—*Tensile properties of steel used for* J-R *curve testing.*

Specimen Diameter		Displacement Rate		Ultimate Tensile Strength		Yield Strength (0.2% Offset)		Elongation, %	Reduction of Area, %
mm	(in.)	mm/min	(in./min)	MPa	(ksi)	MPa	(ksi)		
12.8	(0.505)	0.25	(0.01)	731	(106)	614	(89)	23	63
6.4	(0.252)	0.25	(0.01)	724	(105)	621	(90)	24	69
6.4	(0.252)	50	(2)	765	(111)	669	(97)	23	69
6.4	(0.252)	0.125 m/s	(3.2 in./s)	779	(113)	669	(97)	22	61

[a]All specimens have transverse orientation.

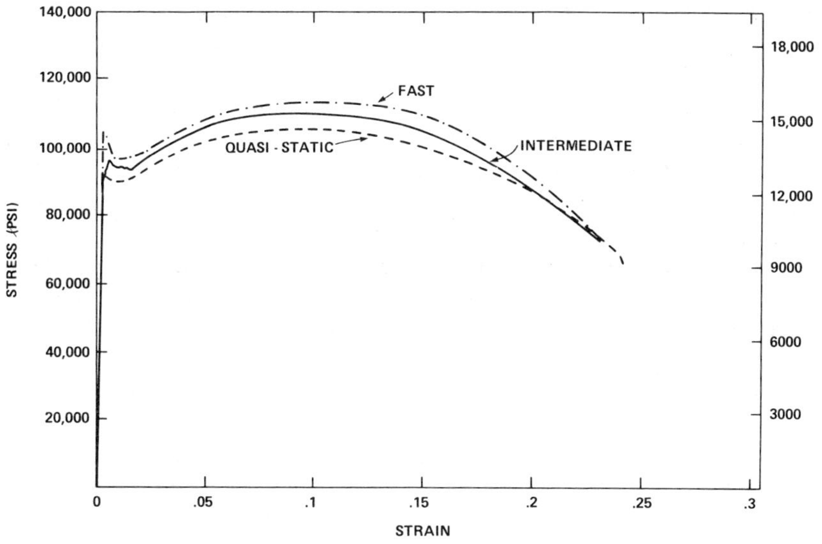

FIG. 4—*Engineering stress-strain curves for 3-Ni steel at three loading rates.*

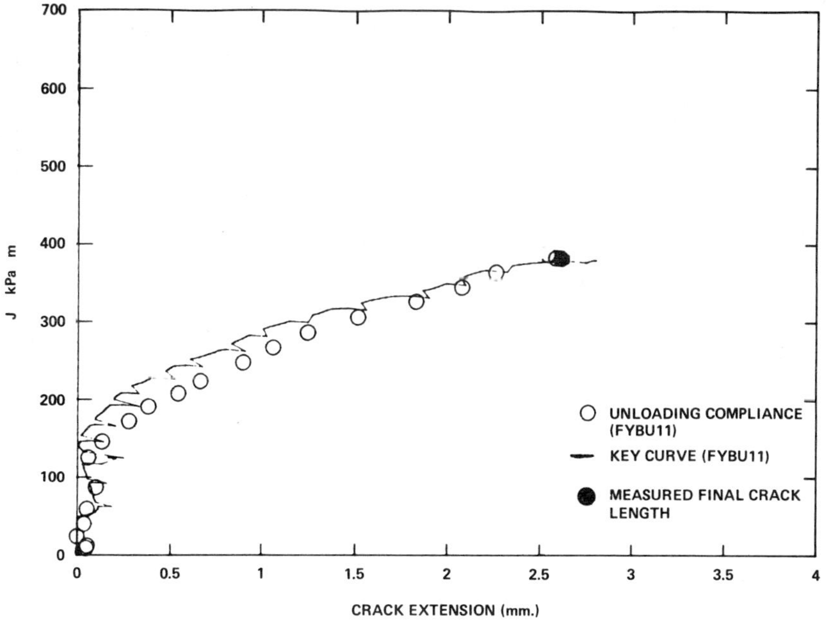

FIG. 5—*Unloading compliance and key curve* J-R *curves for a quasi-static 1T three-point bend specimen.*

a measured final crack length result as long as side-grooved specimens are used. This type of result has been shown previously for compact specimens of several materials [*19,21*], but for the bend bar case used here some modification of the method was required. As before, additional ½T specimens were tested without unloadings as shown by the two examples in Fig. 6. These curves are very different because the a/W values were 0.55 for the upper curve and 0.61 for the lower curve. Putting these curves into key curve form gives the result shown in Fig. 7 where now the two curves are nearly coincident giving a key curve for the bend geometry which is independent of a/W.

Figure 8 compares a ½T bend specimen and a 1T bend specimen both plotted in key curve form. It is clear from this plot that the two geometries behave similarly until a Δ/W value of about 0.024 at which point the larger specimen begins to crack and its normalized load displacement curve begins to fall. Problems are present here as far as application of the key curve analysis technique is concerned in that at about $\Delta/W = 0.038$ the load begins to fall on the ½T specimen, which indicates that crack extension is also occurring in this specimen by this time. Thus the key curve technique as used previously is applicable here only for $\Delta/W = 0.038$, which gives a very limited *J-R* curve on this material. To extend the key curve, additional ½T specimens were tested that had machined notch roots, but these were found to delay

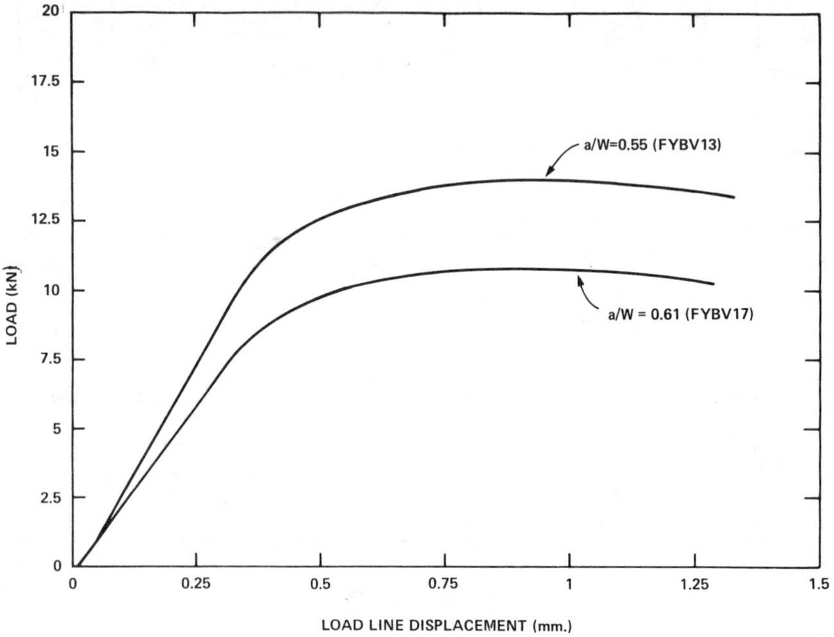

FIG. 6—*Typical load/load-line displacement records for two ½T quasi-static bend bars with different initial crack lengths.*

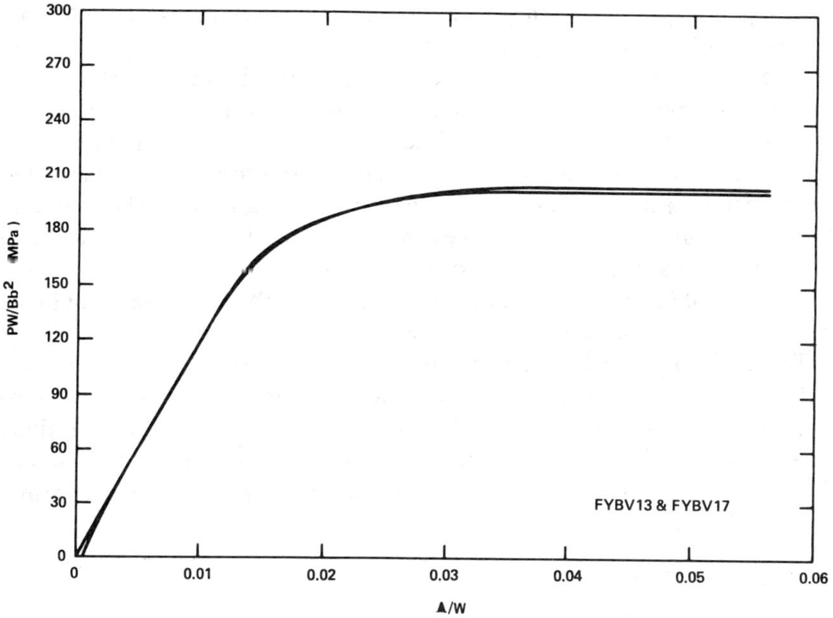

FIG. 7—*Load/load-line displacement records for two 1/2T quasi-static bend bars after key curve scaling showing* a/w *independence of the load displacement records.*

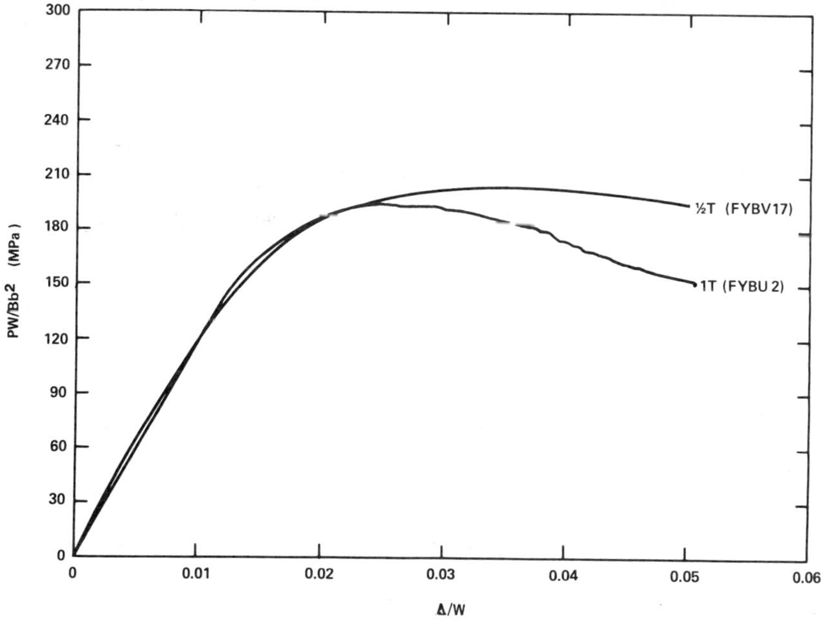

FIG. 8—*Key curve comparison of quasi-static 1/2T and 1T bend bars.*

cracking only slightly and they gave poor correspondence between the 1T and ½T key curve shapes near crack initiation.

To avoid this difficulty the key curve was extended in two ways (Fig. 9). In the first method the key curve was assumed to be constant at the maximum value reached by the experimental data. In the second method the key curve was extended as a straight line with a slope chosen to match the measured final crack length to the result of the key curve calculations. As shown in Fig. 10, the *J-R* curve resulting from the zero slope extension gives a considerable underestimate of the true crack extension, while the choice of a larger slope, 830 MPa (120 000 psi) in this case, allows matching the key curve result to the measured result found optically after completion of the test.

The use of a power law extension to the key curve with coefficients chosen to fit the experimental curve could also be used but to date has not been attempted. The particular slope of 830 MPa (120 000 psi) has been determined experimentally for static tests on this particular structural steel, but as shown in subsequent sections this same slope appears equally applicable at intermediate or drop tower test rates.

Intermediate Rate J-R *Curve Testing*

Since there is about six orders of magnitude difference in loading rate between quasi-static tests (0.25 mm/min) and drop tower rates (2.54 m/s), an

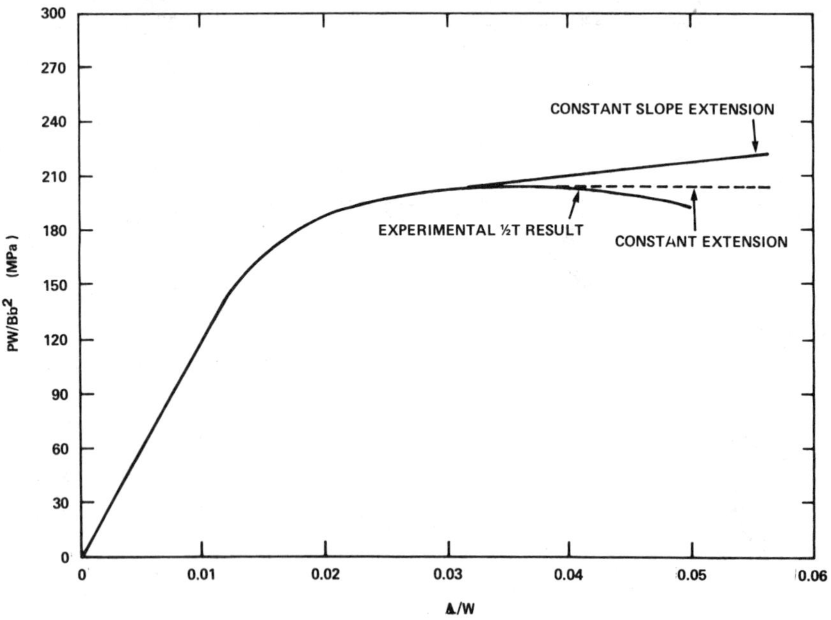

FIG. 9—*Extensions of ½T bend bar records showing both a constant value extension and a constant slope extension.*

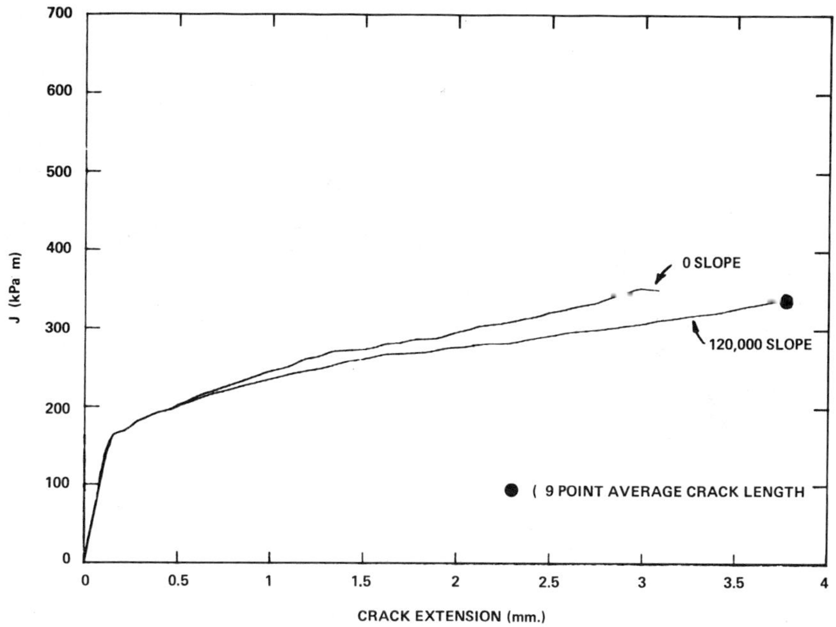

FIG. 10—*Effect of key curve extensions on quasi-static* J-R *curves.*

intermediate rate has been used in this research. The rate chosen was 25 mm/s, since this is a rate easily obtained by standard servohydraulic equipment and can be followed by standard COD clip gages and LVDT transducers.

Both multispecimen and key curve methods were used to obtain *J-R* curve results from intermediate rate tests. Multispecimen results were obtained by using the "stop at breakpoint" feature on the test machine controller to generate a series of tests run at a constant rate to different final load-line displacement values. The *J* integral values were then calculated using Eq 8 and average crack length measurements were obtained in the standard fashion to give a series of J-Δa pairs.

Key curve results were obtained by running ½T bend specimens at the intermediate test rate, generating a key curve file, extended as discussed above at a constant slope of 830 MPa (120 000 psi), and used together with the load displacement data and Eqs 6 and 7 to give *J-R* curves as shown in Fig. 11. Comparison with measured final crack lengths is good though a growing deviation at larger crack extension implies that a power law based key curve extension might give better results.

Comparison of multispecimen and key curve results is shown on Fig. 12 for which the key curve results have been redone without the crack growth corrections to *J* which are present in the standard calculation. This modification is

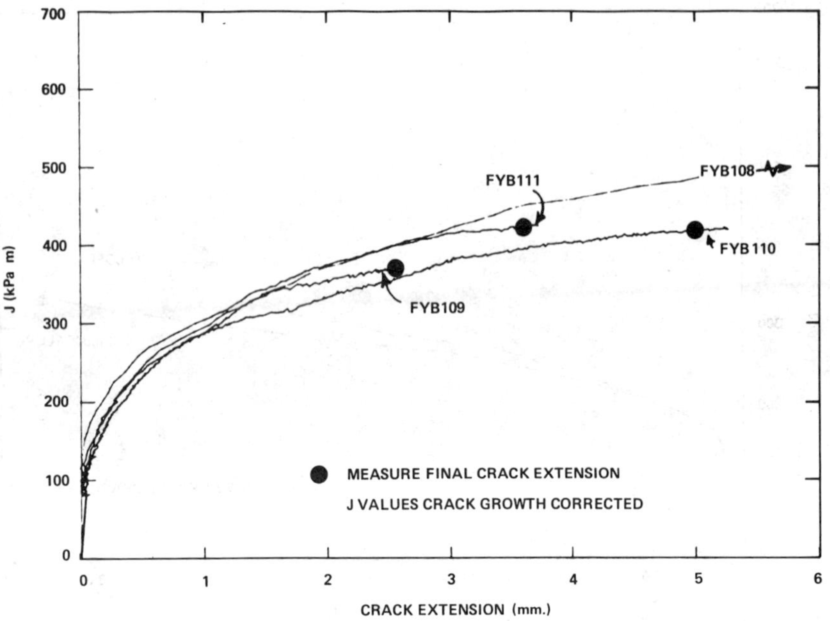

FIG. 11—*Key curve* J-R *curves for four intermediate rate tests using crack growth corrected* J *values.*

accomplished simply by dropping the second term of Eq 6 during the calculation. This produces the close correspondence between multispecimen and key curve *J* values shown in Fig. 12, but it must be emphasized that the resulting *J-R* curves are not an accurate measure of deformation theory *J* beyond the point of crack initiation.

Drop Tower J-R *Curve Testing*

The most straightforward way to conduct a drop tower test is to utilize the manufacturer's calibrated striker or tup signal for load and to assume a constant drop velocity validated by a simple velocometer system to give the load point displacement. From the start of this program two additional data signals were monitored, the first being a fiber optic load-line displacement transducer, the second a "transmitted load" signal obtained from strain gages attached to the test specimen. Typical signal outputs are shown in Figs. 13*a* to 13*c* for a ½T drop tower bend bar loaded at 1.25 m/s. Figure 13*a* shows clearly that the specimen outruns the striker shortly after impact, then bounces against the striker about three times before settling against the striker after about 1 ms of elapsed time. During this same 1 ms interval the striker load signal shown in Fig. 13*b* oscillates considerably, while the transmitted load signal of Fig. 13*c* oscillates but in a less extreme fashion.

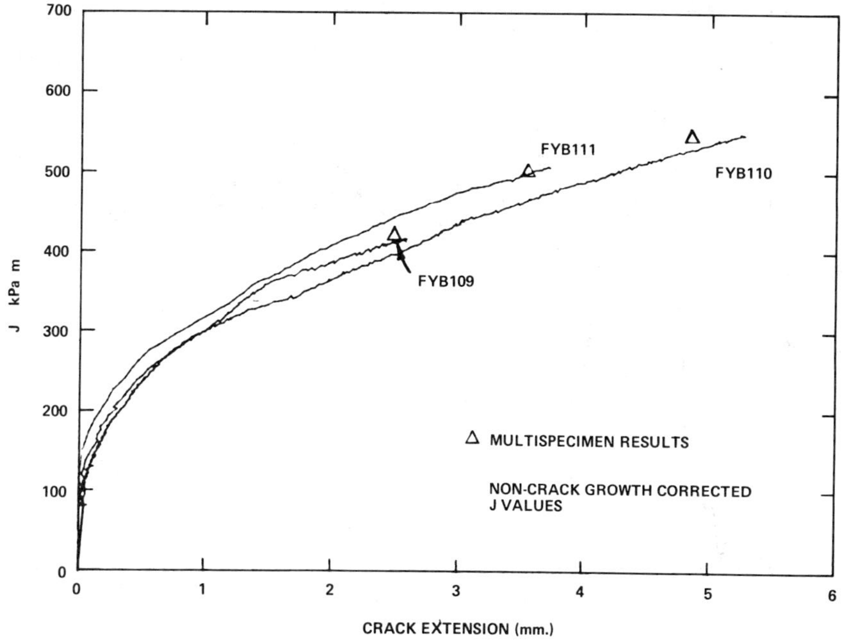

FIG. 12—*Comparison of multispecimen and key curve* J-R *curves at intermediate rate using non-crack-growth corrected* J *values.*

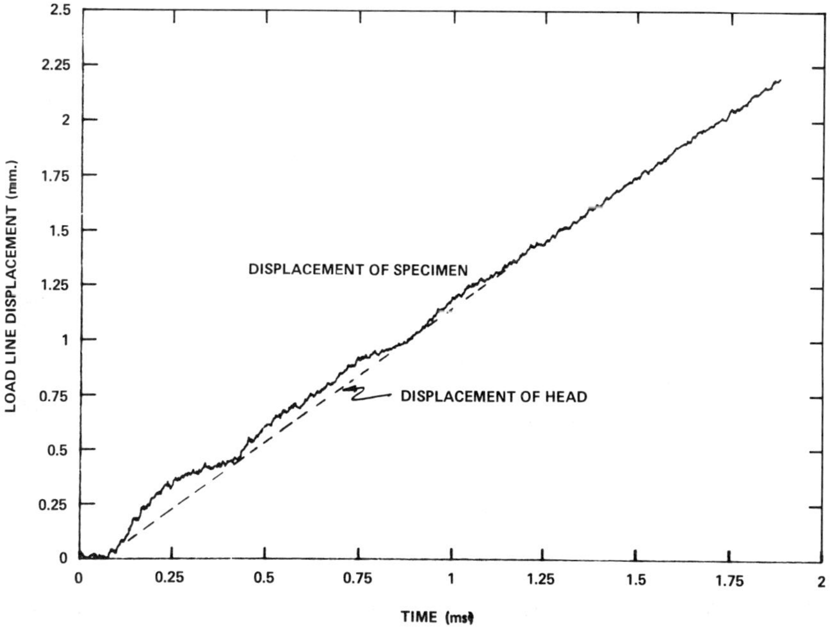

FIG. 13*a*—*Drop tower records for a 1/2T bend bar showing load-line displacement versus time.*

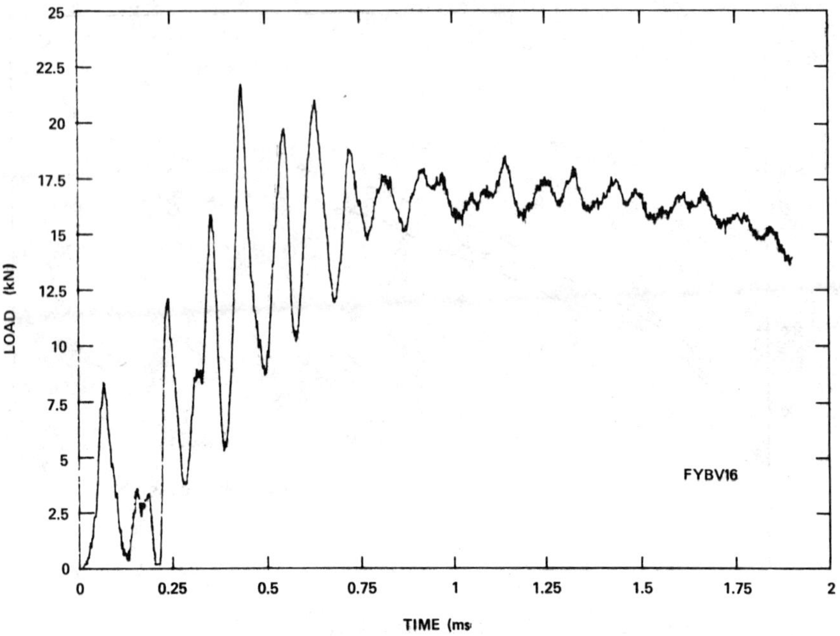

FIG. 13*b*—*Drop tower records for a* ½*T bend bar showing tup load versus time.*

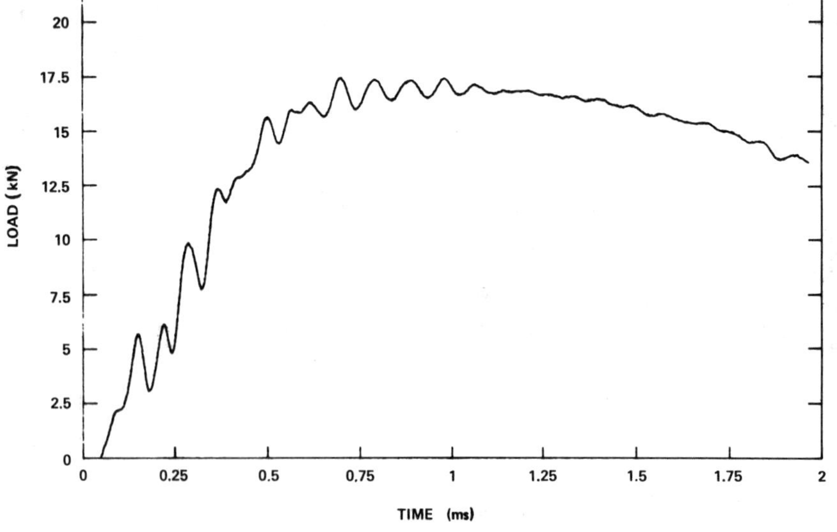

FIG. 13*c*—*Drop tower records for a* ½*T bend bar showing integral strain gage load versus time.*

Since the specimen and striker are only occasionally in contact over this 1 ms interval, it seems unreasonable to place much confidence in the striker load signal, and it is quite satisfying to see the smoother result measured by the specimen mounted strain gages of Fig. 13*c*.

Taking the smoother load signal, obtained from the specimen strain gage system, gives a plot of load versus load-line displacement that is not adequately smooth to be used to construct a key curve file for crack growth estimation, since the key curve technique is very sensitive to the slope of the contributing load displacement records. An iterated numerical polynomial smoothing routine has been developed which smooths both the displacement time and load time records. Basically, segments of the load time curve (for example) are fit with a low-order polynomial, usually a third-order polynomial, and smoothed values are obtained. The polynomial is fit in turn to a series of overlapping regions until a smoother resulting curve is obtained; then, if necessary, the process is iterated to eliminate (or reduce substantially) any resulting steps present between adjacent segments. Smoothed load time and displacement time curves for this specimen are shown in Figs. 14*a* and 14*b* and the resulting smoothed load displacement plot is shown in Fig. 14*c*. Except for the load elevation present at the beginning of this curve, which is a residual of the initial impact spike, the curve in Fig. 14*c* is now adequate for use in the key curve analysis routines.

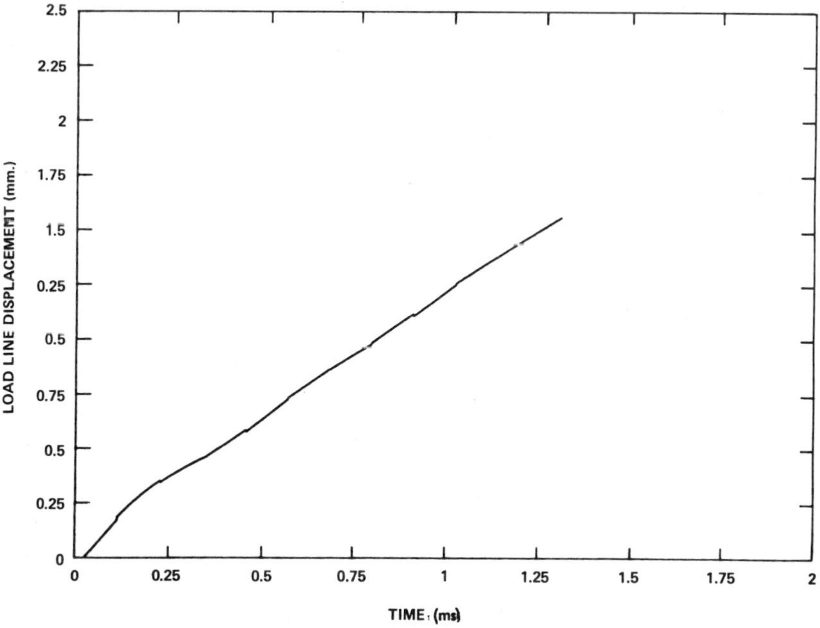

FIG. 14*a*—*Smoothed drop tower results for a* ½*T bend bar showing load-line displacement versus time.*

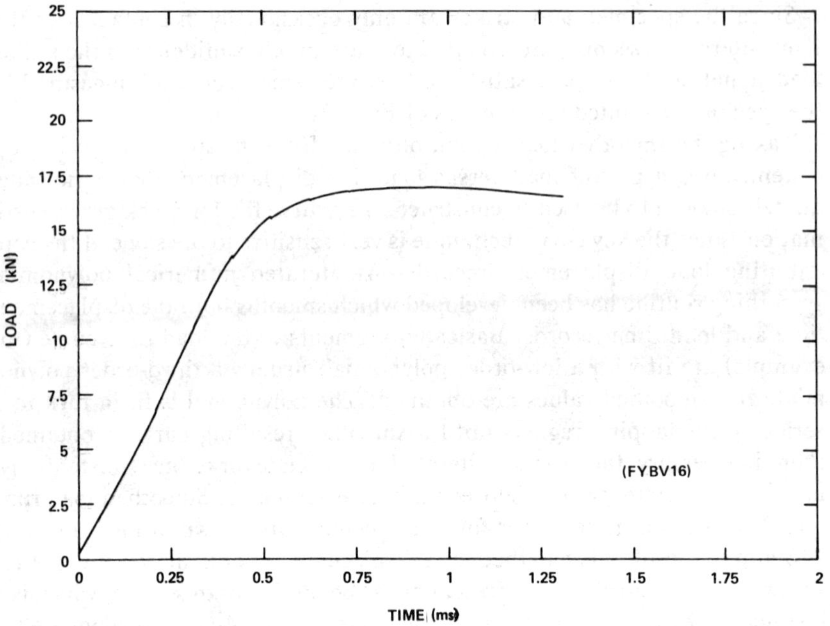

FIG. 14*b*—*Smoothed drop tower results for a* ½*T bend bar showing integral strain gage load versus time.*

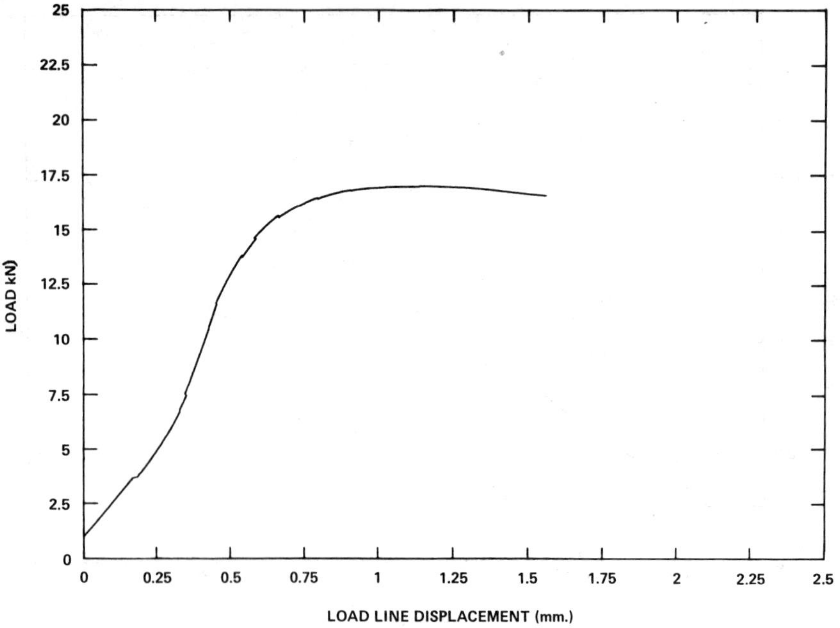

FIG. 14*c*—*Smoothed drop tower results for a* ½*T bend bar showing load versus load-line displacement.*

Data obtained on 1T scale bend bars also show the transient inertial oscillations. Raw data obtained on a typical 1T bend bar are shown in Figs. 15*a* to 15*c*. Figure 15*b* shows that for this specimen geometry even larger load oscillations occur than found for the ½T specimen, making the need for a smoothing operation more apparent and also making the smoothing operation more difficult. The load signal used in Fig. 15*b* is the one obtained from the strain gages mounted directly on the specimen arms. The signal phase for the strain gage signal was much more in agreement with the displacement signal, however, than was the tup signal and was free of the impact spike which dominates the initial stages of the tup load time record. As before, the tup signal must be treated as suspect in regions where tup and specimen are not in continuous contact. For these reasons the strain gage load signal is used in the results which follow.

The displacement versus time curve again shows that the specimen outruns the average striker velocity, and this time it appears that the striker does not catch the specimen until just before the stop block fixture was struck.

Smoothing operations were applied to the load-time and displacement time records as described previously and the smoothed results for this 1T case are shown in Figs. 16*a* to 16*c*. The smoothed load displacement record is compared with the static load displacement curve in Fig. 17. Since the specimen

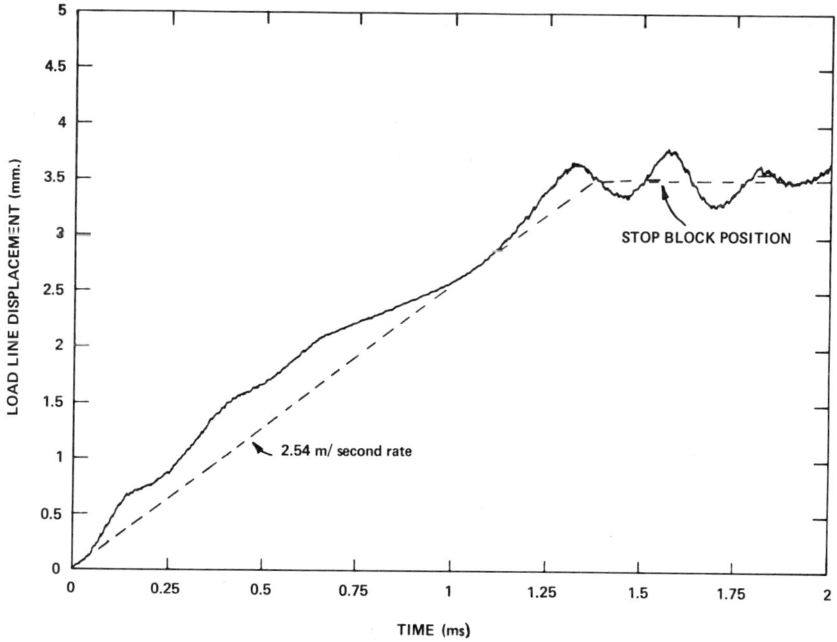

FIG. 15*a*—*Unsmoothed 1T data records showing load-line displacement versus time.*

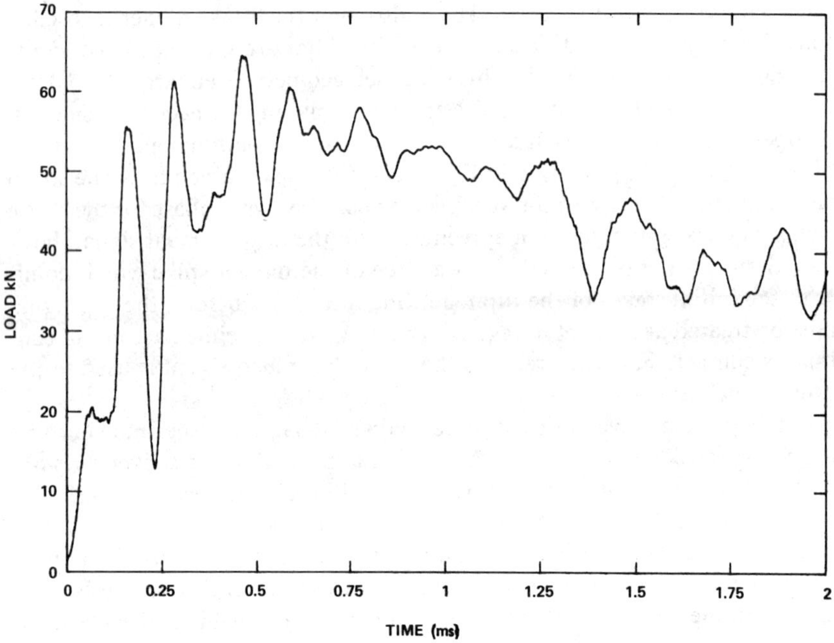

FIG. 15*b*—*Unsmoothed 1T data records showing load versus time.*

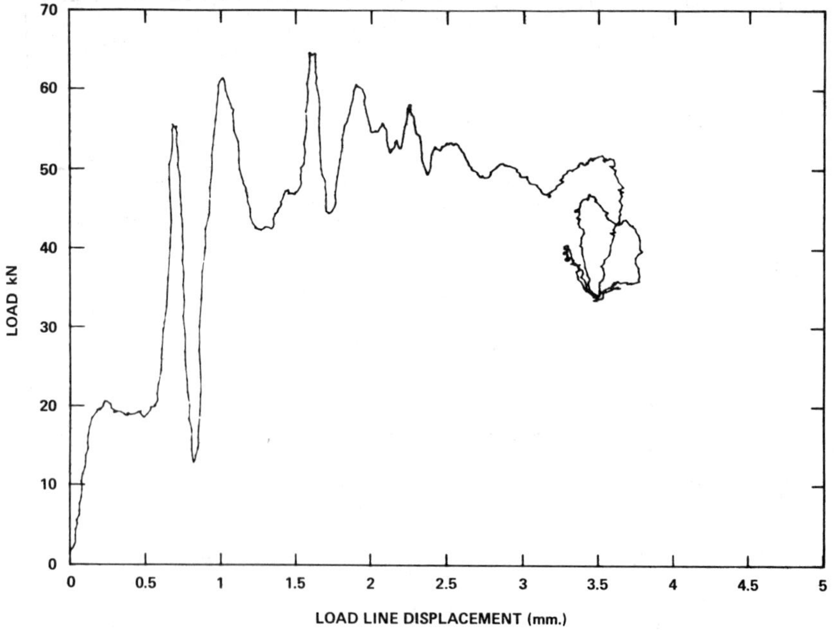

FIG. 15*c*—*Unsmoothed 1T data records showing load versus load-line displacement.*

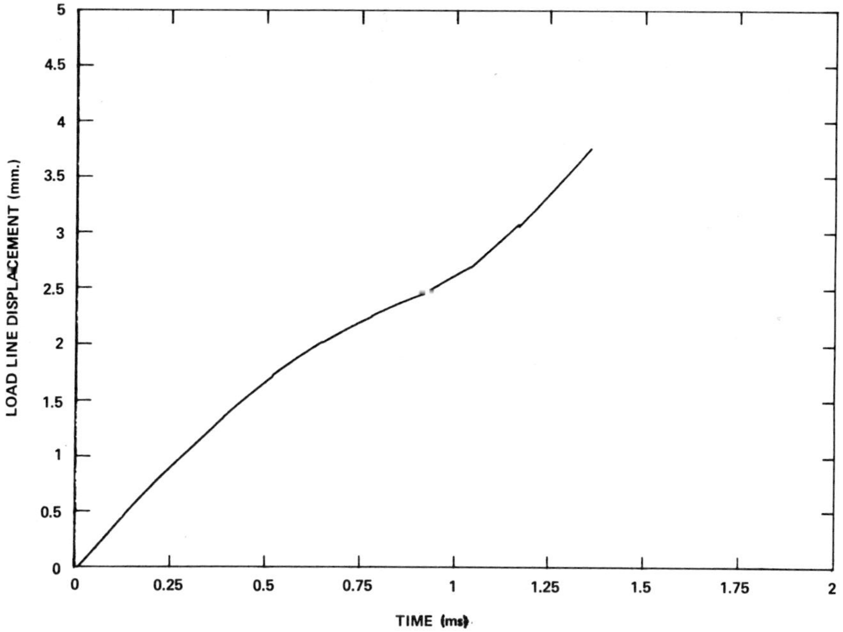

FIG. 16*a*—*Numerically smooth 1T data records showing load-line displacement versus time.*

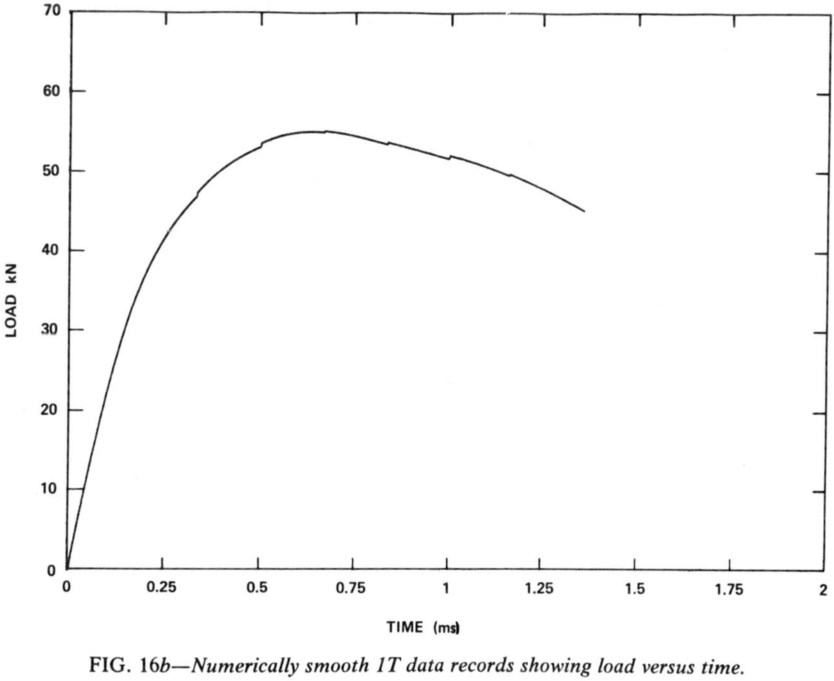

FIG. 16*b*—*Numerically smooth 1T data records showing load versus time.*

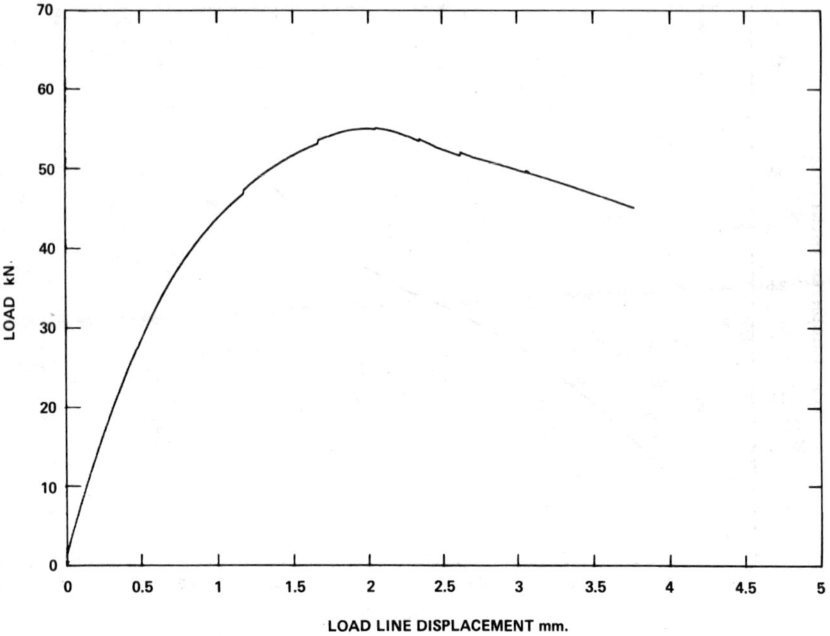

FIG. 16c—*Numerically smooth 1T data records showing load versus load-line displacement.*

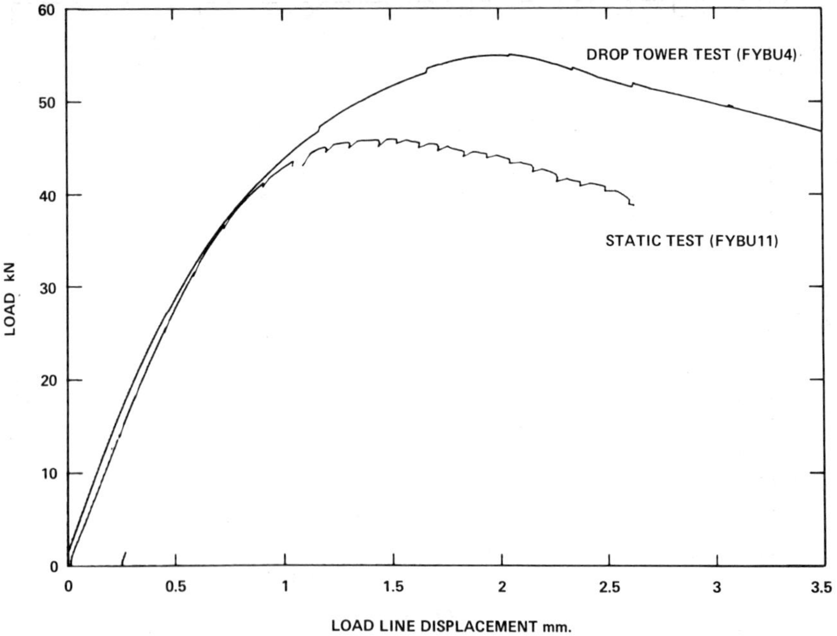

FIG. 17—*Comparison of typical quasi-static and drop tower load/load-line displacement records.*

geometries were identical, the load elevation present on the drop weight specimen is a result of the loading rate. The initial stiffnesses of the two specimens are, however, nearly identical. This is taken here as implying that the smoothing operation has given, in this region, an accurate representation of the real load displacement record.

Figure 18 shows a plot of the smoothed and normalized load displacement records of ½T and 1T drop weight specimens. For the static and intermediate rate tests discussed above, the key curves for ½T and 1T specimens were shown to correspond well until crack initiation occurred. For the drop weight results, correspondence is present in the early portion and again in the vicinity of maximum load but a distinct separation is found in the region between. This appears to result from the large amount of smoothing which was applied to the 1T load time record to remove the inertial oscillations shown in Fig. 15*b*. In the process the rather distinct yield bend demonstrated by the material was also smoothed out.

The result obtained from the ½T specimen is therefore taken as the more accurate key curve result because of the smaller amount of smoothing needed to reduce the original data. In order to apply the key curve method to this case it was assumed that no crack extension was present in the 1T specimen geometry before maximum load. As in the earlier cases the ½T key curve result must be extended to give a *J-R* curve beyond a Δ/W value at which crack

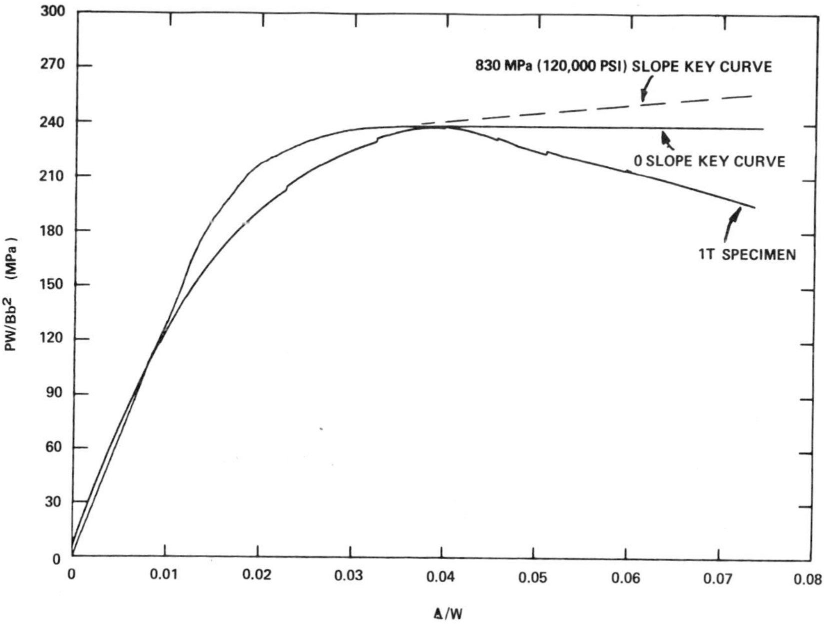

FIG. 18—*Key curve comparison of drop tower ½T and 1T bend bar records after numerical smoothing.*

extension occurs in the ½T specimen. The key curve was extended here as done in the previous two sections by using a constant positive slope, with the slope chosen to give agreement between the key curve estimated final crack length and the measured value.

Since crack initiation is not expected to occur in this material until maximum load, accurate *J-R* curves are expected here in spite of the smoothing applied. Future applications of the key curve method to drop tower testing could be enhanced by using an alternative technique for the determination of crack initiation or by finding experimental methods to smooth the load time record.

Three drop weight *J-R* curves are presented in Fig. 19 along with final measured crack lengths. The straight "blunting line" portion has been produced artificially in the region before maximum load where the 1T specimens load displacement curve deviates from the key curve as discussed above.

As done previously in the intermediate rate tests, these three specimens and a series of five additional identical specimens tested to smaller total displacement values at the same drop tower rate were analyzed in the standard multispecimen technique using the formula for *J* given by Eq 8. The results are shown in Fig. 20 where they are compared with both crack growth corrected and un-crack-growth corrected key curve results. The comparison between the multispecimen results and the uncorrected key curve results is very good,

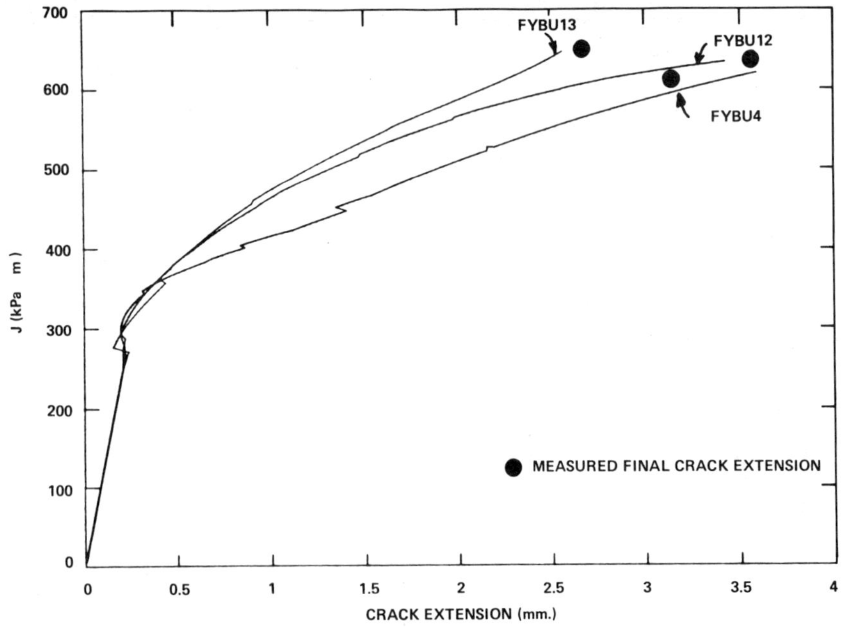

FIG. 19—*Drop tower* J-R *curves obtained by the key curve method.*

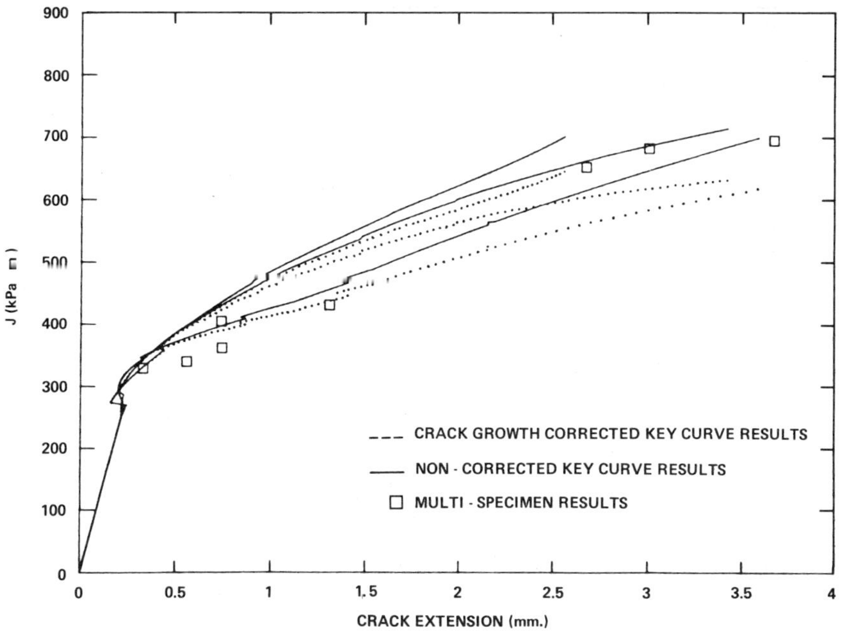

FIG. 20—*Comparison of multispecimen and key curve* J-R *curves obtained from the drop tower tests.*

even in the zone near crack initiation, justifying the key curve analysis assumption that crack initiation does not occur until maximum load for this material.

Applicability of Numerical Smoothing to *J-R* Curve Evaluation

An important question that needs to be answered to support this work is to determine what has been removed from the load time record as a result of the numerical smoothing procedure described in an earlier section.

To study the unsteady transient it was first evaluated by taking the difference between the smoothed and unsmoothed data. This was done for several specimens, and the most important observation was that the result was very repeatable. Shifting all transients to a new time scale with zero at the center of the first tensile spike and doing a simple average at each time instant gave the results shown in Fig. 21.

A separate test was then run by placing a similar specimen in the test machine and loading by dropping the head from a low height of about 0.5 cm. This produced an elastic loading giving a load time trace as shown in Fig. 22. This load time curve was analyzed using the polynomial smoothing procedure to extract the oscillatory transient. After an adjustment of scale and a time

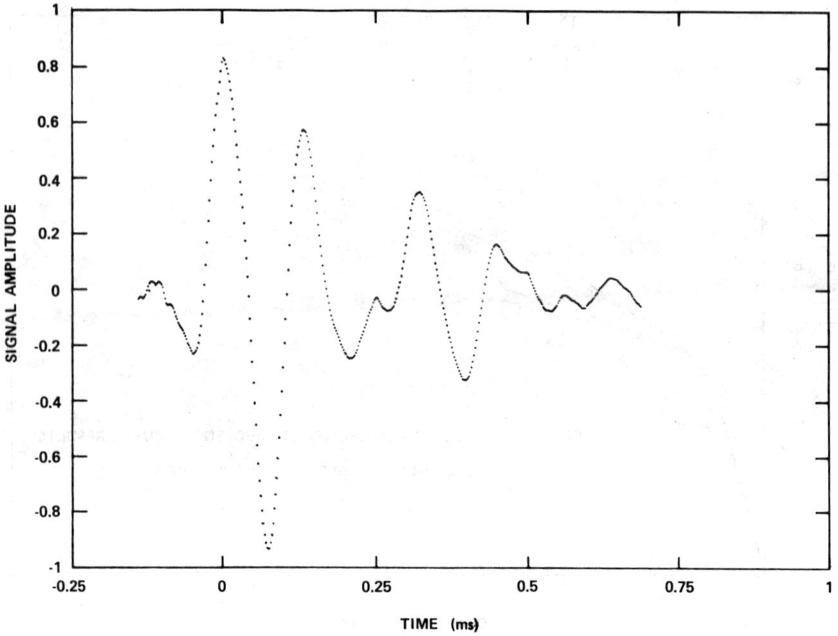

FIG. 21—*Difference between smoothed and unsmoothed load-time records for 1T drop tower tests (average of six tests).*

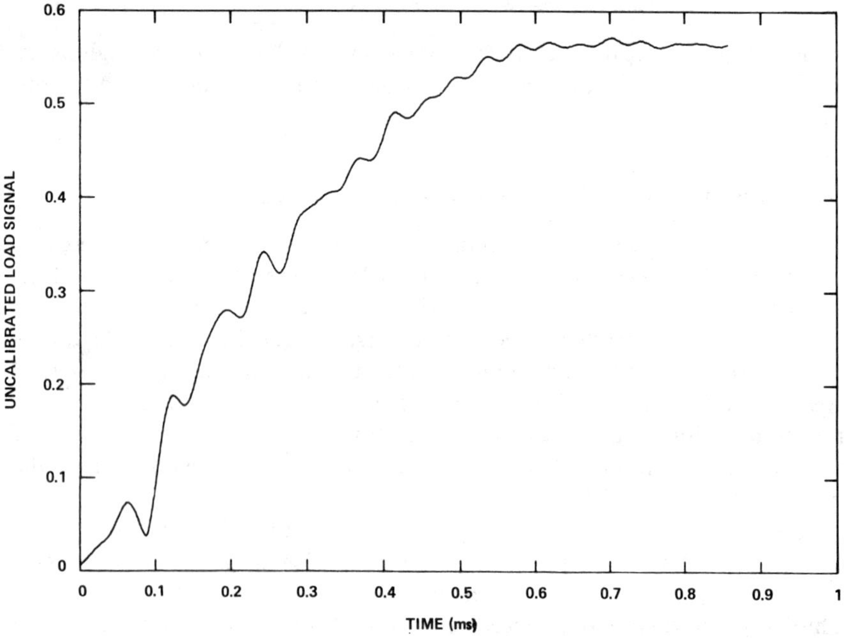

FIG. 22—*Load-time record for a low drop height elastic test of a 1T bend bar.*

shift the plot shown in Fig. 23 was produced. Though the amplitudes of the transients are much different, when scaled by the maximum value as shown, the similarity of the elastic and elastic-plastic transients are very similar even though the impact velocities are different by a factor of 100 and the material properties are very different. This result conclusively shows that the transient being removed by the smoothing operations is an elastic transient which is a function of the specimen geometry and test machine stiffness but not a strong function of material toughness or loading rate.

Comparison of *J-R* Curves at the Three Test Rates

Results of all three test rates are compiled in Table 3. Tearing modulus values are not presented for the multispecimen tests, since J values are calculated for these tests without being corrected for crack growth. The J_{Ic} values are calculated in accordance with ASTM E 813 for all tests, while the T values used are obtained from the slope of data within the ASTM E 813 exclusion lines. These results show how the elevation of test rate for this material increases both J_{Ic} and T. Figure 24 shows the general trend exhibited by 3-Ni steel R-curves with increasing loading rate as determined by both the key curve and multispecimen techniques.

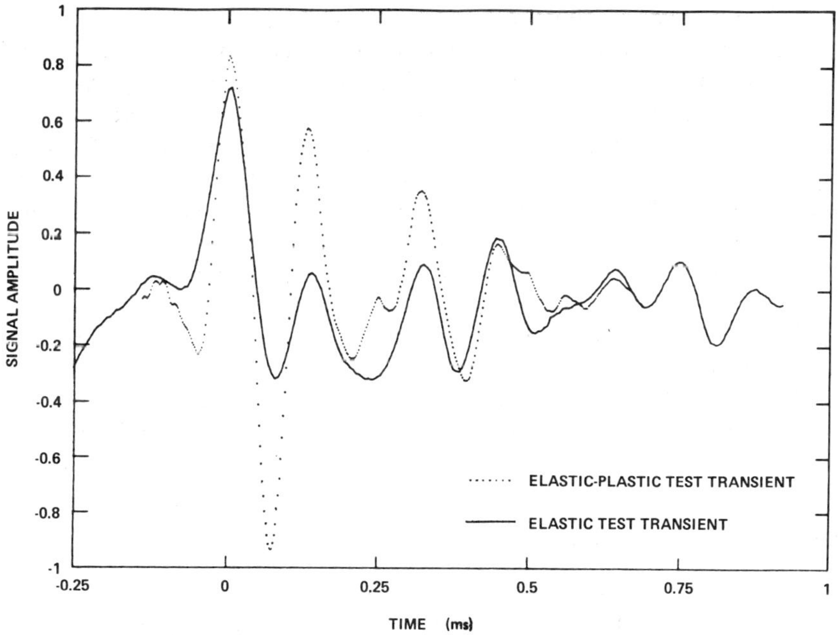

FIG. 23—*Comparison of high frequency amplitude components obtained from an elastic drop and the average of six elastic plastic tests.*

TABLE 3—J_{Ic} *and* T *values for specimens tested at three different loading rates.*

	Loading Rate					
	Quasi-Static		Intermediate		Drop Tower	
Technique/ Specimen	J_{Ic}, MPa·m	T	J_{Ic}, MPa·m	T	J_{Ic}, MPa·m	T
Multispecimen	...	...	290[a]	...	321-323	...
Unloading compliance	168	12.7	...	...	...	...
Key curve						
FYB-U2	209	22.2	...	...	...	...
FYB-U11	244	28.3	...	...	...	...
FYB-108	...	...	229-230	34-39	...	...
FYB-109	...	...	214-215	38-44	...	...
FYB-110	...	...	241-244	22-25	...	...
FYB-111	...	...	241-243	32-37	...	...
FYB-U4	...	...	...	...	344-346	40-46
FYB-U12	...	...	...	...	383-388	48-54
FYB-U13	...	...	...	...	383-391	48-60

[a]Not valid as per ASTM E 813.

To calculate values for J_{Ic} and T flow stresses are required, and a method for relating tensile bar loading rates to bend bar loading rates is at present not available. As an approximation, a two dimensional, plane strain, elastic plastic finite element analysis of the 1T three-point bend specimen used in this study shows that stresses on the order of yield occur at integration points near the crack tip after load-line displacements of 0.25 mm. This corresponds to strain rates in drop tests of 10 mm/mm/s at 0.5 mm radius from the crack tip, a strain rate roughly equivalent to the high speed tensile strain rate generated in the servohydraulic test and presented in Table 2. For this 3-Ni steel material the flow stress is only moderately dependent on strain rate; the J_{Ic} and T values shown in Table 3 show the range obtained by using static flow stress (larger J_{Ic} and T) and servohydraulic flow stress (smaller J_{Ic} and T).

The numbers show a close similarity in J_{Ic} values for static and intermediate test results, with a tendency for intermediate rate T values to be elevated above static T values. Drop tower tests show a clear elevation in J_{Ic} over the other test rates, but T now appears to be only slightly elevated in comparison with intermediate rate results.

Discussion of Quasi-Static *J* Analysis Applied to Dynamic Loading

The elevation in the *J-R* curve fracture behavior shown above can be due to omission of inertial input to the *J* equation, inclusion of an inappropriate load in the *J*-equation, or to a true elevation in material fracture toughness.

Recent investigations [*25-27*] have shown dynamic stress intensity (K) to

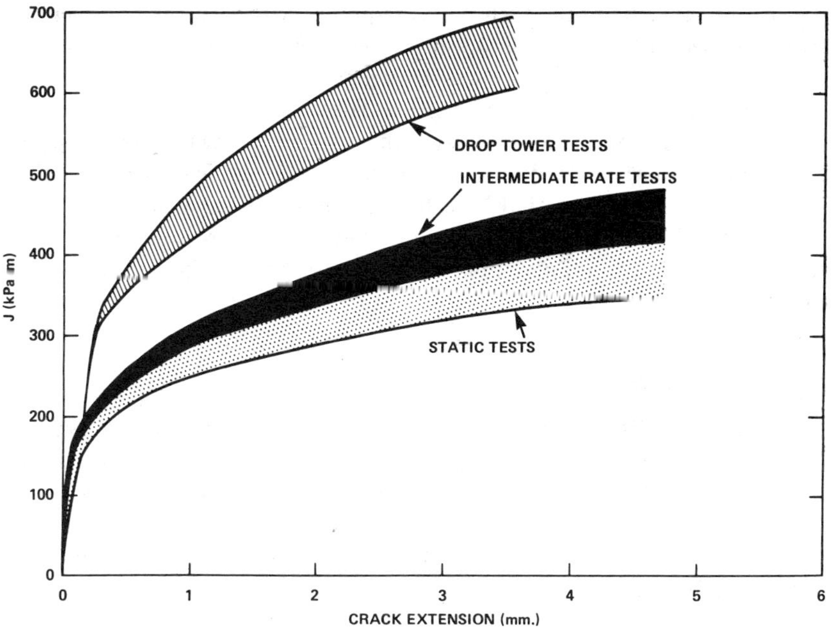

FIG. 24—*Summary of* J-R *curve results at the three test rates for 1T bend bars using the key curve, multispecimen, and unloading compliance techniques.*

be an increasing function of crack velocity only when the crack velocity approaches a significant fraction of the material shear wave speed. The drop weight tests on the 3-Ni steel bend specimens used in this investigation generated ductile crack velocities on the order of 3 m/s and, since the shear wave velocity in steels is approximately 3000 m/s, little dynamic effect would be expected for these tests.

Nakamura et al [*28*] stress the importance of the inclusion of the correct load in the *J* formulation. For the static case the remote load and the transmitted ligament load are equivalent. In the dynamic case the remote and transmitted ligament loads may not be equivalent. For the tests in this investigation, integral strain gages were used to obtain a time-averaged estimate of the transmitted ligament load.

The elevation in *J-R* curves found in this investigation seems therefore to be a material rate effect resulting at least in part from the elevation of material yield stress properties.

Conclusions

- Multispecimen and key curve tests give upper shelf *J-R* curves which are in substantial agreement at elevated loading rates of 0.025 and 2.5 m/s.

• Numerical smoothing techniques required to apply a key curve method appear to separate the oscillatory high frequency component from the load displacement record. This oscillatory component is a function of specimen geometry and elastic material properties. The oscillation amplitude is a function of test speed, but many features of the high frequency transient appear to be unchanged across a wide range of loading rates and load magnitudes.

• For the 3-Ni steel tested here both J_{Ic} and J-R curve slopes (T's) are elevated at higher loading rates, with J_{Ic} and T values being increased by a factor of roughly two for a loading rate increase of about six orders of magnitude.

Acknowledgments

The authors wish to acknowledge the support of Milt Vagins and Jack Strosnider of the U.S. Nuclear Regulatory Commission. The authors also wish to express thanks to Mr. M. G. Vassilaros of the David Taylor Naval Ship R&D Center for his help and advice during the course of this research and to Mr. John P. Gudas, also of the David Taylor Center, who supervised this research program.

References

[*1*] Hutchinson, J. W., "Singular Behavior at the End of a Tensile Crack in Hardening Material," *Journal of the Mechanics and Physics of Solids*, Vol. 16, 1968, pp. 13-31.

[*2*] Rice, J. R. and Rosengren, G. F., "Plane Strain Deformation Near a Crack Tip in a Power-Law Hardening Material," *Journal of the Mechanics and Physics of Solids*, Vol. 16, 1968, pp. 1-12.

[*3*] Begley, J. A. and Landes, J. D. in *Fracture Toughness*, *ASTM STP 514*, American Society for Testing and Materials, Philadelphia, 1972, pp. 1-20.

[*4*] Clarke, G. A., Andrews, W. R., Paris, P. C., and Schmidt, D. W., "Single Specimen Tests for J_{Ic} Determination," in *Mechanics of Crack Growth*, *ASTM STP 590*, American Society for Testing and Materials, Philadelphia, 1976, pp. 27-42.

[*5*] Joyce, J. A. and Gudas, J. P., "Computer Interactive J_{Ic} Testing of Navy Alloys," in *Elastic-Plastic Fracture*, *ASTM STP 668*, American Society for Testing and Materials, Philadelphia, 1979, pp. 451-468.

[*6*] Paris, P. C., Tada, H., Zahoor, A., and Ernst, H., "The Theory of Instability of the Tearing Mode of Elastic Plastic Crack Growth," in *Elastic-Plastic Fracture*, *ASTM STP 668*, American Society for Testing and Materials, Philadelphia, 1979, pp. 5-36.

[*7*] Tada, H., Paris, P., and Gamble, R., "Stability Analysis of Circumferential Cracks in Reactor Piping Systems," NUREG CR-0838, U.S. Nuclear Regulatory Commission, Washington, D.C., June 1979.

[*8*] Tada, H., Paris, P. C., and Gamble, R. M. in *Fracture Mechanics (Twelfth Conference)*, *ASTM STP 700*, American Society for Testing and Materials, Philadelphia, 1980, pp. 296-313.

[*9*] Zahoor, A. and Kanninen, M. F., "A Plastic Fracture Mechanics Prediction of Fracture Instability in a Circumferentially Cracked Pipe in Bending: Part I—*J*-Integral Analysis," *Journal of Pressure Vessel Technology*, Vol. 103, No. 4, Nov. 1981.

[*10*] Paris, P. C., Tada, H., Ernst, H., and Zahoor, A. in *Elastic-Plastic Fracture*, *ASTM STP 668*, American Society for Testing and Materials, Philadelphia, 1979, pp. 251-265.

[*11*] Vassilaros, M. G., Joyce, J. A., and Gudas, J. P. in *Fracture Mechanics: Fourteenth Symposium—Vol. I: Theory and Analysis*, *ASTM STP 791*, American Society for Testing and Materials, Philadelphia, 1983, pp. I-65-I-83.

[*12*] Joyce, J. A. and Vassilaros, M. G., in *Fracture Mechanics (Thirteenth Conference), ASTM STP 743*, American Society for Testing and Materials, Philadelphia, 1980, pp. 525–542.
[*13*] Joyce, J. A. in *Elastic-Plastic Fracture: Second Symposium—Vol. II: Fracture Curves and Engineering Applications*, *ASTM STP 803*, American Society for Testing and Materials, Philadelphia, 1983, pp. II-439–II-463.
[*14*] Vassilaros, M. G., Hays, R. A., Gudas, J. P., and Joyce, J. A., "*J*-Integral Tearing Instability Analysis for 8-Inch Diameter ASTM A106 Steel Pipe," NUREG/CR-3740, U.S. Nuclear Regulatory Commission, Washington, D.C., April 1984.
[*15*] "Response of Metals and Metallic Structures to Dynamic Loading," Report NMAB-341, National Materials Advisory Board, National Academy of Sciences, Washington, D.C., May 1978.
[*16*] Albrecht, P. et al, "Tentative Test Procedure for Determining the Plane Strain *J-R* Curve," *Journal of Testing and Evaluation*, Vol. 10, No. 6, Nov. 1982, pp. 245–251.
[*17*] Ernst, H. A., Paris, P. C., and Landes, J. D. in *Fracture Mechanics (Thirteenth Conference)*, *ASTM STP 743*, American Society for Testing and Materials, Philadelphia, 1980, pp. 476–502.
[*18*] Ernst, H. A., Paris, P. C., Rossour, M., and Hutchinson, J. W. in *Fracture Mechanics (Eleventh Conference)*, *ASTM STP 677*, American Society for Testing and Materials, Philadelphia, 1979, pp. 581–599.
[*19*] Joyce, J. A., Ernst, H. A., and Paris, P. C. in *Fracture Mechanics (Twelfth Conference)*, *ASTM STP 700*, American Society for Testing and Materials, Philadelphia, 1980, pp. 222–236.
[*20*] Kumar, V., German, M. D., and Shih, C. F., "An Engineering Approach for Elastic-Plastic Fracture Analysis," Project 1237-1, EPRI NP-1931, Electric Power Research Institute, Palo Alto, Calif., July 1981.
[*21*] Joyce, J. A. in *Fracture Mechanics: Fourteenth Symposium—Vol. I: Theory and Analysis*, *ASTM STP 791*, American Society for Testing and Materials, Philadelphia, 1983, pp. I-543–I-560.
[*22*] Joyce, J. A. and Czyryca, E. J., "Dynamic *J-R* Curve Testing of HY-130 Steel," DTNSRDC/SME-81-57, Oct. 1981.
[*23*] Rice, J. R., *Journal of Applied Mechanics*, *Transactions of ASME*, Vol. 35, 1968, pp. 379–386.
[*24*] Logsdon, W. A. in *Elastic-Plastic Fracture*, *ASTM STP 668*, American Society for Testing and Materials, Philadelphia, 1979, pp. 515–536.
[*25*] Freund, L. B. and Douglas, A. S. in *Elastic-Plastic Fracture: Second Symposium—Vol. I: Inelastic Crack Analysis*, *ASTM STP 803*, American Society for Testing and Materials, Philadelphia, 1983, pp. I-5–I-20.
[*26*] Achenbach, J. D. and Li, Z. L., "Plastic Deformations Near a Rapidly Propagating Crack Tip," Report NU-SML-TR-No. 84-1, Office of Naval Research, Jan. 1984.
[*27*] Rosakis, A. J., Freund, L. B., and Duffy, J., "Determination of the Fracture Initiation and Crack Propagation Resistance of Structural Steels by the Optical Method of Caustics," Report MRL E-142, Materials Research Laboratory, Brown University, Providence, R.I., Sept. 1982.
[*28*] Nakamura, T., Shih, C. F., and Freund, L. B., "Elastic Plastic Analysis of a Dynamically Loaded Circumferentially Notched Round Bar," Report MRL-152, Materials Research Laboratory, Brown University, Providence, R.I., March 1984.

DISCUSSION

J. H. Giovanola[1] *(written discussion)*—This discussion addresses two questions: (1) the origin of the load oscillations as recorded by the strain gages

[1]SRI International, Menlo Park, CA 94025.

mounted on the impact face of the bend specimen, and (2) the influence of these oscillations on the measured *J-R* curve.

The dynamics of the impacted three-point bend specimen have been studied extensively and interpreted by Kalthoff and co-workers. They have shown that crack tip loading results from two types of superimposed oscillations. The first type of oscillation, with a period of about a millisecond or more, is the vibration of the system hammer-specimen-supports. It can be analyzed with a simple spring-mass model. The second type of oscillation is due to the vibration of the specimen alone and has a period about an order of magnitude smaller than the hammer-specimen-support system. The oscillations observed on the load record presented in this paper are in all likelihood of this second type. Böhme and Kalthoff have shown that proper choice of specimen size—specifically, the length of specimen overhang beyond the supports—can greatly reduce the amplitude of undesired oscillations.

The effect of these load oscillations can be assessed in the following way. The first few oscillations occur in the elastic regime, and their amplitude appears to remain below the yield load. Thus they can be regarded as a few additional fatigue cycles experienced by the precracked specimen. These additional cycles should not have any significant influence on the resistance curve of a tough ductile material. As the average load approaches yield, the load oscillations appear to be considerably damped and do not exceed the maximum load level later reached during the impact test. These oscillations in the near yield region can be regarded as a few partial unloadings similar to the unloadings used to record the elastic compliance in conventional *J-R* curve testing. It has also been recognized that these unloadings do not influence the measurement of the *J-R* curve. Therefore the procedure of obtaining *J* from the average dynamic load-displacement curve, neglecting the effect of the oscillations, seems legitimate.

J. A. Joyce and E. M. Hackett (authors' closure)—We basically agree with the above comment. We certainly feel that both the test machine oscillations and specimen vibrations are superimposed on the oscilloscope data traces. Many important aspects of work in this area remain unsettled, including determining whether the *J* integral remains a controlling parameter of crack tip behavior during rapid loading.

C. W. Smith,[1] J. S. Epstein,[1] and O. Olaosebikan[1]

Boundary Layer Effects in Cracked Bodies: An Engineering Assessment

REFERENCE: Smith, C. W., Epstein, J. S., and Olaosebikan, O., **"Boundary Layer Effects in Cracked Bodies: An Engineering Assessment,"** *Fracture Mechanics: Seventeenth Volume, ASTM STP 905*, J. H. Underwood, R. Chait, C. W. Smith, D. P. Wilhem, W. A. Andrews, and J. C. Newman, Eds., American Society for Testing and Materials, Philadelphia, 1986, pp. 775–788.

ABSTRACT: Experimental methods of analysis and corresponding algorithms for converting data into fracture parameters were reviewed. Results obtained from applying the methods to an analysis of boundary layer effects in compact bending specimens and in moderately deep surface flaws under Mode I loading were presented. Finally, a means for incorporating the new results into linear elastic fracture mechanics (LEFM) design rationale was suggested.

KEY WORDS: stress intensity factors, boundary layer effects, linear elastic fracture mechanics, photoelastic and moiré analysis, eigenvalue problems

Recent analyses [*1–4*] have suggested that the inverse square root singularity commonly associated with linear elastic fracture mechanics (LEFM) is lost when a straight front crack intersects a free boundary at right angles. Benthem's analysis of a quarter infinite crack in a half space [*2*] is generally regarded in the technical community as providing the proper value of the exponent of r at the free surface for the elastic case as a lowest eigenvalue from a series obtained from a variables-separable three-dimensional approach. His results have been verified by Bazant and Estenssoro [*5*] and by Solecki and Swedlow [*6*] using finite element analyses.

The present authors and their associates have developed, over a period of years, an experimental method which integrates the approaches of frozen stress photoelasticity and high density moiré interferometry in order to yield

[1]Alumni Professor, Assistant Professor, and Assistant Professor, respectively, Department of Engineering Science and Mechanics, Virginia Polytechnic Institute and State University, Blacksburg, VA 24061. Dr. Epstein is currently a Research Engineer at the Idaho National Engineering Laboratory.

experimentally determined stress intensity factor (SIF) distributions for three-dimensional problems [*7–8*]. The experimental procedure involves the following steps:

1. Prepare a scale model of the body using di-phase (stress freezing) photoelastic material.
2. Insert a natural or artificial crack into the model at the desired location. Natural starter cracks are made by striking a sharp blade held normal to the surface of the body. The starter crack emanates from the blade tip normal to the blade, propagates a short distance, and arrests. Artificial cracks are introduced by machining with a vee-notched tip with a notch angle not greater than 30°.
3. Place cracked model in a loading frame without restraint in a programmed oven and heat to critical temperature. Then apply live load. Natural cracks may be extended without controlling their shape by using a sufficiently high load and then reducing the load when the desired crack size is reached to stop crack growth.
4. Cool under load to room temperature and remove load. Stress fringes and deformations produced above critical temperature will be retained.
5. Remove thin slices from the model which are mutually orthogonal to the crack border and the crack surface.
6. Analyze each slice photoelastically and use the extracted optical data in a suitably chosen algorithm to estimate the stress intensity factor photoelastically.
7. Deposit a high density linear grid (1200 lines/mm) on a surface of each slice. Anneal the slices so that the grid senses the recovered deformation field.
8. View the deformed gratings through a "virtual" grating produced by recombining a split laser beam in space to obtain moiré displacement fringes around the crack tip. From suitable algorithms, convert the moiré data into a second estimate of the stress intensity factor. By using an algorithm not conforming to linear elastic fracture mechanics, the authors have been able to achieve good correlation with Benthem's result at the crack border free surface intersection [*9*].

After briefly reviewing the features of the method, the authors present results obtained by applying the above approach to two classes of problems in order to measure the variation in the first eigenvalue through the boundary layer within which a crack intersects a free surface. Then, on the basis of these results, the authors suggest an approach for correlating the results with linear elastic fracture mechanics.

Experimental Methods

As noted above, two experimental methods, frozen stress photoelasticity and high density moiré interferometry, have been utilized in studying the boundary layer effect as well as the variation in the SIF distribution in three-dimensional cracked body problems. These methods have been described in detail elsewhere [*8, 10*] so will only be briefly reviewed here.

Frozen Stress Photoelasticity

This method capitalizes on the fact that the transparent model material exhibits di-phase mechanical and optical behavior. At room temperature, it responds to load in a Kelvin-like manner (Fig. 1). When the model's temperature is raised to its "critical" value, however, the viscous coefficient vanishes and the mechanical response becomes linearly elastic with a corresponding reduction in material modulus of some two and a half orders of magnitude and an increase in optical sensitivity to stress fringes of about twenty five times the room temperature sensitivity. To take advantage of this behavior, we heat the cracked model to critical temperature, load, and cool under load. Upon load removal, and even after slicing through the body after load removal, the deformations and stress fringes produced above critical temperature are retained and can be analyzed photoelastically in the usual way. The photoelastic fringes are proportional to the maximum shearing stress in the plane of the slice, and the algorithms for converting the data into stress intensity information are referred to a local Cartesian coordinate system *tnz* (Fig. 2) with slices always taken parallel to the *nz* plane for Mode I loading [*10, 11*].

High Density Moiré Interferometry

By assuming that the annealing of stress frozen slices reverses the effect of stress freezing (which has been verified experimentally), one can deposit a

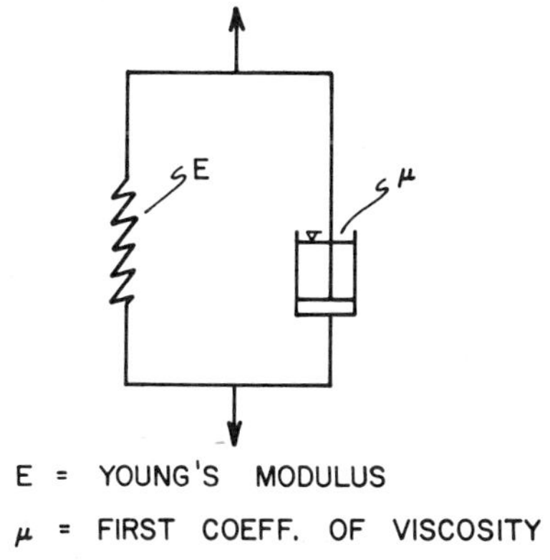

FIG. 1—*Kelvin material.*

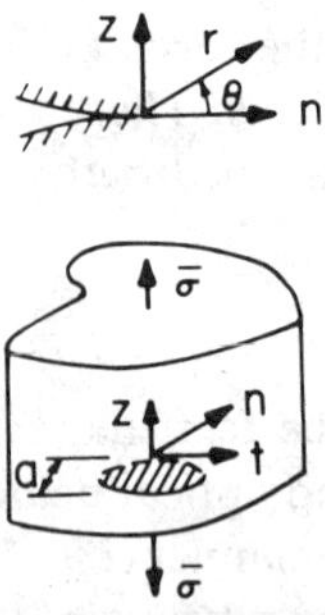

FIG. 2—*General problem geometry and notation.*

grating of 1200 lines/mm on a slice surface and then anneal the slice, deforming the grating. By viewing this grating through a "virtual" master grating developed by Post [*12*], moiré fringes which are proportional to the in-plane displacement components normal to the master grating can be obtained. The virtual grating is formed by splitting and recombining rays from a laser light source so as to form an optical grating of lines of constructive and destructive interference. It is important that rigid body motions be eliminated from the data used to compute SIF values. By following the photoelastic analysis of a slice with a moiré analysis, two separate estimates of the SIF for that slice may be obtained.

With the measuring techniques described above, we measure maximum shearing stress magnitude and displacement components in the slice plane. Generally, these data are collected close to the crack tip, but must exclude a very near-tip non-linear zone. For problems in which boundary layer effects are not present (such as embedded flaws) or are neglected, algorithms are constructed from classical LEFM for converting optical data into SIF values. These algorithms have been discussed elsewhere [*8,10*] and will not be repeated here. In order to study the boundary layer effect, however, algorithms must be constructed which are outside the realm of LEFM.

Moiré Displacement Algorithm

Benthem's solution [*2*] was a variables-separable result which, for the displacement components, took the form of a series of eigenfunctions for a quarter infinite crack in a half space. Focusing on the lowest order eigenvalue term as is customary in such problems and noting that, for Mode I, moiré fringes (Fig. 3) are readily measurable along $\theta = \pm\pi/2$ we have, near the crack tip,

$$u_z = Cr^{\lambda_u} \tag{1}$$

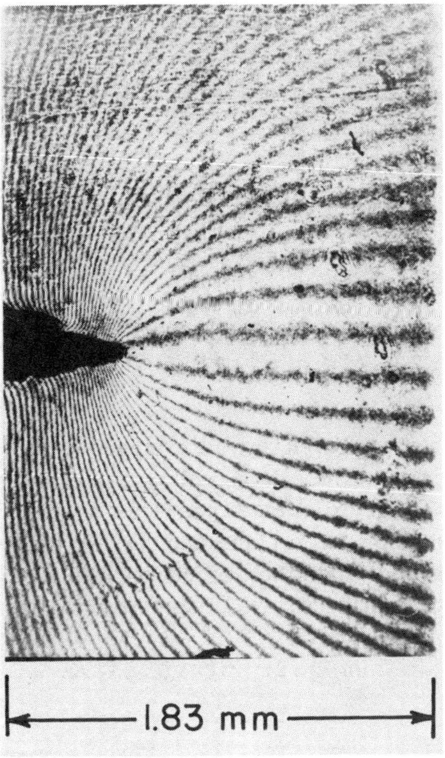

FIG. 3—*Moiré pattern for displacement components normal to crack plane due to Mode I.*

where

u_z = component of displacement normal to crack plane,
C = coefficient of leading term in displacement series, constant for $\theta = \pm \pi/2$, and
λ_u = first or lowest order eigenvalue in the solution for the displacement.

Therefore

$$\log u_z = \log C + \lambda_u \log r \quad (2)$$

and λ_u may be determined as the slope of a plot of log u_z versus log r. Figure 4 shows a typical result from a slice removed from a compact bending specimen. The log r location for the slope measurement is in the linear range with the smallest slope and is nearly common to all slices for a given crack border. In general, the zone lies between $r = 0.5$ and 1.5 mm from the crack tip along $\theta = \pm \pi/2$.

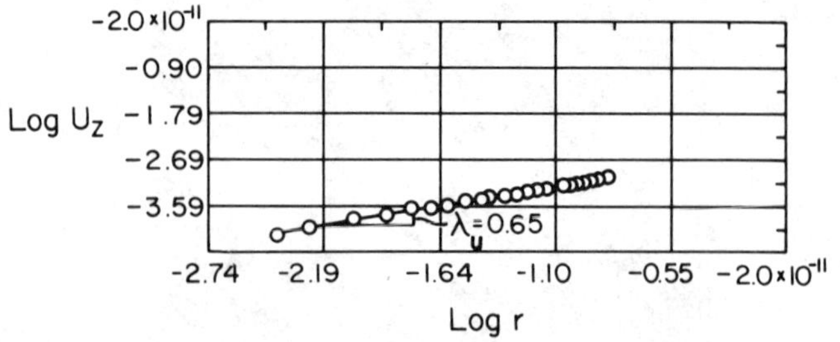

FIG. 4—*Determination of* λ_u *from moiré data.*

Photoelastic Algorithm

Since data from a stress frozen slice are averaged optically through the slice thickness, we can describe these data locally with a two-dimensional function analogous to a Westergaard stress function for a small enough data zone taken so as to focus upon the in-plane singular effect. Using this approach, and again focusing upon the lowest order eigenvalue, we have, along $\theta = \pm \pi/2$ where the stress fringes spread the most for Mode I (Fig. 5),

$$\lim_{|\zeta| \to 0} \{\sqrt{2\pi}\, [Z(\zeta)\,]\, \zeta^{\lambda_\sigma}\} = K_\lambda \tag{3}$$

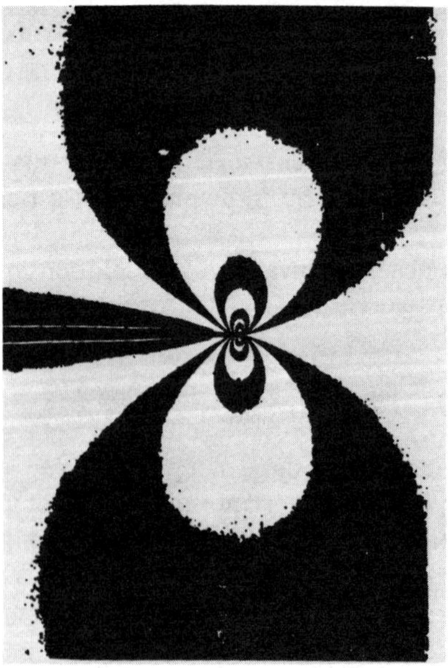

FIG. 5—*Mode I photoelastic stress pattern.*

where Z is the stress function, ζ is as defined in Fig. 6, K_λ is a "stress eigenfactor," and λ_σ is the lowest order eigenvalue in the stress equations. By computing σ_{ij} $(i,j = n,z)$ from Eq 3 and $\tau_{nz}^{\max}$ from these stresses, one arrives at the approximate expression [*13*]

$$\tau_{nz}^{\max} = \frac{K_\lambda f(\lambda_\sigma)}{r^{\lambda_\sigma}} + \sigma_{\text{on}} \tag{4}$$

where σ_{on} represents the contribution of the nonsingular stresses in the measurement zone. By defining

$$(K_\lambda)_{\text{AP}} = \tau_{nz}^{\max}\, r^{\lambda_\sigma} \qquad (\text{AP} = \text{apparent}) \tag{5}$$

then

$$\log\,(\tau_{nz}^{\max}) = \log\,(K_\lambda)_{\text{AP}} - \lambda_\sigma \log r \tag{6}$$

Thus, by plotting $\log\,(\tau_{nz}^{\max})$ versus $\log r$, λ_σ can be determined. We note that λ_σ as defined by Eq 3 equals $1 - \lambda_u$ where λ_u is as defined by Eq 1. However, they can be determined independently of this relation.

Experiments and Results

The foregoing methods were applied to two different cracked body geometries:

1. The four-point loaded compact bending specimen (Fig. 7) containing a straight front crack which intersected the free boundaries at right angles. This geometry was used to benchmark the experimental results against Benthem's value at the free surface using the moiré method and to confirm that the photoelastic and moiré methods yielded essentially the same results. Table 1 shows the results leading to the former observation, and Fig. 8 confirms the latter statement.
2. Wide flat plates containing natural semi-elliptic surface flaws (Fig. 9).

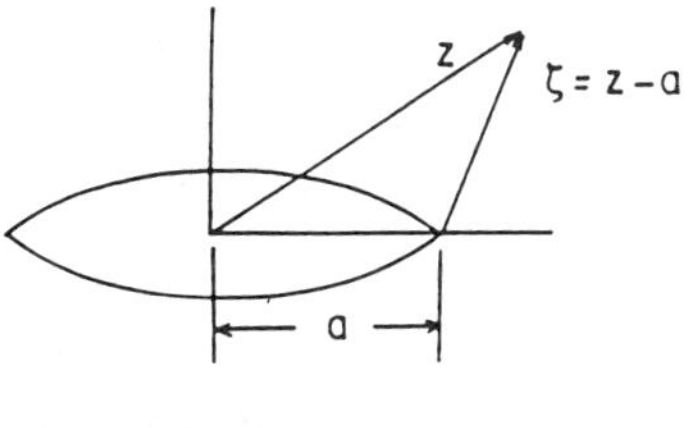

FIG. 6—*Crack tip coordinates.*

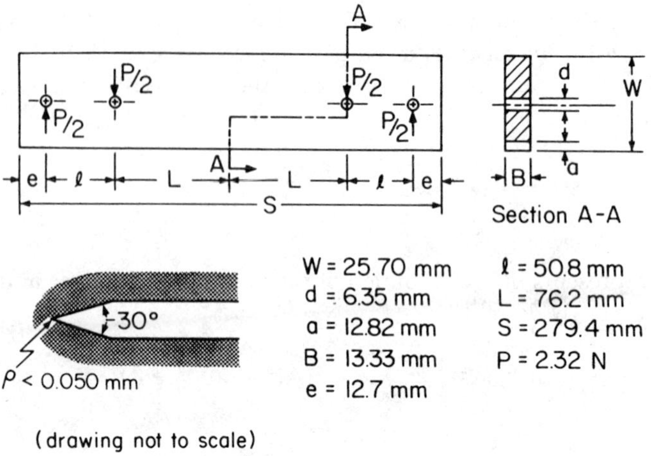

FIG. 7—*Compact bending specimen.*

TABLE 1—*Experimental and analytical values of* λ_u.

Poisson's Ratio	λ_u (Experimental)	λ_u [2]
0.40	0.58	0.59
0.48	0.63	. . .
0.50	. . .	0.65

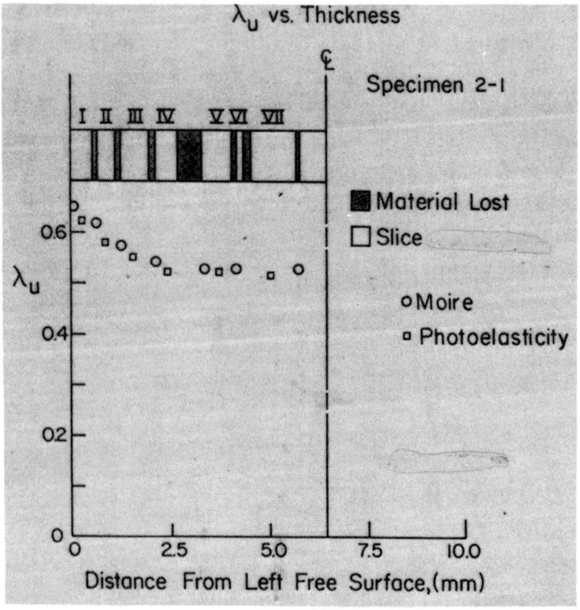

FIG. 8—λ_u *distribution in compact bending specimen from moiré and photoelastic data.*

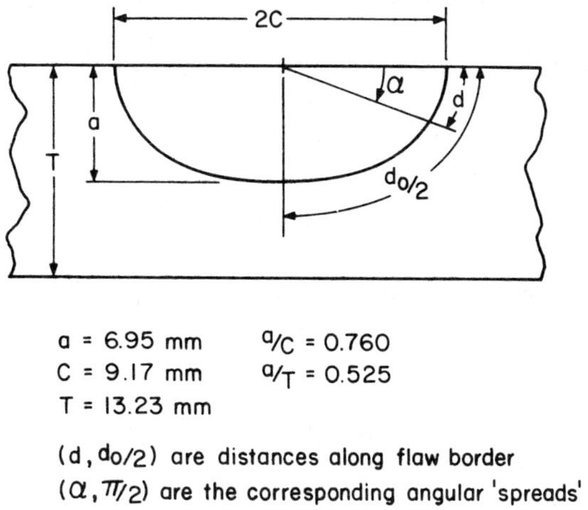

FIG. 9—*Natural semi-elliptic surface crack geometry.*

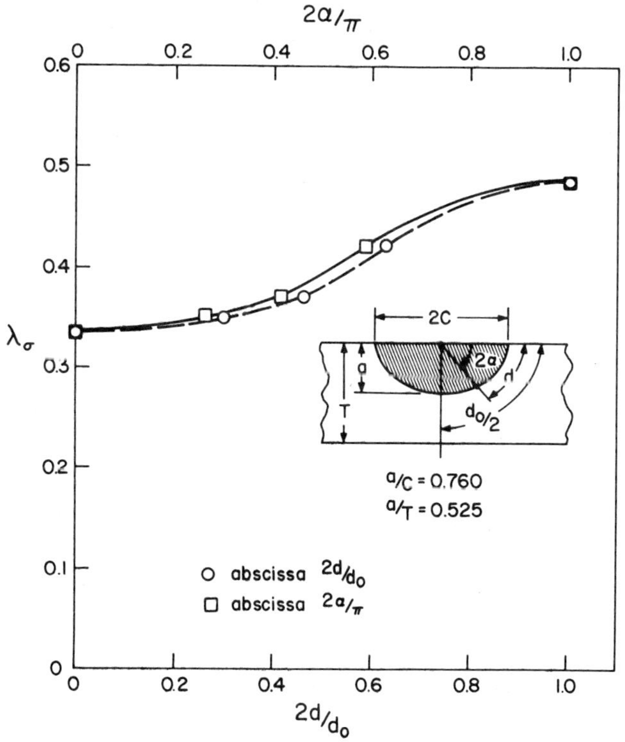

FIG. 10—λ_σ *distribution for surface crack.*

This geometry was used because of its prevalence in structural elements in service in order to investigate the extent to which the aforementioned concepts might apply in a technologically important strongly three-dimensional configuration. Moreover, in these experiments, only standard photoelastic methods were used so that boundary surface values of λ_σ were obtained only by extrapolation. However, this approach results in a significant reduction (over 50%) in analysis time and, as seen in Fig. 8, would not appear to yield significant error.

The distribution of λ_σ along the flaw border for a semi-elliptic crack of moderate depth is shown in Fig. 10 for both scales of $2d/d_0$ and $2\alpha/\pi$. This distribution is compared with the λ_σ distribution for a compact bending specimen of about the same crack depth in Fig. 11. These results agree at the free surface and at the crack midpoint. However, there is a substantial divergence between these points suggesting that, for the surface flaw, the loss of the in-

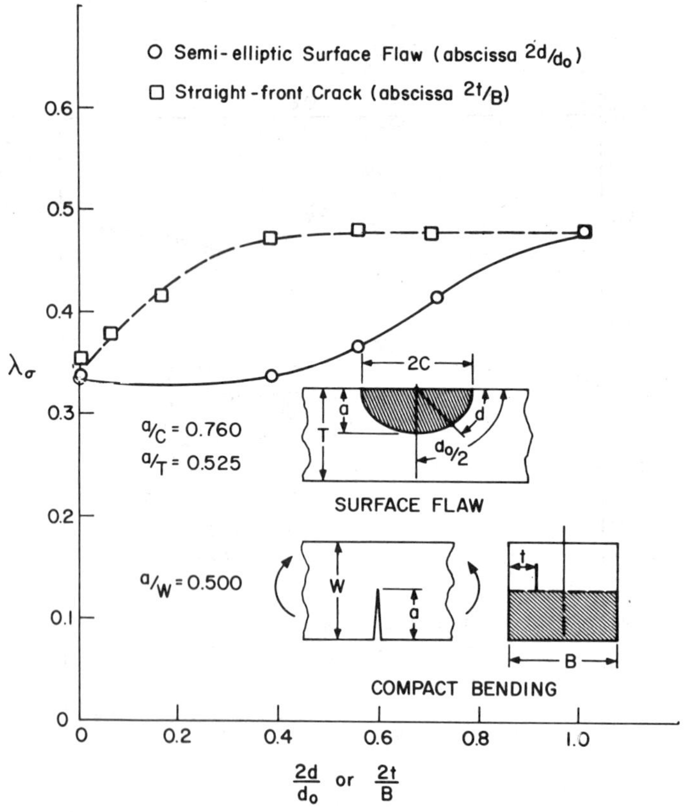

FIG. 11—*Comparison of λ_σ distributions for straight front and surface cracks.*

verse square root singularity extends much further from the free surface than for the straight front crack. This latter trend was also observed for both shallower and deeper flaws. On the basis of these preliminary studies, we conjecture that the extended boundary layer effect is geometric, and results from the fact that a slice taken normal to the crack border in the surface flaw sees a much smaller a/T value than the compact bending slices and thus approaches the free surface condition.

Interpretation of Results for Design

Benthem's analysis predicted that the boundary layer effect would be strongest in materials with high values of Poisson's ratio (rubbers, some plastics and adhesives, rocket motor propellants) and, for such materials, the issue is raised as to how these effects should be accounted for in designing against fracture. Since K_λ, or $K_\lambda f(\lambda)$, do not have the dimensions of the classical SIF, additional interpretations of our results would appear to be needed here.

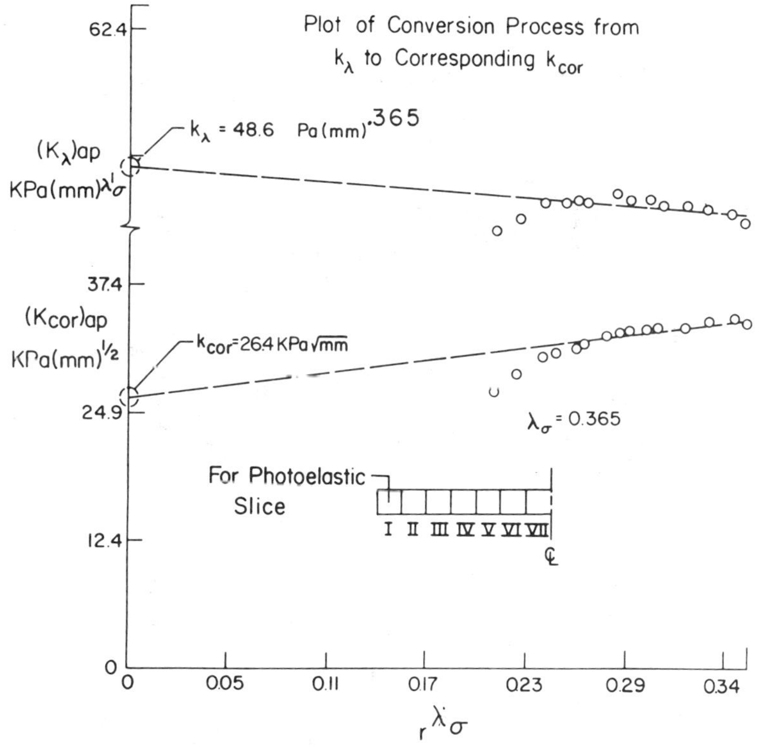

FIG. 12—*Determination of* K_{cor} *from* $(K_\lambda)_{AP}$ *data from photoelastic data from a surface slice from compact bending specimen.*

When the frozen stress method is employed, the quantity measured is proportional to a stress averaged through the thickness in a local region near the crack tip. In this region we have obtained expressions for an apparent stress intensity value from both LEFM [3] and from Eq 5 for the boundary layer algorithm. Since each describes the same quantity (in-plane maximum shear stress), we equate them in the measurement zone to obtain

$$\tau_{nz}^{max} = \frac{K_{AP}}{r^{1/2}} = \frac{(K_\lambda)_{AP}}{r^{\lambda_\sigma}} \tag{7}$$

where K_{AP} "corresponds" to $(K_\lambda)_{AP}$ so we call it $(K_{cor})_{AP}$. Thus

$$(K_{cor})_{AP} = (K_\lambda)_{AP}\, r^{1/2-\lambda_\sigma} \tag{8}$$

If we then use Eq 8 on the boundary layer data, we can convert to a "corresponding" classical LEFM value. Once this is done, $(K_{cor})_{AP}$ values may be extrapolated to the origin on a plot of K_{AP} versus r^{λ_σ} to obtain K_{cor} values as is done to obtain K_I in LEFM [*10,11*]. Figure 12 shows an example of this proce-

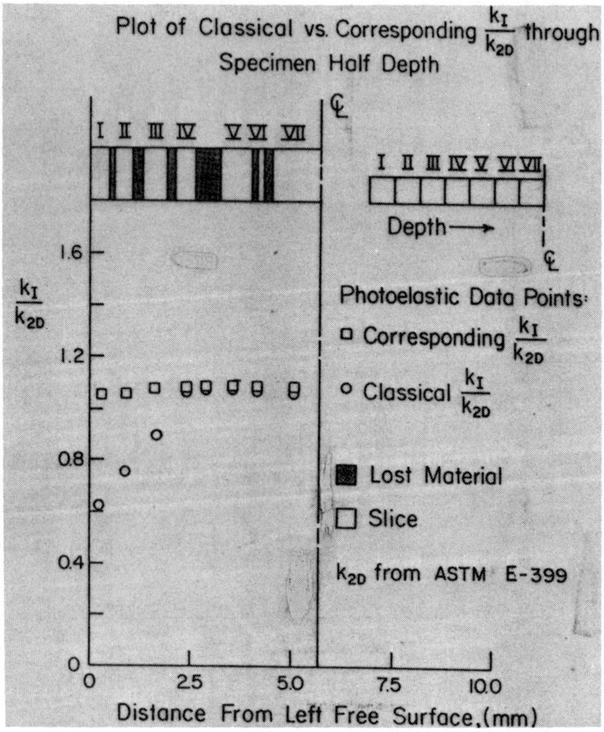

FIG. 13—K_I *and* K_{cor} *distributions in compact bending specimen.*

dure. In Fig. 13 we compare the photoelastic data using an LEFM algorithm with the results obtained using the boundary layer algorithm and Eq 8 for the compact bending specimen. We note that use of Eq 8 essentially eliminates the influence of the free surface, although the result is slightly higher than the two-dimensional result. Thus the new algorithm, when modified by Eq 8, yields a conservative result which could be employed within a two-dimensional context by the designer. Figure 14 shows a similar result for a semi-elliptic crack. We again see an elevation of K_{cor} above the classical SIF as before with greatest elevation again occurring near the free surface, and resulting in a variation in K_{cor} of less than 10% around the flaw border. This result again suggests that the problem could be greatly simplified by basing LEFM design rationale on the maximum value of K_{cor} with a conservative result.

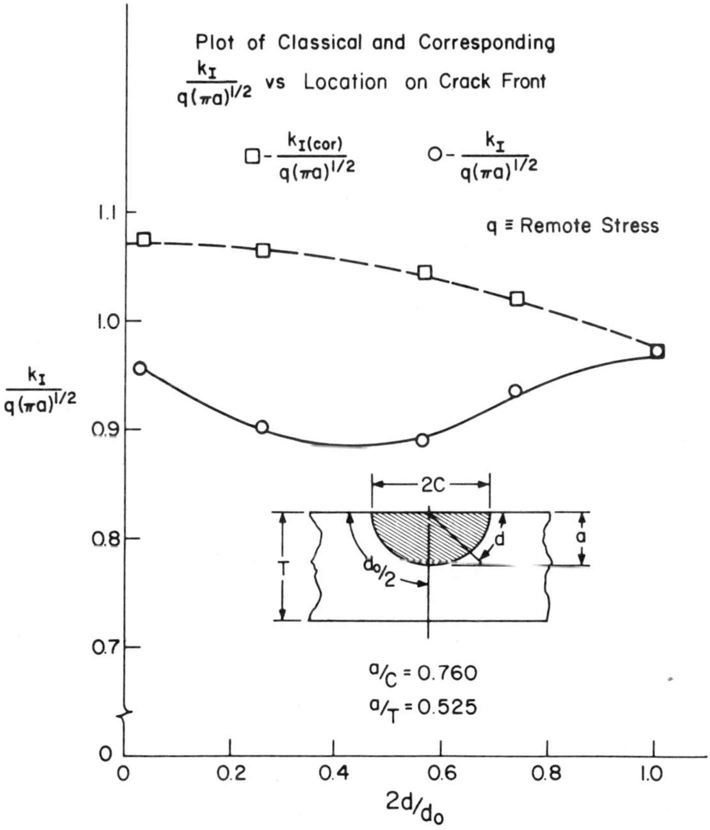

FIG. 14—K_I *and* K_{cor} *distributions for surface crack.*

Summary

After briefly reviewing an optical experimental method for analyzing the stress intensity distributions in cracked bodies, the method was applied to a study of boundary layer effects in such problems where crack fronts intersect free boundaries. Results suggest that such effects are strong for surface flaws in nearly incompressible elastic materials. Based upon limited data, a technique was proposed for accounting for such effects through the use of corresponding SIF together with conventional LEFM design rationale. Studies in this direction are continuing.

Acknowledgments

The authors wish to acknowledge the contributions of their colleagues as noted in the text in providing foundations for this work. The support of the National Science Foundation under Grant MEA-832-0252 is also gratefully acknowledged.

References

[*1*] Villarreal, G., Sih, G. C., and Hartranft, R. J., *Journal of Applied Mechanics*, March 1975, pp. 9–14.

[*2*] Benthem, J. P., *International Journal of Solids and Structures*, Vol. 16, 1980, pp. 119–130.

[*3*] Folias, E. S., *International Journal of Fracture*, Vol. 16, No. 4, 1980, pp. 335–348.

[*4*] Smith, C. W., Post, D., and Epstein, J. S. in *Proceedings*, 40th Anniversary Conference of Society for Experimental Stress Analysis, May 1983, pp. 1–6.

[*5*] Bazant, Z. P. and Estenssoro, L. F. in *Fracture*, Vol. 3, 1977, ICF4, University of Waterloo Press, Ontario, Canada, pp. 371–385.

[*6*] Solecki, J. S. and Swedlow, J. L., "On the Three Dimensional Applications of LEFM II—Finite Element Analysis of Straight and Curved Through Cracks in a Plate," Report SM83-9, Department of Mechanical Engineering, Carnegie Mellon University, Pittsburgh, Aug. 1983, 46 pp.

[*7*] Smith, C. W., *Journal of Optical Engineering*, Vol. 71, No. 4, 1982, pp. 696–703.

[*8*] Smith, C. W., Post, D., and Nicoletto, G. in *Proceedings*, 1982 Joint Conference on Experimental Mechanics, Society for Experimental Stress Analysis—Japan Society of Mechanical Engineers, Part I, 1982, pp. 196–200.

[*9*] Smith, C. W., Epstein, J. S., and Olaosebikan, O., *Advances in Aerospace Structures, Materials and Dynamics*, ASME-Aero Div.-06, 1983, pp. 119–126.

[*10*] Smith, C. W., *Experimental Evaluation of Stress Concentration and Intensity Factors, Mechanics of Fracture*, Vol. 7, G. C. Sih, Ed., Martinus-Nijhoff, Leyden, The Netherlands, Ch. 2, 1981, pp. 162–187.

[*11*] Smith, C. W. in *Experimental Techniques in Fracture Mechanics 2*, SESA Monograph No. 2, A. S. Kobayashi, Ed., Ch. 1, 1975, pp. 3–58.

[*12*] Nicoletto, G., Post, D., and Smith, C. W. in *Proceedings*, 1982 Joint SESA-JSME Conference on Experimental Mechanics, 1982, pp. 258–262.

[*13*] Smith, C. W. and Epstein, J. S., "Measurement of Three Dimensional Effects in Cracked Bodies," in *Proceedings*, 5th International Congress on Experimental Stress Analysis, June 1984.

I. S. Raju[1] *and J. C. Newman*[2]

Stress-Intensity Factors for Circumferential Surface Cracks in Pipes and Rods under Tension and Bending Loads

REFERENCE: Raju, I. S. and Newman, J. C., **"Stress-Intensity Factors for Circumferential Surface Cracks in Pipes and Rods under Tension and Bending Loads,"** *Fracture Mechanics: Seventeenth Volume, ASTM STP 905*, J. H. Underwood, R. Chait, C. W. Smith, D. P. Wilhem, W. A. Andrews, and J. C. Newman, Eds., American Society for Testing and Materials, Philadelphia, 1986, pp. 789-805.

ABSTRACT: Stress-intensity factors for a wide range of nearly semi-elliptical surface cracks in pipes and rods are presented. The surface cracks were oriented on a plane normal to the axis of pipes or rods. The configurations were subjected to either remote tension or bending loads. For pipes, the ratio of crack depth to crack length (a/c) ranged from 0.6 to 1, the ratio of crack depth to wall thickness (a/t) ranged from 0.2 to 0.8, and the ratio of internal radius to wall thickness (R/t) ranged from 1 to 10. For rods, the ratio of crack depth to crack length also ranged from 0.6 to 1, and the ratio of crack depth to rod diameter (a/D) ranged from 0.05 to 0.35. These particular crack configurations were chosen to cover the range of crack shapes (a/c) that have been observed in experiments conducted on pipes and rods under tension and bending fatigue loads. The stress-intensity factors were calculated by a three-dimensional finite-element method. The finite-element models employed singularity elements along the crack front and linear-strain elements elsewhere. The models had about 6500 degrees of freedom. The stress-intensity factors were evaluated using a nodal-force method.

The present results were compared with other analytical and experimental results for some of the crack configurations. The results generally agreed within 10%.

These results should be useful in predicting crack growth rates and fracture strengths, designing structural components, and establishing inspection intervals for pipes and rods.

KEY WORDS: cracks, surface cracks, crack propagation, fracture, stress analysis, fatigue (materials), stress intensity factors, finite elements

[1]Senior Scientist, Analytical Services & Materials, Inc., Tabb, VA 23666.
[2]Senior Scientist, Materials Division, NASA Langley Research Center, Hampton, VA 23665.

Nomenclature

- a Depth of surface crack
- c Half-arc length of surface crack
- D Outer diameter of pipe and rod
- F Stress-intensity boundary-correction factor
- F_A Boundary-correction factor at maximum depth point on surface crack ($\phi = \pi/2$)
- F_B Boundary-correction factor at intersection point of crack and outer surface
- h Half-length of pipe and rod
- K Stress-intensity factor (Mode I)
- K_A Stress-intensity factor at maximum depth point on surface crack ($\phi = \pi/2$)
- Q Shape factor for elliptical crack
- R Internal radius of pipe
- S_b Remote bending stress on outer fiber
- S_t Remote uniform-tension stress
- t Wall thickness of pipe
- x,y Cartesian coordinate system
- ν Poisson's ratio
- ϕ Parametric angle of ellipse

Superscript

Primes denote quantities associated with surface crack in flat plate

Surface cracks can occur in many structural components. Circumferential surface cracks can cause premature failure of landing gear of aircraft, piping, bolts, pins, and reinforcements that employ cylindrically shaped components. Accurate stress analyses of these surface-cracked components are needed for reliable prediction of crack growth rates and fracture strengths. Because of the complexities of such problems, however, exact solutions are not available.

Some investigators have used experimental or approximate analytical methods to obtain stress-intensity factors for surface cracks in rods under tension and bending loads. The stress-intensity factors for a circumferential surface crack growing in a rod under remote uniform tension was obtained experimentally by Wilhem et al [*1*] using the James-Anderson procedure [*2*]. Their results show that the surface cracks intersect the outer surface of the rod at nearly right angles. Athanassiadis et al [*3*] (boundary-integral method) and Nezu et al [*4*] (finite-element method) used analytical methods to obtain stress-intensity factors for circumferential surface cracks of various shapes in rods. Their surface-crack shapes were chosen to agree with their experimental observations. In some of the experiments of Ref *4*, however, the surface cracks did not intersect the outer surface at right angles. Residual stresses at the outer surface could have altered the crack shape near the free surface. Trantina et al [5] conducted an elastic and an elastic-plastic finite-element analysis of a small surface crack in a rod under tension loading. They assumed that cracks intersected the outer surface at right angles.

Much less research has been conducted on circumferential surface cracks in hollow cylinders or pipes than on rods. Delale and Erdogan [*6*] and German et al [*7*] obtained stress-intensity factors for interior and exterior circumferential surface cracks using the line-spring model. Recently, Forman and Shivakumar [*8*] studied the fatigue crack growth behavior of circumferential surface cracks in rods and pipes. Most of their experimental results did indicate that surface cracks intersect the free surfaces at about right angles.

This paper presents stress-intensity factors calculated with a three-dimensional finite-element analysis for a wide range of nearly semi-elliptical surface cracks in pipes and rods. The surface cracks were oriented on a plane normal to the axis of pipes or rods (Fig. 1). The crack configurations were assumed to be such that the crack fronts intersect the free surface at right angles. The pipe and rod were subjected to either remote tension or bending loads. For pipes, the ratio of crack depth to crack length ranged from 0.6 to 1, the ratio of crack depth to wall thickness ranged from 0.2 to 0.8, and the ratio of internal radius to wall thickness was 1 to 10. For rods, the ratio of crack depth to crack length also ranged from 0.6 to 1, and the crack configurations were chosen to cover the range of crack shapes that have been observed in experiments conducted on pipes and rods under tension and bending fatigue loads. The stress-intensity factors were calculated by a nodal-force method [*9–11*]. The present results were compared with other analytical and experimental results from the literature for some of the crack configurations.

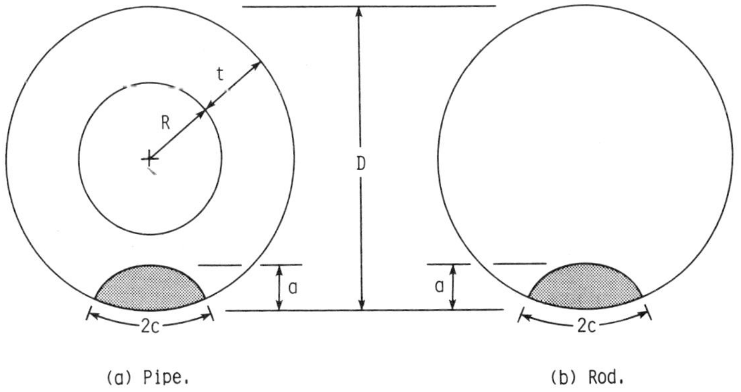

FIG. 1—*Surface crack in a pipe and rod.*

Three-Dimensional Finite-Element Analysis

A three-dimensional finite-element analysis was used to calculate the Mode I stress-intensity factor variations along the crack front for a circumferential surface crack in the pipe and rod shown in Figs. 1*a* and 1*b*, respectively. The pipe and rod were subjected to either remote tension or bending loads. In this analysis, Poisson's ratio (ν) was assumed to be 0.3. The shapes of the surface cracks were nearly but not exactly semi-elliptical. These crack shapes were generated using a conformal transformation as described in the Appendix.

Figures 2*a* and 2*b* show typical finite-element models for a circumferential surface crack in a pipe and rod, respectively. The finite-element models employ singularity elements along the crack front and linear-strain elements elsewhere. The models had about 6500 degrees of freedom. Stress-intensity factors were evaluated from a nodal-force method. Details of the formulation of these types of elements and of the nodal-force method are given in Refs *9* to *11* and are not repeated here. Details on the development of the finite-element models are given in the Appendix.

Loading

Two types of loads were applied to the finite-element models of the surface-cracked pipe and rod: remote uniform-tension and remote bending. The remote uniform-tension stress is S_t and the remote outer-fiber bending stress is S_b. The bending stress S_b in Fig. 3 is calculated at the origin of the surface crack ($x = y = 0$ in Fig. 4) without the crack being present.

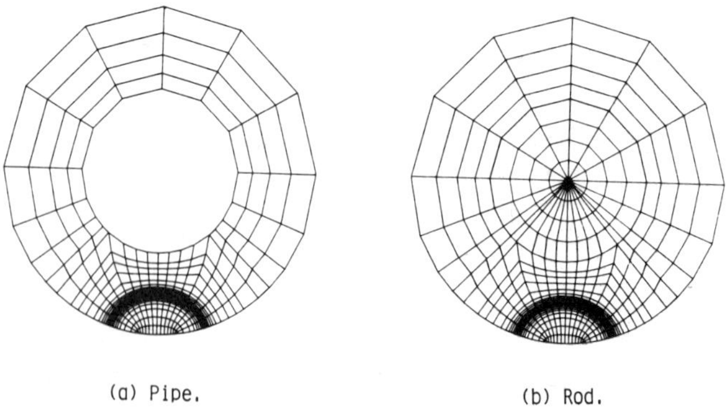

FIG. 2—*Finite-element idealizations of a surface crack in a pipe and rod.*

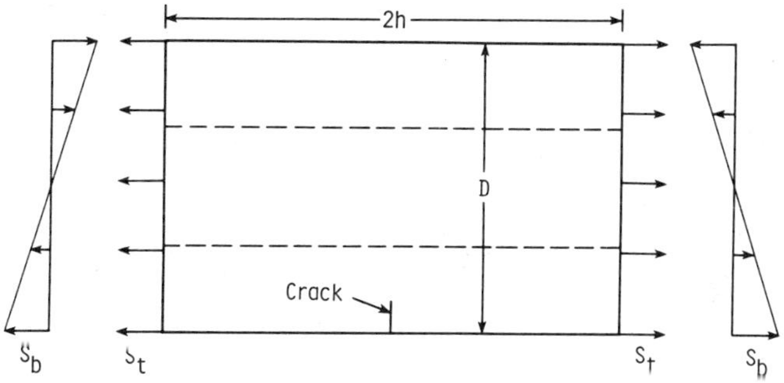

FIG. 3—*Loading applied to the pipe and rod.*

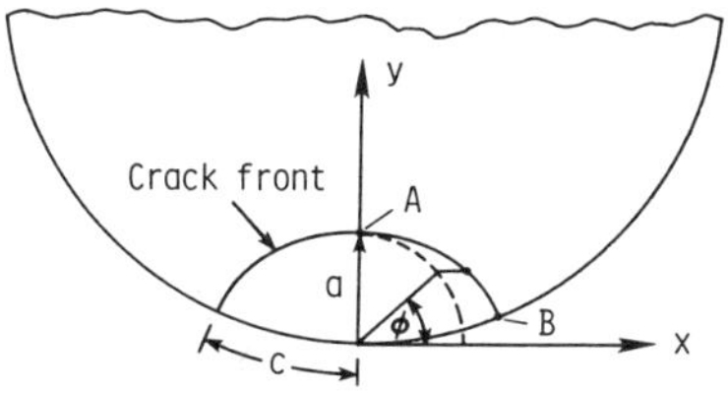

FIG. 4—*Coordinate system and dimensions for surface crack.*

Stress-Intensity Factor

The tension and bending loads only cause Mode I deformations. The Mode I stress-intensity factor K for any point along the surface-crack was taken to be

$$K = S_i \sqrt{\pi \frac{a}{Q}}\, F \tag{1}$$

where the subscript i denotes either tension load ($i = t$) or bending load ($i = b$), and Q, the shape factor for an ellipse, is given by the square of the complete elliptic integral of the second kind. The half-length of the pipe and rod, h, was chosen large enough to have a negligible effect on stress intensity ($h/D \geqslant 10$). Values for F, the boundary-correction factor, were calculated along the crack front for various combinations of parameters (a/t, a/c, R/t, and ϕ for a crack in a pipe; a/c, a/D, and ϕ for a crack in a rod). The crack dimensions and parametric angle, ϕ, are defined in Fig. 4. The range of crack shapes (a/c) and of sizes (a/t or a/D) analyzed are shown in Figs. 5 and 6 for the pipe and rod, respectively.

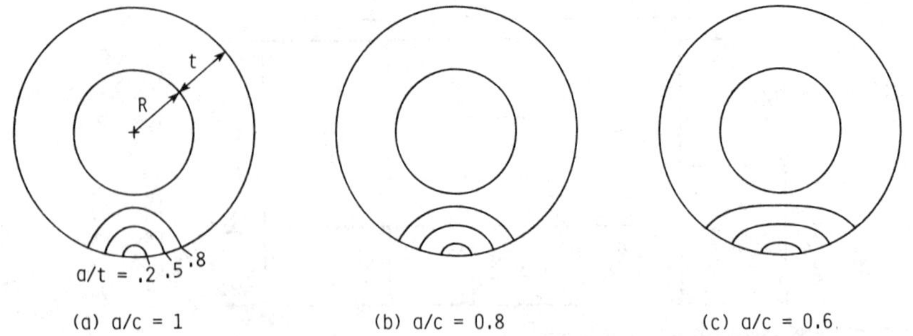

FIG. 5—*Surface-crack shapes and sizes analyzed for the pipe.*

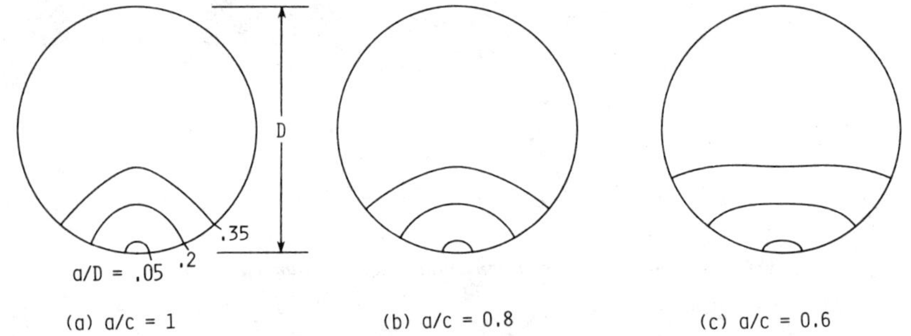

FIG. 6—*Surface-crack shapes and sizes analyzed for the rod.*

The empirical expressions for Q used in this paper are

$$Q = 1 + 1.464(a/c)^{1.65} \text{ for } a/c \leq 1 \tag{2a}$$

$$Q = 1 + 1.464(c/a)^{1.65} \text{ for } a/c > 1 \tag{2b}$$

Results and Discussion

In the following sections, stress-intensity factors for various shape surface cracks in pipes and rods subjected to tension and bending loads are presented. Tables 1 to 3 give the normalized stress-intensity factors, $K/(S\sqrt{\pi a/Q})$, at the maximum depth point (A) and at the point which the crack intersects the free surface (B). Figures 7 to 12 show the variation in normalized stress-intensity factors as a function of the parametric angle ($2\phi/\pi$) for various crack shapes (a/c), crack size (a/t or a/D), and radius of the pipe (R/t).

TABLE 1—*Normalized stress-intensity factor,* $K/(S_t\sqrt{\pi a/Q})$, *for surface crack in a pipe subjected to tension loads* (A = *maximum depth point,* B = *free surface point*).

	a/t = 0.2		*a/t* = 0.5		*a/t* = 0.8	
R/t	*A*	*B*	*A*	*B*	*A*	*B*
			***a/c* = 1.0**			
1	1.015	1.160	1.036	1.237	1.076	1.385
2	1.017	1.157	1.041	1.235	1.072	1.383
4	1.019	1.154	1.046	1.234	1.072	1.381
10	1.020	1.152	1.049	1.233	1.074	1.380
			***a/c* = 0.8**			
1	1.061	1.060	1.114	1.161	1.202	1.354
2	1.059	1.056	1.106	1.155	1.174	1.331
4	1.058	1.053	1.103	1.156	1.157	1.333
10	1.057	1.051	1.101	1.156	1.144	1.335
			***a/c* = 0.6**			
1	1.113	0.943	1.226	1.080	1.455	1.327
2	1.105	0.937	1.194	1.070	1.342	1.285
4	1.101	0.933	1.178	1.071	1.285	1.285
10	1.097	0.930	1.167	1.070	1.247	1.290

TABLE 2—*Normalized stress-intensity factor,* $K/(S_b\sqrt{\pi a/Q})$, *surface crack in a pipe subjected to bending loads* (A = *maximum depth point,* B = *free surface point*).

	a/t = 0.2		*a/t* = 0.5		*a/t* = 0.8	
R/t	*A*	*B*	*A*	*B*	*A*	*B*
			***a/c* = 1.0**			
1	0.943	1.136	0.856	1.162	0.777	1.233
2	0.966	1.137	0.919	1.188	0.870	1.287
4	0.981	1.133	0.971	1.204	0.950	1.327
10	0.995	1.131	1.012	1.212	1.019	1.348
			***a/c* = 0.8**			
1	0.989	1.037	0.931	1.079	0.885	1.162
2	1.007	1.037	0.984	1.107	0.966	1.224
4	1.021	1.033	1.028	1.126	1.033	1.276
10	1.032	1.032	1.064	1.136	1.088	1.303
			***a/c* = 0.6**			
1	1.042	0.919	1.034	0.980	1.094	1.078
2	1.056	0.919	1.069	1.015	1.118	1.152
4	1.065	0.916	1.102	1.039	1.155	1.220
10	1.071	0.913	1.130	1.051	1.188	1.257

TABLE 3—*Normalized stress-intensity factor,* $K/(S_i\sqrt{\pi a/Q})$, *for surface crack in a rod subjected to tension or bending loads* (A = *maximum depth point,* B = *free surface point*).

	$a/c = 1.0$		$a/c = 0.8$		$a/c = 0.6$	
a/D	A	B	A	B	A	B
			Tension Loads			
0.05	1.012	1.156	1.056	1.054	1.107	0.933
0.125	1.015	1.189	1.083	1.101	1.176	0.999
0.20	1.038	1.260	1.131	1.200	1.316	1.129
0.275	1.087	1.356	1.227	1.335	1.565	1.329
0.35	1.175	1.475	1.387	1.509	1.835	1.516
			Bending Loads			
0.05	0.938	1.129	0.984	1.029	1.035	0.907
0.125	0.836	1.114	0.901	1.019	0.987	0.903
0.20	0.749	1.112	0.830	1.028	0.985	0.909
0.275	0.683	1.109	0.795	1.040	1.041	0.924
0.35	0.629	1.106	0.782	1.039	1.056	0.876

Pipes under Tension Loads

Figure 7 shows the normalized stress-intensity factors as a function of the parametric angle (ϕ) for a pipe ($R/t = 2$) subjected to remote tension with a semi-circular crack ($a/c = 1$) for various values of a/t. For this crack shape, the maximum normalized stress-intensity factor occurred at the point where

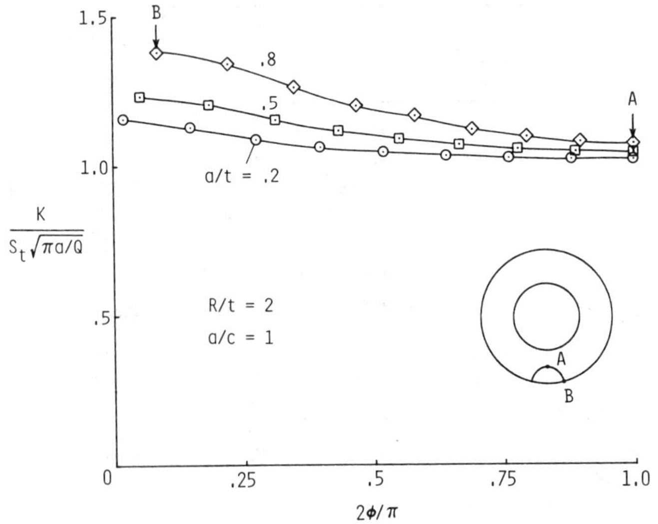

FIG. 7—*Normalized stress-intensity factors along the front of a surface crack* (a/c = 1) *in a pipe under tension.*

the crack meets the free surface (Point B). For all the a/c ratios considered (from 1 to 0.6), larger values of a/t always gave larger normalized stress-intensity factors.

For an a/c ratio of 0.8 and a/t less than or equal to 0.5, the normalized stress-intensity factors at the deepest point and at the free surface are nearly the same (Table 1). On the other hand, for an a/c ratio of 0.6, the maximum normalized stress-intensity factors occurred at the point of maximum depth ($\phi = \pi/2$).

Figure 8 shows the normalized stress-intensity factors for a surface crack with $a/c = 0.8$ and $a/t = 0.5$ in pipes with R/t ratios of 1 and 10. (The results for $R/t = 2$ and 4 lie in between those for $R/t = 1$ and 10 and are not shown for clarity). For this configuration, the effect of varying R/t is insignificant. However, for cracks with $a/c = 0.6$ and $a/t = 0.8$, R/t has a significant effect on the normalized stress-intensity factors (Fig. 9). The values at the deepest point ($\phi = \pi/2$) are affected more than those at the free surface. Figure 9 shows that lower R/t values gave higher stress-intensity factors. However, the differences between the stress-intensity factors are less for larger values of R/t. Thus the effect of R/t diminishes for pipes with larger R/t values.

In summary, the effect of the curvature of the pipe (R/t) is to elevate the stress-intensity factors compared with those of a flat plate ($R/t = \infty$). The effect is more pronounced at the deepest point than at the free surface point.

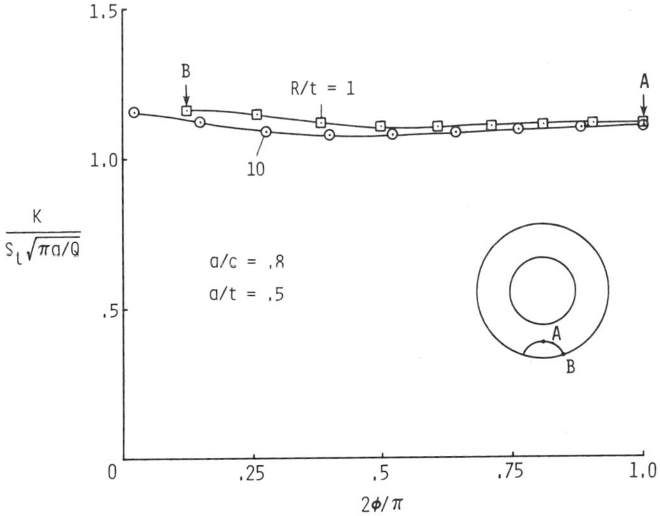

FIG. 8—*Normalized stress-intensity factors along the front of a surface crack* (a/c = 0.8) *in a pipe under tension.*

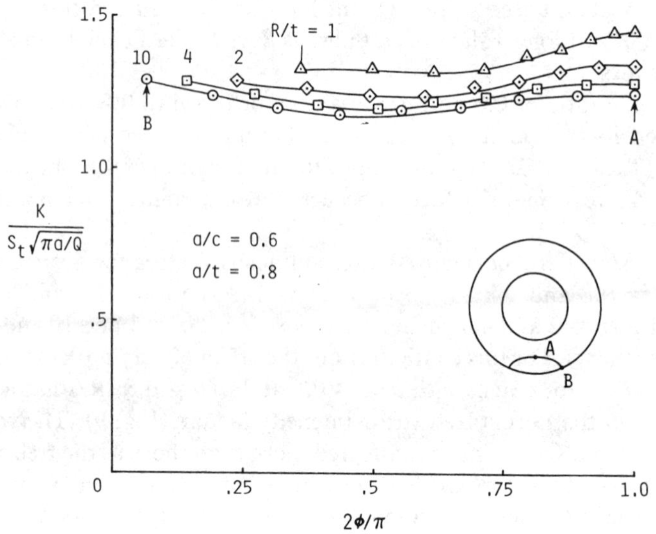

FIG. 9—*Normalized stress-intensity factors along the front of a deep surface crack* (a/c = 0.6) *in a pipe under tension.*

Pipes under Bending Loads

The normalized stress-intensity factors as a function of the parametric angle (ϕ) are shown in Fig. 10 for a pipe ($R/t = 2$) subjected to remote bending with a semi-circular crack for various values of a/t. For this crack shape, the deeper cracks (larger a/t ratios) produced larger normalized stress-intensity factors where the crack meets the free surface but smaller values at the maximum depth point ($\phi = \pi/2$).

Rods under Tension Loads

Figure 11 shows the normalized stress-intensity factors as a function of the parametric angle (ϕ) for a rod with various shape surface cracks with $a/D = 0.2$. When a/c was unity, the maximum normalized stress-intensity factor occurred at the free surface. When a/c was equal to 0.6, the maximum was at the deepest point. For surface cracks with an a/c ratio of 0.8, however, the normalized stress-intensity factors are nearly constant, much like the pipe.

Rods under Bending Loads

The normalized stress-intensity factors as a function of the parametric angle (ϕ) for a rod with various shape surface cracks with $a/D = 0.2$ are shown in Fig. 12. The maximum normalized stress-intensity factor occurred at the free surface for $a/c = 1$ and $a/c = 0.8$. For surface cracks with $a/c = 0.6$,

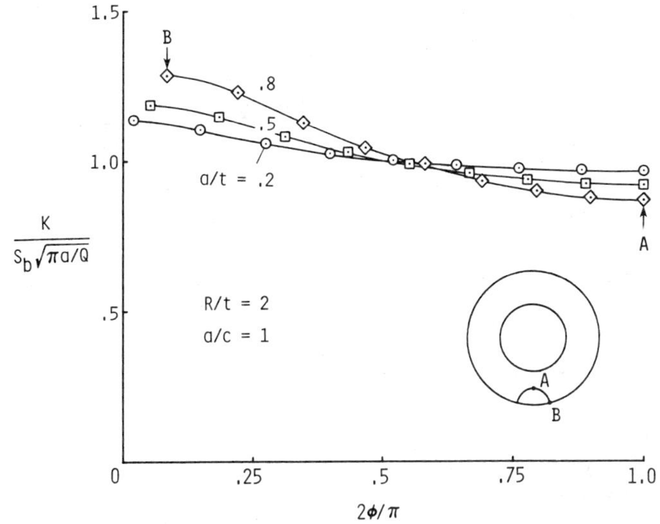

FIG. 10—*Normalized stress-intensity factors along the front of a surface crack* (a/c = 1) *in a pipe under bending.*

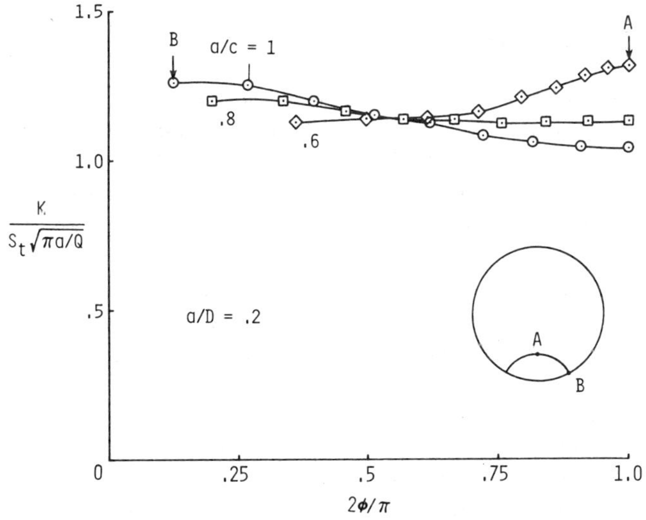

FIG. 11—*Normalized stress-intensity factors along the front of various shaped surface cracks in a rod under tension.*

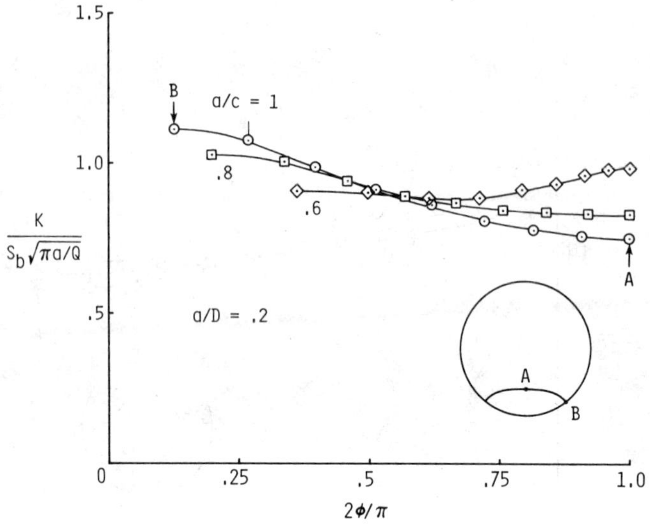

FIG. 12—*Normalized stress-intensity factors along the front of various shaped surface cracks in a rod under bending.*

however, the normalized stress-intensity factors all along the crack front are nearly constant.

Comparisons with Other Solutions

The comparison of the present results with those from the literature are difficult because, at least, three definitions of crack shapes have been used. All definitions differ in how the crack length c is measured. In this paper, c is measured as the arc length (Fig. 4). In some reports, c is measured as the horizontal projection of Point B on the x-axis, while in other reports, c is defined as the intersection point of an ellipse with the x-axis. In the latter case, the crack front will not intersect the free surface at a right angle. The crack front shape, in the latter case, therefore, will be different from that used in this paper. In view of these difficulties, only a few comparisons can be made.

As previously mentioned, stress-intensity factor analysis of circumferential surface cracks in pipes have received very little attention in the literature. Delale and Erdogan [*6*] obtained stress-intensity factors for interior and exterior circumferential surface cracks, and German et al [*7*] obtained stress-intensity factors for interior circumferential surface cracks by using the line-spring model. Most of the external circumferential surface crack configurations presented in Ref *6* were vastly different from the configurations presented in this paper. However, one configuration with $a/c = 0.775$,

$a/t = 0.8$, and $R/t = 5.374$ falls within the range of parameters considered in this paper. (In the notation of Ref *6*, this configuration has $L_0/h = 0.8$, $a/h = 1$ and $\lambda_2 = 0.75$.) The normalized stress-intensity factor at the deepest point ($\phi = \pi/2$) of the crack was computed from the results of Ref *6* as 1.276. Interpolating the present results in Table 1, the normalized stress-intensity factor F_A for this configuration was found to be 1.161. The result from the line-spring model [*6*] is about 10% higher than the present result.

The rod configurations with surface cracks subjected to remote tension have received more attention in the literature than the pipe configurations. These configurations were analyzed by Wilhem et al [*1*], Athanassiadis et al [*3*], and Nezu et al [*4*]. The present results are compared with the results from Refs *1* and *3*. Comparisons with the results from Nezu et al [*4*] could not be made because the crack shapes analyzed in Ref *4* and those in present analysis (Fig. 6) were very different.

Figure 13 compares the normalized stress-intensity factors at the free surface (F_B) and the maximum depth point (F_A) for a surface crack with $a/c = 0.6$ from the present finite-element analysis with those from a boundary integral equation (BIE) method [*3*]. The results of Athanassiadis et al [*3*] were interpolated and plotted in Fig. 13 as open symbols. The present results are shown by solid symbols. The normalized stress-intensity factors obtained by the BIE method were 0 to 10% lower than the present results.

Figure 14 compares the normalized stress-intensity factors at the maximum depth point for surface cracks with various shapes (a/c) and sizes (a/D)

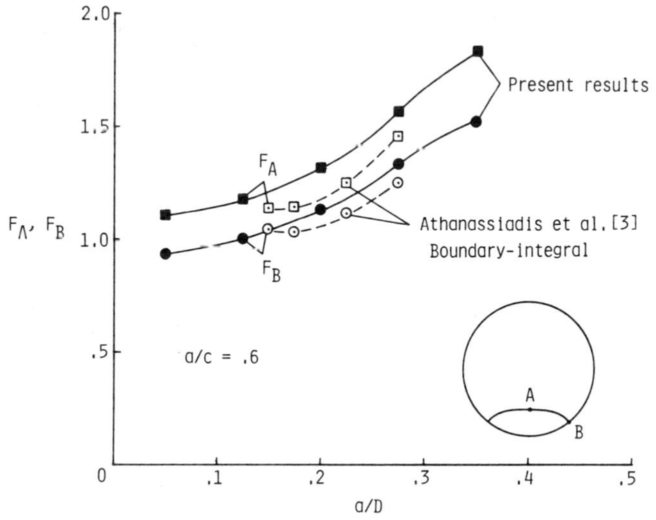

FIG. 13—*Comparison of boundary-correction factors for a surface crack in a rod under tension.*

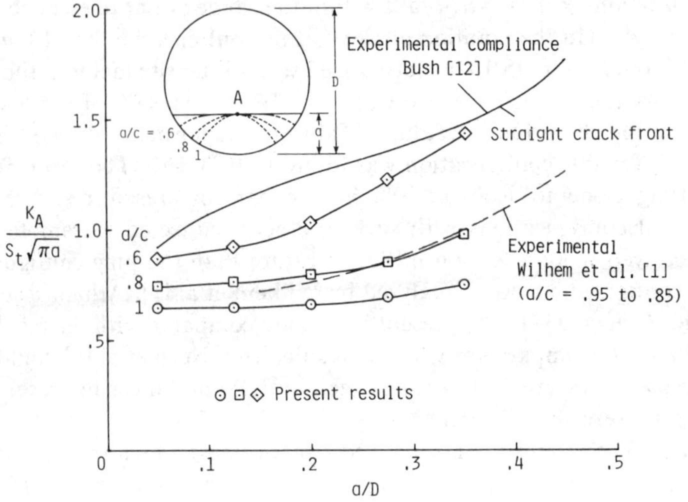

FIG. 14—*Comparison of experimental and calculated stress-intensity factors for a surface crack in a rod under tension.*

from the present analyses and from experimental results. Wilhem et al [*1*] obtained an experimental stress-intensity factor solution using the James-Anderson procedure [*2*]. These results are shown by the dashed curve in Fig. 14. For the surface cracks in their tests, the a/c ratios varied from 0.95 to 0.85. The present results (symbols) for $a/c = 0.8$ and 1.0 bound the experimental results for $a/D < 0.25$ and are a little below for $a/D > 0.25$.

Bush [*12*] considered cracks with straight fronts (see insert in Fig. 14) in rods subjected to remote tension. He obtained stress-intensity factors from experimental compliance for these straight through-cracks of various depths. His results are shown in Fig. 14 by a solid curve. For a surface crack with an a/c ratio of 0.6 and an a/D ratio of 0.35 (Fig. 6*c*), the crack configuration is very nearly the same as that for a crack with a straight front. For this configuration, the present results for $a/c = 0.6$ are a little below (about 2%) the straight through-crack results. The present results need not necessarily agree with the experimental results because the crack shapes are not identical, as noted previously.

Concluding Remarks

Stress-intensity factors for circumferential surface cracks in pipes and rods have been obtained by a three-dimensional finite-element analysis. The pipes and rods were subjected to either remote tension or remote bending loading. The surface cracks were nearly semi-elliptical and were oriented on a plane normal to the axis of pipes or rods. A wide range of crack shapes, crack sizes, and internal radius-to-wall thickness ratios have been considered. For each of these crack configurations and loadings, the stress-intensity factors calculated by the finite element analysis are presented.

Stress-intensity factors for surface cracks in a pipe were found to be insensitive to internal radius-to-wall thickness (R/t) ratios ranging from 1 to 10, for crack depth-to-length (a/c) ratios ranging from 0.8 to 1.0 with crack depth-to-wall thickness (a/t) ratios less than 0.8. For $a/c = 0.6$ and $a/t = 0.8$, however, the stress-intensity factors showed significant variation with R/t. The effect of the curvature of the pipe (R/t) is to elevate the stress-intensity factors compared with those of a flat plate ($R/t = \infty$). This effect is more pronounced at the deepest point than at the free surface point.

Stress-intensity factors for a surface crack in a rod were 0 to 10% higher than those calculated from a boundary-integral analysis. The stress-intensity factors agreed well with experimental results for surface cracks in rods and approached the experimental results for cracks with straight fronts.

The stress-intensity factors obtained here should be useful in predicting fatigue crack growth and fracture of surface cracks in cylinders and rods.

Acknowledgments

The authors gratefully acknowledge the support of this effort by Mr. Royce Forman of the NASA Lyndon B. Johnson Space Center (Shuttle Operations)-Engineering and Test Support) and by Dr. J. H. Crews, Jr., of the NASA Langley Research Center (Materials Division). This work was performed as a part of Contracts NAS1-17090 and NAS1-17683.

APPENDIX

The purpose of this Appendix is to present the procedure used to develop finite-element models for surface cracks in pipe and rod configurations through a conformal transformation.

A cylinder with a surface crack is shown in Fig. 15*a*. The stress-intensity factors for this configuration were evaluated from a nodal-force method [9]. In this method, the nodal forces normal to the crack plane (x,y plane) and ahead of the crack front are used. The nodal-force method also requires that these forces be evaluated at nodes which are very near the crack front and which lie on lines in the x,y plane that are normal to the crack front. Therefore the finite-element model should be such that the normality at the crack front is maintained. This is achieved through a conformal transformation as follows.

First, a finite-element model for a semi-elliptical surface crack with semi-minor and semi-major axes, a' and c', respectively, in a plate of width w' and a thickness of t' (Fig. 15*b*) is developed such that

$$a' = \ln\left(\frac{R + t}{R + t - a}\right) \tag{3}$$

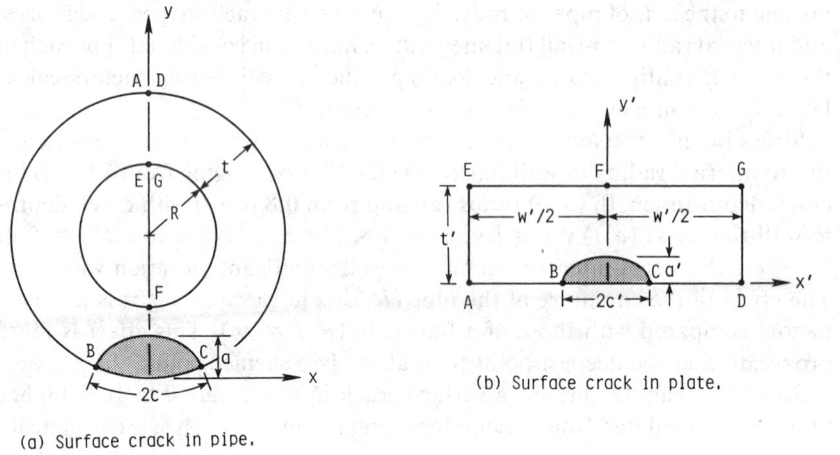

FIG. 15—*Conformal transformation of a surface crack in a plate to a surface crack in a pipe.*

$$c' = \frac{c}{R + t} \tag{4}$$

and

$$t' = \ln \frac{R + t}{R} \tag{5}$$

To obtain the desired configuration in Fig. 15*a*, a conformal transformation

$$x = (R + t)\, e^{-y'} \cos(w'/4 - x') \tag{6}$$

$$y = (R + t)\,[1 - e^{-y'} \sin(w'/4 - x')] \tag{7}$$

$$z = z'$$

and

$$w' = 2\pi \tag{8}$$

is used. This transformation transforms every point in the x',y',z' system to a unique point in the x,y,z system and maintains normality. Because the finite-element model of a surface crack in a flat plate (Fig. 15*b*) has nodes along hyperbolas near the crack front [*11*] (and, hence, normality to the semielliptical crack front in the x',y' plane is assured), the conformal transformation gives nodes along curves in the x,y plane which are also normal to the crack front in the pipe configuration (Fig. 15*a*).

The finite-element models for the surface crack in the rod configuration were obtained from the pipe models by idealizing the inside core with finite elements (Fig. 2).

References

[1] Wilhem, D., Fitzgerald, J., Carter, J., and Dittmer, D., "An Empirical Approach to Determining *K* for Surface Cracks," in *Advances in Fracture Research*, International Conference of Fracture (ICF-5), D. Francois, Ed., Vol. 1, March 1981, pp. 11-21.

[2] James, L. A. and Anderson, W. E., "A Simple Procedure for Stress Intensity Factor Calibration," *Engineering Fracture Mechanics*, Vol. 1, 1969, pp. 565-568.

[3] Athanassiadis, A., Boissenot, J. M., Brevet, P., Francois, D., and Raharinaivo, A., "Linear Elastic Fracture Mechanics Computations of Cracked Cylindrical Tensioned Bodies," *International Journal of Fracture*, Vol. 17, No. 6, Dec. 1981, pp. 553-566.

[4] Nezu, K., Machida, S., and Nakamura, H., "SIF of Surface Cracks and Fatigue Crack Propagation Behavior in a Cylindrical Bar," in *Proceedings*, 25th Japan Congress on Materials Research - Metallic Materials, Society of Materials Science, March 1982, pp. 87-92.

[5] Trantina, G. G., deLorenzi, H. G., and Wilkening, W. W., "Three-Dimensional Elastic-Plastic Finite Element Analysis of Small Surface Cracks," *Engineering Fracture Mechanics*, Vol. 18, No. 5, 1983, pp. 925-938.

[6] Delale, F. and Erdogan, F., "Application of Line-Spring Model to a Cylindrical Shell Containing a Circumferential or Axial Part-Through Crack," *Journal of Applied Mechanics*, Vol. 49, March 1982, pp. 97-102.

[7] German, M. D., Kumar, V., and deLorenzi, H. G., "Analysis of Surface Cracks in Plates and Shells Using the Line-Spring Model and Adina," *Computers and Structures*, Vol. 17, No. 5-6, 1983, pp. 881-890.

[8] Forman, R. G. and Shivakumar, V., "Growth Behavior of Surface Cracks in the Circumferential Plane of Solid and Hollow Cylinders," this publication, pp. 59-74.

[9] Raju, I. S. and Newman, J. C., Jr., "Three-Dimensional Finite-Element Analysis of Finite-Thickness Fracture Specimens," NASA TN D-8414, National Aeronautics and Space Administration, Washington, DC, 1977.

[10] Raju, I. S. and Newman, J. C., Jr., "Improved Stress-Intensity Factors for Semi-Elliptical Surface Cracks in Finite-Thickness Plates," NASA TM X-72825, National Aeronautics and Space Administration, Washington, DC, 1977.

[11] Raju, I. S. and Newman, J. C., Jr., "Stress-Intensity Factors for a Wide Range of Semi-Elliptical Surface Cracks in Finite-Thickness Plates," *Engineering Fracture Mechanics*, Vol. 11, No. 4, 1979, pp. 817-829.

[12] Bush, A. J., "Stress-Intensity Factors for Single-Edge-Crack Solid and Hollow Round Bars Loaded in Tension," *Journal of Testing and Evaluation*, Vol. 9, No. 4, July 1981, pp. 216-223.

Summary

Summary

There were 55 presentations on the program at the Seventeenth National ASTM Symposium on Fracture Mechanics. For a variety of reasons, all related to the technical and time pressures of preparing a submission to this sort of meeting and publication, 44 papers appear in this volume. At the symposium the presentations were divided into five categories: Applications, Subcritical Crack Growth, Fracture Testing, Ductile Fracture, and Analysis and Mechanisms. These categories, although somewhat arbitrary because of the broad scope of many of the papers, are used in this volume and summary. Only a few terms are defined here; the reader can easily refer to the appropriate paper.

Applications

The papers in this section are concerned with the application of fracture mechanics concepts to the analysis of fatigue crack growth and fracture behavior of metallic materials and to the analysis of fracture behavior of ceramic and composite materials.

Hilton et al used fracture mechanics analyses to establish the material fracture toughness requirements to avoid loss of a ship propeller blade. Using engineering approximations and a simple finite-element analysis of a particular propeller crank ring, a toughness value from a J_R resistance curve was established as a minimum requirement for 4150H steel material. The fact that a full-scale laboratory test on a 4150H crank ring did not experience unstable fracture showed that the analysis was conservative.

The paper by *Tanaka et al* presented a new wide-plate short-crack arrest (SCA) test specimen for testing steel weld joints at low temperatures. The basic idea for the development of the SCA test is that brittle fracture should be initiated at the center of a wide plate to eliminate the effects of high compressive residual stresses that exist in welded crack-arrest specimens with edge cracks. One advantage of the SCA test is that it simulates a surface defect which may exist in actual structures.

Yang and Bamford characterized the response of semi-elliptical surface cracks under thermal shock conditions which may result from safety injection actuation in nuclear reactor vessels. The authors developed a methodology to predict the growth behavior of such cracks under simulated thermal shock conditions. Results from the study showed that cracks tend to elongate along the vessel inside surface.

Fatigue crack growth tests on circumferential surface cracks in solid and hollow cylinders under remote tension and bending loads were conducted by *Forman and Shivakumar.* Tests were conducted on both aluminum and titanium alloy specimens. Results show that surface-crack shapes in solid cylinders can be accurately represented by a circular arc, whereas crack shapes for internal or external surface cracks in hollow cylinders can be represented by a transformed semi-ellipse. Stress-intensity factor expressions for surface cracks in hollow cylinders were presented.

The paper by *Bradley et al* studied the dynamic fracture behavior of ductile cast iron and cast steel using either blunt-notched or fatigue precracked (side-grooved) Charpy specimens. The blunt-notched specimens gave erroneous indications of the relative toughness of the two materials (ductile iron being quite inferior to cast steel), whereas the precracked and side-grooved specimens showed that the ductile iron was actually superior to cast steel at temperatures below ambient. The differences in these results were attributed to variations in constraint through the thickness.

The effect of loading rate on the dynamic fracture of reaction-bonded silicon nitride (RBSN) was presented by *Liaw et al.* A novel experimental-numerical procedure, where the experimentally determined crack extension history drives a finite-element code in its generation mode, was used in the study. One of the major findings was that a crack in RSBN can continue to propagate at its terminal velocity under a low dynamic stress-intensity factor.

The last three papers in this section dealt with the fracture behavior of composite materials. *Ratwani and Deo* used the resistance curve approach to characterize the delamination growth resistance of various composite material systems. They found that the delamination growth resistance curves were a function of the resin material as well as the fiber used. *Harris and Morris* compared the fracture behavior of thick graphite-epoxy laminates using standard fracture toughness specimens (compact, three-point bend, and center-cracked tension). Fracture toughness values computed using the load at the intersection of the 5% secant line with the load-COD (crack-opening displacement) record were found to be independent of laminate thickness and specimen configuration. The paper by *Simonds* studied the residual strength of five boron/aluminum laminates with sharp notches with and without prior fatigue loading. Although the fatigue loading (60 to 80% of the static tensile strength) for about 100 000 cycles caused some matrix and fiber cracking, the residual strength was not significantly affected by the prior fatigue loading.

Subcritical Crack Growth

Most of the attention on the subject of fatigue crack propagation rates (FCPR) has been devoted to behavior at ambient temperature under constant-frequency, constant-amplitude conditions. This session delves deeper into the factors that influence FCPR. It also evaluates some interesting techniques for monitoring crack growth behavior.

Nicholas and Weerasooriya examined hold time effects on FCPR at elevated temperatures. Using Inconel 718 and studying the constant K behavior at 650°C, the authors showed that the hold times at maximum load displayed the greatest FCPR. Also, at maximum load hold times greater than 5 s, FCPR was time dependent. A linear cumulative damage model based solely on fatigue and sustained load data was found to be adequate for spectrum loading as long as the hold times were at maximum load. Also evaluating Inconel 718 at 650°C were *Petrovich et al.* They were able to detail the interactive effects of low and high frequency cycling on FCPR. It is interesting that as high-frequency ΔK is combined with constant low-frequency ΔK, two regions of FCPR behavior are noted: one where low frequency cycles dominate and the other where high frequency cycling governs. In the former, a retardation effect which is strongly dependent on low-cycle ΔK is noted.

Saxena studied crack growth under high temperature non-steady-state conditions. He described a crack parameter C_t which correlated well with da/dt under conditions that range from small scale creep to the steady-state creep regime. The correlation appears to be independent of specimen geometry. It was shown that for A470 Class 8 steel, da/dt as a function of C_t is independent of temperature in the range of 482 to 538°C to a first-order approximation.

Soya et al addressed the fatigue behavior of welded Invar sheet in the cryogenic temperature region. An equivalent stress intensity factor, K_{eq}, was calculated from $(K_I^2 + K_{II}^2)^{1/2}$ and was based on a finite element analysis of elastic and elastic-plastic conditions that exist at the weld root. K_{eq}, plotted versus the number of cycles to failure, can be utilized in the same manner as the traditional *S-N* curve. In the temperature range between room temperature and −162°C, for several types of fillet joints and one type of seam weld, it was shown that K_{eq} can be used to normalize the cycles-to-failure data in the region where large plastic deformation is not a factor ($<10^4$ cycles).

Several novel techniques for monitoring crack growth behavior complement the efforts noted above. *Larsen* reported the development of an automated photomicroscope system for monitoring the initiation and growth of small surface cracks. Behavior of surface cracks in the 25 to 2000 μm range was studied. Ti-6Al-2Sn-4Zr-6Mo (Ti-6246) was used to evaluate the system, which had a precision of about 1 μm for cracks of about 25 μm in length. It is noted that this method can provide a cost-effective means of monitoring growth of small cracks.

The growth and coalescence of small cracks was the subject of a paper by *Grandt et al.* A multiple degree of freedom algorithm is utilized to predict the growth of separate cracks. The crack shapes and sizes are then allowed to develop naturally as they join into a simple flow. The analysis was confirmed by the use of heat-tinting techniques on Waspaloy and Tl-6246 alloys.

Sharpe and Lee described an experimental study of crack tip displacement during high-temperature cyclic loading of Inconel 718. At the heart of the technique is a laser-based interferometer that detects fringe patterns produced from two microhardness indentations. Tests conducted between 23 and 650°C showed that compliances were in reasonable agreement with analytical predictions. Opening load ratios were found to be independent of temperature, crack length, and precracking level when measured at positions more than half the specimen thickness away from the crack tip.

The paper by *Wilson and Palazotto* also concentrated on crack-tip behavior utilizing IN-100 compact tension specimens cycled at 732°C. Viscoplastic constitutive equations are utilized in an analysis of the stress and strain field around the crack tip. It was interesting to note that most of the plastic straining occur within the first 3 cycles and that after about 23 cycles the material will no longer undergo any more plastic strain increase. After 1 to 3 cycles, the strain field ahead of the crack remains relatively constant.

Fracture Testing

The following papers were included in Session III of the symposium, held to honor Dr. John E. Srawley and Mr. William F. Brown, Jr. In their long association with the NASA-Lewis Research Center, these two researchers have made truly significant contributions to the area of fracture mechanics test methods. This session, which they chaired, was a tribute to them as they ended their full time work in the field of fracture.

The first three papers described opening mode fracture toughness testing using new test geometries. *Underwood et al* give recommendations for practical arc bend specimen geometries. Solutions are presented for stress intensity, crack mouth displacement, and load-line displacement for fracture specimens that have been compared using finite element and boundary collocation techniques. *Kapp and Bilinsky* described J_{Ic} tests in aluminum and steel alloys using the new arc-tension specimen and the existing compact specimen. A Merkle-Corten type analysis was presented for calculating J for the new specimen. The results showed that the specimen and methods of J analysis are suitable for accurate determination of J_{Ic}, and it is suggested that this information be added to the ASTM J_{Ic} test method. The paper by *Giovanola* shows the influence of specimen dimensions and impact velocity on dynamic fracture toughness of an edge-cracked coupon loaded in bending and impacted at midsection. This new procedure was used to measure the toughness of 4340 steel at three loading rates. The relative advantages of this one-point-bend test in dynamic fracture testing are indicated.

Two papers were presented on Mode II fracture testing. *Buzzard et al* described a test specimen that was developed to obtain fatigue crack propagation data under Mode II shear loading. Stress intensity factor and displacement analyses were performed and compared with photoelastic stress analysis. These results and the nature of the observed fatigue crack growth data suggested that the specimen and analysis would be adequate for Mode II fatigue testing. *Banks-Sills and Arcan* used a finite element analysis of a compact K_{II} specimen and load frame geometry to determine Mode II critical stress intensities. Mode II toughness tests of polymethyl methacrylate (PMMA) indicated that K_{Ic} was greater than K_{IIc} and that the average Mode II crack extension angle occurred along the direction of maximum tangential stress, which was between 63 to 70°.

The last three papers dealt with *J-R* curve testing. *Sutton and Vassilaros* presented comparative data for *J*-integral resistance curves for ASTM Class C and 3-Ni steels, using two reference techniques, multispecimen and d-c potential drop, for comparison with elastic compliance. The *J-R* curves established using elastic compliance showed no significant difference from the reference curve data. In the paper by *Davies and Stearns* the d-c potential drop technique was shown to be unacceptable for measuring the fracture toughness of Zircaloy-2 in the brittle-ductile fracture transition. The potential drop underestimated crack extension by 60%, increasing to 100% at higher test temperatures. Reproducible results were obtained using individual specimen calibrations.

Kapp and Jolles performed *J-R* tests of two aluminum and three steel alloys using Charpy size bend specimens and larger, standard compact specimens. Load-drop and electric-potential methods of crack growth were used. Based on comparisons of results from the two types of specimens and crack growth, the load-drop method with small bend specimens resulted in approximate *J-R* curves which would be suitable for quality control fracture toughness tests.

Ductile Fracture

There were two general groups of papers in this session of the symposium, one emphasizing test methods and results and the other emphasizing the mechanisms of ductile fracture.

Test Methods and Results

Three papers dealt primarily with ductile fracture test methods and results. *Hirano et al* described a single-specimen ultrasonic method for obtaining *J* versus crack growth curves. Measurements of J_{Ic} and tearing modulus in A533B steel were found to be independent of specimen geometry. It was also observed that the crack tip opening displacement (CTOD) remained nearly constant during stable crack growth. *Vassilaros et al* obtained *J-R* curves using compact specimens from A106 steel pipe and compared the results with those from four-point bend tests of 4-ft lengths of 8-in.-diameter pipe. The general result of the comparison was that the small specimen tests do not directly predict the *J-R* curve behavior of full size pipe because of the small amount of crack growth available in the specimen tests. *Towers and Garwood* described J-integral and CTOD analysis of bend tests of HY130 steel. Since these are the two most accepted methods of analysis for ductile fracture, this comparison of results is particularly interesting. In general, the two methods agree in their description of resistance to ductile tearing. Both methods indicate greater tearing resistance for shallower cracks (i.e., for a/w of 0.1 to 0.2 compared with 0.5).

Mechanisms

Four papers dealt primarily with ductile fracture mechanisms. *Etemad and Turner* investigated ductile tearing mechanisms using experiments with HY130 steel and analyses of energy rate, I, and tearing modulus, dJ/da. In the authors' words: "Subject to the choice of appropriate compliance terms the experimental behavior is predicted satisfactorily by both I and dJ/da methods." The authors also stated that satisfactory predictions are restricted to the geometry for which the R-curve was obtained. *Carifo et al* discussed finite element computation of stable crack growth using techniques of element node release. Conditions for nodal release are described and demonstrated in a series of finite element models of stable crack growth. *Hackett et al* investigated environmentally assisted crack growth of 4340 steel in seawater using techniques developed for the study of stable elastic-plastic crack growth. The authors found that J-R curves could be used for this different purpose, provided the rate dependence of the environmentally controlled process is considered. They observed a significant (four-fold) decrease in the energy required for crack initiation, J_{Ic}, due to cathodic polarization of the specimens. *Watson and Jolles* investigated global plastic energy dissipation for crack initiation and growth using experiments with HY130 steel for a variety of specimen configurations. Specimen size, specimen type, crack length, and side grooving were studied. R-curves using plastic energy dissipation (rather than J) showed some geometry dependence, which was minimized using side grooves.

Analysis and Mechanisms

The papers presented on the last day of the symposium focused primarily upon determination of loading and geometric (body shape) effects upon fracture parameters or upon determination of near tip material response. Included in this session was an overview lecture by Dr. George Irwin on Progressive Fracture Mechanics, in which he traced developments over several decades up to the present time. The range of time and technical topic in the lecture covered all aspects of fracture, including analyses, experiments, and applications.

The first paper of the session was that of *Pu*, which addressed the problem of an array of unequal depth radial cracks at the inner radius of a pressurized cylinder. The author used quadrilateral finite elements and collapsed singular elements around the crack tip to calculate stress intensity factor for a number of cases. The general result was that a crack which was slightly deeper than others in the array had a significantly higher K value, so that the deeper crack quickly dominated.

Sha and Yang described a method for computing stress intensity factors by combining the uncracked stress field with explicit crack face weight functions through superposition. The method was illustrated by applying it to the problem of symmetric radial cracks emanating from a circular hole in a plate. *Yau* presented a surface crack solution for fatigue crack propagation analysis of notched components. The solution was based on the weight function technique and was compared with a wide range of test results. Good agreement was observed between experimental results and analytical predictions, including such key information as stress intensity factor, crack propagation rate, residual life, and variation in crack shape.

A paper by *Müller et al* presented results of an analytical and experimental study of surface crack growth under cyclic loading. The authors found that use of local ΔK values gave a poor prediction of the crack aspect ratio, while use of a weighted average ΔK or a crack closure factor improved predictions. *Reuter* presented results of a combined analytical-experimental study of crack growth initiation under elastic and plastic conditions. A modified center cracked panel equation was used in the predictions, and acoustic emission methods were used to detect the load corresponding to crack initiation. The author presented evidence which questions the usefulness of J_{Ic} in structures where J-controlled fields are not attainable. Next, *Balladon and Heritier* compared the fracture toughness and crack growth properties of three grades (316L, 347, and austero-ferritic) of steel at room and elevated temperatures. Comparisons were based on the J-integral concept. Variations observed could not be explained solely on the basis of inclusion content and crack plane orientation.

Homma et al performed experiments in steel and aluminum to measure the minimum time duration at load which is required before a crack grows in an unstable manner. The relation between the minimum times obtained for the materials and their mechanical properties was discussed. The experimental results showed that the minimum time is longer for the more ductile materials. *Shukla and Anand* reported on the results of dynamic photo-elastic experiments under remote biaxial stress fields. They found that while the normal stress parallel to the crack surface had negligible influence on the branching stress intensity factor and crack velocity, a strong influence of the parallel stress was observed on the branching angle.

Next, *Anderson and Williams* used the results of a large number of crack-tip opening displacement (CTOD) measurements on carbon-manganese steels (using single edge notch bending and tensile tests) to assess the dominant mechanism for size effects on CTOD values in the ductile-brittle transition region. The authors concluded that, under conditions of small-scale yielding, a high constraint statistical sampling model works well for explaining size effects. When net section yielding occurs, however, statistical effects are suppressed and size effects are dominated by constraint effects. *Joyce and Hackett* described a series of J-integral R-curve tests on three-point-bend steel specimens at three different load rates. Through the use of multispecimen and key curve procedures at the higher rates, the authors found that J_{Ic} values and tearing modulus (T) values were elevated at the higher loading rates. The J_{Ic} and T values were increased by a factor of approximately two for a loading rate increase of six orders of magnitude.

Smith et al presented a quantitative evaluation of the loss of the inverse square root singularity when a crack intersects a free surface at right angles in nearly incompressible materials. Compact bending and surface flaw specimens were tested using frozen stress photoelasticity and moiré interferometry. After correlating free surface results with analytical results, the authors measured the variation of the lowest eigenvalue through the thickness and found thicker transition zones than previously suspected. The session was closed with a paper by *Raju and Newman* in which they utilized a refined three-dimensional finite element model to study surface flaws. The authors employed singularity elements along the crack front and linear stress elements elsewhere to obtain the stress intensity distribution along flaws in pipes and rods under extension and bending. Results from the models, which contained 6500 degrees of freedom, compared favorably with analyses and experiments of others.

John H. Underwood

Army Armament Research & Development Center, Watervliet, New York; chairman and co-editor.

Richard Chait

Army Materials and Mechanics Research Center, Watertown, Massachusetts; co-chairman and co-editor.

C. William Smith

Virginia Polytechnic Institute & State University, Blacksburg, Virginia; co-chairman and co-editor.

David P. Wilhem

Northrop Aircraft, Hawthorne, California; co-chairman and co-editor.

Wayne R. Andrews

General Electric Company, Schenectady, New York; co-chairman and co-editor.

James C. Newman

NASA Langley Research Center, Hampton, Virginia; co-chairman and co-editor.

Author Index

Subject Index

B

C

D

E

F

S

T

U

V

W

Z